Molecular Genetics of Cancer

Second edition

J. K. Cowell
*Department of Cancer Genetics, Roswell Park Cancer Institute,
Buffalo, NY 14263, USA*

© BIOS Scientific Publishers Limited, 2001
Copyright held by BIOS Scientific Publishers Ltd. For all chapters except that on page 33.

First published in 1995 (1–872748–09–0)
Second edition 2001 (1–85996–169–x)

A CIP catalogue record for this book is available from the British Library.

ISBN 1 85996 169 x

BIOS Scientific Publishers Ltd
9 Newtec Place, Magdalen Road, Oxford OX4 1RE, UK
Tel. +44 (0)1865 726286. Fax +44 (0)1865 246823
World Wide Web home page: http://www.bios.co.uk/

Published in the United States of America, its dependent territories and Canada by Academic Press, Inc., A Harcourt Science and Technology Company, 525 B Street, San Diego, CA 92101–4495. www.academicpress.com

Production Editor: Andrea Bosher
Typeset by Saxon Graphics Ltd, Derby, UK.
Printed by Biddles Ltd, Guildford, UK, www.biddles.co.uk

Contents

Contributors

Ahmed Rasheed, B.K. Department of Pathology, Duke University Medical Center, Durham NC 27710, USA

Barr, F.G. Department of Pathology and Laboratory Medicine, University of Pennsylvania School of Medicine, Philadelphia PA 19104–6–82, USA

Bigner, S.H. Department of Pathology, Duke University Medical Center, Durham NC 27710, USA

Burn, J. School of Biochemistry and Genetics, Department of Genetics, University of Newcastle, Newcastle-upon-Tyne, NE2 4AA, UK

Casey, G. Cleveland Clinic Foundation, Lerner Research Institute, Cancer Biology Department NB40, 9500 Euclid Avenue, Cleveland, OH 44195, USA

Chapman, P.D. School of Biochemistry and Genetics, Department of Genetics, University of Newcastle, Newcastle-upon-Tyne, NE2 4AA, UK

Chumakov, P.M. Department of Molecular Genetics (M/C 669), University of Illinios at Chicago, 900 South Ashland Avenue, Chicago, IL 60607–7170, USA

Cowell, J.K. Department of Cancer Genetics, Roswell Park Cancer Institute, Buffalo, NY 14263, USA

Dagnino, L. Department of Pharmacology and Toxicology and Lawson Health Research Institute, The University of Western Ontario, London, Ontario, Canada

Duckett, A.S. Laboratory of Molecular Signalling, Department of Zoology, University of Cambridge, Cambridge, UK

Eng, C. Director, Clinical Cancer Genetics Program, The Ohio State University, 420 W 12th Avenue, Suite 690 TMRF, Columbus, OH 43210, USA

Faruque, M.U. National Human Genome Research Institute, 9000 Rockville Pike, Bethesda MD 20892–4094, USA

Frevel, M. Cleveland Clinic Foundation, Lerner Research Institute, Cancer Biology Department NB40, 9500 Euclid Avenue, Cleveland, OH 44195, USA

Gallie, B.L. Departments of Opthalmology and Molecular and Medical Genetics and the Divisions of Cancer Informatics and Cellular and Molecular Biology, Ontario Cancer Institute, Princess Margaret Hospital, University of Toronto, Toronto, Canada

Grady, W.M. Vanderbilt University Medical Center, Division of Gastroenterology, 1161 21st Ave. South, Nashville, TN 37232–2279, USA

Gudkov, A.V. Department of Molecular Genetics (M/C 669), University of Illinios at Chicago, 900 South Ashland Avenue, Chicago, IL 60607–7170, USA

Halley, D.J.J. Department of Clinical Genetics, Erasmus University and Academic Hospital Rotterdam, P.O. Box 1738, 3000 DR Rotterdam, The Netherlands

Hui, A.-M. Hepato-Biliary-Pancreatic Surgery Division, Departments of Surgery, Graduate School of Medicine, University of Tokyo, 7-3-1 Hongo, Bunkyo-Ku, Tokyo 113–0033, Japan

Issacs, W. James Buchanan Brady Urological Institute, The John Hopkins Hospital, Baltimore, MD 21287, USA

Kainu, T. National Human Genome Research Institute, National Institutes of Health, Bethesda, MD 20892, USA

Kaye, F.J. Medicine Branch, Bldg. 8, Rm. 5101, National Naval Medical Center, Bethesda MD 20889, USA

Knowles, M.A. Imperial Cancer Research Fund Clinical Centre, St. James's University Hospital, Beckett Street, Leeds LS9 7TF, UK

Komarova, E.A. Department of Molecular Genetics (M/C 669), University of Illinios at Chicago, 900 South Ashland Avenue, Chicago, IL 60607–7170, USA

Kubo, A. Medicine Branch, Bldg. 8, Rm. 5101, National Naval Medical Center, Bethesda MD 20889, USA

Lindhout, D. Department of Clinical Genetics, Erasmus University, P.O. Box 1738, 3000 DR Rotterdam, The Netherlands

MacCollin, M. Neurofibromatosis Clinic, Massachusetts General Hospital CNY-6, Bldg 149, 13th St., Charlestown, MA 02129, USA

Maher, E.R. Department of Paediatrics and Child Health, University of Birmingham, The Medical School, Birmingham B15 2TT, UK

Makuuchi, M. Hepato-Biliary-Pancreatic Surgery Division, Departments of Surgery, Graduate School of Medicine, University of Tokyo, 7–3–1 Hongo, Bunkyo-Ku, Tokyo 113–0033, Japan

Morland, S.J. Cleveland Clinic Foundation, Lerner Research Institute, Cancer Biology Department NB40, 9500 Euclid Avenue, Cleveland, OH 44195, USA

Nellist, M. Department of Clinical Genetics, Erasmus University, P.O. Box 1738, 3000 DR Rotterdam, The Netherlands

Neville, P.J. Cleveland Clinic Foundation, Lerner Research Institute, Cancer Biology Department NB40, 9500 Euclid Avenue, Cleveland, OH 44195, USA

Pressey, J.G. Division of Oncology, Children's Hospital of Philadelphia, Philadelphia PA 19104, USA

Shay, J.W. University of Texas South Western Medical Center, Dallas, TX 75390–9039, USA

Sun, L. Hepato-Biliary-Pancreatic Surgery Division, Departments of Surgery, Graduate School of Medicine, University of Tokyo, 7–3–1 Hongo, Bunkyo-Ku, Tokyo 113–0033, Japan

Teh, B.T. Laboratory of Cancer Genetics, Van Andel Research Institute, 333 Bostwick NE, Grand Rapids MI 49503, USA

Trent, J.M. Division of Intramural Research, National Human Genome Research Institute, 49 Convent Drive, Bethesda MD 20892–4470, USA

Van den Ouweland, A.M.W. Department of Clinical Genetics, Erasmus University and Academic Hospital Rotterdam, P.O. Box 1738, 3000 DR Rotterdam, The Netherlands

Vaziri, S.A.J. Cleveland Clinic Foundation, Lerner Research Institute, Cancer Biology Department NB40, 9500 Euclid Avenue, Cleveland, OH 44195, USA

Verhoef, S. Nederlands Kanker Instituut, Afd. Familiaire Tumoren, Plesmanlaan 121, 1066 CX Amsterdam, The Netherlands

Wallace, M.R. Molecular Genetics and Pediatric Genetics, University of Florida, P.O. Box 100266, 1600 SW Archer Road, Gainesville, FL 32610–0266, USA

Williams, B.R.G. Cleveland Clinic Foundation, Lerner Research Institute, Cancer Biology Department NB40, 9500 Euclid Avenue, Cleveland, OH 44195, USA

Wiltshire, R.N. Department of Pathology, Duke University Medical Center, Durham NC 27710, USA

Wright, W.E.

Abbreviations

A	astrocytoma
AA	anaplastic astrocytoma
ACTH	adrenocorticotropic hormone
ALT	alternative lengthening of telomeres
AML	angiomyolipomas
APC	adenomatous polyposis coli
APKD	adult-type polycystic kidney disease
APUD	amine precursor uptake and decarboxylation
AR	androgen receptor
ARMS	alveolar rhabdomyosarcoma
AT	ataxia telangiectasia
BDNF	brain-derived neurotrophic factor
BLUS	Beckwith–Wiedemann syndrome
BS	Bloom syndrome
Cdk (CDK)	cyclin-dependent kinase
CFCS	cardio-facio-cutaneous syndrome
CGH	comparative genomic hybridization
CIN	chromosomal instability
CIS	carcinoma *in situ*
CRC	colorectal cancer
CT	cranial tomography
DCIS	ductal carcinoma *in situ*
DGGE	denaturing gradient gel electrophoresis
DHPLC	denaturing high performance liquid chromatography
DKC (DC)	dyskeratosis congenital
DMD	Duchenne muscular syndrome
DMs	double minutes
EGFR	epidermal growth factor receptor
ELST	endolymphatic sac tumors
EPV	Epstein–Barr virus
ERMS	embryonal rhabdomyosarcoma
EST	expressed sequence tag
FA	Fanconi anemia
FAP	familial adenomatous polyposis
FGFR2	fibroblast growth factor receptor-2
FIHP	family isolated hyperparathyroidism
FISH	fluorescent *in situ* hybridization
GAP	GTPase-activating protein
GH	growth hormone
GLUT-1	glucose transporter-1
GMB	glioblastoma multiforme

HBV	hepatitis B virus
HCC	hepatocellular carcinoma
HCV	hepatitis C virus
HDGC	hereditary diffuse gastric cancer
HIF	hypoxia-inducible factors
HNPCC	hereditary nonpolyposis colon cancer
HPT	hyperparathyroidism
HPV	human papilloma virus
HSCR	Hirschsprung disease
HSR	homogeneously staining regions
HWE	Hardy–Weinberg equilibrium
IRF1	interferon regulatory factor-1
kb	kilobases
KGFR	keratinocyte growth factor receptor
LAM	lymphangiomyomatosis
LFS	Li-Fraumeni syndrome
LIF	leukemia inhibitory factor
LOH	loss of heterozygosity
LOI	loss of imprinting
MEN	multiple endocrine neoplasia
MMP	matrix metaloproteinase
MMR	mismatch repair
MPNST	malignant peripheral nerve sheath tumor
MRI	magnetic resonance imaging
MRP	multidrug resistance protein
MSI	microsatellite instability
MSS	microsatellite stable
MTC	medullary thyroid carcinoma
NAT	N acetyl transferase
NBCCS	nevoid basal cell carcinoma syndrome
NBS	Nijmegan breakage syndrome
NER	nucleotide excision repair
NES	nuclear export signal
NF1	neurofibromatosis type 1
NF2	neurofibromatosis type 2
NGF	nerve growth factor
NLS	nuclear localization signals
NT-3	neurotrophin-3
NT-4	neurotrophin-4
PC	pheochromocytoma
PCNA	proliferating cell nuclear antigen
PDGF	platelet-derived growth factor
PFGE	pulsed field gel electrophoresis
PJS	Peutz–Jeghers syndrome
PKD	polycystic kidney disease
PML	promyelocytic pigmentosum
PNET	primitive neuroectodermal tumors

PR	progesterone receptor
PSA	prostate-specific antigen
PTT	protein truncation test
RCC	renal cell carcinoma
RFLP	restriction fragment length polymorphism
RMS	rhabdomyosarcoma
RR	relative risk
RTK	receptor tyrosine kinase
RTS	Rothmund–Thomson syndrome
RTS	Rubinstein–Taybi syndrome
SCC	squamous cell carcinoma
SCLC	small cell lung cancer
SHH	sonic hedgehog
SKY	spectral karyotyping
SNP	single nucleotide polymorphism
SSCP	single strand chain polymorphism
TGF	transforming growth factor
TPE	telomere position effect
TSC	tuberous sclerosis complex
TSG	tumor suppressor gene
TSH	thyroid-stimulating hormone
UC	ulcerative colitis
UPD	uniparental disomy
UTR	untranslated region
VEGF	vascular endothelial growth hormone
VHL	von Hippel-Lindau
WS	Werner syndrome
XP	xeroderma pigmentosum

Preface to the second edition

Cancer is a disease of the genes. A genetic predisposition to many forms of cancer can be passed from parent to child and the determining genetic events required for a malignant phenotype are passed from cell to cell. In 1995, when the first edition of this book was published, there were only a handful of cancer predisposition genes that had been identified. Even though it was known that many more existed, the lack of reagents and technologies to isolate them at that time made the cloning process very long and tedious. It could hardly have been imagined then, how the pace of cancer gene discovery would accelerate, due in a large part to the development of enabling technologies and the rapid progress of the human genome project together with the general availability of the information that was being generated. Many of the genes responsible for the common hereditary forms of cancer have now been isolated, with a few notable exceptions such as prostate. The cloning of these genes has been the direct consequence of incremental development of genetic information. It was this rapid expansion in the isolation and characterization of human cancer genes which prompted the production of the second edition of this book. Although the aim was to provide an up-to-date coverage of the underlying details about these genes, in a book of this size it has not been possible to cover every aspect of the subject. For the cancers which were reviewed last time, this volume provides an update on the significant developments in these areas. For the many cancer genes which have been isolated over the past few years, this volume provides a more comprehensive overview of the subject. As before, it was not possible to include all types of hereditary cancer and all of the specific types of tumors where the involvement of specific genes and pathways have been studied in some detail. Consequently, the leukemias have been deliberately excluded since they could produce a volume to themselves. I also chose to exclude syndromes such as Bloom's, Ataxia Telangiectasia and Fanconi's anemia which represent a distinct group of radiation induced cancers which have been the subject of many volumes recently. Deliberately included were tumors of tissues such as liver, lung, bladder and brain where there is no direct evidence for predisposition loci but, because of their relatively high incidence in the population and their importance to public health, a lot is known about the genetic events which occur within these tumors. Also included, and appropriately at the end of the chapters, is a review of the telomerase enzyme which is being shown to have overarching importance to the development of cancer which is of interest to basic biologists, and those involved in cancer therapeutics, alike.

During the course of the analysis of human cancer genes, what had been somewhat surprising was the diversity of the types and functions of genes which are involved in specific types of tumors. We are rapidly approaching the point where the genes for most of the dominantly inherited cancers have been isolated. Certainly most of the genes responsible for the common forms of cancer are now relatively well characterized. From these studies it is becoming clear that many of these genes feed into common pathways which is providing some order within the apparent chaos. The next steps will be to characterize the function of these genes more fully with the hope that they will lead to the development of novel therapeutic approaches. Indeed, in many aspects of cancer genetics research, the realization of the dream that one day this information would be used for treatment of patients, is becoming true. As our understanding of the function of specific genes improves, attention is shifting to the analysis of larger numbers of genes in order to learn more about the interrelationships between genetic events in cancer cells. Although still in its infancy many of the chapters in this book point to the future about the utility of whole genome analysis for improved staging, prognostics and diagnostics.

In keeping with the previous volume, all of the chapters in this book have been written by experts in their respective field. These individuals have active research programs in the particular type of cancers they reviewed and are part of larger teams studying the overall biology of the disease and the application to the clinical management of tumors. I am honored that so many highly talented experts agreed to contribute their time and knowledge to the production of this book, which I feel sure will be an important reference volume in the field of cancer genetics for some time to come. The assembly of such an impressive group of authors guarantees that the details presented are as up-to-date as possible and include insights into tumorigenesis that can only come from the wealth of knowledge that these authors possess. Because of the comprehensive coverage of cancer genetics in this volume it should be of interest to both clinical and basic scientists by providing a single, valuable reference source for human molecular cancer genetics.

John K. Cowell *(2001)*.

Preface to the first edition

In the late 1970s and early 1980s, the analysis of the dominantly transforming oncogenes offered new hopes for the understanding of the genetic basis of cancer. Although much has been learned about signal transduction in tumour and normal cells from these studies, the underlying mechanisms of tumour initiation were not forthcoming. With the development of sophisticated gene cloning strategies, the late 80s and early 90s belong to the analysis of the recessive oncogenes, or 'tumour suppressor genes'. This class of genes play an important role in the normal development and differentiation of cells and have been shown to have diverse roles ranging from extracellular signal recognition to DNA repair. The aim of producing this volume was to provide an up-to-date discussion of the cloning and analysis of genes which are critical in the development and progression of human cancers. It has not been possible to review all the different tumour types which show a hereditary predisposition; in particular, the leukaemias, which could fill a volume on their own, have been deliberately excluded in favour of solid tumours.

Due to the clinical importance of the hereditary cancer genes, their study represents a very rapidly moving area in biomedical research. For example, during the relatively short period in which this volume was compiled, one of the genes responsible for hereditary breast cancer was cloned and the location of a second determined. However, by selecting authors directly involved in the cloning and analysis of tumour suppressor genes, the most up-to-date information possible is presented together with insights into how this information is being used in the clinical management of patients with these tumours. The analysis of mutations in tumour suppressor genes is essential for genetic screening and also provides for a fundamental understanding of the function of many of these genes. Thus, wherever possible, the nature of mutations in these genes is reviewed to provide a valuable reference source for these abnormalities.

Finally, the majority of the chapters are written by individuals who are part of the teams involved in the genetic screening of patients with hereditary forms of cancer, and so discuss the relevance of basic research to the counselling of families. This volume is therefore of interest to clinical as well as research scientists. The book has had an accelerated production time with the aim of providing information both current and topical.

John K. Cowell (*London*)

Basic principles in cancer genetics

John K. Cowell

1. Introduction

It is now well established that the malignant phenotype is genetically determined and that interactions between the environment and the genetic makeup of a cell can influence the course of specific tumors. In certain cases, mutations in specific genes confer a predisposition to cancer, again pointing to the importance of highly specific genetic changes in the development of particular tumors. As our understanding of the series of events which take place during the transformation of a normal cell into a cancer cell improves, it is becoming clear that cancer cells employ a complex repertoire of tricks, both to escape normal cellular growth control as well as overcome the powerful forces of terminal differentiation. The study of the genetic events which underlie this mechanism has been a priority of cancer molecular geneticists for many years, where the expectation is that, through a better understanding of the molecular genetic events that contribute to this commonly fatal collection of diseases, insights into novel therapies may emerge which, if not provide a cure, will hopefully retard the rampant growth of malignant cells that affects 1 in 4 adults and 1 in 800 children during their lifetimes.

Cancer genetics research, like the evolution of species, has progressed through major leaps, followed by consolidation in our understanding of the fundamental processes associated with these diseases. Although some of these major leaps resulted from provocative hypotheses, such as the two-hit theory (Knudson, 1971), more usually they coincided with individual technological advances such as chromosome banding (Caspersson et al., 1968; Seabright, 1971) or the polymerase chain reaction (Mullis and Faloona, 1987), which provided the means to test both new and old hypotheses. An accumulation of data then followed using these technologies in a rigorous, worldwide effort which, in turn, generated new hypotheses. The purpose of this opening chapter is to review some of the critical landmark developments in the field of cancer genetics which have allowed such amazing progress in our understanding of cancer development in such a relatively short time. By presenting the background and principles behind these

Molecular Genetics of Cancer second edition, edited by J.K. Cowell.
© 2001 BIOS Scientific Publishers Ltd, Oxford.

developments, it is anticipated that it will then be possible to put the events leading to the isolation and characterization of human cancer genes in perspective.

When cancer cells divide they produce two cancer cells, clearly demonstrating that the malignant phenotype is genetically determined at the cellular level. The concept that cancer cells arise as a result of genomic instability is attributed to Theodore Boveri in 1914. Long before it was possible to clearly visualize chromosomes and almost half a decade before the structure of DNA was resolved, Boveri noted in his pathology practice that tumor cell nuclei showed mis-division and suggested that this chromosome instability was responsible for the development of cancer cells. He further suggested that random genetic events in a single cell caused chromosome abnormalities which conferred a permanent growth advantage and gave rise to clonal expansion with millions of similarly altered cells forming a malignancy. This hypothesis has proven to be remarkably accurate, and it has since been shown, that cumulative genetic changes give rise to tumor cells and that these changes may either be found in many cancer cells types or may be specific to a particular cell type.

2. Cytogenetics of cancer

One of the earliest developments in cancer research was the ability to be able to adequately separate and view the chromosomes from individual cells. For mammalian cells, analyzing chromosomes from 'squash' preparations was unsatisfying since the chromosomes were difficult to observe and count because of the individual overlaps. This problem was overcome with the development of hypotonic pre-treatments to swell the cells (Hsu, 1952) and chemical agents such as colchicine to disaggregate the spindle microtubules. This combined procedure made the cells fragile and so when dropped onto the microscope slide the cell burst releasing the chromosomes (Tjio and Levan, 1956). Immediately it was possible to count the chromosomes and arrange them not only into pairs of the same size but also to subgroup these chromosomes based on their centromeric position; e.g. telocentric (centromeres at the end), or metacentric (centromeres in the middle). This simple advance gave rise to countless studies of chromosomes in cancer cells describing numerical changes such as trisomies or monosomies for individual chromosomes (some of which are still used in diagnostics today), as well as large changes in chromosome numbers defined as pseudodiploidy, aneuploidy, hyperdiploidy and tetraploidy, depending on the extent of chromosome changes (Sandberg, 1990). Aneuploidy and hyperdiploidy, assessed either by chromosome number or DNA content measurement, is still used as a marker for tumor progression and in some cases may still be an adequate prognostic indicator (*see* Chapter 15). Large deletions within chromosomes could also be identified, although often it was not possible to establish unequivocally which chromosome was involved. An early, and very important, example of this was the report by Lele *et al.* (1963) of heterozygous constitutional chromosome deletions in retinoblastoma patients who showed a range of congenital abnormalities including mental retardation. This chromosome change was in the so-called 'D' group of chromosomes, representing #13, #14 and #15. These studies laid the foundation for the subsequent search for, and cloning of, the retinoblastoma predisposition gene (Friend *et al.*, 1986). A second classic example was the description of a recurrent, tiny marker – the so-called Philadelphia chromosome – in chronic myelogenous leukemia (Nowell and Hungerford, 1960) which turned out to be the consequence of a 9;22 translocation (Rowley, 1973). This observation was the prelude to the demonstration that translocations, in this case in leukemias, led to the generation of chimeric genes with novel functions. The major problem, however, at this point was that consistent chromosome changes in tumors could not be defined because of the inability to distinguish between the individual human chromosomes. This issue was resolved with the advent of chromosome banding techniques (Caspersson *et al.*, 1968; Seabright, 1971). Using intercalating fluorescent dyes or proteases which selectively stripped proteins from fixed chromosomes, a characteristic linear distribution of light and dark bands along the length of the chromosomes could be generated. To the expert it was now possible to identify each of the human chromosomes based on their banding pattern and unequivocally identify each of the 23 different ones (*Figure 1*). The banding patterns also allowed subchromosomal changes such as deletions, translocations

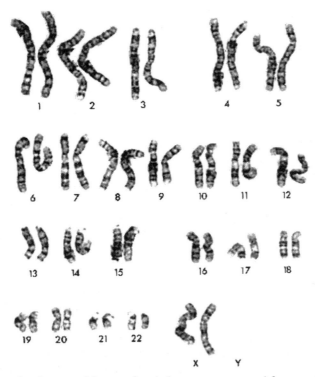

Figure 1. Example of a normal human female karyotype prepared from peripheral blood lymphocytes. Each of the chromosome pairs are arranged side by side based primarily on size and then on centromere position. The individual light-dark linear banding patterns are produced by mild trypsin treatment of acid-methanol fixed chromosome spreads followed by Giemsa staining. The G-band patterns for each of the chromosomes are quite distinct allowing unequivocal identification of each of the 23 different chromosomes (the Y chromosome is not shown).

and inversion to be readily identified. Thus, began the ever increasing effort of characterizing cytogenetic changes in all types of tumor cells which, in turn, formed the basis for much of the molecular cloning era that followed.

Many cancer syndromes had already been defined clinically by the coincidence of a limited set of phenotypes, e.g the association of mental retardation with retinoblastoma (Lele *et al.*, 1963), and aniridia with Wilms tumor (Riccardi *et al.*, 1978). The ability to detect smaller and smaller changes, followed the development of new techniques to generate longer and longer chromosomes, pioneered in the laboratory of Jorge Yunis (Yunis, 1976). The ability to unequivocally identify each of the normal human chromosomes had also opened up the possibility of analyzing the constitutional chromosomes, e.g. in lymphocytes, from patients with cancer. It was subsequently shown that, in some of these cancer syndromes (Riccardi *et al.*, 1978; Yunis and Ramsay, 1978), there were inherited chromosome abnormalities (*Figure 2*). Importantly these rearrangements not only pointed to the location of the genes responsible for the tumor predisposition but also provided a means of isolating them through positional cloning strategies (see below). This proved to be true in many cases and, in fact, the cloning of many cancer predisposition genes has been facilitated by these cytogenetic observations. As cytogenetic data accumulated, there became a need to catalog the various changes in order to identify consistencies between and within tumor types. This data provided a means to identify reliable markers which would not only aid in

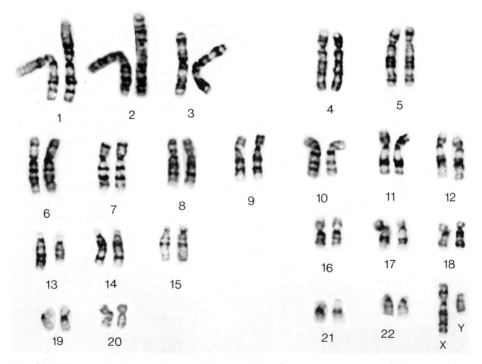

Figure 2. Giemsa-banded karyotype from peripheral blood lymphocytes from a patient with retinoblastoma and mental retardation. The only cytogenetic abnormality is a deletion within the long arm of one of the copies of chromosome 13.

determining diagnosis and predict prognosis, but also to highlight regions within the genome that might contain genes important in the development of a particular malignancy. The task of doing this was adopted by Felix Mitelman and colleagues (Mitelman and Levan, 1981) who, after starting the catalog with an early report of approximately 1900 cytogenetic changes in tumor cells in 1981, now presides over an incredible volume describing over 100 000 changes and which is available for review online (Mitelman, 2000). These studies highlighted several phenomena in addition to the diversity of cytogenetic changes; firstly, that there were clearly consistencies in the regions of the chromosomes involved in abnormalities in specific tumor types. The consistency between tumors indicated that the rearrangements were important in tumorigenesis, especially where these repre-sented the only cytogenetic change in the cells. Secondly, it was clear that the banding procedures could not always resolve the nature of every chromosome change and so valuable genetic information was un-interpretable. These uniden-tified chromosomes were banished to the end of the karyotype and simply described as 'marker' chromosomes to indicate their anonymous nature. Some of these marker chromosomes would turn out to be important and their characteri-zation depended on yet another development in technology which was being developed in the late 1980s by groups such as the one headed by Dan Pinkel (Pinkel *et al.*, 1986, 1988). These developments heralded the beginning of the era of 'molecular cytogenetics'.

3. The evolution of molecular cytogenetics

Although the hybridization of radioactively labeled DNA probes to chromosomes (*in situ* hybridization; ISH) was being applied to gene mapping, its use was mainly confined to repetitive regions of the chromosomes which could produce sufficient signal for detection. The ability to map unique sequences to particular regions of chromosomes was a more controversial issue. This was to change in the late 1980s as a result of several technological advances which proved to be essential for the molecular cytogenetics revolution: (1) the ability to label DNA probes with fluo-rochromes to high density; (2) the development of cameras which collected low levels of fluorescent light; (3) the construction of large insert genomic libraries (see below). The ability to suitably compete out repetitive sequences within these large probes (Landergent, 1987; Pinkel *et al.*, 1988), together with the increased fluorescence signal obtained from large insert clones, provided the opportunity to localize markers directly on chromosomes in metaphase spreads. Thus, mapping by fluorescence *in situ* hybridization (FISH) became the primary way of localizing individual clones (Baldini *et al.*, 1992). This approach also identified chimeric clones, since signal could be seen on two different chromosomes which proved to be an important piece of information for gene cloning experiments.

The technical advance which allowed this area to advance was the development and application of cooled, charge-coupled device (CCD) cameras which were able to capture very low light intensities emitted from the hybridizing clones. These cameras, therefore, which had been developed to stare out at distant galaxies where now being used to focus on some of the smallest structures within a single cell. In combination with software packages which digitally enhanced these images, this technology made visualization of probes relatively straightforward. Filter wheels capturing light from different emission spectra allowed several different probes to be used simultaneously and so their relative order could be determined. The advantage of this automated system was that light emissions were captured with minimal UV exposure and so the bleaching of the image, which is seen in regular fluorescence microscopy, was not a problem. This technique soon resulted in large databases of mapped probes, which made the identification of candidate genes associated with particular cancers much faster. The application of chromosome 'painting', which used complex probes (cDNA libraries, flow sorted chromosomes, Alu-PCR products) derived from single chromosomes or chromosome arms, to highlight specific chromosomes within a metaphase spread, overcame the need to describe the troublesome 'marker' chromosomes, since their origin could now be unequivocally established. With this technology marker chromosomes could be interpreted, subtle chromosome rearrangements could be identified, and the human chromosome complement of somatic cell hybrids could be established (see below).

Cytogenetics has clearly been the source for many hypotheses about cancer and development but its major limitation was that it depended on sufficient numbers of dividing cells in particular tumors to allow chromosome preparation. For many solid tumors this was not always possible. In turn, FISH at this point was mostly used to study single chromosomes or single loci on the chromosomes. FISH probes derived from the centromeres of the individual human chromosomes could be used to detect numerical chromosome changes (Kempski *et al.*, 1995) and gene-specific probes could detect large chromosome deletions (Cowell *et al.*, 1994). Despite this, the need to scale up to look at the whole chromosome complement in non-dividing cells led to the development of comparative genome hybridization (CGH). In this procedure, tumor and normal DNA is competitively hybridized to normal metaphase chromosomes in a single hybridization reaction (Kallioniemi *et al.*, 1992, 1994). Repeat sequences are competed out with Cot-1 DNA and typically the tumor is labeled with a green fluorochrome and the normal DNA is labeled with a red fluorochrome. Copy number differences between tumor and normal are identified by measuring the ratio of green and red fluorescence along the length of normal chromosome spreads. Thus, if a region is deleted in the tumor only the normal red fluorescence is seen at that chromosome site. If a particular region of a chromosome is amplified in the tumor then the signal is more green. Equal hybridization, representing no difference between samples, produces a yellow signal (Getman *et al.*, 1998). In general, this approach is not sensitive enough to identify submicroscopic changes, but it has the advantage of surveying the whole genome without the need for mitotic cells and demonstrate consistent changes in tumors.

Although it has not been possible to discuss all of the modifications that have accompanied the revolution in molecular cytogenetics, it is clear that this field has been an important cornerstone for the isolation of human cancer genes. In the post genome-sequencing era, it is likely that it will continue to provide an important resource which will assist in the implication of genes critical to the development of particular cancers.

4. Genetic linkage and the identification of hereditary cancer genes

Genetic linkage studies have been a vitally important approach for defining the positions of hereditary forms of cancer. For most of the major cancers, approximately 10% will show a family history of disease where there are multiple affected individuals over several generations (*Figure 3*). In these cases it was then possible to use genetic markers to follow the co-inheritance of the marker and the phenotype (*Figure 4*). Statistical analysis of the data provides an estimate of the likelihood that a given allele is co-inherited with a particular disease. The analysis is presented as an odds ratio to the log base 10 – the so-called LOD score. By convention, a LOD score of 3, which is equal to 10^3 or 1:1000, is taken as an indication of linkage., i.e. the possibility of the marker and the disease co-segregating by chance is 1000:1 against (Ott, 1974). Data from individual families can be pooled and the LOD scores combined. This analysis was very important in establishing the location of genes predisposing to malignancy such as retinoblastoma, colon cancer, breast cancer, neurofibromatosis and MEN, where there were sufficiently large pedigrees to obtain the required LOD score. In this way the chromosomal location of the gene could be readily identified and, depending on the markers used, its position could then be refined to within a small region of the chromosome. These analyses could also be used to 'track' the segregation of mutant alleles in families, without any knowledge of the nature of the gene involved (*Figure 4*). The problem with linkage studies was that, because of recombination events within chromosomes, the more distant a marker is from the locus of interest, the greater the chance of recombination which separates the marker from the mutant gene. Recombination, however, can also work in the favor of the researcher, since once the order of specific polymorphic variants of the markers in the vicinity of the predisposition locus was determined, rare recombination events in individual families allowed even further sublocalization of the region

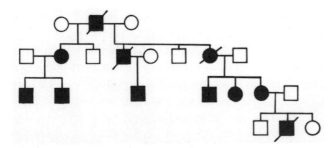

Figure 3. Example of dominant inheritance from a family with hereditary retinoblastoma. Affected individuals are represented by the black symbols. Females are depicted by circles and males are depicted by squares. Individuals with the oblique line drawn through are deceased. Since offspring have a 50% chance of inheriting the chromosome carrying the mutant gene from the founder in the first generation and all of these develop tumors, the tumor phenotype is dominant. There are roughly equal numbers of affected and unaffected members. Individuals who inherit the mutant allele develop the tumor which is the hallmark of dominant inheritance.

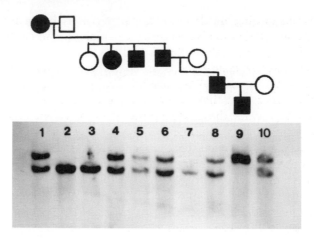

Figure 4. An example of gene tracking using an RFLP in a retinoblastoma pedigree. In this case a chromosome 13q14 specific marker is used as a probe to identify a polymorphic variant band in DNA from family members. The variation is identified by autoradiography following hybridization to DNA from each patient. The presence of two bands in the affected member of the first generation (1) allows each of her copies of chromosome 13 to be distinguished. All patients who inherit the upper band develop the disease thus providing a predictive marker for future generations. The results from patient 9 demonstrate one of the limitations of this approach since this individual acquires another copy of the upper band from an unaffected individual joining the family and so becomes 'uninformative' since his two chromosomes 13 can no longer be distinguished.

which must contain the gene. These critical recombination events proved very important in the final identification of candidate genes, since the smaller the region that could be defined the fewer candidates there were (Kelsell *et al.*, 1993; Simard *et al.*, 1993). Recombination events are defined as a percentage, the unit of which is a centimorgan (cM). Thus, if 100 chromosomes are studied and recombination between two given markers is 5%, then the 2 markers are described as being 5 cM apart. This value does not necessarily equate to a physical distance since recombination is greater in females than males and greater at the ends of chromosomes compared with the central part of the chromosome. The general rule of thumb, however, is that 1 cM is roughly equal to 1000 kilobases which, on average, can be enough DNA to contain 20–30 genes (assuming genes are on average 50 Kb apart). Linkage studies have proved to be the most powerful means of establishing the location of hereditary cancer genes, particularly in the absence of associated constitutional cytogenetic changes such as those seen in retinoblastoma and Wilms tumor. For many common tumors such as lung, brain and liver, however, there have been few reports of extensive families where linkage studies could be applied, although epidemiological data suggest a predisposition in the population may exist. The difficulties of identifying hereditary traits in liver cancer, for example, are that, in places where there is a high incidence, it is often difficult to divorce the environmental factors, such as hepatitis infection, from the hereditary ones. The other complication is that if there is a suggestion of a predisposed population, but the familial incidence is low, this may be due to the lack of penetrance of the specific predisposing mutations. Penetrance is an estimate of the frequency with which a mutant allele will give rise to a particular phenotype (*Figure 5*). If the penetrance is high, e.g. 100%, then all patients with the mutation will develop the tumor which is therefore manifested as an autosomal dominant disorder. If the penetrance is low, however, e.g. 40%, then within a given family only 2/5 people will develop the tumor even though they carry the mutation. In small families, therefore, it may appear that the tumor represents a sporadic case even though it is inherited. One problem with pooling families for linkage studies is that, if several different genes are responsible for the tumor predisposition, then pooling the data will be affected by this heterogeneity. This has clearly been a problem in identi-fying loci important in the development of prostate cancer, for example (*see* Chapter 11). This may be important during the establishment of linkage, but once the linkage to a particular locus has been determined, families can subsequently be classified into one linkage group or another, even without any specific knowledge of the gene in question. These types of studies, as in breast cancer, have then allowed specific correlations to emerge such as the association of ovarian cancer in carriers of BRCA1 mutations and male breast cancer in carriers of BRCA2 mutations (*see* Chapter 3).

5. The two-hit hypothesis

The analysis of inherited cancers such as retinoblastoma clearly showed that, although the tumor phenotype segregated in families as a dominant phenotype, at the cellular level it was recessive, otherwise every cell in the retina would be expected to give rise to tumors which is not the case. Clearly other genetic events were required. To investigate this question, Knudson (1971) analyzed patients

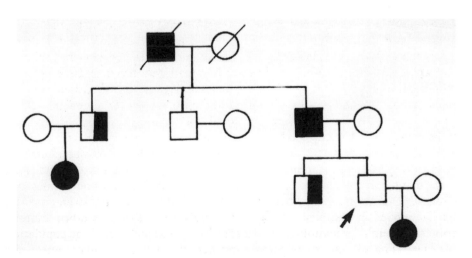

Figure 5. Example of incomplete penetrance in a family with hereditary retinoblastoma. Inheritance generally follows an autosomal dominant pattern except for the individual shown by the arrow. Although not affected himself he is clearly a carrier of the mutation since he has an affected daughter and an affected father.

with the childhood eye tumor, retinoblastoma (Rb). Pedigree analysis clearly identified patients with familial Rb, as well as those which occur apparently sporadically. It was also known that familial Rb cases presented earlier than sporadic cases and with a greater number of tumors (multifocal), which typically affected both eyes in the hereditary cases. The sporadic tumors were more often unifocal and unilateral. Knudson (1971) used data from families with Rb which show an autosomal dominant pattern of inheritance (*Figure 3*) of the tumor phenotype. Knudson analyzed the data mathematically in terms of number of tumors and age of onset in hereditary vs. sporadic cases. From this study the hypothesis was made – it takes only two genetic events for the development of Rb. Similar analyses also predicted that the same might be the case for neuroblastoma, another childhood tumor (Knudson and Strong, 1972). Thus, the 'two hit hypothesis' was formed. In hereditary cases one mutation is inherited and the second is acquired in a susceptible cell type, whereafter the tumor develops. The chances of a single sporadic event occurring in a susceptible population of tumor precursor cells is relatively high and so hereditary cases develop more tumors than sporadic cases, where both mutations must arise in the same cell during the window of development before the cells are terminally differentiated. The chances of two mutations arising in the same cell by chance is relatively low and apparently accounts for the fact that sporadic cases usually have a unilateral, unifocal tumor. Furthermore, since only a single genetic event is required in hereditary cases for tumor formation, these tumors have an earlier age of onset. In sporadic cases, cells experiencing two mutations are considerably older before the tumors begin to develop and so present later. Since random mutations follow a Poisson distribution, there is a liklihood that, in some patients, no cells will experience the second hit and so escape tumor development – so-called 'incomplete penetrance' (*Figure 4*). The original two-hit hypothesis assumed that the inactivating mutational events occur in the same gene (Comings, 1973) but that was not formally proven until over a decade later (see below). An important implication of Knudson's analysis, however, was that if, in a given population, there is evidence of a bimodal distribution of the age of onset, or there are distinct subgroups of patients who develop multifocal disease, then there may be evidence for a genetic predisposition.

6. Loss of heterozygosity

The two-hit hypothesis implied that the critical initiating events in Rb were mutations in the two homologous genes. Subsequent work by Gallie and colleagues (Godbout *et al.*, 1983) and Cavenee and colleagues (Cavenee *et al.*, 1983) provided strong support for this suggestion with their pioneering 'loss of heterozygosity' (LOH) analysis. The principle was simple – find a locus which is constitutionally heterozygous and then analyze the tumor from the same patient to establish whether heterozygosity was retained or not (*Figure 6*). The problem up to this time was that there were insufficient polymorphic markers available around the RB1 locus in the 13q14 region to do the analysis. Gallie and colleagues (Godbout *et al.*, 1983), therefore, used the protein polymorphism associated with the Esterase-D

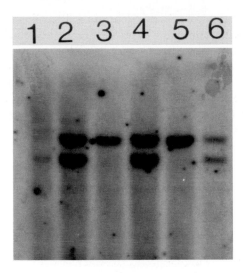

Figure 6. Example of LOH analysis in Wilms tumors using an RFLP from the short arm of chromosome 11. Pairs of tumor and normal DNA respectively are arranged side by side and this autoradiograph demonstrates two bands in the normal cells of all three patients (lanes 2, 4 and 6, respectively). Although the result for the first patient is ambiguous, for the other two patients it is clear that their tumors (lanes 3 and 5) have lost one of the alleles seen in the constitutional cells. The prediction is that the chromosome which is retained carries a mutant allele for the WT gene on chromosome 11.

gene which had previously shown close genetic linkage to the Rb phenotype in familial cases of Rb (Connelly *et al.*, 1983). Cavenee *et al.* (1983) used polymorphic molecular markers available for chromosome 13, which were identified using restriction enzymes and Southern blotting. In both cases, even where tumors could be shown to have two copies of chromosome 13, they were homozygous in 13q14 in 60–70% of cases. In tumors from hereditary cases it was the allele which was derived from the transmitting parent which was retained in the tumor, demonstrating that it was the chromosome region carrying the normal gene from the unaffected parent which had been lost (Cavenee *et al.*, 1985). Loss of the corresponding region containing the normal gene supposedly resulted in the 'exposure' of an otherwise recessive mutation which, in turn, contributes to the development of the malignant phenotype. This work, together with somatic cell hybrid studies (see below), gave rise to the term 'tumor suppressor gene', since the Rb work demonstrated that the presence of a single normal copy of the RB1 gene could protect the cell from malignant transformation. The implications were more far-reaching, however, and, as more and more types of cancer were studied it emerged that, when hereditary tumors for which the chromosomal regions known to harbor tumor predisposition genes were analyzed, LOH was detected in a large proportion of them (Solomon *et al.*, 1987). The LOH phenomenon was so consistent that it spawned the expectation that identifying regions of consistent LOH in sporadic tumors would identify chromosome regions harboring tumor suppressor genes and facilitate their eventual cloning. The work of Cavenee and colleagues (1983) also demonstrated some of the mechanisms that led to the development of LOH. The most frequent were (1) loss of the entire copy of a single chromosome; (2) an interstitial deletion of the critical region; (3) mitotic recombination between sister chromatids which, with the appropriate segregation of the resultant chromosomes at mitosis, resulted in cells homozygous for a given region (*Figure 7*). It was not until the cloning and characterization of the *RB1* gene (see Chapter 13) that it was demonstrated that the 70% of tumors showing LOH carried a homozygous mutation, whereas tumors which had retained heterozygosity carried two distinct mutations in different exons of the gene (Hogg *et al.*, 1993; Yandell *et al.*, 1989).

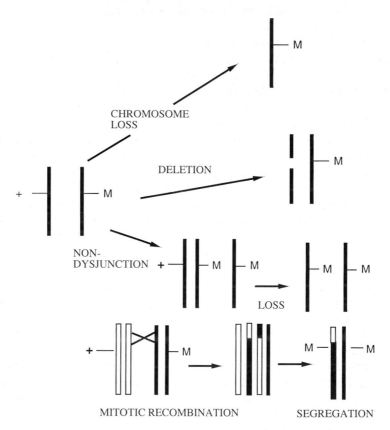

Figure 7. Schematic representation of the mechanisms which most commonly give rise to LOH. From a normally heterozygous cell carrying a mutation (M) in one of the tumor suppressor gene (TSG) alleles (left) whole chromosome loss of the chromosome carrying the normal gene (+) results in the 'exposure' of the recessive mutation. Similarly, an interstitial deletion of the region carrying the TSG results in a cell with two copies of the chromosome but no functional TSG. Reduplication of a chromosome carrying the mutant TSG, as a result of non-dysjunction at mitosis, results in an unstable cell carrying three copies of the particular chromosome. Subsequent loss (followed by selection) of the chromosome carrying the normal gene again produces a cell which is null for normal TSG function. Finally, mitotic recombination between chromatids of homologous chromosomes, when resolved at mitosis generates cells which are null for TSG activity. In all cases loss of TSG function results in a growth advantage which allows these cells to divide uncontrollably.

In some cases, searches for LOH region in tumors has involved whole genome scans made possible by the generation of comprehensive chromosome maps (see later). Others have involved more selected regions (*Figure 6*) based on previous evidence of involvement such as the short arm of chromosome 11 in Wilms tumors (Wadey *et al.*, 1990) following the observation of constitutional deletions of 11p13 in WAGR patients. LOH held great promise for the isolation of tumor suppressor genes specific for the particular tumor type and many of the chapters in this volume address this issue. LOH studies originally depended on using restriction fragment length polymorphisms (RFLPs) which were time consuming to identify and often showed low frequencies of heterozygosity in the population (*Figure 6*). The fact that individuals were constitutionally homozygous made them 'uninformative' in LOH studies. The ability to perform LOH studies was given a boost, however, with the development of the microsatellite repeat markers which, with their large numbers of potential alleles, meant that up to 80% of tumors would be informative in LOH assays (Grundy *et al.*, 1998) compared with the, at best, 50% for two-allele polymorphisms. The density of markers on the microsatellite maps is also much higher so, if one marker is troublesome, then all that was needed was to use another (see below). The other advantage was the fact that the position of the mitotic recombination events could be defined more exactly because the density of useful markers was greater. New discoveries and characterization of single nucleotide polymorphisms (SNPs) through high throughput sequencing now makes it possible to refine this system even further (Sherry *et al.*, 2001). The other opportunity presented by this technology was that it was possible to apply PCR analysis to LOH using tissues which had been tradi-tionally prepared for histopathology. Using this approach, either by scraping the slides containing the tumor and normal tissue, or using laser capture microdis-sectors which are capable of selecting specific small areas within tissues, it was now possible to analyze both large series of tumors available in pathology archives, as well as collections of rare tumors from many different centers.

The combined studies describing LOH paved the way for the era of the recessive oncogene – the 'tumor suppressor gene'. The fact that a single normal copy of a gene could protect a cell from malignant transformation meant that it was loss of function for a critical gene that was needed for tumorigenesis. However, there was clearly another class of cancer gene – the dominantly trans-forming oncogene – which was originally identified from studies of the role of RNA viruses in the transformation of mammalian cells.

7. Oncogenes

Evidence supporting the idea that there were specific genes which could domi-nantly transform normal cells essentially came from studies of the RNA tumor viruses in the 1970s. RNA viruses carry the genes for their own propagation in the host cells but those RNA viruses associated with malignant transformation carry an extra gene, called a viral oncogene (*v-onc*) capable of inducing tumors *in vivo* in animals and in immortalized human cell lines *in vitro* (Varmus, 1984). Surprisingly, *v-onc* DNA sequences were not found to be unique to viruses, but shared broad homology with genes carried in all mammalian cells (Bishop, 1987). These cellular counterparts were termed cellular oncogenes (c-onc) or proto-oncogenes. However, although mammalian cells carry these 'oncogenic' sequences they obviously do not all undergo malignant transformation. Fine structure analysis of the v-oncs of acutely transforming retroviruses showed that they were abnormal in some way. In some cases, the v-oncs were expressed as fusion proteins, containing both proto-oncogene and viral sequences, whereas in other cases the structure of the v-onc differed from that of its cellular counterpart (Varmus, 1984). The role of oncogenes as active instigators of malignant transformation was established by a number of lines of evidence. Infecting immortalized cells in culture with Rous sarcoma virus (RSV) was known to cause malignant transformation and the oncogene in RSV was called v-src. Temperature sensitive variants of RSV occur and their ability to cause transformation can be abolished by culturing infected cells at a non-permissive temperature for v-src expression, suggesting that the oncogene was acting dominantly to induce transformation (Martin, 1970). When immortalized cells in tissue culture were transfected with DNA from malignant human cells, e.g. from bladder tumors, morphological transformation could also be induced (Shih and Weinberg, 1982). Cells from these transformed foci were found to contain DNA sequences that shared homology with the v-onc of the Harvey murine sarcoma virus (H-ras) and Kirsten sarcoma virus (K-ras), both of which are retroviruses causing tumors in animals (Reddy et al., 1982; Tabin et al., 1982). The cellular homologues of the viral genes formed the RAS family of oncogenes and the difference between the transforming oncogene and the normal proto-oncogene was due to a single amino acid substitution (Reddy et al., 1982; Santos et al., 1982; Tabin et al., 1982). The discovery that mutations in the RAS gene could be induced by treatment of cells with carcinogens provided further evidence that oncogenes were directly involved in malignant transformation (Sukumar et al., 1983). Since then, many different oncogenes have been described in human malignancy and their involvement in malignant transformation is now beyond doubt.

While the evidence above is a compelling argument for the role of oncogenes in the malignant process, elucidation of the effects of c-oncs on cell growth and development and the discovery of mechanisms by which c-onc function can become deranged added further weight to the argument. The hallmark of the malignant phenotype is the abnormal growth of cells. Hence, if abnormalities in the structure and function of oncogenes are important in producing the cancer phenotype, it would seem logical that proto-oncogenes are involved in the control of cellular growth and differentiation. The control of cell growth and development is a complex process which is poorly understood. Cells are under the influence of endogenously or exogenously produced growth factors which bind to growth factor receptors on the cell surface. Changes in conformation of the receptor activates a variety of messages, including the membrane-bound tyrosine kinases, and the signals are transduced by cytoplasmic tyrosine kinases. Within the nucleus, nuclear binding proteins and transcriptional regulators further modulate this growth signal. Cellular oncogenes have been associated with each of these stages of signal transduction. The number of dominantly transforming oncogenes has increased over the years and their involvement in cancer development is described throughout the chapters that follow. Oncogene influences are exerted through amplification resulting in overexpression, gene rearrangements resulting in abnormal transcription and altered/acquired gene function and altered expression/stability resulting from intragenic mutations. Induction of morphological transformation in NIH 3T3 cells has become the initial method of choice when establishing whether any particular gene has oncogenic potential or not (Still *et al.*, 1999). In these assays, instead of forming a monolayer in the culture dish, transformed cells lose contact inhibition of growth and pile up, forming foci in the dish (*Figure 8*).

Although oncogenes can often transform cells *in vitro* in NIH 3T3 assays (Lowy *et al.*, 1978) these cells do not, however, represent normal cells since they are already immortalized and serve as convenient assays. In general, increased activity of a single oncogene is not sufficient to induce the malignant transformation of an otherwise 'normal' cell. Normal cells can be transformed by *RAS* but only when immortalized by another oncogene, *MYC* (Land *et al.*, 1983), which has led to the concept that multiple oncogenes work together to transform cells (Weinberg, 1985). As emphasized throughout this volume, oncogenes work in concert with a wide variety of other genetic changes to eventually produce the malignant phenotype, and the ingenious ways that this is achieved is a testament to the power of evolution and natural selection.

8. Somatic cell hybrids

Despite the action of dominantly transforming oncogenes, the concept of cancer as a recessive phenotype, at least in solid tumors, was further supported by somatic cell hybrid studies. It was shown, originally using Sendai virus (Barski *et al.*, 1969), although later using the far more convenient agent, polyethylene glycol (Pontecorvo, 1975), that the membranes of two cells from completely different species can be fused to generate a heterokaryon (Cowell, 1995). Subsequently,

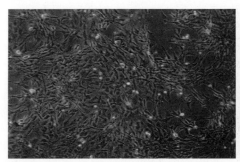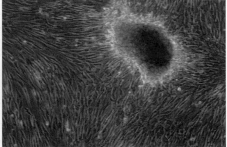

Figure 8. An example of the NIH 3T3 assay for dominantly transforming oncogenes. Cells transfected with an empty vector retain a flat morphology in tissue culture dishes and are contact inhibited at high density (left). When the gene of interested has oncogenic activity (right), following transfection the cells are no longer contact inhibited and continue to grow and pile up on top of a confluent monolayer of cells.

chromosome condensation within these cells, followed by mitosis, generates a hybrid containing the genetic material from both parental cells. The most typical hybrids for human gene analysis have involved human and rodent cells, due largely to the fact that, for as yet still unknown reasons, the human chromosomes are progressively eliminated from these hybrid cells. Eventually, a point is reached where the human chromosome complement in individual hybrids is more or less stable, sometimes retaining only a single human chromosome (*Figure 9*). The application of this simple technology has been extensive. Using whole cell hybrids it was soon shown that, when normal and tumor cells were fused the resulting hybrids were not tumorigenic (Harris *et al.*, 1969), further demonstrating that the malignant phenotype is recessive. From within these suppressed hybrids variants arose which were now tumorigenic and could be shown to have lost additional human chromosomes (Stanbridge, 1990). Careful analysis of the chromosomes which were being lost identified ones which, by their exclusion, allowed the specific cell types used in the original fusion to again become tumorigenic. This type of analysis implicated different chromosomes with suppression of the malignant phenotype in cancers of different cellular origin. This technology gave rise to the microcell mediated chromosome transfer technique. In this case cells are treated with agents that result in the generation of micronuclei containing one or a few human chromosomes (Cowell, 1995). When each chromosome was tagged with a selectable marker it was then possible to fuse the microcells to a tumor cell line and effectively select hybrids which contained the particular chromosome of interest. In this way, it was repeatedly possible to demonstrate more directly that specific chromosomes had the ability to suppress the malignant phenotype of certain cancer cells. By fragmenting the chromosome of interest before the fusion

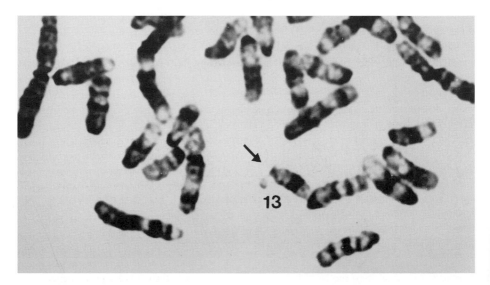

Figure 9. G-banded chromosomes from a mouse–human partial metaphase chromosome spread from cell line PGME. In this hybrid, chromosome 13 is the only human chromosome to be retained (arrow). All of the mouse chromosomes, are characterized by the dark staining telocentric centromere.

(Dowdy *et al.*, 1990) it was possible to define specific regions of that chromosome that carried the tumor suppressor gene as shown, for example in breast cancer cells with chromosome 17 (Plummer *et al.*, 1997). Similarly, chromosome 10 was able to suppress the malignant phenotype in gliomas (Pershouse *et al.*, 1993). Interestingly, no sub-region of chromosome 10 could do this alone suggesting that a set of genes along the length of the chromosome was required. This observation supported LOH data implicating several regions of chromosome 10 in tumorigenicity.

Perhaps an equally important role for somatic cell hybrids in the analysis of cancer was their application to gene mapping. Originally, hybrids which contained overlapping complements of chromosomes were assembled into panel such that analysis of the presence/absence of a gene/marker of interest assigned it to an individual chromosome. Eventually it was possible to build up a panel of somatic cell hybrids (Kelsell *et al.*, 1995) each of which (with minor exceptions) contained only a single, different human chromosome. These panels were very important in assigning large numbers of markers quickly to specific chromosomes (Houlgatte *et al.*, 1995). The challenge was then to sublocalize these markers in order for them to become candidates for particular tumor suppressor genes. This requirement gave rise to panels of somatic cell hybrids (*Figure 10*) which carried deletions and rearrangements of specific chromosomes and allowed fine detailed maps to be constructed (Hawthorn and Cowell 1995; Roberts *et al.*, 1996). Thus, the characterization of large numbers of deletions and translocations involving chromosome 13 resulted in a panel of somatic cell hybrids which allowed the chromosome to be divided into 17 sub-regions and eventually facilitated the integration of the linkage

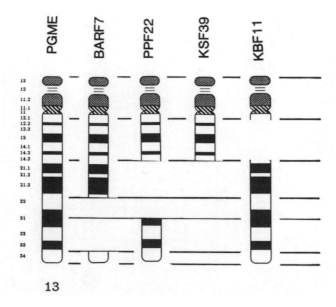

Figure 10. Diagrammatic representation of an early somatic cell hybrid mapping panel for human chromosome 13. In this example the presence of copies of chromosome 13 which carry different and overlapping deletions allow the chromosome to be subdivided into six different regions.

maps and the physical maps (Hawthorn and Cowell, 1995). Although there are many examples of chromosome deletions in somatic cell hybrids contributing to the isolation of cancer genes, this technology usually depended on the availability of hybrids carrying sufficient numbers of overlapping deletions in order to define a small enough region to pursue positional cloning strategies. Chromosome translocations, on the other hand, provided the means to directly identify the gene(s) involved (Roberts *et al.*, 1998).

In many cases chromosome translocations have been identified in patients who are predisposed to a particular tumor type (*Figure 11*). In these patients the assumption is that the chromosome translocation interrupts the critical predisposing gene or a gene in a pathway relevant to the pathogenesis of the tumor. In this case cloning the translocation breakpoint leads directly to the identification of the gene(s) involved. In almost all of the hereditary cancers, and some of the sporadic ones, these rare translocation cases have been instrumental in the isolation of the predisposition gene. Examples include the neurofibromatosis genes, *NF1* and *NF2* (Viskochil *et al.*, 1990) and the *Rb* gene (Mitchell and Cowell,

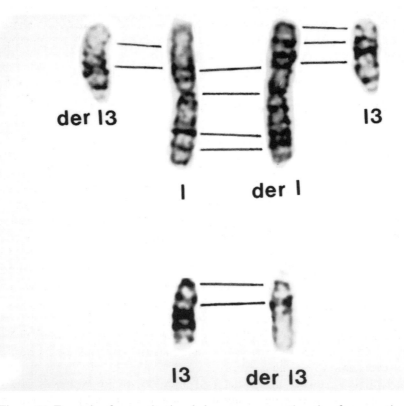

Figure 11. Example of a constitutional chromosome translocation from a patient with retinoblastoma. The G-banded chromosomes reveal a reciprocal exchange between chromosomes 1 and 13 which are lined up appropriately alongside the derivative (der) translocated chromosomes. Molecular analysis subsequently showed that the breakpoint on chromosome 13 occurred within the retinoblastoma predisposition gene (Mitchell and Cowell, 1989).

1989). In sporadic cases, genes implicated in neuroblastoma (Roberts *et al.*, 1998) and familial renal cell carcinoma (Ohta *et al.*, 1996) have also been isolated from patients carrying reciprocal chromosome translocations. These observations have also been extended to ideopathic chromosome translocations from tumors where there is an implication of particular chromosome regions in the development of a specific tumor type. Although, the specific changes seen in leukemias have provided the classic examples such as the 8;13 translocation in myeloproliferative disease (Kempski *et al.*, 1995; Still and Cowell, 1998) and the bcr/abl and myc/Ig rearrangements in AML and CML (Rabbitts, 1994), there are emerging examples of reciprocal translocations to identify critical genes in solid tumors as well. In brain tumors, for example, we have shown that a 10;19 translocation occurs in the T98G cell line (Chernova *et al.*, 1998). The translocation breakpoint in this case lies exactly within a region of frequent LOH in gliomas. Positional cloning of the breakpoint identified the LGI1 gene on chromosome 10q24 which appears to be inactivated during the transition from low grade to high grade tumors (Chernova *et al.*, 1998). The application of FISH to the analysis of these translocation break-points now allows for the rapid pinpointing of the genomic clone which crosses the breakpoint, clearly defining the location of the gene of interest. Subsequent isolation of these chromosome translocations in somatic cell hybrids then allows the exact molecular position of the breakpoint to be defined and so implicate only a single gene.

9. HTF islands and long range maps

Before the advent of the large insert genomic clones, construction of long range physical genome maps was very difficult, and estimating distance between markers nearly impossible. This would change with the observation that certain restriction enzymes were sensitive to methylation at CpG dinucleotides. The best example is the pair of enzymes, MspI and HpaII. Both of these enzymes recognize the CCGG tetranucleotide and cleave DNA at this site. If the internal cytosine is methylated, however, HpaII will not cut the DNA. Thus, it was possible to analyze the methylation status of specific regions in the genome, and in particular the promotors of genes which are generally unmethylated, but where there was some suggestion that methylation could regulate gene expression. This analysis has proved very useful in the analysis of the *P16* gene, for example, which frequently shows inactivation through promotor methylation. Misincorporation of thymidine at the site of a methyl-C during DNA replication is a common event which has resulted in the conversion for mammals into an A-T rich genome throughout evolution. Such changes only become important when they affect a critical nucleotide (resulting in a functional mutation if not repaired) or are involved in the regulation of genes. Since methylation occurs in CpGs, which are frequently seen in relatively high concentrations in the promotors of genes, there has been a strong selection pressure to maintain these regions of high GC. During early experiments with HpaII digestions of genomic DNA, Jack Miller noticed that a subpopulation of tiny fragments were generated in the 200–1000 bp range. These were called HpaII tiny fragments (HTF) and were shown to cluster at the beginnings of genes (Bird, 1986). In fact as other enzymes were identified which also had a high GC content in their recognition sequence, such as NotI, it became clear that the promotors of genes would also be restricted by these into small fragments compared with the rest of the genome, where these enzyme sites were ordinarily hundreds of kilobases apart. The frequent digestion by these so called 'rare-cutter' enzymes within the promotors of genes led to the description of HTF-islands in the genome. It has been estimated that up to 50% of all genes carry HTF islands in their promotors and this was often used to predict the approximate location of genes in large genomic regions and facilitate their cloning (Call *et al.*, 1990).

Accompanying this technology was the ability to separate large fragments of genomic DNA (up to 2 megabases) on agarose gels using the development pioneered by Cantor and colleagues (Olson, 1989; Schwart and Cantor, 1984), called pulse field gel electrophoresis (PFGE). By preparing DNA in agarose plugs to prevent shearing and then digesting them in the same plugs with restriction enzymes, large fragments could be separated when subjected to an alternating current sending the DNA first in one direction and then in another. Many parameters, the interpulse frequency in particular, were important for successful separation, but the net effect was that it was possible to build up long range restriction enzyme digestion maps of particular chromosome regions which allowed good estimates of the distance between probes and HTF islands. This early long range mapping approach proved very important in many early attempts to clone cancer genes and gave rise to the techniques to prepare and analyze the cloning of large pieces of genomic DNA in vectors such as YACs (see below).

10. Construction of physical maps of the genome

Much of the analysis of the genetics of human cancer would not have been possible without the world-wide efforts of numerous laboratories in the compilation of detailed physical and genetic maps for each of the individual chromosomes. Before the advent of technologies to generate libraries of clones from either expressed genes (cDNA libraries) or total genomic DNA, the human gene maps were restricted to phenotypes (particularly on the X chromosome) and mostly included genes where there was an assay for the protein product, such as the esterase D gene on chromosome 13 (Cowell *et al.*, 1987). Somatic cell hybrid mapping panels had been generated in the 1970s and so limited assignments could be made using these panels. The development of Southern blotting (Southern, 1975), incorporation of radioactivity into DNA probes (Feinberg and Vogelstein, 1983) and the widespread availability of enzymes which digested DNA at highly specific recognition sequences, allowed the extension of the maps to include any marker for which there was a probe. This mapping effort tended to cluster genes into regions where the availability of somatic cell hybrids made it possible to subdivide the chromosome (*Figure 9*).

The exploitation of the observation that random single base pair changes which abound throughout the genome could be identified as restriction enzyme site polymorphisms (*Figure 4*) provided the first real opportunities to generate more comprehensive linkage maps for each of the human chromosomes (Botstein *et al.*, 1980). Large cohorts of reference families were collected at the University of Utah and at CEPH in Paris, and these were used to measure recombination frequencies between specific markers. Although physical distances were not accurate in this approach it was very effective in establishing the linear order of markers along the chromosomes and so did not depend on using chromosome breakpoints to define intervals along the chromosome. This is the basis of all linkage studies and, although the discovery of the RFLP was a major breakthrough for genetic research, it was limited by the number of different polymorphisms that could be detected and, because some of the individual alleles would be rare, the often low frequency of heterozygotes in the population. Nonetheless very valuable maps were established for the human chromosomes which provided the framework for subsequent efforts (NIH-CEPH consortium, 1992; Weissenbach *et al.*, 1992).

The development of the polymerase chain reaction (PCR) was perhaps one of the most significant advances for all genetics research (Mullis and Faloona, 1987). The ability to rapidly amplify specific regions of the genome, without cloning (*Figure 12*), for a variety of uses now made it possible to undertake high throughput ventures of mapping and sequencing. The subsequent discovery that thermo-stable polymerases could be isolated from bacteria such as *Thermobacillus*

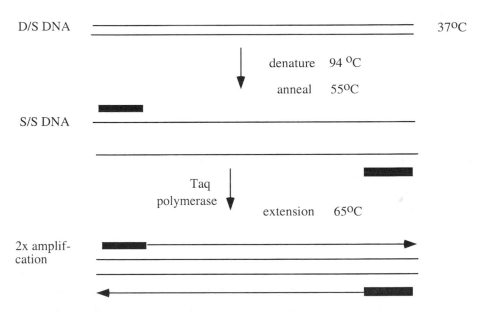

Figure 12. Schematic diagram of the PCR. Double strand DNA is denatured at high temperature in the presence of short (18–25 mer) primers (dark boxes) with a sequence complementary to a known sequence on both strands but in opposite orientations. When the temperature is lowered the primers anneal very specifically to their target sites in the whole genome. This short region of double stranded DNA is recognized by the polymerase and used to prime an extension reaction copying the single stranded template. This cycle generates two double stranded molecules. By repeating this cycle 30–35 times huge amplification of the specific region of interest can be generated in a matter of hours.

aquaticus (Taq) led to a much more efficient automation of the PCR reaction. The applications of PCR have been diverse and only limited by the imagination of the science community. This fact was recognized with the recent award of the Nobel Prize to its inventor, Dr Carey Mullis. For mapping, PCR proved invaluable. It was now possible to assign markers to chromosomes using somatic cell hybrids since human-specific sequences could be selectively amplified in a mouse background. It was also possible to assign markers to the yeast artificial chromosomes (YACs) which were being generated through the work of a number of groups (see below). PCR could also amplify polymorphic regions of the genome for linkage analysis, largely removing the need for Southern blotting. With the observation that small repeats throughout the genome were hypervariable within the population (Jeffreys *et al.*, 1985), the opportunity to create more detailed linkage maps arose. Additionally, microsatellite repeats, which varied in length by 2, 3 or 4 bp were discovered (Wyman and White, 1980) and so different alleles could be resolved on acrylamide gels using radioactive probes or using sequencing technology and fluorescent probes. Because of the high density of these markers throughout the genome maps soon appeared from laboratories of Dr Donnis Keller (1987) and Dr Jean Weissenbach (Gyapay *et al.*, 1994). Throughout the 1990s these maps got increasingly more detailed and replaced many of the earlier maps (Buetow *et al.*, 1994). These markers were first used to create linkage maps of the chromosomes with the anticipation of providing a marker at 1 cM intervals along the length of every chromosome. With a known physical order of markers it was now possible to screen YAC libraries and demonstrate overlapping contigs through the coincident presence of adjacent markers, so providing a crude physical map of the genome. With this physical map it was then possible to incorporate any of the thousands of markers being isolated worldwide onto the framework maps. The availability of these maps made linkage analysis, LOH analysis and positional cloning a more simple procedure than before and this has been largely responsible for the explosion in gene cloning successes over the past 10 years. The production of these maps was also an important stage in the preparation for the sequencing of the human genome.

11. Specialized genomic libraries and positional cloning

As a result of very careful linkage studies in hereditary cancers, cytogenetic abnormalities in constitutional cells and tumor cells and LOH studies in sporadic tumors, the position of cancer genes was being defined to relatively small regions in the genome. Knowing the position of a gene meant that it was then feasible to isolate it based on this information alone and then to confirm it by demonstrating frequent mutation in patients and tumors with the particular disease. This concept led to the era of 'positional cloning' (Collins, 1992) which was the term used to differentiate from functional cloning where the availability of a protein made it possible to use the protein sequence (and hence the mRNA sequence) to identify a cDNA containing the coding sequence. Positional cloning, on the other hand was defined as the identification of a gene based on its chromosomal localization. In this approach, the location of a disease gene was known to within a small (<1 Mbp) region of the genome. If no obvious candidate gene could be identified in this region, cloning in many cases followed the construction of a contiguous overlapping series of DNA clones which spanned the critical region. This contig then provided the substrate in which to use a variety of approaches such as exon trapping (Auch and Reth, 1990; Buckler *et al.*, 1991) or hybrid cDNA capture (Chen-Liu *et al.*, 1995) to identify potential candidate genes. The traditional cloning vectors used to make human genomic DNA libraries, however, were bacteriophage and cosmids where, although improving the size of the inserts considerably over plasmids, still only offered inserts which were 20–50 Kb. This made building up contigs of overlapping clones very time consuming. Clearly, what was needed was a means of generating large insert clones containing megabases of DNA so that chromosome regions could be spanned with as few clones as possible.

The major development in this area came from the suggestion that all that was needed for the creation of an artificial chromosome was a centromere to ensure correct segregation at mitosis and a telomere to ensure completion of replication at the ends of the chromosome. Thus, several groups began to investigate the possibility of creating artificial chromosomes (Burke *et al.*, 1987), the first generation of which were in YACs. In this strategy, large pieces of human DNA could be inserted between the centromeres and telomeres carried in appropriate vectors, which would then be stable in yeast cells. The main technical challenge was to develop a method of keeping mammalian genomic DNA intact enough during its isolation in order to produce these large artificial chromosomes. This complication was not easy to overcome since the solvents commonly used to isolate DNA tended to fragment it. Thus, the early YACs contained inserts which were only between 250 and 500 kb (Anand *et al.*, 1990) which, albeit up to ten times greater than the insert present in cosmids, still posed problems with 'walking' large distance of many megabases. Despite this, given that the positions of genes were being characterized molecularly using somatic cell hybrids and linkage analysis, the position of inherited cancer gene loci was being defined to within 1 cM (generally accepted as 1 Mbp). It was, therefore, becoming possible to construct physical contig maps in YACs which spanned these regions. These contigs, in turn, provided the substrate in which to look for candidate genes. By embedding the genomic DNA in agarose and performing all of the lysis and enzyme manipulations in this medium, random shearing of DNA was reduced and so it became possible to construct YAC libraries with even larger inserts. The next generation of libraries carried inserts of up to 1000 Kb (Larin *et al.*, 1991) and ultimately the final series of libraries made by CEPH had inserts which averaged 1.5–2 megabases. With the availability of these reagents it was becoming possible to build up almost complete contigs of whole chromosomes based on the presence or absence of markers which had come to make up the physical maps of each of the human chromosomes with relatively few gaps. Efforts such as the ones by the Whitehead Institute, Sanger Center and the Universities of Stanford and St Louis, began to create physical contigs in this way and, since the clones were commercially available, it was possible for anyone to obtain their contigs of interest. Where there were gaps in the YAC contigs, merely sequencing the ends of YACs flanking the gaps using a variety of ingenious techniques meant that probes could be generated to screen the YAC libraries for novel overlapping clones (Roberts *et al.*, 1998). The availability of these libraries and the information about the markers they contained led to the construction of contigs of minimal regions between flanking markers of regions known to house cancer genes. This approach was proving particularly valuable to the definition of chromosome translocation breakpoints, since YACs were also ideal for FISH mapping to chromosomes (Baldini *et al.*, 1992). Their size meant that the fluorescence signal obtained was large enough to visualize easily and so the members of the contig which crossed the breakpoints could be identified. At this point however, the size of the YACs became a problem since 2 Mbp of DNA could carry a lot of genes and so a further refinement of the position of the breakpoint was needed. This was achieved in a number of ways exemplified by our attempts to clone genes at a translocation breakpoint present constitutionally in a patient with stage 4S neuoblastoma (Roberts *et al.*, 1998). YAC fragmentation vectors were developed (Lewis *et al.*, 1992) which carry an Alu element and a yeast telomere. By introducing this vector into the yeast cell containing a particular YAC recombination would occur between Alu elements in the YAC and the fragmentation vector. This event would add a new telomere to the end of the truncated YAC whilst still maintaining the centromeric end (*Figure 13*). Thus, the YAC was being fragmented at Alu sites and merely sizing the YACs by PFGE after this procedure gave a series of YACs anchored at one end. These YACs were then used in FISH experiments to establish which crossed the breakpoint and which did not. The relative overlap between these experiments then defined the position of the breakpoint often to within 100 kb or less.

The second development was the creation of artificial chromosome in P1 vectors (PACs) and bacterial artificial chromosomes (BACs). The PACs were on average 90 Kb long and the final generation of BACs were up to 200 Kb long with

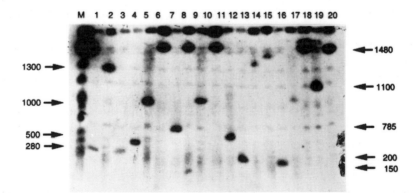

Figure 13. Example of the YAC fragmentation procedure. A 1480 Kb long YAC was fragmented with the pBCL8.1 fragmentation vector. DNA from individual colonies which were identified on selective growth media was then separated using PFGE. It can be seen that the individual colonies carried YACs which were mostly shorter than the original parent YAC. The YACs are discriminated from the yeast host chromosomes by hybridization with a radioactive probe for human repetitive DNA.

an average in the 150 kb size range. BACs (Shizuya *et al.*, 1992) have been the most extensively used and were generally adopted for mapping and eventually sequencing, since they were easy to isolate using standard plasmid lysis procedures (unlike YACs which required preparative gel electrophoresis and only gave very low yields of DNA) and rarely showed rearrangement. BACs could be isolated from commercially available libraries using PCR and primers from the vicinity of the region of interest (e.g. a breakpoint) and then a combination of somatic cell hybrids and FISH would position the breakpoint in one particular BAC. In this positional cloning strategy the position of the gene is reduced from 2 Mbp to 100 Kb or so which has greatly facilitated the cloning of human cancer genes. It was also possible to sequence the ends of BACs directly, unlike YACs which required subcloning of the ends, which made isolating overlapping BACs using PCR relatively easy. The ease of manipulation of BACs made these clones the clones of choice in systematic high throughput sequencing efforts. These positional cloning approaches have been very important in the cloning of many hereditary cancer genes as discussed in the following chapters.

12. cDNAs and ESTs

Complementary to the construction of the large physical maps of the chromosomes was the positioning of large numbers of genes on these maps which would then act as candidates once a specific region of the genome was implicated in a particular cancer. The conversion of RNA into complementary DNA (cDNA) and the subsequent cloning of these cDNAs has been the primary method of sequencing genes, although it has not been without its problems. In order to select the processed genes, cDNA has traditionally been prepared by priming the reverse transcription with an oligo-dT molecule complementary to the poly-A tail of processed mRNA. The competence of the reverse transcription process, however, is such that the resultant cDNA is invariably shorter than the original mRNA, except for the smaller genes (< 2 kb). The result is that, even when a cDNA of interest has been isolated from a cDNA library, considerable effort is still required to be able to reconstruct the full length gene sequence. PCR based methods for the rapid amplification of cDNA ends (RACE), which allows extension of the gene sequence to recover the 5′ end of the gene, has been successful on a gene-by-gene basis (Roberts *et al.*, 1998; Su *et al.*, 1999) although this is a tedious method and cannot usually extend over long distances. Another method employed by some groups was to size select the mRNA using preparative gels before carrying out the cDNA reaction. This approach has resulted in libraries which contain fuller length sequences but, even so, there is often still the requirement to use RACE to find the start of transcription site (Still *et al.*, 1999). Construction of cDNA libraries which were prepared using random primers has generated cDNA clones which contain overlapping fragments of cDNAs, although reconstructing the full-length sequence can still involve multiple rounds of screening using a variety of different libraries. Part of the genome sequencing project designed to investigate only the expressed sequences in the genome involved generating expressed sequence tags (ESTs). This approach involved single pass sequencing from each end of cDNA clones (generating approximately 500 bp of sequence) from tissue specific libraries (Adams *et al.*, 1993). The rationale was that these ESTs would represent the majority of genes expressed in a given tissue. By repeating the same analysis in large numbers of different cDNA libraries it was expected that a tag would be generated for the majority of human genes. These data, representing over 1.8 million sequences, are now deposited in public databases.

13. Mutation analysis

Cloning potential cancer genes was becoming more and more commonplace in the 1990s but the challenge was still to demonstrate their involvement in the development of the malignant phenotype. Whilst it was possible to show altered expression levels in some cells compared with others, with a few exceptions, exactly what this meant was not always clear. The ultimate proof of involvement was to demonstrate function-modifying genetic changes in individual genes in tumors or in patients who developed those tumors. The ultimate proof was to sequence the gene which was relatively straightforward if RNA was available and if the mutation was such that a stable RNA was made. The other approach was to sequence the individual exons of a gene directly from the DNA template after PCR amplification. For genes such as *NF1* and *BRCA2*, however, with over 60 exons, this represented a formidable task not only to establish the exon–intron boundaries (Hogg *et al.*, 1992) but to prepare to sequence many exons for large cohorts of patients using the technologies available at that time. Inevitably, methods of prescreening these exons were developed, and probably the most widely used is the single strand conformation analysis (SSCA) originally described as single strand conformation polymorphism (SSCP). Essentially (*Figure 14*), within small lengths of DNA, variation in the DNA sequence potentially produces a different conformation in each of the DNA strands. This conformation affects the way that the single strand molecules migrate through acrylamide gels (Orita *et al.*, 1989). Thus, as little as a single base pair change can be sufficient to change the conformation (*Figure 15*) and so the result is a band shift on the gel. Although variable, depending on the gene in question, it has been estimated that over 70% of mutations will be detected (Hogg *et al.*, 1993). Thus, depending on the specific conditions used (e.g. low temperature, inclusion of DMSO, etc.) incriminating a given gene in tumorigenesis can be achieved if there are sufficient tumor or patient samples available (Cowell *et al.*, 1994). The advantage of a pre-screening approach is that many exons from many tumors can then be analyzed relatively quickly and sequencing is restricted to those showing band changes. There have been many modifications of this approach and the development of other similar techniques (such as denaturation gradient gel electrophoresis). Most, with the exception of RNAse protection, depend on heteroduplex analysis whereby the melting temperature of homoduplexes and heteroduplexes differs so that sequence changes can be identified (Lerman and Silverstein, 1987; Myers *et al.*, 1987). In combination, these pre-screening approaches have proved very important not only in the demonstration that a particular gene was involved in the development of cancer but also for the clinical

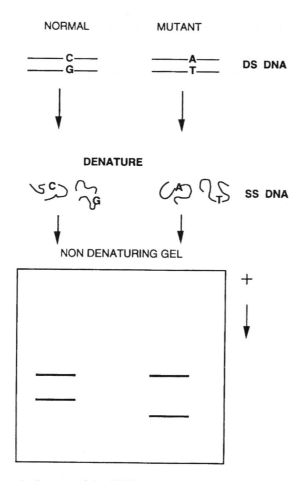

Figure 14. Schematic diagram of the SSCP technique. A comparison between DNA samples which differ by only a single base pair and made radioactive by PCR incorporation of radioactive nucleotides. These DNA molecules are made single stranded by boiling after which they each assume a characteristic secondary conformation. When these single stranded molecules are subjected to acrylamide electrophoresis, their conformation partially determines their electrophoretic mobility. Thus, if a mutation is present, different band patterns are seen when normal and tumor DNA are compared following autoradiography.

management of patients where detection of the predisposing mutation allowed meaningful counseling of the families involved.

14. Summary

From the preceding discussion it will have become clear that the improvement of our understanding of the genetic basis of tumorigenesis has depended on the development of critical technologies and the dedication of the scientific

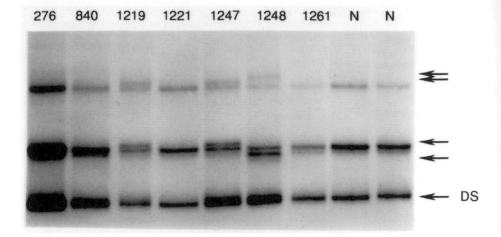

Figure 15. Example of an SSCP gel analyzing exon 9 of the WT1 tumor suppressor gene. PCR products carrying exon 9 from a series of tumors and normal (N) samples were radiolabeled and subjected to acrylamide electrophoresis under conditions to maintain the DNA as single stranded molecules. Although some DNA strands reform duplexes (DS) the positions of the two single strands can be clearly seen. Changes in the banding patterns (arrows) in tumors such as 1247 and 1248 indicate the presence of mutations. In fact, sequencing reveals a very common exon 9 mutation in these patients which have a WT-predisposing condition know as Denys-Drash syndrome (Baird *et al.*, 1992).

community worldwide. These approaches have led not only to the isolation of many genes which are associated with an inherited predisposition to cancer but also to a better understanding of the genetic events that take place in cells during the initiation, progression and metastasis of tumors from many different cell origins. The ingenuity of scientists from many different areas of expertise has come together to lay the foundation for the ultimate goal of identifying all of the genes in the human genome. The challenge now will shift to obtaining a better understanding of how these genes contribute to the malignant transformation of normal cells and the following chapters address this issue.

References

Adams, M.D., Soares, M.B., Kerlavage, A.R., Fields, C. and Venter, J.C. (1993) Rapid cDNA sequencing (expressed sequence tags) from a directionally cloned human infant brain cDNA library. *Nature Genet.* **4**: 373–380.

Anand, R., Riley, J.H., Butler, R., Smith, J.C. and Markham, A.F. (1990) A 3.5 genome equivalent multi access YAC library: Construction, characterisation, screening and storage. *Nucl. Acids Res.* **18**: 1951–1956.

Auch, D. and Reth, M. (1990) Exon trap cloning: using PCR to rapidly detect and clone exons from genomic DNA fragments. *Nucl. Acids Res.* **18**: 6743–6744.

Baird, P.N., Santos, A., Groves, N., Jadresic, L. and Cowell, J.K. (1992) Constitutional mutations in the WT1 gene in patients with Denys-Drash syndrome. *Hum. Molec. Genet.* **1**: 301–305.

Baldini, A., Ross, M., Nizetic, D., Vatcheva, R., Lindsay, E.A., Hehrach, H. and Siniscalco. M. (1992) Chromosomal assignment of human YAC clones by fluorescence in situ hybridisation: use of single-yeast-colony PCR and multiple labeling. *Genomics* **14**: 181–184.

Barski, G., Sorieul, S. and Cornefert, F. (1969) Hybrid type cells in combined cultures of two different mammalian cell strains. *J. Nat. Cancer Inst.* **26**: 1269–1291.

Bird, A.P. (1986) CpG-rich islands and the function of DNA methylation. *Nature* **321**: 209–213.

Bishop, J.M. (1987) The molecular genetics of cancer. *Science* **235**: 305–311.

Botstein, D., White, R.L., Skolnick, M. and Davis, R.W. (1980) Construction of a genetic linkage map in man using restriction fragment length polymorphisms. *Am. J. Hum. Genet.* **32**: 314–331.

Boveri, T. (1914) Zur frage der Entstehung Malinger tumoren. Jena: Fisher.

Buckler, A.J., Chang, D.D., Graw, S.L., Brook, D.J., Haber, D.A., Sharp, P.A. and Houseman, D.E. (1991) Exon amplification: A strategy to isolate mammalian genes based on RNA splicing. *Proc. Natl. Acad. Sci.* **88**: 4005–4009.

Buetow, K.H., Weber, J.L., Ludwigsen, S., Scherpbier-Heddema, T., Duyk, G.M., Sheffield, V.C., Wang, Z. and Murray, J.C. (1994) Integrated human genome wide maps constructed using the CEPH reference panel. *Nature Genet.* **6**: 391–393.

Call, K.M., Glaser, T., Ito, C.Y. *et al.* (1990). Isolation and characterization of a zinc finger polypeptide gene at the human chromosome 11 Wilms' tumour locus. *Cell* **60**: 509–520.

Cassperson, T., Farber, S., Foley, G.E., Kudynowski, J., Modest, E.J., Simonsson, E., Wagh, U. and Zech, L. (1968) Chemical differentiation along metaphase chromosomes. *Exp. Cell Res.* **49**: 219–222.

Cavenee, W.K., Dryja, T.P., Phillips, R.A., Benedict, W.F., Godbout, R., Gallie, B.L., Murphree, A.L., Strong, L.C. and White, R.L. (1983) Expression of recessive alleles by chromosomal mechanisms in retinoblastoma. *Nature* **305**: 779–784.

Cavenee, W.K., Hansen, M.F., Nordenskjold, M., Kock, E., Maumenee, I., Squire, J.A., Phillips, R.A. and Gallie, B.L. (1985) Genetic origin of mutations predisposing to retinoblastoma. *Science* **228**: 501–503.

Chen-Liu, L.W., Huang, B.C., Scalzi, J.M., Hall, B.K., Sims, K.R., Davis, L.M., Siebert, P.D. and Hozier, J.C. (1995) Selection of hybrids by affinity capture (SHAC): a method for the generation of cDNAs enriched in sequences from a specific chromosome region. *Genomics* **30**: 388–392.

Chernova, O., Somerville, R.P.T. and Cowell, J.K. (1998) A novel gene, *LGI1*, from region 10q24, is rearranged and downregulated in malignant brain tumors. *Oncogene* **17**: 2873–2881.

Collins, F.S. (1992) Positional cloning: lets not call it reverse anymore. *Nature Genet.* **1**: 3–6.

Comings, D.E. (1973) A general theory of carcinogenesis. *Proc. Natl Acad. Sci., USA* **70**: 3324–3328.

Connolly, M.J., Payne, R.H., Johnson, G., Gallie, B.L., Allerdice, P.W., Marshall, W.H. and Lawton, R.D. (1983) Familial, ESD-linked, retinoblastoma with reduced penetrance and variable expressivity. *Hum. Genet.* **65**: 122–124.

Cowell, J.K. (1995) Preparation and manipulation of somatic cell hybrids. In: Verma, R.S. and Babu (eds) *Human Chromosomes: Principles and Techniques*, 2nd edn. pp. 31–42.

Cowell, J.K., Jay, M., Rutland, P. and Hungerford, J. (1987) An assessment of the usefulness of electrophoretic variants of esterase-D in the antenatal diagnosis of retinoblastoma in the United Kingdom. *Brit. J. Cancer* **55**: 661–664.

Cowell, J.K., Jaju, R. and Kempski, H. (1994a) Isolation and characterisation of a panel of cosmids which allow unequivocal identification of chromosome deletions involving the RB1 gene using fluorescence in situ hybridisation. *J. Med. Genet.* **31**: 334–337.

Cowell, J.K., Smith, T. and Bia, B. (1994b) Frequent constitutional C—> T mutations in CGA-argenine codons in the RB1 gene produce premature stop codons in patients with bilateral (hereditary) retinoblastoma. *Eur. J. Hum. Genet.* **2**: 281–290.

Donis-Keller, H., Green, P., Helms, C. *et al.* (1987) A genetic linkage map of the human genome. *Cell* **51**: 319–337.

Dowdy, S.F., Scanlon, D.J., Fasching, C.L., Casey, G. and Stanbridge, E.J. (1990) Irradiation microcell-mediated chromosome transfer (XMMCT): the generation of specific chromosomal arm deletions. *Genes Chromosomes Cancer* **2**: 318–327.

Feinberg, A.P. and Vogelstein, B. (1983) A technique for radiolabelling DNA restriction endonuclease fragments to high specific activity. *Anal. Biochem.* **132**: 6–13.

Friend, S.H., Bernards, R., Rogelj, S., Weinberg, R.A., Rapaport, J.M., Albert, D.M. and Dryja, T.P. (1986) A human DNA segment with properties of the gene that predisposes to retinoblastoma and osteosarcoma. *Nature* **323**: 643–646.

Getman, M.E., Houseal, T.W., Miler, G.A., Grundy, P., Cowell, J.K. and Landesc G.M. (1998) Comparative genomic hybridization and its application to Wilms tumorigenesis. *Cytogenet. Cell Genet.* **82**: 284–290.

Godbout, R., Dryja, T.P., Squire, J., Gallie, B.L. and Phillips, R.A. (1983) Somatic inactivation of genes on chromosome 13 is a common event in retinoblastoma. *Nature* **304**: 451–453.

Grundy, R.G., Pritchard, J., Scambler, P. and Cowell, J.K. (1998) Loss of heterozygosity on chromosome 16 in sporadic Wilms tumour. *Br. J. Cancer* **78**: 1181–1187.

Gyapay, G., Morisette, J., Vignal, A. *et al.* (1994) The 1993–94 Genethon human genetic linkage map. *Nature Genetics* **7**: 246–307.

Harris, H., Miller, O.J., Klein, G., Worst, P. and Tachibana, T. (1969) Suppression of malignancy by cell fusion. *Nature* **223**: 363–368.

Hawthorn, L.A. and Cowell, J.K. (1995) Integration of the physical and genetic linkage map for human chromosome 13. *Genomics* **27**: 399–404.

Hogg, A., Onadim, Z., Baird, P.N. and Cowell, J.K. (1992) Detection of heterozygous mutations in the RB1 gene in retinoblastoma patients using single-strand conformation polymorphism analysis and polymerase chain reaction sequencing. *Oncogene* **7**: 1445–1451.

Hogg, A., Bia, B., Onadim, Z. and Cowell, J.K. (1993) Molecular mechanisms of oncogenic mutations in tumours from patients with bilateral and unilateral retinoblastoma. *Proc. Natl Acad. Sci., USA* **90**: 7351–7355.

Houlgatte, R., Mariage-Samson, R., Duprat, S., Tessier, A., Bentolila, S., Lamy, B. and Auffray, C. (1995) The Genexpress Index: a resource for gene discovery and the genic map of the human chromosomes. *Genome Res.* **5**: 272–304.

Hsu, T.C. (1952) Mammalian chromosomes in vitro. 1. The karyotype of man. *J. Hered.* **43**: 167–172.

Jeffreys, A.J., Wilson, V. and Thein, S.L. (1985) Hypervariable minisatellite regions in human DNA. *Nature* **314**: 67–73.

Kallioniemi, A., Kallioniemi, O.P., Sudar, D., Rutovitz, D., Gray, J.W., Waldman, F. and Pinkel, D. (1992) Comparative genomic hybridization for molecular cytogenetic analysis of solid tumours. *Science* **258**: 818–821.

Kallioniemi, A., Kallioniemi, O.P., Piper, J. *et al.* (1994) Detection and mapping of amplified DNA sequences in breast cancer by comparative genomic hybridisation. *Proc. Natl Acad. Sci., USA* **91**: 2156–2160.

Kelsell, D.P., Black, D.M., Bishop, D.T. and Spurr, N.K. (1993) Genetic analysis of the BRCA1 region in a large breast/ovarian family: refinement of the minimal region containing BRCA1. *Hum. Mol. Genet.* **2**: 1823–1828.

Kelsell, D.P., Rooke, L., Warne, D., Bouzyk, M., Cullin, L., Cox, S., West, L., Povey, S. and Spurr, N.K. (1995) Development of a panel of monochromosomal somatic cell hybrids for rapid gene mapping. *Ann. Hum. Genet.* **59**: 233–241.

Kempski, H., McDonald, D., Michalski, A.J., Roberts, T., Goldman, J.M., Cross, N.C.P. and Cowell, J.K. (1995) Localization of the 8;13 translocation breakpoint associated with myeloproliferative disease to a 1.5 Mbp region of chromosome 13. *Genes Chroms. Cancer* **12**: 283–287.

Knudson, A.G. (1971) Mutation and cancer: statistical study of retinoblastoma. *Proc. Natl Acad. Sci., USA* **68**: 820–823.

Knudson, A.G. and Strong, L.C. (1972) Mutation and cancer: neuroblastoma and pheochromocytoma. *Am. J. Hum. Genet.* **24**: 514–532.

Land, H., Parada, L.F. and Weinberg, R.A. (1983) Tumorigenic conversion of primary embryo fibroblasts requires at least two cooperating oncogenes. *Nature* **300**: 596–602.

Landegent, J.E., Jansen in de Wal, N., Dirks, R.W., Baao, F. and van der Ploeg, M. (1987) Use of whole cosmid cloned genomic sequences for chromosomal localization by non-radioactive in situ hybridization. *Hum. Genet.* **77**: 366–370.

Larin, Z., Monaco, T. and Lehrach, H. (1991) Yeast artificial chromosome libraries containing large inserts from mouse and human DNA. *Proc. Natl Acad. Sci., USA* **88**: 412–417.

Lerman, L.S. and Silverstein, K. (1987) Computational simulation of DNA melting and its application to denaturing gradient gel electrophoresis. *Methods Enzymol.* **155**: 482–501.

Lewis, B.C., Shah, N.P., Braun, B.S. and Denny, C.T. (1992) Creation of a yeast artificial chromosome fragmentation vector based on lysine-2. *GATA* **9**: 86–90.

Lele, K.P., Penrose, L.S. and Stallard, H.B. (1963) Chromosomal deletion in a case of retinoblastoma. *Ann. Hum. Genet.* **27**: 171–174.

Lowy, D.R., Rands, E. and Scolnick, E.M. (1978) Helper independent transformation by unintegrated Harvey sarcoma virus DNA. *J. Virol.* **26**: 291–298.

Martin, G.S. (1970) Rous sarcoma virus: a function required for the maintenance of the transformed state. *Nature* **227**: 1021–1023.

Mitchell, C.D. and Cowell, J.K. (1989) Predisposition to retinoblastoma due to a translocation within the 4.7R locus. *Oncogene* **4**: 253–257.

Mitelman, F. (2000) Catalog of chromosome aberrations in cancer. *http://www.wiley.com/products/subject/life/mitelman/*.

Mitelman, F. and Levan, G. (1981) Clustering of aberrations to specific chromosome in human neoplasms. IV. A survey of 1871 cases. *Hereditas* **95**: 79–139.

Mullis, K.B. and Faloona, F.A. (1987) Specific synthesis of DNA in vitro via a polymerase-catalysed chain reaction. *Methods Enzymol.* **155**: 335–350.

Myers, R.M., Maniatis, T. and Lerman, L.S. (1987) Detection and localization of single base changes by denaturing gradient gel electrophoresis. *Methods Enzymol.* **155**: 501–527.

NIH/CEPH Collaborative Mapping Group (1992) A comprehensive genetic linkage map of the Human Genome. *Science* **258**: 67–86.

Nowell, P.C. and Hungerford, D.A. (1960) A minute human chromosome in human granulocytic leukaemia. *Science* **132**: 1497

Ohta, M., Inoue, H., Cotticelli, M. *et al.* (1996) The FHIT gene, spanning the chromosome 3p14.2 fragile site and renal carcinoma-associated t(3;8) breakpoint, is abnormal in digestive tract cancers. *Cell* **84**: 587–597.

Olson, M.V. (1989) Separation of large DNA molecules by pulsed-field gel electrophoresis. A review of the basic phenomenology. *J. Chromatog.* **470**: 377–383.

Orita, M., Iwahana, H., Kanazawa, H., Hayashi, K. and Sekiya, T. (1989) Detection of polymorphisms of human DNA by gel electrophoresis as single-strand conformation polymorphisms. *Proc. Natl Acad. Sci., USA* **86**: 2766–2770.

Ott, J. (1974) Estimation of the recombination fraction in human pedigrees: Efficient computation of the likelihood for human linkage studies. *Am. J. Hum. Genet.* **26**: 588–597.

Pershouse, M.A., Stubblefield, E., Hadi, A., Killary, A.M., Yung, W.K. and Steck, P.A. (1993) Analysis of the functional role of chromosome 10 loss in human glioblastomas. *Cancer Res.* **53**: 5043–5050.

Plummer, S.J., Adams, L., Simmons, J.A. and Casey, G. (1997) Localization of a growth suppressor activity in MCF7 breast cancer cells to chromosome 17q24-q25. *Oncogene* **14**: 2339–2345.

Pinkel, D., Straume, T. and Gray, J.W. (1986) Cytogenetic analysis using quantitative high-sensitivity fluoresence hybridization. *Proc. Natl Acad. Sci., USA* **83**: 2934–2938.

Pinkel, D., Landegent, J., Collins, C., Fuscoe, J., Segraves, R., Lucas, J. and Gray, J. (1988) Fluorescence in situ hybridization with human chromosome-specific libraries; detection of trisomy 21 and translocations of chromosome 4. *Proc. Natl Acad. Sci., USA* **85**: 9138–9142.

Pontecorvo, G. (1975) Production of somatic cell hybrids by means of polyethylene glycol treatment. *Somatic Cell Genet.* **1**: 397–400.

Rabbitts, T.H. (1994) Chromosomal translocations in human cancer. *Nature* **372**: 143–149.

Reddy, E.P., Reynolds, R.K., Santos, E. and Barbacid, M. (1982) A point mutation is responsible for the acquisition of transforming properties by the T24 human bladder carcinoma oncogene. *Nature* **300**: 6447–6458.

Riccardi, V.M., Sujansky, E., Smith, A.C. and Francke, U. (1978) Chromosome imbalance in the aniridia-Wilms' tumour association: 11p interstitial deletion. *Pediatrics* **61**: 604–610.

Roberts, T., Mead, R.S. and Cowell, J.K. (1996) Characterisation of a human chromosome 1 somatic cell hybrid mapping panel and regional assignment of 6 novel STS. *Ann. Human Genet.* **60**: 213–220.

Roberts, T., Chernova, O. and Cowell, J.K. (1998). NB4S, a member of the TBC-1 domain family of genes is truncated as a result of a constitutional t(1;10)(p22;q21) chromosome translocation in a patient with stage 4S neuroblastoma. *Hum Molec. Genet.* **7**: 1169–1178.

Rowley, J.D. (1973) A new consistent chromosome abnormality in chronic myelogenous leukemia identified by quinacrine fluorescence and Giemsa staining. *Nature* **243**: 290–293.

Sandberg, A.A. (1990) *The Chromosomes in Human Cancer and Leukemia*. 2nd edn. New York: Elsevier.

Santos, E., Tronick, S.R., Aaronson, S.A., Pulciani, S. and Barbacid, M. (1982) T24 human bladder carcinoma oncogene is an activated form of the normal human homologue of BALB- and Harvey-MSV transforming genes. *Nature* **298**: 343–347.

Schwartz, D.C. and Cantor, C.R. (1984) Separation of yeast chromosome-sized DNAs by pulsed field gradient gel electrophoresis. *Cell* **37**: 67–75.

Seabright, M. (1971) A rapid banding technique for human chromosomes. *Lancet* **2**: 971–972.

Sherry, S.T., Ward, M.H., Kholodov, M., Baker, J., Phan, L., Smigielski, E.M. and Sirotkin, K. (2001) dbSNP: the NCBI database of genetic variation. *Nucleic Acids Res.* **29**: 308–311.

Shih, C. and Weinberg, R.A. (1982) Isolation of a transforming sequence from a human bladder carcinoma cell line. *Cell* **29**: 161–169.

Shizuya, H., Birren, B., Kim, U.J., Mancino, V., Slepak, T., Tachiiri, Y. and Simon, M. (1992) Cloning and stable maintenance of 300-kilobase-pair fragments of human DNA in *Escherichia coli* using an F-factor-based vector. *Proc. Natl Acad. Sci., USA* **89**: 8794–8797.

Simard, J., Feunteun, J., Lenoir, G. *et al.* (1993) Genetic mapping of the breast-ovarian cancer syndrome to a small interval on chromosome 17q12–21: exclusion of candidate genes EDH17B2 and RARA. *Hum. Mol. Genet.* **2**: 1193–1199.

Solomon, E., Voss, R., Hall, V., Bodmer, W.F., Jass, J.R., Jeffreys, A.J., Lucibello, F.C., Patel, I. and Rider, S.H. (1987) Chromosome 5 allele loss in human colorectal carcinomas. *Nature* **32**: 616–619.

Southern, E. (1975) Detection of specific sequences among DNA fragments separated by gel electrophoresis. *J. Mol. Biol.* **98**: 503.

Stanbridge, E.J. (1990) Human tumor suppressor genes. *Ann. Rev. Genet.* **24**: 615–657.

Still, I.H. and Cowell, J.K. (1998) The t(8;13) atypical myeloproliferative disorder: Further analysis of the ZNF198 gene and lack of evidence for multiple genes disrupted on chromosome 13. *Blood* **92**: 1456–1458.

Still, I.H., Vince, P. and Cowell, J.K. (1999) Identification of a novel gene (ADPRTL1) encoding a poly(ADP-ribosyl) transferase protein. *Genomics* **62**: 533–536.

Su, G., Roberts, T. and Cowell, J.K. (1999) *TTC4*, a novel human gene containing the tetratricopeptide repeat and which maps to the region of chromosome 1p31 which is frequently deleted in sporadic breast cancer. *Genomics* **55**: 157–163.

Sukumar, S., Notario, V., Martin-Zanca, D. and Barbacid, M. (1983) Induction of mammary carcinomas in rats by nitroso-methylurea involves malignant activation of H-ras -1 locus by single point mutations. *Nature* **306**: 658–661.

Tabin, C.J., Bradley, S.M., Barymann, C.I., Weinberg, R.A., Papageorge, A.G., Scdnick, E.M., Dhar, R., Lowy, D.R. and Chang, E.H. (1982) Mechanisms of activation of a human oncogene. *Nature* **300**: 143–149.

Tjio, J.H. and Levan, A. (1956) The chromosome number in man. *Hereditas* **42**: 1–6.

Varmus, H.E. (1984) The molecular genetics of cellular oncogenes. *Ann. Rev. Genet.* **18**: 553–612.

Viskochil, D., Buchberg, A.M., Xu, G. *et al.* (1990) Deletions and a translocation interrupt a cloned gene at the neurofibromatosis type 1 locus. *Cell* **62**: 187–192.

Wadey, R.B., Pal, N.P., Buckle, B., Yeomans, E., Pritchard, J. and Cowell, J.K. (1990). Loss of heterozygosity in Wilms' tumour involves two distinct regions of chromosome 11. *Oncogene* **5**: 901–907.

Weinberg, R.A. (1985) The action of oncogenes in the cytoplasm and nucleus. *Science* **230**: 770–776.

Weissenbach, J., Gyapay, G., Dib, C., Vignal, A., Morissette, J., Millasseau, P., Vaysseix, G. and Lathrop, M. (1992) A second-generation linkage map of the human genome. *Nature* **359**: 794–801.

Wyman, A.R. and White, R. (1980) A highly polymorphic locus in human DNA. *Proc. Natl Acad. Sci., USA* **77**: 6754–6758.

Yandell, D.W., Campbell, T.A., Dayton, S.H., Petersen, R., Walton, D., Little, J.B., McConkie-Rosell, A., Buckley, E.G. and Dryja, T.P. (1989) Oncogenic point mutations in the human retinoblastoma gene: their application to genetic counselling. *New Eng. J. Med.* **321**: 1689–1695.

Yunis, J.J. (1976) High resolution of human chromosomes. *Science* **191**: 1268–1270.

Yunis, J.J. and Ramsay, N. (1978) Retinoblastoma and subband deletion of chromosome 13. *Am. J. Dis. Child.* **132**: 161–163.

2

Defining genetic changes associated with cutaneous malignant melanoma

Mezbah U. Faruque and Jeffrey M. Trent

1. Introduction

Dissection of the genetic contribution of a common complex disease like melanoma is a daunting task. For over 150 years, the medical literature has referred to the hereditary occurrence of melanoma. To date, only two genes have been identified, which are thought to be etiologic in the familial forms of this disease. In addition, genetic changes responsible for melanoma clinical progression in sporadic cases will be critical to further our understanding of disease stages. This brief overview will highlight both genetic changes associated with hereditary cutaneous melanoma, as well as highlight some aspects of the genetic role of disease progression.

2. Development and biology of melanocytes

Melanoma is derived from genetic changes affecting melanocytes which arise from neural crest-derived progenitor cells that migrate from the developing central nervous system into the skin. Melanocytes are homogeneously distributed at the junction between the epidermal and dermal layers, where they synthesize melanin, a brown–black pigment that is distributed to surrounding keratinocytes in the skin by way of dendritic projections (Jimbo et al., 1993; Norris et al., 1998). Mammalian melanocytes synthesize at least three chemically related classes of pigments (Porta, 1988) eumelanins, phenomelanins and trichochromes. Lerch (1981) and Porta (1988) have reviewed the detailed biochemistry of melanin synthesis. Eumelanin is a polyquinoid pigment synthesized from tyrosine. Phenomelanins are polymers containing cysteine and tyrosine, incorporated via the intermediates cysteinyldops and benzothiazines. Trichochromes are dimeric benzothiazine derivatives. All these pigments essentially need tyrosinase for their synthesis (Porta, 1988; Silvers, 1979).

Melanin has a photoprotective function in the skin, directly absorbing ultraviolet photons as well as reactive oxygen species generated by the interaction of ultraviolet photons with membrane lipids and other cellular chromatophores (Pathak, 1995; Riely, 1997). Within cells, melanin tends to be distributed in supranuclear 'caps' that protect the nuclei from injury caused by ultraviolet radiation (Kobayashi et al., 1998). Melanocytes are present in the skin at roughly equal densities in all races (Clark, 1988). In dark-skinned people, each melanocyte produces on average more melanin pigment than does each melanocyte in light-skinned people. Consequently, individuals differ little in the number of precursor cells that can give rise to melanoma.

3. Epidemiology

3.1 Incidence of melanoma

The incidence of cutaneous malignant melanoma (CMM) in Caucasian populations has been rising steadily at a rate of 5% per year (Muir, 1987). In the United States, there were an estimated 34 100 new cases of CMM diagnosed in 1995 (Cancer Facts and Figures, 1995) and 7200 melanoma-related deaths. Furthermore, incidence rates for melanoma in the United States have been increasing more rapidly than for any other except lung cancer (Frey and Hartman, 1991). The lifetime risk of developing melanoma varies in different populations, but is typically 1 in 90 (Rigel et al., 1996; Sober et al., 1991). In England and Wales, approximately 4000 new cases of melanoma are diagnosed each year and there are approximately 1200 deaths (OPCS, 1990). Since the 1930s, the incidence of melanoma has increased nearly 20-fold (Rigel et al., 1996). Moreover, malignant melanoma is one of the most common cancers in young adults (Kosary et al., 1996).

3.2 Risk factors

Sun exposure. Melanoma is a complex disorder with involvement of both host and environmental factors, especially exposure to ultraviolet radiation (Green and Swerdlow, 1989). Epidemiological studies have shown UV radiation as a major cause of CMM in light-pigmented populations and for an increased incidence of this malignancy. Anatomic distribution of melanoma lesions by sex, geographic and racial differences, and migration studies are the basic lines of evidence that support this relationship (Langley and Sober, 1997). In women, the lower legs and the upper back are the most common sites for melanoma, whereas in men, the trunk is the most frequent site. Areas of the body intermittently exposed to sun appear to develop melanoma more frequently than areas of maximal UV exposure (Crombie, 1981; Elwood, 1996; Elwood and Gallagher, 1983; Green et al., 1993; Holman et al., 1980).

Melanoma is nearly three times more common in the southern latitudes of the United States than in the northern United States (Fitzpatrick et al., 1987). The incidence of melanoma among whites is inversely related to latitude of residence (Mack and Floderus, 1991), with the world's highest incidence in Australia (MacLennan et al., 1992), a subtropical country with a largely Celtic population. Conversely, melanomas are uncommon in darker-skinned people; in the United States, the incidence among Blacks is only one-tenth (Parkin et al., 1992) to one-twentieth (Reintgen et al. 1982) of that among Caucasians.

Migration to areas with higher levels of ambient UV radiation has been demonstrated to increase the risk of developing melanoma (Holman et al., 1980, 1984b; Langley and Sober, 1997). Younger immigrants are more prone than adults to develop melanoma. A four-fold increase in the risk of melanoma has been described in immigrants who moved to Western Australia prior to age 10 as compared with those arriving after 15 years of age (Holman and Armstrong, 1984b).

The epidemiological evidence implicating sun exposure in the causation of melanoma is supported by biologic evidence, which suggests that damage caused by UV radiation, particularly damage to DNA, plays a central part in the pathogenesis of these tumors. For example, patients with xeroderma pigmentosum, a family of diseases characterized by grossly deficient repair of DNA photoproducts induced by UV radiation, have a greatly increased risk of melanoma (Cleaver, 1968; Setlow et al., 1969) as compared with the general population. Moreover, melanoma can be induced by exposure to UV radiation in certain animals (Kusewitt et al., 1991; Setlow et al., 1989). Indeed, melanomas were recently induced in human skin that had been grafted onto immunologically tolerant mice by a single exposure to a chemical carcinogen followed by UV radiation (Atillasoy et al., 1998). Much research addresses the relative contribution of ultraviolet B wavelengths (290–320 nm) and ultraviolet A wavelengths (320–400 nm) to photocarcinogenesis, particularly to the development of melanoma (Bentham and Aase, 1996; Langford et al., 1998; Schmitz et al., 1994). UVB radiation is overwhelmingly responsible for the formation of the principal DNA lesions, cyclobutane pyrimidine dimers and pyrimidine (6-4) pyrimidone photoproducts (Freeman et al., 1989; Mitchell et al., 1989, 1991), whose incorrect repair leads to mutations (Kochever et al., 1993; Mitchell et al., 1991; Moriwaki et al., 1996) and induces squamous cell carcinoma in mice (Mitchell et al., 1989, 1991). However, UVA radiation is far more abundant in sunlight than is UVB radiation, and causes oxidative DNA damage that is also potentially mutagenic (Kochever et al., 1993). UV radiation has been demonstrated to induce melanoma in opossums (Kusewitt et al., 1991) and certain fish (Setlow et al., 1989).

Age. Incidence of both melanoma and nonmelanoma skin cancer increases exponentially with age (Kosary et al., 1996; Scotto et al., 1983). Furthermore, although the absolute incidence of skin cancer depends strongly on geographic differences [e.g. the incidence is far higher in New Mexico than in Detroit (Scotto et al., 1983)], the rate of increase with age is independent of the magnitude of risk due to the environmental component (Gilchrest, 1984). This finding may imply that age itself plays a major part in vulnerability to photocarcinogenesis. In particular, there is an age-associated decrease in the capacity to repair DNA (Gad et al., 1998; Wei et al., 1993) and a consequent increase in the rate of mutations in DNA (Langford et al., 1998). Moreover, the removal rate of UV radiation-induced DNA photoproducts from UV irradiated skin decreases with age, particularly during the first two decades of life (Gad et al., 1998).

Precursor lesions. A number of precursor lesions may eventually turn into or increase the risk of malignant melanoma. These include dysplastic nevi, benign nevi, lentigo maligna and congenital nevi.

Clark *et al.* (1978) reported on two families with a high incidence of melanoma and a large number of atypical moles. These nevi, called dysplastic or atypical nevi, are now largely recognized as precursor lesions to melanoma, and as markers for increased melanoma risk (Halpern *et al.*, 1991, 1993; Kruger *et al.*, 1992; Rigel *et al.*, 1989; Tucker *et al.*, 1998).

Increased number of benign nevi has also been observed to increase the risk of developing melanoma (Tucker *et al.*, 1998). It has been reported that individuals with more than 100 common nevi have an odds ratio of 7.6 for developing melanoma (Skender-Kalnenas *et al.*, 1995).

Lentigo maligna, a preinvasive melanoma is considered to be a definite precursor to melanoma (Cohen, 1995). Studies estimate a ten-fold risk of melanoma development in individuals with lentigo maligna before the age of 75 (Weinstock and Sober, 1987).

Giant congenital nevi have a 6–7% lifetime risk of malignant transformation (Kaplan, 1974). Although the exact estimated risk of melanoma development is not known, small to medium sized congenital nevi have also been suggested to be precursor lesions (Kopf *et al.*, 1979).

Cutaneous phenotype and other risk factors. Individuals with blond or red hair, blue or green eyes, many freckles, pale or light skin and an inability to tan have an increased risk of melanoma (Cristofolini *et al.*, 1987; Elwood *et al.*, 1984; Goldstein and Tucker, 1995; Rhodes *et al.*, 1987). Individuals with a history of basal cell or squamous cell carcinoma are at increased risk of developing melanoma (Marghoob *et al.*, 1995).

Hereditary melanoma. The earliest report of melanoma and possible involvement of a hereditary component, may be that by Norris (Norris, 1820). In describing a case of malignant melanoma, W. Norris wrote: 'It is remarkable that this gentleman's father, about thirty years ago, died of a similar disease. A surgeon of this town attended him (the father), and he (the surgeon) informed me that a number of small tumors appeared between the shoulders. ... This tumor..., originated in a mole (in his patient) and it is worth mentioning, that not only my patient and his children had many moles on various parts of their bodies, but also his own father and brothers had many of them. The youngest son had one of these marks exactly in the same place where the disease in his father first manifested itself'. Norris concluded 'These facts, together with a case that has come under my notice, rather similar, would incline me to believe that this disease is hereditary.'

The influence of heredity on melanoma is clearly evident from the clinical observation that cases of melanoma tend to cluster in families (Anderson, 1971; Bergman *et al.*, 1986; Cawley, 1952; Clark *et al.*, 1978; Lynch and Krush, 1968; Lynch *et al.*, 1978, 1980; Reimer *et al.*, 1978). Current estimates indicate that 5–10% of all melanoma cases may have a genetic basis (Cawley, 1952; Greene and Fraumeni, 1979). Increased risks of 2.0 for first-degree relatives of melanoma cases (Wallace *et al.*, 1971) and 3.0 for relatives of one or more cases have been reported (Holman and Armstrong, 1984). Analysis of the Utah population database

(Skolnick, 1980) indicates a risk of 6.5 (Goldgar *et al.*, 1994) for relatives of melanoma cases diagnosed before age 50 and 13.9 for relatives of two melanoma cases in the same family diagnosed at any age. Another large study using the Utah genealogy database also demonstrated familial aggregation of melanoma (Bishop and Skolnick, 1984), showing the fourth highest level of familial aggregation (after ovary, prostate and lip cancer). Genetic susceptibility to melanoma has also been observed in several family cancer syndromes. For example, melanoma accounts for approximately 7% of second malignancies in familial retinoblastoma patients (Traboulsi *et al.*, 1988). An increased number of melanoma cases has been observed in Li-Fraumeni families (Little *et al.*, 1987) and, although less well established, in families with the Lynch type II family cancer syndrome (Lynch *et al.*, 1988). Familial clustering of melanoma, breast and pancreatic cancer has also been reported (Borg *et al.*, 2000; Hemminki and Vaittinen, 1997; Lynch *et al.*, 1975).

4. Clinical characteristics and pathology

Melanoma can be classified into four distinct subtypes based upon clinical and histological features: superficial spreading melanoma, lentigo maligna melanoma, acral lentigious melanoma and nodular melanoma (Clark *et al.*, 1986; Johnson *et al.*, 1995; Langley *et al.*, 1999). Lentigo maligna melanoma and superficial spreading melanoma may exist for several years in a preinvasive state. Superficial spreading melanoma is the most common form, which accounts for approximately 70–80% of all reported cases. It is observed in both familial and sporadic melanomas (Elder and Murphy, 1991; Fountain *et al.*, 1990; Haluska and Housman, 1995; Kraehn *et al.*, 1995) and is the most common type in light-skinned populations (Mihm *et al.*, 1971). This type of melanoma is clinically defined by the presence of putative precursor lesions, common acquired nevi (CAN), dysplastic nevi (DN) or congenital nevi (CN) associated with primary melanoma lesions, radial growth phase (RGP) and vertical growth phase (VGP). Lentigo maligna melanomas grow much more slowly and remain in the RGP stage for much longer, rarely proceeding to the VGP stage. Acral lentigious form remains mainly in indolent RGP lesions. Finally, nodular melanoma is the least common but most aggressive form of this disease. The survival of melanoma patients correlates strongly with the thickness of the primary lesion and dermal invasion. If it is clinically localized, the 5-year survival rate is 85% (Fitzpatrick *et al.*, 1987). The transition from normal melanocytes to metastatic melanoma occurs in a series of defined stages. Distant metastasis is almost invariably fatal. The most common metastatic sites are brain, bone, lung and liver.

5. Genetics of melanoma

Genetic analysis of melanoma has largely focused on two aspects: (1) somatic genetic alterations that occur in tumors revealed by cytogenetic and loss of heterozygosity (LOH) studies; (2) germline susceptibility genes identified by family based studies and genetic linkage analysis. Cytogenetic and LOH studies are used mainly for sporadic melanoma tumors to identify loci involved in the evolution of melanoma (Cowan *et al.*, 1988; Dracopoli *et al.*, 1987; Fountain *et al.*, 1990, 1992; Millikin *et al.*, 1991; Pederson and Wang, 1989; Petty *et al.*, 1993a). Linkage studies use genetic epidemiologic techniques to examine familial melanoma kindreds.

5.1 Somatic genetic alterations of melanoma

Melanomas, like all tumors, progressively accumulate DNA abnormalities including chromosomal losses, duplication, translocations and deletions, as they progress (Fountain *et al.*, 1990; Healy *et al.*, 1995; Su and Trent, 1995). In addition to gross cytogenetic abnormalities, subtle somatic changes, such as microsatellite variability and point mutations are also observed in melanoma (Gruis *et al.*, 1995b; Peris *et al.*, 1995; Quinn *et al.*, 1995; Walker *et al.*, 1994). Cytogenetic alterations frequently pinpoint the location of genes involved in the malignant phenotype. Detailed cytogenetic studies of malignant melanoma have identified a number of recurring chromosomal aberrations. The most frequently observed changes in sporadic melanomas involve chromosomes 1, 6 and 7 followed by chromosomes 9, 2, 11 and 10 (Fountain *et al.*, 1990; Thompson *et al.*, 1995). Chromosomal abnormalities including 9p LOH have been observed as common features in primary melanoma tumors (Kamb and Herlyn, 1998). Allelic loss may be the basis for progression models of melanoma tumorigenesis (Walker *et al.*, 1995). Loss of chromosomal material from 9p has been observed as a common somatic change in melanomas (Fountain *et al.*, 1992; Healy *et al.*, 1995; Olopade *et al.*, 1990). This change has been suggested to be a relatively early event in melanoma development, occurring before the primary lesion matures (Healy *et al.*, 1995). However, more recent studies have demonstrated that a large proportion of the 9p abnormalities involves homozygous deletions of 9p21 in advanced malignancies (Weaver-Feldhaus *et al.*, 1994), suggesting a late event (Reed *et al.*, 1995). Homozygous deletion on 3p involving the *FHIT* gene has also been observed, which is suggestive of a tumor suppressor locus in this region (Sozzi *et al.*, 1996). Although point mutations of the *p53* tumor-suppressor gene on 17p are relatively uncommon, they may account for a portion of 17p LOH (Levin *et al.*, 1995; Volkenandt *et al.*, 1991). Abnormalities on 1p, 3q, and 17q have been described in melanoma cell lines and metastases (Anderson *et al.*, 1993; Balban *et al.*, 1986; Cowan *et al.*, 1988; Dracopoli *et al.*, 1987; Horsman and White, 1993). Chromosome 17 contains the metastasis-suppressor gene *NM23* and the *NF1* tumor-suppressor gene, found to have mutations in some melanoma cell lines (Anderson *et al.*, 1993; Johnson *et al.*, 1993; Weich *et al.*, 1994). The terminal region of the long arm of chromosome 10 has been proposed to harbor one or more genes involved in the early stage of melanocytic neoplasia (Parmiter *et al.*, 1988). LOH studies in melanoma have shown a high frequency of loss of 10q (Healy *et al.*, 1995, 1996; Herbst *et al.*, 1994; Parmiter *et al.*, 1988).

Chromosome 6 is second only to chromosome 1 in its frequency of rearrangements in human cutaneous malignant melanoma (Thompson *et al.*, 1995). Two out of every three melanoma karyotypes have shown abnormalities of chromosome 6 (Fountain *et al.*, 1990; Thompson *et al.*, 1995). Preferential loss of the 6q region was described in one subset of melanoma cases (Trent *et al.*, 1989). The existence of a tumor suppressor gene on chromosome 6 was further substantiated by a microcell-mediated chromosome transfer, where a normal copy of chromosome 6 was introduced into the highly tumorigenic human melanoma cell line UACC 903, resulting in suppression of the tumorigenic phenotype (Trent *et al.*, 1990). Introduction of chromosome 6 has also been shown to suppress melanoma metastasis (Welch *et al.*, 1994; You *et al.*, 1995). The results of LOH, CGH (comparative genomic hybridization), cytogenetic alterations and microcell-transfer all support an important role for chromosome 6 in melanoma development and progression. However, a gene associated with melanoma on chromosome 6 has not yet been identified.

5.2 Linkage analysis

The purpose of linkage analysis is to evaluate the cosegregation of alleles at a particular locus with disease within families; that is, one examines the tendency for alleles, which are close to each other on the same chromosome to be transmitted together. Genetic linkage analysis requires identification and collection of individuals with histopathologically verified melanoma from melanoma-prone families and typing polymorphic markers at a chosen interval throughout the genome to identify the chromosomal locations of genes predisposing individuals to melanoma. The linkage of a putative disease locus and a marker locus is evaluated by measuring the number of recombination events between them with the recombination fraction (θ), which is the probability of an exchange of genetic material, i.e., cross-over between two loci. Linkage is often measured in terms of a LOD score, defined as the $\log_{10}$ of the ratio of two likelihoods: the (relative) probability of observing a given sibship at some particular recombination value relative to the probability of the family given no linkage.

5.3 Melanoma susceptibility loci

Chromosome 1. In 1983, melanoma and dysplastic nevus syndrome were described to be linked to the Rh (Rhesus) blood group locus on the short arm of chromosome 1 (1p34–36) by analyzing 14 melanoma kindreds (θmax = 0.3, Zmax = 1.56; Greene *et al.*, 1983), a region also demonstrating LOH in melanomas (Dracopoli *et al.*, 1987). Bale *et al.* refined the linkage analysis using six melanoma/dysplastic nevi kindreds (Bale *et al.*, 1989). Their data showed linkage to the pronatiodilatin locus (PND) on 1p36 with a maximum LOD score of 3.09 at a recombination fraction of 0.08 and to an anonymous DNA segment, D1S47, with a LOD score of 3.62 at a recombination fraction of 0.107. Subsequently, several laboratories reported evidence against linkage to 1p region for CMM alone, or in families with the combined trait CM/DNS (dysplastic nevus syndrome; Cannon-Albright *et al.*, 1990; Gruis *et al.*, 1990; Kefford *et al.*, 1991; Nancarrow *et al.*, 1992; Van Haeringen *et al.*, 1989). Reanalysis of the updated data on the CMM/DN trait in the families of the Bale model (Bale *et al.*, 1989) by Goldstein *et al.* (1993) found a similar LOD score for D1S47 and a somewhat decreased score for PND. However, when the families for melanoma alone were analyzed, the maximum

LOD score for D1S47 decreased to 1.38 at a recombination fraction of 0.2 and for PND was 0.09 at a recombination fraction of 0.3. Seven additional unpublished NCI families were included in the same study for 1p linkage and gave a LOD score for linkage to D1S47 of 1.99 at a recombination fraction of 0.05 for melanoma alone, and to PND of 0.06 at a recombination fraction of 0.4. When the Bale model (Bale *et al.*, 1989) and a combined CMM/DNS trait was used, the LOD scores were negative for both loci at all recombination values. The results may suggest a locus for CMM or CMM/DN on 1p36 in a subset of the families, indicating genetic locus heterogeneity for this disease.

Chromosome 9. Several independent studies suggested another region on the short arm of chromosome 9 (9p) involved in familial melanoma. Cytogenetically detectable loss or rearrangement of 9p was found in approximately 46% of all melanomas, and 9p rearrangements were identified in both dysplastic nevi and primary lesions, suggestive of a 9p locus involvement in the early events in melanoma development (Cowan *et al.*, 1988). LOH and tumor progression studies were also supportive of 9p as a melanoma candidate region (Cowan *et al.*, 1988; Dracopoli *et al.*, 1987; Fountain *et al.*, 1992; Olopade *et al.*, 1990; Pederson and Wang, 1989; Petty *et al.*, 1993a). About 86% of metastatic melanoma tumors and cell lines were found to have a hemizygous deletion and 10% of melanoma cell lines showed a homozygous deletion on 9p21, a defined region of 2–3 Mb, indicating the presence of a tumor suppressor gene within this interval (Fountain *et al.*, 1992). A 34-year-old Caucasian woman with multiple atypical moles and eight primary cutaneous melanomas was described to have *de novo* germline cytogenetic rearrangement involving 5p and 9p (Petty *et al.*, 1993b). Detailed molecular studies of this germline cytogenetic abnormality showed loss of material from the 9p21 region. Two genetic markers mapping to the critical interval, D9S126 and IFNα were used in comparison studies between the patient's genotype and her unaffected parents. Homozygosity was found at both loci by gene dosage studies in the affected, suggesting that germline loss of the putative gene predisposed to melanoma (Petty *et al.*, 1993b). This feature is in accordance to Knudson's two-hit hypothesis that was based originally on studies of the retinoblastoma (*RB1*) tumor-suppressor gene (Knudson, 1971).

Meanwhile, a linkage analysis was conducted on 10 Utah kindreds and one Texas kindred with multiple cases of invasive melanoma (Cannon-Albright *et al.*, 1992). Strong evidence of linkage was observed for a familial melanoma locus (MLM) to the chromosomal region 9p13–9p22 near IFNα and D9S126. For IFNα, $Z_{max} = 10.57$ at $\theta = 0.01$ and for D9S126, $Z_{max} = 6.12$ at $\theta = 0.01$ were observed. A combined CMM/DN trait was not examined and no evidence for heterogeneity was observed. Another study of invasive melanoma in 26 Australian kindreds supported 9p-linkage (Nancarrow *et al.*, 1993). The peak LOD score was 4.43, 15 cM centromeric to D9S126, although a LOD score of 4.13 was also found 15 cM telomeric of *IFNA*. Gruis *et al.* (1993) examined evidence for linkage to a panel of 9p markers using seven Dutch FAMMM (familial atypical multiple mole-melanoma) kindreds. Suggestive evidence for linkage with two markers was observed, showing a maximum LOD score greater than 2.2. With a somewhat different model in which the phenotype was defined as melanoma and/or >10 atypical nevi, strong linkage to 9p marker D9S171 with $Z_{max} = 3.64$ at $\theta = 0.06$

was observed. Examination of the relationship between CMM/DN and 9p in 13 kindreds previously tested for linkage to 1p (Goldstein *et al.*, 1993) showed significant evidence for linkage of both CMM alone, (Zmax = 4.36 at θ = 0.10) and for CMM/DN(Zmax = 3.05 at θ = 0.20; Goldstein *et al.*, 1994). In addition, a homogeneity test allowing for linkage to 9p, 1p or neither region indicated significant evidence of heterogeneity. Heterogeneity analysis estimated the proportion of families linked to both loci to be 50% each. Two out of 11 families for which data were presented showed conditional probabilities greater than 90% of being linked to 1p. Three of the original families were found to show conditional probabilities greater than 90% for 9p-linkage (Goldstein *et al.*, 1994). These families and an additional 6 CMM/DN families showed evidence of linkage to IFNA for both CMM alone (Zmax = 5.27 at θ = 0.10), and CMM/DN (Zmax = 3.83 at θ = 0.20) (Hussussian *et al.*, 1994). Further analysis placed MLM in a 2cM region flanked by D9S736 and D9S171 (Cannon-Albright *et al.*, 1994). Homozygous deletions of the 9p21 region were found in 56% of 84 melanoma cell lines, and a putative tumor suppressor gene was localized to a region of less than 40 kb that lies centromeric to the α-interferon gene cluster (Weaver-Feldhaus *et al.*, 1994).

6. Genes and melanoma

6.1 p16

Initial attempts to identify MLM were made by using melanoma cell lines. Refinement of the region was obtained by mapping deletion breakpoint in about 100 melanoma cell lines (Weaver-Feldhaus *et al.*, 1994). About 60% of them with a detectable homozygous deletion clustered around a small region on 9p21. This chromosomal region contained two genes, one encoding cyclin dependent kinase (CDK) inhibitor p16 also designated as *P16INK4A*, *MTS1* and *CDKN2A*, and a second gene that was subsequently shown to encode the related CDK inhibitor p15 (Hannon and Beach, 1994; Jen *et al.*, 1994; Kamb *et al.*, 1994a; Nobori *et al.*, 1994; Serrano *et al.*, 1993). CDKN2A was identified by its ability to bind and inhibit cyclin dependent kinases (CDKs) *in vitro* (Serrano *et al.*, 1993). CDKs along with their positive regulatory factors, the cyclins, are principal determinants of the decision to initiate DNA replication and mitosis (Sher, 1993). By inhibiting the phosphorylation of the retinoblastoma protein by CDK4 and CDK6, p16 induces a G_1 cell cycle arrest (Stott *et al.*, 1998). Mutational analysis on the cell lines pointed to *CDKN2A* as the MLM locus (Kamb *et al.*, 1994a). Melanoma cell lines that carried at least one copy of *CDKN2A* frequently carried nonsense, missense and frameshift mutations. Inactivating point mutations were found in the *p16* coding sequence but not in *p15* in cell lines and tumors (Kamb *et al.*, 1994a; Liu *et al.*, 1995; Stone *et al.*, 1995). In no case was homozygous deletion of *p15* observed with *p16* left intact (Kamb *et al.*, 1994). In contrast, there were several examples of *p16* deletions with *p15* left intact.

In the next step, 9p21 linked families were studied for *p16* involvement. *p16* sequence variants were found in many, but not all of these kindreds (Borg *et al.*, 1996; Gruis *et al.*, 1995; Holland *et al.*, 1995; Hussussian *et al.*, 1994; Kamb *et al.*, 1994c; Liu *et al.*, 1995). Segregation of *p16* germline mutations in 9p21-linked kindreds was found consistent with the pattern of codominant Mendelian inheritance (Hussussian *et al.*, 1994). Analysis of *CDKN2A* coding sequences in 15 Dutch familial atypical multiple mole-melanoma (FAMMM) syndrome pedigrees identified a 19 bp germline deletion in 13 of them, all of them originating from an endogamous population (Gruis *et al.*, 1995a). The deletion caused a reading frame shift, predicted to result in a severely truncated p16 protein. Other studies on different melanoma kindreds have also reported mutations in the *CDKN2A* gene (Borg *et al.*, 1996; Ciotti *et al.*, 1996, 2000; Harland *et al.*, 1997; Koh *et al.*, 1995; Liu *et al.*, 1995; Monzon *et al.*, 1998; Ranade *et al.*, 1995; Soufir *et al.*, 1998; Yang *et al.*, 1995).

6.2 CDK4

The *CDK4* gene, located on the long arm of chromosome 12 (12q13) encodes the protein CDK, to which p16 binds. Complexes formed by CDK4 and D-type cyclins are involved in the control of cell proliferation. *CDKN2* mutations in which functional ability of p16 to bind to CDK4 is affected, have been demonstrated by several groups (Lilischkis *et al.*, 1996; Liu *et al.*, 1995). A mutated *CDK4* has been described as a tumor-specific antigen recognized by cytolytic T lymphocytes in a human melanoma (Wolfel *et al.*, 1995). This somatic mutation prevented binding of the CDK4 to *p16*. p16 inhibits phosphorylation of retinoblastoma protein by the G1 cyclin-dependent kinases CDK4 and CDK6, thereby negatively regulating progression through G1 into S phase of the cell cycle. A proportion of melanoma-prone families that are not linked to 9p21 and without any *P16* mutation were found to segregate germline mutations in *CDK4* (Zuo *et al.*, 1996). The authors proposed that the germline Arg24Cys mutation in *CDK4* generated a dominant oncogene resistant to normal physiological inhibition by p16. Another germline mutation of *CDK4* has also been reported (Soufir *et al.*, 1998).

6.3 p15

Kindreds with strong evidence for 9p linkage (Hussussian *et al.*, 1994; Kamb *et al.*, 1994b) but without *p16* mutations, raised the possibility of *p15* as the second melanoma susceptibility gene on 9p21. p15 expression is induced approximately 30-fold in human keratinocytes by treatment with TGF-beta, suggesting that p15 may act as an effector of TGF-beta-mediated cell cycle arrest (Hannon and Beach, 1994). Biochemical behavior of the p15 protein is nearly identical to that of p16 and the two proteins show sequence homology of about 77% (Hannon and Beach, 1994; Jen *et al.*, 1994). Physically, they are about 20 kb apart, and *p15* is thought to be derived from *p16* by gene duplication, divergence and, in human lineage, by a gene conversion event (Jiang *et al.*, 1995). No germ-line mutation has yet been observed in *p15*, although ectopic expression was described to inhibit the growth of tumor-derived cell lines (Stone *et al.*, 1995b). Homozygous deletions as well as mutations of *p16/p15* genes have been observed in primary as well as in metastatic sporadic melanomas (Matsumura *et al.*, 1998).

6.4 PTEN

PTEN, a putative tumor suppressor gene also known as *MMAC1*, was recently cloned and mapped to the long arm of chromosome 10 (10q23.3) (Li *et al.*, 1997; Steck *et al.*, 1997). The product of this gene is a phosphatase that shares structural homology with the protein tyrosine phosphatase family, yet exhibits strong activity toward phosphatidylinositol 3,4,5-triphosphate (PtdIns(3,4,5)P3; Furnari *et al.*, 1998; Maehama and Dixon, 1998; Myers *et al.*, 1998). Protein kinase B (PKB/Akt), one of the direct targets of PtdIns(3,4,5)P3, is constitutively activated by phosphorylation in cells lacking functional PTEN/MMAC1, and the elevated levels of activation can be reduced to normal levels by expression of wild-type PTEN/MMAC1 (Haas-Kogan *et al.*, 1998; Stambolic *et al.*, 1998). These data suggest that the tumor-suppressive properties of PTEN/MMAC1 relate at least in part to its ability to regulate negatively the PtdIns(3,4,5)P3/PKB/Akt signaling pathway. High frequency of deletions and mutations of *PTEN/MMAC1* in some sporadic cancers, in particular glioblastomas, prostate cancers, and endometrial cancers has been reported (Cairns *et al.*, 1997; Kong *et al.*, 1997; Li *et al.*, 1997; Risinger *et al.*, 1997; Steck *et al.*, 1997; Vlietstra *et al.*, 1998; Wang *et al.*, 1997). Deletion or mutation of *PTEN/MMAC1* has been observed in 30–40% of metastatic melanoma cell lines (Guldberg *et al.*, 1997; Robertson *et al.*, 1998; Teng *et al.*, 1997; Tsao *et al.*, 1998). Mutations and allelic loss of *PTEN/MMAC1* in a panel of uncultured primary and metastatic melanomas has also been described, suggesting involvement of this gene in development and progression of melanoma (Birck *et al.*, 2000). An epigenetic mechanism leading to biallelic functional inactivation of PTEN in melanoma tumorigenesis has also been proposed (Zhou *et al.*, 2000).

References

Andersen, L.B., Fountain, J.W., Gutman, D.H., Tarle, S.A., Glover, T.W., Dracopoli, N.C., Housman, D.E. and Collins, F.S. (1993) Mutations in the neurofibromatosis 1 gene in sporadic melanoma cell lines. *Nat. Genet.* **3**: 118.

Anderson, D.E. (1971) Clinical characteristics of the genetic variety of cutaneous malignant melanoma in man. *Cancer* **28**: 721–725.

Atillasoy, E.S., Seykora, J.T., Soballe, P.W. *et al.* (1998) UVB induces atypical melanocytic lesions and melanoma in human skin. *Am. J. Pathol.* **152**: 1179–1186.

Balban, G.B., Herlyn, M., Clark, W.H. Jr. and Nowell, P.C. (1986) Karyotypic evolution in human malignant melanoma. *Cancer Genet. Cytogenet.* **19**: 113.

Bale, S.J., Dracopoli, N.C., Tucker, M.A. *et al.* (1989) Mapping the gene for hereditary cutaneous malignant melanoma-dysplastic nevus to chromosome 1p. *N. Engl. J. Med.* **320**: 1367–1372.

Bentham, G. and Aase, A. (1996) Incidence of malignant melanoma of the skin in Norway, 1955–1989: association with solar ultraviolet radiation, income and holidays abroad. *Int. J. Epidemiol.* **25**: 1132–1138.

Bergman, W., Palan, A. and Went, L.N. (1986) Clinical and genetic studies in six Dutch kindreds with the dysplastic naevus syndrome. *Ann. Hum. Genet.* **50**: 249–258.

Birck, A., Ahrenkiel, V., Zeuthen, J., Hou-Jensen, K. and Guldberg, P. (2000) Mutation and allelic loss of the PTEN/MMAC1 gene in primary and metastatic melanoma biopsies. *J. Invest. Dermatol.* **114**(2): 277–280.

Bishop, D.T. and Skolnick, M.H. (1984) Genetic epidemiology of cancer in Utah geneologies: a prelude to the molecular genetics of common cancer. In: T.W. Mack, I. Tannock (eds). *Cellular and Molecular Biology of Neoplasia. J. Cell Physiol.* (Suppl 3); 63–77.

Borg, A., Johannsson, U., Johannsson, O., Hakansson, S., Westerdahl, J., Masback, A., Olsson, H. and Ingvar, C. (1996) Novel germline p16 mutation in familial malignant melanoma in southern Sweden. *Cancer Res.* **56**: 2497.

Borg, A., Sandberg, T., Nilsson, K., Johannsson, O., Klinker, M., Masback, A., Westerdahl, J., Olsson, H. and Ingvar, C. (2000) High frequency of multiple melanomas and breast and pancreas carcinomas in CDKN2A mutation-positive melanoma families. *J. Natl. Cancer Inst.* **92(15)**: 1260–1266.

Cairns, P., Okami, K., Halachmi, S., Halachmi, N., Esteller, M., Herman, J.G., Jen, J., Isaacs, W.B., Bova, G.S. and Sidransky, D. (1997) Frequent inactivation of PTEN/MMAC1 in primary prostate cancer. *Cancer Res.* **57**: 4997–5000.

Cancer Facts and Figures–1995. American Cancer Society, Inc, Atlanta, GA, 1995.

Cannon-Albright, L.A., Goldgar, D.E., Wright, E.C. *et al.* (1990) Evidence against the reported linkage of the cutaneous melanoma-dysplastic nevus syndrome locus to chromosome 1p36. *Am. J. Hum. Genet.* **46**: 912–918.

Cannon-Albright, L.A., Goldgar, D.E., Meyer, L.J. *et al.* (1992) Assignment of a locus for familial melanoma, MLM, to chromosome 9p13-p22. *Science* **258**: 1148.

Cannon-Albright, L.A., Goldgar, D.E., Neuhausen, S. *et al.* (1994) Localization of the melanoma susceptibility locus to a 2 cM region between D9S736 and D9S171. *Genomics* **23**: 265–268.

Cawley, E.P. (1952) Genetic aspects of malignant melanoma. *Arch. Dermatol. Syph.* **65**: 440–450.

Ciotti, P., Strigini, P. and Bianchi-Scarra, G. (1996) Familial melanoma and pancreatic cancer. (Letter) *New Engl. J. Med.* **334**: 469–470.

Ciotti, P., Struewing, J.P., Mantelli, M. *et al.* (2000) A single genetic origin for the G101W CDKN2A mutation in 20 melanoma-prone families. *Am. J. Hum. Genet.* **67**: 311–319.

Clark, W.H. (1988) The skin. In: Rubin, E. and Farber, J.L. (eds) *Pathology.* Philadelphia: Lippincott.

Clark, W.H. Jr., Reimer, R.R., Greene, M. *et al.* (1978) Origin of familial malignant melanoma from heritable melanocytic lesions: 'the B-K mole syndrome.' *Arch. Dermatol.* **114**: 732–738.

Clark, W.H. Jr., Elder, D.E. and Van Horn, M. (1986) The biologic forms of malignant melanoma. *Hum. Pathol.* **17**: 443.

Cleaver, J.E. (1968) Defective repair replication of DNA in xeroderma pigmentosum. *Nature* **218**: 652–656.

Cohen, L.M. (1995) Lentigo maligna and lentigo maligna melanoma. *J. Am. Acad. Dermatol.* **33**: 923.

Cowan, J.M., Halaban, R. and Francke, U. (1988) Cytogenetic analysis of melanocytes from premalignant nevi and melanomas. *J. Natl. Cancer Inst.* **80**: 1159–1164.

Cristofolini, M. *et al.* (1987) Risk factors for cutaneous malignant melanoma in a northern Italian population. *Int. J. Cancer* **39**: 150.

Crombie, I.K. (1981) Distribution of malignant melanoma on the body surface. *Br. J. Cancer* **43**: 842.

Dracopoli, N.C., Alhadeff, B., Houghton, A.N. *et al.* (1987) Loss of heterozygosity at autosomal and X-linked loci during tumor progression in a patient with melanoma. *Cancer Res.* **47**: 3995–4000.

Elder, D.E. and Murphy, G.F. (1991) Melanocytic Tumors of the Skin. Bethesda: Armed Forces Institutes of Pathology.

Elwood, J.M. (1996) Melanoma and sun exposure. *Semin. Oncol.* **23**: 650.

Elwood, J.M. and Gallagher, R.P. (1983) Site distribution of malignant melanoma. *Can. Med. Assoc. J.* **128**: 1400.

Elwood, J.M., Gallagher, R.P., Hill, G.B., Spinelli, J.J., Pearson, J.C. and Threlfall, W. (1984) Pigmentation and skin reaction to sun as risk factors for cutaneous melanoma: Western Canada Melanoma Study. *BMJ* **288**: 99.

Fitzpatrick, T.B., Sober, A.M. and Mihm, M.C. Jr. (1987) Malignant melanoma of the Skin. In: Braunwald, E., Isselbacher, K.J., Petersdorf, R.G., Wilson, J.D., Martin, J.B. and Fauci, A.S. (eds) *Principles of Internal Medicine*, 13/e, New York: McGraw-Hill, pp. 1595.

Fountain, J.W., Bale, S.J., Housman, D.E. and Dracopoli, N.C. (1990) Genetics of melanoma. *Cancer Surv.* **9**: 645–671.

Fountain, J.W., Karayiorgou, M., Graw, S.L. *et al.* (1991) Chromosome 9p involvement in melanoma. *Am. J. Hum. Genet* **49**: A223, (suppl) (abstr).

Fountain, J.W., Karayiorgou, M., Ernstoff, M.S. *et al.* (1992) Homozygous deletions within human chromosome band 9p21 in melanoma. *Proc. Natl. Acad. Sci. USA* **89**: 10557–10561.

Freeman, S.E., Hacham, H., Gange, R.W., Maytum, D.J., Sutherland, J.C. and Sutherland, B.M. (1989) Wavelength dependence of pyrimidine dimer formation in DNA of human skin irradiated in situ with ultraviolet light. *Proc. Natl. Acad. Sci. USA* **86**: 5605–5609.

Frey, C. and Hartman, J. (1991) National Cancer Institute; 1987 Annual Cancer Statistics Review. NIH publication No. 88-2789, 1988. *J. Natl. Cancer Inst.* **83**: 170.

Furnari, F.B., Huang, H.J. and Cavenee, W.K. (1998) The phosphoinositol phosphatase activity of PTEN mediates a serum-sensitive G1 growth arrest in glioma cells. *Cancer Res.* **58**: 5002–5008.

Gad, F., Yaar, M., Eller, M.S. and Gilchrest, B.A. (1998) The DNA repair capacity of human fibroblast declines with donor age. *J. Invest. Dermatol.* **110**: 690 (abstr.).

Gilchrest, B.A. (1984) Aging and skin cancer. In: Gilchrest, B.A. (ed.) *Skin and Aging Processes.* Boca Raton, FL: CRC Press, pp. 67–81.

Goldgar, D.E., Easton, D.F., Cannon-Albright, L.A. *et al*. (1994) A systematic population-based assessment of cancer risk in first degree relatives of cancer probands. *J. Natl. Cancer Inst.* **86**: 1600–1608.

Goldstein, A.M. and Tucker, M.A. (1995) Genetic Epidemiology of Familial Melanoma. *Dermatol. Clin.* **13(3)**: 605–612.

Goldstein, A.M., Dracopoli, N.C., Ho, E.C. *et al*. (1993) Further evidence for a locus for cutaneous malignant melanoma-dysplastic nevus (CMM/DN) on chromosome 1p, and evidence for genetic heterogeneity. *Am. J. Hum. Genet.* **52**: 537–550.

Goldstein, A.M., Dracopoli, N.C., Engelstein, M. *et al*. (1994) Linkage of cutaneous malignant melanoma/dysplastic nevi to chromosome 9p, and evidence for genetic heterogeneity. *Am. J. Hum. Genet.* **54**: 489–496.

Goldstein, A.M., Goldein, L.R., Dracopoli, N.C., Clark, W.H. Jr. and Tucker, M.A. (1996) Two-locus linkage analysis of cutaneous malignant melanoma/dysplastic nevi. *Am. J. Hum. Genet.* **58**: 1050–1056.

Green, A. and Swerdlow, A.J. (1989) Epidemiology of melanocytic nevi. *Epidemiol. Rev.* **11**: 204.

Green, A. *et al*. (1993) Site distribution of cutaneous melanoma in Queensland. *Int. J. Cancer* **53**: 232.

Greene, M.H. and Fraumeni, J.F. Jr. (1979) The hereditary variant of malignant melanoma. In: Clark, W.H. Jr, Goldman, L.I., Mastrangelo, M.J. (eds) *Human Malignant Melanoma*, New York: Grune and Stratton, pp. 139–166.

Greene, M.H., Goldin, L.R., Clark, W.H. *et al*. (1983) Familial malignant melanoma: Autosomal dominant trait possibly linked to the Rhesus locus. *Proc. Natl. Acad. Sci. USA* **80**: 6071–6075.

Gruis, N.A., Bergman, W. and Frants, R.R. (1990) Locus for susceptibility to melanoma on chromosome 1p. *N. Engl. J. Med.* **322**: 853–854.

Gruis, N.A., Sandkuijl, L.A., Weber, J.L. *et al*. (1993) Linkage analysis in Dutch familial atypical multiple mole-melanoma (FAMMM) syndrome families. Effect of naevus count. *Melanoma Res.* **3**: 271–277.

Gruis, N.A., van der, P.A., Sandkuijl, L.A., Prins, D.E., Weaver-Feldhaus, J., Kamb, A., Bergman, W. and Frants, R.R. (1995a) Homozygotes for CDKN2 (P16) germline mutation in Dutch familial melanoma kindreds. *Nat. Genet.* **10**: 351–353.

Gruis, N.A., Weaver-Feldhaus, J., Liu, Q. *et al*. (1995b) Genetic evidence in melanoma and bladder cancers that p16 and p53 function in separate pathways of tumor suppression. *Am. J. Pathol.* **146**: 1199.

Guldberg, P., Thor Straten, P., Birck, A., Ahrenkiel, V., Kirkin, A.F. and Zeuthen, J. (1997) Disruption of the MMAC1/PTEN gene by deletion or mutation is a frequent event in malignant melanoma. *Cancer Res.* **57**: 3660–3663.

Haas-Kogan, D., Shalev, N., Wong, M., Mills, G., Yount, G. and Stokoe, D. (1998) Protein kinase B (PKB/Akt) activity is elevated in glioblastoma cells due to mutation of the tumor suppressor PTEN/MMAC. *Curr. Biol.* **8**: 1195–1198.

Halpern, A.C., Guerry, D. IV, Elder D.E. *et al*. (1991) Dysplastic nevi as risk markers of sporadic (nonfamilial) melanoma. *Arch. Dermatol.* **127**: 995.

Halpern, A.C., Guerry, D. IV, Elder D.E. *et al*. (1993) A cohort study of melanoma in patients with dysplastic nevi. *J. Invest. Dermatol.* **100**: 346S.

Haluska, F.G. and Housman, D.E. (1995) Recent advances in the molecular genetics of malignant melanoma. *Cancer Surv.* **25**: 277–292.

Hannon, G.J. and Beach, D. (1994) p15 INK4B is a potential effector of TGF-β-induced cell cycle arrest. *Nature* **371**: 257.

Harland, M., Meloni, R., Gruis, N. *et al*. (1997) Germline mutations of the CDKN2 gene in UK melanoma families. *Hum. Molec. Genet.* **6**: 2061–2067.

Healy, E., Rehman, I., Angus, B. and Rees, J.L. (1995) Loss of heterozygosity in sporadic primary cutaneous melanoma. *Genes Chromosome Cancer* 12: 152.

Healy, E., Belgaid, C.E., Takata, M., Vahlquist, A., Rehman, I., Rigby, H. and Rees, J.L. (1996) Allelotypes of primary cutaneous melanoma and benign melanocytic nevi. *Cancer Res.* 56: 589–593.

Hemminki, K. and Vaittinen, P. (1997) Interaction of breast cancer and melanoma genotypes. *Lancet* 350: 931–932.

Herbst, R.A., Weiss, J., Ehnis, A., Cavenee, W.K. and Arden, K.C. (1994) Loss of heterozygosity for 10q22–10qter in malignant melanoma progression. *Cancer Res.* 54: 3111–3114.

Holland, E.A., Beaton, S.C., Becker, T.M., Grulet, T.M., Peters, B.A., Rizos, H., Kefford, R.F. and Mann, G.J. (1995) Analysis of the p16 gene, CDKN2, in 17 Australian melanoma kindreds. *Oncogene* 11: 2289.

Holman, C.D. and Armstrong, B.K. (1984a) Pigmentary traits, ethnic origin, benign nevi, and family history as risk factors for cutaneous malignant melanoma. *J. Natl. Cancer Inst.* 72: 257–266.

Holman, C.D. and Armstrong, B.K. (1984b) Cutaneous malignant melanoma and indicators of total accumulated exposure to the sun: An analysis separating histogenetic types. *J. Natl. Cancer. Inst.* 73: 75.

Holman, C.D. *et al.* (1980) Epidemiology of pre-invasive and invasive malignant melanoma in Western Australia. *Int. J. Cancer* 25: 317.

Horsman, D.E. and White, V.A. (1993) Cytogenetic analysis of uveal melanoma: Consistent occurrence of monosomy 3 and trisomy 8q. *Cancer* 71: 811.

Hussussian, C.J., Struewing, J.P., Goldstein, A.M. *et al.* (1994) Germline p16 mutation in familial melanoma. *Nature Genet.* 8: 15–21.

Jen, J., Harper, J.W., Bigner, S.H., Bigner, D.D., Papadopoulos, N., Markowitz, S., Wilson, J.K.V., Kinzler, K.W. and Vogelstein, B. (1994) Deletion of p16 and p15 genes in brain tumors. *Cancer Res.* 54: 6353.

Jimbo, K., Quevado W.C., Fitzpatric T.B. and Szabo, G. (1993) Biology of melanocytes. In: Fitzpatric, T.B., Eisen, A.Z., Wolff, K., Freedberg, I.M. and Austen, K.F. (eds) *Dermatology in General Medicine.* 4th edn. Vol. 1. New York: McGraw-Hill, pp. 261–289.

Johnson, M.R., Look, A.T., DeClue, J.E., Valentine, M.B. and Lowy, D.R. (1993) Inactivation of the NF1 gene in human melanoma and neuroblastoma cell lines without impaired regulation of GRP.Ras. *Proc. Natl. Acad. Sci. USA* 90: 5539.

Johnson, T.M., Smith, J.W., Nelson, B.R. *et al.* (1995) Current therapy for cutaneous melanoma. *Dermatology* 32: 689.

Kamb, A. and Herlyn, M. (1998) Malignant melanoma. In: Vogelstein, B. and Kinzler, K.W. (eds) *The Genetic Basis of Human Cancer,* New York: McGraw-Hill, pp. 510.

Kamb, A., Gruis, N.A., Weaver-Feldhaus, J. *et al.* (1994a) A cell cycle regulator potentially involved in genesis of many tumor types. *Science* 264: 436.

Kamb, A., Liu, Q., Harshman, K. and Tavtigian, S.V. (1994b) Response to rate of p16 (MTS1) mutations in primary tumors with 9p loss. *Science* 265: 416.

Kamb, A., Shattuck-Eidens, D., Eeles, R. *et al.* (1994c) Analysis of the p16 gene (CDKN2) as a candidate for the chromosome 9p melanoma susceptibility locus. *Nat. Genet.* 8: 22.

Kaplan, E.N. (1974) The risk of malignancy in large congenital nevi. *Plast. Reconstr. Surg.* 53: 421.

Kefford, R.F., Salmon, J., Shaw, H.M. *et al.* (1991) Hereditary melanoma in Australia: Variable association with dysplastic nevi and absence of genetic linkage to chromosome 1p. *Cancer Genet. Cytogenet.* 51: 45–55.

Kobayashi, N., Nakagawa, A., Muramatsu, T. *et al.* (1998) Supranuclear melanin caps reduce ultraviolet induced DNA photoproducts in human epidermis. *J. Invest. Dermatol.* 110: 806–810.

Kochever, I.E., Pathak, M.A. and Parrish, J.A. (1993) Photophysics, photochemistry, and photobiology. In: Fitzpatrick, T.B., Eisen, A.Z., Wolf, K., Freedberg, I.M. and Austen, K.F. (eds) *Dermatology in General Medicine.* 4th edn. Vol. 1. New York: McGraw-Hill, pp. 1627–1638.

Koh, H., Enders, G.H., Cynlacht, B.D. and Harlow, E. (1995) Tumor-derived p16 alleles encoding proteins defective in cell cycle inhibition. *Nature* 375: 506.

Kong, D., Suzuki, A., Zou, T.T. *et al.* (1997) PTEN1 is frequently mutated in primary endometrial carcinomas. *Nat. Genet.* 17: 143–144.

Kopf, W., Bart, R.S. and Hennessey, P. (1979) Congenital nevocytic nevi and malignant melanomas. *J. Am. Acad. Dermatol.* 1: 123.

Kosary, C.L., Ries, L.A.G., Miller, B.A., Hankey, B.F., Harras, A. and Edwards, B.K. (eds) (1996) SEER cancer statistics review, 1973–1992: tables and graphs. Bethesda, MD: National Cancer Institute (NIH publication no. 96-2789).

Kraehn, G.M., Schartl, M. and Peter, R. U. (1995) Human malignant melanoma. A genetic disease? *Cancer* 75: 1228–1237.

Kruger, S., Garbe, C., Buttner, P. *et al.* (1992) Epidemiologic evidence for the role of melanocytic nevi as risk markers and direct precursors of cutaneous malignant melanoma. *J. Am. Acad. Dermatol.* 26: 920.

Kundson, A.G. (1971) Mutation and cancer: Statistical study of retinoblastoma. *Proc. Natl. Acad. Sci. USA* 68: 820.

Kusewitt, D.F., Applegate, L.A. and Ley, R.D. (1991) Ultraviolet radiation-induced skin tumors in a South American opossum (*Monodelphis domestica*). *Vet. Pathol.* 28: 55–65.

Langford, I.H., Bentham, G. and McDonald, A.L. (1998) Multi-level modelling of geographically aggregated health data: a case study on malignant melanoma mortality and UV exposure in the European community. *Stat. Med.* 17: 41–57.

Langley, R.G.B. and Sober, A.J. (1997) A clinical review of the evidence for the role of ultraviolet radiation in the etiology of cutaneous melanoma. *Cancer Invest.* 15: 561.

Langley, R.G.B., Barnhill, R.L., Mihm, M.C. Jr., Fitzpatrick, T.B. and Sober, A.J. (1999) Neoplasm: cutaneous melanoma. In: Freedberg, I.M., Eisen, A.Z., Wolf, K., Austen, K.F., Goldsmith, L.A., Katz, S.I. and Fitzpatrick, T.B. (eds), *Fitzpatrick's Dermatology In General Medicine.* vol. 1, McGraw-Hill, pp. 1085–1089.

Lerch, K. (1981). *Metal Ions Biol. Syst.* 13: 143–186.

Levin, D.B., Wilson, K., Valadares de Amorim, G., Webber, J., Kenny, P. and Kusser, W. (1995) Detection of p53 mutations in benign and dysplastic nevi. *Cancer Res.* 55: 4278.

Li, J., Yen, C., Liaw, D. *et al.* (1997) PTEN, a putative protein tyrosine phosphatase gene mutated in human brain, breast, and prostate cancer. *Science* 275: 1943–1947.

Lilischkis, R., Sarcevic, B., Kennedy, C. *et al.* (1996) Cancer associated missense and deletion mutations impair p16INK4 CDK inhibitory activity. *Int. J. Cancer* 66: 249–254.

Little, J.B., Nove, J., Dahlberg, W.K. *et al.* (1987) Normal cytotoxic response of skin fibroblasts from patients with Li-Fraumeni familial cancer syndrome to DNA-damaging agents in vitro. *Cancer Res.* 47: 4229–4234.

Liu, L., Lassam, N.J.U., Slingerland, J.M., Bailey, D., Cole, D., Jenkins, R. and Hogg, D. (1995a) Germline p16INK4A mutation and protein dysfunction in a family with inherited melanoma. *Oncogene* 11: 405.

Liu, Q., Neuhausen, S., McClure, M. *et al.* (1995b) CDKN2 (MTS1) tumor suppressor gene mutations in human tumor cell lines. *Oncogene* 10: 1061.

Lynch, H.T. and Krush, A.J. (1968) Hereditary and malignant melanoma: implications for early cancer detection. *Can. Med. Assoc. J.* 99: 17–21.

Lynch, H.T., Frichot, B.C., Lynch, P., Lynch, J. and Guirgis, H.A. (1975) Family studies of malignant melanoma and associated cancer. *Surg. Gynecol. Obstet.* 141: 517–522.

Lynch, H.T., Frichot, B.C. and Lynch, J.F. (1978) Familial atypical multiple mole–melanoma syndrome. *J. Med. Genet.* 15: 352–356.

Lynch, H.T., Fusaro, R.M., Pester, J. *et al.* (1980) Familial atypical multiple mole melanoma (FAMMM) syndrome: genetic heterogeneity and malignant melanoma. *Br. J. Cancer* 42: 58–70.

Lynch, H.T., Watson, P., Kriegler, M. *et al.* (1988) Differential diagnosis of hereditary nonpolyposis colorectal cancer (Lynch syndrome I and Lynch syndrome II). *Dis. Colon. Rectum* 31: 372–377.

Mack, T.M. and Floderus, B. (1991) Malignant melanoma risk by nativity, place of residence at diagnosis, and age at migration. *Cancer Causes Control* 2: 401–411.

MacLennan, R., Green, A.C., Mcleod, G.R.C. and Martin, N.G. (1992) Increasing incidence of cutaneous melanoma in Queensland, Australia. *J. Natl. Cancer Inst.* 84: 1427–1432.

Maehama, T. and Dixon, J.E. (1998) The tumor suppressor, PTEN/MMAC1, dephosphorylates the lipid second messenger, phosphatidylinositol 3,4,5-trisphosphate. *J. Biol. Chem.* 273: 13375–13378.

Matsumura, Y., Chikako, N., Yagi, T., Imamura, S. and Takebe, H. (1998) Mutations of p16 and p15 tumor suppressor genes and replication errors contribute independently to the pathogenesis of sporadic malignant melanoma. *Arch. Dermatol. Res.* 290: 175–180.

Marghoob, A., Slade, J., Salopek, T. *et al.* (1995) Basal cell and squamous cell carcinomas are important risk factors for cutaneous malignant melanoma. Screening implications. *Cancer* 75: 707.

Mihm, M.C., Jr. *et al.* (1971) The clinical diagnosis, classification and histogenetic concepts of the early stages of cutaneous malignant melanomas. *N. Engl. J. Med.* **284**: 1078–1082.

Millikin, O., Meese, E. and Vogelstein, B. (1991) Loss of heterozygosity for loci on the long arm of chromosome 6 in human malignant melanoma. *Cancer Res.* **51**: 5449–5453.

Mitchell, D.L. and Nairn, R.S. (1989) The biology of the (6-4) photoproduct. *Photochem. Photobiol.* **86**: 5605–5609.

Mitchell, D.L., Jen, J. and Cleaver, J.E. (1991) Relative induction of cyclobutane dimers and cytosine photohydrates in DNA irradiated in vitro and in vivo with ultraviolet-C and ultraviolet-B light. *Photochem. Photobiol.* **54**: 741–746.

Monzon, J., Liu, L., Brill, H., Goldstein, A.M., Tucker, M.A., From, L., McLaughlin, J., Hogg, D. and Lassam, N.J. (1998) CDKN2A mutations in multiple primary melanomas. *New Engl. J. Med.* **338**: 879–887.

Moriwaki, S., Ray, S., Tarone, R.E., Kraemer, K.H. and Grossman, L. (1996) The effect of donor age on the processing of UV-damaged DNA by cultured human cells: reduced DNA repair capacity and increased DNA mutability. *Mutant Res.* **364**: 117–123.

Muir, C. (1987) *Cancer Incidence in Five Continents*, Volume 5. Lyon: International Agency for Research on Cancer.

Myers, M.P., Pass, I., Batty, I.H., Van der Kaay, J., Stolarov, J.P., Hemmings, B.A., Wigler, M.H., Downes, C.P. and Tonks, N.K. (1998) The lipid phosphatase activity of PTEN is critical for its tumor suppressor function. *Proc. Natl. Acad. Sci. USA* **95**: 13513–13518.

Nancarrow, D.J., Palmer, J.M., Walters, N.K. *et al.* (1992) Exclusion of the familial melanoma locus (MLM) from the PND/D1S47 and MYCL1 regions of chromosome arm 1p in 7 Australian pedigrees. *Genomics* **12**: 18–25.

Nancarrow, D.J., Mann, G.J., Holland, E.A. *et al.* (1993) Confirmation of chromosome 9p linkage in familial melanoma. *Am. J. Human Genet.* **53**: 936–942.

Nobori, T., Miura, K., Wu, D.J., Lois, A., Takabayashi, K. and Karson, D.A. (1994) Deletions of the cyclin-dependent kinase-4 inhibitor gene in multiple human cancers. *Nature* **368**: 753.

Norris, W. (1820) A case of fungoid disease. *Edinb. Med. Surg. J.* **16**: 562–565.

Norris, D.A., Morelli, J.G. and Fuzita, M. (1998) Melanocyte interaction in the skin. In: Nordlund, J.J., Boissy, R.E., Hearing, V.J., King, R.A. and Ortonne, J.-P. (eds) *The Pigmentary System: Physiology and Pathophysiology*. New York : Oxford University Press, pp. 123–133.

Olopade, O.I., Jenkins, R., Linnenbach, A.J. *et al.* (1990) Molecular analysis of Chromosome 9p deletion in human solid tumors. *Proc. Am. Assoc. Cancer Res.* **21**: 318.

OPCS (1990) Office Population of Censuses and Surveys Cancer Statistics: Registration 1985. London: HMSO; Series MB1 No 18.

Parkin, D.M., Muir, C.S., Whelan, S.L., Gao, Y.T. and Ferlay, J. (eds) (1992) *Cancer Incidence in Five Continents*. Vol. 6. Lyon, France; International Agency for Research on Cancer.

Parmiter, A.H., Balaban, G., Clark, W.H. Jr. and Nowell, P.C. (1988) Possible involvement of the chromosome region 10q24-q26 in early stages of melanocytic neoplasia. *Cancer Genet. Cytogenet.* **30**(2): 313–317.

Pathak, M.A. (1995) Functions of melanin and protection by melanin. In: Zeise, L., Chedekel, M.R., Fitzpatric, T.B. (eds). *Melanin: Its Role in Photoprotection*. Overland Park, Kans.: Valdenmar Publishing, pp. 125–134.

Pederson, M.I. and Wang, N. (1989) Chromosomal evolution in the progression and metastasis of human malignant melanoma. *Cancer Genet. Cytogenet.* **41**: 185–201.

Peris, K., Keller, G., Chimenti, S., Amantea, A., Derl, H. and Hofler, H. (1995) Microsatellite instability and loss of heterozygosity in melanoma. *J. Invest. Dermatol.* **105**: 625.

Petty, E.M., Bolognia, J., Bale, A.E. and Yang-Feng, T. (1993a) Cutaneous malignant melanoma and atypical moles associated with a constitutional rearrangement of chromosome 5 and 9. *Am. J. Med. Genet.* **45**: 77–80.

Petty, E.M., Gibson, L.H., Fountain, J.W. *et al.* (1993b) Molecular definition of a chromosome 9p21 germ-line deletion in a woman with multiple melanomas and a plexiform neurofibroma: Implications for 9p tumor suppressor gene(s). *Am. J. Hum. Genet.* **53**: 96–104.

Porta, G. (1988). *Med. Res. Rev.* **8**: 525–556.

Quelle, D.E., Ashmun, R.A., Hannon, G.J. *et al.* (1995) Cloning and characterization of murine p16(INK4a) and p15(INK4b) genes. *Oncogene* **11**: 635–645.

Quinn, A.G., Healy, E., Rehman, I., Sikkink, S. and Rees, J.L. (1995) Microsatellite instability in human non-melanoma and melanoma skin cancer. *J. Invest. Dermatol.* **104**: 309.

Ranade, K., Hussussian, C.J., Sikorski, R.S. *et al.* (1995) Mutation associated with familial melanoma impair p16 INK4 function. *Nat. Genet.* **10**: 114.

Reed, J.A., Loganzo, F. Jr., Shea, C.R. *et al.* (1995) Loss of expression of the p16/cyclin-dependent kinase inhibitor 2 tumor suppressor gene in melanocytic lesions correlates with invasive stage of tumor progression. *Cancer Res.* **55**: 2713.

Reimer, R.R., Clark, W.H. Jr., Greene, M.H. *et al.* (1978) Precursor lesions in familial melanoma: a new genetic preneoplastic syndrome. *JAMA* **239**: 744–746.

Reintgen, D.S. *et al.* (1982) Malignant melanoma in black American and white American populations: Comparative review. *JAMA* **248**: 1856.

Rhodes, A.R. *et al.* (1987) Risk factors for cutaneous melanoma: A practical method for recognizing predisposed individuals. *JAMA* **258**: 3146.

Riely, P.A. (1997) Melanin. *Int. J. Biochem. Cell. Biol.* **29**: 1235–1239.

Rigel, D.S., Rivers J.K., Kopf, A.W. *et al.* (1989) Dysplastic nevi: markers for increased risk for melanoma. *Cancer* **63**: 386.

Rigel, D.S., Fridman, R.J. and Kopf, A.W. (1996) The incidence of malignant melanoma in the United States: Issues as we approach the 21st century. *J. Am. Acad. Dermatol.* **34**: 839.

Risinger, J.I., Hayes, A.K., Berchuck, A. and Barrett, J.C. (1997) PTEN/MMAC1 mutations in endometrial cancers. *Cancer Res.* **57**: 4736–4738.

Robertson, G.P., Furnari, F.B., Miele, M.E., Glendening, M.J., Welch, D.R., Fountain, J.W., Lugo, T.G., Huang, H.J. and Cavenee, W.K. (1998) In vitro loss of heterozygosity targets the PTEN/MMAC1 gene in melanoma. *Proc. Natl. Acad. Sci. USA* **95**: 9418–9423.

Schmitz, S., Garbe, C., Tebbe, B. and Orfanos, C.E. (1994) Langwellige ultraviolet Strahlung (UVA) und Hautkrbs. *Hautarzt* **45**: 517–525.

Scotto, J., Fears, T.R. and Fraumeni, J.F. Jr. (April 1983) Incidence of nonmelanoma skin cancers in the United States. Washington D.C.: Government Printing Office (NIH publication no. 83-2433).

Serrano, M., Hannon, G.J. and Beach, D. (1993) A new regulatory motif in cell cycle control causing specific inhibition of cyclin D/CDK4. *Nature* **366**: 704.

Setlow, R.B., Regan, J.D., German, J. and Carrier, W.L. (1969) Evidence that xeroderma pigmentosum cells do not perform the first step in the repair of ultraviolet damage to their DNA. *Proc. Natl. Acad. Sci. USA* **64**: 1035–1041.

Setlow, R.B., Woodhead, A.D. and Grist, E. (1989) Animal model for ultraviolet radiation-induced melanoma: platyfish-swordtail hybrid. *Proc. Natl. Acad. Sci. USA* **86**: 8922–8926.

Sher, C.J. (1993) Mammalian G1 cyclins. *Cell* **73**: 1059–1065.

Silvers, W.K. (1979) *The Coat Colors of Mice*. New York: Springer-Verlag.

Skender-Kalnenas, T.M., English D.R. and Heenan, P.J. (1995) Benign melanocytic lesions: risk markers or precursors of cutaneous melanoma? *J. Am. Acad. Dermatol.* **33**: 1000.

Skolnick, M.H. (1980) The Utah genealogical database: A resource for genetic epidemiology. In: Cairns, J., Lyon, J.L. and Skolnick, M. (eds) Banbury Report No. 4: *Cancer Incidence in Defined Populations*. New York: NY, Cold Spring Harbor Laboratory, pp. 285–297.

Sober, A.J., Lew, R.A., Koh, H.K. and Barnhill, R.L. (1991) Epidemiology of cutaneous melanoma. *Dermatol. Clin. North Am.* **9**: 617.

Soufir, N., Avril, M., Chompret, A. *et al.* (1998) Prevalence of p16 and CDK4 germline mutations in 48 melanoma-prone families in France. *Hum. Molec. Genetics* **7**: 209–216.

Sozzi, G., Veronese, M.L., Negrini, M. *et al.* (1996) The FHIT gene 3p14.2 is abnormal in lung cancer. *Cell* **85**: 17.

Stambolic, V., Suzuki, A., de la Pompa, J.L. *et al.* (1998) Negative regulation of PKB/Akt-dependent cell survival by the tumor suppressor PTEN. *Cell* **95**: 29–39.

Steck, P.A., Pershouse, M.A., Jasser, S.A. *et al.* (1997) Identification of a candidate tumour suppressor gene, MMAC1, at chromosome 10q23.3 that is mutated in multiple advanced cancers. *Nat. Genet.* **15**: 4: 356–362.

Stone, S., Jiang, P., Dayannanth, P., Tavtigian, S.V., Katcher, H., Parry, D., Peters, G. and Kamb, A. (1995a) Complex structure and regulation of the p16 (MTS1) locus. *Cancer Res.* **55**: 2988.

Stone, S., Dayananth, P., Jiang, P., Weaver-Feldhaus, J.M., Tavtigian, S.V., Skolnick, M.H. and Kamb, A. (1995b) Genomic structure, expression and mutational analysis for the p15 (MTS2) gene. *Oncogene* **11**: 987.

Stott, F.J., Bates, S., James, M.C. *et al.* (1998) The alternative product from the human CDKN2A locus, p14(ARF), participates in a regulatory feedback loop with p53 and MDM2. *EMBO J.* 17(17): 5001–5014.

Su, Y.A. and Trent, J.M. (1995) Genetics of cutaneous malignant melanoma. *Cancer Control* 2(5): 392–397.

Teng, D.H., Hu, R., Lin, H. *et al.* (1997) MMAC1/PTEN mutations in primary tumor specimens and tumor cell lines. *Cancer Res.* 57: 5221–5225.

Thompson, F.H., Emerson, J., Olson, S. *et al.* (1995) Cytogenetics in 158 patients with regional and disseminated melanoma: subset analysis of near-diploid and simple karyotypes. *Cytogenet. Cell. Genet.* 83: 93–104.

Traboulsi, E.I., Zimmerman, L.E. and Manz, H.J. (1988) Cutaneous malignant melanoma in survivors of heritable retinoblastoma. *Arch. Opthalmol.* 109: 1059–1066.

Trent, J.M., Thompson, F.H. and Meyskens, F.L. Jr. (1989) Identification of a recurring translocation site involving chromosome 6 in human malignant melanoma. *Cancer Res.* 49: 420–423.

Trent, J.M., Stanbridge, E.J., McBride, H.L., Meese, E.U., Casey, G., Araujo, D.E., Witkowski, C.M. and Nagle, R.B. (1990) Tumorigenicity in human melanoma cell lines controlled by introduction of human chromosome 6. *Science* 247: 568–571.

Tsao, H., Zhang, X., Benoit, E. and Haluska, F.G. (1998) Identification of PTEN/MMAC1 alterations in uncultured melanomas and melanoma cell lines. *Oncogene* 16: 3397–3402.

Tucker, M.A., Halpern, A., Holly, E. *et al.* (1998) Clinically recognized dysplastic nevi: a central risk factor for cutaneous melanoma. *JAMA* 277: 1439.

Van Haeringen, A., Bergman, W., Nelen, M.R. *et al.* (1989) Exclusion of the dysplastic nevus syndrome (DNS) locus from the short arm of chromosome 1 by linkage studies in Dutch families. *Genomics* 5: 61–64.

Vlietstra, R.J., van Alewijk, D.C., Hermans, K.G., van Steenbrugge, G.J. and Trapman, J. (1998) Frequent inactivation of PTEN in prostate cancer cell lines and xenografts. *Cancer Res.* 58: 2720–2723.

Volkenandt, M., Schlegel, U., Nanus, D.M. and Albino, A.P. (1991) Mutational analysis of the human p53 gene in malignant melanoma cell lines. *Pigment Cell. Res.* 4: 35.

Walker, G.J., Palmer, J.M., Walters, M.K., Nancarrow, D.J. and Hayward, N.K. (1994) Microsatellite instability in melanoma. *Melanoma Res.* 4: 267.

Walker, G.J., Palmer, J.M., Walters, M.K. and Hayward N.K. (1995) A genetic model of melanoma tumorigenesis based on allelic losses. *Genes Chrom. Cancer* 12: 134–141.

Wallace, D.C., Exton, L.A. and McLeod, G.R. (1971) Genetic factors in malignant melanoma. *Cancer* 1262–1266.

Weaver-Feldhaus, J., Gruis, N.A., Neuhausen, S., Le Paslier, D., Stockert, E., Skolnick, M.H. and Kamb, A. (1994) Localization of a putative tumor suppressor gene by using homozygous deletions in melanomas. *Proc. Natl. Acad. Sci. USA* 91: 7563–7567.

Wang, S.I., Puc, J., Li, J., Bruce, J.N., Cairns, P., Sidransky, D. and Parsons, R. (1997) Somatic mutations of PTEN in glioblastoma multiforme. *Cancer Res.* 57: 4183–4186.

Wei, Q., Matanoski, G.M., Farmer, E.R., Hedayati, M.A. and Grossman, L. (1993) DNA repair and aging in basal cell carcinoma: a molecular epidemiology study. *Proc. Natl. Acad. Sci. USA* 90: 1614–1618. [Erratum, *Proc. Natl. Acad. Sci. USA* 1993; 90: 5378.]

Weich, D.R., Chen, P., Miele, M.E., McGary, C.T., Brower, J.M., Stanbridge, E.J. and Weissman, B.E. (1994) Microcell-mediated transfer of chromosome 6 into metastatic human C8161 melanoma cells suppresses metastasis but does not inhibit tumorigenicity. *Oncogene* 9: 255.

Weinstock, M.A. and Sober, A.J. (1987) The risk of progression of lentigo maligna to lentigo maligna melanoma. *Br. J. Dermatol.* 116: 303.

Welch, D.R., Chen, P., Miele, M.E., McGary, C.T., Bower, J.M., Stanbridge, E.J. and Weissman, B.E. (1994) Microcell-mediated transfer of chromosome 6 into metastatic human C8161 melanoma cells suppresses metastasis but does not inhibit tumorigenicity. *Oncogene* 9: 255–262.

Wolfel, T., Hauer, M., Schneider, J. *et al.* (1995) A p16INK4a- insensitive CDK4 mutant targeted by cytotoxic T lymphocytes in human melanoma. *Science* 269: 1281.

Yang, R., Gombart, A.F., Serrano, M. and Koeffler, P. (1995) Mutational effects on the p16[INK4a]. *Cancer Res.* 55: 2503.

You, J., Miele, M.E., Dong, C. and Welch, D.R. (1995) Suppression of human melanoma metastasis by introduction of chromosome 6 may be partially due to inhibition of motility, but not to inhibition of invasion. *Biomed. Biophys. Res. Comm.* 208: 476–484.

Zhou, X.P., Gimm, O., Hampel, H., Niemann, T., Walker, M.J. and Eng, C. (2000) Epigenetic PTEN silencing in malignant melanomas without PTEN mutation. *Am. J. Pathol.* **157(4)**: 1123–1128.

Zuo, L., Weger, J., Yang, Q., Goldstein, A.M., Tucker, M.A., Walker, G.J., Hayward, N. and Dracopoli, N.C. (1996) Germline mutations in the p16[INK4A] binding domain of CDK4 in familial melanoma. *Nat. Genet.* **12**: 97.

3

The genetics of breast and ovarian cancer

Phillippa J. Neville, Sarah J. Morland, Susan A. J. Vaziri and Graham Casey

1. Introduction

Breast and ovarian cancers rank as the first and fifth most common female malignancies, respectively, in the United States (*Figure 1*). Breast cancer is second only to lung cancer as the leading cause of cancer death in women (American Cancer Society, www.cancer.org, 2000). Unlike colorectal cancer, in which a systematic analysis of pathologically defined stages has suggested a model of progressive genetic changes, genetic models for breast and ovarian cancers have not been developed. Major obstacles to developing such models include a relatively poor understanding of the biology of breast and ovarian cancer initiation and progression, and the heterogeneous nature of these diseases.

Current concepts suggest that cancers arise due to deregulation of critical cellular pathways, enabling the cancer cell to evade normal control of cell growth and death. Consistent with this, primary tumors of the breast and ovary frequently show alterations in genes involved in cell cycle control, signaling and cell survival pathways. Breast and ovarian cancer patients who succumb to their disease generally die of complications associated with metastatic disease rather than their primary cancer, and primary tumors often show altered expression of genes associated with extracellular matrix interactions and motility which may be indicative of poor prognosis. In this chapter, we will review the more commonly described genetic changes associated with hereditary and nonhereditary forms of breast and ovarian cancer and their relationship to prognosis.

2. Hereditary breast and ovarian cancer syndromes

Between 5 and 10% of all breast and ovarian cancers are hereditary in nature (Claus *et al.*, 1996). Although germline mutations in *BRCA1* and *BRCA2* (breast cancer 1 and 2) are associated with a significant proportion of inherited breast and ovarian cancers, an increased risk for developing these cancers is also associated

Molecular Genetics of Cancer second edition, edited by J.K. Cowell.

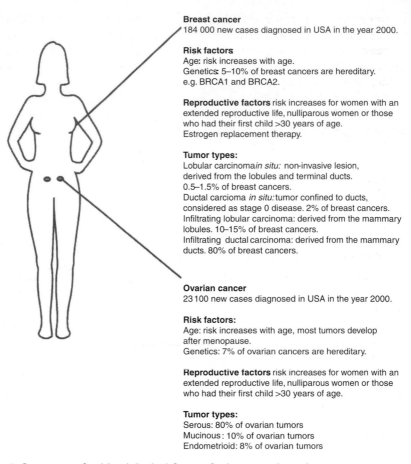

Breast cancer
184 000 new cases diagnosed in USA in the year 2000.

Risk factors
Age: risk increases with age.
Genetics 5–10% of breast cancers are hereditary.
e.g. BRCA1 and BRCA2.

Reproductive factors risk increases for women with an extended reproductive life, nulliparous women or those who had their first child >30 years of age.
Estrogen replacement therapy.

Tumor types:
Lobular carcinoma *in situ:* non-invasive lesion, derived from the lobules and terminal ducts.
0.5–1.5% of breast cancers.
Ductal carcioma *in situ:* tumor confined to ducts, considered as stage 0 disease. 2% of breast cancers.
Infiltrating lobular carcinoma: derived from the mammary lobules. 10–15% of breast cancers.
Infiltrating ductal carcinoma: derived from the mammary ducts. 80% of breast cancers.

Ovarian cancer
23 100 new cases diagnosed in USA in the year 2000.

Risk factors:
Age: risk increases with age, most tumors develop after menopause.
Genetics: 7% of ovarian cancers are hereditary.

Reproductive factors risk increases for women with an extended reproductive life, nulliparous women or those who had their first child >30 years of age.

Tumor types:
Serous: 80% of ovarian tumors
Mucinous: 10% of ovarian tumors
Endometrioid: 8% of ovarian tumors

Figure 1. Summary of epidemiological factors for breast and ovarian cancer.

with other hereditary cancer syndromes. These include Li-Fraumeni syndrome (LFS), ataxia telangiectasia, Cowden disease and hereditary nonpolyposis colon cancer (HNPCC).

2.1 BRCA1 *and* BRCA2-*associated syndromes*

Germline mutations in *BRCA1* or *BRCA2* are associated with an increased risk for developing breast and ovarian cancers and, to a lesser extent, colon and prostate cancers for *BRCA1* and prostate, male breast and pancreatic cancers for *BRCA2* (Ford *et al.*, 1994; Phelan *et al.*, 1996a). Early linkage studies of large, well-defined breast/ovarian cancer families suggested that mutations in *BRCA1* contributed to the majority of hereditary cases (Easton *et al.*, 1995; Narod *et al.*, 1995). Following the identification of *BRCA1* on 17q21 (Miki *et al.*, 1994) and *BRCA2* on 13q12–13 (Tavtigian *et al.*, 1996; Wooster *et al.*, 1995), more accurate estimates of mutation carriers could be determined. Data now suggest that 16–36% of cases with a family history of breast/ovarian cancer are associated with germline mutations in *BRCA1* (Couch *et al.*, 1997; Gayther *et al.*, 1999), and an even lower proportion are associated with *BRCA2* germline mutations (Gayther *et al.*, 1999; Phelan *et al.*, 1996a). *BRCA1* or *BRCA2* mutations account for few early onset breast cancer cases without a family history of cancer (Krainer *et al.*, 1997; Malone *et al.*, 2000). These and other data imply that there may be additional hereditary breast and ovarian cancer predisposing genes yet to be identified (Ford *et al.*, 1998; Rebbeck *et al.*, 1996).

Both *BRCA1* and *BRCA2* are large genes. *BRCA1* encodes a protein of 1863 amino acids, whereas *BRCA2* encodes a protein of 3418 amino acids. *BRCA1* and *BRCA2* share little homology with each other or with other known genes. However, both genes contain recognizable structural motifs (*Figure 2*). BRCA1 contains an amino terminus RING zinc finger domain and two carboxyl terminus BRCT (BRCA1 carboxyl terminus) repeats (Callebaut and Mornon, 1997; Koonin *et al.*, 1996; Paterson, 1998). The BRCT domain is a poorly conserved domain found in several proteins many of which are involved in DNA repair, such as RAD9 and XRCC1 (Callebaut and Mornon, 1997; Koonin *et al.*, 1996). The protein BARD1 (BRCA1-associated ring domain) has been shown to associate with the RING finger domain of BRCA1 (Wu *et al.*, 1996). A conserved feature of BRCA2 is the presence of eight copies of a 30–80 amino acid repeat (BRC repeat) (Bignell *et al.*, 1997; Bork *et al.*, 1996).

Several hundred different germline mutations in *BRCA1* and *BRCA2* have been reported in breast and ovarian cancer patients and families (Breast Information Core (BIC) database; http://www.nhgri.nih.gov/Intramural_research/Lab/transfer/Bic). The majority of mutations reported have been protein truncating nonsense or frameshift mutations that occur throughout both genes. Such mutations can be

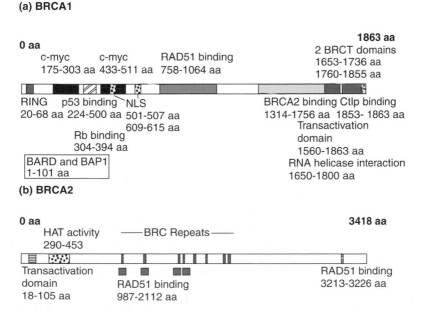

(a) BRCA1

(b) BRCA2

Figure 2. Structure of the proteins derived from the *BRCA1* and *BRCA2* genes indicating the positions of the key domains relative to the amino acid sequence.

identified using a variety of approaches, but a commonly used approach is that of the protein truncation test (PTT) (*Figure 3*). Many missense changes have also been reported, some of which are disease causing, most notably within the RING finger domain (Shattuck-Eidens *et al.*, 1995). While there is evidence that other missense mutations may be functional (Barker *et al.*, 1996), the relevance of the majority of missense changes on increased cancer risk remains uncertain in the absence of functional assays (Dunning *et al.*, 1997; Durocher *et al.*, 1996; Janezic *et al.*, 1999). The recent publication of a functional assay for BRCA1 missense mutations may help resolve this issue (Scully *et al.*, 1999). Large deletions or regulatory mutations may account for up to 20% of *BRCA1* mutations and remain undetected using conventional tests (Petrij-Bosch *et al.*, 1997; Puget *et al.*, 1999; Swensen *et al.*, 1997). The occurrence of similar mutations in *BRCA2* has not been reported.

There are often strong differences in cancer incidence in families carrying the same mutation, which suggests that modifier genes and/or environmental factors may influence cancer risk in hereditary cases. Polymorphisms in two candidate modifier genes for BRCA1 have been reported, the *HRAS1* variable number of tandem repeat rare alleles, and the androgen receptor CAG repeat (Phelan *et al.*, 1996b; Rebbeck *et al.*, 1999a). *BRCA1* mutation carriers harboring one or two rare *HRAS1* alleles were reported to have a 2 times greater risk for developing ovarian cancer compared to carriers with only common alleles (Phelan *et al.*, 1996b). In addition, breast cancer diagnosis occurred at an earlier age in *BRCA1* mutation carriers with very long androgen receptor CAG repeats (Rebbeck *et al.*, 1999a). The influence of hormonal factors has also been noted. *BRCA1*-mutation carriers who underwent a bilateral prophylactic ovariectomy had a statistically significant reduction in breast cancer risk compared to women who did not have this procedure (Rebbeck *et al.*, 1999b).

Several founder mutations in *BRCA1* and *BRCA2* have been reported. Three founder mutations have been described in the Ashkenazi Jewish population (*BRCA1* 185delAG, 5382insC and *BRCA2* 6174delT) with a combined frequency of over 2% in the population (Oddoux *et al.*, 1996; Struewing *et al.*, 1995). Approximately 20% of Ashkenazi women diagnosed with breast cancer under the age of 40 carry the 185delAG mutation (FitzGerald *et al.*, 1996), and nearly 12% of Ashkenazi Jewish breast cancer cases of all ages may be attributable to founder mutations in the *BRCA1* or *BRCA2* genes (Warner *et al.*, 1999). Different founder mutations have been reported in other populations, including Iceland (Thorlacius *et al.*, 1997), Finland (Vehmanen *et al.*, 1997) southern Sweden (Johannsson *et al.*, 1996) and French Canadians (Tonin *et al.*, 1998).

BRCA-associated breast tumors display unique clinico-pathological features compared to breast tumors of nonfamilial cases. There is a higher frequency of medullary carcinomas associated with *BRCA1* mutations (Consortium 1997; Eisinger *et al.*, 1998), and lobular carcinomas associated with *BRCA2* mutations compared to nonfamilial cases (Armes *et al.*, 1998; Marcus *et al.*, 1996). *BRCA1*-mutation carriers generally present with high grade, highly proliferative, aneuploid cancers (Eisinger *et al.*, 1996; Marcus *et al.*, 1996). These features have also been reported for *BRCA2*-related tumors, but the association appears weaker (Chappuis *et al.*, 2000). *BRCA1*-related tumors are generally estrogen and progesterone receptor negative whereas this does not appear to be the case for *BRCA2*-related tumors (Johannsson *et al.*, 1997; Robson *et al.*, 1998).

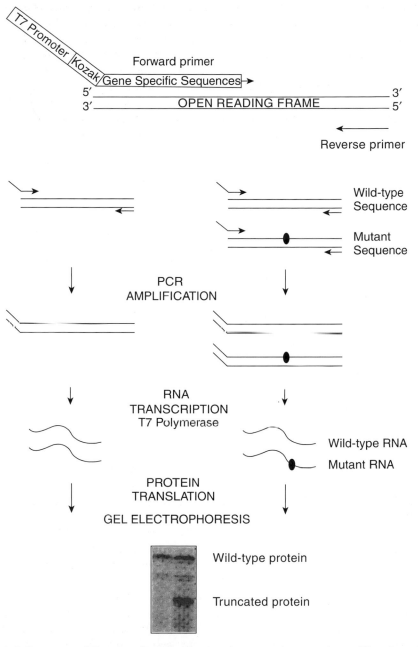

Figure 3. Summary of the procedure for the protein truncation test. A specific primer is designed to include a T7 promotor and Kozak ribosome binding consensus sequence as well as specific sequences from the gene fragment to be analyzed. Following PCR amplification, a coupled *in-vitro* transcription/translation is performed. Where a nonsense mutation is present the translation of the protein will terminate prematurely giving rise to a smaller protein which is then visualized on a SDS-PAGE gel. In this example the RNA sample contained a heterozygous mutation which results in the generation not only of the truncated protein but also the full length, wild-type protein.

A higher frequency of *TP53* mutations in *BRCA1*- and *BRCA2*-related breast and ovarian tumors has been reported (Crook *et al.*, 1997; Sobol *et al.*, 1997). However, this was not reported in other studies (Robson *et al.*, 1998; Schlichtholz *et al.*, 1998), implying that a single model of BRCA-associated cancer development may not exist. The *HER2* receptor tyrosine kinase gene is amplified in approximately 25% of nonfamilial breast and ovarian cancers (Slamon *et al.*, 1989), and amplification is often associated with tumor features similar to those observed in *BRCA*-related tumors. However, studies consistently report a significantly lower frequency of HER2 overexpression in *BRCA1*-related tumors compared to nonfamilial cases and no difference in frequency between *BRCA2*-related tumors and nonfamilial cases (Armes *et al.*, 1999; Noguchi *et al.*, 1999; Robson *et al.*, 1998).

Alterations in a number of other genes implicated in nonfamilial forms of breast and ovarian cancers have also been examined in BRCA-related tumors but no consistent associations have been reported. Those genes include *cyclin D1*, *epidermal growth factor receptor (EGFR)*, *BCL-2, MYC, cathepsin D, P21, P27, E-cadherin* and *β-catenin* (Armes *et al.*, 1999; Jacquemier *et al.*, 1999; Robson *et al.*, 1998). A difference in the patterns of loss of heterozygosity has been reported in *BRCA1*- and *BRCA2*-related tumors compared to nonfamilial tumors (Ingvarsson *et al.*, 1998; Tirkkonen *et al.*, 1997). However, a study using comparative genome hybridization revealed few differences between BRCA-related and nonfamilial tumors (Tapper *et al.*, 1998). These data imply that there are few molecular differences between *BRCA*-related and nonfamilial tumors.

Several studies have evaluated outcomes of patients with *BRCA*-associated cancers. Some studies have reported an improved prognosis for *BRCA1*-mutation carriers compared with nonfamilial breast cancer patients (Marcus *et al.*, 1996; Porter *et al.*, 1994; Rubin *et al.*, 1996). However, the majority of published studies report either no difference (Lee *et al.*, 1999; Robson *et al.*, 1998; Verhoog *et al.*, 1998) or slightly worse survival in carriers (Foulkes *et al.*, 1997; Johannsson *et al.*, 1998; Pharoah *et al.*, 1999). Despite the pathologic features of BRCA-related tumors that are associated with poor prognosis, the absence of a consistent difference in prognosis may suggest that BRCA-related tumors have a greater sensitivity for conventional treatment. In support of this, *BRCA1*- and *BRCA2*-mutant cells are highly sensitive to agents that induce double strand breaks such as radiation, and the chemotherapeutic agents etoposide and cisplatin (Abbott *et al.*, 1998; Foray *et al.*, 1999; Gowen *et al.*, 1998). Furthermore replacement of wild-type *BRCA1* in BRCA1-deficient cells reverses this sensitivity (Abbott *et al.*, 1999).

These data provide insight into possible functions of BRCA1 and BRCA2. Despite the fact that there are few similarities between the BRCA1 and BRCA2 proteins, they appear to co-operate in one or more DNA damage response pathways (Chen *et al.*, 1998; Patel *et al.*, 1998; Scully *et al.*, 1997). Both *Brca1*- and *Brca2*-null mice are embryonic lethals, exhibiting defects in cell division and chromosomal instability (Connor *et al.*, 1997; Hakem *et al.*, 1996; Sharan *et al.*, 1997). Furthermore both proteins bind to the DNA repair gene *TP53*, and *Brca1*-deficient mice can be partially rescued from early embryonic lethality by the presence of *TP53* or *P21* null mutations (Brugarolas and Jacks, 1997; Hakem *et al.*, 1997; Ludwig *et al.*, 1997). A role in DNA repair is further supported by the observation that both proteins bind to the ATM protein product (Cortez *et al.*, 1999;

Marmorstein *et al.*, 1998; Yuan *et al.*, 1999). ATM belongs to a family of protein kinases, homologous to the catalytic subunit of phosphoinositide 3-kinase, that signal cell cycle checkpoint arrest, following DNA damage or incomplete DNA replication (Venkitaraman, 1999).

An important unanswered question regarding BRCA1 and BRCA2 remains. Why do mutations in these two genes contribute towards an increased risk for developing tumors primarily of the breast and ovary, particularly as both genes are ubiquitously expressed and involved in fundamental processes in the cell? One answer to this question may be related to the recent discovery that BRCA1 can inhibit estrogen receptor-α (ER-α) transcription. This raises the possibility that BRCA1 (and possibly BRCA2) may inhibit estrogen-dependent pathways related to mammary epithelial cell proliferation, and that loss of this ability contributes to tumorigenesis (Fan *et al.*, 1999).

2.2 Li-Fraumeni syndrome

Li-Fraumeni syndrome (LFS) is a dominantly inherited syndrome associated with germline mutations in the *TP53* gene (Malkin *et al.*, 1990). LFS is typified by multiple cancers in childhood, including soft tissue sarcomas, brain tumors, leukemia and adrenocortical carcinomas, and also displays early onset and, frequently, bilateral breast cancer in women (Li and Fraumeni, 1969). The type of *TP53* mutation may correlate with disease severity, as germline missense mutations in the core DNA-binding domain show a more highly penetrant cancer phenotype than families with other *TP53* mutations or no mutation (Birch *et al.*, 1998; Plummer *et al.*, 1994). Germline mutations in *TP53* have not been identified in a significant number of LFS families (Bell *et al.*, 1999), raising the possibility that inactivation of genes other than *TP53* may be responsible for the remaining cases. Recently, germline mutations in the *CHK2* gene, a human homolog of the yeast Cds1 and Rad53 G2 checkpoint kinases, have been reported in LFS families (Bell *et al.*, 1999). In response to DNA damage, CHK2 has been shown to phosphorylate BRCA1 (Lee *et al.*, 2000), and stabilize p53 protein leading to cell cycle arrest (Chehab *et al.*, 2000; Hirao *et al.*, 2000). The proportion of LFS families associated with mutations in *CHK2* remains to be determined.

2.3 Ataxia telangiectasia

Ataxia telangiectasia is inherited in an autosomal recessive manner. Epidemiological data suggests that heterozygous carriers of mutations in the ataxia telangiectasia gene (mutated in ataxia telangiectasia, *ATM*) are at increased risk for developing cerebellar ataxia, immune deficiency, leukemia, lymphoma, occulocutaneous telangiectasia and breast cancer (Swift *et al.*, 1987). ATM has been shown to play an essential role in a diverse range of cellular processes, including meiosis, DNA repair, cell cycle checkpoint control and immune development (Xu *et al.*, 1996). It has been estimated that approximately 1% of the population are *ATM* carriers (FitzGerald *et al.*, 1997), and that female carriers may be at a 5-fold increased risk for developing breast cancer (Swift *et al.*, 1991). Most studies have failed to provide any evidence of a role for ATM in breast cancer

(FitzGerald *et al.*, 1997; Izatt *et al.*, 1999). However, recent studies suggest that *ATM* carriers may have a 9-fold increased risk of developing early onset bilateral breast cancer that is associated with long-term survival (Broeks *et al.*, 2000). ATM is a protein kinase involved in the regulation of BRCA1 phosphorylation and double strand DNA repair following damage by ionizing radiation. This association with BRCA1 provides a possible molecular mechanism for ATM in breast cancer development (Cortez *et al.*, 1999; Suzuki *et al.*, 1999).

2.4 Cowden disease

Cowden disease is an autosomal dominant cancer predisposing syndrome associated with an elevated risk for developing breast, thyroid and skin cancers and is characterized by multiple hamartoma lesions of the skin (Eng, 1998; Mallory, 1995; Starink *et al.*, 1986). Cowden disease is associated with germline mutations in the *PTEN* (*p*hosphatase and *ten*sin homolog) gene (Li *et al.*, 1997; Liaw *et al.*, 1997; Steck *et al.*, 1997). PTEN, also known as MMAC1 and TEP1, is a dual specificity phosphatase capable of dephosphorylating serine, threonine and tyrosine residues (Myers *et al.*, 1997). Its targets include the inositol lipid phosphatidylinositol (3,4,5) triphosphate (Maehama and Dixon, 1998), Fak (Tamura *et al.*, 1999a) and AKT (Stambolic *et al.*, 1998; Wu *et al.*, 1998), and it is implicated in the negative regulation of extracellular matrix interactions and the maintenance of cell sensitivity to apoptosis.

2.5 Hereditary non-polyposis colorectal cancer

Hereditary nonpolyposis colorectal cancer (HNPCC) is associated with germ-line mutations in genes involved in mismatch repair (MMR), including *MLH1*, *MSH2*, *MSH6* and others (Peltomaki and Vasen, 1997). The MMR proteins are involved in DNA repair, and loss of these proteins leads to genome instability and an accumulation of DNA replication errors. HNPCC families typically include cancers of the gastrointestinal and genitourinary tracts with basal cell carcinomas, keratocanthomas and colonic diverticula. Female carriers of mutations in the MMR genes also have an increased risk for breast cancer (Risinger *et al.*, 1996). DNA replication errors may be visualized by microsatellite instability (MSI) which is associated with genomic instability in HNPCC, and occurs in approximately 15% of sporadic cases of colon cancer, it is a rare somatic event in nonfamilial breast carcinogenesis (Huiping *et al.*, 1999).

3. Nonhereditary breast and ovarian cancer

The majority (90–95%) of breast and ovarian cancers are nonhereditary and are not associated with dominantly inherited mutations in genes such as *BRCA1* and *BRCA2*. As with dominantly inherited forms of breast and ovarian cancer, aberrant expression of a number of genes has been implicated in the development of nonhereditary breast and ovarian cancers. In the following sections we summarize the data available concerning genes for which there is strongest evidence for a role in the development of nonhereditary breast and ovarian cancers. A growing number of genes have also been implicated in the development of metastasis in breast and ovarian cancers, and those are also described.

3.1 Hormonal involvement – estrogen receptor

Estrogen exposure is an important factor in breast tumorigenesis. Risk for developing breast cancer increases with extended lifetime exposure to estrogens, but decreases following ovariectomy. Estrogen receptor (ESR1) status is a useful prognostic marker in breast cancer, as patients with ER-positive tumors exhibit lower risk of disease recurrence, favorable response to endocrine therapy and improved survival rates (Clark and McGuire, 1988). A role for ESR1 in ovarian cancer development is less well documented (Fischer-Colbrie *et al.*, 1997). Some inactivating mutations of ESR1 have been described in breast tumors and an ovarian cancer cell-line, but generally loss of ESR1 function appears to be caused primarily by methylation (Ottaviano *et al.*, 1994).

3.2 Oncogenes

The HER2/neu *gene family.* The *HER2/neu* proto-oncogene (*ERBB2*), located on 17q21, encodes a transmembrane glycoprotein of 185 kDa with intrinsic tyrosine kinase activity (Coussens *et al.*, 1985). *HER2/neu* belongs to a subfamily of growth factor receptors, which include the epidermal growth factor receptor (EGFR, or HER1) and the receptors HER3 and HER4 (Carraway and Cantley, 1994; Coussens *et al.*, 1985). HER2 appears to play a central role in HER signaling, as it amplifies the signal provided by other HER receptors. Following ligand-dependent activation of HER1, HER3, and HER4 by EGF or heregulin, HER2 is activated by heterodimerization with these receptors (Menard *et al.*, 2000).

Amplification and overexpression of *HER2/neu* has been reported in 20–30% of primary breast tumors and a similar proportion of ovarian tumors (Marks *et al.*, 1994; Slamon *et al.*, 1989). Overexpression is strongly associated with a poor clinical response in both cancers (Slamon *et al.*, 1989). The highest frequency (60%) of *HER2/neu* amplification or overexpression has been reported in large cell ductal carcinoma *in situ* (DCIS) of the breast (Barnes *et al.*, 1992; Liu *et al.*, 1992). Approximately 30% of infiltrating ductal breast carcinomas overexpress *HER2/neu* suggesting that either *HER2/neu* is down-regulated during progression, or that there is a large subset of ductal breast carcinomas which do not develop from DCIS.

The epidermal growth factor receptor gene (*EGFR* or *HER1*) maps to chromosome 7p13-p12. It is overexpressed in approximately 35–60% of breast tumors (Carraway and Cantley, 1994; Prigent and Lemoine, 1992) and up to 61% of ovarian tumors (Fischer-Colbrie *et al.*, 1997). In both diseases protein overexpression correlates with poor prognosis and long-term recurrence (Fischer-Colbrie *et al.*, 1997; Fox *et al.*, 1994; Klijn *et al.*, 1994). Overexpression of EGFR in breast cancer appears independent of *HER2/neu* overexpression (Harris *et al.*, 1989; Tsutsumi *et al.*, 1990). However, overexpression of both EGFR and *HER2/neu* are associated with ER-negative tumors (Adnane *et al.*, 1989; Harris *et al.*, 1989; Press *et al.*, 1989;

Tsutsumi *et al.*, 1990; Zeillinger *et al.*, 1989) and together may be predictive of poor response to hormone therapy (Klijn *et al.*, 1994; Leitzel *et al.*, 1995).

MYC. MYC is a member of the helix-loop-helix/leucine zipper superfamily of genes. It is located on chromosome 8q24, and encodes a DNA-binding protein which, upon heterodimerization with the MAX protein, binds to target DNA sequences and induces transcription of a multitude of genes (Dang *et al.*, 1999). MYC plays a role in many cellular mechanisms including cell cycle control, differentiation, adhesion and apoptosis (Dang 1999).

Reported frequencies of *MYC* amplification in primary breast tumors vary from 4–41% (Bieche and Lidereau, 1995), with higher rates occurring in invasive tumors compared to noninvasive tumors. The level of amplification is also highly variable, ranging from 3 to 18 copies of the gene (Berns *et al.*, 1992a). *MYC* amplification correlates with both tumor size and metastatic spread to the lymph nodes, early disease recurrence and poor prognosis in breast cancer patients (Berns *et al.*, 1992b). Although overexpression of MYC generally occurs as a result of gene amplification, constitutive mRNA overexpression has also been reported (Bieche and Lidereau, 1995). Between 28–50% of ovarian tumors have been reported to show amplification of the *MYC* locus (Arnold *et al.*, 1996; Baker *et al.*, 1990; Bauknecht *et al.*, 1993; Katsaros *et al.*, 1995). Unlike in breast cancer, no clinical associations with *MYC* amplification have been reported for ovarian cancer (Bauknecht *et al.*, 1993).

Cyclin D1 and cyclin E. The G_1 cyclin proteins regulate the progression of a cell through the G_1-S phase of the cell cycle. The sequential activation of cyclin D1 followed by cyclin E leads to the inactivation of pRB1 via phosphorylation and permits the cell to enter the S phase of the cycle (Fernandez *et al.*, 1998).

Several studies have reported overexpression of cyclin D1 protein in approximately 50% of breast tumors, but the clinical importance of this finding remains uncertain (Gillett *et al.*, 1996; McIntosh *et al.*, 1995; Michalides *et al.*, 1996). Up to 70% of ovarian tumors, primarily those of late stage and serous histology, show cyclin D1 protein overexpression, and are associated with poor prognosis (Diebold *et al.*, 2000).

Cyclin E protein overexpression has also been reported in both breast and ovarian cancers. In breast tumors cyclin E overexpression is usually associated with low cyclin D1 expression and ER negative status. Consistent with this, cyclin E overexpression also correlates with poor prognosis in breast cancer patients (Dutta *et al.*, 1995; Keyomarsi *et al.*, 1994; Porter *et al.*, 1997). The importance of cyclin E in ovarian cancer is less clear where protein overexpression has been observed in 12–18% of tumors, although no prognostic associations were reported (Courjal *et al.*, 1996).

3.3 Tumor suppressor genes

BRCA1 *and* BRCA2. Despite the fact that germline mutations in *BRCA1* and *BRCA2* occur at high frequency in hereditary forms of breast and ovarian cancers, few mutations have been reported in nonhereditary forms (Futreal *et al.*, 1994;

Merajver *et al.*, 1995; Stratton *et al.*, 1997). Decreased levels of *BRCA1* mRNA has been reported in association with breast and ovarian cancer progression (Magdinier *et al.*, 1998; Zheng *et al.*, 2000) and decreased BRCA1 protein expression in high-grade breast tumors has also recently been reported (Wilson *et al.*, 1999). These data suggest that BRCA1 is involved in the progression of non-familial forms of breast and ovarian cancer. The identification of aberrant methyl-ation of the BRCA1 promoter suggests a possible mechanism for the reduced expression of *BRCA1* in tumors (Mancini *et al.*, 1998; Rice *et al.*, 1998). Hypermethylation of the *BRCA1* promoter has also been reported in ovarian carcinomas (Catteau *et al.*, 1999; Esteller *et al.*, 2000). In both cancers hypermethyl-ation was strongly associated with negative ESR1 and progesterone receptor (PR) status. *BRCA2* methylation has not been reported and a role for *BRCA2* inacti-vation in nonfamilial cancer remains unclear (Collins *et al.*, 1997).

TP53. The *TP53* tumor suppressor gene maps to 17p13.1 and encodes a nuclear phosphoprotein. The ability of *TP53* to activate the transcription of genes involved in growth arrest or apoptosis following DNA damage has led to *TP53* being described as the 'guardian of the genome' (Lane, 1992). *TP53* mutations are believed to be the most common genetic aberration in human cancer. Approximately 80% of the *TP53* mutations observed in human cancers, most of which are missense alterations, are grouped within an intensively studied and highly conserved region comprising exons 5–8 (Casey *et al.*, 1996; Greenblatt *et al.*, 1994). Missense mutations in p53 generally result in a protein that can be detected by immunohistochemical analysis, presumably due to increased stability (Casey *et al.*, 1996). As a result, immunohistochemical analyses have been used extensively in the analysis of p53 alterations in cancer.

Overall, 25–50% of breast tumors and up to 60% of ovarian tumors harbor *TP53* mutations or are positive for p53 by immunohistochemical approaches (Casey *et al.*, 1996; Horak *et al.*, 1991; Marks *et al.*, 1991; McManus *et al.*, 1994). Breast tumors with *TP53* mutations are, in general, highly aggressive and associated with negative steroid receptor status and high histological grade (Allred *et al.*, 1993; Dong *et al.*, 1997; Elledge *et al.*, 1993). Similarly *TP53* mutation-positive ovarian tumors are usually poorly differentiated, aggressive tumors with a poor prognosis (Milner *et al.*, 1993).

The frequency of *p53* mutations and/or protein overexpression differs between histological subtypes (Domagala *et al.*, 1993; Eccles *et al.*, 1992; Marchetti *et al.*, 1993; Milner *et al.*, 1993). Positive p53 immunostaining is frequently found in medullary and ductal carcinomas of the breast, but not in lobular, mucinous or papillary tumors, histology generally regarded as having a more favorable prog-nosis (Davidoff *et al.*, 1991; Poller *et al.*, 1993; Thor *et al.*, 1992). Similarly up to 50% of malignant epithelial ovarian carcinomas express mutant p53 (Marks *et al.*, 1991), particularly those of the more aggressive, serous papillary origin (Rohlke *et al.*, 1997). In contrast, benign and borderline ovarian tumors generally show low or absent p53 expression. Unlike epithelial tumors, ovarian germ cell tumors do not commonly show p53 alterations (Liu *et al.*, 1995). Positive p53 staining has been seen in only a very small proportion of germ cell tumors studied, implying that aberrations of p53 are not a critical event in the development of these particular tumors (Liu *et al.*, 1995). The majority of *TP53* mutations observed in breast and ovarian tumors are transitions, which are generally believed to arise spontaneously through endogenous mechanisms (Casey *et al.*, 1996; Greenblatt *et al.*, 1994). The low frequency of transversion mutations implies that neither cancer commonly arises through exposure to exogenous carcinogens (Berchuck *et al.*, 1994).

RB1. The RB1 (retinoblastoma) protein (pRb) acts as a critical cell cycle regulator. In its hypophosphorylated, active form, pRb binds to the E2F transcription factor, preventing the cell from entering the S phase of the cell cycle (Gillett and Barnes, 1998). However, following phosphorylation by cyclin D1 and cyclin-dependent kinases (CDKs) and the subsequent release of E2F, the inhibitory effect of pRb is removed allowing progression from the G1 to S phase of the cell cycle.

Between 15–20% of breast tumors show either *RB1* mutations (Berns *et al.*, 1995; Varley *et al.*, 1989) or reduced protein expression (Borg *et al.*, 1992). Between 10–45% of infiltrating breast carcinomas show reduced expression of *RB1* which inversely correlates with tumor proliferation (Jares *et al.*, 1997). Although breast tumors demonstrate an LOH frequency of approximately 25% around the *RB1* locus, LOH does not appear to correlate with observed protein levels (Borg *et al.*, 1992). Similarly, ovarian cancers show between 30–52% LOH at the *RB1* locus, but express wild-type protein (Dodson *et al.*, 1994; Kim *et al.*, 1994; Li *et al.*, 1991). These data suggest that either *RB1* is inactivated without affecting protein levels, or that a second suppressor gene lies close to the *RB1* gene. A definitive role for RB1 in either breast or ovarian cancer therefore remains to be determined.

INK4A/p19^ARF. The INK4A cyclin-dependent kinase inhibitor binds to and inhibits the association between cyclin-dependent kinases (CDK4 and CDK6) and the D cyclins. The resulting hypophosphorylation of pRb causes inhibition of the cell cycle at the G1 phase (Serrano *et al.*, 1993). An alternatively spliced variant of the INK4A (*CDKN2A*) gene encodes a second protein p19^ARF (*A*lternative *R*eading *F*rame), also capable of inducing cell cycle arrest (Quelle *et al.*, 1995). However, loss of p16 expression in tumor cells is more common than p19^ARF (Brenner *et al.*, 1996). Such loss of p16 function can occur through deletions, point mutations or promoter hypermethylation, the frequencies of which vary between tumor types.

Mutations of *INK4 (p16)* are rare in primary breast tumors (Campbell *et al.*, 1995; Shih *et al.*, 1997) although LOH studies have shown that deletions involving the 9p21 locus are associated with tumor development (Brenner and Aldaz, 1995; Cairns *et al.*, 1995). Epigenetic studies have demonstrated that the presence of aberrant or *de novo* 5'CpG island methylation in the gene promoter, which occurs in a third of breast cancer cell lines studied, may be responsible for a proportion of observed *p16* activity loss (Herman *et al.*, 1995). Homozygous deletions of *p16* have been reported in approximately 12–16% of ovarian tumors, particularly those of late stage, although not mucinous subtypes (Ichikawa *et al.*, 1996). There is conflicting evidence of *p16* methylation in ovarian cancer. Studies report methylation of *p16* in primary serous papillary tumors (Niederacher *et al.*, 1999) or borderline tumors (McCluskey *et al.*, 1999). However, others have failed to find any methylation of the gene in these cancers (Ichikawa *et al.*, 1996; Ryan *et al.*, 1998). Thus the *p16* locus does not appear to play a major role in the development of either breast or ovarian cancers.

PTEN. Located on 10q23, *PTEN*, also known as *MMAC1* (*m*utated in *m*ultiple *a*dvanced *c*ancers) or *TEP1* (*T*GFβ-regulated and *e*pithelial cell-enriched *p*hosphatase), encodes a protein that resembles and functions as a dual-specificity phosphatase *in vitro* (Myers *et al.*, 1997), recognizing and dephosphorylating tyrosine and serine/threonine residues. PTEN appears to play a role in phosphatidylinositol 3-kinase (P1 3-K)-mediated growth signaling pathway (Maehama and Dixon, 1998) and anoikis (Di Cristofano and Pandolfi, 2000), and has been implicated in the negative regulation of cell adhesion, migration and tumor invasion (Gu *et al.*, 1999; Tamura *et al.*, 1999b, 1999c).

LOH at the *PTEN* locus has been reported in 29–50% of high-grade, invasive breast carcinomas (Bose *et al.*, 1998) and breast cancer cell lines (Bose *et al.*, 1998; Li *et al.*, 1997). However, *PTEN* mutations are rarely seen in these tumors (Feilotter *et al.*, 1999; Rhei *et al.*, 1997), suggesting the presence of at least one other suppressor gene in this region (Feilotter *et al.*, 1999). LOH at the PTEN locus has been reported in 43% of endometrioid and 28% of serous ovarian tumors, although mutations in the *PTEN* gene have only consistently been reported in endometrioid tumors (21%), particularly those of early stage and low grade (Obata *et al.*, 1998). These data suggest PTEN may play a role in the development of a subset of ovarian tumors.

OVCA1. *OVCA1* maps to 17p13.3, a region of frequent LOH in breast and ovarian cancer. No somatic alterations of OVCA1 have been reported in either breast or ovarian tumors (Bruening *et al.*, 1999). However, *OVCA1* mRNA expression is reduced or lost in both ovarian tumors and ovarian cell lines compared to normal ovarian epithelium (Schultz *et al.*, 1996). Reduced OVCA1 protein expression has also been reported in some breast cancer samples (Bruening *et al.*, 1999). The introduction of exogenous OVCA1 into an ovarian cancer cell line led to a significant reduction in growth compared to the parental cells. Cyclin D1 expression was also reduced in the cells expressing OVCA1, possibly suggesting that OVCA1 could influence tumorigenesis through cell cycle deregulation (Bruening *et al.*, 1999). However, a role for OVCA1 in the development of breast and ovarian cancer remains to be further elucidated.

P21. P21 maps to chromosome 6q21, and has been implicated in the mediation of G_1–S cell arrest, as well as the G_1–M phase transition (Dulic *et al.*, 1998; Niculescu *et al.*, 1998). *P21* gene mutations are rare in both breast and ovarian cancers. Overexpression of the gene has been reported in high-grade breast tumors (Fernandez *et al.*, 1998), where high levels of *p21* expression are associated with favorable prognosis (Schmider *et al.*, 2000). However, low p21 protein expression is consistently associated with ovarian tumors of high grade, late stage and the presence of residual tumor following therapy (Anttila *et al.*, 1999). Expression of *P21* can be induced by wild-type *TP53*. Consistent with this, patients with p21 negative tumors in conjunction with mutant p53 have poorer survival rates and a greater risk of tumor recurrence (Anttila *et al.*, 1999).

P27KIP. P27, also known as KIP1 (cd*k* inhibitory *p*rotein), is a cyclin-dependent kinase inhibitor capable of inducing cell cycle arrest and mediating the assembly of CDK4 and cyclin D1 to the nucleus. Located on 12p13, the *P27KIP* gene shows striking similarities to *P21* demonstrating 42% homology at their N-termini (Polyak *et al.*, 1994; Toyoshima and Hunter, 1994). Although *P27KIP* mutations are rare (Spirin *et al.*, 1996), the expression of *P27KIP* appears to be an important prognostic marker in breast cancer, with low expression significantly associated with a poor prognosis, particularly in patients with node-negative disease (Tan *et al.*, 1997). There is evidence in mouse models that P27KIP may play a role in cancer development through haploinsufficiency (Fero *et al.*, 1998). Co-expression of low levels of *P27KIP* and high levels of cyclin E may be associated with a high mortality rate in breast and ovarian cancer patients (Porter *et al.*, 1997; Sui *et al.*, 1999). In ovarian cancer, 30–50% of tumors show loss of P27KIP expression (Baekelandt *et al.*, 1999a; Sui *et al.*, 1999), although there are no reports of any clinical associations.

Regions of loss of heterozygosity. Many regions of loss of heterozygosity (LOH) have been reported in breast and ovarian tumors, and are suggestive of the location of tumor suppressor genes. In *Figure 4* we summarize available LOH data for each chromosome arm in these cancers. The specific relevance to cancer development and prognosis of specific chromosomal regions identified through loss of

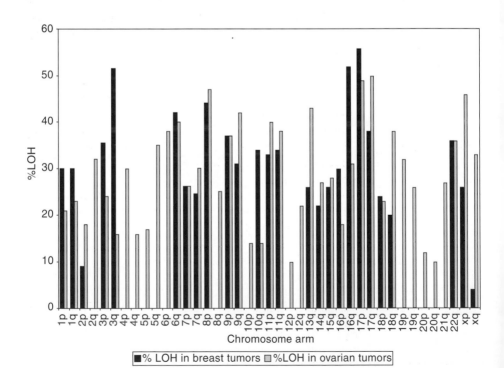

Figure 4. Summary of the frequency of LOH for the various chromosome arms in breast and ovarian cancer.

heterozygosity studies will only clearly be defined once the underlying tumor suppressor genes have been identified.

3.4 Metastasis suppressor genes

E-cadherin (CDH1). E-cadherin is a transmembrane glycoprotein that functions as a cell adhesion molecule (Fujimoto *et al.*, 1997) through its attachment to the actin cytoskeleton via α, β and γ-catenin. In order for a tumor cell to invade and metastasize it must detach from the primary tumor, therefore loss of functional E-cadherin or catenin is believed to promote metastatic progression.

A high frequency (50%) of LOH has been reported in breast tumors at the CDH1 locus on 16q24 (Berx *et al.*, 1995). In addition, approximately 50% of invasive ductal breast carcinomas show reduced expression of CDH1 (Oka *et al.*, 1993). Complete loss of expression has been reported in infiltrative lobular breast carcinomas suggesting that inactivation of CDH1 is of particular relevance to this tumor subtype (Berx *et al.*, 1995; Vos *et al.*, 1997). Consistent with these observations, inactivating mutations of *CDH1* have mainly been reported in lobular carcinomas (Berx *et al.*, 1995). *CDH1* mutations are rare in ovarian cancer (Risinger *et al.*, 1994). Protein expression is generally reduced in poorly differentiated ovarian tumors and ovarian cancer metastases compared with primary tumors implying that loss of CDH1 correlates with acquired invasive and metastatic potential (Davies *et al.*, 1998). The CDH1 pathway may potentially be disrupted via mutations in the catenin genes, and missense mutations of *β-catenin* (CTNNB1) have been reported in endometrioid ovarian tumors but rarely in other histological subtypes (Gamallo *et al.*, 1999; Wright *et al.*, 1999). Reduction or absence of either β or α-catenin expression in ovarian tumors is associated with increased metastatic potential and poor prognosis (Anttila *et al.*, 1998; Davies *et al.*, 1998).

Nm23 *(NME1 and 2).* The *nm23-H1 (NME1)* and *nm23-H2 (NME2)* genes are located on 17q21.3 and encode nucleoside diphosphate kinase A and B respectively. These proteins function to catalyse the phosphorylation of nucleoside diphosphates using ATP, resulting in the production of the corresponding nucleoside triphosphates (Lombardi *et al.*, 2000). Reduced expression of NME1 and NME2 is associated with a highly metastatic phenotype in some tumors and cell lines, including those of the breast and ovary (Iizuka *et al.*, 1995; Keim *et al.*, 1992; Stahl *et al.*, 1991; Steeg *et al.*, 1988). Expression of *NME1* mRNA in breast tumors correlates with lymph-node negative status, less aggressive disease and longer disease-free and overall survival rates (Bevilacqua *et al.*, 1989). Consistent with these observations, the metastatic potential of MDA-MB-435 breast cancer cells can be reduced *in vivo* following overexpression of NME1 (Leone *et al.*, 1993). NME1 may also be important in the progression of ovarian tumors as reduced NME1 expression is associated with late stage disease, lymph node involvement and distant metastasis (Mandai *et al.*, 1994).

3.5 Cell survival and cell death pathways

Telomerase. Telomeres are long tandem repeat sequences at the ends of all eukaryotic chromosomes that function to maintain chromosomal integrity during cell division (see Chapter 21). Incomplete replication of DNA molecules results in the progressive loss of telomeric sequences and, ultimately, cell death. Since the majority of normal somatic cells do not express telomerase, the ribonucleo-protein DNA polymerase responsible for telomere synthesis, they are unable to maintain their telomere length. However, a number of cancers have demonstrated high levels of telomerase, and recent evidence has confirmed telomerase activation as one of the mechanisms by which cancer cells evade death and achieve immortalization (Hahn *et al.*, 1999; Hiyama *et al.*, 1996).

Telomerase activity has been reported in over 90% of breast carcinomas (Landberg *et al.*, 1997). This activity is also associated with cell cycle deregulation, such as up-regulated cyclin D and E levels and increased cell proliferation (Landberg *et al.*, 1997). Similarly, telomerase activity has been reported in almost all malignant ovarian carcinomas, the majority of borderline ovarian tumors and up to 28% of benign ovarian lesions studied, with the intensity of telomerase activity higher in the malignant carcinomas than the other tumor stages (Park *et al.*, 1999). Both *TERC* (*hTR*), the gene encoding the telomerase RNA component (Kyo *et al.*, 1999) and *TERT*, the telomerase catalytic subunit, mRNAs are reported to be up-regulated in benign, borderline and malignant ovarian tumors (Oishi *et al.*, 1998). This may suggest that the altered expression of these genes may also play an important role in the malignant progression of ovarian tumors.

Apoptosis-related genes. The inhibition of apoptosis or programmed cell death plays an important role in cancer development (Lowe and Lin, 2000; Lundberg and Weinberg, 1999). A large number of genes have now been implicated in the regulation of apoptosis. The BCL2 family is comprised of at least 17 pro-apoptotic and anti-apoptotic members that mediate the release of mitchondrial Cytochrome c. Some members of the BCL2 family, including BCL2, BAX, BCLx, among others, are expressed in normal breast and ovarian tissue. Several studies have shown that the levels of several of the BCL2 family proteins are altered in breast cancer. BCL2, an anti-apoptotic protein, is frequently expressed in breast tumor tissue (Krajewski *et al.*, 1995), albeit at decreased levels compared to normal tissue (Zhang *et al.*, 1997). Loss of BAX expression, a pro-apoptotic member of the BCL2 family, has been reported in breast tumors and is associated with poor response to chemotherapy and shorter survival in women with metastatic breast disease (Krajewski *et al.*, 1995).

BCL2-positive and BCL2-negative breast tumors have distinct phenotypes. BCL2 expression levels strongly correlate with ER and PR status (Leek *et al.*, 1994; Silvestrini *et al.*, 1994). Related to this, BCL-2 protein can be up-regulated by 17β-estradiol (Perillo *et al.*, 2000). BCL2 protein expression inversely correlates with tumor grade, apoptotic index, proliferative index and TP53 overexpression (Castiglione *et al.*, 1999; Ellis *et al.*, 1998). TP53 up-regulates *BAX* and down-regulates *BCL2* (Haldar *et al.*, 1994; Miyashita *et al.*, 1994). These data suggest that BCL2-positive tumors may be indicative of a less aggressive phenotype. Indeed, BCL2 protein expression positively correlates with improved survival following treatment with hormone or chemotherapy (Gasparini *et al.*, 1995). In ovarian cancer, BCL2 protein expression is decreased in malignant tumors compared to benign tumors, and this expression inversely correlates with TP53 expression and apoptotic index (Ben-Hur *et al.*, 1999; Chan *et al.*, 2000). The absence of TP53 expression and presence of BCL2 is associated with increased survival (Baekelandt *et al.*, 1999b).

4. Conclusions

At present, it is not possible to develop a genetic progression model of changes associated with breast or ovarian cancer development. However, screening for mutations in hereditary genes such as *BRCA1*, *BRCA2* and *TP53*, can be used in a clinical setting to identify individuals at high-risk for developing these diseases. Psychosocial issues not withstanding, patient benefit will be in early detection and subsequent improved therapies and cure, as well as the identification of women willing to participate in prevention trials. Studies suggest that estrogen modulation may impact cancer risk. For non-hereditary patients, prognostic indicators such as estrogen and progesterone receptor status are currently being used to tailor treatment for patients with tumors responsive to endocrine therapies, as well as molecular targets such as Herceptin, used for patients with tumors over-expressing HER2. The current phase of the molecular revolution is enabling the high throughput analysis of gene expression changes in thousands of genes in tumors. The likely result of this gene profiling and identification of molecular markers for cancer progression will be improvements in the early detection of disease, as well as the identification of markers that will predict disease progression and dissemination. An eventual goal of molecular profiling of tumors will be the ability to individualize treatment options for patients. This will in turn have a significant positive impact on quality of life and patient cure rates.

References

Abbott, D.W., Freeman, M.L. and Holt, J.T. (1998) Double-strand break repair deficiency and radiation sensitivity in BRCA2 mutant cancer cells. *J. Natl Cancer Inst.* **90**: 978–985.

Abbott, D.W., Thompson, M.E., Robinson-Benion, C., Tomlinson, G., Jensen, R.A. and Holt, J.T. (1999) BRCA1 expression restores radiation resistance in BRCA1-defective cancer cells through enhancement of transcription-coupled DNA repair. *J. Biol. Chem.* **274**: 18808–18812.

Adnane, J., Gaudray, P., Simon, M., Simony-Lafontaine, J., Jeanteur, P. and Theillet, C. (1989) Proto-oncogene amplification and human breast tumor phenotype. *Oncogene* **4**: 1389–1395.

Allred, D.C., Clark, G.M., Elledge, R. *et al.* (1993) Association of p53 protein expression with tumor cell proliferation rate and clinical outcome in node-negative breast cancer. *J. Natl Cancer Inst.* **85**: 200–206.

Anttila, M., Kosma, V.M., Ji, H. (1998) Clinical significance of alpha-catenin, collagen IV, and Ki-67 expression in epithelial ovarian cancer. *J. Clinical Oncol.* **16**: 2591–2600.

Anttila, M.A., Kosma, V.M., Hongxiu, J., Puolakka, J., Juhola, M., Saarikoski, S. and Syrjanen, K. (1999) p21/WAF1 expression as related to p53, cell proliferation and prognosis in epithelial ovarian cancer. *Br. J. Cancer* **79**: 1870–1878.

Armes, J.E., Egan, A.J., Southey, M.C. *et al.* (1998) The histologic phenotypes of breast carcinoma occurring before age 40 years in women with and without BRCA1 or BRCA2 germline mutations: a population-based study. *Cancer* **83**: 2335–2345.

Armes, J.E., Trute, L., White, D. *et al.* (1999) Distinct molecular pathogeneses of early-onset breast cancers in BRCA1 and BRCA2 mutation carriers: a population-based study. *Cancer Res.* **59**: 2011–2017.

Arnold, N., Hagele, L., Walz, L., Schempp, W., Pfisterer, J., Bauknecht, T. and Kiechle, M. (1996) Over representation of 3q and 8q material and loss of 18q material are recurrent findings in advanced human ovarian cancer. *Genes Chromosomes Cancer* **16**: 46–54.

Baekelandt, M., Holm, R., Trope, C.G., Nesland, J.M. and Kristensen, G.B. (1999a) Lack of independent prognostic significance of p21 and p27 expression in advanced ovarian cancer: an immunohistochemical study. *Clin. Cancer Res.* **5**: 2848–2853.

Baekelandt, M., Kristensen, G.B., Nesland, J.M., Trope, C.G. and Holm, R. (1999b) Clinical significance of apoptosis-related factors p53, mdm2, and bcl-2 in advanced ovarian cancer. *J. Clin. Oncol.* **17**: 2061.

Baker, V.V., Borst, M.P., Dixon, D., Hatch, K.D., Shingleton, H.M. and Miller, D. (1990) c-myc amplification in ovarian cancer. *Gynecol. Oncol.* **38**: 340–342.

Barker, D.F., Almeida, E.F.A., Casey, G. *et al.* (1996) BRCA1 R841W: a strong candidate for a common mutation with moderate phenotype. *Genet. Epidemiol.* **13**: 595–604.

Barnes, D.M., Bartkova, J., Camplejohn, R.S., Gullick, W.J., Smith, P.J. and Millis, R.R. (1992) Overexpression of the c-erbB-2 oncoprotein: why does this occur more frequently in ductal carcinoma in situ than in invasive mammary carcinoma and is this of prognostic significance? *Eur. J. Cancer* **28**: 644–648.

Bauknecht, T., Angel, P., Kohler, M., Kommoss, F., Birmelin, G., Pfleiderer, A. and Wagner, E. (1993) Gene structure and expression analysis of the epidermal growth factor receptor, transforming growth factor-alpha, myc, jun, and metallothionine in human ovarian carcinomas. Classification of malignant phenotypes. *Cancer* **71**: 419–429.

Bell, D.W., Varley, J.M., Szydlo, T.E. *et al.* (1999) Heterozygous germ line hCHK2 mutations in Li-Fraumeni syndrome. *Science* **286**: 2528–2531.

Ben-Hur, H., Gurevich, P., Huszar, M. *et al.* (1999) Apoptosis and apoptosis-related proteins in the epithelium of human ovarian tumors: immunohistochemical and morphometric studies. *Eur. J. Gynaecol. Oncol.* **20**: 249–253.

Berchuck, A., Kohler, M.F., Marks, J.R., Wiseman, R., Boyd, J. and Bast, R.C., Jr. (1994) The p53 tumor suppressor gene frequently is altered in gynecologic cancers. *Am. J. Obstet. Gynecol.* **170**: 246–252.

Berns, E.M., Klijn, J.G., van Putten, W.L., van Staveren, I.L., Portengen, H. and Foekens, J.A. (1992a) c-myc amplification is a better prognostic factor than HER2/neu amplification in primary breast cancer. *Cancer Res.* **52**: 1107–1113.

Berns, E.M., Klijn, J.G., van Staveren, I.L., Portengen, H., Noordegraaf, E. and Foekens, J.A. (1992b) Prevalence of amplification of the oncogenes c-myc, HER2/neu, and int-2 in one thousand human breast tumours: correlation with steroid receptors. *Eur. J. Cancer* **28**: 697–700.

Berns, E.M., de Klein, A., van Putten, W.L., van Staveren, I.L., Bootsma, A., Klijn, J.G. and Foekens, J.A. (1995) Association between RB-1 gene alterations and factors of favourable prognosis in human breast cancer, without effect on survival. *Int. J. Cancer* **64**: 140–145.

Berx, G., Cleton-Jansen, A.M., Nollet, F., de Leeuw, W.J., van de Vijver, M., Cornelisse, C. and van Roy, F. (1995) E-cadherin is a tumour/invasion suppressor gene mutated in human lobular breast cancers. *EMBO J.* **14**: 6107–6115.

Bevilacqua, G., Sobel, M.E., Liotta, L.A. and Steeg, P.S. (1989) Association of low nm23 RNA levels in human primary infiltrating ductal breast carcinomas with lymph node involvement and other histopathological indicators of high metastatic potential. *Cancer Res.* **49**: 5185–5190.

Bieche, I. and Lidereau, R. (1995) Genetic alterations in breast cancer. *Genes Chromosomes Cancer* **14**: 227–251.

Bignell, G., Micklem, G., Stratton, M.R., Ashwork, A. and Wooster, R. (1997) The BRC repeats are conserved in mammalian BRCA2 proteins. *Hum. Mol. Genet.* **6**: 53–58.

Birch, J.M., Blair, V., Kelsey, A.M., Evans, D.G., Harris, M., Tricker, K.J. and Varley, J.M. (1998) Cancer phenotype correlates with constitutional TP53 genotype in families with the Li-Fraumeni syndrome. *Oncogene* **17**: 1061–1068.

Borg, A., Zhang, Q.X., Alm, P., Olsson, H. and Sellberg, G. (1992) The retinoblastoma gene in breast cancer: allele loss is not correlated with loss of gene protein expression. *Cancer Res.* **52**: 2991–2994.

Bork, P., Blomberg, N. and Nilges, M. (1996) Internal repeats in the BRCA2 protein sequence. *Nat. Genet.* **13**: 22–23.

Bose, S., Wang, S.I., Terry, M.B., Hibshoosh, H. and Parsons, R. (1998) Allelic loss of chromosome 10q23 is associated with tumor progression in breast carcinomas. *Oncogene* **17**: 123–127.

Brenner, A.J. and Aldaz, C.M. (1995) Chromosome 9p allelic loss and p16/CDKN2 in breast cancer and evidence of p16 inactivation in immortal breast epithelial cells. *Cancer Res.* **55:** 2892–2895.

Brenner, A.J., Paladugu, A., Wang, H., Olopade, O.I., Dreyling, M.H. and Aldaz, C.M. (1996) Preferential loss of expression of p16(INK4a) rather than p19(ARF) in breast cancer. *Clin. Cancer Res.* **2:** 1993–1998.

Broeks, A., Urbanus, J.H., Floore, A.N. *et al.* (2000) ATM-heterozygous germline mutations contribute to breast cancer-susceptibility. *Am. J. Hum. Genet.* **66:** 494–500.

Bruening, W., Prowse, A.H., Schultz, D.C., Holgado-Madruga, M., Wong, A. and Godwin, A.K. (1999) Expression of OVCA1, a candidate tumor suppressor, is reduced in tumors and inhibits growth of ovarian cancer cells. *Cancer Res.* **59:** 4973–4983.

Brugarolas, J. and Jacks, T. (1997) Double indemnity: p53, BRCA and cancer. p53 mutation partially rescues developmental arrest in Brca1 and Brca2 null mice, suggesting a role for familial breast cancer genes in DNA damage repair. *Nat. Med.* **3:** 721–722.

Cairns, P., Polascik, T.J., Eby, Y. *et al.* (1995) Frequency of homozygous deletion at p16/CDKN2 in primary human tumours. *Nat. Genet.* **11:** 210–212.

Callebaut, I. and Mornon, J.P. (1997) From BRCA1 to RAP1: a widespread BRCT module closely associated with DNA repair. *FEBS Lett.* **400:** 25–30.

Campbell, I.G., Foulkes, W.D., Beynon, G., Davis, M. and Englefield, P. (1995) LOH and mutation analysis of CDKN2 in primary human ovarian cancers. *Int. J. Cancer* **63:** 222–225.

Carraway, K.L., 3rd and Cantley, L.C. (1994) A neu acquaintance for erbB3 and erbB4: a role for receptor heterodimerization in growth signaling. *Cell* **78:** 5–8.

Casey, G., Lopez, M.E., Ramos, J.C. *et al.* (1996) DNA sequence analysis of exons 2 through 11 and immunohistochemical staining are required to detect all known p53 alterations in human malignancies. *Oncogene* **13:** 1971–1981.

Castiglione, F., Sarotto, I., Fontana, V. *et al.* (1999) Bcl2, p53 and clinical outcome in a series of 138 operable breast cancer patients. *Anticancer Res.* **19:** 4555–4563.

Catteau, A., Harris, W.H., Xu, C.F. and Solomon, E. (1999) Methylation of the BRCA1 promoter region in sporadic breast and ovarian cancer: correlation with disease characteristics. *Oncogene* **18:** 1957–1965.

Chan, W.Y., Cheung, K.K., Schorge, J.O. *et al.* (2000) Bcl-2 and p53 protein expression, apoptosis, and p53 mutation in human epithelial ovarian cancers. *Am. J. Pathol.* **156:** 409–417.

Chappuis, P.O., Nethercot, V. and Foulkes, W.D. (2000) Clinico-pathological characteristics of BRCA1- and BRCA2-related breast cancer. *Semin. Surg. Oncol.* **18:** 287–295.

Chehab, N.H., Malikzay, A., Appel, M. and Halazonetis, T.D. (2000) Chk2/hCds1 functions as a DNA damage checkpoint in G(1) by stabilizing p53. *Genes Dev.* **14:** 278–288.

Chen, J., Silver, D.P., Walpita, D. *et al.* (1998) Stable interaction between the products of the BRCA1 and BRCA2 tumor suppressor genes in mitotic and meiotic cells. *Mol. Cell* **2:** 317–328.

Clark, G.M. and McGuire, W.L. (1988) Steroid receptors and other prognostic factors in primary breast cancer. *Semin. Oncol.* **15:** 20–25.

Claus, E.B., Schildkraut, J.M., Thompson, W.D. and Risch, N.J. (1996) The genetic attributable risk of breast and ovarian cancer. *Cancer* **77:** 2318–2324.

Collins, N., Wooster, R. and Stratton, M.R. (1997) Absence of methylation of CpG dinucleotides within the promoter of the breast cancer susceptibility gene BRCA2 in normal tissues and in breast and ovarian cancers. *Br. J. Cancer* **76:** 1150–1156.

Connor, F., Bertwistle, D., Mee, P.J. *et al.* (1997) Tumorigenesis and a DNA repair defect in mice with a truncating Brca2 mutation. *Nat. Genet.* **17:** 423–430.

Consortium, B.C.L. (1997) Pathology of familial breast cancer: differences between breast cancers in carriers of *BRCA1* or *BRCA2* mutations and sporadic cases. *Lancet* **349:** 1505–1510.

Cortez, D., Wang, Y., Qin, J. and Elledge, S.J. (1999) Requirement of ATM-dependent phosphorylation of brca1 in the DNA damage response to double-strand breaks. *Science* **286:** 1162–1166.

Couch, F.J., DeShano, M.L., Blackwood, M.A. *et al.* (1997) BRCA1 mutations in women attending clinics that evaluate the risk of breast cancer. *N. Engl. J. Med.* **336:** 1409–1415.

Courjal, F., Louason, G., Speiser, P., Katsaros, D., Zeillinger, R. and Theillet, C. (1996) Cyclin gene amplification and overexpression in breast and ovarian cancers: evidence for the selection of cyclin D1 in breast and cyclin E in ovarian tumors. *Int. J. Cancer* **69:** 247–253.

Coussens, L., Yang-Feng, T.L., Liao, Y.C. *et al.* (1985) Tyrosine kinase receptor with extensive homology to EGF receptor shares chromosomal location with neu oncogene. *Science* **230:** 1132–1139.

Crook, T., Crossland, S., Crompton, M.R., Osin, P. and Gusterson, B.A. (1997) *p53* mutations in *BRCA1*-associated familial breast cancer. *Lancet* **350**: 638–639.

Dang, C.V. (1999) c-Myc target genes involved in cell growth, apoptosis, and metabolism. *Mol. Cell Biol.* **19**: 1–11.

Dang, C.V., Resar, L.M., Emison, E. *et al.* (1999) Function of the c-Myc oncogenic transcription factor. *Exp. Cell Res.* **253**: 63–77.

Davidoff, A.M., Kerns, B.J., Iglehart, J.D. and Marks, J.R. (1991) Maintenance of p53 alterations throughout breast cancer progression. *Cancer Res.* **51**: 2605–2610.

Davies, B.R., Worsley, S.D. and Ponder, B.A. (1998) Expression of E-cadherin, alpha-catenin and beta-catenin in normal ovarian surface epithelium and epithelial ovarian cancers. *Histopathology* **32**: 69–80.

Di Cristofano, A. and Pandolfi, P.P. (2000) The multiple roles of PTEN in tumor suppression. *Cell* **100**: 387–390.

Diebold, J., Mosinger, K., Peiro, G. *et al.* (2000) 20q13 and cyclin D1 in ovarian carcinomas. Analysis by fluorescence in situ hybridization. *J. Pathol.* **190**: 564–571.

Dodson, M.K., Cliby, W.A., Xu, H.J. *et al.* (1994) Evidence of functional RB protein in epithelial ovarian carcinomas despite loss of heterozygosity at the RB locus. *Cancer Res.* **54**: 610–613.

Domagala, W., Harezga, B., Szadowska, A., Markiewski, M., Weber, K. and Osborn, M. (1993) Nuclear p53 protein accumulates preferentially in medullary and high-grade ductal but rarely in lobular breast carcinomas. *Am. J. Pathol.* **142**: 669–674.

Dong, Y., Walsh, M.D., McGuckin, M.A. *et al.* (1997) Reduced expression of retinoblastoma gene product (pRB) and high expression of p53 are associated with poor prognosis in ovarian cancer. *Int. J. Cancer* **74**: 407–415.

Dulic, V., Stein, G.H., Far, D.F. and Reed, S.I. (1998) Nuclear accumulation of p21Cip1 at the onset of mitosis: a role at the G2/M-phase transition. *Mol. Cell Biol.* **18**: 546–557.

Dunning, A.M., Chiano, M., Smith, N.R. *et al.* (1997) Common BRCA1 variants and susceptibility to breast and ovarian cancer in the general population. *Hum. Mol. Genet.* **6**: 285–289.

Durocher, F., Shattuck-Eidens, D., McClure, M., Labrie, F., Skolnick, M.H., Goldgar, D.E. and Simard, J. (1996) Comparison of BRCA1 polymorphisms, rare sequence variants and/or missense mutations in unaffected and breast/ovarian cancer populations. *Hum. Mol. Genet.* **5**: 835–842.

Dutta, A., Chandra, R., Leiter, L.M. and Lester, S. (1995) Cyclins as markers of tumor proliferation: immunocytochemical studies in breast cancer. *Proc. Natl Acad. Sci. USA* **92**: 5386–5390.

Easton, D.F., Ford, D. and Bishop, D.T. (1995) Breast and ovarian cancer incidence in BRCA1-mutation carriers. Breast Cancer Linkage Consortium. *Am. J. Hum. Genet.* **56**: 265–271.

Eccles, D.M., Brett, L., Lessells, A., Gruber, L., Lane, D., Steel, C.M. and Leonard, R.C. (1992) Overexpression of the p53 protein and allele loss at 17p13 in ovarian carcinoma. *Br. J. Cancer* **65**: 40–44.

Eisinger, F., Stoppa-Lyonnet, D., Longy, M. *et al.* (1996) Germ line mutation at BRCA1 affects the histoprognostic grade in hereditary breast cancer. *Cancer Res.* **56**: 471–474.

Eisinger, F., Jacquemier, J., Charpin, C. *et al.* (1998) Mutations at *BRCA1*: the medullary breast carcinoma revisited. *Cancer Res.* **58**: 1588–1592.

Elledge, R.M., Fuqua, S.A., Clark, G.M., Pujol, P., Allred, D.C. and McGuire, W.L. (1993) Prognostic significance of p53 gene alterations in node-negative breast cancer. *Breast Cancer Res. Treat.* **26**: 225–235.

Ellis, P.A., Smith, I.E., Detre, S. *et al.* (1998) Reduced apoptosis and proliferation and increased Bcl-2 in residual breast cancer following preoperative chemotherapy. *Breast Cancer Res. Treat.* **48**: 107–116.

Eng, C. (1998) Genetics of Cowden syndrome: through the looking glass of oncology. *Int. J. Oncology* **12**: 701–710.

Esteller, M., Silva, J.M., Dominguez, G. *et al.* (2000) Promoter hypermethylation and BRCA1 inactivation in sporadic breast and ovarian tumors. *J. Natl. Cancer Inst.* **92**: 564–569.

Fan, S., Wang, J., Yuan, R. *et al.* (1999) BRCA1 inhibition of estrogen receptor signaling in transfected cells. *Science* **284**: 1354–1356.

Feilotter, H.E., Coulon, V., McVeigh, J.L. *et al.* (1999) Analysis of the 10q23 chromosomal region and the PTEN gene in human sporadic breast carcinoma. *Br. J. Cancer* **79**: 718–723.

Fernandez, P.L., Jares, P., Rey, M.J., Campo, E. and Cardesa, A. (1998) Cell cycle regulators and their abnormalities in breast cancer. *Mol. Pathol.* **51**: 305–309.

Fero, M.L., Randel, E., Gurley, K.E., Roberts, J.M. and Kemp, C.J. (1998) The murine gene p27Kip1 is haplo-insufficient for tumour suppression. *Nature* **396**: 177–180.

Fischer-Colbrie, J., Witt, A., Heinzl, H., Speiser, P., Czerwenka, K., Sevelda, P. and Zeillinger, R. (1997) EGFR and steroid receptors in ovarian carcinoma: comparison with prognostic parameters and outcome of patients. *Anticancer Res.* **17**: 613–619.

FitzGerald, M.G., MacDonald, D.J., Krainer, M. *et al.* (1996) Germ-line BRCA1 mutations in Jewish and non-Jewish women with early-onset breast cancer. *N. Engl. J. Med.* **334**: 143–149.

FitzGerald, M.G., Bean, J.M., Hegde, S.R. *et al.* (1997) Heterozygous ATM mutations do not contribute to early onset of breast cancer. *Nat. Genet.* **15**: 307–310.

Foray, N., Randrianarison, V., Marot, D., Perricaudet, M., Lenoir, G. and Feunteun, J. (1999) Gamma-rays-induced death of human cells carrying mutations of BRCA1 or BRCA2. *Oncogene* **18**: 7334–7342.

Ford, D., Easton, D.F., Bishop, D.T., Narod, S.A. and Goldgar, D.E. (1994) Risks of cancer in BRCA1-mutation carriers. Breast Cancer Linkage Consortium. *Lancet* **343**: 692–695.

Ford, D., Easton, D.F., Stratton, M. *et al.* (1998) Genetic heterogeneity and penetrance analysis of the BRCA1 and BRCA2 genes in breast cancer families. The Breast Cancer Linkage Consortium. *Am. J. Hum. Genet.* **62**: 676–689.

Foulkes, W.D., Wong, N., Brunet, J.-S. *et al.* (1997) Germ-line *BRCA1* mutation is an adverse prognostic factor in Ashkenazi Jewish women with breast cancer. *Clin. Cancer Res.* **3**: 2465–2469.

Fox, S.B., Smith, K., Hollyer, J., Greenall, M., Hastrich, D. and Harris, A.L. (1994) The epidermal growth factor receptor as a prognostic marker: results of 370 patients and review of 3009 patients. *Breast Cancer Res. Treat.* **29**: 41–49.

Fujimoto, J., Ichigo, S., Hirose, R., Sakaguchi, H. and Tamaya, T. (1997) Expression of E-cadherin and alpha- and beta-catenin mRNAs in ovarian cancers. *Cancer Lett.* **115**: 207–212.

Futreal, P.A., Liu, Q., Shattuck-Eidens, D. *et al.* (1994) BRCA1 mutations in primary breast and ovarian carcinomas. *Science* **266**: 120–122.

Gamallo, C., Palacios, J., Moreno, G., Calvo de Mora, J., Suarez, A. and Armas, A. (1999) Beta-catenin expression pattern in stage I and II ovarian carcinomas: relationship with beta-catenin gene mutations, clinicopathological features, and clinical outcome. *Am. J. Pathol.* **155**: 527–536.

Gasparini, G., Barbareschi, M., Doglioni, C. *et al.* (1995) Expression of bcl-2 protein predicts efficacy of adjuvant treatments in operable node-positive breast cancer. *Clin. Cancer Res.* **1**: 189–198.

Gayther, S.A., Russell, P., Harrington, P., Antoniou, A.C., Easton, D.F. and Ponder, B.A. (1999) The contribution of germline BRCA1 and BRCA2 mutations to familial ovarian cancer: No evidence for other ovarian cancer-susceptibility genes. *Am. J. Hum. Genet.* **65**: 1021–1029.

Gillett, C., Smith, P., Gregory, W., Richards, M., Millis, R., Peters, G. and Barnes, D. (1996) Cyclin D1 and prognosis in human breast cancer. *Int. J. Cancer* **69**: 92–99.

Gillett, C.E. and Barnes, D.M. (1998) Demystified ... cell cycle. *Mol. Pathol.* **51**: 310–316.

Gowen, L.C., Avrutskaya, A.V., Latour, A.M., Koller, B.H. and Leadon, S.A. (1998) BRCA1 required for transcription-coupled repair of oxidative DNA damage. *Science* **281**: 1009–1012.

Greenblatt, M.S., Bennett, W.P., Hollstein, M. and Harris, C.C. (1994) Mutations in the p53 tumor suppressor gene: clues to cancer etiology and molecular pathogenesis. *Cancer Res.* **54**: 4855–4878.

Gu, J., Tamura, M., Pankov, R., Danen, E.H., Takino, T., Matsumoto, K. and Yamada, K.M. (1999) Shc and FAK differentially regulate cell motility and directionality modulated by PTEN. *J. Cell Biol.* **146**: 389–403.

Hahn, W.C., Counter, C.M., Lundberg, A.S., Beijersbergen, R.L., Brooks, M.W. and Weinberg, R.A. (1999) Creation of human tumour cells with defined genetic elements. *Nature* **400**: 464–468.

Hakem, R., de la Pompa, J.L., Sirard, C. *et al.* (1996) The tumor suppressor gene Brca1 is required for embryonic cellular proliferation in the mouse. *Cell* **85**: 1009–1023.

Hakem, R., de la Pompa, J.L., Elia, A., Potter, J. and Mak, T.W. (1997) Partial rescue of Brca1[5–6] early embryonic lethality by *p53* or *p21* null mutation. *Nat. Genet.* **16**: 298–302.

Haldar, S., Negrini, M., Monne, M., Sabbioni, S. and Croce, C.M. (1994) Down-regulation of bcl-2 by p53 in breast cancer cells. *Cancer Res.* **54**: 2095–2097.

Harris, A.L., Nicholson, S., Sainsbury, J.R., Farndon, J. and Wright, C. (1989) Epidermal growth factor receptors in breast cancer: association with early relapse and death, poor response to hormones and interactions with neu. *J. Steroid Biochem.* **34**: 123–131.

Herman, J.G., Merlo, A., Mao, L. *et al.* (1995) Inactivation of the CDKN2/p16/MTS1 gene is frequently associated with aberrant DNA methylation in all common human cancers. *Cancer Res.* **55**: 4525–4530.

Hirao, A., Kong, Y.Y., Matsuoka, S. *et al.* (2000) DNA damage-induced activation of p53 by the checkpoint kinase Chk2. *Science* **287**: 1824–1827.

Hiyama, E., Gollahon, L., Kataoka, T. *et al.* (1996) Telomerase activity in human breast tumors. *J. Natl. Cancer Inst.* **88**: 116–122.

Horak, E., Smith, K., Bromley, L., LeJeune, S., Greenall, M., Lane, D. and Harris, A.L. (1991) Mutant p53, EGF receptor and c-erbB-2 expression in human breast cancer. *Oncogene* **6**: 2277–2284.

Huiping, C., Johannsdottir, J.T., Arason, A., Olafsdottir, G.H., Eiriksdottir, G., Egilsson, V. and Ingvarsson, S. (1999) Replication error in human breast cancer: comparison with clinical variables and family history of cancer. *Oncol. Rep.* **6**: 117–122.

Ichikawa, Y., Yoshida, S., Koyama, Y. *et al.* (1996) Inactivation of p16/CDKN2 and p15/MTS2 genes in different histological types and clinical stages of primary ovarian tumors. *Int. J. Cancer* **69**: 466–470.

Iizuka, N., Oka, M., Noma, T., Nakazawa, A., Hirose, K. and Suzuki, T. (1995) NM23-H1 and NM23-H2 messenger RNA abundance in human hepatocellular carcinoma. *Cancer Res.* **55**: 652–657.

Ingvarsson, S., Geirsdottir, E.K., Johannesdottir, G. *et al.* (1998) High incidence of loss of heterozygosity in breast tumor from carriers of the BRCA2 999del5 mutation. *Cancer Res.* **58**: 4421–4425.

Izatt, L., Greenman, J., Hodgson, S. *et al.* (1999) Identification of germline missense mutations and rare allelic variants in the ATM gene in early-onset breast cancer. *Genes Chromosomes Cancer* **26**: 286–294.

Jacquemier, J., Eisinger, F., Nogues, C. *et al.* (1999) Histological type and syncytial growth pattern affect E-cadherin expression in a multifactorial analysis of a combined panel of sporadic and BRCA1-associated breast cancers. *Int. J. Cancer* **83**: 45–49.

Janezic, S.A., Ziogas, A., Krumroy, L.M. *et al.* (1999) Germline BRCA1 alterations in a population-based series of ovarian cancer cases. *Hum. Mol. Genet.* **8**: 889–897.

Jares, P., Rey, M.J., Fernandez, P.L. *et al.* (1997) Cyclin D1 and retinoblastoma gene expression in human breast carcinoma: correlation with tumour proliferation and oestrogen receptor status. *J. Pathol.* **182**: 160–166.

Johannsson, O., Ostermeyer, E.A., Hakansson, S. *et al.* (1996) Founding BRCA1 mutations in hereditary breast and ovarian cancer in southern Sweden. *Am. J. Hum. Genet.* **58**: 441–450.

Johannsson, O.T., Idvall, I., Anderson, C., Borg, A., Barkardottir, R.B., Egilsson, V. and Olsson, H. (1997) Tumour biological features of BRCA1-induced breast and ovarian cancer. *Eur. J. Cancer* **33**: 362–371.

Johannsson, O.T., Ranstam, J., Borg, A. and Olsson, H. (1998) Survival of BRCA1 breast and ovarian cancer patients: a population-based study from southern Sweden. *J. Clin. Oncol.* **16**: 397–404.

Katsaros, D., Theillet, C., Zola, P. *et al.* (1995) Concurrent abnormal expression of erbB-2, myc and ras genes is associated with poor outcome of ovarian cancer patients. *Anticancer Res.* **15**: 1501–1510.

Keim, D., Hailat, N., Melhem, R. *et al.* (1992) Proliferation-related expression of p19/nm23 nucleoside diphosphate kinase. *J. Clin. Invest.* **89**: 919–924.

Keyomarsi, K., O'Leary, N., Molnar, G., Lees, E., Fingert, H.J. and Pardee, A.B. (1994) Cyclin E, a potential prognostic marker for breast cancer. *Cancer Res.* **54**: 380–385.

Kim, T.M., Benedict, W.F., Xu, H.J. *et al.* (1994) Loss of heterozygosity on chromosome 13 is common only in the biologically more aggressive subtypes of ovarian epithelial tumors and is associated with normal retinoblastoma gene expression. *Cancer Res.* **54**: 605–609.

Klijn, J.G., Look, M.P., Portengen, H., Alexieva-Figusch, J., van Putten, W.L. and Foekens, J.A. (1994) The prognostic value of epidermal growth factor receptor (EGF-R) in primary breast cancer: results of a 10 year follow-up study. *Breast Cancer Res. Treat.* **29**: 73–83.

Koonin, E.V., Altschul, S.F. and Bork, P. (1996) BRCA1 protein products ... Functional motifs ... *Nat. Genet.* **13**: 266–267.

Krainer, M., Silva-Arrieta, S., FitzGerald, M.G. *et al.* (1997) Differential contributions of BRCA1 and BRCA2 to early-onset breast cancer. *N. Engl. J. Med.* **336**: 1416–1421.

Krajewski, S., Blomqvist, C., Franssila, K. *et al.* (1995) Reduced expression of proapoptotic gene BAX is associated with poor response rates to combination chemotherapy and shorter survival in women with metastatic breast adenocarcinoma. *Cancer Res.* **55**: 4471–4478.

Kyo, S., Kanaya, T., Takakura, M., Tanaka, M., Yamashita, A., Inoue, H. and Inoue, M. (1999) Expression of human telomerase subunits in ovarian malignant, borderline and benign tumors. *Int. J. Cancer* **80**: 804–809.

Landberg, G., Nielsen, N.H., Nilsson, P., Emdin, S.O., Cajander, J. and Roos, G. (1997) Telomerase activity is associated with cell cycle deregulation in human breast cancer. *Cancer Res.* **57**: 549–554.

Lane, D.P. (1992) Cancer. p53, guardian of the genome. *Nature* **358**: 15–16.

Lee, J.S., Wacholder, S., Struewing, J.P. *et al.* (1999) Survival after breast cancer in Ashkenazi Jewish BRCA1 and BRCA2 mutation carriers. *J. Natl. Cancer Inst.* **91**: 259–263.

Lee, J.S., Collins, K.M., Brown, A.L., Lee, C.H. and Chung, J.H. (2000) hCds1-mediated phosphorylation of BRCA1 regulates the DNA damage response. *Nature* **404**: 201–204.

Leek, R.D., Kaklamanis, L., Pezzella, F., Gatter, K.C. and Harris, A.L. (1994) bcl-2 in normal human breast and carcinoma, association with oestrogen receptor-positive, epidermal growth factor receptor-negative tumours and in situ cancer. *Br. J. Cancer* **69**: 135–139.

Leitzel, K., Teramoto, Y., Konrad, K. *et al.* (1995) Elevated serum c-erbB-2 antigen levels and decreased response to hormone therapy of breast cancer. *J. Clin. Oncol.* **13**: 1129–1135.

Leone, A., Flatow, U., VanHoutte, K. and Steeg, P.S. (1993) Transfection of human nm23-H1 into the human MDA-MB-435 breast carcinoma cell line: effects on tumor metastatic potential, colonization and enzymatic activity. *Oncogene* **8**: 2325–2333.

Li, F.P. and Fraumeni, J.F., Jr. (1969) Soft-tissue sarcomas, breast cancer, and other neoplasms. A familial syndrome? *Ann. Intern. Med.* **71**: 747–752.

Li, J., Yen, C., Liaw, D. *et al.* (1997) PTEN, a putative protein tyrosine phosphatase gene mutated in human brain, breast, and prostate cancer. *Science* **275**: 1943–1947.

Li, S.B., Schwartz, P.E., Lee, W.H. and Yang-Feng, T.L. (1991) Allele loss at the retinoblastoma locus in human ovarian cancer. *J. Natl. Cancer Inst.* **83**: 637–640.

Liaw, D., Marsh, D.J., Li, J. *et al.* (1997) Germline mutations of the PTEN gene in Cowden disease, an inherited breast and thyroid cancer syndrome. *Nat. Genet.* **16**: 64–67.

Liu, E., Thor, A., He, M., Barcos, M., Ljung, B.M. and Benz, C. (1992) The HER2 (c-erbB-2) oncogene is frequently amplified in in situ carcinomas of the breast. *Oncogene* **7**: 1027–1032.

Liu, F.S., Ho, E.S., Chen, J.T., Shih, R.T., Yang, C.H. and Shih, A. (1995) Overexpression or mutation of the p53 tumor suppressor gene does not occur in malignant ovarian germ cell tumors. *Cancer* **76**: 291–295.

Lombardi, D., Lacombe, M.L. and Paggi, M.G. (2000) nm23: unraveling its biological function in cell differentiation. *J. Cell Physiol.* **182**: 144–149.

Lowe, S.W. and Lin, A.W. (2000) Apoptosis in cancer. *Carcinogenesis* **21**: 485–495.

Ludwig, T., Chapman, D.L., Papaioannou, V.E. and Efstratiadis, A. (1997) Targeted mutations of breast cancer susceptibility gene homologs in mice: lethal phenotypes of Brca1, Brca2, Brca1/Brca2, Brca1/p53, and Brca2/p53 nullizygous embryos. *Genes Dev.* **11**: 1226–1241.

Lundberg, A.S. and Weinberg, R.A. (1999) Control of the cell cycle and apoptosis. *Eur. J. Cancer* **35**: 531–539.

Maehama, T. and Dixon, J.E. (1998) The tumor suppressor, PTEN/MMAC1, dephosphorylates the lipid second messenger, phosphatidylinositol 3,4,5-trisphosphate. *J. Biol. Chem.* **273**: 13375–13378.

Magdinier, F., Ribieras, S., Lenoir, G.M., Frappart, L. and Dante, R. (1998) Down-regulation of BRCA1 in human sporadic breast cancer; analysis of DNA methylation patterns of the putative promoter region. *Oncogene* **17**: 3169–3176.

Malkin, D., Li, F.P., Strong, L.C. *et al.* (1990) Germ line p53 mutations in a familial syndrome of breast cancer, sarcomas, and other neoplasms. *Science* **250**: 1233–1238.

Mallory, S.B. (1995) Cowden syndrome (multiple hamartoma syndrome). *Dermatol. Clin.* **13**: 27–31.

Malone, K.E., Daling, J.R., Neal, C. *et al.* (2000) Frequency of BRCA1/BRCA2 mutations in a population-based sample of young breast carcinoma cases. *Cancer* **88**: 1393–1402.

Mancini, D.N., Rodenhiser, D.I., Ainsworth, P.J., O'Malley, F.P., Singh, S.M., Xing, W. and Archer, T.K. (1998) CpG methylation within the 5′ regulatory region of the *BRCA1* gene is tumor specific and includes a putative CREB binding site. *Oncogene* **16**: 1161–1169.

Mandai, M., Konishi, I., Koshiyama, M. *et al.* (1994) Expression of metastasis-related nm23-H1 and nm23-H2 genes in ovarian carcinomas: correlation with clinicopathology, EGFR, c-erbB-2, and c-erbB-3 genes, and sex steroid receptor expression. *Cancer Res.* **54**: 1825–1830.

Marchetti, A., Buttitta, F., Pellegrini, S. *et al.* (1993) p53 mutations and histological type of invasive breast carcinoma. *Cancer Res.* **53**: 4665–4669.

Marcus, J.N., Watson, P., Page, D.L. *et al.* (1996) Hereditary breast cancer: pathobiology, prognosis, and BRCA1 and BRCA2 gene linkage. *Cancer* 77: 697–709.

Marks, J.R., Davidoff, A.M., Kerns, B.J. *et al.* (1991) Overexpression and mutation of p53 in epithelial ovarian cancer. *Cancer Res.* 51: 2979–2984.

Marks, J.R., Humphrey, P.A., Wu, K., Berry, D., Bandarenko, N., Kerns, B.J. and Iglehart, J.D. (1994) Overexpression of p53 and HER-2/neu proteins as prognostic markers in early stage breast cancer. *Ann. Surg.* 219: 332–341.

Marmorstein, L.Y., Ouchi, T. and Aaronson, S.A. (1998) The BRCA2 gene product functionally interacts with p53 and RAD51. *Proc. Natl Acad. Sci. USA* 95: 13869–13874.

McCluskey, L.L., Chen, C., Delgadillo, E., Felix, J.C., Muderspach, L.I. and Dubeau, L. (1999) Differences in p16 gene methylation and expression in benign and malignant ovarian tumors. *Gynecol. Oncol.* 72: 87–92.

McIntosh, G.G., Anderson, J.J., Milton, I. *et al.* (1995) Determination of the prognostic value of cyclin D1 overexpression in breast cancer. *Oncogene* 11: 885–891.

McManus, D.T., Yap, E.P., Maxwell, P., Russell, S.E., Toner, P.G. and McGee, J.O. (1994) p53 expression, mutation, and allelic deletion in ovarian cancer. *J. Pathol.* 174: 159–168.

Menard, S., Tagliabue, E., Campiglio, M. and Pupa, S.M. (2000) Role of HER2 gene overexpression in breast carcinoma. *J. Cell Physiol.* 182: 150–162.

Merajver, S.D., Pham, T.M., Caduff, R.F. *et al.* (1995) Somatic mutations in the BRCA1 gene in sporadic ovarian tumours. *Nat. Genet.* 9: 439–443.

Michalides, R., Hageman, P., van Tinteren, H., Houben, L., Wientjens, E., Klompmaker, R. and Peterse, J. (1996) A clinicopathological study on overexpression of cyclin D1 and of p53 in a series of 248 patients with operable breast cancer. *Br. J. Cancer* 73: 728–734.

Miki, Y., Swensen, J., Shattuck-Eidens, D. *et al.* (1994) A strong candidate for the breast and ovarian cancer susceptibility gene BRCA1. *Science* 266: 66–71.

Milner, B.J., Allan, L.A., Eccles, D.M. *et al.* (1993) p53 mutation is a common genetic event in ovarian carcinoma. *Cancer Res.* 53: 2128–2132.

Miyashita, T., Krajewski, S., Krajewska, M. *et al.* (1994) Tumor suppressor p53 is a regulator of bcl-2 and bax gene expression in vitro and in vivo. *Oncogene* 9: 1799–1805.

Myers, M.P., Stolarov, J.P., Eng, C. *et al.* (1997) P-TEN, the tumor suppressor from human chromosome 10q23, is a dual-specificity phosphatase. *Proc. Natl Acad. Sci. USA* 94: 9052–9057.

Narod, S.A., Ford, D., Devilee, P. *et al.* (1995) An evaluation of genetic heterogeneity in 145 breast–ovarian cancer families. Breast Cancer Linkage Consortium. *Am. J. Hum. Genet.* 56: 254–264.

Niculescu, A.B., 3rd, Chen, X., Smeets, M., Hengst, L., Prives, C. and Reed, S.I. (1998) Effects of p21(Cip1/Waf1) at both the G1/S and the G2/M cell cycle transitions: pRb is a critical determinant in blocking DNA replication and in preventing endoreduplication. *Mol. Cell Biol.* 18: 629–643.

Niederacher, D., Yan, H.Y., An, H.X., Bender, H.G. and Beckmann, M.W. (1999) CDKN2A gene inactivation in epithelial sporadic ovarian cancer. *Br. J. Cancer* 80: 1920–1926.

Noguchi, S., Kasugai, T., Miki, Y., Fukutomi, T., Emi, M. and Nomizu, T. (1999) Clinicopathologic analysis of BRCA1- or BRCA2-associated hereditary breast carcinoma in Japanese women. *Cancer* 85: 2200–2205.

Obata, K., Morland, S.J., Watson, R.H., Hitchcock, A., Chenevix-Trench, G., Thomas, E.J. and Campbell, I.G. (1998) Frequent PTEN/MMAC mutations in endometrioid but not serous or mucinous epithelial ovarian tumors. *Cancer Res.* 58: 2095–2097.

Oddoux, C., Struewing, J.P., Clayton, C.M. *et al.* (1996) The carrier frequency of the BRCA2 6174delT mutation among Ashkenazi Jewish individuals is approximately 1%. *Nat. Genet.* 14: 188–190.

Oishi, T., Kigawa, J., Minagawa, Y., Shimada, M., Takahashi, M. and Terakawa, N. (1998) Alteration of telomerase activity associated with development and extension of epithelial ovarian cancer. *Obstet. Gynecol.* 91: 568–571.

Oka, H., Shiozaki, H., Kobayashi, K. *et al.* (1993) Expression of E-cadherin cell adhesion molecules in human breast cancer tissues and its relationship to metastasis. *Cancer Res* 53: 1696–1701.

Ottaviano, Y.L., Issa, J.P., Parl, F.F., Smith, H.S., Baylin, S.B. and Davidson, N.E. (1994) Methylation of the estrogen receptor gene CpG island marks loss of estrogen receptor expression in human breast cancer cells. *Cancer Res.* 54: 2552–2555.

Park, T.W., Riethdorf, S., Riethdorf, L., Loning, T. and Janicke, F. (1999) Differential telomerase activity, expression of the telomerase catalytic sub-unit and telomerase-RNA in ovarian tumors. *Int. J. Cancer* 84: 426–431.

Patel, K.J., Yu, V.P., Lee, H. *et al*. (1998) Involvement of Brca2 in DNA repair. *Mol. Cell* 1: 347–357.

Paterson, J.W. (1998) BRCA1: a review of structure and putative functions. *Dis. Markers* 13: 261–274.

Peltomaki, P. and Vasen, H.F. (1997) Mutations predisposing to hereditary nonpolyposis colorectal cancer: database and results of a collaborative study. The International Collaborative Group on Hereditary Nonpolyposis Colorectal Cancer. *Gastroenterology* 113: 1146–1158.

Perillo, B., Sasso, A., Abbondanza, C. and Palumbo, G. (2000) 17beta-estradiol inhibits apoptosis in MCF-7 cells, inducing bcl-2 expression via two estrogen-responsive elements present in the coding sequence. *Mol. Cell Biol.* 20: 2890–2901.

Petrij-Bosch, A., Peelen, T., van Vliet, M. *et al*. (1997) BRCA1 genomic deletions are major founder mutations in Dutch breast cancer patients. *Nat. Genet.* 17: 341–345.

Pharoah, P.D., Easton, D.F., Stockton, D.L., Gayther, S. and Ponder, B.A. (1999) Survival in familial, BRCA1-associated, and BRCA2-associated epithelial ovarian cancer. United Kingdom Coordinating Committee for Cancer Research (UKCCCR) Familial Ovarian Cancer Study Group. *Cancer Res.* 59: 868–871.

Phelan, C.M., Lancaster, J.M., Tonin, P. *et al*. (1996a) Mutation analysis of the BRCA2 gene in 49 site-specific breast cancer families. *Nat. Genet.* 13: 120–122.

Phelan, C.M., Rebbeck, T.R., Weber, B.L. *et al*. (1996b) Ovarian cancer risk in BRCA1 carriers is modified by the HRAS1 variable number of tandem repeat (VNTR) locus. *Nat. Genet.* 12: 309–311.

Plummer, S.J., Santibanez-Koref, M., Kurosaki, T. *et al*. (1994) A germline 2.35 kb deletion of p53 genomic DNA creating a specific loss of the oligomerization domain inherited in a Li-Fraumeni syndrome family. *Oncogene* 9: 3273–3280.

Poller, D.N., Roberts, E.C., Bell, J.A., Elston, C.W., Blamey, R.W. and Ellis, I.O. (1993) p53 protein expression in mammary ductal carcinoma in situ: relationship to immunohistochemical expression of estrogen receptor and c-erbB-2 protein. *Hum. Pathol.* 24: 463–468.

Polyak, K., Kato, J.Y., Solomon, M.J., Sherr, C.J., Massague, J., Roberts, J.M. and Koff, A. (1994) p27Kip1, a cyclin-Cdk inhibitor, links transforming growth factor-beta and contact inhibition to cell cycle arrest. *Genes Dev.* 8: 9–22.

Porter, D.E., Cohen, B.B., Wallace, M.R. *et al*. (1994) Breast cancer incidence, penetrance and survival in probable carriers of *BRCA1* gene mutation in families linked to *BRCA1* on chromosome 17q12–21. *Brit. J. Surg.* 81: 1512–1515.

Porter, P.L., Malone, K.E., Heagerty, P.J. *et al*. (1997) Expression of cell-cycle regulators p27Kip1 and cyclin E, alone and in combination, correlate with survival in young breast cancer patients. *Nat. Med.* 3: 222–225.

Press, M.F., Xu, S.H., Wang, J.D. and Greene, G.L. (1989) Subcellular distribution of estrogen receptor and progesterone receptor with and without specific ligand. *Am. J. Pathol.* 135: 857–864.

Prigent, S.A. and Lemoine, N.R. (1992) The type 1 (EGFR-related) family of growth factor receptors and their ligands. *Prog. Growth Factor Res.* 4: 1–24.

Puget, N., Stoppa-Lyonnet, D., Sinilnikova, O.M., Pages, S., Lynch, H.T., Lenoir, G.M. and Mazoyer, S. (1999) Screening for germ-line rearrangements and regulatory mutations in *BRCA1* led to the identification of four new deletions. *Cancer Res.* 59: 455–461.

Quelle, D.E., Zindy, F., Ashmun, R.A. and Sherr, C.J. (1995) Alternative reading frames of the INK4a tumor suppressor gene encode two unrelated proteins capable of inducing cell cycle arrest. *Cell* 83: 993–1000.

Rebbeck, T.R., Couch, F.J., Kant, J. *et al*. (1996) Genetic heterogeneity in hereditary breast cancer: role of BRCA1 and BRCA2. *Am. J. Hum. Genet.* 59: 547–553.

Rebbeck, T.R., Kantoff, P.W., Krithivas, K. *et al*. (1999a) Modification of BRCA1-associated breast cancer risk by the polymorphic androgen-receptor CAG repeat. *Am. J. Hum. Genet.* 64: 1371–1377.

Rebbeck, T.R., Levin, A.M., Eisen, A. *et al*. (1999b) Breast cancer risk after bilateral prophylactic oophorectomy in BRCA1 mutation carriers. *J. Natl Cancer Inst* 91: 1475–1479.

Rhei, E., Kang, L., Bogomolniy, F., Federici, M.G., Borgen, P.I. and Boyd, J. (1997) Mutation analysis of the putative tumor suppressor gene PTEN/MMAC1 in primary breast carcinomas. *Cancer Res.* 57: 3657–3659.

Rice, J.C., Massey-Brown, K.S. and Futscher, B.W. (1998) Aberrant methylation of the *BRCA1* CpG island promoter is associated with decreased BRCA1 mRNA in sporadic breast cancer cells. *Oncogene* 17: 1807–1812.

Risinger, J.I., Berchuck, A., Kohler, M.F. and Boyd, J. (1994) Mutations of the E-cadherin gene in human gynecologic cancers. *Nat. Genet.* 7: 98–102.

Risinger, J.I., Barrett, J.C., Watson, P., Lynch, H.T. and Boyd, J. (1996) Molecular genetic evidence of the occurrence of breast cancer as an integral tumor in patients with the hereditary nonpolyposis colorectal carcinoma syndrome. *Cancer* **77**: 1836–1843.

Robson, M., Rajan, P., Rosen, P.P. *et al.* (1998) BRCA-associated breast cancer: absence of a characteristic immunophenotype. *Cancer Res.* **58**: 1839–1842.

Rohlke, P., Milde-Langosch, K., Weyland, C., Pichlmeier, U., Jonat, W. and Loning, T. (1997) p53 is a persistent and predictive marker in advanced ovarian carcinomas: multivariate analysis including comparison with Ki67 immunoreactivity. *J. Cancer Res. Clin. Oncol.* **123**: 496–501.

Rubin, S.C., Benjamin, I., Behbakht, K. *et al.* (1996) Clinical and pathological features of ovarian cancer in women with germ-line mutations of BRCA1. *N. Engl. J. Med.* **335**: 1413–1416.

Ryan, A., Al-Jehani, R.M., Mulligan, K.T. and Jacobs, I.J. (1998) No evidence exists for methylation inactivation of the p16 tumor suppressor gene in ovarian carcinogenesis. *Gynecol. Oncol.* **68**: 14–17.

Schlichtholz, B., Bouchind'homme, B., Pages, S. *et al.* (1998) p53 mutations in BRCA1-associated familial breast cancer. *Lancet* **352**: 622.

Schmider, A., Gee, C., Friedmann, W., Lukas, J.J., Press, M.F., Lichtenegger, W. and Reles, A. (2000) p21 (WAF1/CIP1) protein expression is associated with prolonged survival but not with p53 expression in epithelial ovarian carcinoma. *Gynecol. Oncol.* **77**: 237–242.

Schultz, D.C., Vanderveer, L., Berman, D.B., Hamilton, T.C., Wong, A.J. and Godwin, A.K. (1996) Identification of two candidate tumor suppressor genes on chromosome 17p13.3. *Cancer Res.* **56**: 1997–2002.

Scully, R., Chen, J., Ochs, R.L., Keegan, K., Hoekstra, M., Feunteun, J. and Livingston, D.M. (1997) Dynamic changes of BRCA1 subnuclear location and phosphorylation state are initiated by DNA damage. *Cell* **90**: 425–435.

Scully, R., Ganesan, S., Vlasakova, K., Chen, J., Socolovsky, M. and Livingston, D.M. (1999) Genetic analysis of BRCA1 function in a defined tumor cell line. *Mol. Cell* **4**: 1093–1099.

Serrano, M., Hannon, G.J. and Beach, D. (1993) A new regulatory motif in cell-cycle control causing specific inhibition of cyclin D/CDK4. *Nature* **366**: 704–707.

Sharan, S.K., Morimatsu, M., Albrecht, U. *et al.* (1997) Embryonic lethality and radiation hypersensitivity mediated by Rad51 in mice lacking *BRCA2*. *Nature* **386**: 804–810.

Shattuck-Eidens, D., McClure, M., Simard, J. *et al.* (1995) A collaborative survey of 80 mutations in the BRCA1 breast and ovarian cancer susceptibility gene. Implications for presymptomatic testing and screening. *JAMA* **273**: 535–541.

Shih, Y.C., Kerr, J., Liu, J. *et al.* (1997) Rare mutations and no hypermethylation at the CDKN2A locus in epithelial ovarian tumours. *Int. J. Cancer* **70**: 508–511.

Silvestrini, R., Veneroni, S., Daidone, M.G. *et al.* (1994) The Bcl-2 protein: a prognostic indicator strongly related to p53 protein in lymph node-negative breast cancer patients. *J. Natl Cancer Inst.* **86**: 499–504.

Slamon, D.J., Godolphin, W., Jones, L.A. *et al.* (1989) Studies of the HER-2/neu proto-oncogene in human breast and ovarian cancer. *Science* **244**: 707–712.

Sobol, H., Stoppa-Lyonnet, D., Bressac-de Paillerets, B. *et al.* (1997) *BRCA1*–p53 relationship in hereditary breast cancer. *Int. J. Oncol.* **10**: 349–353.

Spirin, K.S., Simpson, J.F., Takeuchi, S., Kawamata, N., Miller, C.W. and Koeffler, H.P. (1996) p27/Kip1 mutation found in breast cancer. *Cancer Res.* **56**: 2400–2404.

Stahl, J.A., Leone, A., Rosengard, A.M., Porter, L., King, C.R. and Steeg, P.S. (1991) Identification of a second human nm23 gene, nm23-H2. *Cancer Res* **51**: 445–449.

Stambolic, V., Suzuki, A., de la Pompa, J.L. *et al.* (1998) Negative regulation of PKB/Akt-dependent cell survival by the tumor suppressor PTEN. *Cell* **95**: 29–39.

Starink, T.M., van der Veen, J.P., Arwert, F., de Waal, L.P., de Lange, G.G., Gille, J.J. and Eriksson, A.W. (1986) The Cowden syndrome: a clinical and genetic study in 21 patients. *Clin. Genet.* **29**: 222–233.

Steck, P.A., Pershouse, M.A., Jasser, S.A. *et al.* (1997) Identification of a candidate tumour suppressor gene, MMAC1, at chromosome 10q23.3 that is mutated in multiple advanced cancers. *Nat. Genet.* **15**: 356–362.

Steeg, P.S., Bevilacqua, G., Pozzatti, R., Liotta, L.A. and Sobel, M.E. (1988) Altered expression of NM23, a gene associated with low tumor metastatic potential, during adenovirus 2 Ela inhibition of experimental metastasis. *Cancer Res.* **48**: 6550–6554.

Stratton, J.F., Gayther, S.A., Russell, P. *et al.* (1997) Contribution of *BRCA1* mutations to ovarian cancer. *New. Engl. J. Med.* **336**: 1125–1130.

Struewing, J.P., Abeliovich, D., Peretz, T., Avishai, N., Kaback, M.M., Collins, F.S. and Brody, L.C. (1995) The carrier frequency of the BRCA1 185delAG mutation is approximately 1 percent in Ashkenazi Jewish individuals. *Nat. Genet.* **11**: 198–200.

Sui, L., Tokuda, M., Ohno, M., Hatase, O. and Hando, T. (1999) The concurrent expression of p27(kip1) and cyclin D1 in epithelial ovarian tumors. *Gynecol. Oncol.* **73**: 202–209.

Suzuki, K., Kodama, S. and Watanabe, M. (1999) Recruitment of ATM protein to double strand DNA irradiated with ionizing radiation. *J. Biol. Chem.* **274**: 25571–25575.

Swensen, J., Hoffman, M., Skolnick, M.H. and Neuhausen, S.L. (1997) Identification of a 14 kb deletion involving the promoter region of *BRCA1* in a breast cancer family. *Hum. Mol. Genet.* **6**: 1513–1517.

Swift, M., Reitnauer, P.J., Morrell, D. and Chase, C.L. (1987) Breast and other cancers in families with ataxia-telangiectasia. *N. Engl. J. Med.* **316**: 1289–1294.

Swift, M., Morrell, D., Massey, R.B. and Chase, C.L. (1991) Incidence of cancer in 161 families affected by ataxia-telangiectasia. *N. Engl. J. Med.* **325**: 1831–1836.

Tamura, M., Gu, J., Danen, E.H., Takino, T., Miyamoto, S. and Yamada, K.M. (1999a) PTEN interactions with focal adhesion kinase and suppression of the extracellular matrix-dependent phosphatidylinositol 3 kinase/Akt cell survival pathway. *J. Biol. Chem.* **274**: 20693–20703.

Tamura, M., Gu, J., Takino, T. and Yamada, K.M. (1999b) Tumor suppressor PTEN inhibition of cell invasion, migration, and growth: differential involvement of focal adhesion kinase and p130Cas. *Cancer Res.* **59**: 442–449.

Tamura, M., Gu, J., Tran, H. and Yamada, K.M. (1999c) PTEN gene and integrin signaling in cancer. *J. Natl. Cancer Inst.* **91**: 1820–1828.

Tan, P., Cady, B., Wanner, M. *et al.* (1997) The cell cycle inhibitor p27 is an independent prognostic marker in small (T1a,b) invasive breast carcinomas. *Cancer Res.* **57**: 1259–1263.

Tapper, J., Sarantaus, L., Vahteristo, P. *et al.* (1998) Genetic changes in inherited and sporadic ovarian carcinomas by comparative genomic hybridization: extensive similarity except for a difference at chromosome 2q24-q32. *Cancer Res.* **58**: 2715–2719.

Tavtigian, S.V., Simard, J., Rommens, J. *et al.* (1996) The complete BRCA2 gene and mutations in chromosome 13q-linked kindreds. *Nat. Genet.* **12**: 333–337.

Thor, A.D., Moore, D.H., II, Edgerton, S.M. *et al.* (1992) Accumulation of p53 tumor suppressor gene protein: an independent marker of prognosis in breast cancers. *J. Natl. Cancer Inst.* **84**: 845–855.

Thorlacius, S., Sigurdsson, S., Bjarnadottir, H. *et al.* (1997) Study of a single BRCA2 mutation with high carrier frequency in a small population. *Am. J. Hum. Genet.* **60**: 1079–1084.

Tirkkonen, M., Johannsson, O., Agnarsson, B.A. *et al.* (1997) Distinct somatic genetic changes associated with tumor progression in carriers of *BRCA1* and *BRCA2* germ-line mutations. *Cancer Res.* **57**: 1222–1227.

Tonin, P.N., Mes-Masson, A.M., Futreal, P.A. *et al.* (1998) Founder BRCA1 and BRCA2 mutations in French Canadian breast and ovarian cancer families. *Am. J. Hum. Genet.* **63**: 1341–1351.

Toyoshima, H. and Hunter, T. (1994) p27, a novel inhibitor of G1 cyclin-Cdk protein kinase activity, is related to p21. *Cell* **78**: 67–74.

Tsutsumi, Y., Naber, S.P., DeLellis, R.A., Wolfe, H.J., Marks, P.J., McKenzie, S.J. and Yin, S. (1990) neu oncogene protein and epidermal growth factor receptor are independently expressed in benign and malignant breast tissues. *Hum. Pathol.* **21**: 750–758.

Varley, J.M., Armour, J., Swallow, J.E. *et al.* (1989) The retinoblastoma gene is frequently altered leading to loss of expression in primary breast tumours. *Oncogene* **4**: 725–729.

Vehmanen, P., Friedman, L.S., Eerola, H. *et al.* (1997) Low proportion of *BRCA1* and *BRCA2* mutations in Finnish breast cancer families: evidence for additional susceptibility genes. *Hum. Mol. Genet.* **6**: 2309–2315.

Venkitaraman, A.R. (1999) Breast cancer genes and DNA repair. *Science* **286**: 1100–1102.

Verhoog, L.C., Brekelmans, C.T.M., Seynaeve, C. *et al.* (1998) Survival and tumour characteristics of breast-cancer patients with germline mutations of *BRCA1*. *Lancet* **351**: 316–321.

Vos, C.B., Cleton-Jansen, A.M., Berx, G. *et al.* (1997) E-cadherin inactivation in lobular carcinoma in situ of the breast: an early event in tumorigenesis. *Br. J. Cancer* **76**: 1131–1133.

Warner, E., Foulkes, W., Goodwin, P. *et al.* (1999) Prevalence and penetrance of BRCA1 and BRCA2 gene mutations in unselected Ashkenazi Jewish women with breast cancer. *J. Natl Cancer Inst.* **91**: 1241–1247.

Wilson, C.A., Ramos, L., Villasenor, M.R. *et al.* (1999) Localization of human BRCA1 and its loss in high-grade, non-inherited breast carcinomas. *Nat. Genet.* **21**: 236–240.

Wooster, R., Bignell, G., Lancaster, J. *et al.* (1995) Identification of the breast cancer susceptibility gene BRCA2. *Nature* **378**: 789–792.

Wright, K., Wilson, P., Morland, S. *et al.* (1999) Beta-catenin mutation and expression analysis in ovarian cancer: exon 3 mutations and nuclear translocation in 16% of endometrioid tumours. *Int. J. Cancer* **82**: 625–629.

Wu, L.C., Wang, Z.W., Tsan, J.T. *et al.* (1996) Identification of a RING protein that can interact in vivo with the BRCA1 gene product. *Nat. Genet.* **14**: 430–440.

Wu, X., Senechal, K., Neshat, M.S., Whang, Y.E. and Sawyers, C.L. (1998) The PTEN/MMAC1 tumor suppressor phosphatase functions as a negative regulator of the phosphoinositide 3-kinase/Akt pathway. *Proc. Natl. Acad. Sci. USA* **95**: 15587–15591.

Xu, Y., Ashley, T., Brainerd, E.E., Bronson, R.T., Meyn, M.S. and Baltimore, D. (1996) Targeted disruption of ATM leads to growth retardation, chromosomal fragmentation during meiosis, immune defects, and thymic lymphoma. *Genes Dev.* **10**: 2411–2422.

Yuan, S.S., Lee, S.Y., Chen, G., Song, M., Tomlinson, G.E. and Lee, E.Y. (1999) BRCA2 is required for ionizing radiation-induced assembly of Rad51 complex in vivo. *Cancer Res.* **59**: 3547–3551.

Zeillinger, R., Kury, F., Czerwenka, K. *et al.* (1989) HER-2 amplification, steroid receptors and epidermal growth factor receptor in primary breast cancer. *Oncogene* **4**: 109–114.

Zhang, G.J., Kimijima, I., Abe, R. *et al.* (1997) Correlation between the expression of apoptosis-related bcl-2 and p53 oncoproteins and the carcinogenesis and progression of breast carcinomas. *Clin. Cancer Res.* **3**: 2329–2335.

Zheng, W., Luo, F., Lu, J.J., Baltayan, A., Press, M.F., Zhang, Z.F. and Pike, M.C. (2000) Reduction of BRCA1 expression in sporadic ovarian cancer. *Gynecol. Oncol.* **76**: 294–300.

4

Genetics of von Hippel-Lindau disease

Eamonn R. Maher

1. Introduction

Insights into the genetic basis of human cancers often follow from the identification of familial cancer syndrome genes. Thus, rare disorders often thought to be only of esoteric interest may provide unique indicators of the role of specific biochemical pathways in tumorigenesis. Von Hippel-Lindau (VHL) disease, a multisystem familial cancer syndrome, is the most common cause of familial renal cell carcinoma. First described more than 100 years ago, VHL disease was considered to be a rare disease of little general interest for most of the last century. However recent insights into the function of the *VHL* gene product have prompted an explosion of interest in VHL research.

2. Clinical aspects

VHL disease is dominantly inherited with variable expression and age-dependent penetrance. Most cases present in the second and third decades, and non-penetrance at age 60 years is uncommon. The minimum birth incidence was estimated at 3 per 100 000 persons (Maher *et al.*, 1991). The major features of VHL disease are retinal angiomatosis, cerebellar and spinal hemangioblastomas and clear cell renal cell carcinoma (RCC). Penetrance is tumor-specific with, on average, an earlier onset for retinal angioma (mean 24.5 years at symptomatic diagnosis) and cerebellar hemangioblastomas (29 years) than for renal cell carcinoma (44 years). Nevertheless, by age 60 years the overall risk of retinal, cerebellar and renal tumors has been estimated at ~70% for each tumor (Maher *et al.*, 1990a). However, the frequency of particular tumors within families is influenced by allelic heterogeneity (see later). This is particularly evident for phaeochromocytoma, which occurs in ~10% of cases but is the most common feature of VHL disease in some kindreds and absent in others. As with other familial cancer syndromes, patients with VHL disease not only have a greatly increased risk of RCC but, in addition, tumors develop at an early age and are frequently multiple (Maher *et al.*, 1990b; Richard *et al.*, 1994).

Molecular Genetics of Cancer second edition, edited by J.K. Cowell.

Ten years ago the median survival of VHL disease patients was estimated at 49 years, however a more recent estimate revealed an improvement to 57 years (Maher *et al.*, 1990; Webster *et al.*, unpublished). Improved survival in VHL disease has resulted from the widespread adoption of a policy of ascertaining affected individuals and at-risk relatives and ensuring they are entered into surveillance programs so that tumors are detected early (Hodgson and Maher, 1999; Maher *et al.*, 1990a). Untreated retinal angiomas enlarge progressively and may produce retinal detachment and hemorrhage resulting in visual impairment. Nevertheless, early detection of these tumors enables treatment by laser- or cryotherapy and reduces the risk of visual loss. Approximately 30% of all patients with cerebellar hemangioblastoma have VHL disease. While results of surgery for an isolated cerebellar hemisphere tumor are often excellent, the treatment of multiple CNS hemangioblastomas and the management of brain stem and spinal tumors may be hazardous and result in significant morbidity. In the future, effective antiangiogenic therapy (e.g. VEGF antagonists) may provide a medical approach to the treatment of CNS and retinal hemangioblastomas.

The major renal manifestations of VHL disease are cysts and clear cell RCC. Although renal cysts are frequent and may mimic the appearance of autosomal dominant polycystic kidney disease, renal impairment occurs rarely. Histopathologic examination of VHL renal cysts may show a continuum from simple benign cysts to frank RCC (Solomon and Schwartz, 1988), and complex cysts are followed carefully as they can develop into solid lesions (Choyke *et al.*, 1992). Although RCC was overlooked as a major feature of VHL disease for many years, recognition that RCC had emerged as a leading cause of death in VHL disease led to a policy of annual renal imaging in adult VHL patients and at-risk relatives. Such surveillance (usually by ultrasound or MRI scanning) allows early presymptomatic diagnosis and treatment of renal tumors so preventing death from metastatic RCC. Most small solid renal tumors enlarge slowly, and the risk of distant metastasis from a solid lesion <3 cm appears to be very remote. Consequently small solid lesions detected presymptomatically by scanning are kept under surveillance until they reach 3 cm in size and then removed by nephron-sparing surgery (partial nephrectomy) (Steinbach *et al.*, 1995). Whereas, 25% of VHL patients with a RCC >3 cm (treated by nephron-sparing surgery or nephrectomy) developed metastatic disease (Walther *et al.*, 1999), metastasis from tumors <3 cm has not been reported. As RCC in VHL disease is frequently multiple and bilateral, patients may undergo repeated partial nephrectomies before becoming anephric. Results from renal transplantation in VHL disease are encouraging as graft and patient survival appear to be comparable to that of other groups requiring transplantation (Goldfarb *et al.*, 1997).

Multiple cystadenomas are the most frequent pancreatic manifestation of VHL disease although these rarely cause significant impairment of pancreatic function. In addition, pancreatic islet cell tumors occur in up to 10% of cases, and a high frequency of malignancy has been reported in these (Binkovitz *et al.*, 1990). Recently it has been suggested that pancreatic tumors >3 cm should be resected (Libutti *et al.*, 1998).

Although the major features of VHL disease were described in the first half of the twentieth century, endolymphatic sac tumors (ELSTs) were only recognized as a complication of VHL disease towards the end of the century (Manski *et al.*,

1997). Many ELSTs are asymptomatic but their presence should be suspected in VHL patients with hearing loss, tinnitus and vertigo. Benign epididymal cysts are very frequent in males with VHL disease and can, if bilateral, impair fertility.

3. VHL disease and isolated features of VHL disease

The standard clinical diagnostic criteria for VHL disease require the presence of a typical VHL tumor (e.g. retinal angioma, cerebellar hemangioblastoma, clear cell RCC, pheochromocytoma, pancreatic islet cell tumor or ELST) with a positive family history of VHL disease. However in isolated cases without a family history, a diagnosis of VHL disease can only be made when two tumors (e.g. two hemangioblastomas or a hemangioblastoma and a visceral tumor) have developed. Accordingly the diagnosis of VHL disease is later in *de novo* isolated cases. The identification of the *VHL* gene provided a molecular test for VHL disease. Thus in patients with a single cerebellar hemangioblastoma, the detection of a germline *VHL* gene mutation (*Figure 1*) allows early diagnosis prior to the development of other features of VHL disease. In a cohort of patients with apparently isolated cerebellar hemangioblastoma and no family history or clinical or subclinical (e.g. visceral cysts, asymptomatic tumors) evidence of VHL disease, 4% has a germline *VHL* gene mutation (Hes *et al.*, 2000). For patients with apparently isolated retinal angioma, the risk of VHL disease can be estimated according to patient age, level of screening and results of mutation analysis (Webster *et al.*, 2000). The incidence of germline mutations in unselected cases of clear cell RCC and in familial cases is

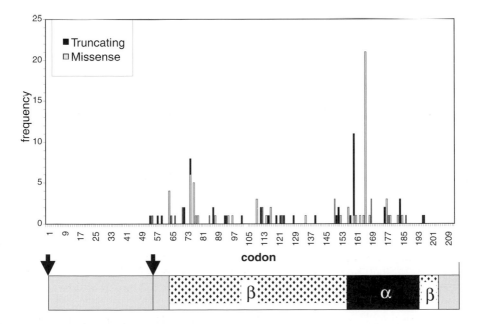

Figure 1. Distribution of germline *VHL* gene mutations in patients with VHL disease (Maher *et al.*, 1996 and unpublished observations).

likely to be small (Neumann *et al.*, 1998, Woodward *et al.*, 2000b). However, ~50% of patients with apparently isolated familial pheochromocytoma or bilateral pheochromocytoma have germline *VHL* gene mutations (Woodward *et al.*, 1997).

4. Molecular genetics of VHL disease

In 1993, five years after the initial mapping to chromosome 3p25, an international collaborative grouping identified the VHL tumor suppressor gene by a positional cloning approach (Latif *et al.*, 1993). The *VHL* coding sequence is represented in three exons and germline mutations have been detected in up to 100% of classical familial VHL disease cases (Chen *et al.*, 1995; Crossey *et al.*, 1994; Maher *et al.*, 1996; Richards *et al.*, 1994; Stolle *et al.*, 1998; Zbar *et al.*, 1996 and see *VHL* mutation database at http://www.umd.necker.fr). This enables reliable predictive testing within families (non-gene carriers can be spared unnecessary surveillance) and a molecular diagnosis of VHL disease to be made in patients with atypical or incomplete forms of VHL disease (see above). Although the detection of large deletions in rare patients by pulsed-field gel electrophoresis was instrumental in mapping the *VHL* gene (Richards *et al.*, 1993; Yao *et al.*, 1993), recent studies suggest that large deletions are relatively common. In total, ~40% of VHL patients have germline deletions which in most cases are detected by Southern analysis, but very large whole gene deletions may be best detected by FISH studies (Pack *et al.*, 1999). Intragenic mutations are detected by direct sequencing and these are predominantly missense or truncating mutations (Maher *et al.*, 1996; Zbar *et al.*, 1995). False negative mutation analysis may reflect mosaicism for a germline mutation (Sgambati *et al.*, 2000).

Two alternatively spliced *VHL* transcripts have been detected reflecting the presence (isoform I) or absence (isoform II) of exon 2. No endogenous isoform II-associated protein product has been reported to date and the identification of VHL patients with germline deletions of exon 2, resulting in only the expression of isoform II from the mutant allele, suggests that isoform II does not encode a functional gene product (Richards *et al.*, 1994). The long 3′ untranslated region (UTR) has been characterized (Renbaum *et al.*, 1996) and the 5′ UTR and its flanking sequence were defined by Kuzmin *et al.* (1995). A minimal promoter region of 106 bp has been defined and there is evidence for the presence of functional SP1 and AP2 binding sites within the minimal promoter ((Kuzmin *et al.*, 1995; Zatyka *et al*, manuscript in preparation). However, to date, no germline mutations have been described in the promoter or 3′ UTR.

Most germline *VHL* mutations occur rarely and recurrent mutations are uncommon. The most frequently reported mutations (e.g. C694T, C712T, G713A) arise from multiple *de novo* events at hypermutable sequences (e.g. CpG dinucleotides), although the T505C (Tyr98His) mutation occurs as a founder effect in patients of German origin (Brauch *et al.*, 1995; Richards *et al.*, 1995).

A striking feature of germline *VHL* mutations is the absence of mutations in the first 53 amino acids (Maher *et al.*, 1998; Zbar *et al.*, 1996). The *VHL* gene encodes two proteins: $pVHL_{30}$ a full length 213 amino acid (~30 kDa) product and a second protein, $pVHL_{19}$, which does not contain the first 53 amino acids of

$pVHL_{30}$. The functional effects of $pVHL_{19}$ and $pVHL_{30}$ are reported to be similar (Iliopoulos *et al.*, 1998; Schoenfeld *et al.*, 1998) and presumably, mutations in the first 53 amino acids would not cause VHL disease as a functional $pVHL_{19}$ protein would be produced. Comparative sequence analysis for human, primate and rodent *VHL* genes and for a putative *C elegans VHL* homolog demonstrates significant sequence conservation over most of $pVHL19$, but the N-terminal sequence of $pVHL_{30}$ contains 8 copies of a GXEEX acidic repeat motif in human and higher primates, but only 3 copies are present in the marmoset and one copy in rodent *VHL* genes. Furthermore, evolutionary analysis suggested that the N-terminal repetitive sequence in $pVHL_{30}$ was of less functional importance than those regions present in both $pVHL_{30}$ and $pVHL_{19}$ (Woodward *et al.*, 2000a). Within the $pVHL_{19}$ coding sequence missense mutations occurred preferentially in residues demonstrating evolutionary conservation.

5. Genotype–phenotype

The clinical variability and the heterogeneity of germline *VHL* gene mutations hinders attempts to perform detailed genotype–phenotype correlations. Nevertheless early investigations demonstrated that interfamilial variations in pheochromocytoma frequency reflected allelic heterogeneity and more recent studies have demonstrated complex genotype–phenotype correlations. Thus initial studies demonstrated a strong association between missense mutations and pheochromocytoma susceptibility such that the risk of pheochromocytoma in patients with germline missense mutations was ~50% at 50 years compared to 5% in those with deletions or truncating mutations (Crossey *et al.*, 1994; Maher *et al.*, 1996; Zbar *et al.*, 1996). Nevertheless missense mutations are heterogeneous with some being associated with a low risk of pheochromocytoma. Clinical subtypes of VHL disease have been devised, with families with hemangioblastomas and RCC but no pheochromocytoma designated as Type 1 and families with pheochromocytoma as Type 2 (Zbar *et al.*, 1996). While there are limitations to this approach (e.g. within a large family with 50 affected individuals, the occurrence of only one pheochromocytoma would change the phenotype from Type 1 to Type 2). Nevertheless, this classification has provided a framework for categorizing family phenotypes for genotype–phenotype studies (*Figure 2*). One interesting aspect of VHL genotype–phenotype correlations is the subdivision of Type 2 VHL disease into 2A, 2B and 2C subgroups. The most common type 2 phenotype is the 2B subgroup, which is associated with missense mutations such as Arg167Gln. VHL type 2B comprises hemangioblastomas, RCC and pheochromocytoma. Two missense mutations (e.g. Tyr98His) have been associated with a Type 2A phenotype in which there is susceptibility to hemangioblastomas and pheochromocytoma, but RCC is uncommon (Brauch *et al.*, 1995; Chen *et al.*, 1996). Intriguingly, among missense mutations described in familial pheochromocytoma-only kindreds (this phenotype is designated type 2C), in addition to mutations described in type 2B, there are apparently unique mutations (e.g. Leu188Val) which appear to only cause pheochromocytoma susceptibility (Crossey *et al.*, 1995; Neumann *et al.*, 1995; Woodward *et al.*, 1997). The complex genotype–phenotype

Modified VHL NCI Classification Scheme

Type	HAB	RCC	PC
1	Yes	Yes	No
2A	Yes	No	Yes
2B	Yes	Yes	Yes
2C	No	No	Yes

Figure 2. NCI classification of VHL disease for genotype–phenotype studies.

correlations suggest that the *VHL* gene product has multiple and tissue-specific functions.

Evidence for genetic modifiers of the phenotypic consequences of germline tumor suppressor gene (TSG) mutations has been reported for NF1 and APC genes (Dietrich *et al.*, 1993; Easton *et al.*, 1993). In addition, Webster *et al.* (1998) investigated phenotypic variation in number of retinal angiomas by relative-pair analysis and found evidence for genetic modifiers of ocular angiomatosis. Furthermore, patients with retinal involvement had a higher risk of cerebellar hemangioblastoma and RCC than those with no retinal angiomas, suggesting that putative modifiers influenced the risk of both retinal and cerebellar hemangioblastoma and of RCC.

6. *VHL* and sporadic tumorigenesis

Chromosome 3p allele loss is frequent in many common tumors including kidney, lung, breast, ovarian and testicular cancers (reviewed by Kok *et al.*, 1997). Statistical analysis of the age at onset of cerebellar HAB and RCC in VHL disease and sporadic cases was consistent with a one- and two-hit model as for familial and sporadic retinoblastoma (Maher *et al.*, 1990b). Furthermore molecular studies have demonstrated that the mechanism of tumorigenesis in hemangioblastomas, RCC, pheochromocytoma, and pancreatic tumors from VHL disease appears to be analogous to that in bilateral retinoblastoma such that tumors demonstrate loss or inactivation (by mutation or promoter hypermethylation) of the wild-type allele in most cases (Crossey *et al.*, 1994; Prowse *et al.*, 1997).

The identification of the *VHL* TSG enabled the hypothesis that *VHL* inactivation was implicated in the pathogenesis of sporadic cancers to be investigated. Notably, somatic *VHL* mutations and allele loss were reported to be present in up to 60% of clear cell RCC tumors and cell lines (Foster *et al.*, 1994; Gnarra *et al.*, 1994). Furthermore transcriptional silencing by promoter hypermethylation occurs in ~15% of sporadic primary clear cell RCC and cell lines (Clifford *et al.*, 1998; Herman *et al.*, 1994). RCC in VHL patients is invariably of the 'clear-cell' type and somatic *VHL* gene mutation and methylation appears to be limited to sporadic clear cell RCC. The tumor suppressor activity of pVHL was confirmed when clear cell RCC cell lines without wild-type VHL were transfected with wild-type pVHL.

Transfected cells demonstrated suppressed tumorigenicity in nude mice studies although *in vitro* effects have not been found consistently (Chen *et al.*, 1995b; Iliopoulos *et al.*, 1995). Recently, Brauch *et al.* (2000) reported an association between somatic *VHL* mutation/hypermethylation and tumor stage in sporadic clear cell RCC, so linking *VHL* gene inactivation with an established prognostic factor.

Inactivation of the *VHL* TSG is thought to be an early event in clear cell RCC tumorigenesis, consistent with the *VHL* TSG having a similar role in RCC to that of APC in colorectal cancer. Nevertheless, it appears that *VHL* inactivation is not sufficient for tumorigenesis and RCC with *VHL* inactivation also show allele loss at chromosome 3p12-p21 (Clifford *et al.*, 1998). Genetic and functional studies suggest further chromosome 3p TSGs at 3p12-p21 (see Kok *et al.*, 1997). Chromosome 3p allele loss patterns suggest that inactivation of TSG(s) in 3p12-p21 occurs in clear cell RCC with and without *VHL* inactivation (Clifford *et al.*, 1998; van den Berg *et al.*, 1997).

Somatic *VHL* mutations are also frequent in sporadic hemangioblastomas (Kanno *et al.*, 1994), but rare in sporadic pheochromocytoma (Eng *et al.*, 1995). Specific germline missense mutations are associated with pheochromocytoma susceptibility in VHL disease (see previously) and somatic *VHL* mutations in sporadic pheochromocytoma are also missense. This suggests that particular missense mutations are required to instigate pheochromocytoma tumorigenesis which would explain the low frequency of somatic *VHL* mutations in sporadic pheochromocytoma.

Chromosome 3p allele loss in common cancers that are not a feature of VHL disease (e.g. lung, breast and gonadal cancers) reflects the involvement of other chromosome 3p TSGs in these cancers as somatic *VHL* mutations have not been described in such tumor types (Foster *et al.*, 1995; Sekido *et al.*, 1994).

7. *VHL* and development

The *VHL* gene encodes a 4.7 kb mRNA which is widely expressed in both fetal and adult tissues, such that expression of the *VHL* transcript and protein is ubiquitous and is not restricted to those organs affected by VHL tumors (Corless *et al.*, 1997; Richards *et al.*, 1996). The pattern of *VHL* mRNA expression in the fetal kidney was considered to be consistent with a role in normal renal tubular development and differentiation (Richards *et al.*, 1996). However, a murine *VHL* gene knockout suggests that the *VHL* gene has a critical role in normal extra-embryonic vascular development. Thus while mice hemizygous for *vhl* (+/−) appeared phenotypically normal, homozygous *vhl* (−/−) knockouts died *in utero* at 10.5 to 12.5 days of gestation. Death resulted from placental dysgenesis caused by a failure of embryonic vasculogenesis of the placenta and development of hemorrhagic lesions in the placenta (Gnarra *et al.*, 1997).

8. *VHL* gene function

The primary sequence of pVHL shows minimal homology to any known protein, so identification of the *VHL* gene did not provide any immediate insights into the likely function of the VHL protein. Although investigations of pVHL function continue and there is growing evidence for multiple tissue-specific actions, recent insights into pVHL function have been derived from two main strategies. Firstly, pVHL-interacting proteins have been sought and characterized. Secondly, the role of pVHL in tumor angiogenesis has been pursued with considerable success.

The first pVHL-interacting proteins to be identified were elongins B and C, which bind to a region of the C-terminal third of the VHL protein that is frequently altered by *VHL*-associated mutations (Duan *et al.*, 1995; Kibel *et al.*, 1995). At that time, elongin B and C were known as regulatory subunits of the general transcriptional elongation complex (composed of elongins A, B and C) known as Elongin or SIII. This led to suggestions that pVHL may regulate tumorigenesis by sequestering elongin B and C so preventing Elongin formation and down-regulating transcriptional elongation of oncogenic target genes. However there is little evidence that pVHL inhibits transcriptional elongation *in vivo*. Subsequently a *CUL2*, a member of the cullin protein family, was shown to associate with pVHL, elongin B and elongin C to form a tetrameric pVHL/elongin C/elongin B/CUL2 (VCBC) complex (Lonergan *et al.*, 1998; Pause *et al.*, 1997). A seminal observation was that elongin C and *CUL2* show significant sequence homology to the yeast proteins Skp1 and Cdc53 which are members of an SCF (Skp1-Cdc53/Cul1-F-box) complex. The SCF complex is part of a family of E3 ubiquitin ligases that regulate multiple cellular processes (e.g. cell cycle, signaling and development) by controlling proteolytic degradation of target proteins (Tyers and Rottapel, 1999). It was predicted that pVHL may be functionally analogous to the SCF F-box protein which recruits the specific protein targets to the core ubiquitylation complex (Lonergan *et al.*, 1998). Further evidence for this model of pVHL function was provided by structural analysis of the VCB complex (Stebbins *et al.*, 1999). The crystal structure of pVHL suggested two protein binding sites: an elongin C binding site in the α-domain and a surface binding site in the β-domain. Stebbins *et al.* (1999) noted that many disease-causing *VHL* mutations map to elongin C binding residues and there were strong structural similarities between the VCBC and SCF complexes. If pVHL were indeed to function as a 'F-box like protein', it would be predicted that the β-domain surface binding site would be involved in targeting proteolytic substrates to the general ubiquitylation complex which is bound via the α-domain. Further evidence for a SCF-like model of pVHL function came with the demonstration Rbx-1 bound to the VCBC complex and was also an essential general component of other SCF complexes (Kamura *et al.*, 1999). In addition, the VCBC complex has been demonstrated to promote ubiquitin ligase activity, in the presence of wild-type pVHL but not mutant pVHL (Iwai *et al.*, 1999; Lisztwan *et al.*, 1999).

Further pVHL binding proteins have been described (see Kaelin and Maher, 1998 and references within). Notably fibronectin co-immunoprecipitates with wild-type pVHL but not mutant pVHL mutants (Ohh *et al.*, 1998). Furthermore formation of an extracellular fibronectin matrix was defective in a VHL null RCC cell line and in mouse embryo fibroblasts from a *VHL* gene knockout mouse. The relationship between the influence of pVHL on fibronectin metabolism and its SCF-like function has not been defined.

In addition to identifying pVHL-interacting proteins, the characterization of pVHL null and pVHL competent cells has been pursued to provide insights into possible pVHL functions. Although VHL-dependent cell cycle effects have been described (VHL null RCC cells failed to exit the cell cycle upon serum withdrawal) (Pause *et al.*, 1998), most attention has focused on the role of pVHL in the regulation of hypoxia-inducible mRNAs such as VEGF.

VHL tumors and their sporadic counterparts (e.g. RCC and hemangioblastomas) are notably hypervascular and express high levels of *VEGF* (Flamme *et al.*, 1998; Siemeister *et al.*, 1996). These observations prompted several groups to investigate the effect of transfecting wild-type pVHL into VHL null RCC cell lines. A consistent finding was that VHL defective RCC constitutively overexpressed *VEGF* mRNA in both normoxic and hypoxic conditions (Gnarra *et al.*, 1996; Iliopoulos *et al.*, 1996; Levy *et al.*, 1996; Siemeister *et al.*, 1996). However, expression of wild-type pVHL restored an appropriate *VEGF* expression pattern with induction of *VEGF* expression by hypoxia.

Maxwell *et al.* (1999) demonstrated pVHL-regulated mRNA expression extended to many hypoxia-inducible mRNAs including VEGF, glucose transporter-1 (GLUT-1) and platelet-derived growth factor (PDGF). Two transcription factors, hypoxia-inducible factors 1 and 2 (HIF-1 and HIF-2/EPAS) play a major role in regulating hypoxia-inducible gene expression responses including genes involved in energy metabolism, angiogenesis and apoptosis (e.g. GLUT-1 and VEGF). HIF-1 and HIF-2 are heterodimeric proteins in which the β-subunits are expressed constitutively, but the α-subunits are normally rapidly degraded by the proteasome in an oxygen-dependent manner. However in the presence of hypoxia the HIF-1α and HIF-2α subunits are stabilized leading to up-regulation of HIF-1 and HIF-2 expression. Maxwell *et al.* (1999) showed that VHL null RCC cell lines demonstrated abnormal up-regulated HIF-1 and HIF-2 expression with loss of hypoxic inducibility. Further experiments revealed that pVHL co-immunoprecipitated with HIF-1α and HIF-2α subunits and pVHL was present in the hypoxic HIF-1 DNA-binding complex and that abnormal up-regulation of HIF-1 and HIF-2 expression in normoxic VHL-defective cells resulted from a failure of oxygen-dependent proteolysis.

The findings of Maxwell *et al.* (1999) could be interpreted as further evidence for an SCF-like function for pVHL. Thus under normoxic conditions the VCBC complex would be predicted to bind HIF-1α and HIF-2α subunits and target these proteins for ubiquitylation and degradation (*Figure 3*). Recently, Cockman *et al.* (2000) confirmed this hypothesis by demonstrating that pVHL bound to HIF-1α and HIF-2α at the pVHL β-domain surface binding site and that extracts from VHL-deficient renal carcinoma cells have a defect in HIF-1α ubiquitylation activity which is restored by exogenous pVHL.

The findings of Cockman *et al.* (2000) link the role of pVHL in VCBC-targeted ubiquitin ligase activity with HIF-1α as a specific pVHL β-domain target substrate and the angiogenic phenotype of VHL defective tumors (see *Figure 3*). However the complex genotype–phenotype correlations observed in VHL disease suggest that pVHL will have multiple functions. Regulation of further pVHL target genes such as transforming growth factor-α (Knebelmann *et al.*, 1998) and the transmembrane carbonic anhydrases, CA9 and CA12 may be mediated via HIF-1 and HIF-2 or by alternative mechanisms (Ivanov *et al.*, 1998). If some pVHL target genes are regulated by HIF-independent mechanisms it will be interesting to know if such regulation involved the targeting of other transcription factors for VCBC-mediated proteolysis or novel mechanisms. The

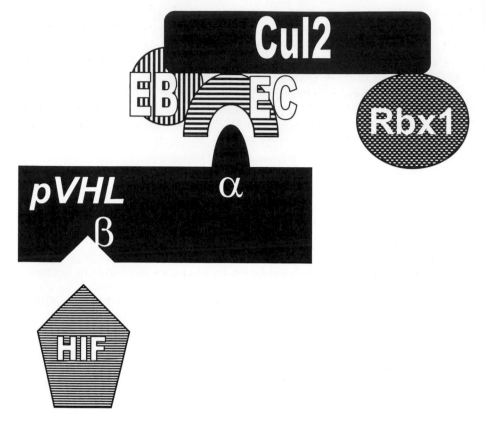

Figure 3. Schematic representation of pVHL function. The pVHL β domain binding site attaches to HIF-1α (or HIF-2α) The pVHL α domain binds directly to elongin C. The pVHL/Elongin C/Elongin B/Cul2/Rbx1 complex ubiquitylates HIF-1 and HIF-2 (and possibly other as yet unidentified substrates).

elucidation of the molecular basis for genotype–phenotype correlations in VHL disease will provide a link between specific pVHL functions and tumorigenesis. Interestingly, missense mutations associated with pheochromocytoma retain partial function in some assays, whereas missense mutations which do not cause pheochromocytoma behave as total loss-of-function mutation (unpublished observations). One possible explanation for the association of missense mutations with pheochromocytoma is that pheochromocytoma cells do not tolerate complete loss of pVHL function.

9. Conclusion

For both retinoblastoma and VHL disease the characterization of a rare familial cancer syndrome gene has led to the identification of a critical gatekeeper gene frequently mutated in more common sporadic cancers. Both *RB1* and *VHL* function as recessive TSGs and the genetics of retinoblastoma have provided a classical model of TSG function. However, the *VHL* TSG model demonstrates well important additional features of TSG function such as the significance of allelic heterogeneity and genotype–phenotype correlations, the existence of genetic modifiers and epigenetic effects in human tumorigenesis. Functional studies of the VHL protein have provided insights into a novel mechanism of TSG function and into cellular responses to hypoxia. Further progress in elucidating pVHL function and the relationship between specific functions and tumor susceptibility will provide further insights into the role of pVHL in tumorigenesis and opportunities to develop novel therapies.

References

Binkovitz, L.A., Johnson, C.D. and Stephens, D.H. (1990) Islet cell tumors in von Hippel-Lindau disease – increased prevalence and relationship to the multiple endocrine neoplasias. *Am. J. Roentgenol.* **155**: 501–505.

Brauch, H., Kishida, T., Glavac, D. *et al.* (1995) Von Hippel-Lindau (VHL) disease with pheochromocytoma in the Black-Forest region of Germany – evidence for a founder effect. *Hum. Genet.* **95**: 551–556.

Brauch, H., Weirich, G., Brieger, J. *et al.* (2000) VHL alterations in human clear cell renal cell carcinoma: Association with advanced tumor stage and a novel hot spot mutation. *Cancer Res.* **60**: 1942–1948.

Chen, F., Kishida, T., Duh, F.M., Renbaum, P., Orcutt, M.L., Schmidt, L. and Zbar, B. (1995a) Suppression of growth of renal-carcinoma cells by the von Hippel-Lindau tumor-suppressor gene. *Cancer Res.* **55**: 4804–4807.

Chen, F., Kishida, T., Yao, M. *et al.* (1995b) Germline mutations in the von Hippel-Hindau disease tumor-suppressor gene – correlations with phenotype. *Hum. Mutat.* **5**: 66–75.

Chen, F., Slife, L., Kishida, T., Mulvihill, J., Tisherman, S.E. and Zbar, B. (1996) Genotype–phenotype correlation in von Hippel-Lindau disease: identification of a mutation associated with VHL type 2A. *J. Med. Genet.* **33**: 716–717.

Choyke, P.L., Glenn, G.M., Walther, M.C.M. *et al.* (1992) The natural-history of renal lesions in von Hippel-Lindau disease – a serial CT study in 28 patients. *Am. J. Roentgenol.* **159**: 1229–1234.

Clifford, S.C., Prowse, A.H., Affara, N.A., Buys, C. and Maher, E.R. (1998) Inactivation of the von Hippel-Lindau (VHL) tumor suppressor gene and allelic losses at chromosome arm 3p in primary renal cell carcinoma: Evidence for a VHL-independent pathway in clear cell renal tumorigenesis. *Gen. Chromosom. Cancer* **22**: 200–209.

Cockman, M.E., Masson, N., Mole, D.R. *et al.* (2000) Hypoxia inducible factor-alpha binding and ubiquitylation by the von Hippel-Lindau tumor suppressor protein. *J. Biol. Chem.* in press.

Corless, C.L., Kibel, A.S., Iliopoulos, O. and Kaelin, W.G. (1997) Immunostaining of the von Hippel-Lindau gene product in normal and neoplastic human tissues. *Hum. Pathol.* **28**: 459–464.

Crossey, P.A., Richards, F.M., Foster, K. *et al.* (1994) Identification of intragenic mutations in the von Hippel-Lindau disease tumor-suppressor gene and correlation with disease phenotype. *Hum. Mol. Genet.* **3**: 1303–1308.

Crossey, P.A., Eng, C., Ginalska-Malinowska, M., Lennard, T.W.J., Wheeler, D.C., Ponder, B.A.J. and Maher, E.R. (1995) Molecular-genetic diagnosis of von Hippel-Lindau disease in familial pheochromocytoma. *J. Med. Genet.* **32**: 885–886.

Dietrich, W.F., Lander, E.S., Smith, J.S., Moser, A.R., Gould, K.A., Luongo, C., Borenstein, N., Dove, W. (1993) Genetic identification of *Mom-1*, a major modifier locus affecting *Min*-induced intestinal neoplasia in the mouse. *Cell* **75**: 631–639.

Duan, D.R., Pause, A., Burgess, W.H. *et al.* (1995) Inhibition of transcription elongation by the VHL tumor-suppressor protein. *Science* **269**: 1402–1406.

Easton, D.F., Ponder, M.A., Huson, S.M., Ponder, B.A.J. (1993) An analysis of variation in expression of neurofibromatosis (NF) type I (NF1): evidence for modifying genes. *Am. J. Hum. Genet.* **53**: 303–313.

Eng, C., Crossey, P.A., Mulligan, L.M. *et al.* (1995) Mutations in the RET proto-oncogene and the von Hippel-Lindau disease tumor suppressor gene in sporadic and syndromic pheochromocytomas. *J. Med. Genet.* **32**: 934–937.

Flamme, I., Krieg, M., Plate, K.H. (1998) Up-regulation of vascular endothelial growth factor in stromal cells of hemangioblastomas is correlated with up-regulation of the transcription factor HRF/HIF-2alpha. *Am. J. Pathol.* **153**: 25–29.

Foster, K., Prowse, A., van den Berg, A. *et al.* (1994) Somatic mutations of the von Hippel-Lindau disease tumor suppressor gene in nonfamilial clear cell renal carcinoma. *Hum. Mol. Genet.* **3**: 2169–2173.

Foster, K., Osborne, R.J., Huddart, R.A., Affara, N.A., Ferguson-Smith, M.A., Maher, E.R. (1995) Molecular genetic analysis of the von Hippel-Lindau disease (VHL) tumor suppressor gene in gonadal tumors. *Eur. J. Cancer* **31A**: 2392–2395.

Gnarra, J.R., Tory, K., Weng, Y. *et al.* (1994) Mutations of the VHL tumor-suppressor gene in renal-carcinoma. *Nat. Genet.* **7**: 85–90.

Gnarra, J.R., Zhou, S.B., Merrill, M.J. *et al.* (1996) Post-transcriptional regulation of vascular endothelial growth factor mRNA by the product of the VHL tumor suppressor gene. *Proc. Natl. Acad. Sci. USA* **93**: 10589–10594.

Gnarra, J.R., Ward, J.M., Porter, F.D. *et al.* (1997) Defective placental vasculogenesis causes embryonic lethality in VHL-deficient mice. *Proc. Natl. Acad. Sci. USA* **94**: 9102–9107.

Herman, J.G., Latif, F., Weng, Y.K. *et al.* (1994) Silencing of the VHL tumor-suppressor gene by DNA methylation in renal-carcinoma. *Proc. Natl. Acad. Sci. USA* **91**: 9700–9704.

Hes, F.J., McKee, S., Taphoorn, M.J.B. *et al.* (2000) Cryptic Von Hippel-Lindau disease: Germline mutations in hemangioblastoma-only patients. *J. Med. Genet.* (in press).

Hodgson, S.V., Maher, E.R. (1999) A practical guide to human cancer genetics. Cambridge University Press, Cambridge. Second Edition.

Iliopoulos, O., Kibel, A., Gray, S. and Kaelin, W.G. (1995) Tumor suppression by the human von Hippel-Lindau gene-product. *Nat. Med.* **1**: 822–826.

Iliopoulos, O., Levy, A.P., Jiang, C., Kaelin, W.G. and Goldberg, M.A. (1996) Negative regulation of hypoxia-inducible genes by the von Hippel-Lindau protein. *Proc. Natl. Acad. Sci. USA* **93**, 10595–10599.

Iliopoulos, O., Ohh, M. and Kaelin, W.G. (1998) pVHL(19) is a biologically active product of the von Hippel-Lindau gene arising from internal translation initiation. *Proc. Natl. Acad. Sci. USA* **95**: 11661–11666.

Ivanov, S.V., Kuzmin, I., Wei, M.H., Pack, S., Geil, L., Johnson, B.E., Stanbridge, E.J. and Lerman, M.I. (1998) Down-regulation of transmembrane carbonic anhydrases in renal cell carcinoma cell lines by wild-type von Hippel-Lindau transgenes. *Proc. Natl. Acad. Sci. USA* **95**: 12596–12601.

Iwai, K., Yamanaka, K., Kamura, T., Minato, N., Conaway, R.C., Conaway, J.W., Klausner, R.D. and Pause, A. (1999) Identification of the von Hippel-Lindau tumor-suppressor protein as part of an active E3 ubiquitin ligase complex. *Proc. Natl. Acad. Sci. USA* **96**: 12436–12441.

Kamura, T., Koepp, D.M., Conrad, M.N. *et al.* (1999) Rbx1, a component of the VHL tumor suppressor complex and SCF ubiquitin ligase. *Science* **284**: 657–661.

Kanno, H., Kondo, K., Ito, S. *et al.* (1994) Somatic mutations of the von Hippel-Lindau tumor suppressor gene in sporadic central nervous system hemangioblastomas. Cancer Res. **54**: 4845–4847.

Kibel, A., Iliopoulos, O., Decaprio, J.A. and Kaelin, W.G. (1995) Binding of the von Hippel-Lindau tumor-suppressor protein to elongin-B and elongin-C. *Science* **269**: 1444–1446.

Knebelmann, B., Ananth, S., Cohen, H.T. and Sukhatme, V.P. (1998) Transforming growth factor-alpha is a target for the von Hippel-Lindau tumor suppressor. *Cancer Res.* **58**: 226–231.

Kok, K., Naylor, S.L. and Buys, C. (1997) Deletions of the short arm of chromosome 3 in solid tumors and the search for suppressor genes. *Adv. Cancer Res.* **71**: 27–92.

Kuzmin, I., Duh, F.M., Latif, F., Geil, L., Zbar, B. and Lerman, M.I. (1995) Identification of the promoter of the human von Hippel-Lindau disease tumor-suppressor gene. *Oncogene* **10**: 2185–2194.

Latif, F., Tory, K., Gnarra, J. *et al.* (1993) Identification of the von Hippel-Lindau disease tumor-suppressor gene. *Science* **260**: 1317–1320.

Levy, A.P., Levy, N.S. and Goldberg, M.A. (1996) Hypoxia-inducible protein binding to vascular endothelial growth factor mRNA and its modulation by the von Hippel-Lindau protein. *J. Biol. Chem.* **271**: 25492–25497.

Libutti, S.K., Choyke, P.L., Bartlett, D.L. *et al.* (1998) Pancreatic neuroendocrine tumors associated with von Hippel-Lindau disease: Diagnostic and management recommendations. *Surgery* **124**: 1153–1159.

Lisztwan, J., Imbert, G., Wirbelauer, C., Gstaiger, M. and Krek, W. (1999) The von Hippel-Lindau tumor suppressor protein is a component of an E3 ubiquitin-protein ligase activity. *Genes Develop.* **13**: 1822–1833.

Lonergan, K.M., Iliopoulos, O., Ohh, M., Kamura, T., Conaway, R.C., Conaway, J.W. and Kaelin, W.G. (1998) Regulation of hypoxia-inducible mRNAs by the von Hippel-Lindau tumor suppressor protein requires binding to complexes containing elongins B/C and Cul2. *Mol. Cell. Biol.* **18**: 732–741.

Los, M., Jansen, G.H., Kaelin, W.G., Lips, C.J.M., Blijham, G.H. and Voest, E.E. (1996) Expression pattern of the von Hippel-Lindau protein in human tissues. *Lab. Invest.* **75**: 231–238.

Maher, E.R., Yates, J.R.W., Harries, R., Benjamin, C., Harris, R., Moore, A.T. and Ferguson-Smith, M.A. (1990a) Clinical-features and natural-history of von Hippel-Lindau disease. *Quart. J. Med.* **77**: 1151–1163.

Maher, E.R., Yates, J.R.W. and Ferguson-Smith, M.A. (1990b) Statistical-analysis of the 2 stage mutation model in von Hippel- Lindau disease, and in sporadic cerebellar hemangioblastoma and renal-cell carcinoma. *J. Med. Genet.* **27**: 311–314.

Maher, E.R., Iselius, L., Yates, J.R.W. *et al.* (1991) Von Hippel-Lindau disease: A genetic study. *J. Med. Genet.* **28**: 443–447.

Maher, E.R., Webster, A.R., Richards, F.M., Green, J.S., Crossey, P.A., Payne, S.J. and Moore, A.T. (1996) Phenotypic expression in von Hippel-Lindau disease: Correlations with germline VHL gene mutations. *J. Med. Genet.* **33**: 328–332.

Maher, E.R., Richards, F.M., Webster, A.R., McMahon, R., Woodward, E.R., Rose, S. (1998) Molecular genetic analysis of von Hippel-Lindau disease. *J. Int. Med.* **243**: 527–533

Manski, T.J., Heffner, D.K., Glenn, G.M. *et al.* (1997) Endolymphatic sac tumors – A source of morbid hearing loss in von Hippel-Lindau disease. *JAMA* **277**: 1461–1466.

Maxwell, P.H., Wiesener, M.S., Chang, G.W. *et al.* (1999) The tumor suppressor protein VHL targets hypoxia-inducible factors for oxygen-dependent proteolysis. *Nature* **399**: 271–275.

Neumann, H.P.H., Eng, C., Mulligan, L., Glavac, D., Ponder, B.A.J., Crossey, P.A., Maher, E.R., Brauch, H. (1995) Consequences of direct genetic testing for germline mutations in the clinical management of families with multiple endocrine neoplasia type 2. *JAMA* **274**: 1149–1151.

Neumann, H.P., Bender, B.U., Berger, D.P. *et al.* (1998) Prevalence, morphology and biology of renal cell carcinoma in von Hippel-Lindau disease compared to sporadic renal cell carcinoma. *J. Urol.* **160**: 1248–1254.

Ohh, M., Yauch, R.L., Lonergan, K.M. *et al.* (1998) The von Hippel Lindau tumor suppressor protein is required for proper assembly of an extracellular fibronectin matrix. *Mol. Cell* **1**: 959–968.

Pack, S.D., Zbar, B., Pak, E. *et al.* (1999) Constitutional von Hippel-Lindau (VHL) gene deletions detected in VHL families by fluorescence in situ hybridization. *Cancer Res.* **59**: 5560–5564.

Pause, A., Lee, S., Worrell, R.A., Chen, D.Y.T., Burgess, W.H., Linehan, W.M. and Klausner, R.D. (1997) The von Hippel-Lindau tumor-suppressor gene product forms a stable complex with human CUL-2, a member of the Cdc53 family of proteins. *Proc. Natl. Acad. Sci. USA* **94**: 2156–2161.

Pause, A., Lee, S., Lonergan, K.M. and Klausner, R.D. (1998) The von Hippel-Lindau tumor suppressor gene is required for cell cycle exit upon serum withdrawal. *Proc. Natl. Acad. Sci. USA* **95**: 993–998.

Prowse, A.H., Webster, A.R., Richards, F.M., Richard, S., Olschwang, S., Resche, F., Affara, N.A. and Maher, E.R. (1997) Somatic inactivation of the VHL gene in von Hippel-Lindau disease tumors. *Am. J. Hum. Genet.* **60**: 765–771.

Renbaum, P., Duh, F.M., Latif, F., Zbar, B., Lerman, M.I. and Kuzmin, I. (1996) Isolation and characterization of the full-length 3′ untranslated region of the human von Hippel-Lindau tumor suppressor gene. *Hum. Genet.* **98**: 666–671.

Richard, S., Chauveau, D., Chretien, Y. *et al.* (1994) Renal lesions and pheochromocytoma in von Hippel-Lindau disease. *Adv. Nephrol. Necker. Hosp.* **23**: 1–27

Richard, S., Resche, F., Vermesse, B. *et al.* (1992) Pheochromocytoma as presenting manifestation of von Hippel-Lindau disease. *Arch. Malad. Coeur Vais.* **85**: 1153–1156.

Richards, F.M., Crossey, P.A., Phipps, M.E. *et al.* (1994) Detailed mapping of germline deletions of the von Hippel-Lindau disease tumor-suppressor gene. *Hum. Mol. Genet.* **3**: 595–598.

Richards, F.M, Phipps, M.E., Latif, F. *et al.* (1993) Mapping the von Hippel-Lindau disease tumor suppressor gene: identification of germline deletions by pulsed field gel electrophoresis. *Hum. Mol. Genet.* **2**: 879–882.

Richards, F.M., Payne, S.J., Zbar, B., Affara, N.A., Ferguson-Smith, M.A., Maher, E.R. (1995a). Molecular analysis of de novo germline mutations in the Von Hippel-Lindau disease gene. *Hum. Mol. Genet.* **4**: 2139–2143.

Richards, F.M., Payne, S.J., Zbar, B., Affara, N.A., Ferguson-Smith, M.A. and Maher, E.R. (1995b) Molecular analysis of de-novo germline mutations in the von Hippel-Lindau disease gene. *Hum. Mol. Genet.* **4**: 2139–2143.

Richards, F.M., Schofield, P.N., Fleming, S. and Maher, E.R. (1996a) Expression of the von Hippel-Lindau disease tumor suppressor gene during human embryogenesis. *Hum. Mol. Genet.* **5**: 639–644.

Schoenfeld, A., Davidowitz, E.J. and Burk, R.D. (1998) A second major native von Hippel-Lindau gene product, initiated from an internal translation start site, functions as a tumor suppressor. *Proc. Natl. Acad. Sci. USA* **95**: 8817–8822.

Sekido, Y., Bader, S., Latif, F., Gnarra, J.R., Gazdar, A.F., Linehan, W.M., Zbar, B., Lerman, M.I., Minna, J.D. (1994) Molecular analysis of the von Hippel-Lindau disease tumor suppressor gene in human lung cancer cell lines. *Oncogene* **9**: 1599–1604.

Sgambati, M.T., Stolle, C., Choyke, P.L., Walther, M.M., Zbar, B., Linehan, W.M., Glenn, G.M. (2000) Mosaicism in von Hippel-Lindau disease: lessons from kindreds with germline mutations identified in offspring with mosaic parents. *Am. J. Hum. Genet.* **66**: 84–91.

Siemeister, G., Weindel, K., Mohrs, K., Barleon, B., MartinyBaron, G. and Marme, D. (1996) Reversion of deregulated expression of vascular endothelial growth factor in human renal carcinoma cells by von Hippel-Lindau tumor suppressor protein. *Cancer Res.* **56**: 2299–2301.

Solomon, D. and Schwartz, A. (1988) Renal pathology in von Hippel-Lindau disease. *Hum. Pathol.* **19**: 1072–1079.

Stebbins, C.E., Kaelin, W.G. and Pavletich, N.P. (1999) Structure of the VHL-ElonginC-ElonginB complex: Implications for VHL tumor suppressor function. *Science* **284**: 455–461.

Steinbach, F., Novick, A.C., Zincke, H. *et al.* (1995) Treatment of renal-cell carcinoma in von Hippel-Lindau disease – a multicenter study. *J. Urol.* **153**: 1812–1816.

Stolle, C., Glenn, G., Zbar, B. *et al.* (1998) Improved detection of germline mutations in the von Hippel Lindau disease tumor suppressor gene. *Hum. Mutat.* **12**: 417–423.

Tyers, M. and Rottapel, R. (1999) VHL: A very hip ligase. *Proc. Natl. Acad. Sci. USA* **96**: 12230–12232.

van-den-Berg, A., Dijkhuizen, T., Draaijers, T.G., Hulsbeek, M.M., Maher, E.R., van-den-Berg-E, Storkel, S., Buys, C.H.C. (1997) Analysis of multiple renal cell adenomas and carcinomas suggests allelic loss at 3p21 to be a prerequisite for malignant development. *Genes Chrom. Cancer* **19**: 228–232.

Walther, M.M., Choyke, P.L., Glenn, G., Lyne, J.C., Rayford, W., Venzon, D. and Linehan, W.M. (1999) Renal cancer in families with hereditary renal cancer: Prospective analysis of a tumor size threshold for renal parenchymal sparing surgery. *J. Urol.* **161**: 1475–1479.

Webster, A.R., Richards, F.M., MacRonald, F.E., Moore, A.T. and Maher, E.R. (1998) An analysis of phenotypic variation in the familial cancer syndrome von Hippel-Lindau disease: Evidence for modifier effects. *Am. J. Hum. Genet.* **63**: 1025–1035.

Webster, A.R., Maher, E.R. and Moore, A.T. (1999) Clinical characteristics of ocular angiomatosis in von Hippel-Lindau disease and correlation with germline mutation. *Arch. Ophthalmol.* **117**: 371–378.

Webster AR, Maher ER, Bird AC, Moore AT (2000) Risk of multisystem disease in isolated ocular angioma. *J. Med. Genet.* **37**: 62–63.

Woodward, E.R., Eng, C., McMahon, R., Voutilainen, R., Affara, N.A., Ponder, B.A.J. and Maher, E.R. (1997) Genetic predisposition to pheochromocytoma: Analysis of candidate genes GDNF, RET and VHL. *Hum. Mol. Genet.* **6**: 1051–1056.

Woodward, E.R., Hurst, L., Clifford, S.C., Affara, N.A. and Maher, E.R. (2000a) Comparative sequence analysis of the VHL tumor suppressor gene. *Genomics*, in press.

Woodward, E.R., Clifford, S.C., Astuti, D., Affara, N.A. and Maher, E.R. (2000b) Familial clear cell renal cell carcinoma (FCRC): clinical features and mutation analysis of the VHL, MET, and CUL2 candidate genes. *J. Med. Genet.* **37**: 348–353.

Yao, M., Latif, F., Orcutt, M.L. *et al.* (1993) von Hippel-Lindau disease – identification of deletion mutations by pulsed-field gel-electrophoresis. *Hum. Genet.* **92**: 605–614.

Zbar, B., Kishida, T., Chen, F. *et al.* (1996) Germline mutations in the von Hippel-Lindau disease (VHL) gene in families from North America, Europe, and Japan. *Hum. Mutat.* **8**: 348–357.

5

Genetics of NF1 and NF2

Margaret R. Wallace and Mia MacCollin

1. History and nomenclature of NF1 and NF2

In 1882, Friedrich von Recklinghausen first described a clinical syndrome characterized by nerve-derived tumors called neurofibromas. This syndrome was subsequently called 'von Recklinghausen neurofibromatosis', and is now referred to as
neurofibromatosis 1 (NF1) (Riccardi, 1992). NF1 is one of the most frequently
occurring human autosomal dominant diseases (1/3500 individuals). The hereditary nature of NF1 was not fully recognized until Borberg (1951) and Crowe *et al.*
(1956) determined that the mode of inheritance is autosomal dominant.
Henneberg and Koch (1903) observed that many cases of 'neurofibromatosis'
lacked skin alterations, but included bilateral eighth cranial nerve tumors. They
referred to this distinct disorder as 'central' neurofibromatosis in contrast with the
'peripheral' features of the disorder von Recklinghausen described. Confusion
persisted in the literature, however, especially with regard to patients with many
spinal or skin tumors. In 1987 and 1990, consensus conferences were held at the
National Institutes of Health to clarify the various clinical types of NF1 (Mulvihill
et al., 1990). The terms neurofibromatosis 1 (NF1) and neurofibromatosis 2 (NF2)
were recommended over all previous designations, and formal diagnostic criteria
were created, which have since been reviewed and updated (Gutmann *et al.*, 1997).
Together, NF1 and NF2 represent the vast majority of the neurofibromatoses.

2. Clinical features of NF1

2.1 Diagnostic criteria and diagnosis

Clinical and pathological findings of NF1, briefly summarized below, have been
reviewed in detail elsewhere (Friedman and Riccardi, 1999). NF1 diagnosis is
currently based on clinical examination, if two or more of the following seven
criteria are met (Gutmann *et al.*, 1997):

(i)   Six or more café-au-lait spots, 1.5 cm or larger in postpubertal individuals, or
      0.5 cm or larger in prepubertal individuals.
(ii) Two or more neurofibromas of any type, or at least one plexiform neurofibroma.
(iii) Freckling of the armpits or groin.

Molecular Genetics of Cancer second edition, edited by J.K. Cowell.
© 2001 BIOS Scientific Publishers Ltd, Oxford.

(iv) Optic pathway glioma.

(v) Two or more Lisch nodules (hamartomata of the iris).

(vi) Dysplasia of the sphenoid bone, or dysplasia or thinning of long bone cortex.

(vii) First degree relative with NF1.

NF1 diagnosis is often unclear, especially in young children with a negative family history, due to phenotypic variability and development of symptoms with age. The first feature to appear is usually café-au-lait spots (*Figure 1*), and young children with multiple café-au-lait spots are at high risk for eventually meeting NF1 diagnostic criteria. Lisch nodules are thought to be nearly pathognomonic for NF1. Additional features which contribute to the variable expressivity include macrocephaly, learning disabilities (rarely outright mental retardation), short stature, scoliosis, pseudoarthrosis, seizures, hypertension, pruritus, headaches, and malignancy (especially tumors of the central nervous system (Friedman and Riccardi, 1999)). Overall, the average life span in NF1 is somewhat reduced due to the increased occurrence of malignancy and other life-threatening complications, and many patients ultimately die of NF1-related conditions, in particular the following malignancies: malignant peripheral nerve sheath tumors (MPNSTs), pheochromocytomas, neuronal tumors, rhabdomyosarcomas, and leukemias (Rasmussen and Friedman, 2000).

The diagnostic criteria are widely accepted as the best diagnostic measure, rather than molecular analysis of the *NF1* gene. In families with NF1, DNA linkage studies using intragenic microsatellite markers are very accurate. For sporadic cases, the protein truncation test (PTT, based on analysis of RNA to detect truncating mutations such as nonsense and frameshifts) is commercially available, but is only about 70% sensitive in most labs (Heim *et al.*, 1995; Park and Pivnick, 1998), and is not used for prenatal diagnosis in the United States.

2.2 Neurofibromas

The hallmark feature of NF1 is the neurofibroma (*Figure 1*), a benign Schwann cell tumor which arises along peripheral nerves. These tumors often appear

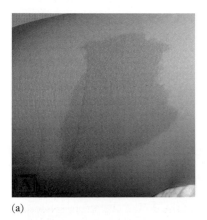 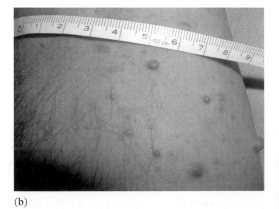

(a) (b)

Figure 1. Typical cutaneous findings in NF1. (a) Prominent cafe au lait macule in an affected child. (b) Multiple cutaneous neurofibromas on the forearm of an affected adult.

beginning in adolescence and increase in size and number with age and hormonal fluctuations such as with pregnancy (Friedman and Riccardi, 1999). Nearly all patients develop cutaneous or subcutaneous neurofibromas in their lifetime, and may have only a few or may have thousands covering their body. Rarely do these become larger than a centimeter in diameter. Somewhat less common, but often more serious, are plexiform neurofibromas, which typically develop along larger peripheral nerves and which may become quite large. These involve multiple nerve fascicles and may extend across a long length of nerve and nerve branches, leading to a diffuse mass of thickened nerves which may infiltrate around normal structures (Korf, 1999). The most evident and difficult ones appear to be congenital in origin, becoming obvious in early childhood. A computerized tomography study found that up to 40% of NF1 patients have thoracic and/or abdominal plexiform tumors, in many cases asymptomatic (Tonsgard et al., 1998). All types of neurofibromas are similar in that they contain Schwann cells, neural components, fibroblasts and mast cells. However, plexiform tumors have a risk of progressing to malignancy (to MPNST in an estimated 6% or more of NF1 patients (Riccardi, 1992)). Plexiform neurofibromas may also lead to severe disfigurement and/or functional impairment of adjacent tissues.

3. Genetics of NF1

NF1 is estimated to have a mutation rate of 1 per 10 000 alleles per generation – about 10-fold higher than most genes (Crowe et al., 1956). The frequency of this disorder, one of the highest among genetic diseases, is likely due to this high mutation rate. In fact, NF1 appears to be the result of new mutation in about half of all patients (Riccardi, 1992). It is possible that the large size of this gene (see 3.1) is responsible for the unusually high mutation rate. Similar to observations in achondroplasia and retinoblastoma, the majority of NF1 mutations arise in the paternal germline, although there does not seem to be a strong paternal age effect (Jadayel et al., 1990; Stephens et al., 1992). However, patients who have large deletions spanning the gene are more likely to have a maternal origin for such mutations, for reasons yet unclear (Leppig et al., 1997; Upadhyaya et al., 1998). NF1 is considered to be completely penetrant, with apparent cases of non-penetrance likely due to either multiple independent mutations in a family (Klose et al., 1999) or somatic/germline mosaicism (reviewed by Carey and Viskochil, 1999).

3.1 The NF1 gene

The NF1 gene (Figure 2), at chromosome 17q11.2, was cloned in 1990 (Cawthon et al., 1990; Viskochil et al., 1990; Wallace et al., 1990). It spans 340 kilobases (kb) of genomic DNA and has 57 exons (plus three that are alternatively spliced) (Viskochil, 1999). The transcript is 12–14 kb in size, with a 3.5 kb 3′ untranslated region. The NF1 mRNA encodes a 2818-amino acid hydrophilic cytoplasmic protein called neurofibromin. The middle eighth of the deduced protein (encoded by exons 21–27b) encodes a GTPase-activating protein (GAP) activity that has been well-established in a number of reports (reviewed by Kim and Tamanoi,

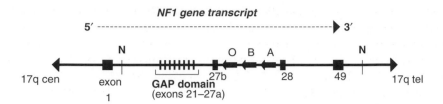

Figure 2. Schematic diagram (not to scale) showing the genomic structure of the *NF1* gene and its direction of transcription on chromosome 17q. N, *Not*I restriction sites found in the CpG islands that flank the gene, with the left-hand one being at the 5′ end of the *NF1* gene. The entire gene except for the extreme 5′ end is contained within the 350-kb *Not*I fragment indicated. Specific exons delineating regions of the gene are shown by black boxes. The three embedded genes in intron 27b, encoded on the opposite strand, are indicated by the shaded arrows: O, *OMGP*; B, *EVI2B*; A, *EVI2A*. The exons encoding the GAP-related domain include 21 through 27a, as indicated by the bracket. No other regions have had functions specifically assigned, although some portions of the 3′ half of the gene show protein-level homology to the yeast GAP proteins IRA1 and IRA2.

1998; Weiss *et al.*, 1999). GAPs inactivate ras-GTP via hydrolysis to the GDP-bound, inactive state (*Figure 3*).

The first discovered alternatively spliced *NF1* exon is 23a, which encodes a 21-amino sequence that is conserved across species (Nishi *et al.*, 1991). Both forms (including and excluding 23a) are present in most adult tissues. The presence of 23a results in slightly decreased GAP activity, although this isoform has higher affinity for ras-GTP (Andersen *et al.*, 1993). Another alternatively spliced exon, 48a, encodes 18 amino acids very near the carboxy end of the molecule. This isoform is expressed primarily in muscle and is proposed to be involved in myocyte development (Gutmann *et al.*, 1993). Exon 9br is a 10-amino acid sequence included only in the brain, particularly in embryogenesis, and its function is unknown (Danglot *et al.*, 1995; Geist and Gutmann, 1996). There are also three small genes (whose functions are not well understood) embedded within intron 27b, encoded on the opposite strand as well as a pseudogene near the 3′ end (Viskochil, 1999). Two of these embedded genes (*EVI2A* and *EVI2B*) are

Function of the NFI protein product

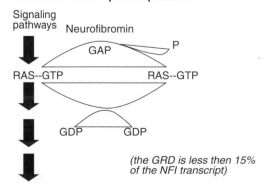

(the GRD is less then 15% of the NFI transcript)

Figure 3. THE NF1 pathway (ras-GAP activity). The NF1 protein product is known to interact with the proto-oncogene Ras. Addition functions are implied, since the majority of the protein product lies outside of the interacting domain, and these other regions are also highly conserved.

predominantly lymphoid-specific, with the other (*OMGP*) involved in oligoden-drocyte myelination. It is unknown whether these genes contribute to the NF1 phenotype, particularly in patients with deletions spanning these genes. The entire *NF1* gene region has been sequenced except for intron 1, which is thought to be approximately 100 kb; it is possible that more embedded genes lie here. The *NF1* gene has a large CpG island at its promoter, with a single transcription start site (Marchuk *et al.*, 1991). Recent studies have identified regions of hyper- and hypo-methylation in the normal promoter that coincide with transcription factor binding (Mancini *et al.*, 1999). A number of *NF1* homologs (unprocessed pseudo-genes) exist in pericentromeric regions of several chromosomes, which do not encode an active protein and mostly include just portions of the 5′ half of *NF1* (Cummings *et al.*, 1996; Purandare *et al.*, 1995; Regnier *et al.*, 1997).

3.2 Neurofibromin

NF1 mRNA is present ubiquitously (Wallace *et al.*, 1990), however neurofibromin is in greatest abundance in neurons, Schwann cells, and oligodendrocytes (Daston *et al.*, 1992; Golubic *et al.*, 1992; Gutmann and Collins, 1993). A number of muta-genesis/*in vitro* expression studies have examined the importance of several amino acid residues within the neurofibromin GAP-related domain, all of which showed that alteration of these conserved residues altered GAP function (reviewed in Colman and Wallace, 1995). Thus, neurofibromin appears to be an essential negative regulator of RAS (*Figure 3*) at least in neural-crest-derived tissues.

Other possible activities/properties of neurofibromin are currently being inves-tigated, with data suggesting possible cAMP regulatory activities (Guo *et al.*, 1997), tubulin-binding (Bollag *et al.*, 1993), co-localization with microtubules (Gregory *et al.*, 1993), and regulation of cytoskeletal organization (Koivunen *et al.*, 2000). There are several other lines of evidence that this molecule has multiple functions, which may potentially vary in a tissue- or developmental-specific pattern: much of neurofibromin is conserved across species (and kingdoms); *NF1* missense mutations and small in-frame deletions well outside the GAP domain have been found; and the alternatively spliced isoforms suggest tissue- or development-specific roles. Furthermore, neurofibromin may have non-redundant roles in tissues affected in NF1, or may be post-translationally modified in a tissue-specific manner or associate with other proteins expressed specifically in these tissues.

3.3 NF1 *germline mutations*

NF1 germline mutation data have accumulated slowly over the years, due to the technical difficulty in screening such a large gene, interference by homologous loci, and the wide variety of mutations (from complete deletion of the locus to point changes, thus far). Thus, no single mutation detection method can be applied with a high degree of sensitivity. Over 400 different *NF1* germline muta-tions scattered across the gene, in nearly all exons, have been described over the past 10 years (single base substitutions, deletions of all sizes, insertions, alter-ations leading to splicing errors, and translocations/inversions), with the largest data sources being the National Neurofibromatosis Foundation International Genetic Consortium Database (www.nf.org) and recent chapters/papers (Ars *et al.*, 2000; Fahsold *et al.*, 2000; Messiaen *et al.*, 2000; Upadhyaya and Cooper, 1998). Most of the mutations are predicted to result in truncated or missing proteins. It is not yet known if such truncated proteins are stable, although the mutant RNAs appear to be present in variable abundance. In a few cases, missense mutations in the GAP domain have been shown to reduce GAP activity (Klose *et al.*, 1998; Li *et al.*, 1992; Upadhyaya and Cooper, 1998), but the functional effects of other mutations, particularly those 3′ to the GAP domain (especially missense and small in-frame deletions), are unclear.

A few mutations are recurrent, with none having a frequency of more than about 1–2% (Boddrich *et al.*, 1997; Dublin *et al.*, 1995; Messiaen *et al.*, 1999), although some exons seem to show an increased occurrence of mutations relative to their size (Fahsold *et al.*, 2000). There are some interesting reports of mutation mechanisms in NF1. For example, in-frame exons containing premature-truncating codons may be skipped in the mRNA processing, presumably resulting in a protein that only lacks that exon's encoded amino acids (Hoffmeyer *et al.*, 1998; Messiaen *et al.*, 1997). Exons 10b and 37 appear to have frequent mutations (Boddrich *et al.*, 1997; Messiaen *et al.*, 1999), although together these still likely account for much less than 10% of all *NF1* mutations. Short repetitive sequences, homonucleotide tracts, and CpG dinucleotides are also common sites for mutation (Krkljus *et al.*, 1997; Rodenhiser *et al.*, 1997). Also, an *Alu* repetitive element was found to have been transposed *de novo* into intron 32 during spermatogenesis, resulting in an out-of-frame splice error in a new mutation patient (Wallace *et al.*, 1991).

As yet, there are no obvious NF1 phenotype/genotype correlations, although no extensive studies have been done. Such studies will likely require large patient numbers with detailed clinical documentation from different ages, if any patterns are to be discerned amidst the variable expressivity. For the recurrent exon 31 nonsense mutation, all three reported patients show approximately the same phenotype (having the same classic NF1 features), with some variability in the less-common symptoms such as scoliosis (Dublin *et al.*, 1995). Our group has identified two patients with this mutation whose NF1 phenotype is also not out of the ordinary, except that one developed a brain tumor (not optic pathway) (unpublished data). A recent study of five patients with the same exon 10b missense (which causes out-of-frame cryptic splicing) found 1/5 had a plexiform tumor, 3/5 had scoliosis, and the other features were either similar (e.g. café au lait spots) or not comparable due to age (Messiaen *et al.*, 1999). Microdeletions spanning the *NF1* gene and flanking regions are thought to be the germline mutation in 2–22% of patients (Messiaen *et al.*, 2000; Rasmussen *et al.*, 1998; Upadhyaya *et al.*, 1998). Additional features such as facial dysmorphia and mental retardation are not always consistent, but early onset of neurofibromas appears to be fairly common among these patients (Leppig *et al.*, 1997; Tonsgard *et al.*, 1997; Upadhyaya *et al.*, 1998; Wu *et al.*, 1995). Recent advances have characterized common breakpoints at the ends of the 1500-kilobase deletion, apparently due to unequal crossover at repeated sequences (Dorschner *et al.*, 2000). This finding raises the theory that the additional features may be due to loss of genes flanking *NF1*. Wu *et al.* (1999) found that 3/6 patients who developed MPNSTs had large germline gene deletions, raising the theory that flanking genes might also predispose to malignancy. Mutations are now being reported in association with unusual NF1 phenotypes, such as in families with consistent spinal neurofibromas (reduced or absent dermal neurofibromas) (Ars *et al.*, 1998; Poyhonen *et al.*, 1997). An additional level of variability added to the phenotype–genotype correlation is that many *NF1* mutations are splicing errors (Ars *et al.*, 2000; Messiaen *et al.*, 2000). Such errors tend to be leaky and the rate of aberrant splicing may even vary within tissues or between individuals. The fact that identical twins can have different NF1 features, yet overall severity and complications tend to be more consistent within a family than between families, suggests that ultimately phenotype–genotype correlations are likely to be found, and may also relate to alleles at modifying genes (Friedman and Riccardi, 1999).

   The involvement of the *NF1* gene in other disorders showing phenotypic overlap is also under study, including investigations in the 'NF-Noonan syndrome' (Stern *et al.*, 1992), Watson syndrome, and familial café-au-lait spots. One Watson syndrome patient has an *NF1* intragenic gene deletion (Upadhyaya *et al.*, 1992), and another has a 42-bp duplication in exon 28 (Tassabehji *et al.*, 1993), supporting the notion that this disorder is allelic to NF1. While some families with café-au-lait spots (consistently showing no other NF1 features) do not show linkage to the *NF1* gene, one family has shown linkage (Abeliovich *et al.*, 1995) although the causative *NF1* mutation (or mutation in another close gene) remains a mystery. Also, there are two reports of non-NF1 patients with *NF1* mutations: one had LEOPARD syndrome with multiple lentigines, and the other was a child with encephalocraniocutaneous lipomatosis (Legius *et al.*, 1995; Wu *et al.*, 1996). It is unclear whether these might also represent coincidental findings or extremes of the NF1 phenotype. There was also a report of a mutation in a 21-year-old with optic glioma who did not meet diagnostic criteria (Buske *et al.*, 1999). It is possible that this patient is mosaic and thus does not express the full phenotype (despite evidence of the mutation in multiple tissues), or it may be an example of a legitimate NF1 phenotype, muddying the issues of diagnostic criteria and penetrance.

4. NF1 tumor genetics

4.1 NF1 *is a tumor suppressor*

A number of functional and genetic studies have firmly concluded that *NF1* is a tumor suppressor gene, following (in at least some tumors) the cancer syndrome paradigm presented by Knudson (1985), that tumors contain one constitutionally inactivated allele (inherited), with the other allele subsequently inactivated ('second hit') via somatic mutation. Under this paradigm, loss of most or all of the function of the associated protein leads to increased cell proliferation or survival, and is often the first step on the tumorigenesis pathway. The *NF1* gene two-hit hypothesis was first proven for malignancies, through loss of heterozygosity (LOH) studies that often included apparent loss of an entire homolog (Glover *et al.*, 1991; Legius *et al.*, 1993; Lothe *et al.*, 1993; Shannon *et al.*, 1994; Xu *et al.*, 1992; and reviewed by Rasmussen and Wallace, 1998). Later LOH studies examined benign neurofibromas, using multiple *NF1* gene markers, and a number of studies detected LOH and other somatic mutations (Colman *et al.*, 1995; John *et al.*, 2000; Kluwe *et al.*, 1999a; Rasmussen *et al.*, 2000;

Sawada *et al.*, 1996; Serra *et al.*, 1997). In the few studies that tested markers outside of *NF1*, these areas of neurofibroma LOH do not extend into 17p, unlike the malignant tumors. In some cases the LOH is limited to the *NF1* gene. Not all NF1-related tumors have shown LOH, but that may be due to cellular heterogeneity of the material used for DNA preparation, and also does not preclude that there is still an inactivating second hit (as found in Sawada *et al.*, 1996). This supports the notion that loss of neurofibromin is probably necessary, and may be an initial event in the tumorigenesis pathway. Whether such loss is sufficient to cause neurofibroma formation is not clear. In addition to genetic events at the DNA level, it appears that epigenetic events at the RNA level might play a role in tumor formation. For example, Cappione *et al.* (1997) reported that the level of *NF1* mRNA editing (converting an arginine to stop codon on the mRNA in exon 23–2; present at a low level constitutionally) is increased in tumors, particularly MPNSTs. Others have speculated that epigenetic events altering normal gene methylation patterns might contribute to neurofibroma pathogenesis, as seen in other conditions, although this has not yet been found (Mancini *et al.*, 1999). In addition to loss of *NF1*, the MPNSTs clearly have multiple genetic abnormalities, as evidenced by cytogenetics (Glover *et al.*, 1991), comparative genomic hybridization studies (Lothe *et al.*, 1996; Mechtersheimer *et al.*, 1999; Schmidt *et al.*, 2000) and a few specific studies at genes such as *TP53* (Kluwe *et al.*, 1999a; Legius *et al.*, 1994; Nigro *et al.*, 1989; Rasmussen *et al.*, 2000) and *CDKN2A/p16* (Nielsen *et al.*, 1999).

4.2 Genomic stability in neurofibromas

There is one report suggesting the presence of microsatellite instability in at least a portion of neurofibromas, including a marker on chromosome 17 (Ottini *et al.*, 1995). Such instability would be indicative of loss of DNA repair mechanisms, such as mismatch repair genes implicated in colon tumors. However, none of the other LOH or tumor genetics studies (including at least one that examined the same marker) report any such observations, nor have we seen any such instability in our own lab's examination of over 50 neurofibromas with multiple markers. At another level, though, there is a suggestion that chromosomal instability may be involved. Two groups reported an overall increase in levels of spontaneous chromosomal aberrations in various cultured skin cells (and 'neurofibroma-derived' culture) from NF1 patients (Hafez *et al.*, 1985; Kehrer and Krone, 1994), and our group found that 3/6 plexiform neurofibroma Schwann cell cultures had multiple clones with different chromosomal abnormalities (Wallace *et al.*, 2000). Krone and Hogemann (1986) found some chromosome anomalies in dermal neurofibroma cultures, although there was no specific enriching for Schwann cells in that method (Schwann cells do not respond to standard tissue culture mitogens such as phytohemagglutinin). Schwann cells now appear to be the tumoral cell type (harboring genetic abnormalities) based on several levels of evidence (Kluwe *et al.*, 1999b; Rutkowski *et al.*, 2000; Wallace *et al.*, 2000), although our group did not find any chromosomal abnormalities in the six Schwann cultures derived from dermal neurofibromas. However, at least one group has failed to find chromosomal instability in NF1 lymphocyte and fibroblast cultures (Troilo *et al.*, 1992), and so the question about the presence and/or frequency of genomic instability in NF1 tumors remains unresolved.

5. NF2 clinical and genetic information

NF2 is an autosomal dominant disorder with full penetrance which is caused by disruption of a tumor suppressor gene located on 22q12 (Rouleau *et al.*, 1993; Trofatter *et al.*, 1993). Half of all affected individuals are sporadic, with clinically and genetically normal parents, and the majority of new mutations arise on the paternal allele (Kluwe *et al.*, 2000). Bilateral vestibular schwannomas are a universal feature of NF2, while other tumor types such as meningioma, ependymoma, and schwannomas (*Figure 4*) of other nerves are more variable (Mautner *et al.*, 1996). The only non-tumorous manifestations of NF2 are ocular abnormalities and peripheral neuropathy. The age of onset of symptoms, diagnosis, and deafness may range from early childhood to late adulthood, but most patients come to medical attention in their late adolescence or early adulthood. The NIH diagnostic criteria for NF2 are summarized below (adapted from Mulvihill *et al.*, 1990); revised criteria are proposed in Gutmann *et al.* (1997). NF2 is present in an individual with:

(1) Bilateral vestibular schwannomas.
OR
(2) A first degree relative with bilateral vestibular schwannomas, and either
    (a) unilateral vestibular schwannoma
    OR
    (b) two or more of the following: meningioma; glioma; schwannoma; juvenile posterior subcapsular lenticular opacity.

6. The *NF2* gene and protein

Initial work in the localization of the *NF2* gene utilized linkage analysis in a small number of NF2-affected families (Rouleau *et al.*, 1987) and loss of heterozygosity

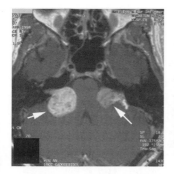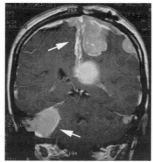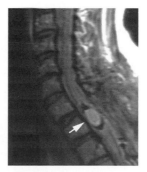

Figure 4. MR imaging appearance of tumors associated with NF2. (a) Bilateral vestibular schwannomas that have exited the internal auditory canals and entered the cerebellar pontine angle (arrows). Axial T1-weighted contrast enhanced image. (b) Multiple meningiomas occurring both super- and infra-tentorially (arrows). Corneal T_1 weighted contrast enhanced image. (c) Intramedullary tumor in at the level of T_1 (arrow). Such tumors are most frequently ependymomas, but may also be astrocytomas. Sagittal T_1-weighted contrast enhanced image.

studies in NF2-related tumors (Seizinger *et al.*, 1987). Both approaches converged on the long arm of human chromosome 22, where the gene was cloned in 1993 (Rouleau *et al.*, 1993; Trofatter *et al.*, 1993). The putative protein product was named 'merlin' (*m*oesin, *e*zrin, *r*adixin-*li*ke prote*in*) for its unexpected relationship with several cytoskeletal elements. The *NF2* gene spans 96.6 kb and comprises 16 constitutive exons and one alternatively spliced exon. Despite its relatively restricted phenotype, *NF2* is widely expressed. *NF2* message can be detected in three different size ranges (7, 4.4 and 2.6 kb) most likely due to variable length of 3′ untranslated sequence. Because exon 16 is both alternatively spliced and contains an in-frame stop codon, two different isoforms of the *NF2* protein product can be made. Isoform 1 is a protein of 595 amino acids produced from exons 1 through 15 and exon 17, and is the dominant form in most tissues. Presence of the alternatively spliced exon 16 substantially changes the carboxy terminus of the protein, replacing 16 amino acids with 11 novel residues in isoform 2. The *NF2* gene is remarkably conserved through evolution. The mouse isoform, isolated by two groups shortly after cloning of the human gene (Haase *et al.*, 1994; Hara *et al.*, 1994), maps to mouse chromosome 11 in a region of synteny with human 22q. It is similarly alternatively spliced in the 3′ end and is 98% identical to human *NF2* at the amino acid level.

On the basis of sequence similarity, the *NF2* protein product was determined to be a member of the protein 4.1 family of cytoskeleton associated proteins (*Figure 5*). This was an unexpected finding, since no other family members are associated with tumor formation and only one, protein 4.1 itself, is associated with human genetic disease (hereditary elliptocytosis). The ERM proteins (ezrin, moesin, radixin), to which the *NF2* protein is most closely related, share 70 to 75% amino acid identity and a common structure. These proteins localize to motile regions of the cell, including microvilli, cellular protrusions and leading and ruffling edges. ERM proteins interact with the Rho family of GTPases in a signaling cascade which controls the organization of the spectrin-actin cytoskeleton and cell adhesion. Although less is currently known about *NF2* protein physiology and function than about other ERM family members, similar interactions with the cytoskeleton are beginning to be identified.

7. Germline NF2 mutations

Over 125 independent disease associated mutations in the *NF2* gene have been reported in the literature to date. Approximately 90% of these events are predicted

Figure 5. Structure of the NF2 protein product. All protein 4.1 family members contain a defining amino terminal end, followed by an alpha helical domain. The 3′ end of the NF2 transcript is alternatively spliced by the inclusion of exon 16 which contains an in-frame stop codon.

to shorten the *NF2* protein product by removal of an exon or introduction of a non-native stop codon. Mutation types are not evenly allotted within the transcript with 'warm spots' at CpG islands in exons 2, 6, 8, and 11. 'Cold spots' with a paucity of mutations are found in exons 5 and 9 and the 3′ half of exon 15; to date no pathogenic mutations have been documented in the alternatively spliced exons of the gene (16 and 17). The underlying *NF2* gene mutation is predictive of the resulting phenotype when the gross variables of mild versus severe disease are examined, with nonsense and frameshift mutation producing the most severe disease and missense mutations producing the most mild disease (Evans *et al.*, 1998a; Kluwe *et al.*, 1996; Parry *et al.*, 1996). Small numbers of large deletions and other mutation types have been reported, and there is growing evidence that mosaicism in founders may account for both typical and atypical NF2 phenotypes (Evans *et al.*, 1998b; Kluwe and Mautner 1998).

8. The *NF2* gene is a tumor suppressor

The finding of acquired, somatic mutations or 'second hits' in NF2-related tumor material is crucial to the confirmation of its role as a tumor suppressor gene. Study of sporadic and NF2-derived schwannomas has shown grossly truncating mutation in the vast majority, with out-of-frame deletion an especially common mechanism of mutation (for example, Jacoby *et al.*, 1996). Many schwannomas also show loss of polymorphic markers on chromosome 22, indicating large deletions or loss of the entire trans chromosome. As would be expected, when the tumor is derived from an NF2-affected individual, the markers lost are always occurs *trans* to the germline mutation, removing the normal copy of the gene. No differences have been reported between mutations detected in vestibular versus non vestibular tumors nor between tumors derived from NF2 patients versus sporadic tumors, although this is an area that deserves further study.

Analysis of other tumor types associated with NF2 has yielded slightly different results. For example, in a comprehensive analysis of 70 sporadic meningiomas, 43 *NF2* mutations were detected in 41 tumors (Wellenreuther *et al.*, 1995). Similar to the results in schwannoma, only 1 of the 43 involved a nontruncating event. Mutational events were much more frequent in tumors which had lost heterozygosity for chromosome 22, supporting the hypothesis that *NF2* is the meningioma suppressor locus on chromosome 22, but that another protein or proteins not on chromosome 22 may also fill this role. *NF2* gene mutations have also been detected in sporadic spinal cord ependymomas (Birch *et al.*, 1996), but not in sporadic astrocytomas (Watkins *et al.*, 1996). Molecular genetic study of glial tumors resected from NF2 patients has not been reported.

Loss of heterozygosity has been observed for chromosome 22q markers in many different types of tumors not characteristic of NF2. Screening for mutations that affect the *NF2* gene in such tumors has yielded mixed results. Only a handful of putative mutations of the *NF2* gene have been found in malignant melanoma, breast adenocarcinoma and colon cancer and none have been seen in ovarian carcinoma, or hepatocellular carcinoma (Arakawa *et al.*, 1994; Bianchi *et al.*, 1994; Englefield *et al.*, 1994; Kanai *et al.*, 1995). Two reports demonstrate a high rate of

NF2 mutation in malignant mesothelioma (Bianchi *et al.*, 1995; Sekido *et al.*, 1995). The significance of these results to patients with NF2 is unclear.

9. Conclusions and future directions

A significant problem in furthering NF1 and NF2 research has been the inability to generate mouse knockout models that adequately recapitulate the human phenotype. Since no naturally occurring animal models exist for either condition, this is especially important. In the *Nf1* knockout (nearly the same mutation made by two independent labs, disrupting the gene in exon 31), the homozygous null embryos die by day 12.5 due to a heart defect/under-developed heart (Brannan *et al.*, 1994; Jacks *et al.*, 1994). The heterozygous mice develop a higher load of neural-crest derived tumors in mid- to late adulthood (e.g. pheochromocytomas and myeloid leukemia), but not benign neurofibromas. These mice do not have pigment problems or show other NF1 diagnostic criteria; they are essentially otherwise normal with one exception – these animals display cognitive deficits in learning and memory similar to those seen in some NF1 patients (Silva *et al.*, 1997). On a p53 null background, *Nf1* heterozygous mice develop what appear to be soft tissue sarcomas (such as MPNSTs) between 3–7 months of age, which appears to be a good model for NF1 malignancy (Vogel *et al.*, 1999). In addition, in mice heterozygous for both *Nf1* and *Tp53* knockouts, but with the mutations in *cis* rather than *trans* (the genes are tightly linked), there was complete mortality by 10 months, again mostly from sarcomas (Cichowski *et al.*, 1999; Vogel *et al.*, 1999). In another experiment, 12/18 chimeric mice containing both normal and null cells were found to develop plexiform neurofibromas (and frequent myeolodysplasia and progressive neuromotor defects) but not dermal neurofibromas (Cichowski *et al.*, 1999). This led to the hypothesis that somatic inactivation of the remaining *Nf1* allele is the rate-limiting step for tumorigenesis, although it is unknown whether additional events at other genes must also take place.

Development of a mouse model of NF2 has met with variable results. Mice homozygous for *Nf2* gene mutation do not survive to term, and those heterozygous develop malignant tumor types not seen in human NF2 (McClatchey *et al.*, 1997, 1998). One group reported that a transgenic mouse line expressing a mutant Nf2 protein (as well as endogenous normal Nf2 protein) displayed Schwann cell hyperplasia and Schwann-cell derived peripheral nerve tumors, suggesting that some mutations might act in a dominant fashion (Giovannini *et al.*, 1999). However, this model did not show any brain tumors, and also had high levels of expression of the mutant protein. Mice engineered to conditionally knock out the Nf2 protein product in specific cell lineages may circumvent these problems; such a knockout targeted to Schwann lineages allowed expression of a phenotype that more clearly mimics the human condition (Giovannini *et al.*, 2000).

Although a great deal has been elucidated about the genetics and cell biology of NF1 and NF2 since the cloning of their respective genes in the past decade, many questions remain to be answered and others have arisen. Ongoing work covers a very broad spectrum of scientific disciplines, including clinical genetics and pathology, human and animal molecular genetics, biochemical analysis of the *NF1* and *NF2* proteins in their signaling pathways, and cognitive/behavioral studies in humans and knockout animals. Hundreds of papers spanning these fields are published each year. Work is also underway to characterize the genomic and cellular changes in NF1 and NF2 tumors, to provide better therapeutic targets and diagnostic markers. The recent advent of NF1 Schwann cell-derived human tumor cultures will be important in many functional studies, perhaps a moderate substitute for the lack of an animal model mimicking NF1 features, although the knockout animals derived so far have proven extremely useful in more detailed studies (such as at the mechanism level). A few clinical trials have begun in both NF1 and NF2, although there is still a need for definitive natural history data to allow best interpretation of the results. Ultimately, the long-term goal of all the research is to contribute to development of therapies for NF1 and NF2, which may also have application in other cancers or benign tumor syndromes.

References

Abeliovich, D., Gelman-Kohan, Z., Silverstein, S., Lerer, I., Chemke, J., Merin, S. and Zlotogora, J. (1995) Familial café-au-lait spots: a variant of neurofibromatosis type 1. *J. Med. Genet.* **32**: 985–986.

Andersen, L.B., Ballester, R., Marchuk, D.A. *et al.* (1993) A conserved alternative splice in the von Recklinghausen neurofibromatosis (*NF1*) gene produces two neurofibromin isoforms, both of which have GTPase-activating protein activity. *Mol. Cell. Biol.* **13**: 487–495.

Arakawa, H., Hayashi, N., Nagase, H., Ogawa, M. and Nakamura, Y. (1994) Alternative splicing of the *NF2* gene and its mutation analysis of breast and colorectal cancers. *Hum. Mol. Genet.* **3**: 565–568.

Ars, E., Kruyer, H., Gaona, A., Casquero, P., Rosell, J., Volpini, V., Serra, E., Lazaro, C. and Estivill, X. (1998) A clinical variant of neurofibromatosis type 1: familial spinal neurofibromatosis with a frameshift mutation in the *NF1* gene. *Am. J. Hum. Genet.* **62**: 834–841.

Ars, E., Serra, E., Garcia, J., Kruyer, H., Gaona, A., Lazaro, C. and Estivill, X. (2000) Mutations affecting mRNA splicing are the most common molecular defects in patients with neurofibromatosis type 1. *Hum. Mol. Genet.* **9**: 237–247.

Bianchi, A.B., Hara, T., Ramesh, V. *et al.* (1994) Mutations in transcript isoforms of the neurofibromatosis 2 gene in multiple human tumor types. *Nat. Genet.* **6**: 185–192.

Bianchi, A.B., Mitsunaga, S.I., Cheng, J.Q., Klein, W.M., Jhanwar, S.C., Seizinger, B., Kley, N., Klein-Szanto, A.J. and Testa, J.R. (1995) High frequency of inactivating mutations in the neurofibromatosis type 2 gene (*NF2*) in primary malignant mesotheliomas. *PNAS USA* **92**: 10854–10858.

Birch, B., Johnson, F., Parsa, A., Desai, R., Yoon, J., Lycette, C., Li, Y.M. and Bruce, J. (1996). Frequent type 2 neurofibromatosis gene transcript mutations in sporadic intramedullary spinal cord ependymomas. *Neurosurgery* **39**: 135–140.

Boddrich, A., Robinson, P.N., Schulke, M., Buske, A., Tinschert, S. and Nurnberg, P. (1997) New evidence for a mutation hotspot in exon 37 of the *NF1* gene. *Hum Mutat* **9**: 374–377.

Bollag, G., McCormick, F. and Clark, R. (1993) Characterization of full-length neurofibromin: tubulin inhibits ras GAP activity. *EMBO J.* **12**: 1923–1927.

Borberg, A. (1951) Clinical and genetic investigations into tuberous sclerosis and Recklinghausen's neurofibromatosis. *Acta Psychiatr. Neurol.* **71 suppl**: 1–239.

Brannan, C., Perkin, A., Vogel, K., *et al.* (1994) Targeted disruption of the Nf1 gene leads to developmental abnormalities in heart and various neural crest-derived tissues. *Genes Dev.* **8**: 1019–1029.

Buske, A., Gewies, A., Lehmann, R., Ruther, K., Algermissen, B., Nurnberg, P. and Tinschert, S. (1999) Recurrent *NF1* gene mutation in a patient with oligosymptomatic neurofibromatosis type 1 (NF1). *Am. J. Med. Genet.* **86**: 328–330.

Cappione, A.J., French, B.L. and Skuse, G.R. (1997) A potential role for *NF1* mRNA editing in the pathogenesis of NF1 tumors. *Am. J. Hum. Genet.* **60**: 305–312.

Carey, J.C. and Viskochil, D.H. (1999) Neurofibromatosis type 1: a model condition for the study of the molecular basis of variable expressivity in human disorders. *Am. J. Med. Genet. (Semin Med Genet)* **89**: 7–13.

Cawthon, R.M., Weiss, R., Xu, G. *et al.* (1990) A major segment of the neurofibromatosis type 1 gene: cDNA sequence, genomic structure, and point mutations. *Cell* **62**: 193–201.

Cichowski, K., Shih, T.S., Schmitt, E., Santiago, S., Reilly, K., McLaughlin, M.E., Bronson, R.T. and Jacks, T. (1999) Mouse models of tumor development in neurofibromatosis type 1. *Science* **286**: 2172–2176.

Colman, S.D. and Wallace, M.R. (1995) Molecular genetics of neurofibromatosis Types 1 and 2. In *Molecular Genetics of Cancer* (ed. Cowell, J.K.), pp. 43–70. BIOS Scientific Publishers, Oxford.

Colman, S.D., Williams, C.A. and Wallace, M.R. (1995) Benign neurofibromas in type 1 neurofibromatosis (NF1) show somatic deletions of the *NF1* gene. *Nat. Genet.* **11**: 90–92.

Crowe, F.W., Schull, W.J. and Neel, J.V. (1956) *A clinical, pathological, and genetic study of multiple neurofibromas.* Charles C. Thomas Publishers, Springfield, IL.

Cummings, L.M., Trent, J.M., and Marchuk, D.A. (1996) Identification and mapping of type 1 neurofibromatosis (*NF1*) homologous loci. *Cytogenet Cell Genet* **73**: 334–340.

Danglot, G., Regnier, V., Fauvet, D., Vassal, G., Kujas, M. and Bernheim, A. (1995) *NF1* mRNAs expressed in the central nervous system are differentially spliced in the 5′ part of the gene. *Hum. Mol. Genet.* **4**: 915–920.

Daston, M.M., Scrable, H., Nordlund, M., Sturbaum, A.K., Nissen, L.M. and Ratner, N. (1992) The protein of the neurofibromatosis type 1 gene is expressed at highest abundance in neurons, Schwann cells, and oligodendrocytes. *Neuron* **8**: 415–428.

Dorschner, M.O., Sybert, V.P., Weaver, M., Pletcher, B.A. and Stephens, K. (2000) *NF1* microdeletion breakpoints are clustered at flanking repetitive sequences. *Hum. Mol. Genet.* **9**: 35–46.

Dublin, S., Riccardi, V.M. and Stephens, K. (1995) Methods for rapid detection of a recurrent nonsense mutation and documentation of phenotypic features in neurofibromatosis type 1 patients. *Hum. Mutat.* **5**: 81–85.

Englefield, P., Foulkes, W.D. and Campbell, I.G. (1994) Loss of heterozygosity on chromosome 22 in ovarian carcinoma is distal to and is not accompanied by mutations in *NF2* at 22q12. *Br. J. Cancer* **70**: 905–907.

Evans, D.G., Trueman, L., Wallace, A., Collins, S., and Strachan, T. (1998a) Genotype/phenotype correlations in type 2 neurofibromatosis (NF2): evidence for more severe disease associated with truncating mutations. *J. Med. Genet.* **35**: 450–455.

Evans, D.G., Wallace, A.J., Wu, C.L., Trueman, L., Ramsden, R.T. and Strachan, T. (1998b) Somatic mosaicism: a common cause of classic disease in tumor-prone syndromes? Lessons from type 2 neurofibromatosis. *Am. J. Hum. Genet.* **63**: 727–736.

Fahsold, R., Hoffmeyer, S., Mischung, C. *et al.* (2000) Minor lesion mutational spectrum of the entire *NF1* gene does not explain its high mutability but points to a functional domain upstream of the GAP-related domain. *Am. J. Hum. Genet.* **66**: 790–818.

Friedman, J.M. and Riccardi, V.M. (1999) Clinical and epidemiological features. In: *Neurofibromatosis: phenotype, natural history, and pathogenesis* (ed. Friedman, J.M., et al.), pp. 29–86. Johns Hopkins Press, Baltimore.

Geist, R. and Gutmann, D. (1996) Expression of a developmentally-regulated neuron-specific isoform of the neurofibromatosis 1 (*NF1*) gene. *Neurosci. Lett.* **211**: 85–88.

Giovannini, M., Robanus-Maandag, E., Niwa-Kawakita, M., van Der Valk, M., Woodruff, J.M., Goutebroze L., Merel, P., Berns A and Thomas G. (1999) Schwann cell hyperplasia and tumors in transgenic mice expressing a naturally occurring mutant NF2 protein. *Genes Dev.* **13**: 978–986.

Giovannini, M., Robanus-Maandag, E., van Der Valk, M., Niwa-Kawakita, M., Abramowski, V., Goutebroze, L., Woodruff, JM., Berns, A. and Thomas, G. (2000) Conditional biallelic Nf2 mutation in the mouse promotes manifestations of human neurofibromatosis type 2. *Genes Dev.* **14**: 1617–1630.

Glover, T.W., Stein, C.K., Legius, E., Andersen, L.B., Brereton, A. and Johnson, S. (1991) Molecular and cytogenetic analysis of tumors in von Recklinghausen neurofibromatosis. *Genes Chrom. Cancer* **3**: 62–70.

Golubic, M., Roudebush, M., Dorbrowolski, S., Wolfman, A. and Stacey, D.W. (1992) Catalytic properties, tissue and intracellular distribution of neurofibromin. *Oncogene* 7: 2151–2159.

Gregory, P.E., Gutmann, D.H., Mitchell, A., Park, S., Boguski, M., Jacks, T., Wood, D.L, Jove, R. and Collins, F.S. (1993) Neurofibromatosis type 1 gene product (neurofibromin) associates with microtubules. *Somat. Cell Mol. Genet.* 19: 265–274.

Guo, H.F., The, I., Hannan, F., Bernards, A. and Zhong, Y. (1997) Requirement of Drosophila NF1 for activation of adenylyl cyclase by PACAP38-like neuropeptides. *Science* 276: 795–798.

Gutmann, D.H. and Collins, F.S. (1993) The neurofibromatosis type 1 gene and its protein product, neurofibromin. *Neuron* 88: 9658–9662.

Gutmann, D.H., Andersen, L.B., Cole, J.L., Swaroop, M. and Collins, F.S. (1993) An alternatively-spliced mRNA in the carboxy terminus of the neurofibromatosis type 1 *(NF1)* gene is expressed in muscle. *Hum. Mol. Genet.* 2: 989–992.

Gutmann, D.H., Aylsworth, A., Carey, J.C., Korf, B., Marks, J., Pyeritz, R.E., Rubenstein, A. and Viskochil, D. (1997) The diagnostic evaluation and multidisciplinary management of neurofibromatosis 1 and neurofibromatosis 2. *JAMA* 278: 51–57.

Haase, V.H., Trofatter, J.A., MacCollin, M., Tarttelin, E., Gusella, J.F. and Ramesh, V. (1994) The murine NF2 homologue encodes a highly conserved merlin protein with alternative forms. *Hum. Mol. Genet.* 3: 407–411.

Hafez, M., Sharaf, L., El-Nabi, S.M.A. and El-Wehedy, G. (1985) Evidence of chromosomal instability in neurofibromatosis. *Cancer* 55: 2434–2436.

Hara, T., Bianchi, A.B., Seizinger, B.R. and Kley, N. (1994) Molecular cloning and characterization of alternatively spliced transcripts of the mouse neurofibromatosis 2 gene. *Cancer Res.* 54: 330–335.

Heim, R.A., Kam-Morgan, L.N.W., Binnie, C.G. *et al.* (1995) Distribution of 13 truncating mutations in the neurofibromatosis type 1 gene. *Hum. Mol. Genet.* 4: 975–981.

Henneberg, K. and Koch, M. (1903) Ueber "Centrale" neurofibromatose und die geschwulste des klein-hirnbruckenwinkels (acusticusneurome). *Arch Psy Nervenkr* (German) 36: 251–304.

Hoffmeyer, S., Nurnberg, P., Ritter, H., Fahsold, R., Leistner, W., Kaufmann, D. and Krone, W. (1998) Nearby stop codons in exons of the neurofibromatosis type 1 gene are disparate splice effectors. *Am. J. Hum. Genet.* 62: 269–277.

Jacks T, Shih S.T., Schmitt., E.M. Bronson, R.T., Bernards, A. and Weinberg, R.A. (1994) Tumour predisposition in mice heterozygous for a targeted mutation in Nf1. *Nature Genet.* 7: 353–361.

Jacoby, L., MacCollin, M., Barone, R., Ramesh, V. and Gusella, J. (1996) Frequency and distribution of *NF2* mutations in schwannomas. *Genes Chrom. Cancer* 17: 45–55.

Jadayel, D., Fain, P., Upadhyaya, M. *et al.* (1990) Paternal origin of new mutations in von Recklinghausen neurofibromatosis. *Nature* 343: 558–559.

John, A.M., Ruggieri, M., Ferner, R. and Upadhyaya, M. (2000) A search for evidence of somatic mutations in the *NF1* gene. *J. Med. Genet.* 37: 44–49.

Kanai, Y., Tsuda, H., Oda, T., Sakamoto, M. and Hirohashi, S. (1995) Analysis of the neurofibromatosis 2 gene in human breast and hepatocellular carcinomas. *Jpn. J. Clin. Oncol.* 25: 1–4.

Kehrer, H. and Krone, W. (1994) Spontaneous chromosomal aberrations in cell cultures from patients with neurofibromatosis 1. *Mut Res* 306: 61–70.

Kim, M.R. and Tamanoi, F. (1998) Neurofibromatosis 1 GTPase activating protein-related domain and its functional significance. In: *Neurofibromatosis Type 1: From genotype to phenotype* (eds Upadhyaya, M., and Cooper, D.N.), pp. 89–112. BIOS Scientific Publishers, Oxford.

Klose, A., Ahmadian, M.R. *et al.* (1998) Selective disactivation of neurofibromin GAP activity in neurofibromatosis type 1. *Hum. Mol. Genet.* 7: 1261–1268.

Klose, A., Peters, H., Hoffmeyer, S., Buske, A., Luder, A., Hess, D., Lehmann, R., Nurnberg, P. and Tinschert, S. (1999) Two independent mutations in a family with neurofibromatosis type 1 (NF1). *Am. J. Med. Genet.* 83: 6–12.

Kluwe, L., and Mautner, V.-F. (1998) Mosaicism in sporadic neurofibromatosis 2 patients. *Hum. Mol. Genet.* 7: 2051–2055.

Kluwe, L. Bayer, S., Baser, M., Hazim, W., Wolfgang, H., Funsterer, C. and Mautner, V.-F. (1996) Identification of *NF2* germ-line mutations and comparison with neurofibromatosis 2 phenotypes. *Hum. Genet.* 98: 534–538.

Kluwe, L., Friedrich, R. and Mautner, V.-F. (1999a) Allelic loss of the *NF1* gene in NF1-associated plexiform neurofibromas. *Cancer Genet. Cytogenet* 113: 65–69.

Kluwe, L., Friedrich, R. and Mautner, V.-F. (1999b) Loss of *NF1* allele in Schwann cells but not fibroblasts derived from an NF1-associated neurofibroma. *Genes Chrom. Cancer* 24: 283–285.

Kluwe, L., Mautner, V., Parry, D., Jacoby, L.B., Baser, M., Gusella, J., Davis, K., Stavrou, D. and MacCollin, M. (2000) The parental origin of new mutations in neurofibromatosis 2. *Neurogenetics* in press.

Knudson, A.G. Jr. (1985) Hereditary cancer, oncogenes, and antioncogenes. *Cancer Res.* 45: 1437–1443.

Koivunen, J., Yla-Outinen, H., Korkiamaki, T., Karvonen, S.L., Poyhonen, M., Laato, M., Karvonen, J., Peltonen, S. and Peltonen, J. (2000) New function for *NF1* tumor suppressor. *J. Invest. Dermatol.* 114: 473–479.

Korf, B.R. (1999) Plexiform neurofibromas. *Am. J. Med. Genet. (Semin. Med. Genet.)* 89: 31–37.

Krkljus, S., Abernathy, C.R., Johnson, J.S. *et al.* (1997) Analysis of CpG C to T mutations in neurofibromatosis type 1. *Hum. Mutat.* Mutation in Brief #129, on-line.

Krone, W. and Hogemann, I. (1986) Cell culture studies on neurofibromatosis (von Recklinghausen) V. Monosomy 22 and other chromosomal anomalies in cultures from peripheral neurofibromas. *Hum. Genet.* 74: 453–455.

Legius, E., Marchuk, D.A., Collins, F.S. and Glover, T.W. (1993) Somatic deletion of the neurofibromatosis type 1 gene in a neurofibrosarcoma supports a tumour suppressor gene hypothesis. *Nat. Genet.* 3: 122–126.

Legius, E., Dierick, H., Wu, R., Hall, B.K., Marynen, P., Cassiman, J.-J. and Glover, T.W. (1994) *TP53* mutations are frequent in malignant NF1 tumors. *Genes Chrom. Cancer* 10: 250–255.

Legius, E., Wu, R., Eyssen, M., Marynen, P., Fryns, J.P. and Cassiman, J.J. (1995) Encephalocraniocutaneous lipomatosis with a mutation in the *NF1* gene. *J Med Genet* 32: 316–319.

Leppig, K., Kaplan, P., Viskochil, D., Weaver, M., Orterberg, J. and Stephens, K. (1997) Familial neurofibromatosis 1 gene deletions: cosegregation with distinctive facial features and early onset of cutaneous neurofibromas. *Am. J. Med. Genet.* 73: 197–204.

Li, Y., Bollag, G., Clark, R. *et al.* (1992) Somatic mutations in the neurofibromatosis 1 gene in human tumors. *Cell* 69: 275–281.

Lothe, R.A., Saeter, G., Danielsen, H.E., Stenwig, A.E., Hoyheim, B., O'Connell, P. and Borresen, A.L. (1993) Genetic alterations in a malignant schwannoma from a patient with neurofibromatosis (NF1). *Path. Res. Pract.* 189: 465–471.

Lothe, R.A., Karhu, R., Mandahl, N., Mertens, F., Saeter, G., Heim, S., Borresen-Dale, A.L. and Kallioniemi, O.P. (1996) Gain of 17q24-qter detected by comparative genomic hybridization in malignant tumors from patients with von Recklinghausen's neurofibromatosis. *Cancer Res.* 56: 4778–4781.

Mancini, D.N., Singh, S.M., Archer, T.K. and Rodenhiser, D.I. (1999) Site-specific DNA methylation in the neurofibromatosis (*NF1*) promoter interferes with binding of CREB and SP1 transcription factors. *Oncogene* 18: 4108–4119.

Marchuk, D.A., Saulino, A.M., Tavakkol, R. *et al.* (1991) cDNA cloning of the type 1 neurofibromatosis gene: complete sequence of the *NF1* gene product. *Genomics* 11: 931–940.

Mautner, V.-F., Lindenau, M., Baser, M., Hazim, W., Tatagiba, M., Haase, W., Samii, M., Wais, R. and Pulst, S.M. (1996). The neuroimaging and clinical spectrum of neurofibromatosis 2. *Neurosurgery* 38: 880–886.

McClatchey, A.I. Saotome, I. Ramesh, V. Gusella, J. and Jacks, T. (1997) The NF2 tumor suppressor gene product is essential for extraembryonic development immediately prior to gastrulation. *Genes Dev.* 11: 1253–1265.

McClatchey, A.I., Saotome, I., Mercer, K., Crowley, D., Gusella, J.F., Bronson, R.T. and Jacks, T. (1998) Mice heterozygous for a mutation at the *NF2* tumor suppressor locus develop a range of highly metastatic tumors. *Genes Dev.* 12: 1121–1133.

Mechtersheimer, G., Otano-Joos, M., Ohl, S. *et al.* (1999) Analysis of chromosomal imbalances in sporadic and NF1-associated peripheral nerve sheath tumors by comparative genomic hybridization. *Genes Chrom. Cancer* 25: 362–369.

Messiaen, L.M., Callens, T., DePaepe, A., Craen, M. and Mortier, G.R. (1997) Characterisation of two different nonsense mutations, C6792A and C6792G, causing skipping of exon 37 in the *NF1* gene. *Hum. Genet.* 101: 75–80.

Messiaen, L.M., Callens, T., Roux,K.J. *et al.* (1999) Exon 10b of the *NF1* gene represents a mutational hot spot and harbors a missense mutation Y489C associated with aberrant splicing in unrelated patients. *Genet. Med* 1: 248–253.

Messiaen, L.M., Callens, T., Mortier, G., Beysen, D., Vandenbroucke, I., Van Roy, N., Speleman, F. and De Paepe, A. (2000) Exhaustive mutation analysis of the *NF1* gene allows identification of 95% of mutations and reveals a high frequency of unusual splicing defects. *Hum. Mutat.* 15: 541–555.

Mulvihill, J.J., Parry, D.M., Sherman, J.L., Pikus, A., Kaiser-Kupfer, M.I. and Eldridge, R. (1990) NIH conference: neurofibromatosis 1 (Recklinghausen disease) and neurofibromatosis 2 (bilateral acoustic neurofibromatosis). An update. *Ann. Intern. Med.* 113: 39–52.

Nielsen, G.P., Stemmer-Rachamimov, A.O., Ino, Y., Moller, M.B., Rosenberg, A.E. and Louis, D.N. (1999) Malignant transformation of neurofibromas in neurofibromatosis 1 is associated with *CDKN2A/p16* inactivation. *Am. J. Pathol.* 155: 1879–1884.

Nigro, J.M., Baker, S.J., Preisinger, A.C. *et al.* (1989) Mutations in the *p53* gene occur in diverse human tumour types. *Nature* 342: 705–708.

Nishi, T., Lee, P.S.Y., Oka, K., Levin, V.A., Tanase, S., Morino, Y. and Saya, H. (1991) Differential expression of two types of the neurofibromatosis type 1 (*NF1*) gene transcripts related to neuronal differentiation. *Oncogene* 6: 1555–1559.

Ottini, L., Esposito, D.L., Richetta, A. *et al.* (1995) Alterations of microsatellites in neurofibromas of von Recklinghausen's disease. *Cancer Res.* 55: 5677–5680.

Park, V.M. and Pivnick, E.K. (1998) Neurofibromatosis type 1 (NF1): a protein truncation assay yielding identification of mutations in 73% of patients. *J. Med. Genet.* 10: 813–820.

Parry, D., MacCollin, M., Kaiser-Kupfer, M., Pulaski, K., Nicholson, H.S., Bolesta, M., Eldridge, R. and Gusella, J. (1996) Germline mutations in the neurofibromatosis 2 (*NF2*) gene: correlations with disease severity and retinal abnormalities. *Am. J. Hum. Genet.* 59: 529.

Poyhonen, M., Leisti, E.-L., Kytola, S. and Leisti, J. (1997) Hereditary spinal neurofibromatosis: a rare form of NF1? *J. Med. Genet.* 34: 184–187.

Purandare, S.M., Huntsman-Breidenbach, H., Li, Y. *et al.* (1995) Identification of neurofibromatosis 1 (*NF1*) homologous loci by direct sequencing, fluorescence in situ hybridization, and PCR amplification of somatic cell hybrids. *Genomics* 30: 476–485.

Rasmussen, S.A. and Friedman, J.M. (2000) *NF1* gene and neurofibromatosis 1. *Am. J. Epidemiol.* 151: 33–40.

Rasmussen, S.A. and Wallace, M.R. (1998) Somatic mutations of the *NF1* gene in type 1 neurofibromatosis and cancer. In: *Neurofibromatosis Type 1: From genotype to phenotype* (eds Upadhyaya, M., and Cooper, D.N.), pp. 153–166. BIOS Scientific Publishers, Oxford.

Rasmussen, S.A., Colman, S.D., Ho, V.T., Abernathy, C.R., Arn, P., Weiss, L., Schwartz, C., Saul, R.A. and Wallace, M.R. (1998) Constitutional and mosaic large *NF1* gene deletions in neurofibromatosis type 1. *J. Med. Genet.* 35: 468–471.

Rasmussen, S.A., Overman, J., Thomson, S.A.M. *et al.* (2000) Chromosome 17 loss of heterozygosity studies in benign and malignant tumors in neurofibromatosis type 1. *Genes Chrom. Cancer* 28: 425–431.

Regnier, V., Meddeb, M., Lecointre, G., Richard, F., Duverger, A., Nguyen, V.C., Dutrillaux, B., Bernheim, A. and Danglot, G. (1997) Emergence and scattering of multiple neurofibromatosis (*NF1*)-related sequences during hominoid evolution suggest a process of pericentromeric interchromosomal transposition. *Hum. Mol. Genet.* 6: 9–16.

Riccardi, V.M. (1992) *Neurofibromatosis. Phenotype, natural history and pathogenesis*. The Johns Hopkins University Press, Baltimore, Maryland.

Rodenhiser, D.I., Andrews, J.D., Mancini, D.N., Jung, J.H. and Singh, S.M. (1997) Homonucleotide tracts, short repeats, and CpG/CpNpG motifs are frequent sites for heterogeneous mutations in the neurofibromatosis type 1 (*NF1*) tumor-suppressor gene. *Mut Res* 373: 185–195.

Rouleau, G.A., Wertelecki, W., Haines, J.L. *et al.* (1987) Genetic linkage of bilateral acoustic neurofibromatosis to a DNA marker on chromosome 22. *Nature* 329: 246–248.

Rouleau, G.A., Merel, P., Lutchman, M. *et al.* (1993) Alteration in a new gene encoding a putative membrane-organizing protein causes neurofibromatosis type 2. *Nature* 363: 515–521.

Rutkowski, J.L., Wu, K., Gutmann, D.H., Boyer, P.J. and Legius, E. (2000) Genetic and cellular defects contributing to benign tumor formation in neurofibromatosis type 1. *Hum. Mol. Genet.* 9: 1059–1066.

Sawada, S., Florell, S., Purandare, S.M., Ota, M., Stephens, K. and Viskochil, D. (1996) Identification of *NF1* mutations in both alleles of a dermal neurofibroma. *Nat Genet* 14: 110–112.

Schmidt, H., Taubert, H., Meye, A., Wurl, P., Bache, M., Bartel, F., Holzhausen, H. and Hinze, R. (2000) Gains in chromosomes 7, 8q, 15q, and 17q are characteristic changes in malignant but not benign peripheral nerve sheath tumors from patients with Recklinghausen's disease. *Cancer Lett.* **155**: 181–190.

Seizinger, B.R., Rouleau, G., Ozelius, L.J., Lane, A.H., St. George-Hyslop, P., Huson, S., Gusella, J.F. and Martuza, R.L. (1987) Common pathogenetic mechanism for three tumor types in bilateral acoustic neurofibromatosis. *Science* **236**: 317–319.

Sekido, Y., Pass, H.I., Bader, S., Mew, D.J., Christman, M.F., Gazdar, A.F. and Minna, J.D. (1995) Neurofibromatosis type 2 (*NF2*) gene is somatically mutated in mesothelioma but not in lung cancer. *Cancer Res.* **55**: 1227–1231.

Serra, E., Puig, S., Otero, D., Gaona, A., Kruyer, H., Ars, E., Estivill, X. and Lazaro, C. (1997) Confirmation of a double-hit model for the *NF1* gene in benign neurofibromas. *Am. J. Hum. Genet.* **61**: 512–519.

Shannon, K.M., O'Connell, P., Martin, G.A., Paderanga, D., Olson, K., Dinndorf, P. and McCormick, F. (1994) Loss of the normal *NF1* allele from the bone marrow of children with type 1 neurofibromatosis and malignant myeloid disorders. *N Engl. J. Med.* **330**: 597–601.

Silva, A.J., Frankland, P.W., Marowitz, Z., Friedman, E., Lazlo, G., Cioffi, D., Jacks, T. and Bourtchuladze, R. (1997) A mouse model for the learning and memory deficits associated with neurofibromatosis type 1. *Nat. Genet.* **15**: 281–284.

Stephens, K., Kayes, L., Riccardi, V.M., Rising, M., Sybert, V.P. and Pagon, R.A. (1992) Preferential mutation of the neurofibromatosis type 1 gene in paternally derived chromosomes. *Hum. Genet.* **88**: 279–282.

Stern, H.G., Saal, H.M., Lee, J.S., Fain, P.R., Goldgar, D.E., Rosenbaum, K.N. and Barker, D.F. (1992) Clinical variability of type 1 neurofibromatosis: is there a neurofibromatosis-Noonan syndrome? *J. Med. Genet.* **29**: 184–187.

Tassabehji, M., Strachan, T., Sharland, M., Colley, A., Donnai, D., Harris, R. and Thakker, N. (1993) Tandem duplication within a neurofibromatosis type 1 (*NF1*) gene exon in a family with features of Watson syndrome and Noonan syndrome. *Am. J. Hum. Genet.* **53**: 90–95.

Tonsgard, J.H., Yelavarthi, K.K., Cushner, S., Short, M.P. and Lindgren, V. (1997) Do *NF1* gene deletions result in a characteristic phenotype? *Am. J. Med. Genet.* **73**: 80–86.

Tonsgard, J.H., Kwak, S.M., Short, M.P. and Dachman, A. (1998) CT imaging in adults with neurofibromatosis-1: frequent asymptomatic plexiform lesions. *Neurology* **50**: 1755–1760.

Trofatter, J.A., MacCollin, M.M., Rutter, J.L. *et al.* (1993) A novel moesin-, ezrin-, radixin-like gene is a candidate for the neurofibromatosis 2 tumor suppressor. *Cell* **72**: 791–800.

Troilo, P., Strong, L.C., Little, J.B. and Nichols, W.W. (1992) Spontaneous and induced levels of chromosomal aberration and sister-chromatid exchange in neurofibromatosis: no evidence of chromosomal hypersensitivity. *Mut. Res.* **283**: 237–242.

Upadhyaya, M. and Cooper, D.N. (1998) The mutational spectrum in neurofibromatosis 1 and its underlying mechanisms. In: *Neurofibromatosis Type 1: From genotype to phenotype* (eds Upadhyaya, M., and Cooper, D.N.), pp. 65–88. BIOS Scientific Publishers, Oxford.

Upadhyaya, M., Shen, M., Cherryson, A., Farnham, J., Maynard, J., Huson, S.M. and Harper, P.S. (1992) Analysis of mutations at the neurofibromatosis 1 (*NF1*) locus. *Hum. Mol. Genet.* **1**: 735–740.

Upadhyaya, M., Ruggieri, M., Maynard, J. *et al.* (1998) Gross deletions of the neurofibromatosis type 1 (*NF1*) gene are predominantly of maternal origin and commonly associated with a learning disability, dysmorphic features and developmental delay. *Hum. Genet.* **102**: 591–597.

Viskochil, D.H. (1999) The structure and function of the *NF1* gene: molecular pathophysiology. In: *Neurofibromatosis: Phenotype, natural history, and pathogenesis* (ed. Friedman, J.M. *et al.*), pp. 119–141. Johns Hopkins Press, Baltimore.

Viskochil, D., Buchberg, A.M., Xu, G. *et al.* (1990) Deletions and a translocation interrupt a cloned gene at the neurofibromatosis type 1 locus. *Cell* **62**: 187–192.

Vogel, K.S. Klesse, L.J. Velasco-Miguel, S. Meyers, K. Rushing, E.J. and Parada, L.F. (1999) Mouse tumor model for neurofibromatosis type 1. *Science* **286**: 2176–2179.

Wallace, M.R., Marchuk, D.A., Andersen, L.B. *et al.* (1990) Type 1 neurofibromatosis gene: identification of a large transcript disrupted in three NF1 patients. *Science* **249**: 181–186.

Wallace, M.R., Andersen, L.B., Saulino, A.M., Gregory, P.E., Glover, T.W. and Collins, F.S. (1991) A *de novo* Alu insertion results in neurofibromatosis type 1. *Nature* **353**: 864–866.

Wallace, M.R., Rasmussen, S.A., Lim, I.T., Gray, B.A., Zori, R.T. and Muir, D. (2000) Culture of cytogenetically abnormal Schwann cells from benign and malignant NF1 tumors. *Genes Chrom. Cancer* **27**: 117–123.

Watkins, D., Ruttledge, M.H., Sarrazin, J., Rangaratnam, S., Poisson, M., Delattre, J.Y. and Rouleau, G.A. (1996) Loss of heterozygosity on chromosome 22 in human gliomas does not inactivate the neurofibromatosis type 2 gene. *Cancer Genet. Cytogenet* **92**: 73–78.

Weiss, B., Bollag, G. and Shannon, K. (1999) Hyperactive ras as a therapeutic target in neurofibromatosis type 1. *Am. J. Med. Genet. (Semin. Med. Genet.)* **89**: 14–22.

Wellenreuther, R., Kraus, J.A., Lenartz, D. *et al.* (1995) Analysis of the neurofibromatosis 2 gene reveals molecular variants of meningioma. *Am. J. Pathol.* **146**: 827–832.

Wu, B.L., Austin, M.A., Schneider, G.H., Boles, R.G. and Korf, B.R. (1995) Deletion of the entire *NF1* gene detected by FISH: four deletion patients associated with severe manifestations. *Am. J. Med. Genet.* **59**: 528–535.

Wu, R., Legius, E., Robberecht, W., Dumoulin, M., Cassiman, J.-J. and Fryns, J.-P. (1996) Neurofibromatosis type 1 gene mutation in a patient with features of LEOPARD syndrome. *Hum. Mutat.* **8**: 51–56.

Wu, R., Lopez-Correa, C., Rutkowski, J.L., Baumbach, L.L., Glover, T.W. and Legius, E. (1999) Germline mutations in NF1 patients with malignancies. *Genes Chrom. Cancer* **26**: 376–380.

Xu, W., Mulligan, L.M., Ponder, M.A., Liu, L., Smith, B.A., Mathew, C.G.P. and Ponder, B.A.J. (1992) Loss of *NF1* alleles in phaeochromocytomas from patients with type 1 neurofibromatosis. *Genes Chrom. Cancer* **4**: 337–342.

6

Genetics of gastric cancer

William M. Grady

1. Introduction

1.1 Epidemiology

Gastric cancer is a leading cause of cancer death worldwide and is responsible for the majority of cancer-related deaths in developing countries. Worldwide in 1999 (Parkin et al., 1999), there were 798 000 new cases of stomach cancer and 628 000 stomach cancer-related deaths (12.1% of all cancer deaths). The incidence of gastric cancer shows considerable geographic variability and tends to be less common in developed countries. In fact, the incidence varies from 78/100 000 people in Japanese males to 2.6/100 000 in North African females (Parkin et al., 1999). Thirty-eight percent of all stomach cancers occur in China where it is the most common form of cancer (Parkin et al., 1999). The geographic differences in risk are assumed to be secondary to environmental factors such as diet. Indeed, epidemiological studies, which include migration and temporal analyses, demonstrate that environmental factors play a role in the etiology of gastric cancer (Correa et al., 1972; Haenszel et al., 1972). In addition to diet, the predominant environmental factors that appear to affect the formation of gastric cancer include chronic gastritis secondary to *Helicobacter pylori* and possibly Epstein–Barr virus (EBV) infection (10–15% of cases) (Imai et al., 1994; Rowlands et al., 1993; Shibata and Weiss, 1992; Wu et al., 2000b).

The incidence of gastric cancer of the distal body and antrum of the stomach has been declining throughout the world and especially in developed countries. A notable exception to this trend is adenocarcinoma of the gastroesophageal junction, which has been increasing dramatically recently. The increase in carcinomas of the gastroesophageal junction is felt to be accounted for primarily by an increase in carcinomas associated with Barrett's esophagus (Blot et al., 1991; Clark et al., 1994; Kabat et al., 1993). The reason for the changes in the incidence of distal stomach gastric cancers and gastroesophageal junction gastric cancers is currently not known but presumably reflects changes in the environmental factors that can influence gastric cancer formation.

Molecular Genetics of Cancer second edition, edited by J.K. Cowell.
© 2001 BIOS Scientific Publishers Ltd, Oxford.

1.2 Pathology

The vast majority of malignant gastric tumors are gastric adenocarcinomas, which are commonly termed gastric cancers or gastric carcinomas. Lymphomas, leiomyosarcomas, and carcinoid tumors also occur in the stomach but account for less than 10% of stomach cancers (Luk, 1998). Gastric adenocarcinomas have been categorized using distinct morphologic, histologic, and biologic features. However, the classification of gastric cancers is complicated by the fact that sporadic gastric adenocarcinomas are a heterogeneous collection of tumors on both a pathological and a genetic basis. Consequently, there are a variety of pathological classification systems that have been developed to describe gastric cancers. The two most commonly used systems are the WHO classification system and the Lauren system. Lauren's system is widely used and classifies gastric adenocarcinomas into histopathological subtypes of either intestinal type or diffuse type. The intestinal type is better differentiated, characterized by cohesive cells that form distinct glands, and resembles colorectal carcinoma histologically. The diffuse type (*Figure 1a*) is often poorly differentiated, is characterized by sheets of cells that fail to form gland-like structures, and often contains mucin and signet ring cells (Luk, 1998). These two subtypes appear to have different biologic behaviors and to arise via different genetic alterations (Correa and Shiao, 1994; Tamura *et al.*, 2000). (*Figure 1b*) The intestinal type more often grows as an exophytic mass and is associated with intestinal metaplasia. In contrast, diffuse gastric cancer grows as an infiltrating lesion into the wall of the stomach and is more clinically aggressive (Bevan and Houlston, 1999). In addition, the intestinal type of cancer tends to predominate in areas where there is a high incidence of gastric cancer, whereas the diffuse type tends to occur with a similar prevalence throughout the world (Craanen *et al.*, 1992; Lauren and Nevalainen, 1993). Thus, at least two distinct types of gastric cancer can be identified on pathological features, and it is likely that these features reflect the effects of at least some of the genetic alterations in the tumors. Consequently, an understanding of the effect of the different genetic alterations that occur in gastric cancer formation is best understood in the context of the pathological changes that accompany it.

2. Genetic alterations and gastric carcinogenesis

2.1 Multistage carcinogenesis of gastric cancer

Similar to the adenoma–carcinoma sequence observed in colon cancer, a multistage tumor progression model for gastric cancer has been proposed. The experimental evidence in support of this model is less well developed than it is for colon cancer (Correa and Shiao, 1994; Maesawa *et al.*, 1995; Rokkas *et al.*, 1991; Wee *et al.*, 1992). Evidence in support of a step-wise progression from normal gastric epithelium to gastric cancer is best developed for an adenoma to carcinoma sequence. The clinical history of gastric adenomas indicates an 11–75% rate of malignant transformation of these polyps over a mean period of observation of 76.4 months (Kamiya *et al.*, 1982; Kolodziejczyk *et al.*, 1994; Ming and Goldman, 1965). Recent studies of microsatellite instability in gastric adenomas and adeno-

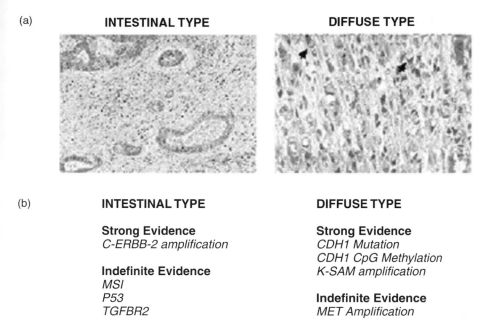

(a)

| INTESTINAL TYPE | DIFFUSE TYPE |

(b)

INTESTINAL TYPE

Strong Evidence
C-ERBB-2 amplification

Indefinite Evidence
MSI
P53
TGFBR2

DIFFUSE TYPE

Strong Evidence
CDH1 Mutation
CDH1 CpG Methylation
K-SAM amplification

Indefinite Evidence
MET Amplification

Figure 1. (a). Photomicrographs of representative examples of intestinal type and diffuse type gastric cancer. Intestinal type gastric cancers are characterized by glandular structures and have a histological appearance similar to colorectal cancer. Diffuse type cancers are poorly differentiated and often contain signet ring cells. Representative signet ring cells are indicated by arrowheads. (b) Genetic and epigenetic alterations associated with the different tumor types. The alterations are divided into those for which there is strong evidence in support of the alterations being associated specifically with that tumor type (Strong Evidence) and those for which the data is suggestive but inconclusive (Indefinite Evidence).

carcinomas have supported the concept that gastric adenomas are potentially premalignant tumors as is the case for colon cancer. Kim *et al* showed that microsatellite-unstable adenomas are more likely to be adjacent to carcinomas and to carry transforming growth factor β receptor type II (*TGFBR2*) mutations (Kim *et al.*, 2000a). *TGFBR2* mutations have been found to occur at the transition of microsatellite-unstable colon adenomas to carcinomas and a similar sequence could be occurring in gastric cancer formation (Grady *et al.*, 1998). Furthermore, others have found *ERBB2* overexpression and p53 immunoreactivity in a subset of gastric adenomas that contained foci of malignant transformation or high-grade dysplasia (Kolodziejczyk *et al.*, 1994; Lauwers *et al.*, 1993; Nogueira *et al.*, 1999). Thus, the evidence supports a model (*Figures 2* and *3*) of adenoma–carcinoma progression but the pre-adenomatous events are currently not understood. Indeed, although dysplasia is not a normal finding in the gastric epithelium and has been clearly demonstrated in stomachs affected by gastric cancer, intestinal metaplasia, or gastritis, its preneoplastic potential is still unknown. In 10% of cases it can progress in severity and even to cancer, however, in the majority of cases it appears to regress or remain stable. In addition, although dysplasia and

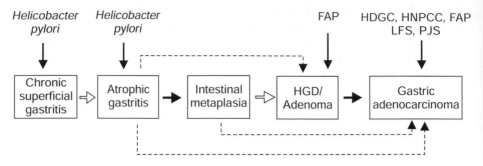

Figure 2. Schematic representation of the multistage process of gastric cancer formation. The bold lines and arrows represent events that are strongly supported by published reports. The thin lines represent events that are less well supported in the literature. The dashed lines and open arrows signify events for which there is inconclusive evidence. The factors listed at the top of the flow diagram are potential contributory factors for these events. *HNPCC = hereditary nonpolyposis colon cancer syndrome, HDGC = hereditary diffuse gastric cancer, PJS = Peutz-Jeghers syndrome, LFS = Li Fraumeni syndrome, FAP = familial adenomatous polyposis, HGD = high-grade dysplasia

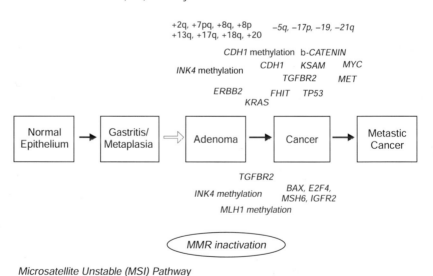

Figure 3. Schematic representation of the genetic and epigenetic alterations observed in two postulated pathways of gastric carcinogenesis. The two pathways, the chromosome instability (CIN) pathway and the microsatellite instability (MSI) pathway, involve different forms of genomic instability. The genetic events that occur in the MSI pathway involve genes that contain microsatellite-like repeats in their exons. Many of the genetic alterations that are observed in the CIN pathway likely occur in the MSI pathway; however, the identification of which specific alterations occur in both pathways remains to be performed. The location of the genes in the figure indicates their probable timing in the multi-stage gastric cancer sequence.

intestinal metaplasia are often found in association with gastric cancers, it is unclear whether they are the end results of a common etiologic process/agent or part of a progression to malignancy (Luk, 1998; Rugge *et al.*, 1994). Consequently, there has been intense interest in the study of genetic alterations in dysplasia and intestinal metaplasia in the hope of resolving these issues.

2.2 Sporadic gastric cancer

The vast majority of gastric cancers form as the consequence of sporadic genetic alterations that occur in the gastric epithelium of individuals who have no inherent predisposition to gastric cancers. These sporadic gastric adenocarcinomas arise as the result of the accumulation of these genetic (and epigenetic) alterations that cause the neoplastic transformation of gastric epithelium. As described above, these alterations appear to underlie a multistage process of carcinogenesis that results in normal gastric epithelial cells transforming into gastric adenocarcinoma cells (*Figures 2* and *3*). These alterations affect an array of oncogenes and tumor suppressor genes. The altered tumor suppressor genes and oncogenes currently identified in gastric carcinogenesis will be discussed in this chapter in further detail. Of note, these alterations appear to occur in the context of genomic instability with subsequent clonal selection of those alterations that favor tumor formation. In a broad sense, it appears that global DNA damage and/or the loss of the normal mechanisms that maintain DNA fidelity is important in initiating gastric carcinogenesis. In support of this model, *Helicobacter pylori*-induced gastric inflammation has been shown to cause significant oxidative DNA damage, as detected by 8-hydroxydeoxyguanosine (8-OHdG) formation, and this damage may play a role in initiating *H. pylori*-associated gastric adenocarcinoma (Farinati *et al.*, 1998). Thus, a plausible model for sporadic gastric cancer formation could entail environmental factors creating a milieu in which genetic alterations favorable for tumor formation occur.

2.3 Familial gastric cancer

Although, most adenocarcinomas of the stomach occur sporadically without a clear inherited predisposition, a small proportion of gastric cancers arises in clearly identified inherited gastric cancer predisposition syndromes (*Figure 2*). These include two recently well-described inherited predisposition syndromes, hereditary nonpolyposis colon cancer syndrome (HNPCC) and the hereditary diffuse gastric cancer syndrome (HDGC), as well as Peutz-Jeghers syndrome (PJS), Cowdens syndrome and subsets of families affected with Li Fraumeni syndrome (LFS) and familial adenomatous polyposis (FAP) (*Figure 2*). HNPCC is a syndrome caused by germline mutations in one of the mutation mismatch repair genes (e.g. *MLH1*, *MSH2*, *PMS1*, *PMS2*) and is characterized by the predisposition to colon cancer, endometrial cancer, and stomach cancer. A form of genomic instability termed microsatellite instability (MSI) characterizes all the tumors that arise in association with HNPCC. Stomach cancers arise in 11% of HNPCC families and have been shown to occur in families with either *MLH1* or *MSH2* germline mutations (Aarnio *et al.*, 1997; Vasen *et al.*, 1996). The majority of these cancers (79% in one study) are intestinal type and they have the same natural history as sporadic gastric cancer (Aarnio *et al.*, 1997). HDGC results from germline mutations in *CDH1*, the gene for E-cadherin (Bevan and Houlston, 1999; Guilford *et al.*, 1998). Originally identified in three Maori families predisposed to diffuse gastric cancer, several families from a variety of ethnic backgrounds have since been described (Gayther *et al.*, 1998; Guilford *et al.*, 1999; Stone *et al.*, 1999). HDGC appears to have 70% penetrance and may also be associated with an increased risk of breast and colon cancer (Keller *et al.*, 1999; Stone *et al.*, 1999). Li-Fraumeni syndrome is caused by germline mutations in *TP53* and results in a marked predisposition to soft-tissue sarcomas, leukemia, brain cancer, and breast cancer as well as gastric cancer in some families. Familial adenomatous polyposis (FAP) families of Japanese descent, who carry germline mutations in *adenomatous polyposis coli* (*APC*), occasionally develop gastric cancer (Bevan and Houlston, 1999). Peutz-Jeghers syndrome, a hamartomatous polyposis syndrome, is discussed in more detail in the section on genetic alterations in *STK11*.

3. Genomic instability and gastric cancer

Although it is now widely appreciated that gastric cancer is a disease that results from genetic alterations, the mechanism(s) responsible for the acquisition and accumulation of these gene alterations is less well understood. The role of genomic instability in the process of gastric cancer tumorigenesis has been an area of intense investigation since 1993 when Aaltonen *et al.* identified microsatellite instability (MSI) in cancers arising in hereditary nonpolyposis colon cancer (HNPCC) families (Aaltonen *et al.*, 1993). This discovery of MSI in HNPCC cancers (*Figure 4*) provided evidence for the mutator phenotype hypothesis proposed by Loeb (Eshleman and Markowitz, 1995; Loeb *et al.*, 1974). This hypothesis argues that, for potential tumor cells to acquire the many mutations needed to attain a malignant state, they must have a higher than normal mutation rate, which would presumably occur through inactivation of the mechanisms that normally regulate DNA fidelity. In fact, if one assumes the known non-germline mutation rate of approximately 10^{-8}, it is impossible for a cell to accumulate sufficient gene mutations during a human lifetime to acquire a malignant phenotype (Eshleman and Markowitz, 1995). However, the role of genomic instability, as an integral mechanism in cancer formation, is not universally accepted. Others have concluded that the genetic events observed in cancer can occur under normal mutation rates, without the need to invoke genomic instability, as long as the normal mutation rates are coupled with rounds of clonal expansion that select for gene alterations that promote tumor formation (Tomlinson *et al.*, 1996). Nonetheless, despite the controversy over the exact role of genomic instability in gastric tumorigenesis, a wealth of data generated over the last 2–3 years provides significant indirect support for the concept that genomic instability is a common mechanism that plays a central role in the formation of most, if not all, stomach cancers. Evidence in support of the concept that genomic instability is common, if not integral, to tumor formation comes from studies that have shown considerable intratumoral genetic heterogeneity in gastric cancers, which is consistent with a loss of the intrinsic control over DNA fidelity that normal cells have (Chung *et al.*,

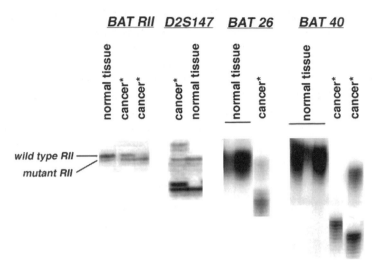

Figure 4. Representative examples of microsatellite instability detected at different loci in gastric cancer cases. The cases marked by an asterisk indicate the presence of shifted allele sizes, which is the hallmark of microsatellite instability. The normal tissue is from tissue adjacent to the tumors and shows the size of the unaltered maternal and paternal alleles. The alleles are detected by PCR amplification using ^{33}P-labeled PCR primers. The PCR products are then subjected to gel electrophoresis and visualized using autoradiography.

1999). In addition, 13–40% of gastric cancers have microsatellite instability, a specific type of genomic instability, demonstrating that at least a subset of gastric cancers clearly lack genomic stability (Mycroff *et al.*, 1995; Strickler *et al.*, 1994; Tamura *et al.*, 1995). Furthermore, many of the tumors that do not display MSI, termed microsatellite stable (MSS) tumors, do show a different form of genomic instability called chromosomal instability (CIN) (*Figure 3*).

3.1 Chromosomal instability

Chromosomal instability (CIN) is recognized by the presence of chromosomal aberrations such as aneuploidy and chromosomal translocations. Our understanding of CIN in gastric cancer is limited secondary to a relative dearth of studies on this particular aspect of gastric carcinogenesis. Fewer cytogenetic studies of gastric cancer have been performed than of other tumor types because of the low incidence of gastric cancer in Western countries and because of technical difficulties involved with culturing gastric tumor cells (Mettlin, 1988). Nonetheless, a small number of studies have been published documenting a number of variable as well as common numerical and structural clonal cytogenetic abnormalities (Cagle *et al.*, 1989; Chun *et al.*, 2000; Espinoza *et al.*, 1999; Ferti-Passantonopoulou *et al.*, 1987; Ochi *et al.*, 1986; Rodriguez *et al.*, 1990; Sekiguchi and Suzuki, 1995; Seruca *et al.*, 1993; Xiao *et al.*, 1992). Using comparative genomic hybridization (CGH), Koizumi *et al.* (1997) detected DNA copy number changes in 94% of primary gastric cancers ($n = 31/33$). The number of changes, and the location of the changes, varied among tumors with some changes showing correlation to histological or clinical features (Koizumi *et al.*, 1997). The most common karyotypic abnormalities seen in gastric cancers are numerical changes in chromosomes 2, 9, 10, 20 and Y and structural changes in chromosomes 1, 3, 5, 7, 9, 11 and 13 (Heim *et al.*, 1995). Consistent chromosomal abnormalities detected to date (*Figure* 3) include gains in 2q, 7pq, 8q, 8p, 13q, 17q, 18q, and 20pq, and loss of 5q, 17p, 19 and 21q (Chun *et al.*, 2000; Koo *et al.*, 2000; Sakakura *et al.*, 1999; Tamura *et al.*, 1996a). The severity of chromosomal instability has been shown to vary among gastric cancer cell lines and primary gastric cancers most likely reflecting different types or degrees of genomic instability as has been demonstrated for colon cancer (Bertoni *et al.*, 1998; Chun *et al.*, 2000; Eshleman *et al.*, 1998; Koizumi *et al.*, 1997). For examples, analysis of two gastric cancer cell lines revealed one, GK2, has a pseudodiploid karyotype with an add(6)(q27) while the other cell line, AKG, is highly rearranged with multiple structural changes (Bertoni *et al.*, 1998). Recently, Lengauer *et al.* (1997) identified somatic mutations in *BUB1* and *BUB1B* that caused CIN in colon cancer. Such findings raise the possibility that mutations in *BUB1* or *BUB1B*, or other genes that are involved in the regulation of chromosome replication, will be found in stomach cancer. However, the underlying mechanism responsible for the chromosomal instability in gastric cancer still remains to be determined.

3.2 Microsatellite instability

Microsatellite instability (MSI) is a form of genomic instability that is characterized by frameshift mutations in microsatellite repeats located throughout the genome. It occurs as the consequence of inactivation of the mutation mismatch repair system (MMR) that normally corrects these errors that arise during DNA replication. It is characteristic of cancers that arise in the Hereditary Nonpolyposis Colon Cancer syndrome (HNPCC), a syndrome defined by the early onset of colon, gastric, and endometrial cancers in affected families. Microsatellite instability has also been found in sporadic gastric cancer in high-risk (Japan, Korea, Portugal, etc.) and low risk populations (Italy, United States). MSI has been identified in 16–39% of sporadic gastric cancers (Chong *et al.*, 1994; Chung *et al.*, 1996; Han *et al.*, 1993; Lin *et al.*, 1995; Myeroff *et al.*, 1995; Nakashima *et al.*, 1994; Ohue *et al.*, 1996; Rhyu *et al.*, 1994a; Semba *et al.*, 1996; Tamura *et al.*, 1995). The incidence of MSI varies between geographic regions tending to be higher in areas where gastric cancer is more prevalent. Sepulveda *et al.* found MSI in 50% of gastric cancers from Korea, 15% of cases from Columbia, and 7% of cases from the U.S. (Sepulveda *et al.*, 1999). Recently, the mechanism responsible for causing MMR inactivation and subsequent MSI in the majority of sporadic gastric cancer has been shown to be aberrant hypermethylation of CpG dinucleotide repeats in the *MLH1* promoter (Fleisher *et al.*, 1999; Leung *et al.*, 1999). Hypermethylation of the promoter of *MLH1* results in its transcriptional silencing with subsequent microsatellite instability in colon cancer and a similar process is presumed to be occurring in gastric cancer (Herman *et al.*, 1998; Veigl *et al.*, 1998).

MSI predisposes the genome to the mutational inactivation of tumor suppressor genes that have microsatellite repeats in their exons, such as transforming growth factor β receptor type II (*TGFBR2*), *BAX*, insulin-like growth factor II receptor (*IGF2R*), and *E2F4*. Duval *et al.* studied a series of 22 gastric cancers and found the following results: *TCF4* mutations 9% (2/22), *TGFBR2* mutations 86% ($n = 19/22$), *BAX* mutations 14% ($n = 3/22$), IGFR2 mutations 32% ($n = 7/22$), *MSH3* mutations 9% ($n = 2/21$), and *MSH6* mutations 54% ($n = 5/23$) (Duval *et al.*, 1999). The inactivation of these genes appears to be part of a unique pathway of gastric carcinogenesis that can be distinguished from tumors that display chromosomal instability (CIN). For example, *TP53* mutations occur uncommonly in MSI gastric cancers but are observed in the majority of microsatellite stable (MSS) cancers consistent with there being two distinct pathways that lead to gastric cancer formation (Luinetti *et al.*, 1998; Renault *et al.*, 1996; Strickler *et al.*, 1994; Yamamoto *et al.*, 1999). In addition, low level MSI (1 microsatellite locus shifted) has been detected in a subset of pre-neoplastic lesions in the stomach or in adenomas (Leung *et al.*, 2000; Ottini *et al.*, 1997; Semba *et al.*, 1996; Tamura *et al.*, 1995). The significance of this finding remains to be demonstrated but it suggests that at least some of these lesions are truly pre-malignant and that they have acquired pre-cancerous genetic alterations. Also of interest, MSI gastric cancers display intratumoral histologic and genetic heterogeneity consistent with a state of increased genomic instability (Chong *et al.*, 1994; Chung *et al.*, 1997, 1999; Ohue *et al.*, 1996). *TGFBR2* and *BAX* mutations occur most frequently throughout these genetically heterogeneous tumors, suggesting they occur early in the development of gastric adenocarcinomas and are important in the acquisition of the malignant phenotype (Chung *et al.*, 1997; Ohue *et al.*, 1996).

A consistent association between MSI and clinicopathologic features of gastric cancers has not been shown to date, although some studies have found an increased association with intestinal type cancer, cancers in the distal body of the stomach, less frequent lymph node or vessel invasion, prominent lymphoid infiltration and *Helicobacter pylori* infection (Chung *et al.*, 1996; Dos Santos *et al.*, 1996; Han *et al.*, 1993; Keller *et al.*, 1995; Luinetti *et al.*, 1998; Wirtz *et al.*, 1998; Wu *et al.*, 1998a). The differences between these studies most likely are a consequence of inconsistent criteria for the definition of MSI (i.e., how many microsatellite loci examined, the nature of the microsatellites assessed), and variability in the histopathologic criteria used to classify the tumors. Studies of the association between MSI and survival have also demonstrated inconsistent results. The prognostic significance of MSI relative to patient survival remains to be demonstrated by a prospective study (Chong *et al.*, 1994; Dos Santos *et al.*, 1996; Ottini *et al.*, 1997; Powell, 1998; Seruca *et al.*, 1995a; Yamamoto *et al.*, 1999).

4. Specific genetic alterations in gastric cancer

Genetic alterations are fundamental to cancer formation. These genetic alterations perturb the function of tumor suppressor genes and oncogenes contributing to the acquisition of biological characteristics that define malignancies. The specific alterations that have been identified in gastric cancers and their biological effects on gastric cancer will be addressed in the next two sections of this chapter (*Table 1*).

Table 1. Specific genetic alterations and epigenetic alterations observed in gastric cancers

Specific genetic and epigenetic alterations in gastric cancer	
Tumor suppressor genes	*Frequency*
TP53	30–60%
TGFBR2	15–40%
SMAD4	3%
CDH1	40–83% (diffuse type cancers)
APC	0–20%
CTNNB1	0–25%
STK11	3%
FHIT	2.5–50%
INK4–5′ CpG DNA methylation	32–42%
p300	
IRF1	
Oncogenes	
CMET	4–26%
KSAM (FGFR2)	21% (diffuse type cancers)
KRAS	0–9%
MYC	16–67%
ERBB2	6–23%
ERBB1	
CYCLIN E1	
DNA caretaker genes	
MLH1–5′ CpG DNA methylation	60% (MSI cancers)

The frequency is obtained from published studies. If there is more than one study available that reports this information and if the frequency varies between studies, the range is provided. The annotation next to the frequency in parentheses indicates that the frequency is only that particular gastric cancer type.

   *DNA caretaker genes is a term that refers to genes that control genomic stability and whose inactivation creates a state that favors the acquisition of genetic alterations.

4.1 Altered genetic loci

A variety of alterations of loci have been demonstrated in sporadic gastric cancer using cytogenetic studies and studies of allelic loss. The identification of these alterations helped establish the concept that gastric cancer is fundamentally a genetic disease. These alterations are predicted to affect specific tumor suppressor genes and oncogenes. The target genes of some of these alterations have been demonstrated in certain instances but the majority of the targets involved in the alterations of genetic loci remain to be determined. Examples of genetic loci associated with tumor suppressor genes that are lost in gastric cancer include those located on the following chromosomes: 17p (over 60% at the *TP53* locus), 18q (over 60% at the *DCC* locus), and 5q (30–40% at or near the *APC* locus) (McKie *et al.*, 1993; Panani *et al.*, 1995; Rhyu *et al.*, 1994b; Sano *et al.*, 1991; Uchino *et al.*, 1992b). Other loci that have been shown to have less frequent but significant allelic losses include 1p, 1q, 7q, 13q, 14q, 18q, 22q (Koo *et al.*, 2000; Sipponen *et al.*, 1985). Furthermore, a variable number of numerical and structural aberrations including those involving chromosome 3 (rearrangements), 6 (deletions distal to 6q21), 8 (trisomy and gain of 8q), 11 (11p13–15 aberrations), 13 (monosomy and translocations), and 20 (gains) have been shown by several different investigators (Koo *et al.*, 2000; Ochi *et al.*, 1986; Panani *et al.*, 1995; Powell, 1998; Rodriguez *et al.*, 1990; Seruca *et al.*, 1993). Further studies are needed to identify the specific genes affected by these genetic changes.

4.2 Tumor suppressor genes

TP53. The p53 protein was originally identified as a protein forming a stable complex with the SV40 large T antigen and was originally suspected to be an oncogene (Ochiai and Hirohashi, 1997). Subsequent studies demonstrated that TP53 is a transcription factor that is located at 17p13.1 and that it is mutated in 50% of primary human tumors, including tumors of the gastrointestinal tract (Somasundaram, 2000). p53 is currently appreciated to be a transcription factor that is involved in maintaining genomic stability through the control of cell cycle progression and apoptosis in response to genotoxic stress (Somasundaram, 2000). The protein encoded by TP53 has been structurally divided into 4 domains: 1) an acidic amino-terminal domain (codons 1–43) required for transcriptional activation; 2) a central core sequence-specific DNA-binding domain (codons 100–300); 3) a tetramerization domain (codons 324–355); and a C-terminal regulatory domain (codons 363–393), rich in basic amino acids and believed to regulate the core DNA-binding domain (Somasundaram, 2000).

The *TP53* gene has been shown to be altered in gastric cancers by a variety of different methods. *TP53* mutation is frequently accompanied by loss of heterozygosity (LOH) which provides the second hit needed to inactivate all p53 function in the tumor cells. LOH has been shown in 60% of gastric cancers and mutational analysis reveals 30–50% of tumors carry *TP53* mutations. The differences in incidence between studies is likely a consequence of the mutation screening technique employed (single strand chain polymorphism (SSCP) or degenerative gradient gel electrophoresis (DGGE)) and sample sizes of the study (Hollstein *et al.*, 1991; Powell, 1998). The spectrum of mutations in *TP53* seen in gastric cancer appears similar to that seen in other tumors with mutations of *TP53* clustering at 4 hot spots in highly conserved regions (domains II–V). The mutations found to occur commonly in gastric carcinoma are G:C to A:T transitions at CpG dinucleotide repeats.

In gastric cancers, *TP53* mutations have been identified to occur commonly and to occur in both early and advanced cancers (Cho *et al.*, 1994; Tamura *et al.*, 1991; Uchino *et al.*, 1993; Yokozaki *et al.*, 1992). In addition, some studies have detected alterations of *TP53* in early dysplastic lesions that are premalignant, such as gastric adenomas, or in potentially preneoplastic lesions such as intestinal metaplasia. However, in general, *TP53* mutations appear to be most frequent in advanced dysplastic lesions (Powell, 1998; Tamura *et al.*, 1991). Mutant *TP53* has a longer half-life than wild-type *TP53* and thus accumulation of p53 in the nucleus of cells can be taken as indirect evidence of a mutated *TP53* allele. The presence of *TP53* mutation as demonstrated by immunohistochemical staining of gastric cancers in one study identified *TP53* mutations in 44/101 (43%) intestinal-type cancers as compared to 2/48 (4%) diffuse-type cancers (Uchino *et al.*, 1992a). Other studies that have employed other mutation detection methods, however, have not identified a correlation between clinicopathologic features and *TP53* mutations (Luinetti *et al.*, 1998; Renault *et al.*, 1996). Furthermore, assessment of the effect of *TP53* mutations on prognosis using immunohistochemistry has not yielded consistent results to date (Gabbert *et al.*, 1995; Hurlimann and Saraga, 1994).

TGF-β receptor type II. Transforming growth factor β (TGF-β) is a multi-functional cytokine that can induce growth inhibition, apoptosis, and differentiation in intestinal epithelial cells (Fynan and Reiss, 1993; Markowitz and Roberts, 1996). TGF-β mediates its effects through a heteromeric receptor complex that consists of type I (RI) and type II (RII) components. The genes for these receptors are called *TGFBR1* and *TGFBR2* respectively. TGFBR1 and TGFBR2 are serine-threonine kinases that phosphorylate downstream signaling proteins, like the Smad proteins, upon activation by TGF-β binding to RI. In addition to its role in tissue homeostasis, the TGF-β signaling pathway has been recently shown to act as a tumor suppressor in a variety of human tumor types, including gastric cancer (Kim *et al.*, 2000b; Markowitz and Roberts, 1996).

Evidence of TGF-β's role in gastric cancer formation came first from studies that demonstrated gastric cancer cell lines were resistant to the normal growth inhibitory effects of TGF-β (Park *et al.*, 1994). One common mechanism through which cancers can acquire TGF-β resistance is through genetic alterations of the *TGFBR2* gene. In fact, functionally significant deletions of *TGFBR2* were identified in those gastric cancer cell lines mentioned above that were found to be resistant to TGF-β (Park *et al.*, 1994). No alterations in *TGFBR1* or the *type III TGF-β receptor* were observed in this study suggesting mutational inactivation of *TGFBR2* is a particularly favorable event that leads to tumor formation (Park *et al.*, 1994). The overall role of *TGFBR2* mutations in cancer formation became clearer after a landmark study by Markowitz *et al.* demonstrated that mutational inactivation of *TGFBR2* is an extremely common event in cancers that display MSI. *TGFBR2* has a microsatellite-like region in exon 3 that consists of a 10 base pair polyadenine tract, which is named *BAT RII*, making it particularly susceptible to mutation in the setting of MSI (Markowitz *et al.*, 1995; Myeroff *et al.*, 1995; Parsons *et al.*, 1995). These mutations are frameshift mutations that result in the insertion or deletion of one or two adenines between nucleotides 709–718 introducing nonsense mutations that encode a truncated TGFBR2 protein. This truncated protein is only 129–161 amino acids in length compared to the wild-type protein (565 amino acids) and lacks the receptor's transmembrane domain and intracellular kinase domain (Markowitz *et al.*, 1995). Approximately 50–91% of these MSI gastric cancers have mutant *TGFBR2* genes (Chung *et al.*, 1996; Myeroff *et al.*, 1995; Ohue *et al.*, 1996; Oliveira *et al.*, 1998; Wu *et al.*, 1998a, 2000a). The role of *TGFBR2* as a tumor suppressor gene in gastric cancer has been further elucidated by studies showing reconstitution of wild-type *TGFBR2* in a gastric cancer cell line with a mutant *TGFBR2* gene suppresses the tumor phenotype of the cell line (Chang *et al.*, 1997). In addition, the specific location of the mutation has been shown to have specific consequences on the behavior of the mutant receptor. Lu *et al* identified a mutation of *TGFBR2* resulting in a substitution of methionine for threonine at codon 315 in the kinase subdomain IV of *TGFBR2* that inactivated the growth inhibitory effects of RII but not its transcriptional activity on plasminogen activator inhibitor type I (*PAI 1*) (Lu *et al.*, 1999). Of note, the overall incidence of *TGFBR2* mutation in both microsatellite stable (MSS) and MSI gastric cancers remains to be determined but appears currently to be between 15 and 40% (Shitara *et al.*, 1998; Yang *et al.*, 1999).

The effect of *TGFBR2* mutations on the clinical behavior of gastric cancers has not been clearly shown. Of interest, colon cancers with MSI and *TGFBR2* mutations display unique clinicopathologic features including an increased incidence in the proximal colon, presentation at an early stage, and better prognosis than microsatellite stable (MSS) colon cancer (Lynch and Smyrk, 1996). Similarly, some investigators have shown gastric cancers with MSI and *TGFBR2* mutations display a higher frequency of antral location, intestinal subtype, *H. pylori* seropositivity, and a lower frequency of lymph node metastases (Wu *et al.*, 2000a). The specific role of mutational inactivation of *TGFBR2* in mediating these clinico-pathologic features, however, remains to be determined.

Smad *genes.* The Smad proteins are a family of proteins that serve as intracellular mediators to regulate TGF-β superfamily signaling. Numerous studies have identified three major classes of Smad proteins: 1) the receptor-regulated Smads (R-Smads) which are direct targets of the TGF-β receptor type I kinase and include Smads1, 2, 3, and 5; 2) the common Smads (Co-Smads: Smad4) which form heteromeric complexes with the R-Smads and propagate the TGF-β mediated signal; and 3) the inhibitory Smads (I-Smads: Smad6 and Smad7) which antagonize TGF-β signaling through the Smad pathway. Mutational inactivation of *SMAD2* and *SMAD4* has been observed in a high percentage of pancreatic cancers and in 5–10% of colon cancers (Hahn *et al.*, 1996; Riggins *et al.*, 1996; Schutte *et al.*, 1996). Mutations in *SMAD4* in stomach cancer occur less frequently and have been demonstrated in approximately 3% of tumors (Powell *et al.*, 1997). Powell *et al.* identified biallelic inactivation in 1/35 primary gastric carcinomas. One allele was found to carry a nonsense mutation at codon 334, which is predicted to truncate the conserved carboxy-terminal domain of the protein, and the other allele was shown to be missing, using adjacent microsatellite markers (Powell *et al.*, 1997). *SMAD2* and *SMAD5* mutations have not been found to date in sporadic gastric carcinomas (Gemma *et al.*, 1998; Shitara *et al.*, 1999). Thus, *SMAD* mutations appear to play a role in tumor formation in a subset of gastric cancers, but are not as common as *TGFBR2* mutations. This finding suggests there are non-Smad TGF-β signaling pathways that play an important role in the tumor suppressor activity of *TGFBR2*.

CDH1. E-cadherin is a 120 kDa calcium dependent cell–cell adhesion molecule that regulates intercellular adhesion and cell polarity in the normal gastric epithelium. E-cadherin is a member of the cadherin family of proteins, a class of cell surface proteins that establish intercellular connections through homophilic interactions, and plays an important role in the organization of epithelial tissues. Furthermore, alpha-, beta-, and gamma-catenin bind to E-cadherin on its intracellular domain. This interaction with E-cadherin plays an important role in the normal cellular localization and function of these proteins. *CDH1*, the gene that encodes E-cadherin, is frequently mutated or inactivated in a variety of cancers including breast cancer, thyroid cancer, prostate cancer, colon cancer, and gastric cancer (Birchmeier and Behrens, 1994). Somatic mutations in *CDH1* have been identified in 40–83% of diffuse gastric cancers and in no intestinal type cancers (Becker and Hofler, 1995; Machado *et al.*, 1999; Muta *et al.*, 1996; Oda *et al.*, 1994; Tamura *et al.*, 1996b). The mutations observed in diffuse gastric cancers consist of missense mutations and predominantly splice site mutations and truncation mutations caused by insertions, deletions, and nonsense mutations (Berx *et al.*, 1998). Of interest, there is a major difference in mutation types between diffuse gastric cancers and infiltrative lobular breast cancers, another tumor type that has a high frequency of *CDH1* mutations (Berx *et al.*, 1996; Vos *et al.*, 1997). In diffuse gastric cancers, the *CDH1* mutations generate exon skipping causing in-frame deletions. However, infiltrating lobular breast cancers have mutations that are out-of-frame mutations that are predicted to generate secreted truncated E-cadherin fragments. This difference may reflect differences in the tumor type etiologies or in the clonal growth advantages the different *CDH1* mutations offer for each particular tumor type. In regards to the timing of these mutations in tumorigenesis, *CDH1* mutations have been found in early stage gastric cancers and lobular breast cancer suggesting they occur early in the multi-stage process of cancer formation (Berx *et al.*, 1998; Muta *et al.*, 1996; Vos *et al.*, 1997).

There is substantial evidence demonstrating the functional significance of inactivation of *CDH1* in gastric carcinoma formation. Inactivation of both alleles through LOH and/or aberrant methylation of the *CDH1* promoter has been shown by several groups and is consistent with Knudson's two-hit hypothesis of biallelic inactivation of tumor suppressor genes in carcinogenesis (Becker and Hofler, 1995; Grady *et al.*, 2000; Tamura *et al.*, 1996b). The biological consequences of E-cadherin inactivation is first suggested by the segregation of *CDH1* mutations in diffuse gastric cancers and not in intestinal type cancers. The segregation of *CDH1* mutations with diffuse gastric cancer suggests that these mutations mediate the histologic appearance of stomach cancers (Machado *et al.*, 1999). More directly, *in-vitro* studies have shown that E-cadherin acts as an invasion suppressor protein in a variety of cancers (Behrens, 1993; Frixen *et al.*, 1991; Handschuh *et al.*, 1999; Vleminckx *et al.*, 1991). Furthermore, studies of primary tumors have demonstrated that loss of E-cadherin is correlated with more aggressive tumor invasion and a poor prognosis (Gabbert *et al.*, 1996; Jawhari *et al.*, 1997). Finally convincing genetic evidence for the role of *CDH1* alterations in gastric cancer formation was provided by Guilford *et al.* when they identified germline mutations in *CDH1* as the cause of a gastric cancer predisposition syndrome in 3 Maori families (Guilford *et al.*, 1998). Additional non-Maori families affected by diffuse gastric cancer have since been identified demonstrating that this syndrome is not ethnically restricted (Gayther *et al.*, 1998; Guilford *et al.*, 1999; Richards *et al.*, 1999; Shinmura *et al.*, 1999; Yoon *et al.*, 1999). Thus, *CDH1* mutations are common in diffuse gastric cancers and have been definitively shown through a number of different lines of investigation to be functionally important in gastric cancer formation.

APC. Allelic loss at 5q was originally identified in gastric cancer a decade ago and suggested at least one tumor suppressor gene important in gastric cancer formation resides at this location (McKie *et al.*, 1993; Powell, 1998; Powell *et al.*, 1996; Rhyu *et al.*, 1994b; Sano *et al.*, 1991; Tamura *et al.*, 1996a). LOH at 5q has been shown in 24–31% of gastric cancers regardless of histologic subtype (Rhyu *et al.*, 1994b; Sano *et al.*, 1991). The exact target of the allelic imbalance is still uncertain but somatic mutations of *APC* have been reported in gastric adeno-carcinomas and adenomas arising in Japan (Nagase and Nakamura, 1993). These mutations occur at low frequency and are most commonly missense mutations (Nagase and Nakamura, 1993). Of note, other studies of somatic *APC* mutations in Japanese gastric cancers have not demonstrated any tumors with *APC* muta-tions, either because of the low frequency of these mutations in combination with small sample sizes or because of differences in mutation screening techniques (Maesawa *et al.*, 1995; Powell *et al.*, 1996). Other evidence that implicates *APC* mutations as playing a role, at least in some gastric cancers, comes from the increased risk of gastric cancer seen in people with germline *APC* mutations in high-risk areas of Asia (Park *et al.*, 1992; Utsunomiya, 1990). However, no risk of gastric cancer has been observed in FAP families in other populations (Burt, 1991; Offerhaus *et al.*, 1992). Thus, the evidence supporting genetic alterations of *APC* as playing a role in gastric cancer formation is not consistent enough to clearly define *APC*'s role in gastric cancer formation. Other loci that have been mapped to commonly deleted regions on 5q in gastric cancer include the interferon regu-latory factor-1 locus and D5S428 (Tamura *et al.*, 1996a). These may be the physio-logically relevant targets on 5q rather than *APC*, but further studies are needed to resolve this issue.

β-catenin. β-catenin is a member of the APC/β-catenin/T cell factor-lymphoid enhancer factor (tcf/lef) pathway that has been recently shown to play an important role in the formation of certain tumors such as colon cancer, melanoma, and gastric cancer. The β-catenin gene (*CTNNB1*) is a homolog of armadillo and its expression is increased by increased wnt signaling (Aberle *et al.*, 1994; Hulsken *et al.*, 1994; Moon *et al.*, 1997). APC interacts with β-catenin and forms a macro-molecular complex with it and glycogen synthase kinase 3β (GSK3B). β-catenin is consequently directed toward degradation as a result of phosphorylation by GSK3B (Munemitsu *et al.*, 1995, 1996; Rubinfeld *et al.*, 1997a). Mutations of *CTNNB1* or *APC* often render β-catenin insensitive to APC/β-catenin/GSK3B mediated degradation (Morin *et al.*, 1997; Rubinfeld *et al.*, 1997b). One of the func-tions of β-catenin is to bind members of the Tcf family of transcription factors and activate gene transcription. Accordingly, cancers with *APC* or *CTNNB1* muta-tions have increased β-catenin/Tcf-mediated transcription, which leads to the overexpression of genes such as *CYCLIN D1 (CCND1)* and *MYC* (He *et al.*, 1998; Shtutman *et al.*, 1999). *CTNNB1* mutations have been identified in 0–25% of gastric cancers (Caca *et al.*, 1999; Candidus *et al.*, 1996; Kawanishi *et al.*, 1995; Park *et al.*, 1999). The majority of these mutations are in a portion of exon 3 encoding for the GSK3B phosphorylation consensus region of β-catenin. These mutations are often missense mutations in the highly conserved aspartic acid 32 and presumably impair the ability of GSK3B to phosphorylate β-catenin (Park *et al.*, 1999). Caca *et al* identified *CTNNB1* mutations in the NH2-terminal phosphoryl-ation sites of β-catenin and found increased Tcf/Lef transcriptional activity in association with this mutation (Caca *et al.*, 1999). Mutations that abolish β-catenin binding with E-cadherin have also been identified and have been shown to impair cell adhesion (Kawanishi *et al.*, 1995).

STK11. *STK11* was recently cloned from the locus linked to Peutz-Jeghers syndrome. This syndrome is characterized most prominently by gastrointestinal hamartomatous polyps and melanocytic macules of the lips, buccal mucosa, and digits. In addition, it is associated with an increased risk of various neoplasms including sex cord tumors, pancreatic cancer, melanoma, and gastric cancer (Itzkowitz and Kim, 1998; Rowan *et al.*, 1999; Su *et al.*, 1999; Wang *et al.*, 1999). Germline mutations in *STK11*, also known as *LKB1*, appear to be responsible for most if not all cases of Peutz-Jeghers syndrome (Hemminki *et al.*, 1998; Jenne *et al.*, 1998; Westerman *et al.*, 1999; Ylikorkala *et al.*, 1999; Yoon *et al.*, 2000). *STK11* maps to 19p13.3 and encodes a serine-threonine kinase. The function of the gene product is presently unknown but the mouse homolog appears to be a nuclear protein (Smith *et al.*, 1999). Inactivating mutations of *STK11* have been found in a subset of sporadic tumors including melanomas, pancreatic cancers, and gastric cancers. Park *et al.* screened a collection of 28 sporadic gastric cancers from Korea for *STK11* mutations using SSCP and DNA sequencing and found 1 (3%) missense mutation and 2 silent mutations. The role of *STK11* mutations in sporadic gastric cancer in low-risk areas remains to be determined (Park *et al.*, 1998).

FHIT. Identified by Ohta *et al.* in 1996, *FHIT* is a human gene located in 3p14.2, a region that is commonly deleted in multiple different tumor types, including gastric cancer. 3p14.2 was originally identified as an area of interest in cancer formation when Cohen *et al.* (1979) observed a constitutional reciprocal t(3;8) translocation associated with familial bilateral multifocal clear cell renal carcinoma. It was later shown that the site of the chromosomal break is 3p14.2 and that this 200–300 kb region is frequently homozygously deleted in many types of cancer (Wang and Perkins, 1984). This region of deletion contains the fragile site locus FRA3B as well as *FHIT*. *FHIT* ('fragile histidine triad') is a member of the histidine triad gene family and encodes a protein similar to the *aph1* gene in *S. pombe*. The gene consists of 10 exons distributed over at least 500 kb with three 5' untranslated exons centromeric to the 3p14.2 translocation associated with clear cell renal carcinoma and the remaining exons telomeric to the translocation breakpoint. Exon 5 is within the homozygously deleted fragile site (Gemma *et al.*, 1997). *FHIT* encodes a protein that is a 147-amino acid AP3A hydrolase. The preferred substrates for FHIT, AP3A (diadenosine 5',5'''P(1),P(3)-triphosphate) and AP4A, appear to have various intracellular functions including the regulation of DNA replication and signaling stress responses (Barnes *et al.*, 1996).

Aberrant transcripts from the *FHIT* locus have been found in 50% of gastric carcinomas (Ohta *et al.*, 1996). Baffa *et al.* (1998) further demonstrated deletions or rearrangements within the *FHIT* gene in four of eight gastric cancer cell lines. None of the cell lines expressed *FHIT* transcript or protein. They also identified rearrangements or aberrant transcripts of *FHIT* in 53% (*n* = 17/32) of primary gastric cancers and loss of *FHIT* protein expression in 67% of cases (*n* = 20/30).

FHIT has been proposed to be particularly susceptible to rearrangements secondary to carcinogen exposure and thus may reflect the degree to which gastric cancer causation is influenced by these agents (Baffa *et al.*, 1998). Tamura *et al* (1997) also observed homozygous *FHIT* deletion in 57% of gastric cancer cell lines ($n = 4/7$) but only found an incidence of homozygous deletion of *FHIT* in primary gastric cancers of 13% ($n = 2/16$). They concluded that the frequent homozygous deletions seen in cell lines may reflect the plasticity of the genome at FRA3B under tissue culture conditions rather than a frequent pathogenetic event in primary gastric cancer formation (Tamura *et al.*, 1997). Somatic missense mutations in *FHIT* and allelic deletions have also been found in primary gastric cancers using SSCP at a frequency of 2.5% ($n = 1/40$) and 42% ($n = 16/38$), respectively (Gemma *et al.*, 1997).

4.3 Oncogenes

MET. *MET* is a proto-oncogene that is a receptor-like tyrosine kinase comprised of disulfide-linked subunits of 50 kDa (alpha) and 145 kDa size (beta). The alpha subunit is extracellular, and the beta subunit has extracellular, transmembrane, and tyrosine kinase domains in addition to sites of tyrosine phosphorylation (Bottaro *et al.*, 1991). Furthermore, the beta subunit of the MET proto-oncogene is the cell surface receptor for the hepatocyte growth factor (Bottaro *et al.*, 1991). Hepatocyte growth factor/scatter factor (HGF/SF) is a known potent mitogen for epithelial cells and many kinds of carcinoma cell lines (Miyazawa *et al.*, 1989; Nakamura *et al.*, 1989; Stoker *et al.*, 1987; Weidner *et al.*, 1990; Zarnegar *et al.*, 1989). Fluorescent *in situ* hybridization (FISH) has assigned the location of *MET* to 7q21-q31 in humans (Dean *et al.*, 1985).

MET was originally cloned from an osteosarcoma cell line (MNNG-HOS) transformed by exposure to N-methyl-N-nitro-N-nitrosoguanidine, a carcinogen that also induces gastric tumors in mice and rats (Cooper *et al.*, 1984). Rearrangements involving *MET* and amplification of this gene have been identified in gastric cancer cell lines and in primary gastric cancers. Soman *et al.* (1991) observed a translocation involving the translocation promoter region (TPR) on chromosome 1 with the 5' region of the *MET* gene on chromosome 7 that produced fusion transcripts in several gastric cancer cell lines. This translocation was also observed in a subset of gastric cancers and potentially pre-neoplastic lesions like gastritis. Amplification of *MET* has been identified by several investigators in both primary gastric cancers and gastric cancer cell lines (Hara *et al.*, 1998; Kuniyasu *et al.*, 1992; Tsujimoto *et al.*, 1997). Kuniyasu *et al.* found amplification of *MET* in 23% ($n = 15/64$) of advanced gastric cancers and 38% ($n = 5/13$) of scirrhous gastric cancers (Kuniyasu *et al.*, 1992). They further demonstrated the expression of two different-sized transcripts from both gastric cancers and non-neoplastic tissue using Northern-blot analysis. The 7.0 kb transcript was overexpressed in 48% of the tumors ($n = 15/31$), which were predominantly of the well-differentiated type (Kuniyasu *et al.*, 1993). In addition, a smaller 6.0 kb transcript was also overexpressed in 52% of the tumors and correlated with tumor stage, the presence of lymph node metastases, and tumor invasion (Kuniyasu *et al.*, 1993). The frequency of *MET* amplification using FISH or Southern blot analysis has demonstrated a lower incidence of *MET* amplification of 4–9% in primary tumors (Hara *et al.*, 1998; Nakajima *et al.*, 1999; Tsujimoto *et al.*, 1997). The primary pathogenetic event in all of the described cases of gastric cancers with abnormal *MET* expression has been amplification as opposed to mutations (Park *et al.*, 2000; Ponzetto *et al.*, 1991).

The amplification of *MET* has been associated with MET tyrosine kinase activation in a cell line providing evidence of the biological significance of *MET* over expression (Ponzetto *et al.*, 1991). The clinical behavior of gastric cancer has also been shown to be affected by *MET* overexpression and to be associated with more aggressive tumor behavior (Nakajima *et al.*, 1999; Tsugawa *et al.*, 1998). Nakajima *et al.* demonstrated amplification of *MET* in 10.2% of gastric carcinomas and correlated this finding with MET overexpression by immunohistochemistry. The tumors with amplified and over-expressed *MET* more frequently were associated with lymph node metastases (2.3% vs. 14%) and had a worse 5-year survival (Nakajima *et al.*, 1999).

KSAM. *KSAM* is identical to the genes for keratinocyte growth factor receptor (*KGFR*) or fibroblast growth factor receptor-2 (*FGFR2*) and is located on chromosome 10q25 (Jaye *et al.*, 1992). It was originally demonstrated to have a role in gastric cancer formation when it was isolated from an amplified sequence in the KATO-III cell line derived from a diffuse-type stomach carcinoma (Hattori *et al.*, 1990; Nakatani *et al.*, 1986, 1990). The keratinocyte growth factor binding motif and carboxyl-terminal portion of the protein encoded by *FGFR2/KSAM* have been demonstrated to play an important role in carcinogenesis (Ishii *et al.*, 1995). In gastric cancer, amplification of *FGFR2/KSAM* has been shown in 21% of diffuse-type gastric cancers and none of intestinal-type cancers (Hattori *et al.*, 1990; Nakatani *et al.*, 1990). Thus, the overexpression of *FGFR2/KSAM* seems to be specific to the formation of gastric cancers with diffuse type histology.

FGFR2/KSAM generates multiple transcripts secondary to alternative splicing. The alternate transcripts, *FGFR2/KSAM I/BEK* and *FGFR2/KSAM III/KGFR*, appear to differ in ligand specificity and thus biological activities (Ishii *et al.*, 1994). Furthermore, some of the transcripts appear to be more abundant in stomach cancers and thus presumably mediate some of the malignant characteristics of the tumors. *FGFR2/KSAM III* can be detected in a secreted form and may serve as a competitor for ligand binding (Katoh *et al.*, 1992). The most abundant transcript of *FGFR2/KSAM* in the KATO-III cell line is *FGFR2/KSAM IIC3*, which has the KGF-binding motif and a short carboxyl terminus lacking a putative phospholipase C-γ1 association site, Tyr-769. This transcript has been shown to have greater transforming activity in NIH-3T3 cells, to induce stronger mitogenic responses to KGF, and to attenuate the differentiation response to differentiation-inducing media (Ishii *et al.*, 1995). The mechanisms controlling the production of the different transcripts are under investigation.

RAS. Alterations of members of the *RAS* oncogene family (*HRAS, KRAS,* and *NRAS*) were originally identified as transforming oncogenes in *in-vitro* experiments employing the transfection of NIH/3T3 cells (Land *et al.*, 1983). The *RAS* family genes encode a highly conserved family of 21 kDa proteins that are involved in signal transduction. The RAS protein binds guanine nucleotides with high affinity and serves as a proximal signaling element in a number of signal pathways that are involved in cell proliferation and differentiation (Barbacid, 1987). *KRAS* mutations and amplification have been observed in some cancers, including cancers of the gastrointestinal tract. In fact, transforming mutations at codon 12, 13, and 61 amino acid residues have been detected commonly in adenocarcinoma of the pancreas, colon, and bile duct (Sukumar, 1990). However, while *KRAS* mutations do occur in stomach cancers, they are much less common than in the lower gastrointestinal tract. *KRAS* mutations have been identified in 0–9% of gastric cancers using PCR-based assays to detect the common mutations present in this gene (Arber *et al.*, 2000; Fujita *et al.*, 1987; Hirohashi and Sugimora, 1991; Hongyo *et al.*, 1995; Jiang *et al.*, 1989; Kihana *et al.*, 1991; Nanus *et al.*, 1990; Victor *et al.*, 1990; Wang, 1995). Most recently an analysis for *c-K-RAS* codon 12 mutations using PCR showed 1/39 (2%) cases of Barrett's esophagus, 1/21 (5%) cases of esophageal adenocarcinoma, 0/27 cases of squamous cell esophageal cancer, and 1/32 (3%) gastric adenocarcinomas with mutations (Arber *et al.*, 2000). Of note, analysis of tumors and preneoplastic lesions in the stomach from Japan and Columbia have demonstrated a higher incidence of *KRAS* mutations indicating a possible difference between tumors that occur in high prevalence vs. low prevalence gastric cancer areas (Gong *et al.*, 1999; Miki *et al.*, 1991). Furthermore, some studies have suggested *KRAS* mutations occur early in the gastric cancer sequence. Gong *et al.* identified mutated *KRAS* in 15% of atrophic gastritis samples from cancers that occurred in Columbia and Kihana *et al.* identified *KRAS* mutations in a subset of gastric adenomas (Gong *et al.*, 1999; Kihana *et al.*, 1991). Thus, while genetic alterations of *KRAS* do appear to play a role in the development of some gastric cancers their incidence is significantly less than it is elsewhere in the gastrointestinal tract.

ERBB2. *ERBB2* is a *v-erbB* related gene that encodes a growth factor receptor highly homologous to the epidermal growth factor receptor. The protein encoded by *ERBB2* is a transmembrane glycoprotein that is 185 kDa in size and is a member of the *ERBB* gene family (Akiyama *et al.*, 1986). Of note, the *ERBB* oncogene was originally called *NGL* after it was identified as a human homolog to the *NEU* oncogene found in a rat neuro/glioblastoma cell line (Yang-Feng *et al.*, 1985). *ERBB2* was identified as an *ERBB* related gene distinct from *ERBB* by Semba *et al.* in 1985 (Semba *et al.*, 1985). Fukushige *et al.* mapped *ERBB2* to 17q21 and found it was amplified in a gastric cancer cell line (Fukushige *et al.*, 1986). Yokota *et al.* identified amplification of *ERBB2* in 23% (*n* = 3/13) of primary gastric tumors. In this particular study all of these tumors were of the intestinal type; 30 tumors of the diffuse type failed to show any detectable *ERBB2* (Yokota *et al.*, 1988). Others have found clear evidence of amplification of *ERBB2* in 6.3–18% of gastric cancers (Ishikawa *et al.*, 1997; Sato *et al.*, 1997; Yonemura *et al.*, 1998). The functional significance of amplification of *ERBB2* has been shown using immunohistochemistry to demonstrate overexpression of the gene product (Ishikawa *et al.*, 1997; Park *et al.*, 1989; Sato *et al.*, 1997). These studies revealed only high-level amplification led to significant overexpression of *ERBB2*

(Ishikawa *et al.*, 1997). The overexpression of *ERBB2* has been associated with a poor prognosis and an increased risk of lymph node metastases (Nakajima *et al.*, 1999; Tsugawa *et al.*, 1993, 1998; Yonemura *et al.*, 1991, 1998).

MYC. Human *MYC* is located on chromosome 8 and encodes a nuclear protein with structural features and DNA-binding properties of transcription factors. Amplification and rearrangements of *MYC* has been observed in many tumor types. Furthermore, *MYC* has been shown to transform primary cell lines and to be transforming in transgenic mice (DePinho *et al.*, 1991). The genetic alterations identified to date that have been associated with aberrant *MYC* expression include gene amplification, chromosomal translocation, and insertional muta-tions (Dang, 1991). Multiple double minute (DMs) have been observed in SNU16, a gastric cancer cell line derived from a poorly differentiated tumor with 50-fold amplified *MYC* by Southern blot analysis. Sublines from SNU16 not only had DMs but homogeneously staining regions by karyotyping (Park *et al.*, 1990). Another gastric cancer cell line, HSC-39, also was found to have 16- to 32-fold amplified *MYC* and numerous DMs (Yanagihara *et al.*, 1991). *MYC* amplification has also been found in primary gastric cancers. *MYC* amplification has been found in 15.9–67% gastric cancers using FISH on both fresh tissue samples and paraffin-embedded, formalin-fixed tissue samples (Hara *et al.*, 1998; Ooi *et al.*, 1998; Suzuki *et al.*, 1997). High level amplification (>10 copies) is only found in 3.4% of cases however (Hara *et al.*, 1998). In a subset of tumors co-amplification of *MYC* and *ERBB2* and *MYC* and *MET* have been observed suggesting there may be clonal selection for the co-amplification of these genes (Hara *et al.*, 1998; Seruca *et al.*, 1995b; Tsuchiya *et al.*, 1989). Amplification of another *MYC* family gene, *MYCN*, has also been observed in a subset of gastric cancers (Nessling *et al.*, 1998).

4.4 Other alterations

HST-1 was originally identified as a transforming gene from studies that involved transfecting NIH-3T3 cells with DNA samples from gastric carcinomas and adjacent non-neoplastic gastric mucosa (Sakamoto *et al.*, 1986). It is homologous to fibroblast growth factor receptor 4 (*FGFR4*) and is located on chromosome 11q13 (Sakamoto *et al.*, 1986). *FGFR4* has been shown to be co-amplified with *INT-2* in a variety of tumor types including urinary bladder carcinomas, melanomas, esophageal carcinomas as well as gastric cancers (Yoshida *et al.*, 1988). The *INT-2* gene, which was subsequently shown to be the *FGF3* gene, has been implicated in mouse mammary carcinogenesis, is activated by viral insertion, and is located adjacent to *FGFR4* on chromosome 11. *FGFR4* amplification in gastric cancer (2%, *n* = 1/43) has been shown in one study (Yoshida *et al.*, 1988). However, Tsuda *et al.* found no amplification of either gene in 42 gastric carcinomas but co-amplification of both genes in 50% of esophageal carcinomas (Tsuda *et al.*, 1989). Finally, amplification of *FGFR4* does not correlate with overexpression of the *FGFR4* gene product. Thus, the role *FGFR4/FGR3* plays in gastric cancer formation is uncertain.

Other infrequently observed alterations found in gastric cancer include the amplification of *AKT-1, CYCLIN E1 (CCNE1), YES1*, and *ERBB1* (Nomura *et al.*, 1986; Ochiai *et al.*, 1997; Seki *et al.*, 1985; Staal, 1987). Mutation analysis of *p300* has found at least one gastric cancer case with a somatic mutation in the functionally important Cys/His region coupled with deletion of the second allele (Muraoka *et al.*, 1996). Of note, *p300* may be the target for deletions of chromosome 22q seen in gastric cancer. Interferon regulatory factor-1 (*IRF1*) also appears to be mutated in a small percentage of gastric cancers. Nozawa *et al.* identified a missense mutation in *IRF1* in 1/9 gastric cancers. This mutation was accompanied by allelic loss of the second allele and decreased transcriptional activity from the mutant *IRF1* (Nozawa *et al.*, 1998). The role these alterations play in sporadic gastric cancer formation remains to be determined.

5. Epigenetic alterations in gastric cancer

The identification of oncogenes and tumor suppressor genes in the early 1980s gave rise to the fundamental concept that carcinogenesis is a process driven by genetic alterations. However, it has become increasingly appreciated that there are also fundamental epigenetic events occurring during the formation of some tumor types, including cancers of the gastrointestinal tract. These epigenetic changes are changes in DNA other than changes in nucleotide structure that mediate heritable traits. Furthermore, it appears that the epigenetic and genetic changes cooperate to promote cancer formation (Baylin and Herman, 2000). The type of cancer-related epigenetic changes that are currently best understood involve aberrant methylation of CpG dinucleotides in the 5′ region of tumor suppressor genes. CpG dinucleotides are relatively enriched in the human genome in the 5′ region of many genes. These 5′ CpG islands are normally unmethylated except in certain situations such as imprinting, X-inactivation, aging, and carcinogenesis (Baylin and Herman, 2000). Specific loci that are subject to CpG methylation only in cancers (cancer specific) have been identified and many of these contain tumor suppressor genes (Toyota *et al.*, 1999a, 1999b). In addition, analysis of the methylation status of specific CpGs in the *MLH1* promoter has shown that the methylation status of small clusters of CpGs in the 5′ region of the gene dictates the transcriptional status of the gene (Deng *et al.*, 1999). Importantly, hypermethylation of *MLH1* has been shown to associate with loss of *MLH1* expression and loss of mutation mismatch repair activity demonstrating the functional consequences of this epigenetic event (Herman *et al.*, 1998; Veigl *et al.*, 1998). The mechanisms that control the CpG methylation status of these regions are still incompletely understood and under intense investigation.

The aberrant hypermethylation of 5′ CpG dinucleotides has been demonstrated to silence a variety of tumor suppressor genes including *CDH1*, *p16*, *thrombospondin-1* (*TSP1*), *MLH1*, and *GSTP1* (Baylin and Herman, 2000; Herman *et al.*, 1995, 1998; Veigl *et al.*, 1998). In gastric cancer, the hypermethylation of *INK4* (*P16*) (32%, *n* = 18/56), *MLH1* (5%, 3/56), *CDH1* (51%, *n* = 27/53), *PS2*, *TGFBRI*, and MINT1, 2, 12, 25, and 31 has been well demonstrated (Fujimoto *et al.*, 2000; Kang *et al.*, 1999; Tamura *et al.*, 2000; Toyota *et al.*, 1999b). The MINT loci are loci that were found to be hypermethylated in a colorectal cancer cell line using methylated CpG island amplification, a PCR-based technique that allows the cloning of differentially methylated sequences of DNA (Toyota *et al.*, 1999b). The functional role of MINT loci hyper-methylation in cancer formation is still being determined. On the other hand, the effect of CpG methylation on *INK4* and *MLH1* is considerably better understood. Epigenetic silencing through DNA hypermethylation is the most common mechanism for silencing *INK4* and *MLH1* in gastric cancer. Interestingly, mutations of either of these genes are uncommon in sporadic gastric cancers; however, methylation of *INK4* is detected in 32–42% of gastric cancers and *MLH1* is detected in 60% of MSI gastric cancers (Igaki *et al.*, 1995; Kane *et al.*, 1997; Kim *et al.*, 1998; Shim *et al.*, 2000; Toyota *et al.*, 1999b; Wu *et al.*, 1998b). In fact, the hypermethylation of *MLH1* is the most common mechanism for inducing microsatellite instability in sporadic gastric cancer (Fleisher *et al.*, 1999; Leung *et al.*, 1999). In contrast to *INK4* and *MLH1*, both mutations and promoter CpG methylation of *CDH1* have been commonly observed in diffuse gastric cancers (Machado *et al.*, 1999; Tamura *et al.*, 2000). This difference suggests there are gene-specific mechanisms mediating the mode of inactivation of tumor suppressor genes during gastric cancer formation.

Of interest, methylation of *INK4* and the *17S5* locus have been found in not only gastric cancer but also in potentially preneoplastic tissue (*Figure 3*). Kanai *et al.* demonstrated hypermethylation of *17S5* in 13% of samples of chronic gastritis and 25% of samples of intestinal metaplasia, and Toyota *et al.* found *INK4* methylation in the normal mucosa adjacent to gastric cancers (Kanai *et al.*, 1998; Toyota *et al.*, 1999b). The significance of these findings is controversial in light of uncertainty about the true preneoplastic potential of these lesions and about the specificity of the assays in these studies for detecting cancer-specific methylation events. Nonetheless, Grady *et al.* recently demonstrated loss of expression of E-cadherin in association with CpG methylation of the wild-type *CDH1* allele in tumors occurring in the setting of the cancer family syndrome hereditary diffuse gastric cancer (Grady *et al.*, 2000). This finding strongly suggests that aberrant hypermethylation of tumor suppressor genes, such as *CDH1*, is involved in the initiation of gastric cancer formation.

6. Conclusions

Gastric cancer is the product of multiple genetic and epigenetic alterations that transform normal gastric epithelial cells into malignant neoplasms. Molecular biology, cell biology, and genetics studies have yielded some insight into the functionally important specific alterations mediating these effects. As with carcinogenesis in general, the alterations appear to arise in the permissive setting of genomic instability and, in general, activate oncogenes or silence tumor suppressor genes. The alterations observed in gastric cancer affect some genes that are commonly known to influence tumor formation in a variety of tumor types, such as *TP53*, *RAS*, and *TGFBR2*, and others that appear to be more specific for gastric cancer formation, such as FGFR2/*KSAM* and *MET*. Furthermore, the genetic alterations that accumulate during gastric carcinogenesis differ depending on the histological type of the tumor suggesting the alterations can influence the histology of the tumor and that there are multiple molecular pathways for gastric carcinogenesis. Finally, and perhaps most importantly, the effect of some of these alterations on the clinical behavior of these cancers has been determined. Thus, the information obtained from the studies of the genetic and epigenetic alterations of gastric cancer conducted to date provide hope that further studies will enhance our understanding of these tumors and will lead to the prevention and treatment of these cancers.

References

Aaltonen, L., Peltomaki, P., Leach, F. *et al.* (1993) Clues to the pathogenesis of familial colorectal cancer. *Science* **260**: 812–816.

Aarnio, M., Salovaara, R., Aaltonen, L.A., Mecklin, J.P. and Jarvinen, H.J. (1997) Features of gastric cancer in hereditary non-polyposis colorectal cancer syndrome. *Int. J. Cancer* **74**: 551–555.

Aberle, H., Butz, S., Stappert, J., Weissig, H., Kemler, R. and Hoschuetzky, H. (1994) Assembly of the cadherin-catenin complex in vitro with recombinant proteins. *J. Cell. Sci.* **107**: 3655–3663.

Akiyama, T., Sudo, C., Ogawara, H., Toyoshima, K. and Yamamoto, T. (1986) The product of the human c-erbB-2 gene: a 185-kilodalton glycoprotein with tyrosine kinase activity. *Science* **232**: 1644–1646.

Arber, N., Shapira, I., Ratan, J., Stern, B., Hibshoosh, H., Moshkowitz, M., Gammon, M., Fabian, I. and Halpern, Z. (2000) Activation of c-K-ras mutations in human gastrointestinal tumors. *Gastroenterology* **118**: 1045–1050.

Baffa, R., Veronese, M.L., Santoro, R., Mandes, B., Palazzo, J.P., Rugge, M., Santoro, E., Croce, C.M. and Huebner, K. (1998) Loss of FHIT expression in gastric carcinoma. *Cancer Res.* **58**: 4708–4714.

Barbacid, M. (1987) ras genes. *Annu. Rev. Biochem.* **56**: 779–827.

Barnes, L.D., Garrison, P.N., Siprashvili, Z., Guranowski, A., Robinson, A.K., Ingram, S.W., Croce, C.M., Ohta, M. and Huebner, K. (1996) Fhit, a putative tumor suppressor in humans, is a dinucleoside 5′,5‴- P1,P3-triphosphate hydrolase. *Biochemistry* **35**: 11529–11535.

Baylin, S.B. and Herman, J.G. (2000) DNA hypermethylation in tumorigenesis: epigenetics joins genetics. *Trends Genet.* **16**: 168–174.

Becker, K.F. and Hofler, H. (1995) Frequent somatic allelic inactivation of the E cadherin gene in gastric carcinomas. *J. Natl. Cancer Inst.* **87**: 1082–1084.

Behrens, J. (1993) The role of cell adhesion molecules in cancer invasion and metastasis. *Breast Cancer Res. Treat.* **24**: 175–184.

Bertoni, L., Zoli, W., Mucciolo, E., Ricotti, L., Nergadze, S., Amadori, D. and Giulotto, E. (1998) Different genome organization in two new cell lines established from human gastric carcinoma. *Cancer Genet. Cytogenet.* **105**: 152–159.

Berx, G., Cleton-Jansen, A.M., Strumane, K., de Leeuw, W.J., Nollet, F., van Roy, F. and Cornelisse, C. (1996) E-cadherin is inactivated in a majority of invasive human lobular breast cancers by truncation mutations throughout its extracellular domain. *Oncogene* **13**: 1919–1925.

Berx, G., Becker, K.-F., Hofler, H. and van Roy, F. (1998) Mutations of the human E-cadherin (CDH1) gene. *Hum. Mutat.* **12**: 226–237.

Bevan, S. and Houlston, R.S. (1999) Genetic predisposition to gastric cancer. *QJM* **92**: 5–10.

Birchmeier, W. and Behrens, J. (1994) Cadherin expression in carcinomas: role in the formation of cell junctions and the prevention of invasiveness. *Biochim. Biophys. Acta.* **1198**: 11–26.

Blot, W.J., Devesa, S.S., Kneller, R.W. and Fraumeni J.F., Jr. (1991) Rising incidence of adenocarcinoma of the esophagus and gastric cardia [see comments]. *JAMA* **265**: 1287–1289.

Bottaro, D.P., Rubin, J.S., Faletto, D.L., Chan, A.M., Kmiecik, T.E., Vande Woude, G.F. and Aaronson, S.A. (1991) Identification of the hepatocyte growth factor receptor as the c-met proto-oncogene product. *Science* **251**: 802–804.

Burt, R. (1991) Polyposis syndromes. In: T. Yamada and T. Alpers (eds). *Textbook of Gastroenterology*. New York. J. B. Lippincott Company, pp. 1674–1696.

Caca, K., Kolligs, F.T., Ji, X., Hayes, M., Qian, J., Yahanda, A., Rimm, D.L., Costa, J. and Fearon, E.R. (1999) Beta- and gamma-catenin mutations, but not E-cadherin inactivation, underlie T-cell factor/lymphoid enhancer factor transcriptional deregulation in gastric and pancreatic cancer. *Cell Growth Differ.* **10**: 369–376.

Cagle, P.T., Taylor, L.D., Schwartz, M.R., Ramzy, I. and Elder, F.F. (1989) Cytogenetic abnormalities common to adenocarcinoma metastatic to the pleura. *Cancer Genet. Cytogenet.* **39**: 219–225.

Candidus, S., Bischoff, P., Becker, K.F. and Hofler, H. (1996) No evidence for mutations in the alpha- and beta-catenin genes in human gastric and breast carcinomas. *Cancer Res.* **56**: 49–52.

Chang, J., Park, K., Bang, Y.-J., Kim, W., Kim, D. and Kim, S.-J. (1997) Expression of transforming growth factor β type II receptor reduces tumorigenicity in human gastric cancer cells. *Cancer Res.* **57**: 2856–2859.

Cho, J.H., Noguchi, M., Ochiai, A., Uchino, S. and Hirohashi, S. (1994) Analysis of regional differences of p53 mutation in advanced gastric carcinoma: relation to heterogeneous differentiation and invasiveness. *Mod. Pathol.* **7**: 205–211.

Chong, J.-M., Fukayama, M., Hayashi, Y., Takizawa, T., Koike, M., Konishi, M., Kikuchi-Yanoshita, R. and Miyaki, M. (1994) Microsatellite instability in the progression of gastric cancer. *Cancer Res.* **54**: 4595–4597.

Chun, Y.H., Kil, J.I., Suh, Y.S., Kim, S.H., Kim, H. and Park, S.H. (2000) Characterization of chromosomal aberrations in human gastric carcinoma cell lines using chromosome painting. *Cancer Genet. Cytogenet.* **119**: 18–25.

Chung, Y.-J., Song, J.-M., Lee, J.-Y., Jung, Y.-T., Seo, E.-J., Choi, S.-W. and Rhyu, M.-G. (1996) Microsatellite instability-associated mutations associate preferentially with the intestinal type of primary gastric carcinomas in a high-risk population. *Cancer Res.* **56**: 4662–4665.

Chung, Y.-J., Park, S.-W., Song, J.-M., Lee, K.-Y., Seo, E.-J., Choi, S.-W. and Rhyu, M.-G. (1997) Evidence of genetic progression in human gastric carcinomas with microsatellite instability. *Oncogene* **15**: 1719–1726.

Chung, Y.J., Kim, K.M., Choi, J.R., Choi, S.W. and Rhyu, M.G. (1999) Relationship between intratumor histological heterogeneity and genetic abnormalities in gastric carcinoma with microsatellite instability. *Int. J. Cancer* **82**: 782–788.

Clark, G.W., Smyrk, T.C., Burdiles, P., Hoeft, S.F., Peters, J.H., Kiyabu, M., Hinder, R.A., Bremner, C.G. and DeMeester, T.R. (1994) Is Barrett's metaplasia the source of adenocarcinomas of the cardia? *Arch. Surg.* **129**: 609–614.

Cohen, A.J., Li, F.P., Berg, S., Marchetto, D.J., Tsai, S., Jacobs, S.C. and Brown, R.S. (1979) Hereditary renal-cell carcinoma associated with a chromosomal translocation. *N. Engl. J. Med.* **301**: 592–595.

Cooper, C.S., Park, M., Blair, D.G., Tainsky, M.A., Huebner, K., Croce, C.M. and Vande Woude, G.F. (1984) Molecular cloning of a new transforming gene from a chemically transformed human cell line. *Nature* **311**: 29–33.

Correa, P. and Haenszel, W. (1972) Epidemiology of gastric cancer. In P. Correa and W. Haenszel (eds). *Epidemiology of Cancer of the Digestive Tract*. The Hague, The Netherlands. Martinus, Nijhoff. **58**.

Correa, P. and Shiao, Y.H. (1994) Phenotypic and genotypic events in gastric carcinogenesis. *Cancer Res.* **54**: 1941s–1943s.

Craanen, M.E., Dekker, W., Blok, P., Ferwerda, J. and Tytgat, G.N. (1992) Time trends in gastric carcinoma: changing patterns of type and location [see comments]. *Am. J. Gastroenterol* **87**: 572–579.

Dang, C.V. (1991) c-myc oncoprotein function. *Biochim. Biophys. Acta.* **1072**: 103–113.

Dean, M., Park, M., Le Beau, M.M., Robins, T.S., Diaz, M.O., Rowley, J.D., Blair, D.G. and Vande Woude, G.F. (1985) The human met oncogene is related to the tyrosine kinase oncogenes. *Nature* **318**: 385–388.

Deng, G., Chen, A., Hong, J., Chae, H. and Kim, Y. (1999) Methylation of CpG in a small region of the hMLH1 promoter invariably correlates with the absence of gene expression. *Cancer Res.* **59**: 2029–2033.

DePinho, R.A., Schreiber-Agus, N. and Alt, F.W. (1991) myc family oncogenes in the development of normal and neoplastic cells. *Adv. Cancer Res.* **57**: 1–46.

Dos Santos, N., Seruca, R., Constancia, M., Seixas, M. and Sobrinho-Simoes, M. (1996) Microsatellite instability at multiple loci in gastric carcinoma: clinicopathologic implications and prognosis. *Gastroenterology* **110**: 38–44.

Duval, A., Gayet, J., Zhou, X.P., Iacopetta, B., Thomas, G. and Hamelin, R. (1999) Frequent frameshift mutations of the TCF-4 gene in colorectal cancers with microsatellite instability. *Cancer Res.* **59**: 4213–4215.

Eshleman, J. and Markowitz, S. (1995) Microsatellite instability in inherited and sporadic neoplasms. *Curr. Opin. Oncol.* **7**: 83–89.

Eshleman, J., Casey, G., Kochera, M., Sedwick, W., Swinler, S., Veigl, M., Willson, J., Stuart, S. and Markowitz, S. (1998) Chromosome number and structure both are markedly stable in RER colorectal cancers and are not destabilized by mutation of p53. *Oncogene* **17**: 719–725.

Espinoza, L.A., Barbieri Neto, J. and Casartelli, C. (1999) Pathological and karyotypic abnormalities in advanced gastric carcinomas. *Cancer Genet. Cytogenet.* **109**: 45–50.

Farinati, F., Cardin, R., Degan, P., Rugge, M., Mario, F.D., Bonvicini, P. and Naccarato, R. (1998) Oxidative DNA damage accumulation in gastric carcinogenesis. *Gut* **42**: 351–356.

Ferti-Passantonopoulou, A.D., Panani, A.D., Vlachos, J.D. and Raptis, S.A. (1987) Common cytogenetic findings in gastric cancer. *Cancer Genet. Cytogenet.* **24**: 63–73.

Fleisher, A.S., Esteller, M., Wang, S. *et al.* (1999) Hypermethylation of the hMLH1 gene promoter in human gastric cancers with microsatellite instability. *Cancer Res.* **59**: 1090–1095.

Frixen, U.H., Behrens, J., Sachs, M., Eberle, G., Voss, B., Warda, A., Lochner, D. and Birchmeier, W. (1991) E-cadherin-mediated cell–cell adhesion prevents invasiveness of human carcinoma cells. *J. Cell. Biol.* **113**: 173–185.

Fujimoto, J., Yasui, W., Tahara, H., Tahara, E., Kudo, Y. and Yokozaki, H. (2000) DNA hypermethylation at the pS2 promoter region is associated with early stage of stomach carcinogenesis. *Cancer Lett.* **149**: 125–134.

Fujita, K., Ohuchi, N., Yao, T., Okumura, M., Fukushima, Y., Kanakura, Y., Kitamura, Y. and Fujita, J. (1987) Frequent overexpression, but not activation by point mutation, of ras genes in primary human gastric cancers. *Gastroenterology* **93**: 1339–1345.

Fukushige, S., Matsubara, K., Yoshida, M., Sasaki, M., Suzuki, T., Semba, K., Toyoshima, K. and Yamamoto, T. (1986) Localization of a novel v-erbB-related gene, c-erbB-2, on human chromosome 17 and its amplification in a gastric cancer cell line. *Mol. Cell. Biol.* **6**: 955–958.

Fynan, T.M. and Reiss, M. (1993) Resistance to inhibition of cell growth by transforming growth factor-beta and its role in oncogenesis. *Crit. Rev. Oncog.* **4**: 493–540.

Gabbert, H.E., Muller, W., Schneiders, A., Meier, S. and Hommel, G. (1995) The relationship of p53 expression to the prognosis of 418 patients with gastric carcinoma. *Cancer* **76**: 720–726.

Gabbert, H.E., Mueller, W., Schneiders, A., Meier, S., Moll, R., Birchmeier, W. and Hommel, G. (1996) Prognostic value of E-cadherin expression in 413 gastric carcinomas. *Int. J. Cancer* **69**: 184–189.

Gayther, S., Gorringe, K., Ramus, S. *et al.* (1998) Identification of germ-line E-cadherin mutations in gastric cancer families of European origin. *Cancer Res.* **58**: 4086–4089.

Gemma, A., Hagiwara, K., Ke, Y., Burke, L.M., Khan, M.A., Nagashima, M., Bennett, W.P. and Harris, C.C. (1997) FHIT mutations in human primary gastric cancer. *Cancer Res.* **57**: 1435–1437.

Gemma, A., Hagiwara, K., Vincent, F., Ke, Y., Hancock, A.R., Nagashima, M., Bennett, W.P. and Harris, C.C. (1998) hSmad5 gene, a human hSmad family member: its full length cDNA, genomic structure, promoter region and mutation analysis in human tumors. *Oncogene* **16**: 951–956.

Gong, C., Mera, R., Bravo, J.C., Ruiz B., Diaz-Escamilla, R., Fontham, E.T., Correa, P. and Hunt, J.D. (1999) KRAS mutations predict progression of preneoplastic gastric lesions. *Cancer Epidemiol. Biomarkers Prev.* **8**: 167–171.

Grady, W., Rajput, A., Myeroff, L., Liu, D., Kwon, K.-H., Willis, J. and Markowitz, S. (1998) Mutation of the type II transforming growth factor-β receptor is coincident with the transformation of human colon adenomas to malignant carcinomas. *Cancer Res.* **58**: 3101–3104.

Grady, W., Willis, J., Guilford, P. *et al.* (2000) E-cadherin gene promoter methylation as the second genetic hit in hereditary diffuse gastric cancer. *Nat. Genet.* in press.

Guilford, P., Hopkins, J., Harraway, J. *et al.* (1998) E-cadherin germline mutations in familial gastric cancer. *Nature* **392**: 402–405.

Guilford, P., Hopkins, J., Grady, W. *et al.* (1999) E-cadherin germline mutations define an inherited cancer syndrome dominated by diffuse gastric cancer. *Hum. Mutat.* **14**: 249–255.

Haenszel, W., Kurihara, M., Segi, M. and Lee, R.K. (1972) Stomach cancer among Japanese in Hawaii. *J. Natl. Cancer Inst.* **49**: 969–988.

Hahn, S., Schutte, M., Shamsul Hoque, A. *et al.* (1996) *DPC4*, a candidate tumor supressor gene at human chromosome 18q21.1. *Science* **271**: 350–353.

Han, J.-H., Yanagisawa, A., Kato, Y., Park, J.-G. and Nakamura, Y. (1993) Genetic instability in pancreatic cancer and poorly differentiated type of gastric cancer. *Cancer Res.* **53**: 5087–5089.

Handschuh, G., Candidus, S., Luber, B. *et al.* (1999) Tumour-associated E-cadherin mutations alter cellular morphology, decrease cellular adhesion and increase cellular motility. *Oncogene* 18: 4301–4312.

Hara, T., Ooi, A., Kobayashi, M., Mai, M., Yanagihara, K. and Nakanishi, I. (1998) Amplification of c-myc, K-sam, and c-met in gastric cancers: detection by fluorescence in situ hybridization. *Lab. Invest.* 78: 1143–1153.

Hattori, Y., Odagiri, H., Nakatani, H. *et al.* (1990) K-sam, an amplified gene in stomach cancer, is a member of the heparin-binding growth factor receptor genes. *Proc. Natl Acad. Sci. USA* 87: 5983–5987.

Heim, S. (1995) Tumors of the digestive tract. In: S. Heim and F. Mitelman (eds), *Cancer Cytogenetics.* New York. Wiley-Liss, pp. 331–332.

He, T.C., Sparks, A.B., Rago, C., Hermeking, H., Zawel, L., da Costa, L.T., Morin, P.J., Vogelstein, B. and Kinzler, K.W. (1998) Identification of c-MYC as a target of the APC pathway [see comments]. *Science* 281: 1509–1512.

Hemminki, A., Markie, D., Tomlinson, I. *et al.* (1998) A serine/threonine kinase gene defective in Peutz-Jeghers syndrome. *Nature* 391: 184–187.

Herman, J.G., Merlo, A., Mao, L., Lapidus, R.G., Issa, J.P., Davidson, N.E., Sidransky, D. and Baylin, S.B. (1995) Inactivation of the CDKN2/p16/MTS1 gene is frequently associated with aberrant DNA methylation in all common human cancers. *Cancer Res.* 55: 4525–4530.

Herman, J., Umar, A., Polyak, K. *et al.* (1998) Incidence and functional consequences of *hMLH1* promoter hypermethylation in colorectal carcinoma. *Proc. Natl Acad. Sci. USA* 95: 6870–6875.

Hirohashi, S. and Sugimura, T. (1991) Genetic alterations in human gastric cancer. *Cancer Cells* 3: 49–52.

Hollstein, M., Sidransky, D., Vogelstein, B. and Harris, C.C. (1991) p53 mutations in human cancers. *Science* 253: 49–53.

Hongyo, T., Buzard, G.S., Palli, D. *et al.* (1995) Mutations of the K-ras and p53 genes in gastric adenocarcinomas from a high-incidence region around Florence, Italy. *Cancer Res.* 55: 2665–2672.

Hulsken, J., Birchmeier, W. and Behrens, J. (1994) E-cadherin and APC compete for the interaction with beta-catenin and the cytoskeleton. *J. Cell. Biol.* 127: 2061–2069.

Hurlimann, J. and Saraga, E.P. (1994) Expression of p53 protein in gastric carcinomas. Association with histologic type and prognosis. *Am. J. Surg. Pathol.* 18: 1247–1253.

Igaki, H., Sasaki, H., Tachimori, Y., Kato, H., Watanabe, H., Kimura, T., Harada, Y., Sugimura, T. and Terada, M. (1995) Mutation frequency of the p16/CDKN2 gene in primary cancers in the upper digestive tract. *Cancer Res.* 55: 3421–3423.

Imai, S., Koizumi, S., Sugiura, M., Tokunaga, M., Uemura, Y., Yamamoto, N., Tanaka, S., Sato, E. and Osato, T. (1994) Gastric carcinoma: monoclonal epithelial malignant cells expressing Epstein-Barr virus latent infection protein. *Proc. Natl Acad. Sci. USA* 91: 9131–9135.

Ishii, H., Hattori, Y., Itoh, H. *et al.* (1994) Preferential expression of the third immunoglobulin-like domain of K-sam product provides keratinocyte growth factor-dependent growth in carcinoma cell lines. *Cancer Res.* 54: 518–522.

Ishii, H., Yoshida, T., Oh, H., Yoshida, S. and Terada, M. (1995) A truncated K-sam product lacking the distal carboxyl-terminal portion provides a reduced level of autophosphorylation and greater resistance against induction of differentiation. *Mol. Cell. Biol.* 15: 3664–3671.

Ishikawa, T., Kobayashi, M., Mai, M., Suzuki, T. and Ooi, A. (1997) Amplification of the c-erbB-2 (HER-2/neu) gene in gastric cancer cells. Detection by fluorescence in situ hybridization. *Am. J. Pathol.* 151: 761–768.

Itzkowitz, S. and Kim, Y. (1998) Colonic polyps and polyposis syndromes. In M. Feldman, B. Scharschmidt and M. Sleisenger (eds). *Sleisenger and Fordtran's Gastrointestinal and Liver Disease.* Philadelphia. W.B. Saunders Co. 1891.

Jawhari, A., Jordan, S., Poole, S., Browne, P., Pignatelli, M. and Farthing, M.J. (1997) Abnormal immunoreactivity of the E-cadherin-catenin complex in gastric carcinoma: relationship with patient survival. *Gastroenterology* 112: 46–54.

Jaye, M., Schlessinger, J. and Dionne, C.A. (1992) Fibroblast growth factor receptor tyrosine kinases: molecular analysis and signal transduction. *Biochim. Biophys. Acta.* 1135: 185–199.

Jenne, D.E., Reimann, H., Nezu, J., Friedel, W., Loff, S., Jeschke, R., Muller, O., Back, W. and Zimmer, M. (1998) Peutz-Jeghers syndrome is caused by mutations in a novel serine threonine kinase. *Nat. Genet.* 18: 38–43.

Jiang, W., Kahn, S.M., Guillem, J.G., Lu, S.H. and Weinstein, I.B. (1989) Rapid detection of ras oncogenes in human tumors: applications to colon, esophageal, and gastric cancer. *Oncogene* 4: 923–928.

Kabat, G.C., Ng, S.K. and Wynder, E.L. (1993) Tobacco, alcohol intake, and diet in relation to adenocarcinoma of the esophagus and gastric cardia. *Cancer Causes Control* **4**: 123–132.

Kamiya, T., Morishita, T., Asakura, H., Miura, S., Munakata, Y. and Tsuchiya, M. (1982) Long-term follow-up study on gastric adenoma and its relation to gastric protruded carcinoma. *Cancer* **50**: 2496–2503.

Kanai, Y., Ushijima, S., Ochiai, A., Eguchi, K., Hui, A. and Hirohashi, S. (1998) DNA hypermethylation at the D17S5 locus is associated with gastric carcinogenesis. *Cancer Lett.* **122**: 135–141.

Kane, M., Loda, M., Gaida, G., Lipman, J., Mishra, R., Goldman, H., Jessup, J. and Kolodner, R. (1997) Methylation of the *hMLH1* promoter correlates with lack of expression of hMLH1 in sporadic colon tumors and mismatch repair-defective human tumor cell lines. *Cancer Res.* **57**: 808–811.

Kang, S.H., Bang, Y.J., Im, Y.H., Yang, H.K., Lee, D.A., Lee, H.Y., Lee, H.S., Kim, N.K. and Kim, S.J. (1999) Transcriptional repression of the transforming growth factor-beta type I receptor gene by DNA methylation results in the development of TGF-beta resistance in human gastric cancer. *Oncogene* **18**: 7280–7286.

Katoh, M., Hattori, Y., Sasaki, H., Tanaka, M., Sugano, K., Yazaki, Y., Sugimura, T. and Terada, M. (1992) K-sam gene encodes secreted as well as transmembrane receptor tyrosine kinase. *Proc. Natl Acad. Sci. USA* **89**: 2960–2964.

Kawanishi, J., Kato, J., Sasaki, K., Fujii, S., Watanabe, N. and Niitsu, Y. (1995) Loss of E-cadherin-dependent cell–cell adhesion due to mutation of the beta-catenin gene in a human cancer cell line, HSC-39. *Mol. Cell. Biol.* **15**: 1175–1181.

Keller, G., Rotter, M., Vogelsang, H., Bischoff, P., Becker, K.F., Mueller, J., Brauch, H., Siewert, J.R. and Hofler, H. (1995) Microsatellite instability in adenocarcinomas of the upper gastrointestinal tract. Relation to clinicopathological data and family history [see comments]. *Am. J. Pathol.* **147**: 593–600.

Keller, G., Vogelsang, H., Becker, I. *et al.* (1999) Diffuse type gastric and lobular breast carcinoma in a familial gastric cancer patient with an E-cadherin germline mutation. *Am. J. Pathol.* **155**: 337–342.

Kihana, T., Tsuda, H., Hirota, T., Shimosato, Y., Sakamoto, H., Terada, M. and Hirohashi, S. (1991) Point mutation of c-Ki-ras oncogene in gastric adenoma and adenocarcinoma with tubular differentiation. *Japan J. Cancer Res.* **82**: 308–314.

Kim, J.R., Kim, S.Y., Kim, M.J. and Kim, J.H. (1998) Alterations of CDKN2 (MTS1/p16INK4A) gene in paraffin-embedded tumor tissues of human stomach, lung, cervix and liver cancers. *Exp. Mol. Med.* **30**: 109–114.

Kim, H.S., Woo, D.K., Bae, S.I., Kim, Y.I. and Kim, W.H. (2000a) Microsatellite instability in the adenoma-carcinoma sequence of the stomach. *Lab. Invest.* **80**: 57–64.

Kim, S.J., Im, Y.H., Markowitz, S.D. and Bang, Y.J. (2000b) Molecular mechanisms of inactivation of TGF-beta receptors during carcinogenesis. *Cytokine Growth Factor Rev* **11**: 159–168.

Koizumi, Y., Tanaka, S., Mou, R. *et al.* (1997) Changes in DNA copy number in primary gastric carcinomas by comparative genomic hybridization. *Clin. Cancer Res.* **3**: 1067–1076.

Kolodziejczyk, P., Yao, T., Oya, M., Nakamura, S., Utsunomiya, T., Ishikawa, T. and Tsuneyoshi, M. (1994) Long-term follow-up study of patients with gastric adenomas with malignant transformation. An immunohistochemical and histochemical analysis. *Cancer* **74**: 2896–2907.

Koo, S.H., Kwon, K.C., Shin, S.Y., Jeon, Y.M., Park, J.W., Kim, S.H. and Noh, S.M. (2000) Genetic alterations of gastric cancer: comparative genomic hybridization and fluorescence in situ hybridization studies. *Cancer Genet. Cytogenet.* **117**: 97–103.

Kuniyasu, H., Yasui, W., Kitadai, Y., Yokozaki, H., Ito, H. and Tahara, E. (1992) Frequent amplification of the c-met gene in scirrhous type stomach cancer. *Biochem. Biophys. Res. Commun.* **189**: 227–232.

Kuniyasu, H., Yasui, W., Yokozaki, H., Kitadai, Y. and Tahara, E. (1993) Aberrant expression of c-met mRNA in human gastric carcinomas. *Int. J. Cancer* **55**: 72–75.

Land, H., Parada, L.F. and Weinberg, R.A. (1983) Cellular oncogenes and multistep carcinogenesis. *Science* **222**: 771–778.

Lauren, P. and Nevalainen, J. (1993) Epidemiology of intestinal and diffuse types of gastric carcinoma: A time-trend study in Finland with comparison between studies in high- and low-risk areas. *Cancer* **71**: 2926.

Lauwers, G.Y., Wahl, S.J., Melamed, J. and Rojas-Corona, R.R. (1993) p53 expression in precancerous gastric lesions: an immunohistochemical study of PAb 1801 monoclonal antibody on adenomatous and hyperplastic gastric polyps [published erratum appears in *Am. J. Gastroenterol* 1994 Feb; 89(2): 300]. *Am. J. Gastroenterol* **88**: 1916–1919.

Lengauer, C., Kinzler, K. and Vogelstein, B. (1997) Genetic instability in colorectal cancers. *Nature* **386**: 623–627.

Leung, S.Y., Yuen, S.T., Chung, L.P., Chu, K.M., Chan, A.S. and Ho, J.C. (1999) hMLH1 promoter methylation and lack of hMLH1 expression in sporadic gastric carcinomas with high-frequency microsatellite instability. *Cancer Res.* **59**: 159–164.

Leung, W.K., Kim, J.J., Kim, J.G., Graham, D.Y. and Sepulveda, A.R. (2000) Microsatellite instability in gastric intestinal metaplasia in patients with and without gastric cancer. *Am. J. Pathol.* **156**: 537–543.

Lin, J.-T., Wu, M.-S., Shun, C.-T., Lee, W.-J., Sheu, J.-C. and Wang, T.-H. (1995) Occurrence of microsatellite instability in gastric carcinoma is associated with enhanced expression of erbB-2 oncoprotein. *Cancer Res.* **55**: 1428–1430.

Loeb, L., Springgate, C. and Battula, N. (1974) Errors in DNA replication as a basis of malignant changes. *Cancer Res.* **34**: 2311–2321.

Lu, S.L., Kawabata, M., Imamura, T., Miyazono, K. and Yuasa, Y. (1999) Two divergent signaling pathways for TGF-beta separated by a mutation of its type II receptor gene. *Biochem. Biophys. Res. Commun.* **259**: 385–390.

Luber, B., Candidus, S., Handschuh, G., Mentele, E., Hutzler, P., Feller, S., Voss, J., Hofler, H. and Becker, K.F. (2000) Tumor-derived mutated E-cadherin influences beta-catenin localization and increases susceptibility to actin cytoskeletal changes induced by pervanadate [In Process Citation]. *Cell Adhes. Commun.* **7**: 391–408.

Luinetti, O., Fiocca, R., Villani, L., Alberizzi, P., Ranzani, G.N. and Solcia, E. (1998) Genetic pattern, histological structure, and cellular phenotype in early and advanced gastric cancers: evidence for structure-related genetic subsets and for loss of glandular structure during progression of some tumors [see comments]. *Hum Pathol* **29**: 702–709.

Luk, G. (1998) Tumors of the stomach. In M. Feldman, B. Scharschmidt and M. Sleisenger (eds). *Sleisenger and Fordtran's Gastrointestinal and Liver Disease*. Philadelphia. W.B. Saunders Company. 733–757.

Lynch, H. and Smyrk, T. (1996) Hereditary nonpolyposis colorectal cancer (Lynch syndrome): an updated review. *Cancer* **78**: 1149–1167.

Machado, J., Soares, P., Carneiro, F., Rocha, A., Beck, S., Blin, N., Berx, G. and Sobrinho-Simoes, M. (1999) E-cadherin gene mutations provide a genetic basis for the phenotypic divergence of mixed gastric carcinomas. *Lab. Invest.* **79**: 459–465.

Maesawa, C., Tamura, G., Suzuki, Y., Ogasawara, S., Sakata, K., Kashiwaba, M. and Satodate, R. (1995) The sequential accumulation of genetic alterations characteristic of the colorectal adenoma-carcinoma sequence does not occur between gastric adenoma and adenocarcinoma [see comments]. *J. Pathol.* **176**: 249–258.

Markowitz, S. and Roberts, A. (1996) Tumor suppressor activity of the TGF-β pathway in human cancers. *Cytokine and Growth Factor Reviews* **7**: 93–102.

Markowitz, S., Wang, J., Myeroff, L. *et al.* (1995) Inactivation of the type II TGF-β receptor in colon cancer cells with microsatellite instability. *Science* **268**: 1336–1338.

McKie, A.B., Filipe, M.I. and Lemoine, N.R. (1993) Abnormalities affecting the APC and MCC tumour suppressor gene loci on chromosome 5q occur frequently in gastric cancer but not in pancreatic cancer. *Int. J. Cancer* **55**: 598–603.

Mettlin, C. (1988) Epidemiologic studies in gastric adenocarcinoma. In H. Douglass (ed). *Gastric Cancer*. New York, Churchill Livingstone. 1–25.

Miki, H., Ohmori, M., Perantoni, A.O. and Enomoto, T. (1991) K-ras activation in gastric epithelial tumors in Japanese. *Cancer Lett.* **58**: 107–113.

Ming, S. and Goldman, H. (1965) Gastric polyps. A histogenetic classification and its relation to carcinoma. *Cancer* **18**: 721–726.

Miyazawa, K., Tsubouchi, H., Naka, D. *et al.* (1989) Molecular cloning and sequence analysis of cDNA for human hepatocyte growth factor. *Biochem. Biophys. Res. Commun.* **163**: 967–973.

Moon, R.T., Brown, J.D., Yang-Snyder, J.A. and Miller, J.R. (1997) Structurally related receptors and antagonists compete for secreted Wnt ligands. *Cell* **88**: 725–728.

Morin, P.J., Sparks, A.B., Korinek, V., Barker, N., Clevers, H., Vogelstein, B. and Kinzler, K.W. (1997) Activation of beta-catenin-Tcf signaling in colon cancer by mutations in beta-catenin or APC [see comments]. *Science* 275: 1787–1790.

Munemitsu, S., Albert, I., Souza, B., Rubinfeld, B. and Polakis, P. (1995) Regulation of intracellular beta-catenin levels by the adenomatous polyposis coli (APC) tumor-suppressor protein. *Proc. Natl Acad. Sci. USA* 92: 3046–3050.

Munemitsu, S., Albert, I., Rubinfeld, B. and Polakis, P. (1996) Deletion of an amino-terminal sequence beta-catenin in vivo and promotes hyperphosphorylation of the adenomatous polyposis coli tumor suppressor protein. *Mol. Cell. Biol.* 16: 4088–4094.

Muraoka, M., Konishi, M., Kikuchi-Yanoshita, R., Tanaka, K., Shitara, N., Chong, J.M., Iwama, T. and Miyaki, M. (1996) p300 gene alterations in colorectal and gastric carcinomas. *Oncogene* 12: 1565–1569.

Muta, H., Noguchi, M., Kanai, Y., Ochiai, A., Nawata, H. and Hirohashi, S. (1996) E-cadherin gene mutations in signet ring cell carcinoma of the stomach. *Japan J. Cancer Res.* 87: 843–848.

Myeroff, L., Parsons, R., Kim, S.-J. *et al.* (1995) A transforming growth factor β receptor type II gene mutation common in colon and gastric but rare in endometrial cancers with microsatellite instability. *Cancer Res.* 55: 5545–5547.

Nagase, H. and Nakamura, Y. (1993) Mutations of the APC (adenomatous polyposis coli) gene. *Hum. Mutat.* 2: 425–434.

Nakajima, M., Sawada, H., Yamada, Y. *et al.* (1999) The prognostic significance of amplification and overexpression of c-met and c-erb B-2 in human gastric carcinomas [see comments]. *Cancer* 85: 1894–1902.

Nakamura, T., Nishizawa, T., Hagiya, M., Seki, T., Shimonishi, M., Sugimura, A., Tashiro, K. and Shimizu, S. (1989) Molecular cloning and expression of human hepatocyte growth factor. *Nature* 342: 440–443.

Nakashima, H., Inoue, H., Mori, M., Ueo, H., Ikeda, M. and Akiyoshi, T. (1994) Microsatellite instability in Japanese gastric cancer. *Cancer* 75: 1503–1507.

Nakatani, H., Tahara, E., Yoshida, T. *et al.* (1986) Detection of amplified DNA sequences in gastric cancers by a DNA renaturation method in gel. *Japan J. Cancer Res.* 77: 849–853.

Nakatani, H., Sakamoto, H., Yoshida, T., Yokota, J., Tahara, E., Sugimura, T. and Terada, M. (1990) Isolation of an amplified DNA sequence in stomach cancer. *Japan J. Cancer Res.* 81: 707–710.

Nanus, D.M., Kelsen, D.P., Mentle, I.R., Altorki, N. and Albino, A.P. (1990) Infrequent point mutations of ras oncogenes in gastric cancers. *Gastroenterology* 98: 955–960.

Nessling, M., Solinas-Toldo, S., Wilgenbus, K.K., Borchard, F. and Lichter, P. (1998) Mapping of chromosomal imbalances in gastric adenocarcinoma revealed amplified protooncogenes MYCN, MET, WNT2, and ERBB2. *Genes Chromosomes Cancer* 23: 307–316.

Nogueira, A.M., Carneiro, F., Seruca, R., Cirnes, L., Veiga, I., Machado, J.C. and Sobrinho-Simoes, M. (1999) Microsatellite instability in hyperplastic and adenomatous polyps of the stomach. *Cancer* 86: 1649–1656.

Nomura, N., Yamamoto, T., Toyoshima, K. *et al.* (1986) DNA amplification of the c-myc and c-erbB-1 genes in a human stomach cancer. *Japan J. Cancer Res.* 77: 1188–1192.

Nozawa, H., Oda, E., Ueda, S., Tamura, G., Maesawa, C., Muto, T., Taniguchi, T. and Tanaka, N. (1998) Functionally inactivating point mutation in the tumor-suppressor IRF-1 gene identified in human gastric cancer. *Int. J. Cancer* 77: 522–527.

Ochi, H., Douglass, H.O., Jr. and Sandberg, A.A. (1986) Cytogenetic studies in primary gastric cancer. *Cancer Genet. Cytogenet.* 22: 295–307.

Ochiai, A. and Hirohashi, S. (1997) Multiple genetic alterations in gastric cancer. In T. Sugimura and M. Sasako (eds). *Gastric Cancer*. New York. Oxford University Press. 87–99.

Oda, T., Kanai, Y., Oyama, T., Yoshiura, K., Shimoyama, Y., Birchmeier, W., Sugimura, T. and Hirohashi, S. (1994) E-cadherin gene mutations in human gastric carcinoma cell lines. *Proc. Natl Acad. Sci. USA* 91: 1858–1862.

Offerhaus, G.J., Giardiello, F.M., Krush, A.J., Booker, S.V., Tersmette, A.C., Kelley, N.C. and Hamilton, S.R. (1992) The risk of upper gastrointestinal cancer in familial adenomatous polyposis [see comments]. *Gastroenterology* 102: 1980–1982.

Ohta, M., Inoue, H., Cotticelli, M.G. *et al.* (1996) The FHIT gene, spanning the chromosome 3p14.2 fragile site and renal carcinoma-associated t(3; 8) breakpoint, is abnormal in digestive tract cancers. *Cell* 84: 587–597.

Ohue, M., Tomita, N., Monden, T. *et al.* (1996) Mutations of the transforming growth factor beta type II receptor gene and microsatellite instability in gastric cancer. *Int. J. Cancer* **68**: 203–306.

Oliveira, C., Seruca, R., Seixas, M. and Sobrinho-Simoes, M. (1998) The clinicopathological features of gastric carcinomas with microsatellite instability may be mediated by mutations of different "target genes": a study of the TGFbeta RII, IGFII R, and BAX genes. *Am. J. Pathol.* **153**: 1211–1219.

Ooi, A., Kobayashi, M., Mai, M. and Nakanishi, I. (1998) Amplification of c-erbB-2 in gastric cancer: detection in formalin-fixed, paraffin-embedded tissue by fluorescence in situ hybridization. *Lab. Invest.* **78**: 345–351.

Ottini, L., Palli, D., Falchetti, M. *et al.* (1997) Microsatellite instability in gastric cancer is associated with tumor location and family history in a high-risk population from Tuscany. *Cancer Res.* **57**: 4523–4529.

Panani, A.D., Ferti, A., Malliaros, S. and Raptis, S. (1995) Cytogenetic study of 11 gastric adenocarcinomas. *Cancer Genet. Cytogenet.* **81**: 169–172.

Park, J.B., Rhim, J.S., Park, S.C., Kimm, S.W. and Kraus, M.H. (1989) Amplification, overexpression, and rearrangement of the erbB-2 protooncogene in primary human stomach carcinomas. *Cancer Res.* **49**: 6605–6609.

Park, J.G., Frucht, H., LaRocca, R.V. *et al.* (1990) Characteristics of cell lines established from human gastric carcinoma. *Cancer Res.* **50**: 2773–2780.

Park, J.G., Park, K.J., Ahn, Y.O., Song, I.S., Choi, K.W., Moon, H.Y., Choo, S.Y. and Kim, J.P. (1992) Risk of gastric cancer among Korean familial adenomatous polyposis patients. Report of three cases. *Dis. Colon. Rectum.* **35**: 996–998.

Park, K., Kim, S.-J., Bang, Y.-J., Park, J.-G., Kim, N.-K., Roberts, A. and Sporn, M. (1994) Genetic changes in the transforming growth factor β (TGF-β) type II receptor gene in human gastric cancer cells: correlation with sensitivity to growth inhibition by TGF-β. *Proc. Natl Acad. Sci. USA* **91**: 8772–8776.

Park, W.S., Moon, Y.W., Yang, Y.M. *et al.* (1998) Mutations of the STK11 gene in sporadic gastric carcinoma. *Int. J. Oncol.* **13**: 601–604.

Park, W.S., Oh, R.R., Park, J.Y. *et al.* (1999) Frequent somatic mutations of the beta-catenin gene in intestinal-type gastric cancer. *Cancer Res.* **59**: 4257–4260.

Park, W.S., Oh, R.R., Kim, Y.S., Park, J.Y., Shin, M.S., Lee, H.K., Lee, S.H., Yoo, N.J. and Lee, J.Y. (2000) Absence of mutations in the kinase domain of the Met gene and frequent expression of Met and HGF/SF protein in primary gastric carcinomas. *Apmis.* **108**: 195–200.

Parkin, D., Pisani, P. and Ferlay, J. (1999) Global cancer statistics. *CA* **49**: 33–64.

Parsons, R., Myeroff, L., Liu, B., Willson, J., Markowitz, S., Kinzler, K. and Vogelstein, B. (1995) Microsatellite instability and mutations of the transforming growth factor β type II receptor gene in colorectal cancer. *Cancer Res.* **55**: 5548–5550.

Ponzetto, C., Giordano, S., Peverali, F., Della Valle, G., Abate, M.L., Vaula, G. and Comoglio, P.M. (1991) c-met is amplified but not mutated in a cell line with an activated met tyrosine kinase. *Oncogene* **6**: 553–559.

Powell, S.M. (1998) Stomach cancer. In B. Vogelstein and K. Kinzler (eds). *The Genetic Basis of Human Cancer*. New York, The McGraw-Hill Companies. 647–652.

Powell, S.M., Cummings, O.W., Mullen, J.A. *et al.* (1996) Characterization of the APC gene in sporadic gastric adenocarcinomas. *Oncogene* **12**: 1953–1959.

Powell, S., Harper, J., Hamilton, S., Robinson, C. and Cummings, O. (1997) Inactivation of *Smad4* in gastric carcinomas. *Cancer Res.* **57**: 4221–4224.

Renault, B., Calistri, D., Buonanti, G., Nanni, O., Amadori, D. and Ranzani, G. (1996) Microsatellite instability and mutations of p53 and TGF-β RII genes in gastric cancer. *Hum. Genet.* **98**: 601–607.

Rhyu, M.-G., Park, W.-S. and Meltzer, S. (1994a) Microsatellite instability occurs frequently in human gastric carcinoma. *Oncogene* **9**: 29–32.

Rhyu, M.G., Park, W.S., Jung, Y.J., Choi, S.W. and Meltzer, S.J. (1994b) Allelic deletions of MCC/APC and p53 are frequent late events in human gastric carcinogenesis. *Gastroenterology* **106**: 1584–1588.

Richards, F., McKee, S., Rajpar, M., Cole, T., Evans, D., Jankowski, J., McKeown, C., Sanders, D. and Maher, E. (1999) Germline E-cadherin gene (CDH1) mutations predispose to familial gastric cancer and colorectal cancer. *Hum. Mol. Genet.* **8**: 607–610.

Riggins, G., Thiagalingam, S., Rozenblum, E. *et al.* (1996) *Mad*-related genes in the human. *Nat. Genet.* **13**: 347–349.

Rodriguez, E., Rao, P.H., Ladanyi, M., Altorki, N., Albino, A.P., Kelsen, D.P., Jhanwar, S.C. and Chaganti, R.S. (1990) 11p13–15 is a specific region of chromosomal rearrangement in gastric and esophageal adenocarcinomas. *Cancer Res.* **50**: 6410–6416.

Rokkas, T., Filipe, M.I. and Sladen, G.E. (1991) Detection of an increased incidence of early gastric cancer in patients with intestinal metaplasia type III who are closely followed up. *Gut* **32**: 1110–1113.

Rowan, A., Bataille, V., MacKie, R., Healy, E., Bicknell, D., Bodmer, W. and Tomlinson, I. (1999) Somatic mutations in the Peutz-Jeghers (LKB1/STKII) gene in sporadic malignant melanomas. *J. Invest. Dermatol.* **112**: 509–511.

Rowlands, D.C., Ito, M., Mangham, D.C. *et al.* (1993) Epstein-Barr virus and carcinomas: rare association of the virus with gastric adenocarcinomas. *Br. J. Cancer* **68**: 1014–1019.

Rubinfeld, B., Albert, I., Porfiri, E., Munemitsu, S. and Polakis, P. (1997a) Loss of beta-catenin regulation by the APC tumor suppressor protein correlates with loss of structure due to common somatic mutations of the gene. *Cancer Res.* **57**: 4624–4630.

Rubinfeld, B., Robbins, P., El-Gamil, M., Albert, I., Porfiri, E. and Polakis, P. (1997b) Stabilization of beta-catenin by genetic defects in melanoma cell lines [see comments]. *Science* **275**: 1790–1792.

Rugge, M., Farinati, F., Baffa, R., Sonego, F., Di Mario, F., Leandro, G. and Valiante, F. (1994) Gastric epithelial dysplasia in the natural history of gastric cancer: a multicenter prospective follow-up study. Interdisciplinary Group on Gastric Epithelial Dysplasia [see comments]. *Gastroenterology* **107**: 1288–1296.

Sakakura, C., Mori, T., Sakabe, T. *et al.* (1999) Gains, losses, and amplifications of genomic materials in primary gastric cancers analyzed by comparative genomic hybridization. *Genes Chromosomes Cancer* **24**: 299–305.

Sakamoto, H., Mori, M., Taira, M., Yoshida, T., Matsukawa, S., Shimizu, K., Sekiguchi, M., Terada, M. and Sugimura, T. (1986) Transforming gene from human stomach cancers and a noncancerous portion of stomach mucosa. *Proc. Natl Acad. Sci. USA* **83**: 3997–4001.

Sano, T., Tsujino, T., Yoshida, K., Nakayama, H., Haruma, K., Ito, H., Nakamura, Y., Kajiyama, G. and Tahara, E. (1991) Frequent loss of heterozygosity on chromosomes 1q, 5q, and 17p in human gastric carcinomas. *Cancer Res.* **51**: 2926–2931.

Sato, T., Abe, K., Kurose, A., Uesugi, N., Todoroki, T. and Sasaki, K. (1997) Amplification of the c-erbB-2 gene detected by FISH in gastric cancers. *Pathol. Int.* **47**: 179–182.

Schutte, M., Hruban, R., Hedrick, L. *et al.* (1996) DPC4 gene in various tumor types. *Cancer Res.* **56**: 2527–2530.

Seki, T., Fujii, G., Mori, S., Tamaoki, N. and Shibuya, M. (1985) Amplification of c-yes-1 proto-oncogene in a primary human gastric cancer. *Japan J. Cancer Res.* **76**: 907–910.

Sekiguchi, M. and Suzuki, T. (1995) Gastric-tumor cell lines. In H. RJ, J.-G. Park and A. Gazdar (eds). *Gastric-tumor Cell Lines*. San Diego. Academic Press. 287–316.

Semba, K., Kamata, N., Toyoshima, K. and Yamamoto, T. (1985) A v-erbB-related protooncogene, c-erbB-2, is distinct from the c-erbB-1/epidermal growth factor-receptor gene and is amplified in a human salivary gland adenocarcinoma. *Proc. Natl Acad. Sci. USA* **82**: 6497–6501.

Semba, S., Yokozaki, H., Yamamoto, S., Yasui, W. and Tahara, E. (1996) Microsatellite instability in precancerous lesions and adenocarcinomas of the stomach. *Cancer* **77**: 1620–1627.

Sepulveda, A.R., Santos, A.C., Yamaoka, Y., Wu, L., Gutierrez, O., Kim, J.G. and Graham, D.Y. (1999) Marked differences in the frequency of microsatellite instability in gastric cancer from different countries. *Am. J. Gastroenterol* **94**: 3034–3038.

Seruca, R., Castedo, S., Correia, C., Gomes, P., Carneiro, F., Soares, P., de Jong, B. and Sobrinho-Simoes, M. (1993) Cytogenetic findings in eleven gastric carcinomas. *Cancer Genet. Cytogenet.* **68**: 42–48.

Seruca, R., Santos, N.R., David, L. *et al.* (1995a) Sporadic gastric carcinomas with microsatellite instability display a particular clinicopathologic profile. *Int. J. Cancer* **64**: 32–36.

Seruca, R., Suijkerbuijk, R.F., Gartner, F. *et al.* (1995b) Increasing levels of MYC and MET co-amplification during tumor progression of a case of gastric cancer. *Cancer Genet. Cytogenet.* **82**: 140–145.

Shibata, D. and Weiss, L.M. (1992) Epstein-Barr virus-associated gastric adenocarcinoma. *Am. J. Pathol.* **140**: 769–774.

Shim, Y.H., Kang, G.H. and Ro, J.Y. (2000) Correlation of p16 hypermethylation with p16 protein loss in sporadic gastric carcinomas. *Lab. Invest.* **80**: 689–695.

Shinmura, K., Kohno, T., Takahashi, M. *et al.* (1999) Familial gastric cancer: clinicopathological characteristics, RER phenotype and germline p53 and E-cadherin mutations. *Carcinogenesis* **20**: 1127–1131.

Shitara, Y., Yokozaki, H., Yasui, W., Takenoshita, S., Nagamachi, Y. and Tahara, E. (1998) Mutation of the transforming growth factor-beta type II receptor gene is a rare event in human sporadic gastric carcinomas. *Int. J. Oncol.* **12**: 1061–1065.

Shitara, Y., Yokozaki, H., Yasui, W., Takenoshita, S., Kuwano, H., Nagamachi, Y. and Tahara, E. (1999) No mutations of the Smad2 gene in human sporadic gastric carcinomas. *Japan J. Clin. Oncol.* **29**: 3–7.

Shtutman, M., Zhurinsky, J., Simcha, I., Albanese, C., D'Amico, M., Pestell, R. and Ben-Ze'ev, A. (1999) The cyclin D1 gene is a target of the beta-catenin/LEF-1 pathway. *Proc. Natl Acad. Sci. USA* **96**: 5522–5527.

Sipponen, P., Kekki, M., Haapakoski, J., Ihamaki, T. and Siurala, M. (1985) Gastric cancer risk in chronic atrophic gastritis: statistical calculations of cross-sectional data. *Int. J. Cancer* **35**: 173–177.

Smith, D.P., Spicer, J., Smith, A., Swift, S. and Ashworth, A. (1999) The mouse Peutz-Jeghers syndrome gene Lkb1 encodes a nuclear protein kinase. *Hum. Mol. Genet.* **8**: 1479–1485.

Soman, N.R., Correa, P., Ruiz, B.A. and Wogan, G.N. (1991) The TPR-MET oncogenic rearrangement is present and expressed in human gastric carcinoma and precursor lesions. *Proc. Natl Acad. Sci. USA* **88**: 4892–4896.

Somasundaram, K. (2000) Tumor suppressor p53: regulation and function. *Front Biosci.* **5**: D424–437.

Staal, S.P. (1987) Molecular cloning of the akt oncogene and its human homologs AKT1 and AKT2: amplification of AKT1 in a primary human gastric adenocarcinoma. *Proc. Natl Acad. Sci. USA* **84**: 5034–5037.

Stoker, M., Gherardi, E., Perryman, M. and Gray, J. (1987) Scatter factor is a fibroblast-derived modulator of epithelial cell mobility. *Nature* **327**: 239–242.

Stone, J., Bevan, S., Cunningham, D., Hill, A., Rahman, N., Peto, J., Marossy, A. and Houlston, R. (1999) Low frequency of germline E-cadherin mutations in familial and nonfamilial gastric cancer. *Brit. J. Cancer* **79**: 1935–1937.

Strickler, J., Zheng, J., Shu, Q., Burgart, L., Alberts, S. and Shibata, D. (1994) *p53* mutations and microsatellite instability in sporadic gastric cancer: when guardians fail. *Cancer Res.* **54**: 4750–4755.

Su, G.H., Hruban, R.H., Bansal, R.K. *et al.* (1999) Germline and somatic mutations of the STK11/LKB1 Peutz-Jeghers gene in pancreatic and biliary cancers. *Am. J. Pathol.* **154**: 1835–1840.

Sukumar, S. (1990) An experimental analysis of cancer: role of ras oncogenes in multistep carcinogenesis. *Cancer Cells* **2**: 199–204.

Suzuki, S., Tenjin, T., Watanabe, H., Matsushima, S., Shibuya, T. and Tanaka, S. (1997) Low level c-myc gene amplification in gastric cancer detected by dual color fluorescence in situ hybridization analysis. *J. Surg. Oncol.* **66**: 173–178.

Tamura, G., Kihana, T., Nomura, K., Terada, M., Sugimura, T. and Hirohashi, S. (1991) Detection of frequent p53 gene mutations in primary gastric cancer by cell sorting and polymerase chain reaction single-strand conformation polymorphism analysis. *Cancer Res.* **51**: 3056–3058.

Tamura, G., Sakata, K., Maesawa, C., Suzuki, Y., Terashima, M.K.S., Sekiyama, S., Suzuki, A., Eda, Y. and Satodate, R. (1995) Microsatellite alterations in adenoma and differentiated adenocarcinoma of the stomach. *Cancer Res.* **55**: 1933–1936.

Tamura, G., Ogasawara, S., Nishizuka, S., Sakata, K., Maesawa, C., Suzuki, Y., Terashima, M., Saito, K. and Satodate, R. (1996a) Two distinct regions of deletion on the long arm of chromosome 5 in differentiated adenocarcinomas of the stomach. *Cancer Res.* **56**: 612–615.

Tamura, G., Sakata, K., Nishizuka, S., Maesawa, C., Suzuki, Y., Iwaya, T., Terashima, M., Saito, K. and Satodate, R. (1996b) Inactivation of the E-cadherin gene in primary gastric carcinomas and gastric carcinoma cell lines. *Japan J. Cancer Res.* **87**: 1153–1159.

Tamura, G., Sakata, K., Nishizuka, S., Maesawa, C., Suzuki, Y., Iwaya, T., Terashima, M., Saito, K. and Satodate, R. (1997) Analysis of the fragile histidine triad gene in primary gastric carcinomas and gastric carcinoma cell lines. *Genes Chromosomes Cancer* **20**: 98–102.

Tamura, G., Yin, J., Wang, S. *et al.* (2000) E-cadherin gene promoter hypermethylation in primary human gastric carcinomas. *J. Natl. Cancer Inst.* **92**: 569–573.

Tomlinson, I., Novelli, M. and Bodmer, W. (1996) The mutation rate and cancer. *Proc. Natl Acad. Sci. USA* **93**: 14800–14803.

Toyota, M., Ahuja, N., Ohe-Toyota, M., Herman, J., Baylin, S. and Issa, J. (1999a) CpG island methylator phenotype in colorectal cancer. *Proc. Natl Acad. Sci. USA* **96**: 8681–8686.

Toyota, M., Ahuja, N., Suzuki, H., Itoh, F., Ohe-Toyota, M., Imai, K., Baylin, S. and Issa, J.-P. (1999b) Aberrant methylation in gastric cancer associated with the CpG island methylator phenotype. *Cancer Res.* **59**: 5438–5442.

Tsuchiya, T., Ueyama, Y., Tamaoki, N., Yamaguchi, S. and Shibuya, M. (1989) Co-amplification of c-myc and c-erbB-2 oncogenes in a poorly differentiated human gastric cancer. *Japan J. Cancer Res.* **80**: 920–923.

Tsuda, T., Tahara, E., Kajiyama, G., Sakamoto, H., Terada, M. and Sugimura, T. (1989) High incidence of coamplification of hst-1 and int-2 genes in human esophageal carcinomas. *Cancer Res.* **49**: 5505–5508.

Tsugawa, K., Fushida, S. and Yonemura, Y. (1993) Amplification of the c-erbB-2 gene in gastric carcinoma: correlation with survival. *Oncology* **50**: 418–425.

Tsugawa, K., Yonemura, Y., Hirono, Y., Fushida, S., Kaji, M., Miwa, K., Miyazaki, I. and Yamamoto, H. (1998) Amplification of the c-met, c-erbB-2 and epidermal growth factor receptor gene in human gastric cancers: correlation to clinical features. *Oncology* **55**: 475–481.

Tsujimoto, H., Sugihara, H., Hagiwara, A. and Hattori, T. (1997) Amplification of growth factor receptor genes and DNA ploidy pattern in the progression of gastric cancer. *Virchows Arch.* **431**: 383–389.

Uchino, S., Noguchi, M., Hirota, T., Itabashi, M., Saito, T., Kobayashi, M. and Hirohashi, S. (1992a) High incidence of nuclear accumulation of p53 protein in gastric cancer. *Japan J. Clin. Oncol.* **22**: 225–231.

Uchino, S., Tsuda, H., Noguchi, M., Yokota, J., Terada, M., Saito, T., Kobayashi, M., Sugimura, T. and Hirohashi, S. (1992b) Frequent loss of heterozygosity at the DCC locus in gastric cancer. *Cancer Res.* **52**: 3099–3102.

Uchino, S., Noguchi, M., Ochiai, A., Saito, T., Kobayashi, M. and Hirohashi, S. (1993) p53 mutation in gastric cancer: a genetic model for carcinogenesis is common to gastric and colorectal cancer. *Int. J. Cancer* **54**: 759–764.

Utsunomiya, J. (1990) The concept of hereditary colorectal cancer and the implications of its study. In J. Utsunomiya and H. Lynch (eds). *Hereditary colorectal cancer.* Tokyo, Springer-Verlag. 3.

Vasen, H.F., Wijnen, J.T., Menko, F.H. *et al.* (1996) Cancer risk in families with hereditary nonpolyposis colorectal cancer diagnosed by mutation analysis [published erratum appears in *Gastroenterology* 1996 Nov; **111**(5): 1402]. *Gastroenterology* **110**: 1020–1027.

Veigl, M., Kasturi, L., Olechnowicz, J. *et al.* (1998) Biallelic inactivation of hMLH1 by epigenetic gene silencing, a novel mechanism causing human MSI cancers. *Proc. Natl Acad. Sci. USA* **95**: 8698–8702.

Victor, T., Du Toit, R., Jordaan, A.M., Bester, A.J. and van Helden, P.D. (1990) No evidence for point mutations in codons 12, 13, and 61 of the ras gene in a high-incidence area for esophageal and gastric cancers. *Cancer Res.* **50**: 4911–4914.

Vleminckx, K., Vakaet, L., Jr., Mareel, M., Fiers, W. and van Roy, F. (1991) Genetic manipulation of E-cadherin expression by epithelial tumor cells reveals an invasion suppressor role. *Cell* **66**: 107–119.

Vos, C.B., Cleton-Jansen, A.M., Berx, G., de Leeuw, W.J., ter Haar, N.T., van Roy, F., Cornelisse, C.J., Peterse, J.L. and van de Vijver, M.J. (1997) E-cadherin inactivation in lobular carcinoma in situ of the breast: an early event in tumorigenesis. *Br. J. Cancer* **76**: 1131–1133.

Wang, T. (1995) Biology of gastric cancer. in A. Rustgi (eds). *Gastrointestinal Cancer*. New York. Lippincott-Raven. 243–259.

Wang, N. and Perkins, K.L. (1984) Involvement of band 3p14 in t(3; 8) hereditary renal carcinoma. *Cancer Genet. Cytogenet.* **11**: 479–481.

Wang, Z.J., Churchman, M., Campbell, I.G., Xu, W.H., Yan, Z.Y., McCluggage, W.G., Foulkes, W.D. and Tomlinson, I.P. (1999) Allele loss and mutation screen at the Peutz-Jeghers (LKB1) locus (19p13.3) in sporadic ovarian tumours. *Br. J. Cancer* **80**: 70–72.

Wee, A., Kang, J.Y. and Teh, M. (1992) *Helicobacter pylori* and gastric cancer: correlation with gastritis, intestinal metaplasia, and tumour histology. *Gut* **33**: 1029–1032.

Weidner, K.M., Behrens, J., Vandekerckhove, J. and Birchmeier, W. (1990) Scatter factor: molecular characteristics and effect on the invasiveness of epithelial cells. *J. Cell. Biol.* **111**: 2097–2108.

Westerman, A.M., Entius, M.M., Boor, P.P. *et al.* (1999) Novel mutations in the LKB1/STK11 gene in Dutch Peutz-Jeghers families. *Hum. Mutat.* **13**: 476–481.

Wirtz, H.C., Muller, W., Noguchi, T., Scheven, M., Ruschoff, J., Hommel, G. and Gabbert, H.E. (1998) Prognostic value and clinicopathological profile of microsatellite instability in gastric cancer. *Clin. Cancer Res.* **4**: 1749–1754.

Wu, M.S., Lee, C.W., Shun, C.T., Wang, H.P., Lee, W.J., Sheu, J.C. and Lin, J.T. (1998a) Clinicopathological significance of altered loci of replication error and microsatellite instability-associated mutations in gastric cancer. *Cancer Res.* **58**: 1494–1497.

Wu, M.S., Shun, C.T., Sheu, J.C., Wang, H.P., Wang, J.T., Lee, W.J., Chen, C.J., Wang, T.H. and Lin, J.T. (1998b) Overexpression of mutant p53 and c-erbB-2 proteins and mutations of the p15 and p16 genes in human gastric carcinoma: with respect to histological subtypes and stages. *J. Gastroenterol Hepatol.* **13**: 305–310.

Wu, M.S., Lee, C.W., Shun, C.T., Wang, H.P., Lee, W.J., Chang, M.C., Sheu, J.C. and Lin, J.T. (2000a) Distinct clinicopathologic and genetic profiles in sporadic gastric cancer with different mutator phenotypes. *Genes Chromosomes Cancer* **27**: 403–411.

Wu, M.S., Shun, C.T., Wu, C.C., Hsu, T.Y., Lin, M.T., Chang, M.C., Wang, H.P. and Lin, J.T. (2000b) Epstein-Barr virus-associated gastric carcinomas: relation to *H. pylori* infection and genetic alterations. *Gastroenterology* **118**: 1031–1038.

Xiao, S., Geng, J.S., Feng, X.L., Liu, X.Q., Liu, Q.Z. and Li, P. (1992) Cytogenetic studies of eight primary gastric cancers. *Cancer Genet. Cytogenet.* **58**: 79–84.

Yamamoto, H., Perez-Piteira, J., Yoshida, T., Terada, M., Itoh, F., Imai, K. and Perucho, M. (1999) Gastric cancers of the microsatellite mutator phenotype display characteristic genetic and clinical features. *Gastroenterology* **116**: 1348–1357.

Yanagihara, K., Seyama, T., Tsumuraya, M., Kamada, N. and Yokoro, K. (1991) Establishment and characterization of human signet ring cell gastric carcinoma cell lines with amplification of the c-myc oncogene. *Cancer Res.* **51**: 381–386.

Yang, H.K., Kang, S.H., Kim, Y.S., Won, K., Bang, Y.J. and Kim, S.J. (1999) Truncation of the TGF-beta type II receptor gene results in insensitivity to TGF-beta in human gastric cancer cells. *Oncogene* **18**: 2213–2219.

Yang-Feng, T., Schechter, A., Weinberg, R. and Francke, U. (1985) Oncogene from rat neuro/glioblastomas (human gene symbol NGL) is located on the proximal long arm of human chromosome 17 and EGFR is confirmed at 7p13-q11.2. *Cytogenet. Cell Genet.* **40**: 784.

Ylikorkala, A., Avizienyte, E., Tomlinson, I.P. *et al.* (1999) Mutations and impaired function of LKB1 in familial and non-familial Peutz-Jeghers syndrome and a sporadic testicular cancer. *Hum. Mol. Genet.* **8**: 45–51.

Yokota, J., Yamamoto, T., Miyajima, N., Toyoshima, K., Nomura, N., Sakamoto, H., Yoshida, T., Terada, M. and Sugimura, T. (1988) Genetic alterations of the c-crbB-2 oncogene occur frequently in tubular adenocarcinoma of the stomach and are often accompanied by amplification of the v-erbA homolog. *Oncogene* **2**: 283–287.

Yokozaki, H., Kuniyasu, H., Kitadai, Y., Nishimura, K., Todo, H., Ayhan, A., Yasui, W., Ito, H. and Tahara, E. (1992) p53 point mutations in primary human gastric carcinomas. *J. Cancer Res. Clin. Oncol.* **119**: 67–70.

Yonemura, Y., Ninomiya, I., Ohoyama, S. *et al.* (1991) Expression of c-erbB-2 oncoprotein in gastric carcinoma. Immunoreactivity for c-erbB-2 protein is an independent indicator of poor short-term prognosis in patients with gastric carcinoma. *Cancer* **67**: 2914–2918.

Yonemura, Y., Ninomiya, I., Tsugawa, K., Fushida, S., Fujimura, T., Miyazaki, I., Uchibayashi, T., Endou, Y. and Sasaki, T. (1998) Prognostic significance of c-erbB-2 gene expression in the poorly differentiated type of adenocarcinoma of the stomach. *Cancer Detect. Prev.* **22**: 139–146.

Yoon, K.A., Ku, J.L., Yang, H.K., Kim, W.H., Park, S.Y. and Park, J.G. (1999) Germline mutations of E-cadherin gene in Korean familial gastric cancer patients. *J. Hum. Genet.* **44**: 177–180.

Yoon, K.A., Ku, J.L., Choi, H.S. *et al.* (2000) Germline mutations of the STK11 gene in Korean Peutz-Jeghers syndrome patients. *Br. J. Cancer* **82**: 1403–1406.

Yoshida, M.C., Wada, M., Satoh, H. *et al.* (1988) Human HST1 (HSTF1) gene maps to chromosome band 11q13 and coamplifies with the INT2 gene in human cancer. *Proc. Natl. Acad. Sci. USA* **85**: 4861–4864.

Zarnegar, R., Muga, S., Enghild, J. and Michalopoulos, G. (1989) NH2-terminal amino acid sequence of rabbit hepatopoietin A, a heparin-binding polypeptide growth factor for hepatocytes. *Biochem. Biophys. Res. Commun.* **163**: 1370–1376.

7

Multiple endocrine neoplasia type 1

Bin Tean Teh

1. Introduction

The molecular genetics of Multiple Endocrine Neoplasia Type 1 (MEN1) fulfills the paradigm of a tumor suppressor gene consistent with Knudson's two-hit mutation theory. From its mapping in 1988, to its cloning in 1997 and since then, all data that has come out supports its role as a tumor suppressor gene. However, despite the wide range of germline and somatic mutations found in MEN1 patients and sporadic tumors, little is known about its functions. This article summarizes recent studies in the characterization of the function of MEN1 as well as a review of mutations in the *MEN1* gene.

2. Clinical manifestations

MEN1 is a familial cancer syndrome that is transmitted in an autosomal dominant pattern with an equal sex distribution and an almost complete penetrance. It is characterized by endocrine tumors of the parathyroid (90–97% of patients), enteropancreas (30–80%), and anterior pituitary (15–50%), the three principal MEN1-related tumors (Brandi *et al.*, 1987; Wermer, 1958). The MEN1-related parathyroid tumors are invariably multiglandular hyperplasia rather than solitary adenoma which is commonly found in sporadic primary hyperparathyroidism. Parathyroid carcinoma is not known to be associated with MEN1. Many of the pancreatic tumours identified in MEN1 patients behave as 'non-functioning' lesions. Gastrinomas are commonly multifocal, arising in both the pancreas and duodenal wall causing Zollinger-Ellison syndrome. Other pancreatic tumors include insulinoma, glucagonoma, somatostatinoma and VIPoma. MEN1-related pituitary tumors in most cases are either prolactin-secreting or nonfunctioning but growth hormone (GH)-secreting, adrenocorticotropic hormone (ACTH)-secreting and very rarely thyroid stimulating hormone (TSH)-secreting tumors have also been described.

Besides the three principal tumors, other endocrine and non-endocrine tumors (*Table 1*) are also found in MEN1 patients (Teh, 1998). Some of these tumors, such

---

Molecular Genetics of Cancer second edition, edited by J.K. Cowell.

Table 1. Endocrine and non-endocrine tumors in MEN1

Endocrine tumors	Non-endocrine tumors
Parathyroid hyperplasia/multiglandular disease	Cutaneous and visceral lipoma
Enteropancreatic endocrine tumors	Skin tumors: angiofibroma, collagenomas
Anterior pituitary tumors	Ependymoma
Adrenocortical tumors	Leiomyoma of gastrointestinal tract and
Carcinoids (thymic, bronchus, gastrointestinal tract)	kidney

as adrenocortical tumors, lipomata, and skin tumors, have been reported to occur in relatively high frequencies. The definition of MEN1 refers to one patient with at least two of the three principal tumors, plus one first-degree relative with one or more of these tumors. However, it is not uncommon to find the so-called MEN1 patients with two or three of the principal tumors but without a family history. These patients have been found to carry *de-novo* germline mutations.

Clinically, MEN1 manifests the hypersecretory effects of endocrine tumors, i.e., signs and symptoms associated with abnormally high levels of the hormones they secrete. For example, because of hyperparathryoidism, patients may present with signs and symptoms of hypercalcemia such as polydipsia, polyuria or renal calculi. The patients may also present with Zollinger-Ellison syndrome or intractable peptic ulcer disease as a result of hypergastrinemia from gastrinomas. If patients have a prolactinoma in the anterior pituitary gland, they may present with galactorrhea, menstrual irregularity or impotence as a result of high prolactin level.

The age of diagnosis is dependent on two factors: (a) if the patient is symptomatic or asymptomatic and diagnosed by biochemial screening, and (b) the type of tumor that the patient develops (Burgess *et al.*, 1998; Marx *et al.*, 1998; Trump *et al.*, 1996). The asymptomatic patients are diagnosed at an earlier age by biochemical screening compared with those who are symptomatic. For example, Trump *et al.* (1996) have shown that, in the symptomatic group, the cumulative percentages of patients who develop MEN1 are 18%, 52% and 78% at the ages of 20, 35 and 50 years, respectively. In the asymptomatic group who are diagnosed by biochemical screening, however, the cumulative percentages increase to 43%, 85% and 94% in the same age groups. In a study of the largest MEN1 kindred, it has been shown that, by the age of 20 years, two-thirds of patients are found to have primary hyperparathyroidism and, by age 30 years, the incidence increased to 95% (Burgess *et al.*, 1998). Both studies show that endocrine pancreatic tumors have two patterns: the gastrinoma occur commonly in the older group of patients (e.g., above 30 or 40) whereas insulinoma tends to occur in the young patients (e.g., below 30 or 40). For anterior pituitary tumors, most patients are diagnosed between their twenties and forties (Burgess *et al.*, 1998; Marx *et al.*, 1998; Trump *et al.*, 1996).

3. Positional cloning of the *MEN1* gene

Based on his epidemiological studies of retinoblastoma, Knudson put forward his two-hit mutation theory in 1971 which translated that familial cancer patients would inherit one mutated gene (first-hit) and acquire the second hit of mutation in the tissues involved, thereby eliminating the tumor suppressive activity of the gene and setting off tumor development. In 1978, Knudson proposed that the second hit could involve a loss of genetic material, which was first proven by Cavenee *et al.* (1983) in retinoblastoma. Their studies compared the constitutional and tumor genotypes and found one allele was lost in the tumor DNA, i.e., loss of heterozygosity (LOH). Subsequently they found that in familial tumors, the LOH invariably involves the wild-type allele, i.e., the allele inherited from the unaffected parent (Cavenee *et al.*, 1985). Based on these concepts, Larsson *et al.* (1988) set out to screen for LOH in MEN1-related tumors by using RFLP markers covering most chromosomes. They identified LOH in chromosome 11 and the ensuing linkage analysis performed on MEN1 families confirmed its linkage to chromosome 11q13. Following the mapping, the search for the gene went through the full process of positional cloning. It involved the development of: a) a reliable physical map, b) new genetic markers for family and tumor studies, and c) contigs (clones containing the whole sequence of a chromosomal region) that cover the MEN1 region. Combined studies of family linkage (to identify critical recombinants) and loss of heterozygosity in tumors (to identify critical deletions) refined the MEN1 region to a few hundred thousand base pairs. From the contigs that covered the MEN1 region, genes were identified by conventional cDNA screening, cDNA direct selection or direct sequencing followed by mutation analysis. A number of these genes were excluded before the *MEN1* gene was finally identified (Chandrasekharappa *et al.*, 1997; European Consortium on MEN1, 1997).

4. The *MEN1* gene

The MEN1 gene (*Figure 1*) contains 10 exons, but only exons 2–9 are transcribed encoding a 610-amino acid protein product which has an estimated weight of 67 kDa. The first exon is noncoding and constitutes most of the 111 nucleotide 5′ UTR. The exon sizes range from 41–1296 nucleotides and the introns from 79–1563 nucleotides. It has no homology to any previously known protein, making its function unpredictable. However, several lines of research have been undertaken to attempt to elucidate its functional roles.

5. The functions of the *MEN1* gene

Northern analysis of the *MEN1* gene has revealed two transcripts: one is a 2.9 kb transcript that is expressed in all tissues and the other a 4.2 kb transcript found in pancreas and thymus. A number of cancer cell lines that have been tested also reveal the expression of this transcript (Debelenko *et al.*, 2000; Srivatsan *et al.*, in press). Recently, Ikeo *et al.* (2000) have studied its expression in different organs by *in-situ* hybridization using a 150 base pair riboprobe containing an antisense *MEN1* cDNA sequence. It is expressed in a variety of adult human organs but most prominent in the proliferative phase of endometrium (as in contrast with

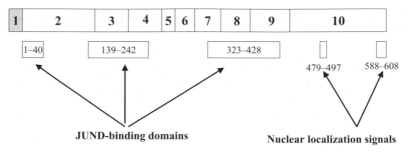

Figure 1. Diagram of the *MEN1* gene showing the three JUND-binding domains and two nuclear-localization signals.

faint expression in secretory phase of endometrium) and in parabasal cells of the esophageal mucosa. In the same study, DNA synthesis in the cultured cells is measured under different chemical exposure. DNA synthesis is inhibited when exposed to a DNA-cross-linking agent, but not to ultraviolet light suggesting that the protein product, called menin, may negatively regulate cell cycle under DNA damage.

Mainly by Western analysis, the expression of menin in relation to cell cycle has also been investigated by several groups. Using synchronized NIH 3T3 and HeLa cell lines, no variation can be found in the amount or size of menin throughout the cell cycle (Guru *et al.*, 1999; Wautot *et al.*, 2000). Using microarray technology, a study of global gene expression including menin by human fibroblasts in response to serum stimulation has not revealed any variation at different time intervals (Iyer *et al.*, 1999). In the pituitary cell line, GH4C1, however, menin expression has been shown to be 30–50% lower at the G_1–S boundary compared with G_0, but expression levels increase once the cell enters S phase. However, at the G_2–M phase, these cells express lower levels of menin. These data suggest that the menin expression is cell-cycle regulated in pituitary cells (Kaji *et al.*, 1999).

Menin has been found to be predominantly a nuclear protein using immunoflu-orescent-staining, subcellular fractionation and Western blotting (Guru *et al.*, 1998; Kaji *et al.*, 1999). Two nuclear localization signals (NLS) or regions involved in nuclear localization are identified in the C-terminus of menin: one as a stretch of 19 amino acids from amino acid position 479 to 497, and the other in 20 amino acids from position 588 to 608. It has been postulated that most of the truncating mutations resulting in the loss of both NLS may cause a shift of the truncated protein from nucleus to cytoplasm.

Using yeast two-hybrid systems, an interacting protein of menin has been isolated which turns out to be the AP1 transcription factor JUND (Agarwal *et al.*, 1999). The AP1 transcription factors have basic leucine zipper (bZip) domains that pair to bind DNA as a Y-shaped heterodimer. Menin has been found to speci-fically bind to JUND via its N-terminal, but not to the other members of JUN or FOS. Menin causes repression on JUND-activated transcription and one recent study has shown that this repression is released by the histone deacetylase inhibitor trichostatin A, suggesting that the repression is dependent on histone deacetylation (Gobl *et al.*, 1999). Some, but not all, MEN1 mutations disrupt its binding to JUND causing an increase in JUND-activated transcription. Based on the binding effects demonstrated by *MEN1* mutations from different locations, three binding domains have been identified: amino acids 1–40, 139–242 and 323–428 (*Figure 1*) and there are a number of disease-causing mutations that are located outside the binding domains. The identification of other interacting proteins, which may constitute parts of the menin-JUND complex or separate functional entities, may further elucidate the molecular mechanisms involved in MEN1 tumorigenesis.

To demonstrate the tumor suppressive role of menin, Kim *et al.* (1999) have studied the effect of menin expression in RAS transformed NIH-3T3 cells. The latter are used because, to date, no menin-null tumor cell line is available. When these transformed NIH-3T3 cells over-express menin, they revert to the more pre-transformed morphology and become less clonogenic. Furthermore, menin expression also reduces growth after injection of cells in nude mice. These data support the tumor suppressive role of the MEN1 gene.

6. Homologous genes

The *MEN1* murine homolog, *Men1*, has been characterized and is located in the mouse synteny region of chromosome 19 and shows 97% homology (Bassett *et al.*, 1999; Guru *et al.*, 1999, Karges *et al.*, 1999; Maruyama *et al.*, 1999; Stewart *et al.*, 1998). Like its human counterpart, it has 10 exons with two transcripts (3.2 and 2.8 kb), a result of alternative splicing in intron 1. During mouse embryogenesis, the *men1* activity is already detected at day 7. It is generally expressed in all tissues, although by day 17 it is most prominent in thymus, skeletal muscles, and the CNS system. In adult tissues, it is also generally expressed although most prominently in the testis (perinuclear spermatogonia), cerebral cortex (nerve cell nuclei), and thymus. Its universal expression pattern, especially in non-MEN1 related tissues, suggests its universal functional role, both in development and housekeeping.

To date, the MEN1 homologs from rat (Karges *et al.*, 1999; Maruyama *et al.*, 1999), zebrafish (Khodaei *et al.*, 1999; Manickam *et al.*, 2000), and drosophila (CG33266 in Flybase:http://flybase.bio.Indiana.edu: 7081/annot/) have also been identified and characterized which share 96.7%, 67%, 50% identity with human menin, respectively. These homologs will definitely facilitate future studies in elucidating their functions and roles in tumorigenesis. Interestingly, from the full sequencing of yeast (*Saccharomyces cerevisiae*) and roundworm (*Caenorhabditis elegans*), no MEN1 homologs could be found.

7. *MEN1* mutations

To date, over 250 germline mutations have been identified distributed throughout the whole of the *MEN1* gene (*Table 2*). More than two thirds of these changes are inactivating mutations which are either nonsense or frameshift. The latter consists of small and large deletions, insertions and splice mutations. The remainder are either missense or in-frame deletions. There are several mutations that recur in unrelated families, including those from different ethnic origin, and geographical locations (Teh *et al.*, 1998a). These 'warm spots' are 359del4, K119del, 734del4 and 1650–1657delC which occur in the CpG/CpNpG direct repeats or single nucleotide repeats (*Table 2*). The detection rate of MEN1 mutations in patients is estimated to be around 80%. Approximately 5–10% are thought to be *de-novo* mutations found in so-called 'sporadic MEN 1' – those with two or more of MEN 1-related endocrine tumors without a family history.

Table 2. Summary of published MEN1 germline mutations

Exon	Codon	Name	Reference
2	12	P12L	Agarwal *et al.*, 1997
2	22	L22R	Chandrasekharappa *et al.*, 1997
2	26	E26K	Bartsch *et al.*, 1998
2	29	R29X	Sato *et al.*, 1998; Matsubara *et al.*, 1998
2	34	221ins15	Bassett *et al.*, 1998
2	38	222delT	Roijers *et al.*, 2000
2	39	L39W	Poncin *et al.*, 1999; Roijers *et al.*, 2000
2	42	G42D	Bassett *et al.*, 1998
2	43	239del3	Cebrián A *et al.*, 1999
2	45	E45G	Sato *et al.*,1998; Miyauchi A *et al.*, 1998
2	45	E45K	Morelli *et al.*, 2000
2	55	275del TinsGG	Mayr *et al.*, 1997; Bartsch *et al.*, 1998
2	63	299ins10	Bassett *et al.*, 1998
2	63	299ins5	Hai *et al.*,1999
2	66	307delGinsAA	Bartsch *et al.*, 1998
2	66	309dup10	Bassett *et al.*, 1998
2	67	310dup5	Giraud *et al.*, 1998
2	68	298ins5	Tanaka *et al.*, 1998a
2	68	311insG	Teh *et al.*, 1998a
2	68	311ins5	Morelli *et al.*, 2000
2	68	313delC	Agarwal *et al.*, 1997
2	69	315del11	Basset *et al.*, 1998
2	69	317ins5	Mayr *et al.*, 1997
2	70	320del2	Cote *et al.*, 1998
2	71	321ins4	Tanaka *et al.*, 1998a
2	76	337delC	Giraud *et al.*, 1998
2	78	341insTA	Hamaguchi *et al.*, 1999
2	79	345delC	Tanaka *et al.*, 1998a
2	82	352insC	Cebrián *et al.*, 1999
2	83	358del4	Cebrián *et al.*, 1999
2	83	359del4	Chandrasekharappa *et al.*, 1997; Agarwal *et al.*, 1997 Poncin *et al.*, 1999; Bassett *et al.*, 1998; Teh *et al.*, 1998b; ECM *et al.*, 1997; Giraud *et al.*, 1998; Sakurai *et al.*, 1999; Hai *et al.*, 1999; Cote *et al.*, 1998; Roijers *et al.*, 2000; Morelli *et al.*, 2000
2	84	84insGT	Giraud *et al.*, 1998; ECM *et al.*, 1997
2	90	379delAT	Bassett *et al.*, 1998; Cebrián *et al.*, 1999
2	96	Q96X	Giraud *et al.*, 1998
2	98	R98X	ECM *et al.*, 1997; Bassett *et al.*, 1998; Giraud *et al.*, 1998; Mayr *et al.*, 1997; Teh *et al.*, 1998a
2	102	416del32	Mayr *et al.*, 1997

Exon	Codon	Name	Reference
2	102	416delC	Chandrasekharappa et al., 1997; Agarwal et al., 1997; Teh et al., 1998a; Mutch et al., 1999
2	103	307delC	Mutch et al., 1999
2	106	319delAT	Mutch et al., 1999
2	108	R108X	ECM et al., 1997; Giraud et al., 1998
2	110	437insGG	Giraud et al., 1998
2	119	K119del	Chandrasekharappa et al., 1997; Agarwal et al., 1997; Bassett et al., 1998; Shimizu et al., 1997; Sakurai et al., 1999; Cote et al., 1998; Roijers et al., 2000
2	120	K120X	Agarwal et al., 1997
2	124	480delG	Evans et al., 1998
2	125	483del2	Aoki et al., 1997; Shimizu et al., 1997; Tanaka et al., 1998a
2	126	W126X	Bassett et al., 1998
2	131	131delC	Bassett et al., 1998
2	133	Y133X	Giraud et al., 1998
2	134	512delC	Chandrasekharappa et al., 1997; Agarwal et al., 1997; Bassett et al., 1998
2	139	H139D	Agarwal et al., 1997; Zhuang et al., 1997
2	139	H139Y	Agarwal et al., 1997
2	139	H139R	Martin-Campos et al., 1999
2	144	F144V	Agarwal et al., 1997
2	149	555delGGinsC	Mutch et al., 1999
Intron 2		556-3C>G	Burgess et al., 2000
3	151	560insA	Teh et al., 1998a
3	153	569delC	Sato et al., 1998
3	155	S155F	Engelbach et al., 1999
3	156	G156D	Mutch et al., 1999
3	160	A160T	Teh et al., 1998a
3	160	A160P	Agarwal et al., 1997; Bassett et al., 1998
3	162	594ins5	Bassett et al., 1998
3	163	597delG	Teh et al., 1998a
3	163	599insA	Teh et al., 1998a
3	164	A164D	Bassett et al., 1998
3	166	Q166X	Tanaka et al., 1998a
3	168	L168P	Bartsch et al., 1998
3	171	621del9	Sakurai et al., 1999
3	172	D172Y	Giraud et al., 1998; Poncin et al., 1999; Fuji et al., 1999
3	174	630delC	Bassett et al., 1998
3	176	A176P	Agarwal et al., 1997
3	179	E179D	Poncin et al., 1999
3	179	E179Q	Roijers et al., 2000
3	183	W183R	Hai et al., 1999
3	183	W183S	Bassett et al., 1998; ECM et al., 1997
3	183	W183X	Agarwal et al., 1997; ECM et al., 1997
3	183	W183X	ECM et al., 1997; Bassett et al., 1998
3	184	V184E	Fujimori et al., 1998
3	191	E191X	Teh et al., 1998; Bassett et al., 1998
3	194	A194D	Lam et al., 1998
3	197	699del3	Giraud et al., 1998
3	198	W198X	Hamaguchi et al., 1999
3	198	W198X	Chandrasekharappa et al., 1997; Agarwal et al., 1997
3	201	711del11	Sato et al., 1998

continued overleaf

Table 2. *continued*

Exon	Codon	Name	Reference
3	201	712delA	Ohye et al., 1999
3	201	713delG	Agarwal et al., 1997
3	208	734delC	Teh et al., 1998a
3	209	Q209X	Cebrián et al., 1999
3	209	735del4	Chandrasekharappa et al., 1997; Agarwal et al., 1997; Engelbach et al., 1999
3	210	738del4	Mutch et al., 1999; Morelli et al., 2000
3	210	739del4	Giraud et al., 1998; Bassett et al., 1998; ECM et al., 1997; Teh et al., 1998a; Sakurai et al., 1999
3	214	214delG	Bassett et al., 1998
3	215	V215M	Morelli et al., 2000
3	218	764G>T	Giraud et al., 1998
intron 3		764+1G>T	Teh et al., 1998a
intron 3		765-6C>T	Roijers et al., 2000
intron 3		765-4delT	Morelli et al., 2000
intron 3		765-1G>T	Mayr et al., 1998
intron 3		765-1G>C	Morelli et al., 2000
4	220	W220X	Teh et al., 1998a
4	220	W220X	Mutch et al., 1999
4	222	776delC	Giraud et al., 1998
4	223	778delT	Giraud et al., 1998
4	223	L223P	Giraud et al., 1998; Roijers et al., 2000
4	225	G225R	Hai et al., 1999
4	227	Y227X	Teh et al., 1998a; Dackiw et al., 1999
4	227	799del6	Ludwig et al., 1999
4	236	817del9	Bassett et al., 1998
4	236	818insT	Bartsch et al., 1998
4	241	C241R	Mutch et al., 1999
4	241	C241Y	Hai et al.,1999
4	242	A242V	Agarwal et al., 1997
4	246	848del4ins9	Cebrián et al., 1999
4	253	S253P	Hai et al., 1999
4	253	879ins7	Poncin et al., 1999
4	258	Q258X	Roijers et al., 2000; Poncin et al., 1999
4	260	Q260X	Agarwal et al., 1997; Shimizu et al., 1997
4	261	R261X	Giraud et al., 1998
intron 4		893+1G>A	Morelli et al., 2000
intron 4		893+1G>C	Poncin et al., 1999
intron 4		893+1G>T	Teh et al., 1998a; Giraud et al., 1998
intron 4		894-9G>A	Mutch et al., 1999; Görtz et al., 1999b; Kishi et al., 1999; Hai et al., 1999; Engelback et al., 1999
intron 4		894-1G>C	Giraud et al., 1998; Poncin et al., 1999
5	262	894delAinsCC	Bassett et al., 1998
5	262	K262X	Bassett et al., 1998
5	263	899del9	Mutch et al., 1999
5	264	L264P	Roijers et al., 2000; Poncin et al., 1999
5	265	W265X	Agarwal et al., 1997; Giraud et al., 1998; Roijers et al., 2000
5	267	L267P	Poncin et al., 1999
5	268	912insT	Chico et al., 1998
5	268	Y268X	Bartsch et al., 1998; Roijers et al., 2000
5	272	924insC	Giraud et al., 1998

Exon	Codon	Name	Reference
6	284	A284E	Bassett et al., 1998
6	286	L286P	Agarwal et al., 1997
6	297	1001delC	Hai et al., 1999
intron 6		1022+1G>A	Mutch et al., 1999
intron 6		1023-1G>C	Mutch et al., 1999
7	305	G305D	Honda et al., 2000
7	308	S308X	Agarwal et al., 1997
7	309	A309P	Agarwal et al., 1997
7	312	Y312X	Agarwal et al., 1997; Sato et al., 2000
7	313	Y313X	Mutch et al., 1999
7	314	R314P	Giraud et al., 1998
7	315	1054insA	Giraud et al., 1998; Poncin et al., 1999
7	317	1059delC	Morell et al., 2000
7	317	H317R	Roijers et al., 2000
7	320	P320L	Tanaka et al., 1998a
7	321	1071delT	Morelli et al., 2000
7	323	Y323X	Agarwal et al., 1997
7	327	1089delT	Teh et al., 1998b; Bassett et al., 1998
7	335	1114delG	Fuji et al., 1999
7	337	A337D	Giraud et al., 1998
7	337	A337P	Roijers et al., 2000
7	341	W341X	Cebrián et al., 1999; Cote et al., 1998
7	341	W341R	Giraud et al., 1998
7	341	1132delG	Chandrasekharappa et al., 1997
7	341	W341X	Bassett et al., 1998; Teh et al., 1998a; Mutch et al., 1999
7	344	T344R	Agarwal et al., 1997; Morelli et al., 2000
7	344	1142delG	Hai et al., 1999
7	345	1143delG	Sakurai et al., 1999
7	348	I348N	Roijers et al., 2000
7	349	Q349X	Bassett et al., 1998; Mutch et al., 1999
intron 7		1159+1G>A	Bassett et al., 1998
intron 7		1159+5G>A	Roijers et al., 2000
intron 7		1160-2A>G	Tanaka et al., 1998a
8	353	Y353X	Shimizu et al., 1997; Roijers et al., 2000
8	358	1182del3	Bassett et al., 1998
8	362	1195ins14	Roijers et al., 2000
8	363	E363del	Chandrasekharappa et al., 1997; Agarwal et al., 1997; Bassett et al., 1998
8	364	1202del2	Agarwal et al., 1997
8	368	A368D	Giraud et al., 1998
8	373	1226delC	Bassett et al., 1998
8	385	1264delC	Morelli et al., 2000
8	385	A385V	Roijers et al., 2000
8	386	1267delG	Morelli et al., 2000
8	388	E388X	Bassett et al., 1998
8	391	1280delG	Agarwal et al., 1997
8	392	E392X	Engelbach et al., 1999
8	393	Q393X	Mutch et al., 1999
9	397	1300delC	Giraud et al., 1998
9	398	Q398X	Roijers et al., 2000
9	405	1325insA	Dackiw et al., 1999
9	405	1325delG	Giraud et al., 1998
9	406	406del5	Bassett et al., 1998
9	414	L414del	Sato et al., 1998; Ohye et al., 1998

continued overleaf

Table 2. *continued*

Exon	Codon	Name	Reference
9	415	R415X	ECM *et al.*, 1997; Giraud *et al.*, 1998; Lam *et al.*, 1998; Morelli *et al.*, 2000; Roijers *et al.*, 2000
9	418	1362del12	Giraud *et al.*, 1998
9	418	D418del	Agarwal *et al.*, 1997
9	418	D418N	Teh *et al.*, 1998a; Bassett *et al.*, 1998; Giraud *et al.*, 1998
9	418	1363delAC	Giraud *et al.*, 1998; Poncin *et al.*, 1999
9	422	1374delA	Giraud *et al.*, 1998
9	425	E425del	Giraud *et al.*, 1998
3	423	W423S	Mayr *et al.*, 1998
9	427	S427R	Cote *et al.*, 1998
9	431	W431X	Dackiw *et al.*, 1999
9	436	W436R	Chandrasekharappa *et al.*, 1997; ECM *et al.*, 1997
9	436	W436X	Chandrasekharappa *et al.*, 1997; ECM *et al.*, 1997
9	437	1419delG	Bartsch *et al.*, 1998
9	438	1422insA	Sakurai *et al.*, 1999
9	438	1424del4	Giraud *et al.*, 1998
9	440	1429del2ins4	Bassett *et al.*, 1998
9	442	Q442X	Shimizu *et al.*, 1997; Roijers *et al.*, 2000
9	444	L444P	Cetani *et al.*, 1990
9	447	1449del11	Giraud *et al.*, 1998
9	447	F447S	Agarwal *et al.*, 1997
9	448	1452del11	Teh *et al.*, 1998a; Dackiw *et al.*, 1999
9	450	Q450X	Hai *et al.*, 1999; Morelli *et al.*, 2000
intron 9		1460+1del12	Mutch *et al.*, 1999
intron 9		1460+1del14insAT	Hai *et al.*, 1999
intron 9		1460+4del2	Giraud *et al.*, 1998
intron 9		1461-2A>C	Martin-Campos *et al.*, 1999
10	452	1466del12	Teh *et al.*, 1998a
10	453	Q453X	Bassett *et al.*, 1998
10	454	1472del5	Tanaka *et al.*, 1998a
10	455	1473del5	Sakurai *et al.*, 1999
10	455	1475delG	Roijers *et al.*, 2000
10	458	1484del8	Agarwal *et al.*, 1997
10	459	1487ins6	Giraud *et al.*, 1998
10	460	R460X	Agarwal *et al.*, 1997; Bassett *et al.*, 1998; Giraud *et al.*, 1998
10	461	1491insCC	Giraud *et al.*, 1998
10	463	1499dup8	Giraud *et al.*, 1998
10	464	1502del4	Mutch *et al.*, 1999
10	467	1508ins	Giraud *et al.*, 1998
10	467	1509insGA	Agarwal *et al.*, 1997
10	471	W471X	Mutch *et al.*, 1999; Valdes *et al.*, 1999
10	472	1526insG	Roijers *et al.*, 2000
10	473	E473X	Cote *et al.*, 1998
10	477	E477X	Cote *et al.*, 1998
10	477	1539insG	Poncin *et al.*, 1999
10	482	1555insG	Morelli *et al.*, 2000
10	493	1587delCCinsG	Bartsch *et al.*, 1998
10	499	1607delA	Giraud *et al.*, 1998
10	503	7736del25	Cebrián *et al.*, 1999
10	507	1630insC	Giraud *et al.*, 1998
10	508	Q508X	Morelli *et al.*, 2000
10	510	1639delCA	Bartsch *et al.*, 1998

Exon	Codon	Name	Reference
10	514	1650delC	Agarwal et al., 1997; Giraud et al., 1998; Bassett et al., 1998
10	514	1650insC	Agarwal et al., 1997; Giraud et al., 1998; Lam et al., 1998; Bassett et al., 1998; Cebrián et al., 1999; Hai et al., 1999
10	516	1657insC	Teh et al., 1998a; Bassett et al., 1998; Sakurai et al., 1999; Roijers et al., 2000
10	519	1666delC	Poncin et al., 1999
10	519	1667insT	Hai et al., 1999
10	527	R527X	Chandrasekharappa et al., 1997; Bassett et al., 1998; Giraud et al., 1998; Teh et al., 1998a; Mutch et al., 1999
10	530	1699del3ins2	Debelenko et al., 1999
10	536	Q536X	Giraud et al., 1998
10	536	1717delA	Hai et al., 1999
10	544	P544S	Roijers et al., 2000
10	550	1768delT	Giraud et al., 1998
10	554	1771ins8	Roijers et al., 2000
10	555	S555N	Giraud et al., 1998
10	557	1780del3	Giraud et al., 1998
10	558	1782delA	Giraud et al., 1998
10	559	1785insG	Mutch et al., 1999
10	571	1823delCT	Hai et al., 1999

Over 100 mutations have also been found in the sporadic counterparts of the MEN1-related tumors (*Table 3*) suggesting that the *MEN1* gene is involved in the genesis of these tumors. Furthermore, the majority of these tumors have been found to have LOH, thus fulfilling Knudson's two-hit hypothesis. However, it is important to appreciate that only a subset of these sporadic tumors harbor *MEN1* mutations. For example, about 25–30% of parathyroid and enteropancreatic endocrine tumors have MEN1 mutations and very few sporadic anterior pituitary or adrenocortical tumors carry mutations at all. This suggests that, in those tumors without *MEN1* mutations, different underlying genetic mechanisms may be involved.

8. Genotype–phenotype correlation

Genotype–phenotype correlation has so far been difficult in MEN1. It is well-known that its clinical presentation, age of onset, and the natural history of the disease vary extensively even among members of the same family (Burgess et al., 1996; Trump et al., 1996). Furthermore, the finding of mutations throughout the gene rather than clustered in regions makes any correlation even more challenging.

Several groups have also tried to establish whether some of the previously described 'endocrine neoplasia syndromes' that are atypical but characterized by MEN1-related tumors, notably family isolated hyperparathyroidism (FIHP) and familial acromegaly/gigantism, are associated with MEN1 mutations or better

Table 3. Summary of published MEN1 somatic mutations in different types of sporadic tumors. NET = neuroendocrine tumor

Exon	Codon	Mutation	Tumor	Reference
2	8	134del13	lung carcinoid	Debelenko et al., 1997
2	13	147del61	gastrinoma	Wang et al., 1988
2	15	154del16	parathyroid adenoma	Carling et al., 1998
2	26	E26K	parathyroid adenoma	Heppner et al., 1997
2	37	L37P	pancreatic VIPoma	Görtz et al., 1999a
2	42	G42S	nonfunctional pancreatic NET	Toliat et al., 1997
2	42	G42S	parathyroid adenoma	Sato et al., 2000
2	45	E45D	parathyroid adenoma	Farnebo et al., 1998
2	49	A49P	pancreatic somatostatinoma	Görtz et al., 1999a
2	50	258del4	parathyroid adenoma	Carling et al., 1998
2	53	V53I	ileal NET	Görtz et al., 1999
2	70	320del16	gastrinoma	Wang et al., 1988
2	70	320del7	neuroendocrine, liver metastasis	Mailman et al., 1999
2	73	328del9	gastrinoma	Wang et al., 1988
2	81	352delC	parathyroid adenoma	Dwight et al., 2000
2	83	357del4	gastric neuroendocrine carcinoma	Fuji et al., 1998
2	83	358del25	gastrinoma, lymph node metastasis	Zhuang et al., 1997a
2	83	358del4	duodenum gastrinoma	Zhuang et al., 1997a; Wang et al., 1988
2	86	I86F	duodenum gastrinoma	Zhuang et al., 1997a
2	89	L89R	pancreatic glucagonoma	Hessman et al., 1998
2	90	378delT	prolactinoma	Wenbiet et al., 1999
2	98	R98X	non-functioning pancreatic endocrine tumor	Fuji et al., 1998
2	99	405del1	parathyroid adenoma	Carling et al., 1998
2	108	R108X	lung NET	Baudin et al., 1999
2	108	R108X	parathyroid adenoma	Heppner et al., 1997
2	108	434ins29	bronchial NET	Görtz et al., 1999a
2	109	E109X	adrenocortical adenoma	Görtz et al., 1999b
2	114	452delC	parathyroid adenoma	Dwight et al., 2000
2	119	K119del	parathyroid adenoma	Farnebo et al., 1998
2	119	K119del	nonfunctioning pancreatic NET	Toliat et al., 1997
2	125	483delAT	pancreas gastrinoma	Zhuang et al., 1997a
2	126	W126G	pancreatic gastrinoma	Zhuang et al., 1997a
2	135	K135I	skin Angiofibroma	Böni et al., 1998
2	138	522delG	pancreatic VIPoma	Wang et al., 1988

continued overleaf

Exon	Codon	Mutation	Phenotype	Reference
2	139	H139D	parathyroid adenoma	Carling et al., 1998
2	141	Q141R	gastrinoma	Wang et al., 1988
2	145	S145R	nonfunctioning pancreatic NET	Wang et al., 1988
2	145	545insT	pancreatic insulinoma	Zhuang et al., 1997a
2	147	550del3	gastrinoma	Wang et al., 1988
3	152	L152W	parathyroid hyperplasia	Carling et al., 1998
3	159	F159C	gastrinoma, lymph node metastasis	Zhuang et al., 1997a
3	162	V162C	gastrinoma	Wang et al., 1988
3	162	V162F	parathyroid adenoma	Dwight et al., 2000
3	172	D172V	bronchial NET	Görtz et al., 1999a
3	175	L175R	parathyroid hyperplasia	Farnebo et al., 1998
3	175	634insC	gastrinoma, peripancreatic node metastasis	Mailman et al., 1999
3	178	S178Y	non-functioning pancreatic NET	Fuji et al., 1998
3	183	W183S	irradiation-related parathyroid tumor	Farnebo et al., 1999
3	196	699insA	parathyroid adenoma	Farnebo et al., 1998
3	198	W198X	pancreatic VIPoma	Shan et al., 1998
3	199	707delCinsGG	parathyroid adenoma	Dwight et al., 2000
3	209	Q209X	duodenal NET	Görtz et al., 1999a
intron 3		764+3A>G	lung carcinoid	Debelenko et al., 1997
4	220	770del9	gastrinoma	Wang et al., 1988
4	224	781del5	parathyroid hyperplasia	Farnebo et al., 1998
4	253	S253L	parathyroid adenoma	Dwight et al., 2000
4	255	875insA	gastrinoma, lymph node metastasis	Zhuang et al., 1997a
5	259	816del22	pituitary adenoma	Tanaka et al., 1998b
5	265	904delG	metastatic pancreatic VIPoma	Toliat et al., 1997
intron 5		934+1G>A	parathyroid hyperplasia	Heppner et al., 1997
intron 5		935-2A>G	pancreatic insulinoma	Toliat et al., 1997
6	284	A284P	parathyroid adenoma	Dwight et al., 2000
7	328	1091insAGC	parathyroid adenoma	Sato et al., 2000
7	329	E329X	pancreatic glucagonoma, liver metastasis	Hessman et al., 1998
7	330	R330P	gastrinoma	Wang et al., 1988
7	340	A340T	parathyroid adenoma	Shan et al.,1998
7	345	1144delC	parathyroid adenoma	Farnebo et al., 1998
7	347	347insI	irraPT	Farnebo et al., 1999
7	349	Q349X	metastatic melanoma	Nord et al., 2000
intron 7		1160-2A>G	pituitary adenoma	Schmidt et al., 1999

Table 3. *continued*

Exon	Codon	Mutation	Tumor	Reference
intron 7	358	1160-5del7	metastatic non-functioning pancreatic NET	Toliat *et al.*, 1997
8	358	E359K	skin angiofibroma	Böni *et al.*, 1998
8	363	E363del	gastrinoma, lymph node metastasis	Zhuang *et al.*, 1997a
8	368	1212del7	gastrinoma, lymph node metastasis	Zhuang *et al.*, 1997a
8	373	1226delC	large-cell neuroendocrine lung carcinoma	Debelenko *et al.*, 1997
8	387	1269delG	gastrinoma	Wang *et al.*, 1988
8	387	1269delG	irradiation-related parathyroid tumor	Farnebo *et al.*, 1999
8	388	E388X	parathyroid adenoma	Sato *et al.*, 2000
8	390	1279ins11	parathyroid adenoma	Heppner *et al.*, 1997
8	392	E392X	pancreatic glucagonoma, liver metastasis	Hessman *et al.*, 1998
8	393	Q393X	thymic carcinoid	Fuji *et al.*, 1998
9	401	1313del19	parathyroid adenoma	Carling *et al.*, 1998
9	410	F410L	pituitary adenoma	Zhuang *et al.*, 1997b
9	415	R415X	non-functioning pancreatic NET, liver metastasis	Hessman *et al.*, 1998
9	418	D418N	parathyroid adenoma	Heppner *et al.*, 1997
9	429	T429K	pancreatic insulinoma	Shan *et al.*, 1998
9	450	Q450X	pancreasVIPoma metastasis	Görtz *et al.*, 1999a
intron 9		1461-4C>T	adrenal hyperplasia	Schulte *et al.*, 1999
10	451	1461delG	lung carcinoid	Debelenko *et al.*, 1997
10	460	R460X	parathyroid adenoma	Sato *et al.*, 2000
10	462	A462P	irradiation-related parathyroid tumor	Farnebo *et al.*, 1999
10	502	K502M	pituitary adenoma	Zhuang *et al.*, 1997b
10	514	1650insC	lung carcinoid	Debelenko et al., 1997
10	522	1674del7	pancreatic insulinoma	Görtz *et al.*, 1999a
10	535	A535V	pancreatic insulinoma	Zhuang *et al.*, 1997a
10	541	A541T	parathyroid adenoma	Shan *et al.*, 1998
10	541	A541T	nonfunctional hepatic NET	Toliat *et al.*, 1997
10	543	S543L	gastrinoma, lymph node metastasis	Zhuang *et al.*, 1997a
10	552	T552S	adreno-cortical adenoma	Schulte *et al.*, 1999
10	561	M561T	parathyroid hyperplasia	Heppner *et al.*, 1997
10	568	1812del5	parathyroid hyperplasia	Heppner *et al.*, 1997
10	568	1814delC	parathyroid adenoma	Dwight *et al.*, 2000
10	580	T580R	parathyroid adenoma	Dwight *et al.*, 2000

still, specific mutations that will allow a genotype–phenotype correlation. Our findings of two missense mutations in close proximity in exon 4 in two of the largest FIHP families (one with 7 affected and the other with 14 affected) – considered a milder form of MEN 1, maybe an example (Kassem *et al.*, 2000; Teh *et al.*, 1998b). These two mutations, E255K and Q260P, interestingly fall outside the sites for nuclear localization signals and JUND binding, suggesting a correlation of functionally 'milder' mutations with a milder form of disease. However, a few smaller FIHP families have been reported to be associated with truncating mutations in other exons. Whether these families represent true FIHP or MEN1 families that are yet to develop pituitary or enteropancreatic tumors is not known. In the studies of familial acromegaly/gigantism, several groups have failed to identify any MEN1 mutation (Ackermann *et al.*, 1999; Gadelha *et al.*, 1999; Teh, 1998), suggesting that it may be a separate entity caused by a separate gene or that the mutations in these families may be unique and lie in the *MEN1* regulatory region which is yet to be identified.

In our genotype–phenotype correlation efforts, two observations have come to our attention: MEN1 phenocopies and MEN1 modifiers. The former refers to family members that have the phenotype but not the genotype. In our study of the largest MEN1 family, 7 of 71 individuals satisfying clinical diagnostic criteria for MEN1 have been found to be genetically negative, confirmed by both mutation and linkage analyses (Burgess *et al.*, 2000). These MEN1 phenocopies comprise of 4 cases of primary hyperparathyroidism, two 'non-secretory' pituitary adenoma and one case of coincident prolactinoma and hyperparathyroidism. This phenomenon can possibly be attributed to the following two factors: 1) a relatively high frequency of both conditions, i.e., hyperparathyroidism and asymptomatic pituitary tumors, in the general populations and 2) stringent biochemical and radiological screening prior to genetic diagnosis in this well-known family. Also, from the studies of this family, it has been noted that gene carriers from different branches have developed differential rates of endocrine neoplasia. For example, more than 50% of gene carriers from two branches develop prolactinomas, whereas in the other branches the prevalence of prolactinoma among gene carriers is less than 10% suggesting the existence of a disease-modifying gene capable of modulating the MEN1 clinical manifestation (Burgess *et al.*, 1996). Besides that, studies of MEN1-related thymic carcinoids also suggest the influence of a sex-related modifier gene. Of the 42 reported MEN1-related thymic carcinoids, only two are found in females (Teh *et al.*, 1998c).

To date, genotype–phenotype correlation has also been attempted in sporadic tumors but has not been very fruitful. Goebel *et al.* (2000) have observed that more than half of the mutations in sporadic gastrinomas clustered in exon 2 but they have not found any correlation either in clinical characteristics or outcome of treatment between mutation-positive and mutation-negative tumors or between tumors carrying different mutations.

9. Mitogenic factors and chromosome instability

For years, an elusive mitogenic factor has been reported in MEN1 patients (Brandi *et al.*, 1986). Plasma of MEN1 patients has been shown to have mitogenic activity on cultured bovine parathyroid cells. This mitogenic factor, estimated to have an apparent molecular weight of 50 000 to 55 000 kD, appears to be a fibroblast growth factor-like factor which might be secreted by the pituitary tumors (Zimering *et al*. 1990, 1993). By all accounts, this mitogenic factor is unlikely to be the protein product of the MEN1 gene but its existence and role in MEN1 cannot be disregarded without further studies. Example of a growth factor playing a role in TSG-related familial cancer has been known. For example, in von Hippel-Lindau disease (VHL), vascular endothelial growth factor (VEGF) is highly expressed in renal cell carcinoma and the VHL gene has been found to suppress its expression (Gnarra *et al*., 1996).

The other phenomenon that has been reported, even before the age of positional cloning, is chromosomal instability in MEN1 (Gustavson *et al*., 1983). Increased frequency of chromosome breakage, numerical and structural abnormalities and more recently premature centrosome division, have been reported in MEN 1 patients (Sakurai *et al*., 1999; Scappaticci *et al*., 1991). These studies suggest that the MEN1 gene may play a role in maintaining DNA integrity and DNA repairing.

10. Conclusions

It is obvious that more studies are needed to characterize the functions of the *MEN1* gene and its molecular mechanisms in MEN1 tumorigenesis. A MEN1 animal model will likely be very useful in achieving these aims and, in addition, may serve as a tool for testing any new therapies. It is also expected that more interacting proteins of MENIN will be identified. Meanwhile, genetic predictive testing and proper clinical surveillance will reduce morbidity and mortality in MEN1 patients. In order to achieve our full understanding of MEN1 tumorigenesis and progression, it is also timely to look both for genetic and epigenetic modifiers that influence its phenotypes and natural history. This information will help in improving the clinical outcome of the disease, which varies between families and even within the same family.

Acknowledgements

I am very grateful to Dr Fung Ki Wong for contributing to the mutation list and Lynn Ritsema, Andrea Brenner and Jane Kao for preparation of the manuscript.

References

Ackermann, F., Krohn, K., Windgassen, M., Buchfelder, M., Fahlbusch, R. and Paschke, R. (1999) Acromegaly in a family without a mutation in the menin gene. *Exp. Clin. Endocrinol. Diabetes* **107**: 93–96.

Agarwal, S.K., Kester, M.B., Debelenko, L.V. *et al*. (1997) Germline mutations of the MEN1 gene in familial multiple endocrine neoplasia type 1 and related states. *Hum. Mol. Genet.* **6**: 1169–1175.

Agarwal, S.K., Guru, S.C., Heppner, C. *et al*. (1999) Menin interacts with the AP1 transcription factor JunD and represses JunD-activated transcription. *Cell* **96**: 143–152.

Aoki, A., Tsukada, T., Yasuda, H. *et al.* (1997) Multiple endocrine neoplasia type 1 presented with manic-depressive disorder: a case report with an identified MEN1 gene mutation. *Jpn J. Clin. Oncol.* **27**: 419–422.

Bartsch, D., Kopp, I., Bergenfelz, A. *et al.* (1998) MEN1 gene mutations in 12 MEN1 families and their associated tumors. *Eur. J. Endocrinol.* **139**: 416–420.

Bassett, J.H., Forbes, S.A., Pannett, A.A. *et al.* (1998) Characterization of mutations in patients with multiple endocrine neoplasia type 1. *Am. J. Hum. Genet.* **62**: 232–244.

Bassett, J.H., Rashbass, P., Harding, B., Forbes, S.A., Pannett, A.A. and Thakker, R.V. (1999) Studies of the murine homolog of the multiple endocrine neoplasia type 1 (MEN1) gene, men1. *J. Bone Miner Res.* **14**: 3–10.

Baudin, E., Bidart, J.M., Rougier, P. *et al.* (1999) Screening for multiple endocrine neoplasia type 1 and hormonal production in apparently sporadic neuroendocrine tumors. *J. Clin. Endocrinol. Metab.* **84**: 69–75.

Böni, R., Vortmeyer, A.O., Pack, S., Park, W.S., Burg, G., Hofbauer, G., Darling, T., Liotta, L. and Zhuang, Z. (1998) Somatic mutations of the MEN1 tumor suppressor gene detected in sporadic angiofibromas. *J. Invest. Dermatol.* **111**: 539–540.

Brandi, M.L., Aurbach, G.D., Fitzpatrick, L.A., Quarto, R., Spiegel, A.M., Bliziotes, M.M., Norton, J.A., Doppman, J.L. and Marx, S.J. (1986) Parathyroid mitogenic activity in plasma from patients with familial multiple endocrine neoplasia type 1. *N. Engl. J. Med.* **314**: 1287–1293.

Brandi, M.L., Marx, S.J., Aurbach, G.D. and Fitzpatrick, L.A. (1987) Familial multiple endocrine neoplasia type 1: A new look at pathophysiology. *Endocr. Rev.* **8**: 391–405.

Burgess, J.R., Shepherd, J.J., Parameswaran, V., Hoffman, L. and Greenaway, T.M. (1996) Prolactinomas in a large kindred with multiple endocrine neoplasia type 1: clinical features and inheritance pattern. *J. Clin. Endocrinol. Metab.* **81**: 1841–1845.

Burgess, J.R., Greenaway, T.M. and Shepherd, J.J. (1998) Expression of the MEN1 gene in a large kindred with multiple endocrine neoplasia type 1. *J. Int. Med.* **243**: 465–470.

Burgess, J.R., Rueben, D., Greenaway, T.M., Larsson, C., Parameswaran, V., Shepherd, J.J. and Teh, B.T. Clinical and genetic correlates in multiple endocrine neoplasia type 1 (MEN1). *Clin. Endocrinol.* **53**: 205–211.

Carling, T., Correa, P., Hessman, O., Hedberg, J., Skogseid, B., Lindberg, D., Rastad, J., Westin, G. and Akerstrom, G. (1998) Parathyroid MEN1 gene mutations in relation to clinical characteristics of nonfamilial primary hyperparathyroidism. *J. Clin. Endocrinol. Metab.* **83**: 2960–2963.

Cavenee, W.K., Dryja, T.P., Phillips, R.A., Benedict, W.F., Godbout, R., Gallie, B.L., Murphree, A.L., Strong, L.C. and White, R.L. (1983) Expression of recessive alleles by chromosomal mechanisms in retinoblastoma. *Nature* **305**: 779–784.

Cavenee, W.K., Hansen, M.F., Nordenskjold, M., Kock, E., Maumenee, I., Squire, J.A., Phillips, R.A. and Gallie, B.L. (1985) Genetic origin of mutations predisposing to retinoblastoma. *Science* **228**: 501–503.

Cebrián, A., Herrera-Pombo, J.L., Diez, J.J. *et al.* (1999) Genetic and clinical analysis in 10 Spanish patients with multiple endocrine neoplasia type 1. *Eur. J. Hum. Genet.* **7**: 585–589.

Cetani, F., Pardi, E., Cianferotti, L., Vignali, E., Picone, A., Miccoli, P., Pinchera, A. and Marcocci, C. (1990) A new mutation of the MEN1 gene in an Italian kindred with multiple endocrine neoplasia type 1. *Eur. J. Endocrinol.* **140**: 429–433.

Chandrasekharappa, S.C., Guru, S.C., Manickam, P. *et al.* (1997) Positional cloning of the gene for multiple endocrine neoplasia type 1. *Science* **276**: 404–407.

Chico, A., Gallart, L., Mato, E. *et al.* (1998) A novel germline mutation in exon 5 of the multiple endocrine neoplasia type 1 gene. *J. Mol. Med.* **76**: 837–839.

Cote, G.J., Lee, J.E., Evans, D.B. *et al.* (1998) Five novel mutations in the familial multiple endocrine neoplasia type 1 (MEN1) gene. *Hum Mutat.* **12**: 219 Mutations in brief no. 188. Online.

Dackiw, A.P., Cote, G.J., Fleming, J.B., Schultz, P.N., Stanford, P., Vassilopoulou-Sellin, R., Evans, D.B., Gagel, R.F. and Lee, J.E. (1999) Screening for MEN1 mutations in patients with atypical endocrine neoplasia. *Surgery* **126**: 1097–1103.

Debelenko, L.V., Brambilla, E., Agarwal, S.K., Swalwell, J.I., Kester, M.B., Lubensky, I.A., Zhuang, Z., Guru, S.C., Manickam, P., Olufemi, S.E., Chandrasekharappa, S.C., Crabtree, J.S., Kim, Y.S., Heppner, C., Burns, A.L., Spiegel, A.M., Marx, S.J., Liotta, L.A., Collins, F.S., Travis, W.D. and Emmert-Buck, M.R. (1997) Identification of MEN1 gene mutations in sporadic carcinoid tumors of the lung. *Hum. Mol. Genet.* **6**: 2285–2290.

Debelenko, L.V., Brambilla, E., Agarwal, S.K. *et al.* (1999) Germline mutations in the MEN1 gene: creation of a new splice acceptor site and insertion of 7 intron nucleotides into the mRNA. *Int. J. Mol. Med.* **4**: 483–485.

Debelenko, L.V., Swalwell, J.I., Kelley, M.J. *et al.* (2000) MEN1 gene mutation analysis of high-grade neuroendocrine lung carcinoma. *Genes Chromosomes Cancer* **28**: 58–65.

Dwight, T., Twigg, S., Delbridge, L. *et al.* Loss of heterozygosity in sporadic parathyroid tumours: involvement of chromosome 1 and the MEN1 gene locus in 11q13. *Clin. Endocrinol.* (in press).

Engelbach, M., Forst, T., Hankeln, T. *et al.* (1999) Germline mutations in the MEN1 gene: creation of a new splice acceptor site and insertion of 7 intron nucleotides into the mRNA. *Int. J. Mol. Med.* **4**: 483–485.

European Consortium on MEN1. (1997) Identification of the Multiple Endocrine Neoplasia Type 1 (MEN1) gene. *Hum. Mol. Genet.* **6**: 1177–1183.

Evans, S., Curtis, D., Dalton, A., Cook, J., Ross, R. and Quarrell, O.W.J. (1998) Identification of a new nt480delG mutation in Multiple Endocrine Neoplasia Type 1. *J. Med. Genet.* **35**: Suppl 1 Poster 14.01.

Farnebo, F., Teh, B.T., Kytola, S. *et al.* (1998) Alterations of the MEN1 gene in sporadic parathyroid tumors. *J. Clin. Endocrinol. Metab.* **83**: 2627–2630.

Farnebo, F., Kytölä, S., Teh, B.T. *et al.* (1999) Alternative genetic pathways in parathyroid tumorigenesis. *J. Clin. Endocrinol. Metab.* **84**: 3775–3780.

Fuji, T., Kawai, T., Saito, K., Hishima, T., Hayashi, Y., Imura, J., Hironaka, M., Hosoya, Y., Koike, M. and Fukayama, M. (1999) MEN1 gene mutations in sporadic neuroendocrine tumors of foregut derivation. *Pathol. Int.* **49**: 968–973.

Fujimori, M., Shirahama, S., Sakurai, A. *et al.* (1998) Novel V184E MEN1 germline mutation in a Japanese kindred with familial hyperparathyroidism. *Am. J. Med. Genet.* **80**: 221–222.

Gadelha, M.R., Prezant, T.R., Une, K.N., Glick, R.P., Moskal, S.F. II, Vaisman, M., Melmed, S., Kineman, R.D. and Frohman, L.A. (1999) Loss of heterozygosity on chromosome 11q13 in two families with acromegaly/gigantism is independent of mutations of the multiple endocrine neoplasia type 1 gene. *J. Clin. Endocrinol. Metab.* **84**: 249–256.

Giraud, S., Zhang, C.X., Serova-Sinilnikova, O. *et al.* (1998) Germ-line mutation analysis in patients with multiple endocrine neoplasia type 1 and related disorders. *Am. J. Hum. Genet.* **63**: 455–467.

Gnarra, J.R., Zhou, S.B., Merrill, M.J., Wagner, J.R., Krumm, A., Papavassiliou, E., Oldfield, E.H., Klausner, R.D. and Linehan, W.M. (1996) Post-transcriptional regulation of vascular endothelial growth factor mRNA by the product of the VHL tumor suppressor gene. *Proc. Natl Acad. Sci. USA* **88**: 8405–8409.

Gobl, A.E., Berg, M., Lopez-Egido, J.R., Oberg, K., Skogseid, B. and Westin, G. (1999) Menin represses JunD-activated transcription by a histone deacetylase-dependent mechanism. *Biochim. Biophys. Acta* **1447**: 51–56.

Goebel, S.U., Heppner, C., Burns, A.L. *et al.* (2000) Genoetype/phenotype correlation of multiple endocrine neoplasia type 1 gene mutations in sporadic gastrinomas. *J. Clin. Endocrinol. Metab.* **85**: 116–123.

Gortz, B., Roth, J., Krahenmann, A., de Krijger, R.R., Muletta-Feurer, S., Rutmann, K., Saremaslani, P., Speel, E.J., Heitz, P.U. and Komminoth, P. (1999a) Mutations and allelic deletions of the MEN1 gene are associated with a subset of sporadic endocrine pancreatic and neuroendocrine tumors and not restricted to foregut neoplasms. *Am. J. Pathol.* **154**: 429–436.

Gortz, B., Roth, J., Speel, E.J. *et al.* (1999b) MEN1 gene mutation analysis of sporadic adrenocortical lesions. *Int. J. Cancer* **80**: 373–379.

Guru, S.C., Crabtree, J.S., Brown, K.D. *et al.* (1999) Isolation, genomic organization, and expression analysis of Men1, the murine homolog of the MEN1 gene. *Mamm. Genome* **10**: 592–596.

Gustavson, K-H, Jansson, R. and Öberg K. (1983) Chromosomal breakage in multiple endocrine adenomatosis (types 1 and II). *Clin. Genet.* **23**: 143–149.

Hai, N., Aoki, N., Matsuda, A., Mori, T. and Kosugi, S. (1999) Germline MEN1 mutations in sixteen Japanese families with multiple endocrine neoplasia type 1 (MEN1). *Eur. J. Endocrinol.* **141**: 475–480.

Hamaguch, K., Nguyen, D.C., Yanase, T. *et al.* (1999) Novel germline mutations of the MEN1 gene in Japanese patients with multiple endocrine neoplasia type 1. *J. Hum. Genet.* **44**: 43–47.

Heppner, C., Kester, M.B., Agarwal, S.K. *et al.* (1997a) Somatic mutation of the MEN1 gene in parathyroid tumours. *Nat. Genet.* **16**: 375–378.

Heppner, C., Burns, A.L., Spiegel, A.M., Marx, S.J., Liotta, L.A., Collins, F.S., Travis, W.D. and Emmert-Buck, M.R. (1997b) Identification of MEN1 gene mutations in sporadic carcinoid tumors of the lung. *Hum. Mol. Genet.* **6**: 2285–2290.

Heppner, C., Reincke, M., Agarwal, S.K., Mora, P., Allolio, B., Burns, A.L., Spiegel, A.M. and Marx, S.J. (1999) MEN1 gene analysis in sporadic adrenocortical neoplasms. *J. Clin. Endocrinol. Metab.* **84**: 216–219.

Hessman, O., Lindberg, D., Skogseid, B., Carling, T., Hellman, P., Rastad, J., Akerstrom, G. and Westin, G. (1998) Mutation of the multiple endocrine neoplasia type 1 gene in nonfamilial, malignant tumors of the endocrine pancreas. *Cancer Res.* **58**: 377–379.

Honda, M., Tsukada, T., Tanaka, H., Maruyama, K., Yamaguchi, K., Obara, T., Yamaji, T. and Ishibashi, M. (2000) A novel mutation of the MEN1 gene in a Japanese kindred with familial isolated primary hyperparathyroidism. *Eur. J. Endocrinol.* **142**: 138–143.

Ikeo, Y., Sakurai, A., Suzuki, R., Zhang, M-X., Koizumi, S., Takeuchi, Y., Yumita, W., Nakayama, J. and Hashizume, K. (2000) Proliferation-associated expression of the *MEN1* gene as revealed by *in situ* hybridization: possible role of the menin as a negative regulator of cell proliferation under DNA damage. *Lab. Invest.* **80**: 797–804

Iyer, V.R., Eisen, M.B., Ross, D.T., Schuler, G., Moore, T., Lee, J.C.F., Trent, J.M., Staudt, L.M., Hudson, J. Jr., Boguski, M.S., Lashkari, D., Shalon, D., Botstein, D. and Brown, P.O. (1999) The transcriptional program in the response of human fibroblasts to serum. *Science* **283**: 83–87.

Kaji, H., Canaff, L., Goltzman, D. and Hendy, G.N. (1999) Cell cycle regulation of menin expression. *Cancer Res.* **59**: 5097–5101.

Karges, W., Maier, S., Wissmann, A., Dralle, H., Dosch, H.M. and Boehm, B.O. (1999) Primary structure, gene expression and chromosomal mapping of rodent homologs of the MEN1 tumor suppressor gene. *Biochim. Biophys. Acta* 3;1446: 286–294.

Kassem, M., Kruse, T.A., Wong, F.K., Larsson, C. and Teh, B.T. (2000) Familial Isolated Hyperparathyroidism – A Variant of MEN1. *J. Clin. Endocrinol. Metab.* **85**: 165–167.

Khodaei, S., O'Brien, K., Dumanski, J., Wong, F. and Weber, G. (1999) Characterization of the Men 1 ortholog in zebrafish. *Biochem. Biophys. Res. Commun.* **264**: 404–408.

Kim, Y.S., Burns, A.L., Goldsmith, P.K., Heppner, C., Park, S.Y., Chandrasekharappa, S.C., Collins, F.S., Spiegel, A.M. and Marx, S.J. (1999) Stable overexpression of MEN1 suppresses tumorigenicity of RAS. *Oncogene* **18**: 5936–5942.

Kishi, M., Tsukada, T., Shimizu, S., Futami, H., Ito, Y., Kanbe, M., Obara, T. and Yamaguchi, K. (1998) A large germline deletion of the MEN1 gene in a family with multiple endocrine neoplasia type 1. *Jpn J. Cancer Res.* **89**: 1–5.

Kishi, M., Tsukada, T., Shimizu, S. *et al.* (1999) A novel splicing mutation (894–9 G → A) of the MEN1 gene responsible for multiple endocrine neoplasia type 1. *Cancer Lett.* **142**: 105–110.

Knudson, A.G. (1971) Mutation and cancer: Statistical study of retinoblastoma. *Proc. Natl Acad. Sci. USA* **68**: 820–823.

Knudson, A.G. (1978) Retinoblastoma: a prototypic hereditary neoplasm. *Semin. Oncol.* **5**: 57–60.

Lam, W.W.K., Deeble, J., Charlton, R., Taylor, G., Belchetz, P., Fitzpatrick, D. and Chu, C.E. (1998) Germline mutations in patients with multiple endocrine neoplasia type I (MEN1) *J. Med. Genet.* **35**: Suppl 1 S64, poster 05.41.

Larsson, C., Skogseid, B., Oberg, K., Nakamura, Y. and Nordenskjold, M. (1988) Multiple endocrine neoplasia type 1 gene maps to chromosome 11 and is lost in insulinoma. *Nature* **332**: 85–87.

Ludwig, L., Schleithoff, L., Kessler, H., Wagner, P.K., Boehm, B.O. and Karges, W. (1999) Loss of wild-type MEN1 gene expression in multiple endocrine neoplasia type 1-associated parathyroid adenoma. *Endocr. J.* **46**: 539–544.

Mailman, M.D., Muscarella, P., Schirmer, W.J., Ellison, E.C., O'Dorisio, T.M. and Prior, T.W. (1999) Identification of MEN1 mutations in sporadic enteropancreatic neuroendocrine tumors by analysis of paraffin-embedded tissue. *Clin. Chem.* **45**: 29–34.

Manickam, P., Vogel, A.M., Agarwal, S.K., Oda, T., Spiegel, A.M., Marx, S.J., Collins, F.S., Weinstein, B.M. and Chandrasekharappa, S.C. (2000) Isolation, characterization, expression and functional analysis of the zebrafish ortholog of MEN1. *Mamm. Genome* **11**: 448–454.

Martin-Campos, J.M., Catasus, L., Chico, A. *et al.* (1999) Molecular pathology of multiple endocrine neoplasia type I: two novel germline mutations and updated classification of mutations affecting MEN1 gene. *Diagn. Mol. Pathol.* **8**: 195–204.

Maruyama, K., Tsukada, T., Hosono, T., Ohkura, N., Kishi, M., Honda, M., Nara-Ashizawa, N., Nagasaki, K. and Yamaguchi, K. (1999) Structure and distribution of rat menin mRNA. *Mol. Cell. Endocrinol.* **156**: 25–33.

Marx, S.J., Spiegel, A.M., Skarulis, M.C., Doppman, J.L., Collins, F.S. and Liotta, A. (1998) Multiple endocrine neoplasia type 1: Clinical and Genetic Topics. *Ann. Intern. Med.* **129**: 484–494.

Matsubara, S., Sato, M., Ohye, H., Iwata, Y., Imachi, H., Yokote, R., Murao, K., Miyauchi, A. and Takahara, J. (1998) Detection of a novel nonsense mutation of the MEN1 gene in a familial multiple endocrine neoplasia type 1 patient and its screening in the family members. *Endocr J.* **45**: 653–657.

Mayr, B., Apenberg, S., Rothamel, T., von zur Muhlen, A. and Brabant, G. (1997) Menin mutations in patients with multiple endocrine neoplasia type 1. *Eur. J. Endocrinol.* **137**: 684–687.

Miyauchi, A., Sato, M., Matsubara, S., Ohye, H., Kihara, M., Matsusaka, K., Nishitani, A. and Takahara, J. (1998) A family of MEN1 with a novel germline missense mutation and benign polymorphisms. *Endocr. J.* **45**: 753–759.

Morelli, A., Falchetti, A., Martineti, V., Becherini, L., Mark, M., Friedman, E. and Brandi, M.L. (2000) MEN1 gene mutation analysis in Italian patients with multiple endocrine neoplasia type 1. *Eur. J. Endocrinol.* **142**: 131–137.

Mutch, M.G., Dilley, W.G., Sanjurjo, F., DeBenedetti, M.K., Doherty, G.M., Wells, S.A. Jr., Goodfellow, P.J., Lairmore, T.C. (1999) Germline mutations in the multiple endocrine neoplasia type 1 gene: evidence for frequent splicing defects. *Hum. Mutat.* **13**: 175–185.

Nord, B., Platz, A., Smoczynski, K. *et al.* Malignant melanoma in patients with multiple endocrine neoplasia type 1 and the involvement of the MEN1 gene in sporadic melanoma. *Int. J. Cancer* (in press).

Ohye, H., Sato, M., Matsubara, S., Miyauchi, A., Imachi, H., Murao, K., Takahara, J. (1998) Germline mutation of the multiple endocrine neoplasia type 1 (MEN1) gene in a family with primary hyperparathyroidism. *Endocr J.* **45**: 719–723.

Ohye, H., Sato, M., Matsubara, S., Miyauchi, A., Kishi-Imai, K., Murao, K., Takahara, J. (1999) A novel germline mutation of multiple endocrine neoplasia type 1 (MEN1) gene in a Japanese MEN1 patient and her daughter. *Endocr. J.* **46**: 325–329.

Poncin, J., Abs, R., Velkeniers, B. *et al.* (1999) Mutation analysis of the MEN1 gene in Belgian patients with multiple endocrine neoplasia type 1 and related diseases. *Hum. Mutat.* **13**: 54–60.

Roijers, J.F., De Wit, M.J., Van Der Luijt, R.B., Ploos Van Amstel, H.K., Hoppener, J.W., Lips, C.J. (2000) Criteria for mutation analysis in MEN 1-suspected patients: MEN 1 case-finding. *Eur. J. Clin. Invest.* **30**: 487–492.

Sakurai, A., Katai, M., Itakura, Y., Nakajima, K., Baba, K. and Hashizume, K. (1996) Genetic screening in hereditary multiple endocrine neoplasia type 1: Absence of a founder effect among Japanese families. *Jpn J. Cancer Res.* **87**: 985–994.

Sakurai, A., Katai, M., Itakura, Y., Ikeo, Y. and Hashizume, K. (1999) Premature centromere division in patients with multiple endocrine neoplasia type 1. *Cancer Genet. Cytogenet.* **109**: 138–140.

Sato, M., Matsubara, S., Miyauchi, A., Ohye, H., Imachi, H., Murao, K. and Takahara, J. (1998) Identification of five novel germline mutations of the MEN1 gene in Japanese multiple endocrine neoplasia type 1 (MEN1) families. *J. Med. Genet.* **35**: 915–919.

Sato, K., Yamazaki, K., Zhu, H. *et al.* (2000) Somatic mutations of the multiple endocrine neoplasia type 1 (MEN1) gene in patients with sporadic, nonfamilial primary hyperparathyroidism. *Surgery* **127**: 337–341.

Scappaticci, S., Maraschio, P., del Ciotto, N., Fassati, G.S., Zonta, A. and Fraccaro, M. (1991) Chromosome abnormalities in lymphocytes and fibroblasts of subjects with multiple endocrine neoplasia type 1. *Cancer Genet. Cytoget.* **52**: 85–92.

Schmidt, M.C., Henke, R.T., Stangl, A.P., Meyer-Puttlitz, B., Stoffel-Wagner, B., Schramm, J. and von Deimling, A. (1999) Analysis of the MEN1 gene in sporadic pituitary adenomas. *J. Pathol.* **188**: 168–173.

Schulte, K.M, Heinze, M., Mengel, M., Simon, D., Scheuring, S., Kohrer, K., Roher, H.D. (1999) MEN I gene mutations in sporadic adrenal adenomas. *Hum. Genet.* **105**: 603–610.

Shan, L., Nakamura, Y., Nakamura, M., Yokoi, T., Tsujimoto, M., Arima, R., Kameya, T. and Kakudo, K. (1998) Somatic mutations of multiple endocrine neoplasia type 1 gene in the sporadic endocrine tumors. *Lab. Invest.* **78**: 471–475.

Schmidt, M.C., Henke, R.T., Stangl, A.P., Meyer-Puttlitz, B., Stoffel-Wagner, B., Schramm, J. and von Deimling, A. (1999) Analysis of the MEN1 gene in sporadic pituitary adenomas. *J. Pathol.* **188**: 168–173.

Shimizu, S., Tsukada, T., Futami, H. *et al.* (1997) Germline mutations of the MEN1 gene in Japanese kindred with multiple endocrine neoplasia type 1. *Jpn J. Cancer Res.* **88**: 1029–1032.

Stewart, C., Parente, F., Piehl, F. *et al.* (1998) Characterization of the mouse Men1 gene and its expression during development. *Oncogene* **17**: 2485–2493.

Tanaka, C., Yoshimoto, K., Yamada, S., Nishioka, H., Ii, S., Moritani, M., Yamaoka, T. and Itakura, M. (1998a) Absence of germ-line mutations of the multiple endocrine neoplasia type 1 (MEN1) gene in familial pituitary adenoma in contrast to MEN1 in Japanese. *J. Clin. Endocrinol. Metab.* **83**: 960–965.

Tanaka, C., Kimura, T., Yang, P., Moritani, M., Yamaoka, T., Yamada, S., Sano, T., Yoshimoto, K. and Itakura, M. (1998b) Analysis of loss of heterozygosity on chromosome 11 and infrequent inactivation of the MEN1 gene in sporadic pituitary adeonomas. *J. Clin. Endocrinol. Metab.* **83**: 2631–2634.

Teh, B.T. (1998) Recent advances in multiple endocrine neoplasia type 1. *Curr. Opin. Diabetes Endocrinol.* **5**: 35–39.

Teh, B.T., Kytölä, S, Farnebo, F. *et al.* (1998a) Mutation analysis of the MEN1 gene in multiple endocrine neoplasia type 1, familial acromegaly and familial isolated hyperparathyroidism. *J. Clin. Endocrinol. Metab.* **83**: 2621–2626.

Teh, B.T., Esapa, C.T., Grandell, U., Houlston, R., Nordenskjöld, M., Larsson, C. and Harris, P.E. (1998b) Familial isolated hyperparathyroidism associated with a constitutional MEN1 mutation. *Am. J. Hum. Genet.* **63**: 1544–1549.

Teh, B.T., Zedenius, J., Kytölä, S. *et al.* (1998c) Thymic carcinoids in multiple endocrine neoplasia type 1. *Ann. Surg.* **228**: 99–105.

Toliat, M.R., Berger, W., Ropers, H.H., Neuhaus, P. and Wiedenmann, B. (1997) Mutations in the MEN I gene in sporadic neuroendocrine tumours of gastroenteropancreatic system. *Lancet* **350**: 1223.

Trump, D., Farren, B., Wooding, C. *et al.* (1996) Clinical studies of multiple endocrine neoplasia type 1 (MEN1). *Q. J. Med.* **89**: 653–669.

Valdes, N., Perez de Nanclares, G., Alvarez, V., Castano, L., Diaz-Cadorniga, F., Aller, J. and Coto, E. (1999) Multiple endocrine neoplasia type 1 (MEN1): clinical heterogeneity in a large family with a nonsense mutation in the MEN1 gene (Trp471Stop). *Clin. Endocrinol.* (Oxf) **50**: 309–313.

Wang, E.H., Ebrahimi, S.A., Wu, A.Y., Kashefi, C., Passaro, E. Jr. and Sawicki, MP. (1988) Mutation of the MENIN gene in sporadic pancreatic endocrine tumors. *Cancer Res.* **58**: 4417–4420

Wautot, V., Khodaei, S., Frappart, L., Buisson, N., Baro, E., Lenoir, G.M., Calender, A., Zhang, C.X. and Weber, G. (2000) Expression analysis of endogenous menin, the product of the multiple endocrine neoplasia type 1 gene, in cell lines and human tissues. *Int. J. Cancer.* **85**: 877–881.

Wenbin, C., Asai, A., Teramoto, A., Sanno, N. and Kirino, T. (1999) Mutations of the MEN1 tumor suppressor gene in sporadic pituitary tumors. *Cancer Lett.* **142**: 43–47.

Wermer, P. (1958) Genetic aspects of adenomatosis of endocrine glands. *Am. J. Med.* **16**: 363–371.

Zhuang, Z., Ezzat, S.Z., Vortmeyer, A.O. *et al.* (1997a) Mutations of the MEN1 tumor suppressor gene in pituitary tumors. *Cancer Res.* **57**: 5446–5451.

Zhuang, Z., Vortmeyer, A.O., Pack, S. *et al.* (1997b) Somatic mutations of the MEN1 tumor suppressor gene in sporadic gastrinomas and insulinomas. *Cancer Res.* **57**: 4682–4686.

Zimering, M.B., Brandi, M.L., deGrange, D.A., Marx, S.J., Steeten, E., Katsumata, N., Murphy, P.R., Sao, Y. and Friesen, H.G. (1990) Circulating fibroblast growth factor-like substance in familial multiple endocrine neoplasia type 1. *J. Clin. Endocrinol. Metab.* **70**: 149–154.

Zimering, M.B., Katsumata, N., Sato, Y., Brandi, M.L., Aurbach, G.D., Marx, S.J. and Friesen, H.G. (1993) Increased basic fibroblast growth factor in plasma from multiple endocrine neoplasia type 1: relation to pituitary tumor. *J. Clin. Endocrinol. Metab.* **76**: 1182–1187.

8

Multiple endocrine neoplasia type 2

Charis Eng

1. Introduction

Multiple endocrine neoplasia type 2 (MEN 2) is an autosomal dominant inherited cancer syndrome whose incidence is 1: 500 000 live births (de la Chapelle and Eng, 1999). MEN 2 is characterized by medullary thyroid carcinoma (MTC), pheochromocytoma (PC) and hyperparathyroidism (HPT). Germline mutations in the *RET* proto-oncogene on 10q11.2, which encodes a receptor tyrosine kinase, cause MEN 2 (Eng, 1999; Eng *et al.*, 1996a). With the discovery of a limited number of hotspot *RET* mutations in MEN 2, molecular diagnostic testing and predictive testing for this syndrome not only became a practical reality but also has become the standard of care for all MEN 2 and MTC cases.

From the fundamental scientific point of view, RET is unusual in that it needs to interact with one of four co-receptors before it can bind one of four ligands. Downstream, it signals down the MAP kinase (MAPK)-RAS-RAF pathway as well as the phosphoinositide 3-kinase (PI3K) and PLCγ pathways. Gain-of-function *RET* mutations have been associated with its neoplastic properties while loss-of-function mutations or haploinsufficiency result in a non-neoplastic, developmental disorder, Hirschsprung disease (HSCR), whose end result is aganglionosis of the gut or absent nervous innervation of hindgut derivatives.

2. Clinical features of MEN 2

MEN 2 is traditionally divided into three sub-types based on the clinical features. MEN 2A, the most common sub-type, is characterized by MTC in >99% of affected individuals, PC in 50% and 15–30% in HPT (reviewed in (Eng, 1999)). MEN 2B, which accounts for 5% of all MEN 2, is similar to MEN 2A except that the average age of tumor onset is 10 years earlier, HPT is not clinically apparent, and such stigmata as marfanoid habitus, ganglioneuromatosis of the gut, and mucosal neuromatosis are present. Familial MTC (FMTC) is operationally defined as two or more related MTC cases in a family in the absence of objective

Molecular Genetics of Cancer second edition, edited by J.K. Cowell.

evidence of PC and HPT in affected members and at-risk members (Farndon *et al.*, 1986).

Like most, if not all, inherited cancer syndromes, the MEN 2 component tumors are multifocal, and if paired organs are involved, e.g. adrenal glands, bilateral involvement is common. Synchronous and metachronous component tumors and multiple primary tumor presentations occur with some frequency. In MEN 2, MTC is almost always the first presenting sign. PC or HPT as the first and only presenting sign is the exception.

3. Genetics of MEN 2

MEN 2 is inherited as an autosomal dominant disorder with age-related pene-trance (Easton *et al.*, 1989b; reviewed in Eng, 1999). Within a single family and certainly among families, there is a range of variation with respect to disease expression such as age of onset and incidence of each type of component tumor. Classically, penetrance is said to be 70% by 70 years of age for MEN 2A (Easton *et al.*, 1989a). Penetrance is higher at a younger age group in MEN 2B and likely lower for FMTC.

3.1 Germline mutations of the RET proto-oncogene in MEN 2

Germline *RET* mutations are associated with >92% of MEN 2 overall (Donis-Keller *et al.*, 1993; Eng *et al.*, 1996a; Mulligan *et al.*, 1993, 1995). In MEN 2A, >98% of cases have an identifiable germline *RET* mutation, all of which are located in the cysteine-rich extra-cellular domain encoded by exons 10 and 11 (*Figure 1*) (Eng *et al.*, 1996a). In MEN 2B, 95% harbor a germline *RET* codon 918 (exon 16) mutation, M918T (*Figure 1*) (Carlson *et al.*, 1994b; Eng *et al.*, 1994, 1996a; Hofstra *et al.*, 1994). Another 2–3% carry the germline A883F (exon 15) mutation (Gimm *et al.*, 1997; Smith *et al.*, 1997). Thus, 97–98% of MEN 2B cases have an identifiable *RET* mutation at one of two hotspots. Approximately 85% of FMTC probands have been found to have a germline *RET* mutation (Eng *et al.*, 1996a). In contrast to the mutations associated with MEN 2A and MEN 2B, *FMTC* mutations are more scattered and varied (*Figure 1*) (Eng *et al.*, 1996a). They not only involve similar cysteine codons as those associated with MEN 2A, but they also affect residues of the intracellular domain, codons 768 (exon 13), 804 (exon 14) and 891 (exon 15) (Bolino *et al.*, 1995; Eng *et al.*, 1995d, 1996a; Hofstra *et al.*, 1997). It is currently unclear if the rare mutations found at codons 790 and 791 in the tyrosine kinase domain, thus far found only in German families (Berndt *et al.*, 1998), are peculiar to a certain geographic region or are more universal.

3.2 De novo MEN2 mutations and parent-of-origin

MEN 2A and FMTC cases are mainly found in families. Between ≤ 1 and 10% of all MEN 2A and FMTC probands have *RET* mutations which arise *de novo* (Mulligan *et al.*, 1994c; Schuffenecker *et al.*, 1997; Wohlik *et al.*, 1996). To date, all (10/10 informative cases) *de novo* MEN 2A-FMTC *RET* mutations have been

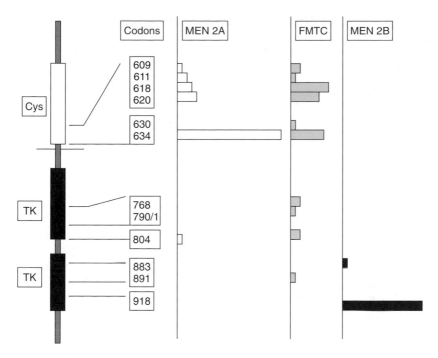

Figure 1. Relative frequency and distribution of germline *RET* mutations in MEN 2A, MEN 2B and FMTC. Cys, cysteine-rich domain; TK, tyrosine kinase domain.

shown to arise on the paternal allele (Mulligan *et al.*, 1994c; Schuffenecker *et al.*, 1997; Wohlik *et al.*, 1996). In *de novo* MEN 2B, which occurs 40% of the time, all but two mutations examined have been shown to arise on the paternal allele as well (Carlson *et al.*, 1994a; Kitamura *et al.*, 1995). In these instances, preferential susceptibility of paternally derived DNA has been invoked. In MEN 2 and several other inherited cancer syndromes, e.g., bilateral retinoblastoma, type 1 neurofibromatosis and Wilms tumor, *de novo* mutations have been associated with increased paternal age (Carlson *et al.*, 1986; reviewed by Sapienza and Hall, 1994). Thus, the preferential susceptibility of paternally derived DNA has been attributed to the increased paternal age during conception.

Among all series comprising *de novo* MEN 2 cases, there was a significant sex ratio distortion. When the mutations arose on the paternal alleles, it would appear that transmission to females was preferred (Carlson *et al.*, 1994a; Schuffenecker *et al.*, 1997). If type I error (simply a chance finding) is not involved, then this type of observation could reflect genomic imprinting.

3.3 RET genotype and its relationship to phenotypic features in MEN 2

There are at least two main reasons for examining genotype–phenotype correlations. The first is to gain insight from genetic clues as to the function of RET with respect to organ-specific development of tumors. The second is to determine if such data could be applicable to clinical practice in this era of data-based, cost-effective medicine. To date, reports of genotype–phenotype associations are based on moderately sized (100 or fewer families) series originating from single referral centers (Frank-Raue *et al.*, 1996; Mulligan *et al.*, 1994b; Schuffenecker *et al.*, 1994) or large (hundreds of families) international collaborative studies where pooled analyses are performed (Eng *et al.*, 1996a; Mulligan *et al.*, 1995). The most recent study of the International *RET* Mutation Consortium (IRMC) pooled 477 unrelated MEN 2 families from 18 centers of excellence around the world for genotype–phenotype analyses (Eng *et al.*, 1996a). Unlike single center analyses, phenotypes and operational phenotypic classifications in the IRMC analysis were made uniform.

Of these 477 MEN 2 families, 42.6% were classic MEN 2A, 16.6% MEN 2B and 7.1% FMTC. The remaining were operationally classified into an 'other' category comprising 'small' FMTC (≤ 3 affected members) and incompletely documented MEN 2A families. Over 98% of the MEN 2A families, 95% MEN 2B and 85% FMTC were found to have germline *RET* mutations (Eng *et al.*, 1996a).

Germline mutations at codons 609, 611, 618, 620 and 634 were associated with MEN 2A (Eng *et al.*, 1996a). Eighty-five percent of *MEN2A* mutations occurred at codon 634. Among codon 634 mutations, the most frequent alteration was a TGC (cysteine) to CGC (arginine) (C634R). The presence of any germline mutation at codon 634 was found to be highly associated with the development of PC and HPT in a given MEN 2A family (Eng *et al.*, 1996a). In the IRMC dataset as a whole, C634R was associated with specific development of HPT. This correlation has been controversial. The original single center analysis from Cambridge, UK first unveiled this association (Mulligan *et al.*, 1994b) but various small studies from continental Europe could not confirm it (Frank-Raue *et al.*, 1996; Schuffenecker *et al.*, 1994). If the original Cambridge dataset (Mulligan *et al.*, 1994b) were removed from the IRMC analysis, the C634R-HPT association was no longer significant (Eng *et al.*, 1996a). Initially explained by geographic differences or differences in criteria for the diagnosis of HPT between the UK families and those of continental Europe (Eng *et al.*, 1996a), another alternative, but not mutually exclusive, explanation is possible. A recent French population-based study examining the risk and penetrance of HPT in MEN 2A families again demonstrated a lack of association between C634R and the development of HPT on a family-as-unit basis but revealed a highly significant association between C634R and the development of HPT on an individual basis (Schuffenecker *et al.*, 1998). Additionally, the risk for HPT development rose sharply after the age of 30. Hence, it is possible that the members of MEN 2 families ascertained in the UK were, on average, older than those from continental Europe. From a clinical point of view, the C634R–HPT correlation is somewhat moot: the presence of any codon 634 mutation should alert the clinician to the risk of both PC and HPT.

Germline *RET* mutations in FMTC were distributed over the codons in a more uniform relative frequency (Eng *et al.*, 1996a). For example, 30% of all mutations found in FMTC families occurred at codon 634 in contrast to 85% for MEN 2A. Interestingly, amongst this modest group of FMTC families, no C634R mutations were noted. Clinicians should, therefore, be alerted if they find an apparent FMTC family with a C634R mutation: they should rigorously pursue PC and HPT screening in all affected or mutation positive members. A novel correlation was found: E768D and V804L seemed to be associated with FMTC and 'small' FMTC only. Whether such families can forego PC and HPT surveillance in the future, as the Consortium accrues larger numbers of families with codon 768 and 804 mutations and has longer follow-up of existing families, is as yet unknown. Indeed, an anecdotal family who carried germline V804L and only had MTC had a family member with MTC also develop a unilateral pheochromocytoma (Nilsson *et al.*, 1999). It is difficult to differentiate whether a unilaternal unifocal pheochromocytoma is a coincidence or truly part of this family's MEN 2. Additionally, the investigators were cautious enough to exclude the coincidental occurrence of von Hippel-Lindau disease in this individual by performing *VHL* gene analysis (Nilsson *et al.*, 1999).

The V804M mutation was first reported in a FMTC family after the 1996 IRMC analysis (Fink *et al.*, 1996). Subsequently, two small FMTC families were ascertained when their probands presented with apparently sporadic MTC after the age of 60 years (Shannon *et al.*, 1999). Both carried V804M mutations. It should be noted that a mutation positive, clinically unaffected, member of one of these families was found to have microcarcinoma in her late twenties at prophylactic surgery.

3.4 Occult or de novo germline RET mutations in apparently sporadic MTC, PC and HPT patients

Clinical epidemiologic observations have revealed that 25% of all MTC presentations are hereditary, thus 75% are sporadic. There are several series examining the frequency of germline *RET* mutations in apparently sporadic MTC. Most series ascertained such patients by relatively stringent criteria, including those with no associated features suggestive of MEN 2 in a potential subject, known family history of MEN 2, and any family history suggestive of MEN 2 (family histories were taken to at least second degree relatives). These three series suggest that between 1.5 and 10% of apparently sporadic MTC cases will carry occult or *de novo* germline *RET* mutations (Eng *et al.*, 1995c; Schuffenecker *et al.*, 1997; Wohlik *et al.*, 1996). Note should be made, however, that because each of these projects spanned several years, the total number of known MEN 2-associated mutations examined increased as time went on. For example, the 1995 study only looked for the known hotspots within exons 10, 11, 13 and 16 (Eng *et al.*, 1995c) while the 1997 one looked for mutations in exons 10, 11, 13, 14 and 16 (Schuffenecker *et al.*, 1997). Because of the putative low penetrance of mutations at codon 804 (exon 14), it is conceivable that many of the so-called isolated MTC cases will carry these mutations (Fink *et al.*, 1996; Shannon *et al.*, 1999). A fourth series essentially took 'all comers' with MTC and found a germline *RET* mutation rate of 25% (Borst *et al.*, 1995), which agrees with the figures obtained in clinical epidemiologic studies. An informal survey of apparently sporadic MTC cases collected by the International *RET* Mutation Consortium revealed an occult or *de novo* germline mutation frequency of approximately 3–4% (Eng C, Mulligan LM, unpublished data). Recently, a particular polymorphic variant within *RET*, S836S (c.2439C>T; exon 15), was found to be over-represented among cases with sporadic MTC compared to region-matched controls (Gimm *et al.*, 1999). This conferred a relative risk (RR) of approximately 2. Of note, approximately 90% of cases with the polymorphic variant also had MTC tumors harboring somatic *RET* M918T mutation (Gimm *et al.*, 2000). Either the variant itself or another locus in linkage disequilibrium appears to be acting as a low penetrance allele conferring suscepti- bility to 'sporadic' MTC.

In contrast to MTC, occult or *de novo* germline mutations in apparently sporadic PC presentations are relatively uncommon, and this is especially true if careful medical and family histories have been obtained. For example, the first series that systematically examined this issue comprised 48 apparently sporadic PC patients, among which only one (2%) was shown to have a germline *RET* mutation (Eng *et al.*, 1995a). In this instance, when the referring clinician was asked to re-examine the patient, who was already a young adult, and first degree relatives, he discovered that the patient had an MTC, and the father had a large neck mass which was found to be MTC as well. Further, a more extensive family history revealed the index case's paternal grandfather dying of 'a goitre' (Eng *et al.*, 1995a). Three other series revealed no occult or *de novo* germline *RET* mutations in apparently isolated PC cases (Beldjord *et al.*, 1995; Hofstra *et al.*, 1996; Lindor *et al.*, 1995). Recently, over- representation of polymorphic alleles of the *CUL2* gene were described in sporadic PC cases originating from Brazil (Duerr *et al.*, 1999). CUL2, a member of the cullin family believed to play a role in the ubiquitination-proteosome degradative pathway, binds the VHL/elongin-B/elongin-C complex (see Chapter 4). The precise mechanism of low penetrance susceptibility by *CUL2* alleles is currently unknown.

No *RET* mutations have been found in apparently sporadic HPT patients (Komminoth *et al.*, 1996).

3.5 Somatic RET mutations in sporadic MTC, PC and HPT

Somatic mutations of the *RET* proto-oncogene have been found in sporadic MTC. Depending on series, and mutation detection technology used, the somatic *RET* mutation frequency can range from 25–70% in sporadic MTC (Blaugrund *et al.*, 1994; Eng *et al.*, 1994, 1995b; Hofstra *et al.*, 1994; Komminoth *et al.*, 1995; Marsh *et al.*, 1996b; Romei *et al.*, 1996; Zedenius *et al.*, 1994). Somatic M918T make up the largest proportion of all somatic mutations in sporadic MTC. Of note, when microdissected subpopulations of MTC were examined, the somatic *RET* mutation status was found to be heterogeneous even within a single MTC and among metastases (Eng *et al.*, 1996b). Interestingly, there was a high concordance rate between subpopulations with *RET* mutation and RET expression as evidenced by immunohistochemistry (Eng *et al.*, 1998). Approximately 80% of all MTC tumors in this manner were found to harbor at least one subpopulation with somatic *RET* mutation, mainly M918T (Eng *et al.*, 1996b). These observations likely reflect clonal evolution within MTC, a relatively slow growing tumor and/or polyclonal origin. The latter seems to be corroborated by an independent study using X-chromosome inactivation (Ferraris *et al.*, 1997). In rare cases, somatic M918T has been detected in MTC subpopulations in MTC from MEN 2A/FMTC patients (Eng *et al.*, 1996b; Marsh *et al.*, 1996a).

In contrast to sporadic MTC, sporadic PC were found to carry somatic *RET* mutations in approximately 10% of tumors (Beldjord *et al.*, 1995; Eng *et al.*, 1995a; Komminoth *et al.*, 1994). While the great majority of somatic *RET* mutations in sporadic MTC are M918T, there seems to be representation of other *RET* mutations in sporadic PC. Interestingly, the non-M918T *RET* mutations in PC are not necessarily those found in the germline in patients with MEN 2. Of note, no somatic *RET* mutations have been found in sporadic HPT (Komminoth *et al.*, 1996).

4. Germline *RET* mutations in Hirschsprung disease

4.1 RET and Hirschsprung disease

Hirschsprung disease (HSCR), characterized by aganglionosis of the gut, was seemingly unrelated to MEN 2. This syndrome is usually sporadic but a proportion are familial as well, although the genetics were felt to be complex (Badner *et al.*, 1990). When a putative gene for HSCR was mapped to 10q11.2 (Angrist *et al.*, 1993; Lyonnet *et al.*, 1993), *RET*, which is expressed in the precursors of the enteric ganglia, became a prime candidate. Indeed, loss-of-function germline *RET* mutations have been found in a proportion of sporadic and familial HSCR (Edery *et al.*, 1994; Romeo *et al.*, 1994). Among various highly selected series, approximately 50% of familial HSCR and 30% of isolated cases were shown to have germline *RET* mutations scattered throughout the 21-exon gene (Angrist *et al.*, 1995; Attié *et al.*, 1994, 1995; Luo *et al.*, 1994; Myers *et al.*, 1995). However, an unselected, population-based HSCR series in the catchment area of Stockholm revealed a *RET* mutation frequency of 3% (Svensson *et al.*, 1998).

When assayed for loss-of-function, *HSCR-RET* mutations result in haploinsufficiency, whether the mechanism is structural, e.g. whole gene deletion, nonsense mutation, frameshift mutation, or functional, e.g. lack of maturation of receptor to cell surface due to missense mutations (see below).

Recent analysis of a population-based series of 64 isolated HSCR cases in Western Andalusia, Spain revealed over-representation of polymorphic *RET* alleles in HSCR cases compared to region-matched controls (Borrego *et al.*, 1999). One specific polymorphism at codon 45 (A45A, c.135G>A) was highly significantly over-represented among HSCR cases compared to controls (Borrego *et al.*, 1999). Subsequently, these data were replicated in HSCR cases from other population bases, namely, Germany and the United Kingdom (Fitze *et al.*, 1999; Sancandi *et al.*, 2000). Taken together, these observations suggest several hypotheses. Either the polymorphic variants themselves affect RET expression, e.g., by introducing a cryptic splice site, or there is another locus in linkage disequilibrium with the polymorphic loci within *RET* which predisposes to isolated HSCR in a low penetrance manner (Borrego *et al.*, 1999, 2000).

In addition to *RET*, other minor susceptibility genes for HSCR exist, including the genes encoding endothelin receptor beta, its ligand endothelin-3, GDNF, NTN (ligands for RET – see below) and SOX10 (Amiel *et al.*, 1996; Doray *et al.*, 1998; Puffenberger *et al.*, 1995; Southard-Smith *et al.*, 1998).

4.2 RET mutations in families segregating MEN 2 and HSCR

There are at least 14 families segregating MEN 2A or FMTC and HSCR and each of these families harbor germline *RET* mutations, mainly affecting codons 618 and 620 (Borrego *et al.*, 1998; Borst *et al.*, 1995; Caron *et al.*, 1996; Decker and Peacock, 1998; Eng *et al.*, 1996a; Inoue *et al.*, 1999; Mulligan *et al.*, 1994a). Most of the mutations are C618R and C620R. All of the earlier papers noted that a second mutation within *RET* could not be found despite complete sequencing of the gene in these families. Recently, however, a single family segregating MEN 2 and HSCR with germline *RET* C620S was described and may lend some insight (Borrego *et al.*, 1998). Only the individual with the C620S as well as homozygosity for the polymophic sequence variant at codon 45 had both HSCR and MEN 2. Taken together with the HSCR-low penetrance allele data (Section 4.1) (Borrego *et al.*, 1999), it is possible that variants within *RET* can act as phenotypic modifiers, although other explanations also exist (below).

5. Clinical management of individuals at risk

5.1 Management of MEN 2 and suspected MEN 2 in the pre-DNA era

First-degree relatives (children, parents, siblings) of affected individuals are at 50% risk of inheriting the mutated gene. Before DNA-based predictive testing, all unaffected individuals at 50% risk were subjected to annual screening for MTC, PC and HPT from the age of 6 to the age of 35 years. This involves pentagastrin-stimulated calcitonin levels, 24-hour urinary levels for catecholamines and serum calcium and parathyroid levels. Especially in the US, many centers advocated prophylactic thyroidectomy in individuals who are first degree relatives of affected individuals prior to the age of 6 for two reasons: firstly, the youngest age at diagnosis reported for MTC in MEN 2A is around age 6 years (Telander and Moir, 1994; Wells *et al.*, 1994), and secondly, MTC can be lethal.

5.2 DNA-based predictive testing in MEN 2

The strategy of early prophylactic thyroidectomy is lifesaving in 50% of all clinically at-risk individuals but would represent a senseless, and not trivial, procedure, condemning a child to lifelong thyroid replacement, in the other 50%. Pentagastrin screening, on the other hand, appears to be less invasive, but the trade-off is lack of sensitivity when it is needed most – prior to the teens. False positives and negatives obtained with pentagastrin screening are well documented (Lips *et al.*, 1994; Marsh *et al.*, 1994; Neumann *et al.*, 1995; Zedenius *et al.*, 1995).

Since mutations of the *RET* proto-oncogene have been identified in >92% of all MEN 2 families, DNA-based testing is possible. This has such distinct advantages as not having age-dependent sensitivity, being useful as a molecular diagnostic test to confirm a clinical diagnosis of MEN 2, and most importantly, being useful as a predictive test for asymptomatic clinically at-risk individuals. In a known MEN 2A or FMTC family, a clinically at-risk individual should undergo DNA testing prior to the age of 6 (Wells *et al.*, 1994) or certainly prior to surgery.

Predictive testing for *RET* mutations in MEN 2 kindreds is performed using PCR-based protocols; target exonic sequences are amplified either for direct sequencing and/or restriction endonuclease digestion, where mutations would create or cause loss of specific sites (Lips *et al.*, 1994; Marsh *et al.*, 1994; McMahon *et al.*, 1994). Such tests are reproducible, accurate and allow presymptomatic identification of at-risk individuals (Lips *et al.*, 1994; Neumann *et al.*, 1995), permitting more effective and timely surgical intervention or initiation of screening, although the course advocated by many is prophylactic thyroidectomy prior to the age of 6 years (Wells *et al.*, 1994).

In a known MEN 2 family with an identified family-specific mutation, therefore, the detection of the same mutation in a clinically at-risk individual indicates that that person has MEN 2. Conversely, if a mutation is not detected, that individual will not have MEN 2. Barring administrative errors, DNA-based predictive testing is 100% accurate. Thus, individuals found to carry a germline *RET* mutation can be subjected to targeted screening for the presence of PC and HPT annually from the age of 6 and offered prophylactic thyroidectomy (reviewed in Learoyd *et al.* (1995) and Eng and Ponder (1998)). The only exception is the discovery of the MEN 2B-specific M918T or A883F mutations: those should be offered prophylactic thyroidectomy at a younger age. Those clinically at-risk individuals who test mutation negative can be reassured, do not have to be subjected to annual biochemical screening and can be spared prophylactic thyroidectomy. In general, there is no need for prophylactic parathyroidectomy. Once primary HPT is diagnosed or when operating on the thyroid gland, all four parathyroids need to be identified and all enlarged parathyroids should be removed. In case of enlargement of all four parathyroids, the least pathologic gland needs to be identified. The least pathologic part should either remain *in situ* or should be autotransplanted into an easily accessible muscle. Similar results should be achieved with either technique (Decker *et al.*, 1995; Wells *et al.*, 1994).

RET testing has been recommended by the American Society of Clinical Oncology (Offit *et al.*, 1996) and is the clinical standard of care for MEN 2 in this and other countries worldwide. This is because the great majority (>92%) of MEN 2 cases have germline *RET* mutations and that the determination of mutation status does influence medical management (Eng, 1996, 1999). In general, in the US, the standard practice is to perform a prophylactic thyroidectomy between ages 5 and 10 years in mutation-positive asymptomatic individuals, except in individuals with the MEN 2B-specific mutations, M918T or A883F, when prophylactic thyroidectomy should occur before the age of 3 years. This strategy is advocated, and seems obvious, for the highly penetrant mutations at codons 918, 883 and 634, or in families where the diagnoses of MTC has occurred at a young age. However, recent clinical genetic and preliminary functional data have suggested that mutations such as the exon 10 mutations, codon 768 and codon 804 mutations may have low penetrance (Carlomagno *et al.*, 1997; Ito *et al.*, 1997; Moers *et al.*, 1996; Pasini *et al.*, 1997; Shannon *et al.*, 1999). These types of mutations tend to occur only in FMTC and 'small' FMTC families, already an indication that the full blown syndrome with PC and HPT is not manifested; secondly, the codon 804 mutations seem to confer a later age of onset and the proband usually presents as an apparently sporadic case (Shannon *et al.*, 1999). If these anecdotes and *in vitro* studies can be proven in larger studies, then clinical management might actually be tailored, such as targeted biochemical surveillance or later prophylactic surgery for individuals who carry low penetrance mutations.

In a syndrome where over 92% of cases have identified germline mutations, it can be uncomfortable when a clinician is faced with a *RET* mutation-negative MEN 2 family. The mutation negative family will most likely tend to be a 'small' FMTC family. Since MTC is rare, the occurrence of even two MTC in a single family is unusual. So, the chances that this has occurred entirely by coincidence are low. It would, therefore, be most conservative to treat these sorts of families as we would any MEN 2 family prior to the era of DNA-based diagnosis.

5.3 Germline RET mutation analysis for apparently sporadic MTC cases

Differentiating between truly sporadic from hereditary MTC is important as it has implications for the clinical care of the individual patient as well as his/her family. Because there is a finite proportion of apparently sporadic MTC cases that have been consistently found to carry occult or *de novo RET* mutation (see Section 3.4) and because *RET* mutation analysis has a high sensitivity, it is the clinical standard of care to perform genetic testing in all presentations of MTC, regardless of age at diagnosis or family history (Eng, 1997, 1999). Currently, this recommendation does not hold for apparently sporadic PC and HPT presentations: if careful medical and family histories are obtained, no occult *RET* mutations should be found in these cases (see Section 3.4).

5.4 Mutation analysis for HSCR in the clinical setting?

Because there are several genes which may be considered HSCR predisposing genes, it is not recommended that genetic testing be instituted for HSCR in the clinical setting.

6. The *RET* proto-oncogene

The *RET* proto-oncogene, a 21-exon gene localized to 10q11.2, encodes a receptor tyrosine kinase (RTK) expressed primarily in cells and lineages derived from the neural crest, branchial arches and kidney (Attié-Bitach *et al.*, 1998; Gardner *et al.*, 1993; Ivanchuk *et al.*, 1997; Myers *et al.*, 1995; Nakamura *et al.*, 1994; Santoro *et al.*, 1990; Takahashi *et al.*, 1985). Expression is developmentally regulated, with highest levels in early embryonic stages (Attié-Bitach *et al.*, 1998; Ivanchuk *et al.*, 1997, 1998). Like other RTKs, RET is comprised of an intracellular kinase domain, a transmembrane domain and a cysteine-rich domain, implicated in tertiary structure formation Takahashi *et al.*, 1985, 1988, 1989, 1991, 1993; Takahashi and Cooper, 1987).

In contrast to other mammalian RTKs, however, RET is unique in that it requires ligand and coreceptor for activation. One of four related ligands, GDNF, NTN, PSP and ART, needs to interact with one of four coreceptors, GFRα-1, 2, 3 and 4 before the heterotetrameric complex can bind RET (*Figure 2*) (Baloh *et al.*, 1997; Buj-Bello *et al.*, 1997; Davies *et al.*, 1997; Enokido *et al.*, 1998; Jing *et al.*,

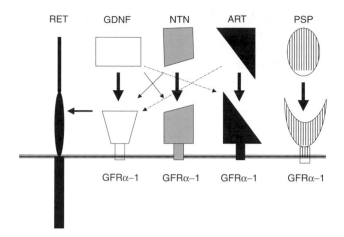

Figure 2. A multicomponent complex comprising RET, a co-receptor and a ligand are required for RET signaling. GDNF preferentially binds GFRα-1 with high affinity, GFRα-2 with lower affinity and GFRα-3 with lowest affinity. NTN binds GFRα-2 with high affinity and GFRα-1 with lower affinity. ART and PSP bind GFRα-3 and GFRα-4, respectively.

1996, 1998; Kotzbauer *et al.*, 1996; Sanicola *et al.*, 1997; Thompson *et al.*, 1998; Treanor *et al.*, 1996; Trupp *et al.*, 1996; Vega *et al.*, 1996). As GFRα family members have no transmembrane domain or cytoplasmic region, they were initially believed not to have direct signaling capabilities until recently, when GDNF treatment of neuronal cell lines not expressing RET resulted in GFRα-1-related Src-like kinase activity (Trupp *et al.*, 1999). Nonetheless, the major role of the GFRα family members in RET activation is to act as adapter molecules. Ligand and GFRα form a complex which then binds to RET and triggers RET dimerization and autophosphorylation as well as phosphorylation of downstream molecules (Baloh *et al.*, 1998). GDNF binds with high affinity to GFRα-1 and to a lesser extent GFRα-2 and with the lowest affinity to GFRα-3 (*Figure 2*) (Sanicola *et al.*, 1997; Trupp *et al.*, 1998). NTN binds GFRα-2 with high affinity but can also use GFRα-1 as its adaptor (Jing *et al.*, 1998). Thus far, PSP has been shown to bind GFRα-4 only (Enokido *et al.*, 1998; Thompson *et al.*, 1998) although it is currently unclear whether GFRα-4 has a human homolog. ART uses GFRα-3 as its high affinity receptor, and to a lesser extent GFRα-1 (Baloh *et al.*, 1998).

6.1 Downstream targets of RET signaling

All the natural downstream targets of RET activation are still unknown despite much work in the last decade. To date, it would appear that activation of RET can result in growth, survival or differentiative signals depending on the downstream pathways triggered, the cell type as well as the developmental stage. Of a total of 18 intracellular tyrosines, all of which are potential phosphorylation sites, 9 have been shown to be phosphorylated as a direct result of RET activation (Liu *et al.*, 1996). Activated (phosphorylated) RET has been shown to interact either directly or indirectly with several adaptor molecules containing src homology (SH2) domains, such as Grb2 (P-Y1096 is the docking site), Grb7 (Y905), Grb 10 (Y905), phospholipase Cγ (Y1015), Crk and Nck and with enigma proteins (*Figure 3*) (Alberti *et al.*, 1998; Bocciardi *et al.*, 1997; Borrello *et al.*, 1994; Durick *et al.*, 1998b; Murakami *et al.*, 1999a; Pandey *et al.*, 1995).

Shc adaptor proteins have been shown to bind RET *in vivo*, which leads to activation of the RAS-MAP kinase (MAPK) pathway (*Figure 3*) (Asai *et al.*, 1995; Santoro *et al.*, 1995). The tyrosine residue Y1062 has been shown to play a significant role in Shc binding, among other adaptor molecules, as it is a multi-functional docking site (Asai *et al.*, 1996). Y1062 appears to be a phosphorylation-dependent docking site for Shc and a phosphorylation-independent site for interaction with the membrane-anchored enigma through its LIM domain, the latter believed to play a role of positioning RET in the cellular membrane (Durick *et al.*, 1996, 1998a). RET activation of the RAS-MAPK pathway is mediated by Shc and Grb2 or Grb2 alone.

Apart from RAS-MAPK, the phosphoinositol 3-kinase (PI3K) pathway is also an important downstream pathway for RET signaling (*Figure 3*). It is believed that RET activation of PI3K occurs indirectly via Shc, Grb2 and GAB1 or via GAB1 alone (Murakami *et al.*, 1999b). Through PI3K, RET activates the PKB/Akt pathway, which is an anti-apoptotic pathway, and it has been shown that the oncogenic potential of RET is dependent on Akt (Murakami *et al.*, 1999b; Segouffin-Cariou and Billaud, 2000). Because there is a panoply of pathways downstream of the Akt pathway, it is conceivable that RET-associated survival and growth signals are also mediated by Akt-dependent apoptotic pathways which include Bad and Bax-Bcl. It is also believed that RasGAP and Nck can be recruited in a PI3K and p62Dok dependent manner, downstream of which cell motility is affected, and phosphorylation of molecules involved in cell–cell communication and cell migration, such as focal adhesion kinase, paxillin and p130Cas (Murakami *et al.*, 1999a, 1999b; Romano *et al.*, 1994).

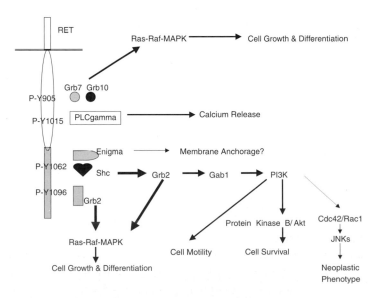

Figure 3. Activated RET signals down several downstream pathways. See text for details.

It would also appear that RET signaling down to the Jun kinases (JNK) and the extracellular signal-regulated protein kinase (ERK), both groups of which are considered MAPKs, diverge (Chiariello *et al.*, 1998). RET activation of JNK may be mediated by Rho/Rac-related small GTPases, specifically cdc42 (Chiariello *et al.*, 1998). To date, it is unclear if JNK is activated in a PI3K-dependent manner.

6.2 Gain-of-function mutations in neuroendocrine neoplasia

In general, the germline mutations that characterize MEN 2A and many FMTC are missense mutations which change a cysteine codon to another non-sulfhydryl-containing amino acid. Like other receptor tyrosine kinases, the cysteines of RET form intramolecular disulfide bonds that help form the ligand-binding pocket. When one of the cysteines is mutated, its respective partner cysteine can no longer form a disulfide bond with it, and hence, a free sulfhydryl group is exposed. Two mutant receptors with free sulfhydryl groups can then form intermolecular disulfide bonds, thus mimicking ligand activation in a constitutive manner (Asai *et al.*, 1995; Borrello *et al.*, 1995; Santoro *et al.*, 1995).

While it is true that a single gain-of-function mutation in RET can cause transformation, at least *in vitro* (Borrello *et al.*, 1995; Santoro *et al.*, 1995), *in vivo*, different cysteine codon mutations result in different phenotypes (Eng *et al.*, 1996a) (Section 3.3). In general, MTC, PC and HPT, i.e., classic MEN 2A, are more prevalent with codon 634 mutations, and it would appear that the ages of onset are relatively young compared to those with mutations at codons 609, 611, 618 and 620, where both MEN 2A and FMTC phenotypes can result. Indeed, early stable transfection studies demonstrated that constructs with the C634R mutation resulted in more transformants on focus assay than those with the C620R mutation (Santoro *et al.*, 1995). Subsequently, it has been shown that missense mutations affecting the extracellular domain result in RET molecules that fail to mature and fail to reach the cell surface (Carlomagno *et al.*, 1997; Ito *et al.*, 1997). Accordingly, mutations of cysteine codons closest to the transmembrane domain (e.g. C634R) result in the greatest fraction of receptors that reach the cell surface while those in cysteine codons farthest from the transmembrane domain (e.g. C609W) result in the lowest fraction of receptors that reach the cell surface. Thus, these observations can explain, at least partially, the relative penetrance of mutations at codon 634 versus those at cysteine codons away from the transmembrane domain. RET Y905 appears to be an important residue with respect to *MEN2A* mutations and P-Y905 is required for MEN 2A-related transformation (Iwashita *et al.*, 1996). Further, the ability for MEN 2A-related mutant RET to transform is dependent on an intact PI3K-Akt pathway (Segouffin-Cariou and Billaud, 2000).

The MEN 2B-specific mutation at codon 918 occurs at a residue which lies in the substrate recognition pocket of the catalytic core of the tyrosine kinase domain (Hanks *et al.*, 1988). This methionine residue is highly conserved among receptor tyrosine kinases as well as across species (Hanks *et al.*, 1988). M918T changes the methionine to a threonine. Comparisons at that equivalent position reveals that cytosolic tyrosine kinases have a threonine in that position (Hanks *et al.*, 1988). *In vitro*, substitution of the methionine for a threonine results in loss of substrate specificity (Santoro *et al.*, 1995; Songyang *et al.*, 1995). M918T-RET constructs 'preferred' substrates specific to cytosolic tyrosine kinases. Y864 and Y952 are critical in MEN 2B-related transformation (Liu *et al.*, 1996). Interestingly, although substrate specificity is altered, M918T-RET appears to result in enhanced activation of the PI3K pathway and its downstream targets (Murakami *et al.*, 1999b). While M918T-RET can be constitutively active (Santoro *et al.*, 1995), ligand stimulation, e.g. by GDNF, of M918T-RET can increase its activity. It is tantalizing to postulate that the severe phenotype seen in most MEN 2B cases results from the high level of PI3K-Akt signaling with consequent increased cell survival, growth and cell motility in addition to the pathways involved downstream of altered substrate specificity. In contrast, MEN 2A could result from activation of both the RAS-MAPK pathway as well as lower level PI3K-Akt signaling.

The phenotypically less penetrant mutations at *RET* codons 768 and 804 have been associated with lower transformation potential (Iwashita *et al.*, 1999; Pasini *et al.*, 1997).

6.3 Loss-of-function mutations in HSCR

A broad spectrum of mutations in a proto-oncogene are capable of causing loss-of-function. Indeed, in HSCR, germline loss-of-function mutations in *RET* include gross deletions and chromosomal aberrations, nonsense mutations, frameshift mutations, splice site mutations and missense mutations (section 4.1) (Angrist *et al.*, 1993, 1995; Attié *et al.*, 1994, 1995; Edery *et al.*, 1994; Luo *et al.*, 1994; Mulligan *et al.*, 1994a). Thus, it would seem obvious that haploinsufficiency, whether structural or functional, plays a large role in causing HSCR. Whole gene deletions and truncated protein secondary to nonsense, frameshift and nonsense mutations result in such haploinsufficiency. Even missense mutations which cause loss-of-function, e.g. in the catalytic core of the tyrosine kinase, would be predicted to cause functional haploinsufficiency. Missense mutations in the extracellular domain of *RET* have been shown to result in lack of maturation of the mutant receptor and hence, lack of receptor at the cell surface with consequent haploinsufficiency surface (Carlomagno *et al.*, 1997; Ito *et al.*, 1997).

6.4 Gain-of-function and loss-of-function mutations in MEN 2-HSCR families: a unifying hypothesis and explanation of penetrance in MEN 2

Initially, it was puzzling that missense mutations at codons 618 and 620, typical of classic MEN 2A and FMTC, are also found in families and individuals segregating both HSCR and MEN 2A or FMTC (Eng *et al.*, 1996a; Mulligan *et al.*, 1994a). However, the functional data which demonstrates that extracellular missense mutations cause lack of maturation of mutant RET, and hence, lack of migration of the mutant receptors to the surface might partially account for the MEN 2-HSCR phenotype.

Non-634 cysteine codon mutations appear to be more weakly transforming in transfection studies as well as in biochemical assays of kinase activity (Carlomagno *et al.*, 1997; Ito *et al.*, 1997). Additionally, mutations of the more 5' cysteine codons seem to result in a decreased proportion of receptor molecules on the cell surface (Ito *et al.*, 1997). These data together appear to suggest that MEN 2A and FMTC are a single genetic entity, whose phenotypic manifestation is a consequence of differences in penetrance of each mutation type. So, codon 634 mutations seem to have the highest penetrance likely due to the fact that all the molecules mature and are on the cell surface. This results in the full manifestation of MTC, PC and HPT, i.e. classic MEN 2A. In contrast, non-codon 634 mutations have a smaller proportion of mutant receptors at the cell surface. This results in some constitutively dimerised, and hence, consitutively activated, receptors but perhaps, only enough for MTCs to form, and to a lesser extent, PC. In addition to the variable penetrance of these mutations, there also appear to be tissue-specific differences in transformation threshold, such that C cells have the lowest threshold and parathyroid cells, the highest.

A similar explanation might solve the dilemma of gain of function mutations, C618R and C620R, that appear causative of both MEN 2A/FMTC and HSCR in a single family, indeed, in a single patient (above). Only a proportion of C618R- and C620R-bearing receptor molecules mature to the cell surface, and hence, are constitutively activated. If this hypothesis were correct, any non-codon 634 MEN 2A/FMTC-type mutation has some potential to be etiologic for the co-occurrence of MEN 2A/FMTC and HSCR. While it is enough for MTC, and perhaps, PC to result from constitutive activation of some receptors at the cell surface, enteric ganglia appear to require a certain threshold number of 'workable' receptors for normal gangliogenesis to occur or perhaps to prevent inappropriate apoptosis. However, because this is not an all or none phenomenon, we find that approximately 1/3 of families with C618R and C620R have both MEN 2A/FMTC and HSCR, while the other 2/3 have MEN 2A or FMTC only (Mulligan *et al.*, 1994a).

Recently, when a very strong association between two germline *RET* sequence variants and isolated HSCR was noted (above) (Borrego *et al.*, 1998, 1999; Fitze *et al.*, 1999; Sancandi *et al.*, 2000), an obvious hypothesis was that the presence of these variants modulated phenotypic expression. For example, in the presence of the *RET* codon 45 variant, HSCR or a decreased phenotype would be observed (Borrego *et al.*, 1998, 1999). More interestingly, there also appeared to be genotypes comprised of *RET* haplotype pairs which occurred only in HSCR cases and not in region-matched controls; conversely, there were also specific genotypes which occurred only in normal controls (Borrego *et al.*, 2000).

7. Animal models

The loss-of-function or ret null animal model has a phenotype similar to that in humans. Homozygous ret–/– mice develop the equivalent of HSCR (Schuchardt *et al.*, 1994). Unlike their human counterparts, however, ret+/– heterozygotes do not show any phenotype. In addition, these null mice have renal agenesis or dysgenesis, which only rarely occurs together with human HSCR. The difference in dosage requirements for phenotypic expression between mouse and human has been attributed to the presence of other modifier genes, which recently, has been shown to be plausible (Bolk *et al.*, 1999; Borrego *et al.*, 1999, 2000). A HSCR-like phenotype, among other features, have been observed in mice null for *gdnf, gfrα-1* and *gfrα-2* (Enomoto *et al.*, 1998; Rossi *et al.*, 1999; Sánchez *et al.*, 1996). While rare germline *GDNF* variants in humans have been found in individuals with HSCR (Angrist *et al.*, 1996; Salomon *et al.*, 1996), no germline mutations in *GFRα-1* have been found in a series of 269 unrelated human HSCR cases (Myers *et al.*, 1999). Human *GFRα-2* has yet to be examined in HSCR.

Mice carrying *men2* mutations were difficult to generate: four years elapsed from the time *RET* was identified as the MEN 2 susceptibility gene to the moment the first mouse was generated. The first model had a transgene with an *men2a* mutation and used the calcitonin promoter (Michiels *et al.*, 1997). The mice which expressed this transgene in C cells developed MTC. Given that the promoter chosen for this model is specific to C cells, it was not surprising that the two other classic components of human MEN 2A, PC and HPT, were not observed. When a transgene carrying the major *men2b* mutation and the dopamine-β-hydroxylase promoter was introduced into mice, they developed benign neuroglial tumors, similar to human ganglioneuromas, and renal malformations similar to those observed in *ret –/–* mice (Sweetser *et al.*, 1999). These manifestations may partially result from the promoter used, and thus, lack of MTC in this instance is not a surprise. Interestingly, when levels of MAPK phosphorylation was assayed, they were found not be different from non-MEN 2B mice. This might support the studies which suggest that the human *MEN2B* M918T mutation results in alteration of substrate specificity and not merely simple supra-activation of wildtype pathways (Songyang *et al.*, 1995). An *men2b* mutation knock-in mouse model resulted in PC, C-cell hyperplasia and ganglioneuromas of the adrenal medulla as well as male reproductive tract defects (Smith-Hicks *et al.*, 2000), all of which can be seen in human MEN 2B.

Interestingly, there exist at least two sets of mouse models that have a MEN-like phenotype which result from the presence of a *c-mos* transgene (Schulz *et al.*, 1992) or double hemizygous null at *rb* and *p53* (Coxon *et al.*, 1998). Why MTC from *rb +/– p53 +/–* mice have somatic cysteine codon mutations equivalent to the germline mutations seen in human MEN 2A and FMTC is unknown.

8. The triumph of molecular oncology and the dilemma of penetrance and expressional modification

RET mutation analysis in the clinical setting for MEN 2 and apparently sporadic MTC is the standard of care for patient management in the world (Eng, 1999; Offit *et al.*, 1996). The syndrome is potentially lethal; the test is highly sensitive; and the management is altered by the results of the test, whether as a molecular diagnostic or a predictive test. These are the traditional *sine qua non* for an effective clinical test. Given the advances in molecular genetics, it has become obvious that other modifying molecular influences come into play (Borrego *et al.*, 1999; Gimm *et al.*, 1999). How should FMTC with the lower penetrance mutations be handled? Surely, they should not undergo prophylactic surgery at the same age as those with the higher penetrance mutations. One day, can our knowledge of germline and somatic profiling of variants help determine these issues?

Acknowledgments

I would like to thank my longtime collaborators in the *RET* field, Lois M. Mulligan and Stanislas Lyonnet, and members of my laboratory, especially Oliver Gimm, Mary Armanios and Heather Dziema, the neuroendocrine enthusiasts, some of whom have contributed to work described in this review and who put up with my hypotheses. The International *RET* Mutation Consortium is >25 member centers strong and is dedicated to the examination of genetic data as it translates to clinical practice. Other centers of excellence in the neuroendocrine field who wish to contribute to such translational research should contact the coordinators and co-chairs, C. Eng (eng-1@medctr.osu.edu or ceng@hgmp.mrc.ac.uk) or L.M. Mulligan (mulligal@post.queensu.ca).

References

Alberti, L., Borrello, M.G., Ghizzoni, S., Torriti, F. and Pierotti, M.A. (1998) Grb2 binding to the different isoforms of Ret tyrosine kinase. *Oncogene* **17**: 1079–1087.

Amiel, J., Attié, T., Jan, D. *et al.* (1996) Heterozygous endothelin receptor B (*EDNRB*) mutations in isolated Hirschsprung disease. *Hum. Mol. Genet.* **5**: 355–357.

Angrist, M., Kauffman, E., Slaugenhaupt, S.A. *et al.* (1993) A gene for Hirschsprung disease (megacolon) in the pericentromeric region of human chromosome 10. *Nature Genet.* **4**: 351–356.

Angrist, M., Bolk, S., Thiel, B., Puffenberger, E.G., Hofstra, R.M., Buys, C.H.C.M., Cass, D.T. and Chakravarti, A. (1995) Mutation analysis of the RET receptor tyrosine kinase in Hirschsprung disease. *Hum. Mol. Genet.* **4**: 821–830.

Angrist, M., Bolk, S., Halushka, M., Lapchak, P.A. and Chakravarti, A. (1996) Germline mutations in glial cell line-derived neurotrophic factor (*GDNF*) and *RET* in a Hirschsprung disease patient. *Nature Genet.* **14**: 341–344.

Asai, N., Iwashita, T., Matsuyama, M. and Takahashi, M. (1995) Mechanism of activation of the *ret* proto-oncogene by multiple endocrine neoplasia 2A mutations. *Mol Cell Biol.* **3**: 1613–1619.

Asai, N., Marukami, H., Iwashita, T. and Takahashi, M. (1996) A mutation at tyrosine 1062 in MEN2A-Rct and MEN2B-Ret impairs their transforming activity and association with shc adaptor proteins. *J. Biol. Chem.* **271**: 17644–17649.

Attié, T., Pelet, A., Sarda, P. *et al.* (1994) A 7 bp deletion of the RET proto-oncogene in familial Hirschsprung's disease. *Hum. Mol. Genet.* **3**: 1439–1440.

Attié, T., Pelet, A., Edery, P. *et al.* (1995) Diversity of *RET* proto-oncogene mutations in familial and sporadic Hirschsprung disease. *Hum. Mol. Genet.* **4**: 1381–1386.

Attié-Bitach, T., Abitbol, M., Gérard, M. *et al.* (1998) Expression of the *RET* proto-oncogene in human embryos. *Am. J. Med. Genet.* **80**: 481–486.

Badner, J.A., Sieber, W.K., Garver, K.L. and Chakravarti, A. (1990) A genetic study of Hirschsprung disease. *Am. J. Hum. Genet.* **46**: 568–580.

Baloh, R.H., Tansey, M.G., Golden, J.P. *et al.* (1997) TrnR2, a novel receptor that mediates neurturin and GDNF signaling through Ret. *Neuron* **18**: 793–802.

Baloh, R.H., Tansey, M.G., Lampe, P.A. *et al.* (1998) Artemin, a novel member of the GDNF ligand family, supports peripheral and central neurons and signals through the GFRalpha-3-RET receptor complex. *Neuron* **21**: 1291–1302.

Beldjord, B., Desclaux-Arramond, F., Raffin-Sanson, M., Corvol, J.-C., de Keyser, Y., Luton, J.-P., Plouin, P.-F. and Bertagna, X. (1995) The *RET* proto-oncogene in sporadic pheochromocytomas: frequent MEN 2-like mutations and new molecular defects. *J. Clin. Endocrinol. Metab.* **80**: 2063–2068.

Berndt, I., Reuter, M., Saller, B., Frank-Raue, K., Groth, P., Grussendorf, M., Raue, F., Ritter, M.M. and Höppner, W. (1998) A new hotspot for mutations in the *RET* proto-oncogene causing familial medullary thyroid carcinoma and multiple endocrine neoplasia. *J. Clin. Endocrinol. Metab.* **83**: 770–774.

Blaugrund, J.E., Johns, M.M., Eby, Y.J., Ball, D.W., Baylin, S.B., Hruban, R.H. and Sidransky, D. (1994) *RET* proto-oncogene mutations in inherited and sporadic medullary thyroid cancer. *Hum. Mol. Genet.* **3**: 1895–1897.

Bocciardi, R., Mograbi, B., Pasini, B., Borrello, M.G., Pierotti, M.A., Bourget, I., Fischer, S., Romeo, G. and Rossi, B. (1997) The multiple endocrine neoplasia type 2B point mutation switches the specificity of the Ret tyrosine kinase towards cellular substrates that are susceptible to interact with Crk and Nck. *Oncogene* **15**: 2257–2265.

Bolino, A., Schuffenecker, I., Luo, Y. *et al.* (1995) *RET* mutations in exons 13 and 14 of FMTC patients. *Oncogene* **10**: 2415–2419.

Bolk, S., Pelet, A., Hofstra, R.M.W., Angrist, M., Salomon, R., Croaker, D., Buys, C.H.C.M., Lyonnet, S. and Chakravarti, A. (1999) A human model for multigenic inheritance: phenotypic expression in Hirschsprung disease requires both the *RET* gene and a new 9q31 locus. *Proc. Natl Acad. Sci. USA* **97**: 268–273.

Borrego, S., Eng, C., Sánchez, B., Sáez, M.-E., Navarro, E. and Antiñolo, G. (1998) Molecular analysis of *RET* and *GDNF* genes in a family with multiple endocrine neoplasia type 2A and Hirschsprung diease. *J. Clin. Endocrinol. Metab.* **83**: 3361–3364.

Borrego, S., Saez, M.E., Ruiz, A., Gimm, O., Lopez-Alonso, M., Antiñolo, G. and Eng, C. (1999) Specific polymorphisms in the *RET* proto-oncogene are over-represented in individuals with Hirschsprung disease and may represent loci modifying phenotypic expression. *J. Med. Genet.* **36**: 771–774.

Borrego, S., Saez, M.E., Ruiz, A. *et al.* (2000) *RET* genotypes comprising specific haplotypes of polymorphic variants predispose to isolated Hirschsprung disease. *J. Med. Genet.* **37**: 572–578.

Borrello, M.G., Pelicci, G., Arighi, E., DeFillippis, L., Greco, A., Bongarzone, I., Rizzetti, M.G., Pelicci, P.G. and Pierotti, M.A. (1994) The oncogenic versions of the ret and trk tyrosine kinases bind to Shc and Grb2 adaptor proteins. *Oncogene* **9**: 1661–1668.

Borrello, M.G., Smith, D.P., Pasini, B. *et al.* (1995) *RET* activation by germline *MEN2A* and *MEN2B* mutations. *Oncogene* **11**: 2419–2427.

Borst, M.J., van Camp, J.M., Peacock, M.L. and Decker, R.A. (1995) Mutation analysis of multiple endocrine neoplasia type 2A associated with Hirschsprung's disease. *Surgery* **117**: 386–389.

Buj-Bello, A., Adu, J., Pinón, L.G.P., Horton, A., Thompson, J., Rosenthal, A., Chinchetru, M., Buchman, V.L. and Davies, A.M. (1997) Neurturin responsiveness requires a GPI-linked receptor and the Ret receptor tyrosine kinase. *Nature* **387**: 721–724.

Carlomagno, F., Salvatore, G., Cirafici, A.M., Devita, G., Melillo, R.M., Defranciscis, V., Billaud, M., Fusco, A. and Santoro, M. (1997) The different *RET*-activating capability of mutations of cysteine 620 or cysteine 634 correlates with the multiple endocrine neoplasia type 2 disease phenotype. *Cancer Res.* **57**: 391–395.

Carlson, H.E., Burns, T.W., Davenport, S.L., Luger, A.M., Spence, M.A., Sparkes, R.S. and Orth, D.N. (1986) Cowden disease: gene marker studies and measurements of epidermal growth factor. *Am. J. Hum. Genet.* **38**: 908–917.

Carlson, K.M., Bracamontes, J., Jackson, C.E., Clark, R., Lacroix, A., Wells Jr., S.A. and Goodfellow, P.J. (1994a) Parent-of-origin effects in multiple endocrine neoplasia type 2B. *Am. J. Hum. Genet.* **55**: 1076–1082.

Carlson, K.M., Dou, S., Chi, D., Scavarda, N., Toshima, K., Jackson, C.E., Wells, S.A., Goodfellow, P.J. and Donis-Keller, H. (1994b) Single missense mutation in the tyrosine kinase catalytic domain of the *RET* protooncogene is associated with multiple endocrine neoplasia type 2B. *Proc. Natl. Acad. Sci. USA* **91**: 1579–1583.

Caron, P., Attié, T., David, D., Amiel, J., Brousset, F., Roger, P., Munnich, A. and Lyonnet, S. (1996) C618R mutation in exon 10 of the *RET* proto-oncogene in a kindred with multiple endocrine neoplasia type 2A and Hirschsprung's disease. *J. Clin. Endocrinol. Metab.* **81**: 2731–2733.

Chiariello, M., Visconti, R., Carlomagno, F. *et al.* (1998) Signalling of the Ret receptor tyrosine kinase through c-Jun NH2-terminal protein kinases (JNKs): evidence for divergence of ERKs and JNKs pathways induced by Ret. *Oncogene* **16**: 2435–2445.

Coxon, A.B., Ward, J.M., Geradts, J., Otterson, G.A., Zajac-Kaye, M. and Kaye, F.J. (1998) RET cooperates with RB/p53 inactivation in a somatic multi-step model for murine thyroid cancer. *Oncogene* **17**: 1625–1628.

Davies, A.M., Dixon, J.E., Fox, G.M. *et al.* (1997) Nomenclature for GPI-linked receptors for the GDNF ligand family. *Neuron* **19**: 485.

de la Chapelle, A, and Eng, C. (1999) Molecular genetic testing in hereditary cancer. in: *ASCO Educational Book* (M.C. Perry, ed.), pp. 445–453. Lippincott, Williams and Wilkins, Baltimore.

Decker, R.A. and Peacock, M.L. (1998) Occurrence of MEN 2a in familial Hirschsprung disease: another new indication for genetic testing of the *RET* proto-oncogene. *J. Pediatr. Surg.* **33**: 207–214.

Decker, R.A., Peacock, M.L., Borst, M.J., Sweet, J.D. and Thompson, N.W. (1995) Progress in genetic screening of multiple endocrine neoplasia type 2A: Is calcitonin testing obsolete? *Surgery* **118**: 257–264.

Donis-Keller, H., Dou, S., Chi, D. *et al.* (1993) Mutations in the *RET* proto-oncogene are associated with MEN 2A and FMTC. *Hum. Mol. Genet.* **2**: 851–856.

Doray, B., Salomon, R., Amiel, J. *et al.* (1998) Mutation of the RET ligands, neurturin, supports multigenic inheritance in Hirschsprung disease. *Hum. Mol. Genet.* **7**: 1449–1452.

Duerr, E.-M., Gimm, O., Kumm, J.B., Neuberg, D.S., Clifford, S.C., Toledo, S.P.A., Maher, E.R., Dahia, P.L.M. and Eng, C. (1999) Differences in allelic distribution of two polymorphisms in the VHL-associated gene *CUL2* in pheochromocytoma patients without somatic *CUL2* mutations. *J. Clin. Endocrinol. Metab.* **84**: 3207–3211.

Durick, K., Wu, R.Y., Gill, G.N. and Taylor, S.S. (1996) Mitogenic signaling by Ret/ptc2 requires association with enigma via a LIM domain. *J. Biol. Chem.* **271**: 12691–12694.

Durick, K., Gill, G.N. and Taylor, S.S. (1998a) Shc and Enigma are both required for mitogenic signaling by Ret/ptc2. *Mol. Cell. Biol.* **18**: 2298–2308.

Durick, K., Gill, G.N. and Taylor, S.S. (1998b) Tyrosines outside the kinase core and dimerization domain are required for the mitogenic activity of RET/ptc2. *J. Biol. Chem.* **270**: 24642–24645.

Easton, D.F., Ponder, M.A., Cummings, T. *et al.* (1989a) The clinical and age-at-onset distribution for the MEN-2 syndrome. *Am. J. Hum. Genet.* **44**: 208–215.

Easton, D.F., Ponder, M.A., Cummings, T. *et al.* (1989b) The clinical and screening age-at-onset distribution for the MEN-2 syndrome. *Am. J. Hum. Genet.* **44**: 208–215.

Edery, P., Lyonnet, S., Mulligan, L.M. *et al.* (1994) Mutations of the *RET* proto-oncogene in Hirschsprung's disease. *Nature* **367**: 378–380.

Eng, C. (1996) The *RET* proto-oncogene in multiple endocrine neoplasia type 2 and Hirschsprung disease. *N. Engl. J. Med.* **335**: 943–951.

Eng, C. (1997) From bench to bedside ... but when? *Genome Res.* **7**: 669–672.

Eng, C. (1999) *RET* proto-oncogene in the development of human cancer. *J. Clin. Oncol.* **17**: 380–393.

Eng, C. and Ponder, B.A.J. (1998) Multiple endocrine neoplasia type 2 and medullary thyroid carcinoma. In: *Clinical Endocrinology* (A. Grossman, ed.) pp. 635–650. Blackwell Science, Oxford.

Eng, C., Smith, D.P., Mulligan, L.M. *et al.* (1994). Point mutation within the tyrosine kinase domain of the *RET* proto-oncogene in multiple endocrine neoplasia type 2B and related sporadic tumours. *Hum. Mol. Genet.* **3**: 237–241.

Eng, C., Crossey, P.A., Mulligan, L.M. *et al.* (1995a) Mutations of the *RET* proto-oncogene and the von Hippel-Lindau disease tumour suppressor gene in sproadic and syndromic phaeochromocytoma. *J. Med. Genet.* **32**: 934–937.

Eng, C., Mulligan, L.M., Smith, D.P. *et al.* (1995b) Mutation in the *RET* proto-oncogene in sporadic medullary thyroid carcinoma. *Genes Chrom. Cancer* **12**: 209–212.

Eng, C., Mulligan, L.M., Smith, D.P. *et al.* (1995c) Low frequency of germline mutations in the *RET* proto-oncogene in patients with apparently sporadic medullary thyroid carcinoma. *Clin. Endocrinol.* **43**: 123–127.

Eng, C., Smith, D.P., Mulligan, L.M. *et al.* (1995d) A novel point mutation in the tyrosine kinase domain of the *RET* proto-oncogene in sporadic medullary thyroid carcinoma and in a family with FMTC. *Oncogene* **10**: 509–513.

Eng, C., Clayton, D., Schuffenecker, I. *et al.* (1996a) The relationship between specific *RET* proto-oncogene mutations and disease phenotype in multiple endocrine neoplasia type 2: International *RET* Mutation Consortium analysis. *JAMA* **276**: 1575–1579.

Eng, C., Mulligan, L.M., Healey, C.S., Houghton, C., Frilling, A., Raue, F., Thomas, G.A. and Ponder, B.A.J. (1996b) Heterogeneous mutation of the *RET* proto-oncogene in subpopulations of medullary thyroid carcinoma. *Cancer Res.* **56**: 2167–2170.

Eng, C., Thomas, G.A., Neuberg, D.S. *et al.* (1998) Mutation of the *RET* proto-oncogene is correlated with RET immunostaining in subpopulations of cells in sporadic medullary thyroid carcinoma. *J. Clin. Endocrinol. Metab.* **83**: 4310–4313.

Enokido, Y., de Sauvage, F., Hongo, J.-A., Ninkina, N., Rosenthal, A., Buchman, V.L. and Davies, A.M. (1998) GFRa-4 and the tyrosine kinase Ret form a functional receptor complex for persephin. *Curr. Biology* 8: 1019–1022.

Enomoto, H., Araki, T., Jackman, A., Heuckeroth, R.O., Snider, W.D., Johnson, E.M. and Milbrandt, J. (1998) GFR alpha-1-deficient mice have deficits in the interic nervous system and kidneys. *Neuron* 21: 317–324.

Farndon, J.R., Leight, G.S., Dilley, W.G., Baylin, S.B., Smallridge, R.C., Harrison, T.S. and Wells, S.A. (1986) Familial medullary thyroid carcinoma without associated endocrinopathies: a distinct clinical entity. *Br. J. Surg.* 73: 278–281.

Ferraris, A.M., Mangerini, R., Gaetani, G.F., Romei, C., Pinchera, A. and Pacini, F. (1997) Polyclonal origin of medullary carcinoma of the thyroid in multiple endocrine neoplasia type 2. *Hum. Genet.* 99: 202–205.

Fink, M., Weinhäusel, A., Niederle, B. and Haas, O.A. (1996) Distinction between sporadic and hereditary medullary thyroid carcinoma (MTC) by mutation analysis of the *RET* proto-oncogene. *Int. J. Cancer* 69: 312–316.

Fitze, G., Schreiber, M., Kuhlisch, E., Schackert, H.K. and Roesner, D. (1999) Association of the *RET* proto-oncogene codon 45 polymorphism with Hirschsprung disease. *Am. J. Hum. Genet.* 65: 1469–1473.

Frank-Raue, K., Höppner, W., Frilling, A. *et al.* (1996) Mutations of the *RET* proto-oncogene in German MEN families: relation between genotype and phenotype. *J. Clin. Endocrinol. Metab.* 81: 1780–1783.

Gardner, E., Papi, L., Easton, D.F. *et al.* (1993) Genetic linkage studies map the multiple endocrine neoplasia type 2 loci to a small interval on chromosome 10q11.2. *Hum. Mol. Genet.* 2: 241–246.

Gimm, O., Marsh, D.J., Andrew, S.D., Frilling, A., Dahia, P.L.M., Mulligan, L.M., Zajak, J.D., Robinson, B.G. and Eng, C. (1997) Germline dinucleotide mutation in codon 883 of the *RET* proto-oncogene in multiple endocrine neoplasia type 2B without codon 918 mutation. *J. Clin. Endocrinol. Metab.* 82: 3902–3904.

Gimm, O., Neuberg, D.S., Marsh, D.J., Dahia, P.L.M., Hoang-Vu, C., Raue, F., Hinze, R., Dralle, H. and Eng, C. (1999) Over-representation of a germline *RET* sequence variant in patients with sporadic medullary thyroid carcinoma and somatic *RET* codon 918 mutation. *Oncogene* 18: 1369–1370.

Gimm, O., Perren, A., Weng, L.P. *et al.* (2000) Differential nuclear and cytoplasmic expression of PTEN in normal thyroid tissue, and benign and malignant epithelial thyroid tumors. *Am. J. Pathol.* 156: 1693–1700.

Hanks, S.K., Quinn, A.M. and Hunter, T. (1988) The protein kinase family: conserved features and deduced phylogeny of the catalytic domain. *Science* 241: 42–52.

Hofstra, R.M.W., Landsvater, R.M., Ceccherini, I. *et al.* (1994) A mutation in the *RET* proto-oncogene associated with multiple endocrine neoplasia type 2B and sporadic medullary thyroid carcinoma. *Nature* 367: 375–376.

Hofstra, R.M.W., Stelwagen, T., Stulp, R.P. *et al.* (1996) Extensive mutation scanning of *RET* in sporadic medullary thyroid carcinoma and of *RET* and *VHL* in sporadic pheochromocytoma reveals involvement of these genes in only a minority of cases. *J. Clin. Endocrinol. Metab.* 81: 2881–2884.

Hofstra, R.M.W., Fattoruso, O., Quadro, L., Wu, Y., Libroia, A., Verga, U., Colantuoni, V. and Buys, C.H.C.M. (1997) A novel point mutation in the intracellular domain of the *RET* proto-oncogene in a family with medullary thyroid carcinoma. *J. Clin. Endocrinol. Metab.* 82: 4176–4178.

Inoue, K., Shimotake, T., Inoue, K., Tokiwa, K. and Iwai, N. (1999) Mutational analysis of the *RET* proto-oncogene in a kindred with multiple endocrine neoplasia type 2A and Hirschsprung's disease. *J. Pediatr. Surg.* 34: 1552–1554.

Ito, S., Iwashita, T., Asai, N., Murakami, H., Iwata, Y., Sobue, G. and Takahashi, M. (1997) Biological properties of Ret with cysteine mutations correlate with multiple endocrine neoplasia type 2A, familial medullary thyroid carcinoma, and Hirchsprung's disease phenotype. *Cancer Res.* 57: 2870–2872.

Ivanchuk, S.M., Eng, C., Cavanee, W.K. and Mulligan, L.M. (1997) The expression of *RET* and its multiple splice forms in developing human kidney. *Oncogene* 14: 1811–1818.

Ivanchuk, S.M., Myers, S.M. and Mulligan, L.M. (1998) Expression of *RET* 3' splicing variants during human kidney development. *Oncogene* 16: 991–996.

Iwashita, T., Asai, N., Murakami, H., Matsuyama, M. and Takahashi, M. (1996) Identification of tyrosine residues that are essential for transforming activity of the *ret* proto-oncogene with MEN2A or MEN2B mutation. *Oncogene* 12: 481–487.

Iwashita, T., Kato, M., Murakami, H. *et al.* (1999) Biological and biochemical properties of Ret with kinase domain mutations identified in multiple endocrine neoplasia type 2B and familial medullary thyroid carcinoma. *Oncogene* **18**: 3919–3922.

Jing, S., Wen, D., Yu, Y. *et al.* (1996) GDNF-induced activation of the Ret protein tyrosine kinase is mediated by GDNFR-a, a novel receptor for GDNF. *Cell* **85**: 1113–1124.

Jing, S.Q., Yu, Y.B., Fang, M. *et al.* (1998) GFR-alpha-2 and GFR-alpha-3 are two new receptors for ligands of the GDNF family. *J. Biol. Chem.* **272**: 33111–33117.

Kitamura, Y., Scavarda, N., Wells, S.A., Jackson, C.E. and Goodfellow, P.J. (1995) Two maternally derived missense mutations in the tyrosine kinase domain of the *RET* protooncogene in a patient with *de novo* MEN 2B. *Hum. Mol. Genet.* **4**: 1987–1988.

Komminoth, P., Kunz, E., Hiort, O., Schöder, S., Matias-Guiu, X., Christiansen, G., Roth, J. and Heitz, P.U. (1994) Detection of *RET* proto-oncogene point mutations in paraffin-embedded pheochromocytoma specimens by nonradioactive single-strand conformation polymorphism analysis and direct sequencing. *Am. J. Pathol.* **145**: 922–929.

Komminoth, P., Kunz, E.K., Matias-Guiu, X., Hiort, O., Christensen, G., Colomer, A., Roth, J. and Heitz, P.U. (1995) Analysis of *RET* proto-oncogene point mutations distinguishes heritable from nonheritable medullary thyroid carcinomas. *Cancer* **76**: 479–489.

Komminoth, P., Roth, J., Muletta-Feurer, S., Saremaslani, P., Seelentag, W.K.F. and Heitz, P.U. (1996) *RET* proto-oncogene point mutations in sporadic neuroendocrine tumors. *J. Clin. Endocrinol. Metab.* **81**: 2041–2046.

Kotzbauer, P.T., Lampe, P.A., Heuckeroth, R.O., Golden, J.P., Creedon, D.J., Johnson, E.M. and Milbrandt, J. (1996) Neurturin, a relative of glial-cell-line-derived neurotrophic factor. *Nature* **384**: 467–470.

Learoyd, D.L., Twigg, S.M., Marsh, D.J. and Robinson, B.G. (1995) The practical management of multiple endocrine neoplasia. *Trend Endocrinol. Metab.* **6**: 273–278.

Lindor, N.M., Honchel, R., Khosla, S. and Thibodeau, S.N. (1995) Mutations in the *RET* protooncogene in sporadic pheochromocytomas. *J. Clin. Endocrinol. Metab.* **80**: 627–629.

Lips, C.J.M., Landsvater, R.M., Höppener, J.W.M. *et al.* (1994) Clinical screening as compared with DNA analysis in families with multiple endocrine neoplasia type 2A. *N. Engl. J. Med.* **331**: 828–835.

Liu, X., Vega, Q.C., Decker, R.A., Pandey, A., Worby, C.A. and Dixon, J.E. (1996) Oncogenic RET receptors display different autophosphorylation sites and substrate binding specificities. *J. Biol. Chem.* **271**: 5309–5312.

Luo, Y., Barone, V., Seri, M. *et al.* (1994) Heterogeneity and low detection rate of *RET* mutations in Hirschsprung disease. *Eur. J. Hum. Genet.* **2**: 272–280.

Lyonnet, S., Bolino, A., Pelet, A. *et al.* (1993) A gene for Hirschsprung disease maps to the proximal long arm of chromosome 10. *Nature Genet.* **4**: 346–350.

Marsh, D.J., Robinson, B.G., Andrew, S., Richardson, A.-L., Pojer, R., Schnitzler, M., Mulligan, L.M. and Hyland, V.J. (1994) A rapid screening method for the detection of mutations in the RET proto-oncogene in multiple endocrine neoplasia type 2A and familial medullary thyroid carcinoma families. *Genomics* **23**: 477–479.

Marsh, D.J., Andrew, S.D., Eng, C. *et al.* (1996a) Germline and somatic mutations in an oncogene: *RET* mutations in inherited medullary thyroid carcinoma. *Cancer Res.* **56**: 1241–1243.

Marsh, D.J., Learoyd, D.L., Andrew, S.D. *et al.* (1996b) Somatic mutations in the *RET* proto-oncogene in sporadic medullary thyroid carcinoma. *Clin. Endocrinol.* **44**: 249–257.

McMahon, R., Mulligan, L.M., Healey, C.S., Payne, S.J., Ponder, M., Ferguson-Smith, M.A., Barton, D.E. and Ponder, B.A.J. (1994) Direct, non-radioactive detection of mutations in multiple endocrine neoplasia type 2A families. *Hum. Mol. Genet.* **3**: 643–646.

Michiels, F.M., Chappuis, S., Caillou, B., Pasini, A., Talbot, M., Monier, R., Lenoir, G.M., Feunteun, J. and Billaud, M. (1997) Development of medullary thyroid carcinoma in transgenic mice expressing the *RET* proto-oncogene altered by a multiple endocrine neoplasia type 2A mutation. *Proc. Natl Acad. Sci. USA* **94**: 3330–3335.

Moers, A.M.J., Landsvater, R.M., Schaap, C. *et al.* (1996) Familial medullary thyroid carcinoma – not a distinct entity – genotype–phenotype correlation in a large family. *Am. J. Med.* **101**: 635–641.

Mulligan, L.M., Kwok, J.B.J., Healey, C.S. *et al.* (1993) Germline mutations of the *RET* proto-oncogene in multiple endocrine neoplasia type 2A. *Nature* **363**: 458–460.

Mulligan, L.M., Eng, C., Attié, T. *et al.* (1994a) Diverse phenotypes associated with exon 10 mutations of the *RET* proto-oncogene. *Hum. Mol. Genet.* **3**: 2163–2167.

Mulligan, L.M., Eng, C., Healey, C.S. *et al.* (1994b) Specific mutations of the *RET* proto-oncogene are related to disease phenotype in MEN 2A and FMTC. *Nature Genet* **6**: 70–74.

Mulligan, L.M., Eng, C., Healey, C.S., Ponder, M.A., Feldman, G.L., Li, P., Jackson, C.E. and Ponder, B.A.J. (1994c) A *de novo* mutation of the *RET* proto-oncogene in a patient with MEN 2A. *Hum. Mol. Genet.* **3**: 1007–1008.

Mulligan, L.M., Marsh, D.J., Robinson, B.G. *et al.* (1995) Genotype-phenotype correlation in MEN 2: Report of the International *RET* Mutation Consortium. *J. Intern. Med.* **238**: 343–346.

Murakami, H., Iwashita, T., Asai, N., Iwata, Y., Narumiya, S. and Takahashi, M. (1999a) Rho-dependent and independent tyrosine phosphorylation of focal adhesion kinase, paxillin and p130Cas mediated by Ret kinase. *Oncogene* **18**: 1975–1982.

Murakami, H., Iwashita, T., Asai, N., Shimono, Y., Iwata, Y., Kawai, K. and Takahashi, M. (1999b) Enhanced phosphoinositol 3-kinase activity and high phosphorylation state of its downstream signalling molecules mediated by RET with the *MEN2B* mutation. *Biochem. Biophys. Res. Comm.* **262**: 68–75.

Myers, S.M., Eng, C., Ponder, B.A.J. and Mulligan, L.M. (1995) Characterization of *RET* proto-oncogene 3′ splicing variants and polyadenylation sites: a novel C-terminus for RET. *Oncogene* **11**: 2039–2045.

Myers, S.M., Salomon, R., Gössling, A., Pelet, A., Eng, C., von Deimling, A., Lyonnet, S. and Mulligan, L.M. (1999) Absence of germline *GFRa-1* mutations in Hirschsprung disease. *J. Med. Genet.* **36**: 217–220.

Nakamura, T., Ishizaka, Y., Nagao, M., Hara, M. and Ishikawa, T. (1994) Expression of the *ret* proto-oncogene product in human normal and neoplastic tissues of neural crest origin. *J. Pathol.* **172**: 255–260.

Neumann, H.P.H., Eng, C., Mulligan, L.M., Glavac, D., Zaüner, I., Ponder, B.A.J., Crossey, P.A., Maher, E.R. and Brauch, H. (1995) Consequences of direct genetic testing for germ-line mutations in the clinical management of families with multiple endocrine neoplasia type 2. *JAMA* **274**: 1149–1151.

Nilsson, O., Tissell, L.-E., Jansson, S., Ahlman, H., Gimm, O. and Eng, C. (1999) Adrenal and extra-adrenal pheochromocytomas in a family with germline *RET* V804L mutation. *JAMA* **281**: 1587–1588.

Offit, K., Biesecker, B.B., Burt, R.W., Clayton, E.W., Garber, J.E. and Kahn, M.J.E. (1996) Statement of the American Society of Clinical Oncology – Genetic testing for cancer susceptibility. *J. Clin. Oncol.* **14**: 1730–1736.

Pandey, A., Duan, H., Di Fiore, P.P. and Dixit, V.M. (1995) The Ret receptor protein tyrosine kinase associates with the SH2-containing adaptor protein Grb10. *J. Biol. Chem.* **270**: 21461–21463.

Pasini, A., Geneste, O., Legrand, P. *et al.* (1997) Oncogenic activation of *RET* by two distinct FMTC mutations affecting the tyrosine kinase domain. *Oncogene* **15**: 393–402.

Puffenberger, E.G., Hosoda, K., Washington, S.S., Nakao, K., de Wit, D., Yanagisawa, M. and Chakravarti, A. (1995) A missense mutation of the endothelin B receptor gene in multigenic Hirschsprung's disease. *Cell* **79**: 1257–1266.

Romano, A., Wong, W.T., Santoro, M., Wirth, P.J., Thorgeissen, S.S. and DiFiore, P.P. (1994) The high transforming potency of erbB-2 and ret is associated with phosphorylation of paxillin and a 23 kDa protein. *Oncogene* **9**: 2923–2933.

Romei, C., Elisei, R., Pinchera, A., Ceccherini, I., Molinaro, E., Mancusi, F., Martino, E., Romeo, G. and Pacini, F. (1996) Somatic mutations of the *RET* proto-oncogene in sporadic medullary thyroid carcinoma are not restricted to exon 16 and are associated with tumor recurrence. *J. Clin. Endocrinol. Metab.* **81**: 1619–1622.

Romeo, G., Ronchetto, P., Luo, Y. *et al.* (1994) Point mutations affecting the tyrosine kinase domain of the *RET* proto-oncogene in Hirschsprung's disease. *Nature* **367**: 377–378.

Rossi, J., Lukko, K., Poteryaev, D. *et al.* (1999) Retarded growth and deficits in the enteric and parasympathetic nervous system in mice lacking GFR alpha2, a functional neurturin receptor. *Neuron* **22**: 243–252.

Salomon, R., Attié, T., Pelet, A. *et al.* (1996) Germline mutations of the RET ligand, GDNF, are not sufficient to cause Hirschsprung disease. *Nature Genet.* **14**: 345–347.

Sancandi, M., Ceccherini, I., Costa, M. *et al.* (2000) Incidence of RET mutations in patients with Hirschsprung's disease. *J. Pediatr. Surg.* **35**: 139–142.

Sánchez, M.P., Silos-Santiago, I., Frisén, J., He, B., Lira, S.A. and Barbacid, M. (1996) Renal agenesis and the absence of enteric neurons in mice lacking GDNF. *Nature* **382**: 70–73.

Sanicola, M., Hession, C., Worley, D. *et al.* (1997) GDNF-dependent RET activation can be mediated by two different cell-surface accessory proteins. *Proc. Natl Acad. Sci. USA* **94**: 6238–6243.

Santoro, M., Rosato, R., Grieco, M., Berlingieri, M.T., Luca-Colucci D'Amato, G., de Franciscis, V. and Fusco, A. (1990) The *ret* proto-oncogene is consistently expressed in human pheochromocytomas and thyroid medullary carcinomas. *Oncogene* **5**: 1595–1598.

Santoro, M., Carlomagno, F., Romano, A. *et al.* (1995) Activation of *RET* as a dominant transforming gene by germline mutations of MEN 2A and MEN 2B. *Science* **267**: 381–383.

Sapienza, C. and Hall, G. (1994) Genome imprinting in human disease. In: *The Metabolic Basis of Inherited Disease, 7th edition* (C.R. Scriver, A.L. Beaudet, W.S. Sly, and D. Valle, eds. McGraw-Hill, Baltimore.

Schuchardt, A., D'Agati, V., Larsson-Blomberg, L., Costantini, F. and Pachnis, V. (1994) The *c-ret* receptor tyrosine kinase gene is required for the development of the kidney and enteric nervous system. *Nature* **367**: 380–383.

Schuffenecker, I., Billaud, M., Calender, A., Chambe, B., Ginet, N., Calmettes, C., Modigliani, E., Lenoir, G.M. and GETC. (1994) *RET* proto-oncogene mutations in French MEN 2A and FMTC families. *Hum. Mol. Genet.* **3**: 1939–1943.

Schuffenecker, I., Ginet, N , Goldgar, D. *et al.* (1997) Prevalence and parental origin of *de novo RET* mutations in MEN 2A and FMTC. *Am. J. Hum. Genet.* **60**: 233–237.

Schuffenecker, I., Virally-Monod, M., Brohet, R. *et al.* (1998) Risk and penetrance of primary hyperparathyroidism in MEN 2A families with codon 634 mutations of the *RET* proto-oncogene. *J. Clin. Endocrinol. Metab.* **83**: 487–491.

Schulz, N., Propst, F., Rosenberg, M.P., Linnoila, R.I., Paules, R.S., Kovatch, R., Ogiso, Y. and Vande Woude, G.F. (1992) Pheochromocytomas and c-cell thyroid neoplasms in transgenic *c-mos* mice: a model for the human multiple endocrine neoplasia type 2 syndrome. *Cancer Res* **52**: 450–455.

Segouffin-Cariou, C. and Billaud, M. (2000) Transforming ability of MEN2A-RET requires activation of the phosphatidylinositol 3-Kinase/AKT signaling pathway. *J. Biol. Chem.* **275**: 3568–3576.

Shannon, K.E., Gimm, O., Hinze, R., Dralle, H. and Eng, C. (1999) Germline V804M in the *RET* proto-oncogene in two apparently sporadic cases of MTC presenting in the seventh decade of life. *J. Endo. Genet.* **1**: 39–46.

Smith, D.P., Houghton, C. and Ponder, B.A.J. (1997) Germline mutation of *RET* codon 883 in two cases of *de novo* MEN 2B. *Oncogene* **15**: 1213–1217.

Smith-Hicks, C.L., Sizer, K.C., Powers, J.F., Tischler, A.S. and Costantini, F. (2000) C-cell hyperplasia, pheochromocytoma and sympathoadrenal malformation in a mouse model of multiple endocrine neoplasia type 2B. *EMBO J.* **19**: 612–622.

Songyang, Z., Carraway III, K.L., Eck, M.J. *et al.* (1995) Catalytic specificity of protein-tyrosine kinases is critical for selective signalling. *Nature* **373**: 536–539.

Southard-Smith, E.M., Kos, L. and Pavan, W.J. (1998) *Sox10* mutation disrupts neural crest development in *Dom* Hirschsprung mouse model. *Nature Genet.* **18**: 60–64.

Svensson, P.-J., Molander, J.-L., Eng, C., Anvret, M. and Nordenskjöld, A. (1998) Low frequency of *RET* mutations in Hirschsprung disease in Sweden. *Clin. Genet.* **54**: 39–44.

Sweetser, D.A., Froelick, G.J., Matsumoto, A.M., Kafer, K.E., Palmiter, R.D. and Kapur, R.P. (1999) Ganglioneuromas and renal anomalies are induced by activated RET (MEN2B) in transgenic mice. *Oncogene* **18**: 877–886.

Takahashi, M. and Cooper, G.M. (1987) *RET* transforming gene encodes a fusion protein homologous to tyrosine kinases. *Mol. Cell. Biol.* **3**: 1378–1385.

Takahashi, M., Ritz, J. and Cooper, G.M. (1985) Activation of a novel human transforming gene, *ret*, by DNA rearrangement. *Cell* **42**: 581–588.

Takahashi, M., Buma, Y., Iwamoto, T., Inaguma, Y., Ikeda, H. and Hiai, H. (1988) Cloning and expression of the *ret* proto-oncogene encoding a receptor tyrosine kinase with two potential transmembrane domains. *Oncogene* **3**: 571–578.

Takahashi, M., Buma, Y. and Hiai, H. (1989) Isolation of ret proto-oncogene cDNA with an amino-terminal signal. *Oncogene* **4**: 805–806.

Takahashi, M., Buma, Y. and Taniguchi, M. (1991) Identification of ret proto-oncogene products in neuroblastoma and leukemia cells. *Oncogene* **6**: 297–301.

Takahashi, M., Asai, N., Iwashita, T., Isomura, T., Miyazaki, K. and Matsuyama, M. (1993) Characterisation of the *ret* proto-oncogene products expressed in mouse L. *Oncogene* **8**: 2925–2929.

Telander, R.L. and Moir, C.R. (1994) Medullary thyroid carcinoma in children. *Sem. Pediatr. Surg.* **3**: 188–193.

Thompson, J., Doxakis, E., Pinon, L.G.P., Strachan, P., Buj-Bello, A., Wyatt, S., Buchman, V.L. and Davies, A.M. (1998) GFRa-4, a new GDNF family receptor. *Mol. Cell. Neurosci.* **11**: 117–126.

Treanor, J.J.S., Goodman, L., de Sauvage, F. *et al.* (1996) Characterization of a multicomponent receptor for GDNF. *Nature* **382**: 80–83.

Trupp, M., Arenas, E., Fainzilber, M. *et al.* (1996) Functional receptor for GDNF encoded by the *c-ret* proto-oncogene. *Nature* **381**: 785–789.

Trupp, M., Raynoschek, C., Belluardo, N. and Ibanez, C.F. (1998) Multiple GPI-anchored receptors control GDNF-dependent and independent activation of the c-Ret receptor tyrosine kinase. *Moll. Cell. Neurobiol.* **11**: 47–63.

Trupp, M., Scott, R., Whittemore, S.R. and Ibáñez, C.F. (1999) Ret-dependent and independent mechanisms of glial cell line-derived neurotrophic factor signaling in neuronal cells. *J. Biol. Chem.* **274**: 20885–20894.

Vega, Q.C., Worby, C.A., Lechner, M.S., Dixon, J.E. and Dressler, G.R. (1996) Glial cell line-derived neurotrophic factor activates the receptor tyrosine kinase RET and promotes kidney morphogenesis. *Proc. Natl Acad. Sci. USA* **93**: 10657–10661.

Wells, S.A., Chi, D.D., Toshima, D. *et al.* (1994) Predictive DNA testing and prophylactic thyroidectomy in patients at risk for multiple endocrine neoplasia type 2A. *Ann. Surg.* **200**: 237–250.

Wohlik, N., Cote, G.J., Bugalho, M.M.J. *et al.* (1996) Relevance of RET proto-oncogene mutations in sporadic medullary thyroid carcinoma. *J. Clin. Endocrinol. Metab.* **81**: 3740–3745.

Zedenius, J., Wallin, G., Hamberger, B., Nordenskjöld, M., Weber, G. and Larsson, C. (1994) Somatic and MEN 2A *de novo* mutations identified in the RET proto-oncogene by screening of sporadic MTCs. *Hum. Mol. Genet.* **3**: 1259–1262.

Zedenius, J., Larsson, C., Bergholm, U. *et al.* (1995) Mutations of codon 918 in the *RET* proto-oncogene correlate to poor prognosis in sporadic medullary thyroid carcinoma. *J. Clin. Endocrinol. Metab.* **80**: 3088–3090.

9

TP53 in cancer origin and treatment

Elena A. Komarova, Peter M. Chumakov and Andrei V. Gudkov

1. Introduction

TP53 is the most well-studied tumor suppressor gene whose function is affected in the majority of malignancies. The high frequency of TP53 inactivation in tumors is a reflection of its unique role in the organism: TP53 is a key component of a cellular emergency response mechanism that converts a variety of intra- and extracellular stress signals, such as DNA damage, as well as conflicts between positive and negative proliferation stimuli, into growth arrest or apoptosis. This protein is also an essential component of control of cellular senescence. This set of properties assigns an important 'social' function to TP53 within the cell society of the organism: its activity results in the elimination of damaged and potentially dangerous cells. At the same time, an intact TP53 pathway is something unbearable for the tumor cell, which unavoidably has to overcome normal growth control mechanisms, including senescence, and to tolerate accumulation of additional mutations. This explains why TP53 function is suppressed in the majority of tumors. Moreover, the unique position of TP53 on the crossroads of several signaling pathways, and its involvement in numerous interactions, results in an unusually frequent acquisition of mutations in the protein itself, turning it not only into an inactive form, but also often changing its properties and converting it from a tumor suppressor into an oncogene.

During the last 10 years, impressive progress has been achieved in understanding the mechanisms of *TP53* function. The goal of this review is to summarize what has been learned about *TP53* in cancer. We will first focus on the cellular mechanisms of TP53 activity. We will analyze the mechanisms of *TP53* inactivation in tumors and the consequences of *TP53* inactivation on tumor cell phenotype. We will then focus on the role of *TP53* in the organism and, finally, will overview potential clinical applications that have stemmed from *TP53* research.

2. *TP53* gene structure and regulation

2.1 The TP53 *gene and protein structure*

Human *TP53* is encoded by a single gene localized on the short arm of chromosome

Molecular Genetics of Cancer second edition, edited by J.K. Cowell.
© 2001 BIOS Scientific Publishers Ltd, Oxford.

17 (Miller *et al.*, 1986) that consists of 11 exons and is transcribed in all tissues, although with different efficiency (see below). This 393 amino acid-long protein can hardly be detected in the majority of normal cells under normal conditions due to its rapid ubiquitin-mediated degradation in proteasomes. The TP53 protein is the subject of complex post-translational modifications and regulation. Fast accumulation of TP53 occurs in response to a variety of stresses, including DNA damage (Kastan *et al.*, 1991; Maltzman and Czyzyk, 1984), deregulation of microtubules (Tishler *et al.*, 1995) and actin microfilaments (Rubtsova *et al.*, 1998), hypoxia (Graeber *et al.*, 1996), hyperthermia (Valenzuela *et al.*, 1997), and activation of dominant oncogenes (Serrano *et al.*, 1997). The description of the mechanisms of TP53 protein activation and function is one of the main topics of current review.

The TP53 protein (*Figure 1*) consists of several functional domains (Ko and Prives, 1996). The N-terminal sequence (amino acids 1–50) contains a transactivation domain, involved in the modulation of transcription of target genes. This locus is also responsible for the interaction with MDM2, a protein with ubiquitin ligase activity that controls proteasomal degradation of *TP53* (see below) (Kussie *et al.*, 1996; Lin *et al.*, 1998). The central core domain between amino acids 100 and 300 participates directly in recognition and binding with specific DNA sequences (Cho *et al.*, 1994). A locus between amino acids 323 and 356 is responsible for oligomerization of *TP53* molecules into tetramers (Kussie *et al.*, 1996; Lin *et al.*, 1994), the form in which this protein functions in the cell (McLure and Lee, 1998).

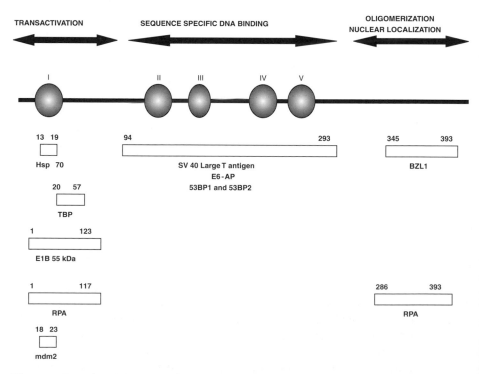

Figure 1. Functional domains of p53. The coordinates correspond to human p53.

A nuclear export signal lies within the tetramerization domain (Stommel *et al.*, 1999), which is positioned between the first and second of three nuclear local-ization signals spanning through amino acids 316–325, 369–375 and 379–384 (Shaulsky *et al.*, 1990). The C-terminal domain (amino acids 363–393) is essential for the regulation of TP53 activity and is a target for modification by kinases, acetylases, glycosylases, and for binding with other interacting proteins. In addition, the C-terminal TP53 fragment is able to bind nonspecifically to single-stranded DNA regions, unpaired bases, and DNA ends (Bakalkin *et al.*, 1995), indicating its possible involvement in the process of recognition of damaged DNA.

TP53 gene orthologs can be found in all studied vertebrates (Soussi *et al.*, 1987, 1988) as well as in insects (Ollmann *et al.*, 2000). This comparison reveals five highly conserved stretches of amino acids, four of which are located in the DNA-binding domain, while one is located at the amino terminal part of the protein. Regions of similarity between *TP53* of Drosophila (DmTP53) and vertebrates are limited to the central DNA-binding domain and include residues identified in human *TP53* as critical for DNA sequence recognition (Ollmann *et al.*, 2000).

2.2 Regulation of TP53 expression at mRNA level

Transcriptional regulation of the *TP53* gene might play an important role since: (i) attenuated transcription of wild type *TP53* can be involved in its inactivation in tumors (Raman *et al.*, 2000); (ii) the level of *TP53* mRNA determines the susceptibility of normal tissues to apoptosis in response to genotoxic stress (Komarova *et al.*, 1997a, 2000; Zhao *et al.*, 2000); and (iii) mutant *TP53* genes are generally expressed at higher levels due in part to increased rates of tran-scription of the gene (Balint and Reisman, 1996). The relatively well-charac-terized mouse *tp53* promoter contains binding sites and is controlled by a number of regulatory transcription factors. The promotor includes an E-box motif that serves as a recognition site for members of the basic-helix-loop-helix family of transcription factors (Ronen *et al.*, 1991). USF and Myc/Max bind to this site in the promoter (Reisman and Rotter, 1993; Reisman *et al.*, 1993) and positively regulate *TP53* transcription. NFkB, NF1, SP1, PF1, PF2, PBF I and II are among the factors regulating transcription activity of *TP53* (Reisman and Loging, 1998). It was recently shown that *HOXA5* is involved in positive regulation of the human *TP53* promoter. Down-regulation of *HOXA5* in breast carcinomas bearing wild-type *TP53* is likely to be the cause of low expression of the *TP53* gene in these tumors; ectopic expression of *HOXA5* in such tumors results in apoptosis that does not occur if *TP53* is inactivated (Raman *et al.*, 2000).

TP53 mRNA is also the subject of alternative splicing. Although alternative transcripts of *TP53* are found both in mouse and human, this results in the gener-ation of a protein with altered structure (with truncation and substitution of its C-terminus) only in the mouse (Flaman *et al.*, 1996; Will *et al.*, 1995). Considering the numerous functions ascribed to the carboxy-terminus of the protein (see below), the product of such splice variants may have some functional role in mouse cells.

3. Cell growth control by *TP53*

3.1 TP53 *induction of growth arrest and apoptosis in response to stress*

The phenotypic consequences of *TP53* activation were determined by comparison of the cells with intact *TP53* or with *TP53* that is experimentally inactivated either by gene knock-out (cells derived from *tp53*-deficient mice) or by expression of proteins acting as TP53 inhibitors (see below). The general conclusion drawn from numerous studies is that TP53 is a negative regulator of cell growth which, upon activation, induces growth arrest at cell cycle checkpoints, or apoptosis. The choice among different types of growth arrest and apoptosis depends on the cell type and growth conditions. For example, thymocytes enter apoptosis after DNA damage (Clarke *et al.*, 1993; Lowe *et al.*, 1993b), while fibroblasts predominantly undergo irreversible growth arrest. Growth conditions determine the TP53-dependent response in gamma irradiated human hematopoietic cell lines Baf-3 and DA-1 that enter growth arrest in the presence of interleukin 3 and undergo apoptosis in the absence of the cytokine (Canman *et al.*, 1995; Gottlieb and Oren, 1998). Different splice variants of JMY, a protein that together with p300/CBP participate in a multiprotein complex that acts as coactivator for diverse transcriptional factors including TP53, may determine whether TP53 mediates growth arrest or apoptosis (Shikama *et al.*, 1999). Modulation of pRB (Haupt *et al.*, 1995a) and E2F (Wu and Levine, 1994) expression may also play a role in the final outcome. In addition, the levels of TP53 mRNA and the TP53 protein itself may also regulate the nature of the TP53 response (Komarova *et al.*, 1997a, 2000; Zhao *et al.*, 2000).

3.2 TP53 *as a transcription factor*

The mechanism of TP53-mediated growth arrest and apoptosis is determined, at least in part, by the activity of TP53 as a transcription factor and is dependent on TP53 binding to specific DNA elements located near the promoter regions of TP53 responsive genes. A modification of the C-terminal part of the molecule, initiated by various stress mechanisms, is followed by a conformational rearrangement and acquisition of the ability of the protein to associate with DNA. A complex interaction of the TP53 N-terminal transcription-activator region with components of the transcription apparatus such as proteins of the TFIID complex: TBP (TATA-Box Binding Protein) and TBP-associated factors TAFII, is essential (Liu *et al.*, 1993; Lu and Levine, 1995; Seto *et al.*, 1992; Thut *et al.*, 1995; Truant *et al.*, 1993). In addition, the transcription coactivators, CREB-binding protein (CBP) and its closely related p300 protein, are required in a complex with TFIID for activation of the specific promoters. p300/CBP displays histone-acetyl-transferase activity. It is able to directly interact with the RNA-polymerase II complex conducting initiation of transcription in response to the association with a transcription factor.

Accumulation of TP53 in the nucleus is associated with altered transcription of a large number of genes, usually containing TP53-binding sites inside or outside of their promoter regions (el-Deiry, 1998). The first known TP53-responsive genes were identified by using differential screening strategies, including a cDNA subtraction technique and yeast two-hybrid screen (el-Deiry *et al.*, 1993; Harper *et al.*, 1993; Jiang *et al.*, 1994). This gene list grew slowly during the early years but has recently been expanded considerably through the application of cDNA microarray hybridization approaches (Komarova *et al.*, 1998; Zhao *et al.*, 2000). Hundreds of genes that are either up- or down-regulated by TP53 have now been identified. Of 6000 genes examined for TP53-regulatory responses by the group of Arnold Levine, 107 were found to be up, and 54 down-regulated by TP53. The TP53-responsive genes fell into the groups involved in apoptosis, growth arrest, cytoskeletal functions, growth factors and their inhibitors, extracellular matrix, adhesion genes and others (Komarova *et al.*, 1998; Zhao *et al.*, 2000). The pattern of genes whose expression is altered by TP53 is different depending on the cell type, the inducing agent, levels of *TP53* expression and time after *TP53* induction.

TP53 not only up-regulates but also down-regulates many genes including topoisomerase II (Wang *et al.*, 1997), MDR1 (Chin *et al.*, 1992), HSP70 (Agoff *et al.*, 1993), microtubule associated protein 4 (Murphy *et al.*, 1996), the transcription factors SPI1 (Bargonetti *et al.*, 1997) and HIF-1 (Blagosklonny *et al.*, 1998).

3.3 TP53-*dependent cell cycle checkpoint control*

Eukaryotic cells have developed a network of surveillance mechanisms (checkpoints) whereby their progression through the cell cycle can be stopped in response to a variety of stresses, in order to repair the damage caused by stress and to prevent damaged cells from entering mitosis. Inactivation of *TP53* often results in abrogation of these checkpoints, indicating that TP53 is involved in their function.

Most of the information regarding the role of TP53 in checkpoint control was obtained using fibroblast models, in which TP53 was inactivated either by experimental or spontaneous gene knockout, or by expression of the proteins suppressing the TP53 pathway (such as oncogenes of DNA tumor viruses, see below). This is a reflection of the lack of apoptosis in fibroblasts which usually react to stress by reversible or irreversible growth arrest. Similar mechanisms are likely to be involved in deregulation of checkpoints in tumor cells of epithelial origin which have an inactivated control of apoptosis.

G_1 *arrest.* The TP53-dependent G_1 arrest results mainly from transactivation of the *WAF1* gene (wild-type TP53-activated factor) (el-Deiry *et al.*, 1993), also known as *CIP1* (Cdk-interacting protein 1) (Harper *et al.*, 1993) or *SDI1* (senescent cell-derived inhibitor) (Noda *et al.*, 1994), that encodes the small protein p21, an inhibitor of cyclin-dependent kinases (Cdks). p21 interferes with cell-cycle progression and prevents S phase entry by blocking the activity of Cdks (Cdk2, Cdk3, Cdk4, and Cdk6). Cdks are catalytic partners of the cyclins, controlling cell-cycle progression (Dulic *et al.*, 1994; el-Deiry *et al.*, 1993). Although *p21* is a TP53-responsive gene, it is a subject of complex transcriptional regulation that is, in part, TP53-independent (Macleod *et al.*, 1995). Up-regulation of *P21* often interferes with the induction of apoptosis (Gorospe *et al.*, 1996; Polyak *et al.*, 1996). Although *p21* knockout cells have impaired G_1 checkpoint control, inactivation of *p21* does not result in any severe phenotype and does not match the consequences of *tp53* knockout mice (Brugarolas *et al.*, 1995; Deng *et al.*, 1995; Komarova *et al.*, 2000). Abrogation of the G1 checkpoint results in a higher proportion of cells arrested at other checkpoints and, in some cases, increases sensitivity to apoptosis (Gartel *et al.*, 1996; Komarova *et al.*, 2000; Wang *et al.*, 1996b).

S phase checkpoint. Another checkpoint response pathway acts in S phase and blocks DNA replication in the event of nucleotide depletion or inactivation of the DNA replication machinery (Taylor *et al.*, 1999). Human and mouse fibroblasts with normal *TP53* fail to enter mitosis when DNA synthesis is blocked by aphidicolin or hydroxyurea. Isogenic *tp53*-null fibroblasts do enter mitosis with incompletely replicated DNA, revealing that TP53 contributes to a checkpoint that prevents mitosis before the completion of DNA synthesis. When treated with N-(phosphonacetyl)-L-aspartate (PALA), which inhibits pyrimidine nucleotide synthesis and thus generates damaged DNA due to highly unbalanced dNTP pools, *tp53*-null cells enter mitosis after they have completed DNA replication, but cells with wild-type *tp53* do not, revealing that TP53 also mediates a checkpoint that monitors the quality of newly replicated DNA.

The mechanism of the TP53-mediated block of DNA replication remains unclear. It might involve P21 binding to the proliferating cell nuclear antigen (PCNA) which results in suppression of the elongation step in DNA replication and interference with cell cycle progression (Waga *et al.*, 1994).

G_2/M *checkpoint.* TP53 has also been implicated in the control of the G_2/M checkpoint, the last line of cell defense before entering mitosis. After DNA damage, many cells appear to enter a sustained arrest in the G_2 phase of the cell cycle. It was shown that this arrest could be sustained only when TP53 was present in the cell and capable of transcriptionally activating the cyclin-dependent kinase inhibitor P21 (Bunz *et al.*, 1998). The mechanism is p21 and retinoblastoma protein (pRB)-dependent and involves an initial inhibition of cyclin B1-Cdc2 activity and a secondary decrease in cyclin B1/Cdc2 complex (Flatt *et al.*, 2000), a major regulatory factor required for entry into mitosis (Elledge, 1996; O'Connor *et al.*, 1997). Abrogation of p21 or pRB function in cells containing wild-type *TP53* prevents down-regulation of the cyclin B1 and Cdc2 expression leading to an accelerated exit from G2 after genotoxic stress (Flatt *et al.*, 2000). Thus, similar to what occurs in p21 and TP53 deficiency, pRB loss can uncouple S phase and mitosis after genotoxic stress. These results indicate that similar molecular mechanisms are required for TP53 regulation of G_1 and G_2 checkpoints.

In addition to this mechanism, cyclin B1 can be directly down-regulated by TP53 (Agarwal *et al.*, 1995; Stewart *et al.*, 1995). The 14–3–3-sigma protein, a product of the TP53-induced gene (Hermeking *et al.*, 1997) can bind to Cdc25C preventing its phosphorylation and the activation of *CDC2* (Peng *et al.*, 1997). These mechanisms may also contribute to TP53-mediated G_2/M arrest.

Abrogation of cell cycle checkpoints in *TP53*-deficient cells leads to two opposite alterations in cell reaction to stress. On one hand, it is often associated with higher resistance of cell populations to a variety of treatments that otherwise would lead to growth arrest. For example, inactivation of G_2 checkpoint control in TP53-deficient cells is believed to be responsible for rapid accumulation of poly- and aneuploid variants in cell populations subjected to treatments with drugs targeting G_2 (May and May, 1999). On the other hand, cells with impaired check-point control lose the capability to pause for necessary repairs and, therefore, continue to proceed through the cell cycle, thus increasing the risk of death from the damage. This latter mechanism is likely to explain why, in some instances, TP53 suppression results in increased tumor cell sensitivity to genotoxic stress (Lowe and Lin, 2000). Thus, under certain circumstances TP53 can serve as a survival factor.

3.4 TP53-dependent apoptosis

Ectopic expression of wild type TP53 induces apoptosis in many tumor cell lines lacking functional TP53 (May and May, 1999; Yonish-Rouach *et al.*, 1991), providing the rationale for numerous attempts to use TP53 as an anticancer agent (see below). Moreover, apoptosis occurring in some cell types and tissues in response to a variety of stimuli was found to be TP53 dependent. There are not that many *in vitro* cell systems that can serve as models for studying TP53-mediated programmed cell death. This is not surprising considering the necessity for any cell line to have the TP53 pathway at least partially suppressed to ensure unconstrained growth and immortalization. Primary thymocytes and mouse embryo fibroblasts transformed with either *E1a+ras* or with *myc* oncogenes are the most commonly used cell models of TP53-dependent apoptosis in normal and tumor cells, respectively (Lowe *et al.*, 1993a, 1993b). Apoptosis in the first two systems can be triggered by a wide variety of treatments, including DNA damage (Lowe *et al.*, 1993b), while *myc*-transformed mouse fibroblasts represent the model for apoptosis caused by growth factor deprivation (Rohn *et al.*, 1998). These cells, together with other models, were used to determine the mechanism of TP53-mediated death and specifically the role of TP53-responsive genes in inducing programmed cell death.

A group of TP53 target genes was identified which encoded the effectors of TP53-mediated apoptosis. A proapoptotic *BAX* gene (Oltvai and Korsmeyer, 1994) contains a TP53-binding sequence in its promoter and is up-regulated after TP53 activation. This promotes the release of cytochrome *c* from mitochondria (Rosse *et al.*, 1998), acting as a trigger of the apoptotic cascade (Srinivasula *et al.*, 1998). A gene encoding the death receptor, *CD95/Fas/Apo1*, was also found to be activated by TP53 in some cell systems (Bennett *et al.*, 1998; Owen-Schaub *et al.*, 1995). The TP53-responsive gene for the KILLER/DR5 protein belongs to another group of 'death domain' receptor proteins (Wu *et al.*, 1997) and mediates cell killing by TRAIL (Ashkenazi and Dixit, 1998). Another mechanism of cell sensitization to apoptosis is exerted by another TP53-responsive gene, encoding the inhibitor of the insulin-like growth factor 3 (*IGFBP3*). It blocks the anti-apoptotic activity of the IGF cytokine (Harrington *et al.*, 1994), thus promoting the induction of death. Interestingly, TP53 simultaneously suppresses tran-scription of IGF (Prisco *et al.*, 1997), leading to an even stronger pro-apoptotic effect.

Besides the genes that are directly involved in the apoptotic machinery, TP53 is likely to promote programmed cell death by modulating expression of a large group of genes (so called PIGs, or TP53-inducible genes), involved in the generation of reactive oxygen species (Polyak et al., 1997). Accumulation of free radicals induces damage of mitochondrial membranes, causing the release of cytochrome c and initiation of apoptosis.

As the list of TP53-responsive genes involved in apoptosis continues to grow (Attardi et al., 2000; Elkeles et al., 1999; Yin et al., 1998) it becomes clear that there is no single major TP53-dependent apoptosis-inducing gene and that the ultimate cell response depends on the combined effect of altered expression of the TP53 gene targets available in a particular cell type. Indeed, the spectra of TP53-responsive genes in the spleen and thymus, both undergoing gamma-radiation-induced apoptosis, were shown to be different (Komarova et al., 1998).

TP53 activity could not be entirely attributed to its function as a transcriptional regulator. The demonstration that TP53-mediated apoptosis can occur in the absence of protein or RNA synthesis (Bissonnette et al., 1997; Caelles et al., 1994; Levine, 1997) implied that TP53 transcription activity is not always obligatory for apoptosis. It was also shown that a TP53 mutant, deficient in transactivation domain, is still capable of inducing apoptosis in some cell systems (Haupt et al., 1995b). It is interesting that TP53 can induce apoptosis through the Fas/Apo1 receptor by a transcription-independent mechanism, leading to a fast translocation of APO1 from the Golgi apparatus to the cell surface (Bennett et al., 1998). Thus, TP53 is involved in modulation of Fas-mediated apoptosis in two ways: by mediating transport of the protein and by transcriptional regulation of the gene encoding the Fas receptor.

TP53 is an integral part of a cellular signaling network interacting with other growth regulatory pathways. For example, TP53 can be involved in determination of cell sensitivity to TNF as demonstrated by comparison of two variants of prostate carcinoma LNCaP cells, differing in their TP53 status (Rokhlin et al., 2000). TP53 also serves within a sensor mechanism reacting to abnormal proliferation associated with activation of dominant oncogenes by promoting growth arrest or apoptosis. In mouse fibroblasts, it determines the high sensitivity to apoptosis of myc-transformed cells and induces irreversible growth arrest and premature senescence in response to ras activation. These functions of TP53 constitute important facets of its tumor suppressor activity and are discussed below.

TP53-dependent apoptosis occurs through the mitochondria route and therefore can be suppressed by BCL2 or AKT. Thus, the activity of the Akt survival pathway is essential for the viability of myc-transformed mouse fibroblasts since its suppression by growth factor deprivation results in rapid activation of TP53-dependent apoptosis (Chiarugi et al., 1995; Kennedy et al., 1999; Sabbatini and McCormick, 1999).

3.5 Role of TP53 in cell sensitivity to dominant oncogenes

In addition to the stress induced by external factors, TP53 was shown to be involved in the cellular reaction to abnormal growth signaling by dominant oncogenes. It was established long ago that cell transformation either by the activated

RAS oncogene or the adenoviral gene *E1a* requires cooperation with other genetic alterations resulting in TP53 inactivation. In the presence of intact TP53, cells react to activated *RAS* by irreversible growth arrest, resembling premature senescence (Lin *et al.*, 1998; Serrano *et al.*, 1997). E1A can stabilize the TP53 protein causing an induction of TP53-dependent apoptosis (Lowe and Ruley, 1993), while TP53-deficient cells are resistant to E1a (Debbas and White, 1993; Lowe *et al.*, 1994b). MYC also sensitizes cells to TP53-dependent apoptosis (Hermeking and Eick, 1994; Wagner *et al.*, 1994). The loss of TP53, therefore, not only makes cells genetically unstable and resistant to a variety of stresses, but also allows them to tolerate abnormal growth stimuli through abrogation of natural growth regulatory control that prevents cancer formation.

3.6 TP53 and control of cellular senescence

TP53 is involved in cell aging resulting in irreversible growth arrest after a certain number of divisions. This process is also known as replicative senescence. This is another function of TP53 that could represent an important mechanism of its anti-tumor activity. Senescence is abrogated by TP53 inactivation since fibro-blasts from *tp53*-knockout mice stay immortal in culture (Jacks and Weinberg, 1996). The establishment of immortal stable cell lines requires TP53 inactivation that is often achieved by expression of the oncoproteins of DNA-containing viruses (see below). Irreversible growth arrest of senescent cells is associated with the transcriptional activation of TP53-responsive genes (Atadja *et al.*, 1995; Bringold and Serrano, 2000; Kulju and Lehman, 1995; Vaziri *et al.*, 1999). The cause of TP53 activation in senescent cells remains unknown although the most obvious hypothesis is that it is induced by chromosomal breaks associated with the telomere shortening in aging cells (Harley and Sherwood, 1997; Vaziri *et al.*, 1999).

3.7 TP53 and control of genomic stability

TP53 deficiency results not only in cell survival or continuous proliferation under conditions when a normal cell would become arrested or commit suicide but is also associated with genomic instability detected by a high frequency of gene amplification (Livingstone *et al.*, 1992; Yin *et al.*, 1992). Consistently, an extremely high incidence of chromosomal aberrations was found in somatic cells of *tp53*-knockout mice (Lee *et al.*, 1994). Tumors from *tp53*-knockout mice are characterized by increased aneuploidy and chromosomal instability (Donehower *et al.*, 1995; Purdie *et al.*, 1994). Based on these observations, *TP53* was termed by David Lane 'a guardian of the genome' (Lane, 1992). It remains unclear, however, whether TP53 exerts its control over genomic stability exclusively through growth control by eliminating mutated cells from proliferation. Additionally, TP53 may be involved in the control of genomic stability by taking a direct part in DNA repair. TP53 can bind and modulate the repair activity of the nucleotide excision repair factors XPB and XPD (components of the multiprotein transcription/reparation complex TFIIH) and RAD51, a proposed member of the mammalian recombination machinery (Albrechtsen *et al.*, 1999). In addition, TP53 binds to the ends of double-strand DNA breaks through its C-terminal domain (Bakalkin *et al.*, 1995) which is a potential mechanism for recognizing DNA damage. TP53 could be involved in the DNA repair process by several additional biochemical activities, including 3' – 5' exonuclease activity (Mummenbrauer *et al.*, 1996).

In summary, inactivation of TP53 dramatically changes the cellular phenotype, making cells more tolerant to environmental or therapy-associated stresses but, at the same time, reducing normal control of genomic stability, reaction on growth proliferation stimuli and cell aging, leading to rapid accumulation of genetically altered variants in cell populations.

4. Mechanisms of TP53 regulation

TP53 function is regulated at different levels, including control of protein accumulation and subcellular localization. This regulation operates via covalent modifications of the TP53 protein and/or its interaction with different counterparts, determining fast activation of TP53 in response to stresses and shutting down TP53-mediated response when the stress is over. The general scheme of regulation of the TP53 pathway is shown in *Figure 2*, with the details described below.

4.1 TP53 degradation: role of MDM2

Regulation of TP53 activity during stress response is controlled mostly at the level of the protein stability and specifically through modulation of TP53 degradation. Under normal growth conditions, TP53 is a subject of rapid ubiquitin-dependent degradation in proteasomes. TP53 degradation involves its interaction with a specific ubiquitin ligase, identified through its interaction with TP53, named *mdm2* (mouse double minute gene) (Momand *et al.*, 2000; Piette *et al.*, 1997). *MDM2* is a key player in the regulation of TP53. It can physically bind to TP53, within the TP53 transactivation domain and inactivate its ability to recruit the components of basal transcription machinery (Lu and Levine 1995; Momand *et al.*, 1992; Thut *et al.*, 1995). Moreover, binding to MDM2 promotes TP53 nuclear export followed by proteasomal degradation (Haupt *et al.*, 1997; Kubbutat *et al.*, 1997; Maki *et al.*, 1996; Wu *et al.*, 1993). The MDM2 protein is encoded by a TP53-responsive gene containing two TP53-responsive elements, which are located in the promoter region and in the first intron (Zauberman *et al.*, 1995). Thus, TP53 and MDM2 form a feedback regulatory loop.

The function of *MDM2* is essential for the control of normal activity of the TP53-dependent response. This is illustrated by the embryonic lethality of *mdm2*-deficient mice, which can be rescued by the inactivation of tp53 (Jones *et al.*, 1995). This observation shows that uncontrolled activity of TP53 is a life-threatening event. On the other hand, hyperactivity of *MDM2* may result in inactivation of the TP53 response and is used by some cancer cells to shut down the TP53 pathway (see below).

4.2 Covalent modifications of TP53

Stabilization of TP53, making it more resistant to ubiquitination and subsequent proteolysis, can be achieved either by conformational changes in the protein, or by modulation of activity of the TP53 degradation machinery. Phosphorylation and acetylation play important roles in TP53 activation by DNA damage. Recent studies have shown that different types of DNA damage induce different site-specific phosphorylation within the N-terminus of TP53, specifically at residues 15, 20, 33, 37 and 46 (Banin *et al.*, 1998; Bulavin *et al.*, 1999; Canman *et al.*, 1998; Kamijo *et al.*, 1998; Sakaguchi *et al.*, 1998; Shieh *et al.*, 1999), and this phosphorylation correlates well with stabilization of the TP53 protein. Phosphorylation at serines 15 and 37 or at serine 20 was shown to reduce the interaction between TP53 and MDM2 *in vitro* (Liu *et al.*, 1999; Shieh *et al.*, 1997), and replacement of both serines 15 and 37 with aspartic acid partially protected TP53 from degradation by MDM2 (Unger *et al.*, 1999). Taken together, accumulated observations indicate that N-terminal phosphorylation at sites in or around the MDM2 binding region of TP53 may regulate TP53 stability.

Several kinases are capable of phosphorylation of TP53 *in vitro*. Most of them increase their activity in response to DNA damage. The list of kinases include ATM kinase (Ser 15), ATR kinase (Ser 15 and 37), DNA-PK (Ser 15 and 37), ERK/p38 kinase (Ser 15), JUN aminoterminal kinase (Ser 34), CHK1 and CHK2 kinases (Ser 20), casein kinase (Tre 18) (Banin *et al.*, 1998; Canman *et al.*, 1998; Cliby *et al.*, 1998; Keegan *et al.*, 1996; Lees-Miller *et al.*, 1992; She *et al.*, 2000; Shieh *et al.*, 1997; Siliciano *et al.*, 1997).

DNA-PK consists of a large catalytic subunit (DNA-PKcs) that is targeted to DNA ends by the Ku heterodimer (polypeptides Ku70 and Ku80) (Dvir *et al.*, 1993; Gottlieb and Jackson, 1993). Binding of DNA-PK to DNA ends *in vitro* is required in order to phosphorylate a number of different substrates, including TP53 (Anderson and Lees-Miller, 1992; Smith and Jackson, 1999). DNA-PK has been shown to be capable of phosphorylating TP53 on serine-15 and serine-37 *in vitro* in a DNA-dependent manner (Lees-Miller *et al.*, 1990, 1992).

The gene mutated in ataxia-telangiectasia patients (exibiting strong specific symptoms including 100-fold increased incidence of cancer, chromosomal instability and radiosensitivity) has been cloned and termed *ATM* (Gatti *et al.*, 1988; Savitsky *et al.*, 1995). ATM kinase can also mediate phosphorylation of serine-15 of TP53 *in vitro*, and this kinase activity is activated in response to ionizing radiation (Banin *et al.*, 1998; Canman *et al.*, 1998).

Acetylation of the TP53 C-terminal domain plays an important role in its activation. Acetylation of TP53 at Lys373/Lys382 and Lys320 by transcriptional co-activators, p300/CBP and PCAF respectively, enhance sequence-specific DNA binding *in vitro*. *In vivo*, acetylation of these sites is increased in response to DNA-damaging agents (Liu *et al.*, 1999) and overexpression of p300/CBP increases transactivation by TP53 (Gu *et al.*, 1997; Lill *et al.*, 1997). Phosphorylation of Ser-15 causes significant activation of Lys-382 acetylation carried out by p300/CBP (Lambert *et al.*, 1998).

4.3 Protein–protein interactions and TP53 activation: role of ARF in TP53 response

While covalent modifications of TP53 determine its activation in response to DNA damage, TP53 stabilization by abnormal proliferative signals mediated by oncogenes is regulated by a separate mechanism involving another tumor suppressor, *ARF* (de Stanchina *et al.*, 1998; Zindy *et al.*, 1998). This small protein (p14[ARF] in humans and p19[ARF] in mice) arises through translation of an alternative reading frame derived from the *INK4A* tumor suppressor gene also encoding the CDK inhibitor p16 (Kamijo *et al.*, 1997). ARF acts by binding to MDM2 and to TP53, preventing nuclear export of MDM2 (Tao and Levine, 1999b; Zhang and Xiong, 1999) and MDM2-mediated TP53 proteolysis (Kamijo *et al.*, 1998; Pomerantz *et al.*, 1998), apparently by blocking the E3 ligase activity of MDM2 (Honda and Yasuda, 1999; Honda *et al.*, 1997). It was shown that ARF reduces the ability of MDM2 to interact with TP53 by changing the subcellular localization of MDM2 (Weber *et al.*, 1999; Zhang *et al.*, 1998) and so driving it to the nucleoli. In this manner ARF inhibits ubiquitinating activity of MDM2 and prevents its escape from the nucleus.

Induction of expression of p19/ARF is an essential factor of oncogene-mediated TP53 activation (Lowe and Lin, 2000). For example, E1A or MYC can increase *ARF* mRNA and protein levels in normal mouse embryo fibroblasts, resulting in activation of TP53 and subsequent apoptosis (Dang *et al.*, 1999). In contrast, the same oncogenes fail to activate TP53 in ARF-null cells, resulting in promotion of proliferation without substantial apoptosis (de Stanchina *et al.*, 1998; Zindy *et al.*, 1998). Together, these studies indicate that p19ARF acts as part of a TP53-dependent fail-safe mechanism to counter hyperproliferative signals. Disruption of ARF should cooperate with mitogenic oncogenes during tumor development.

Initially it has been suggested that the ARF-related mechanism is limited to oncogene-mediated TP53 activation, and is not used in the DNA-damage response (de Stanchina *et al.*, 1998; Zindy *et al.*, 1998). Consistently, gamma-irradiation-induced *TP53* activation was reported to be independent of ARF and conducting through N-terminal phosphorylation and other modifications (Kamijo *et al.*, 1997; Shieh *et al.*, 1997; Siliciano *et al.*, 1997). Recently (Khan *et al.*, 2000) it was demonstrated that ARF is also required for TP53 activation by gamma-radiation. It was shown that *arf*[-/-] mouse fibroblasts are defective in *tp53* accumulation and growth arrest (Khan *et al.*, 2000). They continue cycling after DNA damage, although not to the same extent as *tp53*[-/-] MEF. Also, reduced TP53 levels are detected in *p19*[-/-] fibroblasts. Thus, DNA damage may require ARF to induce the TP53 responses. A possible explanation for the apparent TP53 instability is that the absence of ARF allows a higher proportion of TP53 to be bound by MDM2, which increases TP53 turnover.

Besides ARF, TP53 has other cellular counterparts that may be involved in its function. For example, CBP/p300 (Gu *et al.*, 1997; Lill *et al.*, 1997) and p33 (ING1) (Garkavtsev *et al.*, 1998), a nuclear PHD finger protein, are required for TP53 to function as a transcription activator, while REF1 is involved in TP53-mediated apoptosis (Jayaraman *et al.*, 1997).

4.4 Nuclear transport of TP53

Distribution of TP53 between the nucleus and cytoplasm is a subject of a complex regulation. TP53 nuclear accumulation after gamma irradiation is cell cycle-dependent and is allowed only during a certain 'window', including G_1–early S phase (Komarova *et al.*, 1997b). Analysis of gamma-irradiated cell fusions shows that TP53 nuclear accumulation is determined by the property of the nucleus, indicating that this stress may produce modifications that inhibit TP53 nuclear export rather than accelerate its nuclear import (Komarova *et al.*, 1997b). Nuclear localization of TP53 is very important for its activity, because many TP53 effects are associated with its transcriptional regulation of the target genes. In some cell types, TP53 is completely excluded from the nucleus. For example, in mouse embryonic stem cells wild-type tp53 is sequestered in the cytoplasm, thus explaining the lack of G_1 arrest in these cells in response to genotoxic stresses (Aladjem *et al.*, 1998; Moll *et al.*, 1996). This natural mechanism of TP53 regulation is often activated in neuroblastomas that use it to functionally inactivate the TP53 pathway. The reason for cytoplasmic sequestration of TP53 is probably associated with the accelerated export of TP53 from the nucleus (Moll *et al.*, 1996).

The mechanics of TP53 nuclear shuttling is not completely understood. It is likely to involve MDM2. Both TP53 and MDM2 carry nuclear export signal (NES) regions responsible for export from the nucleus (Freedman and Levine, 1998; Lain *et al.*, 1999; Stommel *et al.*, 1999). It was demonstrated that nucleocytoplasmic shuttling of MDM2 is essential for its ability to promote TP53 degradation (Tao and Levine, 1999a), although it has been shown that TP53 can be exported from the nucleus without binding MDM2. Nuclear export of TP53 is likely to be achieved through its direct interaction with the export receptor CRM1, rather than through a NES-containing binding partner (Stommel *et al.*, 1999). A model has been proposed in which subcellular localization of TP53 is determined through tetramerization-regulated exposure of the NES to the export machinery (Stommel *et al.*, 1999).

Because MDM2 binding to TP53 is necessary for TP53 degradation, it was proposed that MDM2 functions as a NES-containing chaperone that mediates TP53 export to cytoplasmic proteasomes (Freedman and Levine, 1998; Roth *et al.*, 1998). However, there is currently no evidence that MDM2 is the sole mediator of TP53 nuclear export. On the contrary, TP53 has been shown to be fully capable of nuclear export independently of MDM2 (Stommel *et al.*, 1999). On the other hand, it is clear that continuous nuclear import and export of MDM2 are required for degradation of TP53 (Freedman and Levine, 1998; Roth *et al.*, 1998; Tao and Levine, 1999a).

4.5 Cellular functions and regulation of TP53 (summary)

TP53 is a uniquely multifunctional protein that is involved in several processes (reaction to variety of stresses, control of genomic stability, control of senescence, expression of surface receptors, etc.), acts through numerous counterparts and is regulated at different levels. TP53 function is largely exerted, but not limited to, its activity as a transcriptional regulator. Consistently with its multiple functions, TP53 inactivation at the cellular level results in dramatic changes in cell phenotype, including impaired cell cycle checkpoint and apoptosis control in response to DNA damage and other stresses and resulting in resistance to various treatments, facilitated immortalization and rapid accumulation of mutations. In the following sections we will describe how these changes in cell regulation are translated into the alterations of phenotype at the level of the organism.

5. TP53 in the organism

5.1 TP53 deficiency and high risk of spontaneous cancer

The phenotype of *tp53*-knockout mice (*tp53*-KO) is a direct illustration of the efficacy of TP53 as a tumor suppressor: *tp53* deficiency is associated with a dramatic increase in the incidence of spontaneous tumors both in homozygous and heterozygous animals, although the spectrum of tumors is different. While all homozygous knockouts die predominantly from lymphomas during the first 6 months, all heterozygous animals develop different types of tumors of which sarcomas and lymphomas appear more often. The detailed description of tumor formation in *tp53*-KO mice was summarized elsewhere (Donehower *et al.*, 1992; Jacks *et al.*, 1994). The reason for the high incidence of cancer in *tp53*-deficient animals is consistent with the role of TP53 as a 'guardian of the genome', which prevents propagation of sporadically mutated cells leading to tumor formation, presumably through growth arrest and apoptotic mechanisms. In fact, an extremely high incidence of chromosomal aberrations was found in somatic cells of *tp53*-KO mice (Lee *et al.*, 1994).

In humans, germline mutations of *TP53* resulting in its inactivation, are also associated with high incidence of cancer. These have also been linked to Li-Fraumeni syndrome, an inherited susceptibility disorder in which affected individuals are at increased risk of developing a variety of cancers at an earlier age (Eeles, 1995; Malkin and Friend, 1993; Varley *et al.*, 1997). The risk of cancer in these families can reach 90% by the age of 50 years, with a range of tumor types. Some of the mutations (26%) in families with Li-Fraumeni syndrome are located outside the mutation-clustered region of exon 5–8, which is predominant in human tumors, and sometimes in a non-coding region (Barel *et al.*, 1998).

The human wild-type *TP53* exhibits a sequence polymorphism at position 72 containing either a proline (TP53Pro) or an arginine (TP53Arg) residue (Buchman *et al.*, 1988; Matlashewski *et al.*, 1987), in the region of the protein that is necessary for the induction of apoptosis (Sakamuro *et al.*, 1997; Walker and Levine, 1996). Interestingly, the analysis of this allelic polymorphism revealed their unequal distribution among several types of cancer. For example, skin cancer occurs more frequently among people with the *TP53Arg* variant of the gene, suggesting that the *TP53Pro* gene may be more efficient as a protector against skin cancer. This would be consistent with the observation that the *TP53Pro* gene is more common in darker skin human populations living closer to the equator. Moreover, it was reported that the *TP53Arg* variant is preferentially targeted in cervical cancers and skin cancers associated with human papillomavirus infections (Storey *et al.*, 1998), although this issue needs further testing. Based on the above observations it seems likely that the TP53Pro allele is generally more protective against skin and cervical cancers than TP53Arg.

It is interesting that the mutated *TP53Arg* variant had a higher affinity to p73 (TP53 homolog) (Di Como *et al.*, 1999) than mutants originating from *TP53Pro* or wild-type *TP53Arg*, resulting in an increased binding and inactivation of p73. Therefore, the mutations in the *TP53Arg* variant may have a greater gain-of-function activity than mutations in the *TP53Pro* variant with respect to inhibiting p73.

5.2 TP53 and normal development

tp53 knockout in mice, in addition to a high incidence of tumors, is also associated with several developmental abnormalities. Although the majority of *tp53*-null homozygous embryos survive gestation without obvious defects (Donehower *et al.*, 1992; Jacks *et al.*, 1994), a significant portion (23%) of *tp53*-KO female embryos failed to undergo the normal process of neural tube closure, which led to exencephaly (Armstrong *et al.*, 1995; Sah *et al.*, 1995). In addition to a variety of neural tube defects, many of the *tp53*-KO embryos exibited a range of craniofacial malformations, including defects in tooth formation and ocular abnormalities (Armstrong *et al.*, 1995; Pan and Griep, 1995).

In addition to developmental defects, *tp53* deficiency affects the differentiation of B-cells (Almog *et al.*, 1997; Shick *et al.*, 1997), T-cells (Jiang *et al.*, 1996), neurons and oligodendrocytes (Eizenberg *et al.*, 1996), spermatocytes (Hendry *et al.*, 1996; Schwartz *et al.*, 1993). Also, inhibition of TP53 activity in *Xenopus laevis* embryos blocked the ability of early blastomeres to undergo differentiation (Wallingford *et al.*, 1997). The exact mechanisms of TP53-mediated control of cell differentiation remain generally unknown.

5.3 Tissue specificity of TP53 activity in vivo

To directly monitor the activity of TP53 as a transcriptional activator *in vivo*, three independent groups created transgenic mice expressing the *lacZ* reporter gene from *tp53*-responsive promoters (Gottlieb *et al.*, 1997; Komarova *et al.*, 1997a; MacCallum *et al.*, 1996). Analysis of these mice showed that *tp53* is generally inactive under normal conditions. Whole-body gamma irradiation or treatment with high dosages of DNA-damaging chemotherapeutic drugs, led to a pronounced activation of the transgene (indicative of the p53 activity), in spleen, thymus, small intestine, and tissues of early embryos. Remarkably, *tp53*-mediated transgene activation coincided with the most obvious areas of radiation or drug-induced apoptosis that did not develop in *tp53*-deficient mice. These areas, in turn, coincided with the sites known to be affected by anti-cancer chemotherapy. Thus, *tp53*-dependent transgene activation was found to map to the most radiosensitive tissues in the mouse.

Clear correlation between radiosensitivity and *tp53* activity was also found in embryos. It was recognized long ago that early mammalian embryos are charac-terized by hypersensitivity to gamma-radiation that is gradually decreased during organogenesis (8–13 days) in mice (Komarova and Gudkov, 1998). These observa-tions provide a molecular basis for a well known radiological phenomenon of high sensitivity of early mammalian embryos to gamma radiation that, as it became clear recently, is a consequence of *tp53* activity.

The biological significance of differential activity of TP53 in tissues could be a part of the defense mechanism acting in some rapidly proliferating tissues that prevents accumulation of genetically damaged cells. In fact, gamma irradiation results in a fast development of lymphomas in *tp53*-KO mice (Kemp *et al.*, 1994). These lymphomas possibly arise from cells that would otherwise be eliminated by TP53-dependent apoptosis. Similarly, higher survival rate of *tp53*-deficient embryos after irradiation is accompanied with the development of malformations apparently originating from genetically damaged cells that remained alive because of the absence of *tp53* (Norimura *et al.*, 1996). This latter observation inspired David Lane to call TP53 a 'guardian of babies' (Hall and Lane, 1997).

Why does TP53 act in a tissue-specific manner? Analysis of *tp53* in different mouse organs showed that although protein stabilization occurred in every tissue analyzed, the amount of the protein varied dramatically among tissues. Thus, radiosensitive organs expressed much higher levels of the tp53 protein than radioresistant ones (MacCallum *et al.*, 1996). The variation of TP53 protein abundance among normal tissues correlates well with the levels of TP53 mRNA, suggesting that the amount of mRNA templates dictates how much protein will be available in the cell under stress. In fact, radiosensitive organs such as spleen and thymus express many fold more TP53 mRNA than radioresistant organs such as liver, muscle, and brain (Komarova *et al.*, 2000; Rogel *et al.*, 1985). Moreover, a dramatic decrease in *tp53* mRNA expression occurs between days 12 and 15 of embryonal development (Rogel *et al.*, 1985; Schmid *et al.*, 1991) and correlates well with the differences in radiosensitivity of early and late embryos. It seems likely that it is the level of TP53 mRNA that determines whether apoptosis will or will not be activated. Therefore, that developmental and tissue-specific *TP53* gene regulation, at the level of mRNA, may play a critical role in the determination of TP53 effects *in vivo*.

The level of *TP53* mRNA expression is not the only factor determining tissue specificity of TP53 effects. Any members of the TP53 pathway, if expressed in a tissue-specific manner, could contribute to this phenomenon. Indeed, many of the newly identified TP53-responders were found to be tissue specific (Komarova *et al.*, 1998).

In conclusion, the main function of TP53 in the normal organism is consistent with its role as a 'guardian of the genome' which eliminates potentially dangerous cells in embryonic development and in proliferating adult tissues. This function is probably involved in control of cancer prevention and aging. On one hand, TP53 activity reduces the risk of cancer development, while on the other hand it determines high sensitivity of tissues and the entire organism to genotoxic stress. Another function of TP53 is related to its involvement in differentiation and morphogenesis. Effects of TP53 are tissue specific and are determined by tissue specific transcription of the *TP53* gene and other members of the TP53 pathway.

5.4 TP53 homologs

P73, p63 and p40 are recently discovered *TP53* homologs, revealing less or more sequence homology with TP53 (Jost *et al.*, 1997; Kaghad *et al.*, 1997; Osada *et al.*, 1998; Trink *et al.*, 1998). These genes are now grouped together in the TP53 family. They are highly similar to TP53 in the regions corresponding to the TP53 N-terminal transactivation, central DNA-binding, and C-terminal oligomerization domains. Both P73 and P63 can bind to canonical DNA-binding sites, can activate transcription from TP53-responsive promoters, and can induce apoptosis when overproduced in cells (De Laurenzi *et al.*, 1998; Di Como *et al.*, 1999; Jost *et al.*, 1997; Kaghad *et al.*, 1997; Osada *et al.*, 1998; Yang *et al.*, 1998; Zhu *et al.*, 1998). MDM2 can bind to P73 and inhibit its ability to serve as a transcriptional activator, although MDM2 does not target P73 for degradation (Zeng *et al.*, 1999).

P73, unlike TP53, is not induced by DNA damage (Kaghad *et al.*, 1997), also, P73 and P63 undergo complex alternative splicing (De Laurenzi *et al.*, 1998; Kaghad *et al.*, 1997; Osada *et al.*, 1998; Yang *et al.*, 1998). P63 and P73, appear to be rarely mutated in human cancers, in striking contrast to TP53 (Ichimiya *et al.*, 1999; Kaghad *et al.*, 1997; Kovalev *et al.*, 1998; Kroiss *et al.*, 1998; Mai *et al.*, 1998; Nimura *et al.*, 1998; Nomoto *et al.*, 1998; Osada *et al.*, 1998; Takahashi *et al.*, 1998; Yokomizo *et al.*, 1999). There is currently no firm evidence that P63 and P73 should be considered tumor suppressors (Kaelin, 1998). How they are activated and may replace and function when TP53 is mutated is not known.

6. TP53 in the tumor

6.1 TP53 mutations in human cancer

TP53-dependent mechanisms eliminating abnormal cells are highly efficient and provide a potent shield against proliferation and survival of cells that were exposed to a variety of damaging conditions, or suffered faults in cell duplication and interactions with their neighbors. These mechanisms prevent any chances for selection of a cell that would have more growth or metabolic advantages over its competitors. Therefore, without breakage or mitigation of TP53 pathways the carcinogenesis process is virtually impossible. Moreover, once TP53-dependent mechanisms are broken, conditions for the rapid accumulation of genetic changes are established which lead to dramatic destabilization of the genome and the acceleration of carcinogenesis.

Indeed, it was found that the *TP53* gene is the most frequently mutated gene in cancer. Of more then 10 000 different human tumors analyzed so far 45–50% contained inactivating mutations within the *TP53* gene (Soussi *et al.*, 2000). Frequencies of mutations vary from one tumor type to another (Greenblatt *et al.*, 1994). Point mutations within the *TP53* gene were observed in 60–65% of lung and colon cancer, 40–45% of stomach, esophagus and bladder cancer, 25–30% of breast, liver, prostate cancer and in lymphomas, and only in 10–15% of leukemias (Soussi *et al.*, 2000). A total of 80–85% of point mutations are missense and localize predominantly within evolutionarily conserved specific DNA-binding domain of the TP53 protein (exons 4–9) (Hansen and Oren, 1997; Levine, 1997). TP53 mutants lose the ability to recognize appropriate DNA-binding sites and activate transcription of TP53-responsive genes. In addition, mutant TP53 proteins tend to accumulate in the cells due to decreased degradation. The mutated TP53 protein is able to oligomerize with the product of the wild-type allele, thus exerting a dominant-negative effect. Therefore, acquisition of *TP53* missense mutations would result in immediate mitigation of TP53-dependent control. However, in advanced tumors the wild type allele is lost through subsequent selection process.

6.2 New functions of TP53 mutants

At variance with other tumor suppressor genes, cells with *TP53* mutations typically maintain expression of the full-length protein. This may suggest that mutant *TP53* genes can contribute actively to cancer progression through gain of function

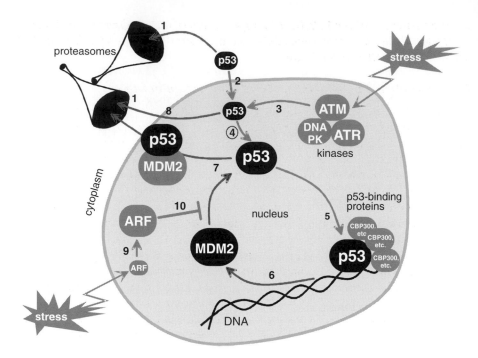

Figure 2. Scheme of p53 activation and regulation in response to genotoxic stress (through kinases) or oncogene activation (through ARF).

1 – proteasomal degradation.
2 – nuclear import.
3 – p53 activation by phosphorylation or acetylation.
4 – p53 nuclear accumulation.
5 – DNA binding.
6 – transactivation of p53-responsive genes, including MDM2.
7 – p53/MDM2 binding.
8 – nuclear export.
9 – activation of ARF.
10 – ARF/MDM2 binding that interferes with p53 degradation.

activity (Dittmer *et al.*, 1993; Michalovitz *et al.*, 1991). Indeed, overexpression of certain mutant *TP53* genes in *tp53*-null cells resulted in enhancement of trans-formed features, increased plating efficiency in culture, increased tumorigenicity and invasiveness *in vivo* (Dittmer *et al.*, 1993; Gloushankova *et al.*, 1997; Hsiao *et al.*, 1994; Lotem and Sachs, 1995; Sun *et al.*, 1993; Wolf *et al.*, 1984). In addition, some *TP53* mutants are capable (i) to further increase genetic instability by abro-gating the mitotic spindle checkpoint (Gualberto *et al.*, 1998); (ii) to change differ-entiation status of cells (Kremenetskaya *et al.*, 1997) and (iii) to interfere with TP53-independent apoptosis (Peled *et al.*, 1996), thus increasing resistance to certain chemotherapeutic drugs (Blandino *et al.*, 1999; Li *et al.*, 1998).

The mechanisms underlying gain of function activities of *TP53* mutants are not quite clear, although it was found that some TP53 mutants are involved in acti-vation of genes that are usually not induced by the wild-type protein. Among these are the anti-apoptotic gene *BAG-1* (Yang *et al.*, 1999), the MYC gene (Frazier *et al.*, 1998), the 15-lipoxygenase gene (Kelavkar and Badr, 1999), the PCNA gene (Deb *et al.*, 1992) as well as several others.

6.3 Alternative mechanisms of TP53 inactivation in tumors

Tumors with wild type *TP53* do not necessarily retain the TP53 pathway active. On the contrary, they often have alterations of other genes that can lead to abrogation of the TP53 pathway. Oncogene products of DNA tumor viruses are among the most efficient inhibitors of TP53 function. SV40 large-T antigen and the 55 kD product of adenovirus E1B gene and the hepatitis B viral X protein can bind to TP53 and interfere with its transcriptional activity (Levine, 1992). The human papilloma virus E6 gene product can also bind to TP53 and target it to rapid proteasome degradation through interaction with the cellular protein, E6-AP (Scheffner *et al.*, 1990). This mechanism of *TP53* gene inactivation is frequently observed in cervical carcinomas (Thomas *et al.*, 1999).

Another widespread mechanism of *TP53* inactivation is through over-expression of its natural inhibitor *MDM2* which binds to the N-terminal part of the TP53 protein and targets TP53 to proteasome degradation (see above). Overall, the frequency of *MDM2* gene amplification in human tumors is 7%, with the highest frequency (Momand *et al.*, 1998) observed in soft tissue tumors (20%) and osteosarcomas (16%).

Alterations in the expression of another modulator of TP53 function, ARF, a protein that binds MDM2 and prevents its ability to inactivate TP53 (see above), are also involved in suppression of TP53 pathways in tumors. Elimination of ARF activity would result in higher MDM2 activity leading to TP53 inactivation. Several human tumors have deletions within the *INK4A* locus that inactivate ARF along with p16-INK4a, an inhibitor of the cyclin D-CDK4/6 complex (Sherr, 1998).

Another mechanism of TP53 inactivation involves aberrant subcellular localization. Constitutive cytoplasmic localization of wild-type TP53 has been reported in inflammatory breast carcinoma (Moll *et al.*, 1992), colorectal adeno-carcinoma (Bosari *et al.*, 1995; Sun *et al.*, 1992), undifferentiated neuroblastoma (Moll *et al.*, 1995, 1996), hepatocellular carcinoma (Ueda *et al.*, 1995), and retinoblastoma (Schlamp *et al.*, 1997) and is associated with tumor metastasis and poor long-term patient survival (Stenmark-Askmalm *et al.*, 1994; Sun *et al.*, 1992). It seems possible that these cancers use the natural mechanism of TP53 cytoplasmic sequestration that is acting in embryonic stem cells (Aladjem *et al.*, 1998).

The above examples do not fully cover the mechanisms of inactivation of the TP53 pathway. It is quite possible that other genes within the TP53 pathway are not functional in tumors that retain the *TP53* gene intact. One of the recent observations indicates that inactivation of the component of apoptotic machinery, Apaf1, might be the reason of tolerance of melanomas to TP53, thus providing the first example of suppression of a pro-apoptotic gene as the mechanism of TP53 pathway inactivation in tumors (Lowe and Lin, 2000).

6.4 TP53 and tumor cell sensitivity to anticancer agents

Abrogation of TP53-dependent apoptosis and growth arrest can be translated into increased resistance of cells to a variety of treatments, including anti-cancer therapy with radiation or drugs. The major impact to the understanding of the role of TP53 in tumor resistance to treatment was made by the group of Scott Lowe who used mouse embryo fibroblasts from wild-type *tp53* and *tp53*-deficient mice transformed with the combination of oncogenes *E1a* and *ras* as experimental models of tumor cells and tumors differing in their *tp53* status (Lowe *et al.*, 1993b, 1994b). These experimental tumor models are believed to represent naturally occurring tumors at early stages of progression when they still retain wild type TP53. The majority of tumor-derived cell lines with wild-type TP53 (such as human breast carcinoma MCF7 or fibrosarcoma HT1080 cell lines) could not be used as adequate models for the analysis of the role of TP53 in tumor behavior, since their TP53 function became strongly suppressed during the long history of maintenance *in vitro*. However, TP53 in freshly prepared *E1a*+*ras*-transformed MEF retains the capability of inducing apoptosis in response to a variety of stresses including gamma radiation and treatment with different chemotherapeutic agents (Lowe *et al.*, 1993). Consistently, similar cells with no *tp53* show strong resistance to the above treatments both *in vitro* and *in vivo*, under the conditions of experimental therapy of tumor-bearing mice (Lowe *et al.*, 1994b). Later, similar conclusions were made by the analysis of experimental mouse hematopoietic malignancies (Schmitt *et al.*, 1999).

6.5 TP53 and control of angiogenesis

Recent studies have indicated that angiogenesis may be regulated, in part, by TP53 tumor suppressor gene function (Bouvet *et al.*, 1998; Nishimori *et al.*, 1997; Van Meir *et al.*, 1994; Yokota, 2000). Neovascularization may be a direct consequence of TP53 inactivation (Van Meir *et al.*, 1994), possibly due to loss of trans-activation of genes regulated by TP53 (Vogelstein and Kinzler, 1992). It is now thought that the tumor angiogenic switch is triggered as a result of a shift in the balance of stimulators to inhibitors. Negative regulators of angiogenesis, including thrombospondin-1 (TSP1), glioma-derived angiogenesis inhibitory factor (GD-AIF) and brain-specific angiogenesis inhibitor 1 (*BAI1*) are TP53-target genes. Overexpression of TP53 may inhibit angiogenesis through the upregulation of these genes (Dameron *et al.*, 1994; Furuhata *et al.*, 1996; Nishimori *et al.*, 1997). Loss of wild-type TP53 results in reduced expression of the inhibitor of angiogenesis thrombospondin 1 gene and can switch fibroblasts to the more angiogenic phenotype (Dameron *et al.*, 1994). Early passage Li-Fraumeni cells which carry one wild type *TP53* allele secrete large amounts of TSP1. However, the late passage cells undergo an angiogenic switch associated with loss or mutation of the wild type *TP53* allele and reduced expression of TSP1 (Dameron *et al.*, 1994; Somasundaram and El-Deiry, 2000). Progression of a glioma to its more malignant form of glioblastoma is usually associated with inactivation of TP53 and striking neovascularization. The expression of BAI1 (containing TSP-type 1 repeats) was absent or significantly reduced in glioblastoma cell lines, suggesting that BAI1 plays a significant role in angiogenesis inhibition, as a mediator of TP53 (Nishimori *et al.*, 1997).

Angiogenesis is naturally regulated by hypoxia through the induction of a transcription factor, named hypoxia-inducible factor 1 (HIF-1a), a positive regulator of the vascular endothelial growth factor (VEGF) gene. At the same time, hypoxic conditions often result in the activation of TP53 occurring together with upregulation of HIF-1 (An *et al.*, 1998; Blagosklonny *et al.*, 1998). It was found that HIF-1a stabilizes TP53 through the formation of hypoxic complex, which in turn enhances the transcription of known TP53 targets (Halterman *et al.*, 1999). It was reported (Ravi *et al.*, 2000) that TP53 can promote MDM2-mediated ubiquitination and proteasomal degradation of HIF-1a leading to the suppression of angiogenic stimulus. Moreover, activation of TP53 in hypoxic regions of tumors may lead to apoptosis (Graeber *et al.*, 1996). Thus, hypoxia that develops within any rapidly growing tumor creates conditions for the selection and clonal expansion of TP53 deficient variants that can avoid apoptotic death and efficiently stimulate angiogenesis.

6.6 TP53 and tumor metastasis

Comparison of isogenic tumor cell lines differing in their TP53 status showed that wild type TP53 activity can delay the natural selection of metastatic tumor variants, possibly through control of genomic stability that attenuates the accumulation of additional mutations.

  Another property of TP53 that may contribute to the control of tumor metastasis is its involvement in cell reaction to detachment from natural substrate and exposure to abnormal microenvironment, the essential events occurring during metastasis. To metastasize, a tumor cell must acquire the ability to survive in the bloodstream and invade a foreign tissue. Normally, this process is prevented by the propensity of epithelial cells to die in suspension, or in the absence of the appropriate survival factors (Frisch and Francis, 1994). Disruption of epithelial cell–matrix interactions induces apoptosis. Clearly, the fact that these processes trigger apoptosis creates selective pressure to mutate apoptotic programs during tumor development. In fact, the status of TP53 can influence cell death in suspension (Nikiforov *et al.*, 1996), and enrichment for *TP53* mutations was observed in metastases (Lowe and Lin, 2000). Hence, loss of apoptosis can impact tumor initiation, progression and metastasis.

6.7 TP53 and the bystander effect (secretion of growth inhibitors)

TP53 response is not limited to the intrinsic effects of stress or injury on proliferation of damaged cells. TP53 activation induces secretion of growth inhibitory factors affecting the proliferation of neighboring cells. In addition to the anti-angiogenic factors, the list of TP53-responsive genes includes many other known growth suppressors, effective against a broad variety of cells (IGBP-3, TGF-beta 2, inhibin-beta, serine protease inhibitors) (Komarova *et al.*, 1998). Thus, TP53 controls the cellular export of stress-dependent growth inhibitory stimuli. This function possibly determines the 'bystander effect' associated with *TP53* gene therapy (Qazilbash *et al.*, 1997) and could be an important factor in determining the efficacy of anti-cancer therapy. In fact, tumor cell death in response to chemotherapeutic drugs was more efficient when cells were co-cultivated with wild-type *TP53* than with *TP53*-null fibroblasts (Komarova *et al.*, 1998).

6.8 Reasons and consequences of TP53 inactivation in tumors (conclusions)

The review of the cell properties that are controlled by TP53 helps to understand the forces stimulating tumors to inactivate this gene. In fact, TP53 activity interferes with the acquisition of most basic features of malignant transformation: abnormal growth regulation caused by the activation of dominant oncogenes and immortalization, thus providing an obstacle for tumor origin. At later stages, TP53 suppresses tumor progression and metastasis through the control of genomic stability (suppression of accumulation of additional mutations), negative regulation of angiogenesis and induction of apoptosis in response to hypoxia and exposure to abnormal microenvironment. Finally, TP53 contributes to tumor sensitivity to anti-cancer treatment by promoting apoptosis in response to stresses caused by radiation or chemotherapy or stimulating a bystander effect by controlling secretion of growth inhibitory factors by damaged cells. All the above explains why mutations in the *TP53* gene itself, or other genetic alterations resulting in inactivation of the TP53 pathway, are so frequently acquired by cancer cells and are associated with more aggressive and less curable forms of the disease.

7. TP53 and cancer treatment

7.1 Prognostic significance of TP53 mutations

Since the loss of TP53 strongly contributes to cancer origin and progression, the TP53 status of the tumor was expected to be a useful prognostic marker. Inactivation of TP53, in addition to its association with a high incidence of cancer, also largely determines the properties of the tumor. Tumors with mutated *TP53* for example can be more anaplastic, have a higher rate of proliferation, and have a more aggressive phenotype than similar tumors with wild-type *TP53*, thereby giving rise to a worse prognosis. Wild type *TP53* is considered a positive prognostic marker in colorectal, breast, bladder, prostatic, hematological malignancies and is usually associated with less aggressive tumors (Cordon-Cardo *et al.*, 1994; Falette *et al.*, 1998; Fresno *et al.*, 1997; Iniesta *et al.*, 1998; Molina *et al.*, 1998; Norberg *et al.*, 1998; Pandrea *et al.*, 1995; Patel *et al.*, 1996; Peller, 1998; Yamaguchi *et al.*, 1992). However, it would be premature to define wild-type *TP53* as a general positive prognostic marker for all types of cancer. Analysis of all relevant information done in a recent review (Nieder *et al.*, 2000) drew the conclusion that the prognostic information of *TP53* is at best marginal, especially when compared to established parameters such as grading, age, etc. Its predictive value, which most likely is rather limited too, can hardly be judged without prospective studies also evaluating other biological factors as well as end-points.

7.2 TP53 activation by small molecules

In the majority of tumors, TP53 protein is present in an inactive form either as a result of mutations altering protein conformation or due to the expression of cellular (MDM2) or viral (E6 of human papilloma virus) inhibitory proteins (see above). Restoration of TP53 function in such tumors is viewed as a therapeutic approach to treatment, potentially applicable to a large proportion of cancers. Chemical compounds capable of restoring active conformation of mutant *TP53* were isolated by screening of chemical libraries for the small molecules that stabilize the DNA-binding domain of TP53 in the active conformation (Foster *et al.*, 1999). A recently reported prototype compound caused the accumulation of conformationally active TP53 in the cells with mutant *TP53*, enabling it to activate transcription and to slow tumor growth in mice.

Numerous attempts have been made to restore TP53 function in tumors, in which it is suppressed by MDM2 or E6 proteins. Since therapeutic targets here are clearly defined, the majority of these studies are focused on generation of either inhibitory antisense oligonucleotides (Beer-Romero *et al.*, 1997; Chen *et al.*, 1999) or small peptides inhibiting TP53 interaction with the inhibitory proteins. Biologically active peptides were isolated from random peptide libraries (Bottger *et al.*, 1997) or generated from the fragments of TP53 inhibitory proteins (Chene *et al.*, 2000; Midgley *et al.*, 2000) or from the fragments of TP53 itself (Hupp *et al.*, 1995; Selivanova *et al.*, 1997). Some of the developed reagents showed tumor suppressor effect in tissue culture and in animal model and have a potential to be converted into useful anti-cancer drugs.

7.3 TP53 gene therapy

Frequent loss of TP53 in tumors opens the possibility of using the wild-type *TP53* gene for gene therapy in cancer treatment. As a general rule, tumor cells are inherently more sensitive to TP53 inhibition than normal cells, perhaps because mitogenic oncogenes can activate TP53 to promote apoptosis (Lowe *et al.*, 1993a). For example, reintroduction of *TP53* into *TP53* mutant tumor cells can directly induce apoptosis or enhance treatment sensitivity in tumor cell lines or in xenographs (Badie *et al.*, 1998; Spitz *et al.*, 1996a, 1996b). Indeed, strategies using this approach are currently in clinical trials (Swisher *et al.*, 1999).

TP53–MDM2 interaction is another prospective target for cancer therapy. Negative regulation of TP53 by MDM2 may limit the magnitude of TP53 activation by DNA damaging agents, thereby limiting their therapeutic effectiveness. *MDM2* is frequently up-regulated in many types of cancer, thus contributing to the suppression of the TP53 function. For example, the *MDM2* gene is amplified in 16–20% of human sarcomas (Beroud and Soussi, 1998) and is strongly expressed in 54% of breast carcinomas (O'Neill *et al.*, 1998). Inhibition of the MDM2–TP53 interaction results in a dramatic increase in TP53 levels followed by subsequent growth inhibition. Several approaches are being extensively tested attempting to use this strategy for selective killing of tumor cells, including polypeptides targeted to the MDM2–TP53-binding domain and antisense oligonucleotides that specifically inhibit *MDM2* expression (Bottger *et al.*, 1997; Zhang and Wang, 2000).

7.4 Viral therapy of TP53-deficient tumors

An elegant approach to the selective killing of *TP53*-deficient cells has been suggested which involves the use of genetically modified adenovirus deficient in its E1B, 55K gene (strain Onyx015) (Bischoff *et al.*, 1996). This gene is essential for efficient virus replication in TP53 wild-type cells since it protects them from TP53-dependent apoptosis induced by another adenoviral protein, E1a (Kirn *et al.*, 1998). Onyx015 virus is unable to multiply in normal cells due to the TP53-dependent apoptotic response occurring faster than the completion of virus replication cycle. However, Onyx15 is able to multiply in the cells with mutated *TP53* genes resulting in completion of its lytic life cycle killing the infected tumor cells. Onyx015 is currently under clinical trials (Ganly *et al.*, 2000).

7.5 Chemoprotection

TP53-dependent apoptosis was found to be a major factor contributing to tissue sensitivity to genotoxic stress and TP53 was defined as a major determinant of radiosensitivity and chemosensitivity of tissues responsible for severe side effects of anti-cancer treatment. As indicated above, the propensity of certain tissues to drug-induced apoptosis may limit the effectiveness of many current therapies. Studies using mouse models have clearly documented the importance of TP53 for apoptosis in thymocytes, bone marrow and intestinal stem cells (see above); consequently, agents which suppress TP53 function may be effective radio- or chemoprotective agents and/or allow dose intensification of current regimens. Chemotherapy-induced hair loss, another well-known side effect of cancer treatment, was also found to be TP53 dependent (Botchkarev *et al.*, 2000). Temporary suppression of TP53 by chemical inhibitors was, therefore, suggested as a therapeutic approach to reduce damage caused by radiation and chemotherapy to normal tissues (Komarov *et al.*, 1999; Komarova and Gudkov, 1998). Since most advanced solid tumors have lost TP53 function, these inhibitors should not interfere with cell death of most tumor cells. Recently, a small molecule (designated pifithrin-alpha) has been identified that inhibits TP53-mediated transcriptional responses and TP53-induced apoptosis in cultured cells (Komarov *et al.*, 1999). In mice, pifithrin-alpha is a potent radio-protective agent, allowing mice to survive otherwise lethal doses of ionizing radiation.

It is likely that suppression of TP53-dependent apoptosis has already been successfully and broadly applied to cancer patients in the form of growth factors supplementing chemotherapy (Johnston and Crawford, 1998). Therapeutic effect of such supplements may be associated with their activity as survival factors suppressing TP53-dependent apoptosis (Bronchud, 1993). The use of TP53 inhibitors may not be limited to protection from chemotherapy-induced side effects but could be applicable to other clinical situations, in which TP53 suppression might be desirable. These include, for example, heart and brain ischemia, both resulting from local hypoxia, known to be a potent activator of TP53 (Graeber *et al.*, 1996).

Acknowledgments

Authors' work in the TP53 field is supported by National Institutes of Health Grants CA60730 and CA75179 to A.V.G. and grants from Quark Biotech, Inc. to A.V.G. and P.M.C.

References

Agarwal, M.L., Agarwal, A., Taylor, W.R. and Stark, G.R. (1995) TP53 controls both the G2/M and the G1 cell cycle checkpoints and mediates reversible growth arrest in human fibroblasts. *Proc. Natl Acad. Sci. USA.* **92**: 8493–8497.

Agoff, S.N., Hou, J., Linzer, D.I. and Wu, B. (1993) Regulation of the human hsp70 promoter by TP53. *Science* **259**: 84–87.

Aladjem, M.I., Spike, B.T., Rodewald, L.W., Hope, T.J., Klemm, M., Jaenisch, R. and Wahl, G.M. (1998) ES cells do not activate TP53-dependent stress responses and undergo TP53-independent apoptosis in response to DNA damage. *Curr. Biol.* 8: 145–155.

Albrechtsen, N., Dornreiter, I., Grosse, F., Kim, E., Wiesmuller, L. and Deppert, W. (1999) Maintenance of genomic integrity by TP53: complementary roles for activated and non-activated TP53. *Oncogene* 18: 7706–7717.

Almog, N., Li, R., Peled, A., Schwartz, D., Wolkowicz, R., Goldfinger, N., Pei, H. and Rotter, V. (1997) The murine C'-terminally alternatively spliced form of TP53 induces attenuated apoptosis in myeloid cells. *Mol. Cell. Biol.* 17: 713–722.

An, W.G., Kanekal, M., Simon, M.C., Maltepe, E., Blagosklonny, M.V. and Neckers, L.M. (1998) Stabilization of wild-type TP53 by hypoxia-inducible factor 1alpha. *Nature* 392: 405–408.

Anderson, C.W. and Lees-Miller, S.P. (1992) The nuclear serine/threonine protein kinase DNA-PK. *Crit. Rev. Eukaryot. Gene. Expr.* 2: 283–314.

Armstrong, J.F., Kaufman, M.H., Harrison, D.J. and Clarke, A.R. (1995) High-frequency developmental abnormalities in TP53-deficient mice. *Curr. Biol.* 5: 931–936.

Ashkenazi, A. and Dixit, V.M. (1998) Death receptors: signaling and modulation. *Science* 281: 1305–1308.

Atadja, P., Wong, H., Garkavtsev, I., Veillette, C. and Riabowol, K. (1995) Increased activity of TP53 in senescing fibroblasts. *Proc. Natl Acad. Sci. USA.* 92: 8348–8352.

Attardi, L.D., Reczek, E.E., Cosmas, C., Demicco, E.G., McCurrach, M.E., Lowe, S.W. and Jacks, T. (2000) PERP, an apoptosis-associated target of TP53, is a novel member of the PMP-22/gas3 family. *Genes. Dev.* 14: 704–718.

Badie, B., Kramar, M.H., Lau, R., Boothman, D.A., Economou, J.S. and Black, K.L. (1998) Adenovirus-mediated TP53 gene delivery potentiates the radiation-induced growth inhibition of experimental brain tumors. *J. Neurooncol.* 37: 217–222.

Bakalkin, G., Selivanova, G., Yakovleva, T., *et al.* (1995) TP53 binds single-stranded DNA ends through the C-terminal domain and internal DNA segments via the middle domain. *Nucleic Acids Res.* 23: 362–369.

Balint, E. and Reisman, D. (1996) Increased rate of transcription contributes to elevated expression of the mutant TP53 gene in Burkitt's lymphoma cells. *Cancer Res.* 56: 1648 1653.

Banin, S., Moyal, L., Shieh, S *et al.* (1998) Enhanced phosphorylation of TP53 by ATM in response to DNA damage. *Science* 281: 1674–1677.

Barel, D., Avigad, S., Mor, C., Fogel, M., Cohen, I.J. and Zaizov, R. (1998) A novel germline mutation in the noncoding region of the TP53 gene in a Li-Fraumeni family. *Cancer Genet Cytogenet.* 103: 1–6.

Bargonetti, J., Chicas, A., White, D. and Prives, C. (1997) TP53 represses Sp1 DNA binding and HIV-LTR directed transcription. *Cell. Mol. Biol. (Noisy-le-grand)* 43: 935–949.

Beer-Romero, P., Glass, S. and Rolfe, M. (1997) Antisense targeting of E6AP elevates TP53 in HPV-infected cells but not in normal cells. *Oncogene* 14: 595–602.

Bennett, M., Macdonald, K., Chan, S.W., Luzio, J.P., Simari, R. and Weissberg, P. (1998) Cell surface trafficking of Fas: a rapid mechanism of TP53-mediated apoptosis. *Science* 282: 290–293.

Beroud, C. and Soussi, T. (1998) TP53 gene mutation: software and database. *Nucleic Acids Res.* 26: 200–204.

Bischoff, J.R., Kirn, D.H., Williams, A. *et al.* (1996) An adenovirus mutant that replicates selectively in TP53-deficient human tumor cells. *Science* 274: 373–376.

Bissonnette, N., Wasylyk, B. and Hunting, D.J. (1997) The apoptotic and transcriptional transactivation activities of TP53 can be dissociated. *Biochem. Cell. Biol.* 75: 351–358.

Blagosklonny, M.V., An, W.G., Romanova, L.Y., Trepel, J., Fojo, T. and Neckers, L. (1998) TP53 inhibits hypoxia-inducible factor-stimulated transcription. *J. Biol. Chem.* 273: 11995–11998.

Blandino, G., Levine, A.J. and Oren, M. (1999) Mutant TP53 gain of function: differential effects of different TP53 mutants on resistance of cultured cells to chemotherapy. *Oncogene* 18: 477–485.

Bosari, S., Viale, G., Roncalli, M., Graziani, D., Borsani, G., Lee, A.K. and Coggi, G. (1995) TP53 gene mutations, TP53 protein accumulation and compartmentalization in colorectal adenocarcinoma. *Am. J. Pathol.* 147: 790–798.

Botchkarev, V.A., Komarova, E.A., Siebenhaar, F., Botchkarev, N.V., Komarov, P.G., Maurer, M., Gilchrest, B.A. and Gudkov, A.V. (2000) TP53 is essential for the chemotherapy-induced hair loss. *Cancer Res.* 60.

Bottger, A., Bottger, V., Sparks, A., Liu, W.L., Howard, S.F. and Lane, D.P. (1997) Design of a synthetic Mdm2-binding mini protein that activates the TP53 response in vivo. *Curr. Biol.* 7: 860–869.

Bouvet, M., Ellis, L.M., Nishizaki, M., Fujiwara, T., Liu, W., Bucana, C.D., Fang, B., Lee, J.J. and Roth, J.A. (1998) Adenovirus-mediated wild-type TP53 gene transfer down-regulates vascular endothelial growth factor expression and inhibits angiogenesis in human colon cancer. *Cancer Res.* 58: 2288–2292.

Bringold, F. and Serrano, M. (2000) Tumor suppressors and oncogenes in cellular senescence. *Exp. Gerontol.* 35: 317–329.

Bronchud, M. (1993) Can hematopoietic growth factors be used to improve the success of cytotoxic chemotherapy? *Anticancer Drugs* 4: 127–139.

Brown, J.M. and Wouters, B.G. (1999) Apoptosis, TP53, and tumor cell sensitivity to anticancer agents. *Cancer Res.* 59: 1391–1399.

Brugarolas, J., Chandrasekaran, C., Gordon, J.I., Beach, D., Jacks, T. and Hannon, G.J. (1995) Radiation-induced cell cycle arrest compromised by p21 deficiency. *Nature* 377: 552–557.

Buchman, V.L., Chumakov, P.M., Ninkina, N.N., Samarina, O.P. and Georgiev, G.P. (1988) A variation in the structure of the protein-coding region of the human TP53 gene. *Gene* 70: 245–252.

Bulavin, D.V., Saito, S., Hollander, M.C., Sakaguchi, K. Anderson, C.W., Appella, E. and Fornace, A.J., Jr. (1999) Phosphorylation of human TP53 by p38 kinase coordinates N-terminal phosphorylation and apoptosis in response to UV radiation. *EMBO J.* 18: 6845–6854.

Bunz, F., Dutriaux, A., Lengauer, C., Waldman, T., Zhou, S., Brown, J.P., Sedivy, J.M., Kinzler, K.W. and Vogelstein, B. (1998) Requirement for TP53 and p21 to sustain G2 arrest after DNA damage [In Process Citation]. *Science* 282: 1497–1501.

Caelles, C., Helmberg, A. and Karin, M. (1994) TP53-dependent apoptosis in the absence of transcriptional activation of TP53-target genes. *Nature* 370: 220–223.

Canman, C.E., Gilmer, T.M., Coutts, S.B. and Kastan, M.B. (1995) Growth factor modulation of TP53-mediated growth arrest versus apoptosis. *Genes Dev.* 9: 600–611.

Canman, C.E., Lim, D.S., Cimprich, K.A., Taya, Y., Tamai, K., Sakaguchi, K., Appella, E., Kastan, M.B. and Siliciano, J.D. (1998) Activation of the ATM kinase by ionizing radiation and phosphorylation of TP53. *Science* 281: 1677–1679.

Chen, L., Lu, W., Agrawal, S., Zhou, W., Zhang, R. and Chen, J. (1999) Ubiquitous induction of TP53 in tumor cells by antisense inhibition of MDM2 expression. *Mol. Med.* 5: 21–34.

Chene, P., Fuchs, J., Bohn, J., Garcia-Echeverria, C., Furet, P. and Fabbro, D. (2000) A small synthetic peptide, which inhibits the TP53-hdm2 interaction, stimulates the TP53 pathway in tumour cell lines. *J. Mol. Biol.* 299: 245–253.

Chiarugi, V., Magnelli, L., Cinelli, M., Turchetti, A. and Ruggiero, M. (1995) Dominant oncogenes, tumor suppressors, and radiosensitivity. *Cell. Mol. Biol. Res.* 41: 161–166.

Chin, K.V., Ueda, K., Pastan, I. and Gottesman, M.M. (1992) Modulation of activity of the promoter of the human MDR1 gene by Ras and TP53. *Science* 255: 459–462.

Cho, Y., Gorina, S., Jeffrey, P.D. and Pavletich, N.P. (1994) Crystal structure of a TP53 tumor suppressor–DNA complex: understanding tumorigenic mutations. *Science* 265: 346–355.

Clarke, A.R., Purdie, C.A., Harrison, D.J., Morris, R.G., Bird, C.C., Hooper, M.L. and Wyllie, A.H. (1993) Thymocyte apoptosis induced by TP53-dependent and independent pathways. *Nature* 362: 849–852.

Cliby, W.A., Roberts, C.J., Cimprich, K.A., Stringer, C.M., Lamb, J.R., Schreiber, S.L. and Friend, S.H. (1998) Overexpression of a kinase-inactive ATR protein causes sensitivity to DNA-damaging agents and defects in cell cycle checkpoints. *EMBO J.* 17: 159–169.

Cordon-Cardo, C., Dalbagni, G., Sarkis, A.S. and Reuter, V.E. (1994) Genetic alterations associated with bladder cancer. *Important Adv. Oncol.* 71–83.

Cui, Y.F., Zhou, P.K., Woolford, L.B., Lord, B.I., Hendry, J.H. and Wang, D.W. (1995) Apoptosis in bone marrow cells of mice with different TP53 genotypes after gamma-rays irradiation in vitro. *J. Environ. Pathol. Toxicol. Oncol.* 14: 159–163.

Dameron, K.M., Volpert, O.V., Tainsky, M.A. and Bouck, N. (1994) Control of angiogenesis in fibroblasts by TP53 regulation of thrombospondin-1. *Science* 265: 1582–1584.

Dang, R.K., Anthony, R.S., Craig, J.I. and Parker, A.C. (1999) A novel 8-bp insertion in codon 281 of TP53 in a patient with acute lymphoblastic leukaemia and 2 separate leukaemic clones. Mutations in brief no. 219. Online. *Hum. Mutat.* 13: 172.

De Laurenzi, V., Costanzo, A., Barcaroli, D., Terrinoni, A., Falco, M., Annicchiarico-Petruzzelli, M., Levrero, M. and Melino, G. (1998) Two new p73 splice variants, gamma and delta, with different transcriptional activity. *J. Exp. Med.* **188**: 1763–1768.

de Stanchina, E., McCurrach, M.E., Zindy, F. *et al.* (1998) E1A signaling to TP53 involves the p19(ARF) tumor suppressor. *Genes Dev.* **12**: 2434–2442.

Deb, S., Jackson, C.T., Subler, M.A. and Martin, D.W. (1992) Modulation of cellular and viral promoters by mutant human TP53 proteins found in tumor cells. *J. Virol.* **66**: 6164–6170.

Debbas, M. and White, E. (1993) Wild-type TP53 mediates apoptosis by E1A, which is inhibited by E1B. *Genes Dev.* **7**: 546–554.

Deng, C., Zhang, P., Harper, J.W., Elledge, S.J. and Leder, P. (1995) Mice lacking p21CIP1/WAF1 undergo normal development, but are defective in G1 checkpoint control. *Cell.* **82**: 675–684.

Di Como, C.J., Gaiddon, C. and Prives, C. (1999) p73 function is inhibited by tumor-derived TP53 mutants in mammalian cells. *Mol. Cell. Biol.* **19**: 1438–1449.

Dittmer, D., Pati, S., Zambetti, G., Chu, S., Teresky, A.K., Moore, M., Finlay, C. and Levine, A.J. (1993) Gain of function mutations in TP53. *Nat. Genet.* **4**: 42–46.

Donehower, L.A., Harvey, M., Slagle, B.L., McArthur, M.J., Montgomery, C.A., Jr., Butel, J.S. and Bradley, A. (1992) Mice deficient for TP53 are developmentally normal but susceptible to spontaneous tumours. *Nature* **356**: 215–221.

Donehower, L.A., Godley, L.A., Aldaz, C.M. *et al.* (1995) Deficiency of TP53 accelerates mammary tumorigenesis in Wnt-1 transgenic mice and promotes chromosomal instability. *Genes Dev.* **9**: 882–895.

Dulic, V., Kaufmann, W.K., Wilson, S.J., Tlsty, T.D., Lees, E., Harper, J.W., Elledge, S.J. and Reed, S.I. (1994) TP53-dependent inhibition of cyclin-dependent kinase activities in human fibroblasts during radiation-induced G1 arrest. *Cell* **76**: 1013–1023.

Dvir, A., Stein, L.Y., Calore, B.L. and Dynan, W.S. (1993) Purification and characterization of a template-associated protein kinase that phosphorylates RNA polymerase II. *J. Biol. Chem.* **268**: 10440–10447.

Eeles, R.A. (1995) Germline mutations in the TTP53 gene. *Cancer Surv.* **25**: 101–124.

Eizenberg, O., Faber-Elman, A., Gottlieb, E., Oren, M., Rotter, V. and Schwartz, M. (1996) TP53 plays a regulatory role in differentiation and apoptosis of central nervous system-associated cells. *Mol. Cell. Biol.* **16**: 5178–5185.

el-Deiry, W.S. (1998) Regulation of TP53 downstream genes. *Semin Cancer Biol.* **8**: 345–357.

el-Deiry, W.S., Tokino, T., Velculescu, V.E *et al.* (1993) WAF1, a potential mediator of TP53 tumor suppression. *Cell.* **75**: 817–825.

Elkeles, A., Juven-Gershon, T., Israeli, D., Wilder, S., Zalcenstein, A. and Oren, M. (1999) The c-fos proto-oncogene is a target for transactivation by the TP53 tumor suppressor. *Mol. Cell. Biol.* **19**: 2594–2600.

Elledge, R.M. (1996) Assessing TP53 status in breast cancer prognosis: where should you put the thermometer if you think your TP53 is sick? [editorial]. *J. Natl Cancer Inst.* **88**: 141–143.

Falette, N., Paperin, M.P., Treilleux, I et al. (1998) Prognostic value of TP53 gene mutations in a large series of node-negative breast cancer patients. *Cancer Res.* **58**: 1451–1455.

Flaman, J.M., Waridel, F., Estreicher, A., Vannier, A., Limacher, J.M., Gilbert, D., Iggo, R. and Frebourg, T. (1996) The human tumour suppressor gene TP53 is alternatively spliced in normal cells. *Oncogene* **12**: 813–818.

Flatt, P.M., Tang, L.J., Scatena, C.D., Szak, S.T. and Pietenpol, J.A. (2000) TP53 regulation of G(2) checkpoint is retinoblastoma protein dependent. *Mol. Cell. Biol.* **20**: 4210–4223.

Foster, B.A., Coffey, H.A., Morin, M.J. and Rastinejad, F. (1999) Pharmacological rescue of mutant TP53 conformation and function. *Science* **286**: 2507–2510.

Frazier, M.W., He, X., Wang, J., Gu, Z., Cleveland, J.L. and Zambetti, G.P. (1998) Activation of c-myc gene expression by tumor-derived TP53 mutants requires a discrete C-terminal domain. *Mol. Cell. Biol.* **18**: 3735–3743.

Freedman, D.A. and Levine, A.J. (1998) Nuclear export is required for degradation of endogenous TP53 by MDM2 and human papillomavirus E6. *Mol. Cell. Biol.* **18**: 7288–7293.

Fresno, M., Molina, R., Perez del Rio, M.J., Alvarez, S., Diaz-Iglesias, J.M., Garcia, I. and Herrero, A. (1997) TP53 expression is of independent predictive value in lymph node-negative breast carcinoma. *Eur. J. Cancer.* **33**: 1268–1274.

Frisch, S.M. and Francis, H. (1994) Disruption of epithelial cell-matrix interactions induces apoptosis. *J. Cell. Biol.* **124**: 619–626.

Furuhata, T., Tokino, T., Urano, T. and Nakamura, Y. (1996) Isolation of a novel GPI-anchored gene specifically regulated by TP53; correlation between its expression and anti-cancer drug sensitivity. *Oncogene* **13**: 1965–1970.

Ganly, I., Kirn, D., Eckhardt, S.G. *et al.* (2000) A phase I study of Onyx-015, an E1B attenuated adenovirus, administered intratumorally to patients with recurrent head and neck cancer. *Clin. Cancer Res.* **6**: 798–806.

Garkavtsev, I., Grigorian, I.A., Ossovskaya, V.S., Chernov, M.V., Chumakov, P.M. and Gudkov, A.V. (1998) The candidate tumour suppressor p33ING1 cooperates with TP53 in cell growth control. *Nature* **391**: 295–298.

Gartel, A.L., Serfas, M.S., Gartel, M., Goufman, E., Wu, G.S., el-Deiry, W.S. and Tyner, A.L. (1996) p21 (WAF1/CIP1) expression is induced in newly nondividing cells in diverse epithelia and during differentiation of the Caco-2 intestinal cell line. *Exp. Cell. Res.* **227**: 171–181.

Gatti, R.A., Berkel, I., Boder, E. *et al.* (1988) Localization of an ataxia-telangiectasia gene to chromosome 11q22–23. *Nature* **336**: 577–580.

Gloushankova, N., Ossovskaya, V., Vasiliev, J., Chumakov, P. and Kopnin, B. (1997) Changes in TP53 expression can modify cell shape of ras-transformed fibroblasts and epitheliocytes. *Oncogene* **15**: 2985–2989.

Gorospe, M., Shack, S., Guyton, K.Z., Samid, D. and Holbrook, N.J. (1996) Up-regulation and functional role of p21Waf1/Cip1 during growth arrest of human breast carcinoma MCF-7 cells by phenylacetate. *Cell Growth Differ.* **7**: 1609–1615.

Gottlieb, E. and Oren, M. (1998) TP53 facilitates pRb cleavage in IL-3-deprived cells: novel pro-apoptotic activity of TP53. *EMBO J.* **17**: 3587–3596.

Gottlieb, E., Haffner, R., King, A., Asher, G., Gruss, P., Lonai, P. and Oren, M. (1997) Transgenic mouse model for studying the transcriptional activity of the TP53 protein: age- and tissue-dependent changes in radiation-induced activation during embryogenesis. *EMBO J.* **16**: 1381–1390.

Gottlieb, T.M. and Jackson, S.P. (1993) The DNA-dependent protein kinase: requirement for DNA ends and association with Ku antigen. *Cell* **72**: 131–142.

Graeber, T.G., Osmanian, C., Jacks, T., Housman, D.E., Koch, C.J., Lowe, S.W. and Giaccia, A.J. (1996) Hypoxia-mediated selection of cells with diminished apoptotic potential in solid tumours. *Nature* **379**: 88–91.

Greenblatt, M.S., Bennett, W.P., Hollstein, M. and Harris, C.C. (1994) Mutations in the TP53 tumor suppressor gene: clues to cancer etiology and molecular pathogenesis. *Cancer Res.* **54**: 4855–4878.

Gu, W., Shi, X.L. and Roeder, R.G. (1997) Synergistic activation of transcription by CBP and TP53. *Nature* **387**: 819–823.

Gualberto, A., Aldape, K., Kozakiewicz, K. and Tlsty, T.D. (1998) An oncogenic form of TP53 confers a dominant, gain-of-function phenotype that disrupts spindle checkpoint control. *Proc. Natl Acad. Sci. USA.* **95**: 5166–5171.

Hall, P.A. and Lane, D.P. (1997) Tumor suppressors: a developing role for TP53? *Curr. Biol.* **7**: R144–147.

Halterman, M.W., Miller, C.C. and Federoff, H.J. (1999) Hypoxia-inducible factor-1alpha mediates hypoxia-induced delayed neuronal death that involves TP53. *J. Neurosci.* **19**: 6818–6824.

Hansen, R. and Oren, M. (1997) TP53; from inductive signal to cellular effect. *Curr. Opin. Genet. Dev.* **7**: 46–51.

Harley, C.B. and Sherwood, S.W. (1997) Telomerase, checkpoints and cancer. *Cancer Surv.* **29**: 263–284.

Harper, J.W., Adami, G.R., Wei, N., Keyomarsi, K. and Elledge, S.J. (1993) The p21 Cdk-interacting protein Cip1 is a potent inhibitor of G1 cyclin-dependent kinases. *Cell.* **75**: 805–816.

Harrington, E.A., Fanidi, A. and Evan, G.I. (1994) Oncogenes and cell death. *Curr. Opin. Genet Dev.* **4**: 120–129.

Hasegawa, M., Zhang, Y., Niibe, H., Terry, N.H. and Meistrich, M.L. (1998) Resistance of differentiating spermatogonia to radiation-induced apoptosis and loss in TP53-deficient mice. *Radiat. Res.* **149**: 263–270.

Haupt, Y., Rowan, S. and Oren, M. (1995a) TP53-mediated apoptosis in HeLa cells can be overcome by excess pRB. *Oncogene* **10**: 1563–1571.

Haupt, Y., Rowan, S., Shaulian, E., Vousden, K.H. and Oren, M. (1995b) Induction of apoptosis in HeLa cells by trans-activation-deficient TP53. *Genes Dev.* **9**: 2170–2183.

Haupt, Y., Maya, R., Kazaz, A. and Oren, M. (1997) Mdm2 promotes the rapid degradation of TP53. *Nature* **387**: 296–299.

Hendry, J.H., Adeeko, A., Potten, C.S. and Morris, I.D. (1996) TP53 deficiency produces fewer regenerating spermatogenic tubules after irradiation. *Int. J. Radiat. Biol.* **70**: 677–682.

Hendry, J.H., Cai, W.B., Roberts, S.A. and Potten, C.S. (1997) TP53 deficiency sensitizes clonogenic cells to irradiation in the large but not the small intestine. *Radiat. Res.* **148**: 254–259.

Hermeking, H. and Eick, D. (1994) Mediation of c-Myc-induced apoptosis by TP53. *Science* **265**: 2091–2093.

Hermeking, H., Lengauer, C., Polyak, K., He, T.C., Zhang, L., Thiagalingam, S., Kinzler, K.W. and Vogelstein, B. (1997) 14-3-3 sigma is a TP53-regulated inhibitor of G2/M progression. *Mol. Cell.* **1**: 3–11.

Honda, R. and Yasuda, H. (1999) Association of p19(ARF) with Mdm2 inhibits ubiquitin ligase activity of Mdm2 for tumor suppressor TP53. *EMBO J.* **18**: 22–27.

Honda, R., Tanaka, H. and Yasuda, H. (1997) Oncoprotein MDM2 is a ubiquitin ligase E3 for tumor suppressor TP53. *FEBS Lett.* **420**: 25–27.

Hsiao, M., Low, J., Dorn, E., Ku, D., Pattengale, P., Yeargin, J. and Haas, M. (1994) Gain-of-function mutations of the TP53 gene induce lymphohematopoietic metastatic potential and tissue invasiveness. *Am. J. Pathol.* **145**: 702–714.

Hupp, T.R., Sparks, A. and Lane, D.P. (1995) Small peptides activate the latent sequence-specific DNA binding function of TP53. *Cell.* **83**: 237–245.

Ichimiya, S., Nimura, Y., Kageyama, H. *et al.* (1999) p73 at chromosome 1p36.3 is lost in advanced stage neuroblastoma but its mutation is infrequent. *Oncogene* **18**: 1061–1066.

Iniesta, P., Vega, F.J., Caldes, T. *et al.* (1998) TP53 exon 7 mutations as a predictor of poor prognosis in patients with colorectal cancer. *Cancer Lett.* **130**: 153–160.

Jacks, T. and Weinberg, R.A. (1996) Cell-cycle control and its watchman [news; comment]. *Nature* **381**: 643–644.

Jacks, T., Remington, L., Williams, B.O., Schmitt, E.M., Halachmi, S., Bronson, R.T. and Weinberg, R.A. (1994) Tumor spectrum analysis in TP53-mutant mice. *Curr. Biol.* **4**: 1–7.

Jayaraman, L., Murthy, K.G., Zhu, C., Curran, T., Xanthoudakis, S. and Prives, C. (1997) Identification of redox/repair protein Ref-1 as a potent activator of TP53. *Genes Dev.* **11**: 558–570.

Jiang, D., Lenardo, M.J. and Zuniga-Pflucker, C. (1996) TP53 prevents maturation to the CD4+CD8+ stage of thymocyte differentiation in the absence of T cell receptor rearrangement. *J Exp Med.* **183**: 1923–1928.

Jiang, H., Lin, J., Su, Z.Z., Collart, F.R., Huberman, E. and Fisher, P.B. (1994) Induction of differentiation in human promyelocytic HL-60 leukemia cells activates p21, WAF1/CIP1, expression in the absence of TP53. *Oncogene* **9**: 3397–3406.

Johnston, E.M. and Crawford, J. (1998) Hematopoietic growth factors in the reduction of chemotherapeutic toxicity. *Semin Oncol.* **25**: 552–561.

Jones, S.N., Roe, A.E., Donehower, L.A. and Bradley, A. (1995) Rescue of embryonic lethality in Mdm2-deficient mice by absence of TP53. *Nature* **378**: 206–208.

Jost, C.A., Marin, M.C. and Kaelin, W.G., Jr. (1997) p73 is a human TP53-related protein that can induce apoptosis. *Nature* **389**: 191–194.

Kaelin, W.G., Jr. (1998) Another TP53 Doppelganger? *Science* **281**: 57–58.

Kaghad, M., Bonnet, H., Yang, A. *et al.* (1997) Monoallelically expressed gene related to TP53 at 1p36, a region frequently deleted in neuroblastoma and other human cancers. *Cell.* **90**: 809–819.

Kamijo, T., Zindy, F., Roussel, M.F., Quelle, D.E., Downing, J.R., Ashmun, R.A., Grosveld, G. and Sherr, C.J. (1997) Tumor suppression at the mouse INK4a locus mediated by the alternative reading frame product p19ARF. *Cell.* **91**: 649–659.

Kamijo, T., Weber, J.D., Zambetti, G., Zindy, F., Roussel, M.F. and Sherr, C.J. (1998) Functional and physical interactions of the ARF tumor suppressor with TP53 and Mdm2. *Proc. Natl Acad. Sci. USA.* **95**: 8292–8297.

Kastan, M.B., Onyekwere, O., Sidransky, D., Vogelstein, B. and Craig, R.W. (1991) Participation of TP53 protein in the cellular response to DNA damage. *Cancer Res.* **51**: 6304–6311.

Keegan, K.S., Holtzman, D.A., Plug, A.W *et al.* (1996) The Atr and Atm protein kinases associate with different sites along meiotically pairing chromosomes. *Genes Dev.* **10**: 2423–2437.

Kelavkar, U.P. and Badr, K.F. (1999) Effects of mutant TP53 expression on human 15-lipoxygenase-promoter activity and murine 12/15-lipoxygenase gene expression: evidence that 15-lipoxygenase is a mutator gene. *Proc. Natl Acad. Sci. USA* **96**: 4378–4383.

Kemp, C.J., Burns, P.A., Brown, K., Nagase, H. and Balmain, A. (1994) Transgenic approaches to the analysis of ras and TP53 function in multistage carcinogenesis. *Cold Spring Harb. Symp. Quant. Biol.* **59**: 427–434.

Kennedy, S.G., Kandel, E.S., Cross, T.K. and Hay, N. (1999) Akt/Protein kinase B inhibits cell death by preventing the release of cytochrome c from mitochondria. *Mol. Cell. Biol.* **19**: 5800–5810.

Khan, S.H., Moritsugu, J. and Wahl, G.M. (2000) Differential requirement for p19ARF in the TP53-dependent arrest induced by DNA damage, microtubule disruption, and ribonucleotide depletion. *Proc. Natl Acad. Sci. USA.* **97**: 3266–3271.

Kirn, D., Hermiston, T. and McCormick, F. (1998) ONYX-015: clinical data are encouraging [letter; comment]. *Nat. Med.* **4**: 1341–1342.

Ko, L.J. and Prives, C. (1996) TP53: puzzle and paradigm. *Genes Dev.* **10**: 1054–1072.

Komarov, P.G., Komarova, E.A., Kondratov, R.V., Christov-Tselkov, K., Coon, J.S., Chernov, M.V. and Gudkov, A.V. (1999) A chemical inhibitor of TP53 that protects mice from the side effects of cancer therapy. *Science* **285**: 1733–1737.

Komarova, E.A. and Gudkov, A.V. (1998) Could TP53 be a target for therapeutic suppression? *Semin. Cancer Biol.* **8**: 389–400.

Komarova, E.A., Chernov, M.V., Franks, R. *et al.* (1997a) Transgenic mice with TP53-responsive lacZ: TP53 activity varies dramatically during normal development and determines radiation and drug sensitivity in vivo. *EMBO J.* **16**: 1391–1400.

Komarova, E.A., Zelnick, C.R., Chin, D., Zeremski, M., Gleiberman, A.S., Bacus, S.S. and Gudkov, A.V. (1997b) Intracellular localization of TP53 tumor suppressor protein in gamma-irradiated cells is cell cycle regulated and determined by the nucleus. *Cancer Res.* **57**: 5217–5220.

Komarova, E.A., Diatchenko, L., Rokhlin, O.W., Hill, J.E., Wang, Z.J., Krivokrysenko, V.I., Feinstein, E. and Gudkov, A.V. (1998) Stress-induced secretion of growth inhibitors: a novel tumor suppressor function of TP53. *Oncogene* **17**: 1089–1096.

Komarova, E.A., Christov, K., Faerman, A. and Gudkov, A.V. (2000) Different impact of TP53 and p21 in determining tissue reaction to gamma-radiation in the mouse. *Oncogene* **18**.

Kovalev, S., Marchenko, N., Swendeman, S., LaQuaglia, M. and Moll, U.M. (1998) Expression level, allelic origin, and mutation analysis of the p73 gene in neuroblastoma tumors and cell lines [In Process Citation]. *Cell Growth Differ.* **9**: 897–903.

Kremenetskaya, O.S., Logacheva, N.P., Baryshnikov, A.Y., Chumakov, P.M. and Kopnin, B.P. (1997) Distinct effects of various TP53 mutants on differentiation and viability of human K562 leukemia cells. *Oncol. Res.* **9**: 155–166.

Kroiss, M.M., Bosserhoff, A.K., Vogt, T., Buettner, R., Bogenrieder, T., Landthaler, M. and Stolz, W. (1998) Loss of expression or mutations in the p73 tumour suppressor gene are not involved in the pathogenesis of malignant melanomas. *Melanoma Res.* **8**: 504–509.

Kubbutat, M.H., Jones, S.N. and Vousden, K.H. (1997) Regulation of TP53 stability by Mdm2. *Nature* **387**: 299–303.

Kulju, K.S. and Lehman, J.M. (1995) Increased TP53 protein associated with aging in human diploid fibroblasts. *Exp. Cell. Res.* **217**: 336–345.

Kussie, P.H., Gorina, S., Marechal, V., Elenbaas, B., Moreau, J., Levine, A.J. and Pavletich, N.P. (1996) Structure of the MDM2 oncoprotein bound to the TP53 tumor suppressor transactivation domain [comment]. *Science* **274**: 948–953.

Lain, S., Midgley, C., Sparks, A., Lane, E.B. and Lane, D.P. (1999) An inhibitor of nuclear export activates the TP53 response and induces the localization of HDM2 and TP53 to U1A-positive nuclear bodies associated with the PODs. *Exp. Cell. Res.* **248**: 457–472.

Lambert, P.F., Kashanchi, F., Radonovich, M.F., Shiekhattar, R. and Brady, J.N. (1998) Phosphorylation of TP53 serine 15 increases interaction with CBP. *J. Biol. Chem.* **273**: 33048–33053.

Lane, D.P. (1992) Cancer. TP53, guardian of the genome [news; comment]. *Nature* **358**: 15–16.

Lee, J.M., Abrahamson, J.L., Kandel, R., Donehower, L.A. and Bernstein, A. (1994) Susceptibility to radiation-carcinogenesis and accumulation of chromosomal breakage in TP53 deficient mice. *Oncogene* **9**: 3731–3736.

Lees-Miller, S.P., Chen, Y.R. and Anderson, C.W. (1990) Human cells contain a DNA-activated protein kinase that phosphorylates simian virus 40 T antigen, mouse TP53, and the human Ku autoantigen. *Mol. Cell. Biol.* **10**: 6472–6481.

Lees-Miller, S.P., Sakaguchi, K., Ullrich, S.J., Appella, E. and Anderson, C.W. (1992) Human DNA-activated protein kinase phosphorylates serines 15 and 37 in the amino-terminal transactivation domain of human TP53. *Mol. Cell. Biol.* **12**: 5041–5049.

Levine, A.J. (1992) The TP53 tumour suppressor gene and product. *Cancer Surv.* **12**: 59–79.

Levine, A.J. (1997) TP53, the cellular gatekeeper for growth and division. *Cell.* **88**: 323–331.

Li, R., Sutphin, P.D., Schwartz, D. *et al.* (1998) Mutant TP53 protein expression interferes with TP53-independent apoptotic pathways. *Oncogene* **16**: 3269–3277.

Lill, N.L., Grossman, S.R., Ginsberg, D., DeCaprio, J. and Livingston, D.M. (1997) Binding and modulation of TP53 by p300/CBP coactivators. *Nature* **387**: 823–827.

Lin, A.W., Barradas, M., Stone, J.C., van Aelst, L., Serrano, M. and Lowe, S.W. (1998) Premature senescence involving TP53 and p16 is activated in response to constitutive MEK/MAPK mitogenic signaling. *Genes Dev.* **12**: 3008–3019.

Lin, J., Chen, J., Elenbaas, B. and Levine, A.J. (1994) Several hydrophobic amino acids in the TP53 amino-terminal domain are required for transcriptional activation, binding to mdm-2 and the adenovirus 5 E1B 55-kD protein. *Genes Dev.* **8**: 1235–1246.

Liu, L., Scolnick, D.M., Trievel, R.C., Zhang, H.B., Marmorstein, R., Halazonetis, T.D. and Berger, S.L. (1999) TP53 sites acetylated in vitro by PCAF and p300 are acetylated in vivo in response to DNA damage. *Mol. Cell. Biol.* **19**: 1202–1209.

Liu, X., Miller, C.W., Koeffler, P.H. and Berk, A.J. (1993) The TP53 activation domain binds the TATA box-binding polypeptide in Holo-TFIID, and a neighboring TP53 domain inhibits transcription. *Mol. Cell. Biol.* **13**: 3291–3300.

Livingstone, L.R., White, A., Sprouse, J., Livanos, E., Jacks, T. and Tlsty, T.D. (1992) Altered cell cycle arrest and gene amplification potential accompany loss of wild-type TP53. *Cell* **70**: 923–935.

Lotem, J. and Sachs, L. (1995) A mutant TP53 antagonizes the deregulated c-myc-mediated enhancement of apoptosis and decrease in leukemogenicity. *Proc. Natl Acad. Sci. USA.* **92**: 9672–9676.

Lowe, S.W. and Ruley, H.E. (1993) Stabilization of the TP53 tumor suppressor is induced by adenovirus 5 E1A and accompanies apoptosis. *Genes Dev.* **7**: 535–545.

Lowe, S.W. and Lin, A.W. (2000) Apoptosis in cancer. *Carcinogenesis* **21**: 485–495.

Lowe, S.W., Ruley, H.E., Jacks, T. and Housman, D.E. (1993a) TP53-dependent apoptosis modulates the cytotoxicity of anticancer agents. *Cell* **74**: 957–967.

Lowe, S.W., Schmitt, E.M., Smith, S.W., Osborne, B.A. and Jacks, T. (1993b) TP53 is required for radiation-induced apoptosis in mouse thymocytes. *Nature* **362**: 847–849.

Lowe, S.W., Jacks, T., Housman, D.E. and Ruley, H.E. (1994a) Abrogation of oncogene-associated apoptosis allows transformation of TP53-deficient cells. *Proc. Natl Acad. Sci. USA.* **91**: 2026–2030.

Lowe, S.W., Bodis, S., McClatchey, A., Remington, L., Ruley, H.E., Fisher, D.E., Housman, D.E. and Jacks, T. (1994b) TP53 status and the efficacy of cancer therapy in vivo. *Science* **266**: 807–810.

Lu, H. and Levine, A.J. (1995) Human TAFII31 protein is a transcriptional coactivator of the TP53 protein. *Proc. Natl Acad. Sci. USA.* **92**: 5154–5158.

MacCallum, D.E., Hupp, T.R., Midgley, C.A. *et al.* (1996) The TP53 response to ionising radiation in adult and developing murine tissues. *Oncogene* **13**: 2575–2587.

Macleod, K.F., Sherry, N., Hannon, G., Beach, D., Tokino, T., Kinzler, K., Vogelstein, B. and Jacks, T. (1995) TP53-dependent and independent expression of p21 during cell growth, differentiation, and DNA damage. *Genes Dev.* **9**: 935–944.

Mai, M., Qian, C., Yokomizo, A., Tindall, D.J., Bostwick, D., Polychronakos, C., Smith, D.I. and Liu, W. (1998) Loss of imprinting and allele switching of p73 in renal cell carcinoma. *Oncogene* **17**: 1739–1741.

Maki, C.G., Huibregtse, J.M. and Howley, P.M. (1996) In vivo ubiquitination and proteasome-mediated degradation of TP53(1). *Cancer Res.* **56**: 2649–2654.

Malkin, D. and Friend, S.H. (1993) Correction: a Li-Fraumeni syndrome TP53 mutation [letter; comment]. *Science* **259**: 878.

Maltzman, W. and Czyzyk, L. (1984) UV irradiation stimulates levels of TP53 cellular tumor antigen in nontransformed mouse cells. *Mol. Cell. Biol.* **4**: 1689–1694.

Matlashewski, G.J., Tuck, S., Pim, D., Lamb, P., Schneider, J. and Crawford, L.V. (1987) Primary structure polymorphism at amino acid residue 72 of human TP53. *Mol. Cell. Biol.* **7**: 961–963.

May, P. and May, E. (1999) Twenty years of TP53 research: structural and functional aspects of the TP53 protein [published erratum appears in *Oncogene* 2000 Mar 23;19(13): 1734]. *Oncogene* **18**: 7621–7636.

McLure, K.G. and Lee, P.W. (1998) How TP53 binds DNA as a tetramer. *EMBO J.* **17**: 3342–3350.

Merritt, A.J., Allen, T.D., Potten, C.S. and Hickman, J.A. (1997) Apoptosis in small intestinal epithelial from TP53-null mice: evidence for a delayed, TP53-independent G2/M-associated cell death after gamma-irradiation. *Oncogene* **14**: 2759–2766.

Michalovitz, D., Halevy, O. and Oren, M. (1991) TP53 mutations: gains or losses? *J. Cell. Biochem.* **45**: 22–29.

Midgley, C.A., Desterro, J.M., Saville, M.K., Howard, S., Sparks, A., Hay, R.T. and Lane, D.P. (2000) An N-terminal p14ARF peptide blocks Mdm2-dependent ubiquitination in vitro and can activate TP53 in vivo. *Oncogene* **19**: 2312–2323.

Miller, C., Mohandas, T., Wolf, D., Prokocimer, M., Rotter, V. and Koeffler, H.P. (1986) Human TP53 gene localized to short arm of chromosome 17. *Nature* **319**: 783–784.

Molina, R., Segui, M.A., Climent, M.A. *et al.* (1998) TP53 oncoprotein as a prognostic indicator in patients with breast cancer. *Anticancer Res.* **18**: 507–511.

Moll, U.M., Riou, G. and Levine, A.J. (1992) Two distinct mechanisms alter TP53 in breast cancer: mutation and nuclear exclusion. *Proc. Natl Acad. Sci. USA* **89**: 7262–7266.

Moll, U.M., LaQuaglia, M., Benard, J. and Riou, G. (1995) Wild-type TP53 protein undergoes cytoplasmic sequestration in undifferentiated neuroblastomas but not in differentiated tumors. *Proc. Natl Acad. Sci. USA* **92**: 4407–4411.

Moll, U.M., Ostermeyer, A.G., Haladay, R., Winkfield, B., Frazier, M. and Zambetti, G. (1996) Cytoplasmic sequestration of wild-type TP53 protein impairs the G1 checkpoint after DNA damage. *Mol. Cell. Biol.* **16**: 1126–1137.

Momand, J., Zambetti, G.P., Olson, D.C., George, D. and Levine, A.J. (1992) The mdm-2 oncogene product forms a complex with the TP53 protein and inhibits TP53-mediated transactivation. *Cell* **69**: 1237–1245.

Momand, J., Jung, D., Wilczynski, S. and Niland, J. (1998) The MDM2 gene amplification database. *Nucleic Acids Res.* **26**: 3453–3459.

Momand, J., Wu, H.H. and Dasgupta, G. (2000) MDM2 – master regulator of the TP53 tumor suppressor protein. *Gene* **242**: 15–29.

Mummenbrauer, T., Janus, F., Muller, B., Wiesmuller, L., Deppert, W. and Grosse, F. (1996) TP53 Protein exhibits 3′-to-5′ exonuclease activity. *Cell* **85**: 1089–1099.

Murphy, M., Hinman, A. and Levine, A.J. (1996) Wild-type TP53 negatively regulates the expression of a microtubule-associated protein. *Genes Dev.* **10**: 2971–2980.

Nieder, C., Petersen, S., Petersen, C. and Thames, H.D. (2000) The challenge of TP53 as prognostic and predictive factor in gliomas. *Cancer Treat. Rev.* **26**: 67–73.

Nikiforov, M.A., Hagen, K., Ossovskaya, V.S., Connor, T.M., Lowe, S.W., Deichman, G.I. and Gudkov, A.V. (1996) TP53 modulation of anchorage independent growth and experimental metastasis. *Oncogene* **13**: 1709–1719.

Nimura, Y., Mihara, M., Ichimiya, S. *et al.* (1998) p73, a gene related to TP53, is not mutated in esophageal carcinomas. *Int. J. Cancer.* **78**: 437–440.

Nishimori, H., Shiratsuchi, T., Urano, T. *et al.* (1997) A novel brain-specific TP53-target gene, BAI1, containing thrombospondin type 1 repeats inhibits experimental angiogenesis. *Oncogene* **15**: 2145–2150.

Noda, A., Ning, Y., Venable, S.F., Pereira-Smith, O.M. and Smith, J.R. (1994) Cloning of senescent cell-derived inhibitors of DNA synthesis using an expression screen. *Exp Cell Res.* **211**: 90–98.

Nomoto, S., Haruki, N., Kondo, M., Konishi, H. and Takahashi, T. (1998) Search for mutations and examination of allelic expression imbalance of the p73 gene at 1p36.33 in human lung cancers. *Cancer Res.* **58**: 1380–1383.

Norberg, T., Lennerstrand, J., Inganas, M. and Bergh, J. (1998) Comparison between TP53 protein measurements using the luminometric immunoassay and immunohistochemistry with detection of TP53 gene mutations using cDNA sequencing in human breast tumors. *Int. J. Cancer* **79**: 376–383.

Norimura, T., Nomoto, S., Katsuki, M., Gondo, Y. and Kondo, S. (1996) TP53-dependent apoptosis suppresses radiation-induced teratogenesis. *Nat. Med.* **2**: 577–580.

O'Connor, P.M., Jackman, J., Bae, I. *et al.* (1997) Characterization of the TP53 tumor suppressor pathway in cell lines of the National Cancer Institute anticancer drug screen and correlations with the growth-inhibitory potency of 123 anticancer agents. *Cancer Res.* **57**: 4285–4300.

Ollmann, M., Young, L.M., Di Como, C.J. *et al.* (2000) Drosophila TP53 is a structural and functional homolog of the tumor suppressor TP53. *Cell* **101**: 91–101.

Oltvai, Z.N. and Korsmeyer, S.J. (1994) Checkpoints of dueling dimers foil death wishes [comment]. *Cell* 79: 189–192.

O'Neill, M., Campbell, S.J., Save, V., Thompson, A.M. and Hall, P.A. (1998) An immunochemical analysis of mdm2 expression in human breast cancer and the identification of a growth-regulated cross-reacting species p170. *J. Pathol.* 186: 254–261.

Osada, M., Ohba, M., Kawahara, C. *et al.* (1998) Cloning and functional analysis of human p51, which structurally and functionally resembles TP53 [In Process Citation]. *Nat. Med.* 4: 839–843.

Owen-Schaub, L.B., Zhang, W., Cusack, J.C. *et al.* (1995) Wild-type human TP53 and a temperature-sensitive mutant induce Fas/APO-1 expression. *Mol. Cell. Biol.* 15: 3032–3040.

Pan, H. and Griep, A.E. (1995) Temporally distinct patterns of TP53-dependent and TP53-independent apoptosis during mouse lens development. *Genes Dev.* 9: 2157–2169.

Pandrea, I.V., Mihailovici, M.S., Carasevici, E., Szekely, A.M., Reynes, M., Tarcoveanu, E. and Dragomir, C. (1995) An immunohistochemical study of TP53 protein on colorectal carcinomas. *Rev. Med. Chir. Soc. Med. Nat. Iasi.* 99: 171–178.

Patel, D.D., Bhatavdekar, J.M., Chikhlikar, P.R., Ghosh, N., Suthar, T.P., Shah, N.G., Mehta, R.H. and Balar, D.B. (1996) Node negative breast carcinoma: hyperprolactinemia and/or overexpression of TP53 as an independent predictor of poor prognosis compared to newer and established prognosticators. *J. Surg. Oncol.* 62: 86–92.

Peled, A., Zipori, D. and Rotter, V. (1996) Cooperation between TP53-dependent and TP53-independent apoptotic pathways in myeloid cells. *Cancer Res.* 56: 2148–2156.

Peller, S. (1998) Clinical implications of TP53: effect on prognosis, tumor progression and chemotherapy response. *Semin Cancer Biol.* 8: 379–387.

Peng, C.Y., Graves, P.R., Thoma, R.S., Wu, Z., Shaw, A.S. and Piwnica-Worms, H. (1997) Mitotic and G2 checkpoint control: regulation of 14–3–3 protein binding by phosphorylation of Cdc25C on serine-216. *Science* 277: 1501–1505.

Piette, J., Neel, H. and Marechal, V. (1997) Mdm2: keeping TP53 under control. *Oncogene* 15: 1001–1010.

Polyak, K., Waldman, T., He, T.C., Kinzler, K.W. and Vogelstein, B. (1996) Genetic determinants of TP53-induced apoptosis and growth arrest. *Genes Dev.* 10: 1945–1952.

Polyak, K., Xia, Y., Zweier, J.L., Kinzler, K.W. and Vogelstein, B. (1997) A model for TP53-induced apoptosis. *Nature* 389: 300–305.

Pomerantz, J., Schreiber-Agus, N., Liegeois, N.J. *et al.* (1998) The Ink4a tumor suppressor gene product, p19Arf, interacts with MDM2 and neutralizes MDM2's inhibition of TP53. *Cell.* 92: 713–723.

Prisco, M., Hongo, A., Rizzo, M.G., Sacchi, A. and Baserga, R. (1997) The insulin-like growth factor I receptor as a physiologically relevant target of TP53 in apoptosis caused by interleukin-3 withdrawal. *Mol. Cell. Biol.* 17: 1084–1092.

Purdie, C.A., Harrison, D.J., Peter, A. et al. (1994) Tumour incidence, spectrum and ploidy in mice with a large deletion in the TP53 gene. *Oncogene* 9: 603–609.

Qazilbash, M.H., Xiao, X., Seth, P., Cowan, K.H. and Walsh, C.E. (1997) Cancer gene therapy using a novel adeno-associated virus vector expressing human wild-type TP53. *Gene Ther.* 4: 675–682.

Raman, V., Martensen, S.A., Reisman, D., Evron, E., Odenwald, W.F., Jaffee, E., Marks, J. and Sukumar, S. (2000) Compromised HOXA5 function can limit TP53 expression in human breast tumours [In Process Citation]. *Nature* 405: 974–978.

Ravi, R., Mookerjee, B., Bhujwalla, Z.M. *et al.* (2000) Regulation of tumor angiogenesis by TP53-induced degradation of hypoxia-inducible factor 1alpha. *Genes Dev.* 14: 34–44.

Reisman, D. and Rotter, V. (1993) The helix-loop-helix containing transcription factor USF binds to and transactivates the promoter of the TP53 tumor suppressor gene. *Nucleic Acids Res.* 21: 345–350.

Reisman, D. and Loging, W.T. (1998) Transcriptional regulation of the TP53 tumor suppressor gene. *Semin Cancer Biol.* 8: 317–324.

Reisman, D., Elkind, N.B., Roy, B., Beamon, J. and Rotter, V. (1993) c-Myc trans-activates the TP53 promoter through a required downstream CACGTG motif. *Cell Growth Differ.* 4: 57–65.

Rogel, A., Popliker, M., Webb, C.G. and Oren, M. (1985) TP53 cellular tumor antigen: analysis of mRNA levels in normal adult tissues, embryos, and tumors. *Mol. Cell. Biol.* 5: 2851–2855.

Rohn, J.L., Hueber, A.O., McCarthy, N.J., Lyon, D., Navarro, P., Burgering, B.M. and Evan, G.I. (1998) The opposing roles of the Akt and c-Myc signalling pathways in survival from CD95-mediated apoptosis. *Oncogene* 17: 2811–2818.

Rokhlin, O.W., Gudkov, A.V., Kwek, S., Glover, R.A., Gewies, A.S. and Cohen, M.B. (2000) TP53 is involved in tumor necrosis factor-alpha-induced apoptosis in the human prostatic carcinoma cell line LNCaP. *Oncogene* 19: 1959–1968.

Ronen, D., Rotter, V. and Reisman, D. (1991) Expression from the murine TP53 promoter is mediated by factor binding to a downstream helix-loop-helix recognition motif. *Proc. Natl Acad. Sci. USA* 88: 4128–4132.

Rosse, T., Olivier, R., Monney, L., Rager, M., Conus, S., Fellay, I., Jansen, B. and Borner, C. (1998) Bcl-2 prolongs cell survival after Bax-induced release of cytochrome c. *Nature* 391: 496–499.

Roth, J., Dobbelstein, M., Freedman, D.A., Shenk, T. and Levine, A.J. (1998) Nucleo-cytoplasmic shuttling of the hdm2 oncoprotein regulates the levels of the TP53 protein via a pathway used by the human immunodeficiency virus rev protein. *EMBO J.* 17: 554–564.

Rubtsova, S.N., Kondratov, R.V., Kopnin, P.B., Chumakov, P.M., Kopnin, B.P. and Vasiliev, J.M. (1998) Disruption of actin microfilaments by cytochalasin D leads to activation of TP53. *FEBS Lett.* 430: 353–357.

Sabbatini, P. and McCormick, F. (1999) Phosphoinositide 3-OH kinase (PI3K) and PKB/Akt delay the onset of TP53-mediated, transcriptionally dependent apoptosis [In Process Citation]. *J. Biol. Chem.* 274: 24263–24269.

Sah, V.P., Attardi, L.D., Mulligan, G.J., Williams, B.O., Bronson, R.T. and Jacks, T. (1995) A subset of TP53-deficient embryos exhibit exencephaly. *Nat. Genet.* 10: 175–180.

Sakaguchi, K., Herrera, J.E., Saito, S., Miki, T., Bustin, M., Vassilev, A. Anderson, C.W. and Appella, E. (1998) DNA damage activates TP53 through a phosphorylation-acetylation cascade. *Genes Dev.* 12: 2831–2841.

Sakamuro, D., Sabbatini, P., White, E. and Prendergast, G.C. (1997) The polyproline region of TP53 is required to activate apoptosis but not growth arrest. *Oncogene* 15: 887–898.

Savitsky, K., Sfez, S., Tagle, D.A., Ziv, Y., Sartiel, A., Collins, F.S., Shiloh, Y. and Rotman, G. (1995) The complete sequence of the coding region of the ATM gene reveals similarity to cell cycle regulators in different species. *Hum Mol Genet.* 4: 2025–2032.

Scheffner, M., Werness, B.A., Huibregtse, J.M., Levine, A.J. and Howley, P.M. (1990) The E6 oncoprotein encoded by human papillomavirus types 16 and 18 promotes the degradation of TP53. *Cell.* 63: 1129–1136.

Schlamp, C.L., Poulsen, G.L., Nork, T.M. and Nickells, R.W. (1997) Nuclear exclusion of wild-type TP53 in immortalized human retinoblastoma cells. *J. Natl. Cancer Inst.* 89: 1530–1536.

Schmid, P., Lorenz, A., Hameister, H. and Montenarh, M. (1991) Expression of TP53 during mouse embryogenesis. *Development* 113: 857–865.

Schmitt, C.A., McCurrach, M.E., de Stanchina, E., Wallace-Brodeur, R.R. and Lowe, S.W. (1999) INK4a/ARF mutations accelerate lymphomagenesis and promote chemoresistance by disabling TP53. *Genes Dev.* 13: 2670–2677.

Schwartz, D., Goldfinger, N. and Rotter, V. (1993) Expression of TP53 protein in spermatogenesis is confined to the tetraploid pachytene primary spermatocytes. *Oncogene* 8: 1487–1494.

Selivanova, G., Iotsova, V., Okan, I., Fritsche, M., Strom, M., Groner, B., Grafstrom, R.C. and Wiman, K.G. (1997) Restoration of the growth suppression function of mutant TP53 by a synthetic peptide derived from the TP53 C-terminal domain. *Nat Med.* 3: 632–638.

Serrano, M., Lin, A.W., McCurrach, M.E., Beach, D. and Lowe, S.W. (1997) Oncogenic ras provokes premature cell senescence associated with accumulation of TP53 and p16INK4a. *Cell* 88: 593–602.

Seto, E., Usheva, A., Zambetti, G.P., Momand, J., Horikoshi, N., Weinmann, R., Levine, A.J. and Shenk, T. (1992) Wild-type TP53 binds to the TATA-binding protein and represses transcription. *Proc. Natl Acad. Sci. USA* 89: 12028–12032.

Shaulsky, G., Goldfinger, N., Ben-Ze'ev, A. and Rotter, V. (1990) Nuclear accumulation of TP53 protein is mediated by several nuclear localization signals and plays a role in tumorigenesis. *Mol. Cell. Biol.* 10: 6565–6577.

She, Q.B., Chen, N. and Dong, Z. (2000) ERKs and p38 kinase phosphorylate TP53 protein at serine 15 in response to UV radiation. *J. Biol. Chem.* 275: 20444–20449.

Sherr, C.J. (1998) Tumor surveillance via the ARF-TP53 pathway. *Genes Dev.* 12: 2984–2991.

Shick, L., Carman, J.H., Choi, J.K. *et al.* (1997) Decreased immunoglobulin deposition in tumors and increased immature B cells in TP53-null mice. *Cell Growth Differ.* 8: 121–131.

Shieh, S.Y., Ikeda, M., Taya, Y. and Prives, C. (1997) DNA damage-induced phosphorylation of TP53 alleviates inhibition by MDM2. *Cell* 91: 325–334.

Shieh, S.Y., Taya, Y. and Prives, C. (1999) DNA damage-inducible phosphorylation of TP53 at N-terminal sites including a novel site, Ser20, requires tetramerization. *EMBO J.* **18**: 1815–1823.

Shikama, N., Lee, C.W., France, S., Delavaine, L., Lyon, J., Krstic-Demonacos, M. and La Thangue, N.B. (1999) A novel cofactor for p300 that regulates the TP53 response. *Mol. Cell.* **4**: 365–376.

Siliciano, J.D., Canman, C.E., Taya, Y., Sakaguchi, K., Appella, E. and Kastan, M.B. (1997) DNA damage induces phosphorylation of the amino terminus of TP53. *Genes Dev.* **11**: 3471–3481.

Smith, G.C. and Jackson, S.P. (1999) The DNA-dependent protein kinase. *Genes Dev.* **13**: 916–934.

Somasundaram, K. and El-Deiry, W. (2000) Tumor suppressor TP53: regulation and function. *Front Biosci.* **5**: D424–437.

Soussi, T., Caron de Fromentel, C., Mechali, M., May, P. and Kress, M. (1987) Cloning and characterization of a cDNA from *Xenopus laevis* coding for a protein homologous to human and murine TP53. *Oncogene* **1**: 71–78.

Soussi, T., Begue, A., Kress, M., Stehelin, D. and May, P. (1988) Nucleotide sequence of a cDNA encoding the chicken TP53 nuclear oncoprotein. *Nucleic Acids Res.* **16**: 11383.

Soussi, T., Dehouche, K. and Beroud, C. (2000) TP53 website and analysis of TP53 gene mutations in human cancer: forging a link between epidemiology and carcinogenesis. *Hum Mutat.* **15**: 105–113.

Spitz, F.R., Nguyen, D., Skibber, J.M., Cusack, J., Roth, J.A. and Cristiano, R.J. (1996a) In vivo adenovirus-mediated TP53 tumor suppressor gene therapy for colorectal cancer. *Anticancer Res.* **16**: 3415–3422.

Spitz, F.R., Nguyen, D., Skibber, J.M., Meyn, R.E., Cristiano, R.J. and Roth, J.A. (1996b) Adenovirus-mediated wild-type TP53 gene expression sensitizes colorectal cancer cells to ionizing radiation. *Clin. Cancer Res.* **2**: 1665–1671.

Srinivasula, S.M., Ahmad, M., Fernandes-Alnemri, T. and Alnemri, E.S. (1998) Autoactivation of procaspase-9 by Apaf-1-mediated oligomerization. *Mol. Cell.* **1**: 949–957.

Stenmark-Askmalm, M., Stal, O., Sullivan, S., Ferraud, L., Sun, X.F., Carstensen, J. and Nordenskjold, B. (1994) Cellular accumulation of TP53 protein: an independent prognostic factor in stage II breast cancer. *Eur. J. Cancer* **30A**: 175–180.

Stewart, N., Hicks, G.G., Paraskevas, F. and Mowat, M. (1995) Evidence for a second cell cycle block at G2/M by TP53. *Oncogene* **10**: 109–115.

Stommel, J.M., Marchenko, N.D., Jimenez, G.S., Moll, U.M., Hope, T.J. and Wahl, G.M. (1999) A leucine-rich nuclear export signal in the TP53 tetramerization domain: regulation of subcellular localization and TP53 activity by NES masking. *EMBO J.* **18**: 1660–1672.

Storey, A., Thomas, M., Kalita, A. *et al.* (1998) Role of a TP53 polymorphism in the development of human papillomavirus-associated cancer. *Nature* **393**: 229–234.

Sun, Y., Hegamyer, G., Cheng, Y.J., Hildesheim, A., Chen, J.Y., Chen, I.H., Cao, Y., Yao, K.T. and Colburn, N.H. (1992) An infrequent point mutation of the TP53 gene in human nasopharyngeal carcinoma. *Proc. Natl Acad. Sci. USA* **89**: 6516–6520.

Sun, Y., Nakamura, K., Wendel, E. and Colburn, N. (1993) Progression toward tumor cell phenotype is enhanced by overexpression of a mutant TP53 tumor-suppressor gene isolated from nasopharyngeal carcinoma. *Proc. Natl Acad. Sci. USA* **90**: 2827–2831.

Swisher, S.G., Roth, J.A., Nemunaitis, J. *et al.* (1999) Adenovirus-mediated TP53 gene transfer in advanced non-small-cell lung cancer. *J. Natl. Cancer Inst.* **91**: 763–771.

Takahashi, H., Ichimiya, S., Nimura, Y., Watanabe, M., Furusato, M., Wakui, S., Yatani, R., Aizawa, S. and Nakagawara, A. (1998) Mutation, allelotyping, and transcription analyses of the p73 gene in prostatic carcinoma. *Cancer Res.* **58**: 2076–2077.

Tao, W. and Levine, A.J. (1999a) Nucleocytoplasmic shuttling of oncoprotein Hdm2 is required for Hdm2-mediated degradation of TP53. *Proc. Natl Acad. Sci. USA* **96**: 3077–3080.

Tao, W. and Levine, A.J. (1999b) P19(ARF) stabilizes TP53 by blocking nucleo-cytoplasmic shuttling of Mdm2. *Proc. Natl Acad. Sci. USA* **96**: 6937–6941.

Taylor, W.R., Agarwal, M.L., Agarwal, A., Stacey, D.W. and Stark, G.R. (1999) TP53 inhibits entry into mitosis when DNA synthesis is blocked. *Oncogene* **18**: 283–295.

Thomas, M., Pim, D. and Banks, L. (1999) The role of the E6-TP53 interaction in the molecular pathogenesis of HPV. *Oncogene* **18**: 7690–7700.

Thut, C.J., Chen, J.L., Klemm, R. and Tjian, R. (1995) TP53 transcriptional activation mediated by coactivators TAFII40 and TAFII60. *Science* **267**: 100–104.

Tishler, R.B., Lamppu, D.M., Park, S. and Price, B.D. (1995) Microtubule-active drugs taxol, vinblastine, and nocodazole increase the levels of transcriptionally active TP53. *Cancer Res.* **55**: 6021–6025.

Trink, B., Okami, K., Wu, L., Sriuranpong, V., Jen, J. and Sidransky, D. (1998) A new human TP53 homologue [letter] [In Process Citation]. *Nat. Med.* **4**: 747–748.

Tron, V.A., Trotter, M.J., Tang, L., Krajewska, M., Reed, J.C., Ho, V.C. and Li, G. (1998) TP53-regulated apoptosis is differentiation dependent in ultraviolet B-irradiated mouse keratinocytes. *Am. J. Pathol.* **153**: 579–585.

Truant, R., Xiao, H., Ingles, C.J. and Greenblatt, J. (1993) Direct interaction between the transcriptional activation domain of human TP53 and the TATA box-binding protein. *J. Biol. Chem.* **268**: 2284–2287.

Ueda, H., Ullrich, S.J., Gangemi, J.D., Kappel, C.A., Ngo, L., Feitelson, M.A. and Jay, G. (1995) Functional inactivation but not structural mutation of TP53 causes liver cancer. *Nat Genet.* **9**: 41–47.

Unger, T., Juven-Gershon, T., Moallem, E., Berger, M., Vogt Sionov, R., Lozano, G., Oren, M. and Haupt, Y. (1999) Critical role for Ser20 of human TP53 in the negative regulation of TP53 by Mdm2. *EMBO J.* **18**: 1805–1814.

Valenzuela, M.T., Nunez, M.I., Villalobos, M., Siles, E., McMillan, T.J., Pedraza, V. and Ruiz de Almodovar, J.M. (1997) A comparison of TP53 and p16 expression in human tumor cells treated with hyperthermia or ionizing radiation. *Int. J. Cancer.* **72**: 307–312.

Van Meir, E.G., Polverini, P.J., Chazin, V.R., Su Huang, H.J., de Tribolet, N. and Cavenee, W.K. (1994) Release of an inhibitor of angiogenesis upon induction of wild type TP53 expression in glioblastoma cells. *Nat. Genet.* **8**: 171–176.

Varley, J.M., Evans, D.G. and Birch, J.M. (1997) Li-Fraumeni syndrome – a molecular and clinical review. *Br. J. Cancer.* **76**: 1–14.

Vaziri, H., Squire, J.A., Pandita, T.K. *et al.* (1999) Analysis of genomic integrity and TP53-dependent G1 checkpoint in telomerase-induced extended-life-span human fibroblasts. *Mol. Cell. Biol.* **19**: 2373–2379.

Vogelstein, B. and Kinzler, K.W. (1992) TP53 function and dysfunction. *Cell* **70**: 523–526.

Waga, S., Hannon, G.J., Beach, D. and Stillman, B. (1994) The p21 inhibitor of cyclin-dependent kinases controls DNA replication by interaction with PCNA. *Nature* **369**: 574–578.

Wagner, A.J., Kokontis, J.M. and Hay, N. (1994) Myc-mediated apoptosis requires wild-type TP53 in a manner independent of cell cycle arrest and the ability of TP53 to induce p21waf1/cip1. *Genes Dev.* **8**: 2817–2830.

Walker, K.K. and Levine, A.J. (1996) Identification of a novel TP53 functional domain that is necessary for efficient growth suppression. *Proc. Natl Acad. Sci. USA* **93**: 15335–15340.

Wallingford, J.B., Seufert, D.W., Virta, V.C. and Vize, P.D. (1997) TP53 activity is essential for normal development in Xenopus. *Curr. Biol.* **7**: 747–757.

Wang, L., Cui, Y., Lord, B.I., Roberts, S.A., Potten, C.S., Hendry, J.H. and Scott, D. (1996a) Gamma-ray-induced cell killing and chromosome abnormalities in the bone marrow of TP53-deficient mice. *Radiat. Res.* **146**: 259–266.

Wang, Y., Okan, I., Pokrovskaja, K. and Wiman, K.G. (1996b) Abrogation of TP53-induced G1 arrest by the HPV 16 E7 protein does not inhibit TP53-induced apoptosis. *Oncogene* **12**: 2731–2735.

Wang, Q., Zambetti, G.P. and Suttle, D.P. (1997) Inhibition of DNA topoisomerase II alpha gene expression by the TP53 tumor suppressor. *Mol. Cell. Biol.* **17**: 389–397.

Weber, J.D., Taylor, L.J., Roussel, M.F., Sherr, C.J. and Bar-Sagi, D. (1999) Nucleolar Arf sequesters Mdm2 and activates TP53. *Nat. Cell. Biol.* **1**: 20–26.

Will, K., Warnecke, G., Bergmann, S. and Deppert, W. (1995) Species- and tissue-specific expression of the C-terminal alternatively spliced form of the tumor suppressor TP53. *Nucleic Acids Res.* **23**: 4023–4028.

Wlodarski, P., Wasik, M., Ratajczak, M.Z. *et al.* (1998) Role of TP53 in hematopoietic recovery after cytotoxic treatment. *Blood* **91**: 2998–3006.

Wolf, D., Harris, N. and Rotter, V. (1984) Reconstitution of TP53 expression in a nonproducer Ab-MuLV-transformed cell line by transfection of a functional TP53 gene. *Cell* **38**: 119–126.

Wu, G.S., Burns, T.F., McDonald, E.R., 3rd *et al.* (1997) KILLER/DR5 is a DNA damage-inducible TP53-regulated death receptor gene [letter]. *Nat Genet.* **17**: 141–143.

Wu, X. and Levine, A.J. (1994) TP53 and E2F-1 cooperate to mediate apoptosis. *Proc. Natl Acad. Sci. USA* **91**: 3602–3606.

Wu, X., Bayle, J.H., Olson, D. and Levine, A.J. (1993) The TP53-mdm-2 autoregulatory feedback loop. *Genes Dev.* **7**: 1126–1132.

Yamaguchi, A., Kurosaka, Y., Fushida, S., Kanno, M., Yonemura, Y., Miwa, K. and Miyazaki, I. (1992) Expression of TP53 protein in colorectal cancer and its relationship to short-term prognosis. *Cancer* **70**: 2778–2784.

Yang, A., Kaghad, M., Wang, Y., Gillett, E., Fleming, M.D., Dotsch, V. Andrews, N.C., Caput, D. and McKeon, F. (1998) p63, a TP53 homolog at 3q27–29, encodes multiple products with transactivating, death-inducing, and dominant-negative activities. *Mol. Cell.* **2**: 305–316.

Yang, X., Pater, A. and Tang, S.C. (1999) Cloning and characterization of the human BAG-1 gene promoter: upregulation by tumor-derived TP53 mutants [In Process Citation]. *Oncogene* **18**: 4546–4553.

Yin, Y., Tainsky, M.A., Bischoff, F.Z., Strong, L.C. and Wahl, G.M. (1992) Wild-type TP53 restores cell cycle control and inhibits gene amplification in cells with mutant TP53 alleles. *Cell.* **70**: 937–948.

Yin, Y., Terauchi, Y., Solomon, G.G., Aizawa, S., Rangarajan, P.N., Yazaki, Y., Kadowaki, T. and Barrett, J.C. (1998) Involvement of p85 in TP53-dependent apoptotic response to oxidative stress. *Nature* **391**: 707–710.

Yokomizo, A., Mai, M., Tindall, D.J., Cheng, L., Bostwick, D.G., Naito, S., Smith, D.I. and Liu, W. (1999) Overexpression of the wild type p73 gene in human bladder cancer. *Oncogene* **18**: 1629–1633.

Yokota, J. (2000) Tumor progression and metastasis. *Carcinogenesis* **21**: 497–503.

Yonish-Rouach, E., Resnitzky, D., Lotem, J., Sachs, L., Kimchi, A. and Oren, M. (1991) Wild-type TP53 induces apoptosis of myeloid leukaemic cells that is inhibited by interleukin-6. *Nature* **352**: 345–347.

Zauberman, A., Flusberg, D., Haupt, Y., Barak, Y. and Oren, M. (1995) A functional TP53-responsive intronic promoter is contained within the human mdm2 gene. *Nucleic Acids Res.* **23**: 2584–2592.

Zeng, X., Chen, L., Jost, C.A. *et al.* (1999) MDM2 suppresses p73 function without promoting p73 degradation. *Mol. Cell. Biol.* **19**: 3257–3266.

Zhang and Wang, H. (2000) MDM2 oncogene as a novel target for human cancer therapy. *Curr. Pharm. Des.* **6**: 393–416.

Zhang, Y. and Xiong, Y. (1999) Mutations in human ARF exon 2 disrupt its nucleolar localization and impair its ability to block nuclear export of MDM2 and TP53. *Mol. Cell.* **3**: 579–591.

Zhang, Y., Xiong, Y. and Yarbrough, W.G. (1998) ARF promotes MDM2 degradation and stabilizes TP53: ARF-INK4a locus deletion impairs both the Rb and TP53 tumor suppression pathways. *Cell* **92**: 725–734.

Zhao, R., Gish, K., Murphy, M., Yin, Y., Notterman, D., Hoffman, W.H., Tom, E., Mack, D.H. and Levine, A.J. (2000) Analysis of TP53-regulated gene expression patterns using oligonucleotide arrays. *Genes Dev.* **14**: 981–993.

Zhu, J., Jiang, J., Zhou, W. and Chen, X. (1998) The potential tumor suppressor p73 differentially regulates cellular TP53 target genes. *Cancer Res.* **58**: 5061–5065.

Zindy, F., Eischen, C.M., Randle, D.H., Kamijo, T., Cleveland, J.L., Sherr, C.J. and Roussel, M.F. (1998) Myc signaling via the ARF tumor suppressor regulates TP53-dependent apoptosis and immortalization. *Genes Dev.* **12**: 2424–2433.

10

Colorectal cancer

P.D. Chapman and J. Burn

1. Introduction

Colorectal cancer (CRC) is the largest cancer killer in non-smokers in Westernized countries. There has been very little change in the death rate of the disease in the last 50 years, partly due to increasing incidence but mainly due to lack of substantial improvement in survival which is still only around 37%. This emphasizes the need for prevention strategies which can only develop when the fundamental processes of carcinogenesis are further understood. The largest step towards such an understanding was in 1987 when the *APC* gene responsible for familial adenomatous polyposis (FAP) was identified on chromosome 5. This was the first colorectal cancer gene to be described, and has led the way to huge research initiatives in the area of CRC genetics. CRC is more sensitive to environmental factors than any other malignancy, and dietary factors play a big part. Another characteristic of CRC is the presence of a premalignant lesion in the form of an adenomatous polyp. Taking all these factors together, primary prevention is likely to be possible, screening for adenomas should save lives, and the study and understanding of underlying genetic and biological processes is extremely important.

2. Family history and CRC risk

Susceptibility to CRC can be predicted on the basis of a family history of the disease, particularly when this involves early age of onset. Other factors of relevance are age and sex, where both increasing age and maleness are associated with increasing risk. Population studies consistently demonstrate a 2-fold increase in CRC in first degree relatives of an individual with CRC (Brown *et al.*, 1998a,b). The cancers are seen at an age comparable to the general population, and have a similar location and age of onset (Lynch and Lynch, 1998).

The risk of CRC in relatives of affected individuals has been shown, in a number of studies, to be related both to the age of onset in any close relative and the number of affected relatives. *Figure 1* shows the relationship between the risk of CRC by age and extent of family history. As a baseline, the risk of CRC in a 45 year old who has no family history is used. As this individual gets older, their risk increases simply on the basis of their own ageing process (at age 55 their risk of

Molecular Genetics of Cancer second edition, edited by J.K. Cowell.
© 2001 BIOS Scientific Publishers Ltd, Oxford.

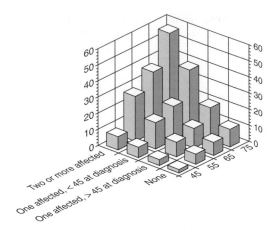

Figure 1. Risk of CRC by age and extent of family history, relative to the risk of a 45 year old with no family history (Burn *et al.*, 2000). The data for this figure are taken from a study of CRC in families in Melbourne, Australia conducted by Dr Jim St John (personal communication).

developing CRC in the next year is five times that of individuals at age 45; at age 65 the risk is seven times that of a 45 year old and at age 75 it is 11 times the risk at 45 years). The risk also increases with the number of affected relatives. Thus, a 45 year old with one affected relative diagnosed over 45 years of age has a risk of 1.8 times that of a 45 year old without a family history. The relative risk is increased to 3.7 when the relative diagnosed was under 45 years and 5.7 if there are two affected first degree relatives. The risk increases across all levels of family history and at all ages such that a 75 year old, for example, with 2 affected first degree relatives has over 50 times the risk of CRC in the next year compared with a 45 year old with no family history.

The causes of this familial risk are largely unknown, but presumably include a contribution from partially penetrant genes conferring susceptibility to colonic neoplasia, common environmental exposures (which are risk factors for CRC and which aggregate in families), and interactions between genetic and environmental factors (Kim, 1997). Colon cancer in these families is not linked to high penetrance genes (described below) implying that these genes are not the major cause of familial CRC. About three quarters of all CRCs are thought to result from somatic mutations. At the present time there is no certain way of identifying the 25% with a predisposition. Approximately 3% of CRC cases have a known, dominantly inherited predisposition, and another 5% of families appear to have highly penetrant predisposing genes which have not yet been identified.

3. Genetic factors predisposing to CRC

CRC provides an excellent system for the study of the genetic changes which occur during the development of a common human cancer. Most CRCs arise from benign adenomas which develop from aberrant colonic crypts, which means that the developing carcinoma can be observed, removed and studied at all stages from an aberrant crypt focus to metastatic carcinoma. Adenomas are usually considered monoclonal compared to the polyclonal composition of colorectal epithelium. There has, however, been one report of polyclonality in FAP in a case with a coincidental mosaic chromosome marker (Novelli *et al.*, 1996).

The study of the stochastic genetic events leading from early adenoma to CRC has led to the identification of three major classes of genes involved in colorectal carcinogenesis; oncogenes, tumor suppressor genes, and DNA mismatch repair (MMR) genes which identify and correct DNA replication errors. Mutations in MMR genes lead to genetic instability and, since defects in somatic cell DNA replication are not corrected, this results in mutations in other genes such as the type-II TGF Beta receptor (Yagi *et al.*, 1997).

CRC development is a multistep process, involving mutations in at least 7 genes, so many genotypes are likely to be involved in susceptibility to CRC. Most, but not all, of these genes exert a biological effect only when both alleles are mutated, i.e. they are recessive at the cellular level. However, some tumor suppressor gene defects, such as *APC*, can exert a phenotypic effect even in the heterozygous state. Many genes are known to be part of the multistep process and some have been shown to be important predictors of CRC when a germline mutation is present.

Although the genetic changes which give rise to CRC tumors often occur in a particular sequence, it is the accumulation of mutations together with the temporal sequence which determines the malignant potential of a tumor (*Figure 2*) (Arends,

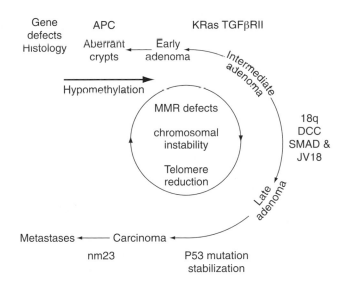

Figure 2. Multi-step interactions in CRC. There is a continuing debate about whether loss of a single *APC* allele can increase proliferation. Whereas mouse studies do not reveal significant changes in proliferation in *Apc* knockout mice, our own work in humans with FAP has shown a significant increase in the number of mitoses per crypt (Mills *et al.*, 2000). In contrast, Wasan and colleagues reported an increase in crypt fission rather than proliferation. DNA hypomethylation is probably not a pivotal event and the role of mismatch repair deficiency early in the process is also equivocal (Wasan *et al.*, 1998)

2000; Fearon and Vogelstein, 1990). In multipotent stem cells, the accumulation of mutations occurs in a stochastic, stepwise manner with each mutation providing a selective advantage for each cell generation, leading to an expanded population of daughter cells (Bodmer et al., 1994). APC mutations occur early (Powell et al., 1992) and are important for initiation. This explains the severe, young colorectal cancer phenotype of familial adenomatous polyposis. In contrast, ras oncogene mutations usually occur in larger adenomas, being present only in 10% of those smaller than 1 cm. It is thought that such mutations are responsible for the development of a small adenoma into a larger lesion. Similarly, allelic losses on chromosomes 5q, 17p and 18q and other chromosomes are frequently observed in malignant tumors, but rarely in adenomas. Such mutations, occurring later in the cascade of mutational events are unlikely to have major relevance in the search for cancer susceptibility as they are largely somatic mutational events, commencing in a single cell. Nevertheless, it is conceivable that variations which predispose to mutational change in any of these genes could influence the nature and rate of cancer development.

Microsatellite instability is seen in about 15% of all CRCs (Bodmer et al., 1994), and at a much higher rate in Lynch syndrome CRC. The latter association probably explains why it is commonly seen in the younger age group. Functional assays for the binding of the mismatch recognition genes can be used to test cells for a mismatch recognition defect (Aquilina et al., 1994). This phenotype is likely to be recessive, reflecting total loss of function of the binding protein that recognizes DNA mismatches, e.g. at CA and CT repeats. This explains the observation that normal cells from Lynch syndrome gene carriers do not exhibit an MSI phenotype, since the cells in a heterozygous individual have one normal, working copy of the mutated, inherited gene.

3.1 Mouse models

Mice bred to have defective copies of the major genes involved in CRC can be of great value in the evaluation of susceptibility factors (Kim et al., 1993). The Min mouse, Apc 1638N and Apc1638T and the Apc [delta 716] are the four mouse models of defective Apc function. The Apc 1638T mouse is interesting since it involves a mutation near the end of the coding sequence but leaving the critical catenin binding function intact. These mice do not develop significant intestinal tumorigenesis and homozygotes can survive to term.

While the phenotypes of the mouse models show important differences from the human with predominance of small gut tumors, the mouse models have, nevertheless, been of great value in studies of the biology of CRC and in investigations of chemopreventive agents. Mouse mutants with defective mismatch repair genes have been less valuable as the phenotype does not include gastrointestinal tumors (Heyer et al., 1999).

4. Inherited CRC predisposition

Although the molecular genetics of most CRCs is unclear at present, some cases can be designated as having a genetic predisposition on the basis of their family history, clinical findings, pathology findings or molecular genetic analysis. Clinical overlap between the syndromes sometimes makes diagnosis difficult, but this is being clarified as genetic and functional histopathological analysis becomes available.

The two major, dominantly inherited forms of CRC are FAP and Lynch syndrome. This nomenclature describes the phenotype of these two conditions at the histological level, but the range of mutations in causative genes is much more disparate. Almost all cases of FAP result from a pathological mutation in a single gene on chromosome 5, *APC*. However, Lynch syndrome results from loss of function of one of at least five separate genes, each one of which encodes part of a protein complex which is responsible for MMR during DNA replication.

4.1 Familial adenomatous polyposis

FAP is the most genetically determined of all inherited predisposing CRC syndromes. A mutation in *APC* results in multiple colorectal adenomatous polyps developing during the teens and early adulthood (*Figure 3*). In the absence of prophylactic colectomy, the large number of adenomas would lead to the almost certain development of CRC at a young age. Much has been learned about the *APC* gene since its localization and cloning in the early 1990s (Bodmer *et al.*, 1987; Nishisho *et al.*, 1991), and the relationship of germline mutations to individual phenotype (*Figure 5*) has been described in more detail than for any other inherited cancer predisposition.

About 80% of FAP families have a different, distinct mutation of *APC*, although almost all of the disease causing mutations so far found inactivate *APC*

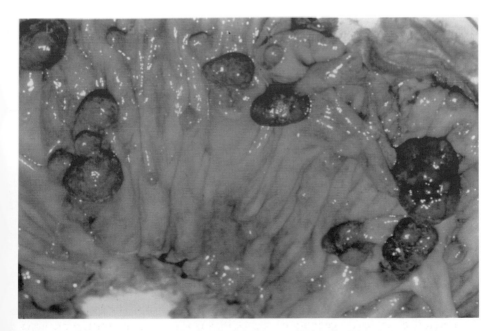

Figure 3. FAP colon showing adenomatous polyps at all stages of development.

and result in protein truncation. *APC* codes for a large 2843 amino acid protein which is involved in cell adhesion and cytoskeleton function, and forms a complex with axin and glycogen synthase kinase 3 beta (*GSK3 beta*) to bind beta catenin, released by the *Wnt* signaling pathway. Free beta catenin crosses into the nucleus to activate the *c myc* oncogene via *Tcf4* (*Figure 4*). Defects in *APC* or beta catenin mutations can lead to abnormal activation in *c myc* (Brown *et al.*, 1999).

There are 15 exons in *APC* of which exons 1 to 14 are short and exon 15 is long. A higher density of polyps occurs in families with a mutation near the center of the gene in exon 15, whereas a sparse pattern of flat adenomas, known as attenuated FAP, is associated with mutations at the extreme 5′ (proximal) end of the gene (*Figure 5*). It is likely that the most severe phenotype is a consequence of mutations which disrupt beta-catenin binding. Events leading to oncogenic activation of beta-catenin, which promotes tumor progression via interaction with a downstream target, can result from the inactivation of tumor suppressor activity of a mutated *APC* gene, from activation of *wnt* receptors, or from direct mutation of the beta catenin gene itself (Polakis, 1999). In the nucleus, beta catenin upregulates the oncogene *c.myc* (He *et al.*, 1998).

Recognition of families and individuals at risk of developing FAP relies on careful pedigree analysis which is aided by a multidisciplinary approach including genetics, surgery, gastroenterology and pathology. Such an approach has been taken in the Northern Region of England, and has been shown to be effective in reducing the burden of CRC in such families (Burn *et al.*, 1991). However, the new mutation rate for germline *APC* mutations is 20–30% and, unfortunately, new mutation cases appear to exhibit a more severe phenotype than familial cases, with mutations more common at codon 1309. It is possible,

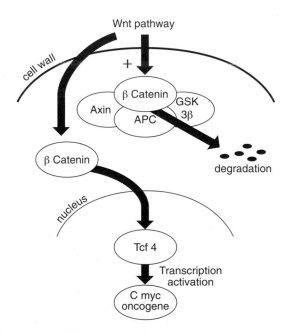

Figure 4. Mechanism of *APC* activation of the *myc* oncogene

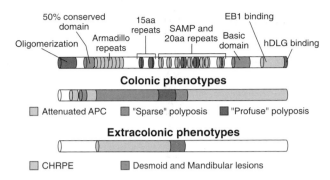

Figure 5. Structure of APC protein and genotype–phenotype correlation.

therefore, that a high new mutation rate, combined with improving survival rates into or beyond reproductive years, will lead to an increase in the incidence of FAP (Gayther *et al.*, 1994).

One explanation for the genotype–phenotype correlation in FAP is that some mutations, such as a truncating mutation at codon 1309, result in a dominant negative effect at the protein level. There is experimental evidence that wild-type *APC* activity is strongly inhibited by a mutant allele with this codon 1309, and this results in a severe phenotype. In contrast, a mutation associated with a mild phenotype (attenuated *APC*, or *AAPC*) produces a gene product which associates only weakly with the wild type product (Dihlmann *et al.*, 1999). In AAPC, CRC occurs at a later age, and extracolonic manifestations are less common (Lynch *et al.*, 1995). It is possible to attribute this mild phenotype to mutations in 3 distinct regions of the gene; at the 5′ end, within exon 9, and at the 3′ distal end of *APC*. When such a mutation is known in a family seeking genetic counseling, it is possible to modify risks according to the known genotype–phenotype relationship, and bowel examination may be less frequent than in a family with a mutation such as *APC*[1309].

Variation in the FAP phenotype even in those with identical mutations presents difficulties in counseling, both within and between families. However, in all cases of FAP, CRC susceptibility remains high, and regular screening with prophylactic surgery are essential components of care in such families (see Genetic modifiers below).

4.2 Lynch syndrome

Lynch syndrome (or hereditary nonpolyposis colon cancer, HNPCC) is a highly penetrant genetic susceptibility gene for CRC, which was referred to in the past as Cancer Family syndrome. The disease was traditionally recognized by the familial clustering of CRC in persons without obvious polyposis, but is now known to be associated with an MMR gene defect. Patients are at risk of many extracolonic malignancies, including cancer of the endometrium, stomach, small bowel, hepatobiliary and urinary tract (Aarnio *et al.*, 1995; Marra and Boland, 1995; Vasen *et al.*, 1995; Watson and Lynch, 1993). For example, gene-carrying women have a 42% risk (to age 70) of developing endometrial cancer, which exceeds CRC risk in some age groups (Dunlop *et al.*, 1997). An individual's family history which includes any of this spectrum of cancers, not just colorectal, could be considered to be at risk of an underlying genetic susceptibility. Lynch syndrome is most often suspected on the basis of family history. The modified Amsterdam criteria (*Table 1*) make use of the pattern of disease in families without access to direct mutation searching (Vasen *et al.*, 1994). 'Amsterdam Criteria' are useful in research to ensure inclusion of high-risk individuals, whereas in clinical use they will only target a proportion of high-risk people. An indirect application of the Amsterdam criteria is their use to select the families most likely to yield positive results when searching for mismatch repair gene defects (Wijnen *et al.*, 1998).

Where there is a family history of CRC, early age of onset in an affected relative and two or more affected generations are the most predictive factors for risk for that individual (Gaglia *et al.*, 1995). However, other independent risk factors discussed earlier, such as age and male sex, greatly modify an individual's risk. For example, Guillem and colleagues found that, at screening colonoscopy, those at greatest risk for harboring an asymptomatic adenoma were males over the age of 50 years with at least one first degree relative with CRC (Guillem *et al.*, 1992).

Clinically, the colorectal neoplastic process in HNPCC appears to follow an adenoma-to-carcinoma progression similar to that described in FAP or other CRC settings, although several aspects of the clinical manifestations, as well as the molecular pathophysiologies underlying them, may be distinctive (Kinzler and Vogelstein, 1996).

Classical Lynch syndrome is caused by inherited mutations in one of the *Mut*-related family of MMR genes, including *MLH1*, and *MSH2*, and *MSH6* (Fishel *et al.*, 1993; Kolodner *et al.*, 1994, 1995; Nicolaides *et al.*, 1994). Three other genes involved in the MMR complex, *PMS1*, *PMS2*, and *MLH3* are rarely or never associated with a mendelian phenotype. This has been attributed to redundancy between them, i.e. some degree of interchangability (Lipkin *et al.*, 2000) (see below). These genes encode protein products that are responsible for recognizing and correcting errors that arise when DNA is replicated (Dunlop *et al.*, 1997; Leach *et al.*, 1993). An early manifestation of this defect *in vivo* is the appearance of microsatellite instability (MSI). A second mutation is required in colorectal cells to inactivate the MMR function. MSI contributes to the progressive accumulation of secondary mutations throughout the genome and thereby affects crucial growth-regulatory genes which ultimately lead to cancer.

Table 1. Revised definition of Lynch syndrome (modified Amsterdam criteria)

There should be at least three relatives with a Lynch syndrome related cancer (CRC, cancer of endometrium, small bowel, ureter or renal pelvis)

- One should be the first degree relative of the other two.
- At least two successive affected generations.
- At least one CRC should be diagnosed <50 years.
- FAP should be excluded.
- Tumors should be verified by pathological examination.

To date, more than 100 different germ-line mutations have been identified in the MMR genes known to be associated with Lynch syndrome. Mutations in *MSH2* and *MLH1* account for roughly equal proportions of Lynch kindreds, and are together responsible for a majority of CRC in these families (Aaltonen *et al.*, 1998). However, germ-line disease-associated mutations are found in only approximately 40–70% of probable Lynch syndrome families. Other germ-line mutations, and/or different classes of genes, may be discovered which play an etiologic role in susceptible individuals. One family has been described in which a probable pathological mutation has occurred in the TGF-beta Receptor II gene (Yagi *et al.*, 1997), a gene known to show altered expression in CRC. This, and the attenuated form of FAP, might be regarded as falling into the broader category of HNPCC (hereditary nonpolyposis colon cancer) but not Lynch syndrome.

One of the recently identified MMR genes, *MLH3*, associates with *MLH1*, *MSH2* and *MSH3* to form a complex involved with repair of insertion–deletion loops of single stranded DNA. Its role in human cancer predisposition is uncertain but it is thought to show functional redundancy with *PMS1* and *PMS2*. This would explain why *PMS1* and *PMS2* mutations are only rarely found in Lynch syndrome families (Lipkin *et al.*, 2000). *MSH6* is involved in repair of single base changes (*Figure 6*). When *MSH6* is defective, errors occur predominantly in mononucleotide repeats (e.g. AAAAAA). Tumors with MLH3 mutations display unstable dinucleotide repeats. This selectivity can explain the 'Microsatellite Instability Low' (MSI(L)) molecular phenotype in such tumors, as MSI status is usually defined on the basis of a mixture of mono- and dinucleotide repeats.

4.3 Rare syndromes

There are several rare genetic syndromes which confer a higher relative CRC risk than expected. The underlying genetic etiology has not yet been defined in all predisposing syndromes, but those most understood are described below. Juvenile polyposis coli is a rare autosomal dominant syndrome which is characterized by multiple polyps in the colon, and occasionally elsewhere in the GI tract. Juvenile polyps arise from the lamina propria, rather than the epithelium. There is a high risk of CRC, probably due to the development of foci of adenomatous change which progresses to dysplasia and adenocarcinoma and the predisposing factor is a mutation in one of two genes; *PTEN* on chromosome 10 (Jacoby *et al.*, 1997), or

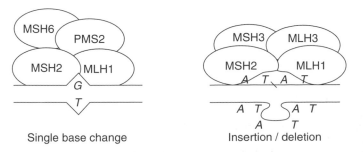

Figure 6. Mismatch repair protein complexes target single and multiple base mutations.

SMAD4 on chromosome 18. It is likely that removal of juvenile polyps will be seen as preventative.

There have been reports of families with atypical juvenile polyps, adenomas and colorectal cancers as well as inflammatory and metaplastic polyps. In one such family, cases of mixed polyps follows an autosomal dominant inheritance pattern, and the putative gene has been localized to the long arm of chromosome 6 (Thomas *et al.*, 1996).

CRC is occasionally seen in Gorlins syndrome (basal cell nevus syndrome), where the polyps are hamartomatous (Schwartz, 1978). There is an association with broad facies, basal cell nevi, ectopic calcification of the falx, and bony abnormalities of sites including the ribs, mandible and maxilla. Other malignancies seen in this condition include medulloblastoma and malignant nevi (Murday and Slack, 1989). Premature termination of the PATCHED protein resulting from germ-line mutations in one of two tumor suppressor genes, *Patched 1* and *2* are thought to be responsible for this syndrome, but no genotype–phenotype correlations have yet been described (Wicking *et al.*, 1997).

In the somewhat rare Peutz-Jeghers syndrome, the predisposing polyps are pathologically discrete from other polyps and are called Peutz-Jeghers polyps. They exhibit some adenomatous features and have an increased malignant potential, appearing in the stomach, small bowel and colon. The distinguishing feature of this syndrome is the mucocutaneous melanin pigment seen around and inside the mouth, and between the fingers. At least some cases of this disorder are due to mutations in the serine/threonine kinase (*STK11*) tumor suppressor gene on chromosome 19 (Jenne *et al.*, 1998).

In Turcot's syndrome, multiple colorectal adenomas are associated with central nervous system tumors, particularly of the brain. This phenotype has been seen with variants of both Lynch syndrome and FAP genotypes. With an *APC* variant the brain tumors are cerebellar and medulloblastomas whereas with *MLH1* or *MSH2* variants, glioblastoma multiforme are more frequently seen.

5. Modifiers and genes of minor effect: allelic variants

There is a strong likelihood that a variety of molecular pathology might result in allelic variants at any of the 'major gene' loci which increase the risk of CRC. The challenge will be to choose candidate genes for detailed sequencing. In future, high-throughput technology will make selection less critical but, in the current era, it is easy to expend large sums of money and achieve little. For example, a splice site mutation in *MSH2* gene has been found in normal individuals with no history of Lynch syndrome. It is therefore possible that some functional effect exists which is associated with increased risk of CRC but which is not sufficiently high to show up as a positive family history.

5.1 Single nucleotide polymorphisms

One method of targeting susceptible individuals is to use single nucleotide polymorphisms (SNPs) to identify haplotypes. *Table 2* shows a series of 5 SNPs at the

Table 2. Single nucleotide polymorphisms (SNPs) at the MSH2 locus.
(SNP data from ICG-HNPCC database@www.nfdht.nl)

Haploytpe	Number	Haplotype frequency
Cgagt	**61**	**0.616**
Ggagt	**14**	**0.141**
Ggaat	**10**	**0.101**
Ggtat	7	0.071
Cgagc	3	0.030
Ggtgt	0	0.000
Ggagc	0	0.000
Cgaat	2	0.020
Ggaac	2	0.020
Total number of chromosomes typed	99	

MSH2 locus. Of the 32 possible haplotypes in our control population, only nine were identified among 99 chromosomes characterized and three (bold in table) accounted for 85 of the 99. In other words, common ancient versions of the gene will occur in different populations. It will now be possible to examine these SNP patterns in people with colon cancer to see if the distribution of haplotypes is different from that in the local population. If, for example, there had been a mutation several hundred years ago in an *MSH2* gene residing on a 'ggtat' chromosome, this would be reflected in an over-representation of that haplotype in the disease population and would focus sequencing studies on *MSH2* in affected people with the 'ggtat' haplotype.

5.2 Modifying genes

In FAP, allelic heterogeneity does not appear to account for all of the observed variation in phenotype and other genes are probably involved in the phenotypic expression of this, the most 'monogenic' of all cancer susceptibility genes. This is an area of ongoing research but there is evidence for a modifier gene on chromosome 1p35–36 which maps to an equivalent locus in the mouse for a known modifying gene called *Mom-1* (Dobbie *et al.*, 1997). The candidate gene for this is type 2 non-pancreatic *PLA2*, a phospholipase gene, but no mutations in the human homolog have been found to date (Spirio *et al.*, 1996). Identification of modifying genes is a powerful tool in the understanding of a gene's function, and the advent of dense genetic linkage maps has made the dissection of polygenic traits, such as CRC susceptibility, more practical.

5.3 The APC I1307K polymorphism

A polymorphism has been identified in *APC* at codon 1307, which is relatively common in the Ashkenazi Jewish population, and increases adenoma formation in this group (Gryfe *et al.*, 1999). This mutation causes a change of isoleucine to lysine (shown as *I1307K*). The underlying mutation changes a thymine to an adenine resulting in a sequence of 8 adenines which is more hypermutable than the wild-type. This leads to mutational susceptibility in somatic colon cells which in turn confers a higher risk of neoplasm. Because the gene variant is dominantly inherited, but the penetrance for the phenotype is low, this form of predisposition could be described as familial rather than dominantly inherited. The lifetime risk of CRC for an individual with the polymorphism is about 10% but, because it is common (6% in New York Ashkenazim and 28% in those with a family history), it is thought to underlie 3–4% of CRC in this population (Laken et al., 1997).

This finding has raised the issue of genetic predictive testing for the APC I1307K polymorphism, which could be targeted towards those of Jewish Ashkenazi descent, to identify those with an increased susceptibility as a prelude to prevention programs. It is thought, for example, that 360 000 polymorphism carriers live in the USA, but the issue of testing is contentious, as a positive test could carry with it unfavorable psychological effects, insurance difficulties and potential sociological problems attendant in selecting out a population on the basis of ethnic descent. Moreover, the predictive value of such a marker in isolation is very small.

5.4 Other genes

A relationship between mucin-producing genes such as MUC2 and the pathogenesis of CRC has been suggested (Sternberg et al., 1999). MUC2 predominates within colorectal goblet cell mucin, and is expressed in adenomas and mucinous carcinomas. Down-regulation of MUC2 is seen in non-mucinous adenocarcinomas arising from adenomas, and cancers that develop de novo do not express MUC2. When more is known about CRC mucin, there will be opportunities to study different cell lineages to further explore the pathogenesis and other susceptibility factors in CRC.

The CDX1 and 2 homeotic genes have the characteristics of transcriptional regulatory proteins and are down-regulated in about 85% of CRCs. In cdx2+/− mice, multiple intestinal polyps are seen in the proximal colon (Chawengsaksophak et al., 1997). However, these polyps are not typical adenomas, and have similar histological characteristics to those seen in the Min mouse. These polyps occasionally contain true metaplasia and occasional large pedunculated tubulovillous colonic adenomas are seen. The fact that polyps are seen mostly in the proximal colon suggests that lowering levels of CDX2 would induce exaggerated cell growth leading to tumor formation, and expression would stimulate cell differentiation and growth arrest (Yagi et al., 1999). However, because the tumors display an unusual histological pattern, and human mutations have not been identified, the role of this gene in human carcinogenesis remains to be elucidated.

5.5 Environmentally sensitive polymorphisms

Environmentally sensitive genetic polymorphisms are some of the most recently described CRC-predisposing factors. A functional polymorphism in the methylenetetrahydrofolate reductase gene (MTHFR) appears to confer a 50% reduction in CRC risk in the US population. This gene plays a key role in providing folate, necessary for DNA replication and methylation, but this polymorphism ([667]C-T, ala-val), found in 10–15% of the study population, only provided protection when adequate folate was present in the diet (Ma *et al.*, 1997). Prior to the report that C677T was protective against CRC in homozygotes it might have been expected that the reverse would be true. The observation could indicate that malignancies are more susceptible to disturbance of the folate pathway and that the homozygotes for the thermolabile variant are relatively protected by the less efficient folate pathway. It is of potential importance that protection against CRC in those homozygous for the thermolabile variant was observed only in study subjects reporting zero to modest alcohol intakes. With high alcohol intakes no protection was evident, indicating an important diet–gene interaction (Ma *et al.*, 1997).

The importance of gene–environment interactions in assessing cancer suscepti-bility was illustrated by the beta carotene trials. Intervention studies designed to test the hypothesis that supplementation with beta carotene would reduce cancer risk showed no protection. However, the trials showed increased risk of lung cancer in smokers given beta carotene. This effect is thought to be due to increase in cell proliferation and squamous metaplasia in the lung, effects that were enhanced by tobacco smoke, and is associated with suppression of retinoic acid receptor beta-gene expression and over expression of *JUN* and *FOS* genes (Wang *et al.*, 1999).

Exposure to environmental carcinogens, such as aromatic amines found in well cooked or preserved meat and cigarette smoking, are associated with an increased risk of CRC. Their metabolism is complex but central to most activation or detoxifi-cation of amylines and heterocyclic amines. Acetylation of heterocyclic amines by N acetyl transferases (NAT) is likely to be of major importance. Acetylation of hetero-cyclic amines by the *NAT1* and *NAT2* gene products can lead to the formation of reactive carcinogenic intermediates or to detoxification. This means that the associ-ation of cancer risk and enzyme activity could go either way. Acetylation activity varies as a result of the sequence polymorphism in the *NAT2* gene, and if two non-functional alleles are inherited ('slow acetylator' alleles) there is no NAT2 activity. Several studies show that NAT2 'rapid acetylation' phenotypes are associated with an increased risk of CRC. An increased CRC risk of 1.9 results from a variation in NAT1, which is again due to a rapid acetylator genotype (Hein *et al.*, 2000). There is an association between fast acetylator status and cancer in those with high intakes of cooked meat (a source of heterocyclic amines). Conversely, slow acetylators who smoke and drink are at increased risk. These genetic polymorphisms provide a good example of how complex the interaction between genotype and environmental factors can be. Again, it must be remembered that 'multiple slices' of a data set might point to apparent interactions which are random events, resulting in claims and counter-claims on their predictive significance.

6. Personal history and risk

Previous colorectal adenomas or cancer, without regular follow up, indicate an overall increase in susceptibility to CRC. Such clinical observations can be further refined by pathology and genetic analysis of the neoplasm, and also by the age and follow up history of the individual. Because there is variability in the normal adenoma–carcinoma sequence, some types of adenomas confer a greater risk than others. Sessile villous lesions behave differently to pedunculated adenomas, and flat adenomas are associated with an increased potential for malignant change (Jass, 1995). Microadenomata, the pathologically detectable precursors for adenomas, are very common, but only a fraction of these will ever progress to malignancy, depending on the genetic status of the individual. For example, the risk is very low for an individual microadenoma in FAP but high in Lynch syndrome. Similarly with adenomas; in FAP the risk of an individual adenoma becoming malignant is very small, whereas the risk is higher in Lynch syndrome.

Together with size and the number of polyps, the pathological examination of a colorectal polyp is crucial in defining recurrence risk (Winawer et al., 1993a). Adenomatous polyps, whether they are tubullovillous, regular or flat in appearance, are the most likely to become malignant whereas metaplastic polyps carry little potential for malignant change. Individuals who have had an adenomatous polyp and followed up regularly, are at low risk of developing cancer (Winawer et al., 1993b). This observation implies that the original adenoma carries the potential for development to a CRC, and once removed, there is little risk of recurrence at that site. Microsatellite instability (MSI) in a colorectal tumor predicts a high risk of recurrence for an individual, regardless of family history (Brown et al., 1999a,b).

Inflammatory bowel disease, and in particular, ulcerative colitis (UC), carries an increased risk for CRC. Those at highest risk have UC throughout the colon, rather than localized disease, are over 40 years of age, and have had UC for more than 10 years. Interestingly this does not depend on continuous manifestation of the disease; those who have a short episode are at the same risk of CRC 10 years later as those who have 10 years without remission. Patients with UC have a small but increasing risk which is approximately equivalent to a 0.5–1% chance of CRC per year of follow up (Ekbom, 1998). For this reason, many gastroenterology units follow up patients for many years, despite lack of definitive evidence that this prevents CRC. However, as in most CRC screening protocols, this intervention has been shown to prevent death from CRC by early detection of small neoplasia. Recent case-control studies show that those with UC who are at the highest risk of CRC are those with a family history of CRC, whereas those with no family history have a risk which is not significantly different from the average person without colitis (Nuako et al., 1998). This finding implies that any inherited genetic risk is not associated with both the development of colitis and colon cancer risk, but that they are separate, discrete risk factors, and that inflammation is not a sufficient risk factor in isolation to have a clinically significant impact in most people.

7. Prevention

There is evidence from many sources indicating that CRC incidence, and also the course of the disease in individuals, can be altered. In different world populations, there are great differences of CRC incidence and this can partly be attributable to dietary differences. The graphs in *Figure 7* show (a) a negative 0.25 correlation coefficient between fibre intake in different countries and prevalence on CRC, and (b) a much closer correlation with starch intake (–0.79) (Cassidy et al., 1994).

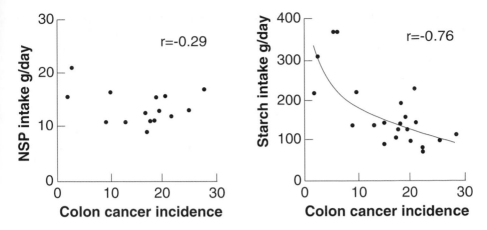

Figure 7. Non-starch polysaccharide intake compared with dietary starch and CRC incidence in different countries.

Nutrition factors are notoriously difficult to investigate due to the complexity of eating behavior and variety in the contents of individual food items. However, nutrition is likely to be a very large factor in CRC so methodology development and collaborative projects are urgently needed so that evidence-based dietary prevention strategies can be developed.

Both observational and interventional studies have shown that regular aspirin intake reduces CRC death rates, and also that NSAIDS reduce polyp size and number in FAP. More recently it has been shown that the immune system can identify mutant proteins by their characteristic tails, which raises the possibility of developing them to prevent cancer in high-risk individuals. These examples highlight the potential to influence CRC prevalence in populations through drug prophylaxis and this is likely to be a growing area of research in coming years.

There have been reports that the NSAID Sulindac reduces the polyp load in FAP, and more recently Celecoxib, a COX2 inhibitor has shown a similar effect (Giardiello et al., 1993; Steinbach et al., 2000). Research into other risk factors and possible interventions can be undertaken in subjects at high risk as opposed to population cohorts. This means that fewer numbers are required as interim or end-point results such as adenomas are likely to be more frequent. Two such trials are underway using patients with FAP and Lynch syndrome gene mutations, in which aspirin and starch which is resistant to digestion are being tested in randomized controlled trials (Burn et al., 1995, 1998). How much relevance the results of these trials will have to the rest of the population remains to be established.

Although primary prevention will always be the preferred strategy, the early detection of colorectal neoplasia will always be important as the adenoma–carcinoma sequence is likely to span many years. Occult blood testing is cheap but fails to detect many lesions whilst identifying many false positive results. Colonoscopy is thorough and has the advantage of allowing biopsy but it is an invasive procedure with some risk of perforation, and sigmoidoscopy excludes examination of the proximal colon. There is therefore no ideal screening method available at present but stool-based assays are being developed, one of which examines stool DNA for microsatellite instability (Laken *et al.*, 1997).

8. Social considerations

FAP has been identifiable in families for over a hundred years because of the presence of multiple polyps, and more recently, by genetic testing for mutations or specific patterns in and around the *APC* gene. In the Northern Region of England, using a registry approach, and by applying clinical and genetic criteria, almost all individuals at high risk have been recruited to screening programes (Burn *et al.*, 1991). One of the biggest advantages of a proactive approach towards offering genetic testing is the relief of uncertainty and the reduction of unnecessary colonoscopic surveillance in those not at high genetic risk. At the level of population healthcare there are benefits in being able to target screening effectively to those at greatest risk.

There are many differences between making a genetic diagnosis of hereditary CRC and a clinical diagnosis of CRC. Firstly, hereditary cancer involves probability statements describing inheritance and penetrance. Because little is known about the role of environmental influences in these families, this contributes to the high levels of uncertainty accompanying predictive testing for late onset disorders. Secondly, there are psychosocial implications for the individual, and the family. How people respond to the information will depend on many variables such as their attribution style, defence mechanisms, and understanding of the disease and its consequences. The effect of a genetic diagnosis can pass through a family like ripples on a pond, and extended families will become more aware of, and discuss, cancer diagnoses. Family beliefs about inheritance may be very different from accepted patterns of Mendelian inheritance. For example, they may believe that only males are affected in Lynch syndrome.

It is easy to assume that those found not to carry a predisposing gene will be unconditionally pleased. However, some individuals react in a negative way to such information. This may be due to life plans built around assumptions that they will become ill having to be altered, or due to guilt about escaping the family disease. It would also be logical to assume that those found to be gene carriers would present themselves for bowel examination. This does not always happen and is a source of frustration to surgical and endoscopy staff when appointments are missed repeatedly. The reasons for this reaction by patients are complex, and deep underlying reasons may include fear of surgery, fear of the examination, or a hedonistic personality which does not easily accept hospitalization and potential illness. These reasons can differ greatly from a non-familial case where the patient is attending with symptoms (Rossi and Srivastava, 1996).

There are many ethical issues surrounding hereditary colon cancer, whether it is a clinical or a genetic diagnosis. Firstly, equitable access to services is a guiding principle for Health Services in the UK, but this is difficult when only some regions have active registers and recall systems with individual counseling services. Bearing in mind the widespread history of eugenics during the early part of the 20th century, there must be caution in pursuing equity by using forms of coercion such as laws and social pressures. Where efforts are made to make contact with at-risk family members in an attempt to offer equitable services, there is a risk of straying over the invisible line between voluntary and enforced testing. Choices for at-risk families should be realistic, and not just be a lack of coercion. For example, whilst DNA testing is undertaken the person should be able to withdraw at any time, or choose not to take any action if the result is positive. Other issues, such as those concerning confidentiality and privacy, must be discussed with family members, so together with the genetic and clinical information and exploration of psychosocial issues, much careful counseling is required. The common thread to all of this discussion is the protection of individual rights, and a multi-professional approach with specialist genetic counseling available for all is the best way to offer such a service.

9. Conclusions

All CRC involves multiple somatic genetic changes. Based on family history data, it is likely that in at least half, and probably three quarters, of CRC in developed countries, this is random acquisition of mutations due to environmental influences. If and when diet improves and people take more exercise, the importance of germline defects will grow. Even now, there are very large numbers of people at significantly elevated risk of cancer who will be identifiable using molecular genetic tests for pathological mutations in *APC* and the mismatch repair genes. It is likely that there will be utility in searching for less penetrant defective alleles in these genes as this might influence treatments and screening strategies. Perhaps more important, knowledge of a specific personal risk factor is liable to stimulate a greater interest in chemoprevention and lifestyle factors which could lead to reduced risk.

On the negative side, this possibility also raises psychosocial issues including making well people 'sick' by describing them as susceptible to cancer and by identifying high-risk ethnic groups such as Ashkenazi Jews. Ethical issues relating to providing equitable screening opportunities across populations, and financial concerns when insurance risks can be stratified by susceptibility factors or medical liaiblity of the clinician become more complex. In the USA, physicians are being pressurized into assuming the additional responsibility of establishing a family history of cancer across several generations. This in turn creates a duty for a health care worker to provide counseling to extended families, and several lawsuits have been instituted claiming negligence when a family history has not been given adequate consideration or has not been communicated to at-risk family members (Nelson, 1996; Severin, 1999). In addition to legal pressures to include genealogy in CRC care, there are economic factors associated with identification and monitoring of susceptible individuals (Bolin, 1996; Brown and Kessler, 1996; Smith and DuBois, 1997).

Where there are advantages to a community or society in providing a health care intervention, there will always be a risk of eugenic policies creeping into practice.

Establishing genetic susceptibility for CRC will soon become a reality, and the advantages to a member of an HNPCC or FAP family member in finding out they are not a gene carrier can be enormous. However, the situation is more complex where the predictability of the genotype is less certain, and this is likely to be the situation for most genetic susceptibility testing (Lynch *et al.*, 1999). The needs and views of the individual must always take precedence over societal needs if maximum uptake of screening alongside freedom of choice is to be assured. Even more important is the recognition that few risk factors will be sufficiently predictive to justify extension from the realm of primary research to the clinic. In many cases, groups of individuals chosen for their susceptible genotype will be used to test which environmental changes might benefit the whole community.

Acknowledgments

PC is supported by Imperial Cancer Research Fund Clinical Cancer Genetics Network and the European Union Biomed Programme.

References

Aaltonen, L.A., Salovaara, R., Kristo, P. *et al.* (1998) Incidence of hereditary nonpolyposis colorectal cancer and the feasibility of molecular screening for the disease. *N. Engl. J. Med.* **338**: 1481–1487.

Aarnio, M., Mecklin, J.P., Aaltonen, L.A., Nystrom-Lahti, M. and Jarvinen, H.J. (1995) Life-time risk of different cancers in hereditary non-polyposis colorectal cancer (HNPCC) syndrome. *Int. J. Cancer* **64**: 430–433.

Aquilina, G., Hess, P., Branch, P., MacGeoch, C., Casciano, I., Karran, P. and Bignami, M. (1994) A mismatch recognition defect in colon carcinoma confers DNA microsatellite instability and a mutator phenotype. *Proc. Natl Acad. Sci. USA* **91**: 8905–8909.

Arends, J.W. (2000) Molecular interactions in the Vogelstein model of colorectal carcinoma. *J. Pathol.* **190**: 412–416.

Bodmer, W.F., Bailey, C.J., Bodmer, J. *et al.* (1987) Localization of the gene for familial adenomatous polyposis on chromosome 5. *Nature* **328**: 614–616.

Bodmer, W., Bishop, T. and Karran, P. (1994) Genetic steps in colorectal cancer. *Nat. Genet.* **6**: 217–219.

Bolin, T.D. (1996) Cost benefit of early diagnosis of colorectal cancer. *Scand. J. Gastroenterol. Suppl* **220**: 142–146.

Brown, M.L. and Kessler, L.G. (1996) Use of gene tests to detect hereditary predisposition to cancer: what do we know about cost effectiveness? *Int. J. Cancer* **69**: 55–57.

Brown, S.R., Finan, P.J. and Bishop, D.T. (1998a) Are relatives of patients with multiple HNPCC spectrum tumours at increased risk of cancer? *Gut* **43**: 664–668.

Brown, S.R., Finan, P.J., Cawkwell, L., Quirke, P. and Bishop, D.T. (1998b) Frequency of replication errors in colorectal cancer and their association with family history. *Gut* **43**: 553–557.

Brown, S.R., Finan, P.J., Cawkwell, L., Quirke, P. and Bishop, D.T. (1999) Frequency of replication errors in colorectal cancer and their association with family history. *Gut* **43**: 553–557.

Burn, J., Chapman, P., Delhanty, J. *et al.* (1991) The UK Northern region genetic register for familial adenomatous polyposis coli: use of age of onset, congenital hypertrophy of the retinal pigment epithelium, and DNA markers in risk calculations. *J. Med. Genet.* **28**: 289–296.

Burn, J., Chapman, P.D., Mathers, J., Bertario, L., Bishop, D.T., Bulow, S., Cummings, J., Phillips, R. and Vasen, H. (1995) The protocol for a European double-blind trial of aspirin and resistant starch in familial adenomatous polyposis: the CAPP study. Concerted Action Polyposis Prevention. *Eur. J. Cancer* **31A**: 1385–1386.

Burn, J., Chapman, P.D., Bishop, D.T. and Mathers, J. (1998) Diet and cancer prevention: the concerted action polyp prevention (CAPP) studies. *Proc. Nutr. Soc.* **57**: 183–186.

Burn, J., Chapman, P.D., Smalley, S., Mickleburgh, I., West, S. and Mathers, J.C. (2000)

Susceptibility markers in colorectal cancer. In: *Biomarkers in Cancer Prevention*. IARC Scientific Press (in press).

Cassidy, A., Bingham, S.A. and Cummings, J.H. (1994) Starch intake and colorectal cancer risk: an international comparison. *Br. J. Cancer* **69**: 937–942.

Chawengsaksophak, K., James, R., Hammond, V.E., Kontgen, F. and Beck, F. (1997) Homeosis and intestinal tumours in Cdx2 mutant mice. *Nature* **386**: 84–87.

Dihlmann, S., Gebert, J., Siermann, A., Herfarth, C. and von Knebel, D.M. (1999) Dominant negative effect of the APC1309 mutation: a possible explanation for genotype–phenotype correlations in familial adenomatous polyposis. *Cancer Res.* **59**: 1857–1860.

Dobbie, Z., Heinimann, K., Bishop, D.T., Muller, H. and Scott, R.J. (1997) Identification of a modifier gene locus on chromosome 1p35–36 in familial adenomatous polyposis. *Hum. Genet.* **99**: 653–657.

Dunlop, M.G., Farrington, S.M., Carothers, A.D., Wyllie, A.H., Sharp, L., Burn, J., Liu, B., Kinzler, K.W. and Vogelstein, B. (1997) Cancer risk associated with germline DNA mismatch repair gene mutations. *Hum. Mol. Genet.* **6**: 105–110.

Ekbom, A. (1998) Risk factors and distinguishing features of cancer in IBD. *Inflamm. Bowel. Dis.* **4**: 235–243.

Fearon, E.R. and Vogelstein, B. (1990) A genetic model for colorectal tumorigenesis. *Cell* **61**: 759–767.

Fishel, R., Lescoe, M.K., Rao, M.R., Copeland, N.G., Jenkins, N.A., Garber, J., Kane, M. and Kolodner, R. (1993) The human mutator gene homolog MSH2 and its association with hereditary nonpolyposis colon cancer [published erratum appears in *Cell* 1994 Apr 8;77(1): 167]. *Cell* **75**: 1027–1038.

Gaglia, P., Atkin, W.S., Whitelaw, S., Talbot, I.C., Williams, C.B., Northover, J.M. and Hodgson, S.V. (1995) Variables associated with the risk of colorectal adenomas in asymptomatic patients with a family history of colorectal cancer. *Gut* **36**: 385–390.

Gayther, S.A., Wells, D., SenGupta, S.B., Chapman, P., Neale, K., Tsioupra, K. and Delhanty, J.D. (1994) Regionally clustered APC mutations are associated with a severe phenotype and occur at a high frequency in new mutation cases of adenomatous polyposis coli. *Hum. Mol. Genet.* **3**: 53–56.

Giardiello, F.M., Hamilton, S.R., Krush, A.J., Piantadosi, S., Hylind, L.M., Celano, P., Booker, S.V., Robinson, C.R. and Offerhaus, G.J. (1993) Treatment of colonic and rectal adenomas with sulindac in familial adenomatous polyposis. *N. Engl. J. Med.* **328**: 1313–1316.

Gryfe, R., Di Nicola, N., Lal, G., Gallinger, S. and Redston, M. (1999) Inherited colorectal polyposis and cancer risk of the APC I1307K polymorphism. *Am. J. Hum. Genet.* **64**: 378–384.

Guillem, J.G., Forde, K.A., Treat, M.R., Neugut, A.I., O'Toole, K.M. and Diamond, B.E. (1992) Colonoscopic screening for neoplasms in asymptomatic first-degree relatives of colon cancer patients. A controlled, prospective study. *Dis. Colon Rectum* **35**: 523–529.

He, T.C., Sparks, A.B., Rago, C., Hermeking, H., Zawel, L., da Costa, L.T., Morin, P.J., Vogelstein, B. and Kinzler, K.W. (1998) Identification of c-MYC as a target of the APC pathway . *Science* **281**: 1509–1512.

Hein, D.W., Doll, M.A., Fretland, A.J., Leff, M.A., Webb, S.J., Xiao, G.H., Devanaboyina, U.S., Nangju, N.A. and Feng, Y. (2000) Molecular genetics and epidemiology of the NAT1 and NAT2 acetylation polymorphisms. *Cancer Epidemiol. Biomarkers Prev.* **9**: 29–42.

Heyer, J., Yang, K., Lipkin, M., Edelmann, W., and Kucherlapati, R. (1999) Mouse models for colorectal cancer. *Oncogene* **18**: 5325–5333.

Jacoby, R.F., Schlack, S., Cole, C.E., Skarbek, M., Harris, C. and Meisner, L.F. (1997) A juvenile polyposis tumor suppressor locus at 10q22 is deleted from nonepithelial cells in the lamina propria. *Gastroenterology* **112**: 1398–1403.

Jass, J.R. (1995) Colorectal adenoma progression and genetic change: is there a link? *Ann. Med.* **27**: 301–306.

Jenne, D.E., Reimann, H., Nezu, J., Friedel, W., Loff, S., Jeschke, R., Muller, O., Back, W. and Zimmer, M. (1998) Peutz-Jeghers syndrome is caused by mutations in a novel serine threonine kinase. *Nat. Genet.* **18**: 38–43.

Kim, S.H., Roth, K.A., Moser, A.R. and Gordon, J.I. (1993) Transgenic mouse models that explore the multistep hypothesis of intestinal neoplasia. *J. Cell Biol.* **123**: 877–893.

Kim, Y.S. (1997) Molecular genetics of colorectal cancer. *Digestion* **58 Suppl 1**: 65–68.

Kinzler, K.W. and Vogelstein, B. (1996) Lessons from hereditary colorectal cancer. *Cell* **87**: 159–170.

Kolodner, R.D., Hall, N.R., Lipford, J. *et al.* (1994) Structure of the human MSH2 locus and analysis of two Muir-Torre kindreds for msh2 mutations [published erratum appears in *Genomics* 1995 Aug 10;28(3): 613]. *Genomics* **24**: 516–526.

Kolodner, R.D., Hall, N.R., Lipford, J. *et al.* (1995) Structure of the human MLH1 locus and analysis of a large hereditary nonpolyposis colorectal carcinoma kindred for mlh1 mutations. *Cancer Res.* 55: 242–248.

Laken, S.J., Petersen, G.M., Gruber, S.B. *et al.* (1997) Familial colorectal cancer in Ashkenazim due to a hypermutable tract in APC. *Nat. Genet.* 17: 79–83.

Leach, F.S., Nicolaides, N.C., Papadopoulos, N. *et al.* (1993) Mutations of a mutS homolog in hereditary nonpolyposis colorectal cancer. *Cell* 75: 1215–1225.

Lipkin, S.M., Wang, V., Jacoby, R., Banerjee-Basu, S., Baxevanis, A.D., Lynch, H.T., Elliott, R.M. and Collins, F.S. (2000) MLH3: a DNA mismatch repair gene associated with mammalian microsatellite instability. *Nat. Genet.* 24: 27–35.

Lynch, H.T. and Lynch, J.F. (1998) Genetics of colonic cancer. *Digestion* 59: 481–492.

Lynch, H.T., Smyrk, T., McGinn, T. *et al.* (1995) Attenuated familial adenomatous polyposis (AFAP). A phenotypically and genotypically distinctive variant of FAP. *Cancer* 76: 2427–2433.

Lynch, H.T., Watson, P., Shaw, T.G. *et al.* (1999) Clinical impact of molecular genetic diagnosis, genetic counseling, and management of hereditary cancer. Part II: Hereditary nonpolyposis colorectal carcinoma as a model. *Cancer* 86: 2457–2463.

Ma, J., Stampfer, M.J., Giovannucci, E. *et al.* (1997) Methylenetetrahydrofolate reductase polymorphism, dietary interactions, and risk of colorectal cancer. *Cancer Res.* 57: 1098–1102.

Marra, G. and Boland, C.R. (1995) Hereditary nonpolyposis colorectal cancer: the syndrome, the genes, and historical perspectives. *J. Natl. Cancer Inst.* 87: 1114–1125.

Mathers, J.C. and Burn, J. (1999) Nutrition in cancer prevention. *Curr. Opin. Oncol.* 11: 402–407.

Mills, S., Mathers, J.C., Chapman, P.D., Burn, J. and Gunn, A. (2000) Aspirin, Sulindac and the rectum in familial adenomatous polyposis. *Dis. Colon Rectum* (in press).

Mills, S.J., Mathers, J.C., Chapman, P.D., Burn, J. and Gunn, A. (2000) Colonic crypt cell proliferation assessed by whole crypt microdissection in sporadic neoplasia and familial adenomatous polyposis. *Gut* 48: 41–46.

Murday, V. and Slack, J. (1989) Inherited disorders associated with colorectal cancer. *Cancer Surv.* 8: 139–157.

Nelson, N.J. (1996) Caution guides genetic testing for hereditary cancer genes. *J. Natl. Cancer Inst.* 88: 70–72.

NHS Executive (1997) Guidance on Commissioning Cancer Services. Improving outcomes in colorectal cancer. The manual, pp. 1–73. Department of Health, London.

Nicolaides, N.C., Papadopoulos, N., Liu, B. *et al.* (1994) Mutations of two PMS homologues in hereditary nonpolyposis colon cancer. *Nature* 371: 75–80.

Nishisho, I., Nakamura, Y., Miyoshi, Y. *et al.* (1991) Mutations of chromosome 5q21 genes in FAP and colorectal cancer patients. *Science* 253: 665–669.

Novelli, M.R., Williamson, J.A., Tomlinson, I.P., Elia, G., Hodgson, S.V., Talbot, I.C., Bodmer, W.F. and Wright, N.A. (1996) Polyclonal origin of colonic adenomas in an XO/XY patient with FAP. *Science* 272: 1187–1190.

Nuako, K.W., Ahlquist, D.A., Mahoney, D.W., Schaid, D.J., Siems, D.M. and Lindor, N.M. (1998) Familial predisposition for colorectal cancer in chronic ulcerative colitis: a case-control study. *Gastroenterology* 115: 1079–1083.

Polakis, P. (1999) The oncogenic activation of beta-catenin. *Curr. Opin. Genet. Dev.* 9: 15–21.

Powell, S.M., Zilz, N., Beazer-Barclay, Y., Bryan, T.M., Hamilton, S.R., Thibodeau, S.N., Vogelstein, B. and Kinzler, K.W. (1992) APC mutations occur early during colorectal tumorigenesis. *Nature* 359: 235–237.

Rossi, S.C. and Srivastava, S. (1996) National Cancer Institute Workshop on Genetic Screening for Colorectal Cancer. *J. Natl. Cancer Inst.* 88: 331–339.

Schwartz, R.A. (1978) Basal-cell-nevus syndrome and gastrointestinal polyposis [letter]. *N. Engl. J. Med.* 299: 49.

Severin, M.J. (1999) Genetic susceptibility for specific cancers. Medical liability of the clinician. *Cancer* 86: 2564–2569.

Smith, B. and DuBois, R.N. (1997) Current concepts in colorectal cancer prevention. *Comprehensive therapy* 23: 184–189.

Spirio, L.N., Kutchera, W., Winstead, M.V. *et al.* (1996) Three secretory phospholipase A(2) genes that map to human chromosome 1P35–36 are not mutated in individuals with attenuated adenomatous polyposis coli. *Cancer Res.* 56: 955–958.

Steinbach, G., Lynch, P.M., Phillips, R.K.S. *et al.* (2000) The effect of celecoxib, a cycloosygenase-2 inhibitor, in familial adenomatous polyposis. *N. Engl. J. Med.* **342**: 1946–1952.

Sternberg, L.R., Byrd, J.C., Yunker, C.K., Dudas, S., Hoon, V.K. and Bresalier, R.S. (1999) Liver colonization by human colon cancer cells is reduced by antisense inhibition of MUC2 mucin synthesis. *Gastroenterology* **116**: 363–371.

Thomas, H.J., Whitelaw, S.C., Cottrell, S.E. *et al.* (1996) Genetic mapping of hereditary mixed polyposis syndrome to chromosome 6q. *Am. J. Hum. Genet.* **58**: 770–776.

Vasen, H.F., Mecklin, J.P., Khan, P.M. and Lynch, H.T. (1994) The International Collaborative Group on HNPCC. *Anticancer Res.* **14**: 1661–1664.

Vasen, H.F., Taal, B.G., Nagengast, F.M., Griffioen, G., Menko, F.H., Kleibeuker, J.H., Offerhaus, G.J., and Meera, K.P. (1995) Hereditary nonpolyposis colorectal cancer: results of long-term surveillance in 50 families. *Eur. J. Cancer* **31A**: 1145–1148.

Wang, X.D., Liu, C., Bronson, R.T., Smith, D.E., Krinsky, N.I. and Russell, M. (1999) Retinoid signaling and activator protein-1 expression in ferrets given beta-carotene supplements and exposed to tobacco smoke. *J. Natl. Cancer Inst.* **91**: 60–66.

Wasan, H.S., Park, H.S., Liu, K.C., Mandir, N.K., Winnett, A., Sasieni, P., Bodmer, W.F., Goodlad, R.A. and Wright, N.A. (1998) APC in the regulation of intestinal crypt fission. *J. Pathol.* **185**: 246–255.

Watson, P. and Lynch, H.T. (1993) Extracolonic cancer in hereditary nonpolyposis colorectal cancer. *Cancer* **71**: 677–685.

Wicking, C., Shanley, S., Smyth, I. *et al.* (1997) Most germ-line mutations in the nevoid basal cell carcinoma syndrome lead to a premature termination of the PATCHED protein, and no genotype–phenotype correlations are evident. *Am. J. Hum. Genet.* **60**: 21–26.

Wijnen, J.T., Vasen, H.F., Khan, P.M., Zwinderman, A.H., van der, K.H., Mulder, A., Tops, C., Moller, P. and Fodde, R. (1998) Clinical findings with implications for genetic testing in families with clustering of colorectal cancer. *N. Engl. J. Med.* **339**: 511–518.

Winawer, S.J., Zauber, A.G., Ho, M.N. *et al.* (1993a) Prevention of colorectal cancer by colonoscopic polypectomy. The National Polyp Study Workgroup. *N. Engl. J. Med.* **329**: 1977–1981.

Winawer, S.J., Zauber, A.G., O'Brien, M.J. *et al.* (1993b) Randomized comparison of surveillance intervals after colonoscopic removal of newly diagnosed adenomatous polyps. The National Polyp Study Workgroup. *N. Engl. J. Med.* **328**: 901–906.

Yagi, O.K., Akiyama, Y., Ohkura, Y., Ban, S., Endo, M., Saitoh, K. and Yuasa, Y. (1997) Analyses of the APC and TGF-beta type II receptor genes, and microsatellite instability in mucosal colorectal carcinomas. *Jpn. J. Cancer Res.* **88**: 718–724.

Yagi, O.K., Akiyama, Y. and Yuasa, Y. (1999) Genomic structure and alterations of homeobox gene CDX2 in colorectal carcinomas. *Br. J. Cancer* **79**: 440–444.

11

Genetics of prostate cancer

Tommi Kainu and William Isaacs

1. Introduction

The genetic basis of cancer development can be delineated into two topics. Genetic susceptibility to cancer is conferred by inherited genetic defects that cause a dramatic increase in the risk of cancer in those carrying the susceptibility allele. Tumor progression is also genetically determined by the accumulation of distinct genetic alterations in tumor cells. Such alterations, including amplification of oncogenes, and loss of tumor suppressor genes, are the cause of the tumorigenic properties of cancer cells. While the known genes involved in tumor progression appear to be to a large degree the same in different tumor types, susceptibility genes, such as the breast and ovarian cancer genes, *BRCA1* and *BRCA2* confer an increased risk of a narrow spectrum of tumors. In this review of the genetic basis of prostate cancer, we will concentrate on inherited predisposition to prostate cancer as the hereditary component of prostate cancer is believed to be the largest of all tumor types (Lichtenstein *et al.*, 2000). First, we present an overview of the somatic changes prostate cancer cells acquire during tumor progession. We then examine the epidemiological and genetic data establishing the presence of a strong hereditary component in prostate cancer. The significance of the susceptibility loci identified so far is evaluated. Finally, we overview the evidence that distinct, phenotypically identifiable groups of families are linked to the different susceptibility loci.

2. Somatic genetic changes in prostate cancer development

Prostate cancer is, after skin cancer, the most common malignancy among men in the United States. One in five men is expected to develop prostate cancer during his lifetime (Feuer, 1997). The advent of prostate-specific antigen (PSA) screening and improvement in other detection methods over the last decade has led to earlier diagnosis – still, however, about a third of all patients are diagnosed at a clinically advanced stage (Kosary *et al.*, 1995). While surgery is usually effective in

Molecular Genetics of Cancer second edition, edited by J.K. Cowell.

localized prostate cancer, advanced forms of the disease can be effectively treated only with androgen withdrawal therapy. Despite a favorable initial response in most patients (Stearns and McGarvey, 1992), the disease will eventually progress. Since no effective form of therapy exists for recurrent disease, the question why prostate cancer progresses from androgen-dependence to androgen-independence is a key-issue in bettering our understanding of the disease.

When considering the molecular genetics of prostate cancer, the somatic changes arising during tumorigenesis play an important role in the progression and development of prostate cancer. The study of such changes has progressed rapidly during the past years benefiting from several novel, molecular cytogenetic technologies, including fluorescence *in situ* hybridization (FISH) and comparative genomic hybridization (CGH). In addition, loss of heterozygosity (LOH) studies and karyotyping have brought about a better understanding of the chromosomal regions involved in prostate tumorigenesis.

2.1 Chromosomal abnormalities in prostate cancer cells

The most common chromosomal abnormalities in prostate cancer cells include losses of 8p, 10q, 13q and 16q as well as gains of 7p, 7q, 8q, and Xq, as detected by CGH (reviewed in Nupponen and Visakorpi, 1999). Allelic loss is seen, furthermore, at 6q, 7q, 17p, 17q and 18q (reviewed in Isaacs and Bova, 1998). In many cases the aberrations seen in chromosomal arms consist of several distinct regions of loss or gain indicating multiple target genes in these regions. For example, allelic loss is seen at three separate regions of chr 13q: 13q14, 13q22, and 13q31 (Hyytinen *et al.*, 1998), and gain of 8q comprises of amplification of sequences at 8q21 and 8q23–24 (Nupponen *et al.*, 1998). The target genes of these gains and losses remain unidentified in most cases. However, alterations in some specific genes have been uncovered, and these studies are described below. Furthermore, a few chromosomal aberrations have been demonstrated to affect clinical outcome. Such aberrations include deletions at 7q31 (Takahashi *et al.*, 1995) as well as losses of 8p/gains of 8q, which are more prevalent in recurrent cancers than in primary tumors (Visakorpi *et al.*, 1995a).

2.2 Oncogenes

MYC. Gain of 8q in prostate cancers was first described by Bova *et al.* (1993). Gain of 8q is more prevalent in recurrent tumors (Visakorpi *et al.*, 1995a) as well as in metastatic lesions (Cher *et al.*, 1996) than in primary tumors. Accordingly, 8q gains are associated with a short progression-free interval (Takahashi *et al.*, 1994), and the presence of lymph-node metastasis (Van Den Berg *et al.*, 1995). The *MYC* oncogene is located at 8q24, the other of the minimally amplified regions at 8q (Cher *et al.*, 1996; Nupponen *et al.*, 1998; Visakorpi *et al.*, 1995a). This well-known oncogene plays an important role in the regulation of cellular proliferation, differentiation, and apoptosis (reviewed in Henrikkson and Lucher, 1996). Both over-expression and amplification of *MYC* have been detected in prostate tumors (Buttyan *et al.*, 1987; Fleming *et al.*, 1986; Nupponen *et al.*, 1998). However, relatively few prostate tumors show high-level amplification of *MYC* (Bubendorf *et al.*, 1999; Nupponen *et al.*, 1998), indicating that there may exist other target genes for the 8q23–24 amplification, in addition to those at the other minimally amplified region 8q21. *MYC* is, however, suggested to play an important role in metastatic prostate cancer (Bubendorf *et al.*, 1999).

RAS. Experimentally, the *ras* family of oncogenes (*HRAS, KRAS,* and *NRAS*) can affect the tumorigenic properties of transfected prostate cancer cells (Partin *et al.*, 1988). Activating point mutations of the *RAS* genes are seen in a variety of solid tumors. In prostate tumors, mutations in the *RAS* gene are relatively uncommon, except in the rare ductal form of the disease (Carter *et al.*, 1990). The role of *ras* in human prostate cancer initiation and progression is, thus, most likely a minor one.

HER-2/neu (ERBB2). In view of the promising therapeutic potential of the commercially available anti-Her-2/neu antibody, the role of this 17q oncogene in prostate cancer is of great interest. Using FISH analysis, several groups have, however, failed to show high level amplification of *ERBB2* (Bubendorf *et al.*, 1999; Mark *et al.*, 1999), even though overexpression of the gene is a frequent event in prostate cancer, as well as an independent prognostic factor for the disease (Morote *et al.*, 1999). An intriguing mechanism for the role of *ERBB2* in hormone-independent prostate cancer was recently presented by Craft *et al.* (1999). In androgen-independent cancer cells, over-expression of *ERBB2* was able to 'superactivate' the androgen receptor pathway, providing a clue to how prostate cancers can circumvent androgen deprivation therapy. Indeed, the commercial HER-2/neu antibody inhibits growth of prostate cancer cells in a xenograft model (Agus *et al.*, 1999).

BCL2. Amplification of chr 18q is present in over a third of prostate tumors (Nupponen and Visakorpi, 1999). The anti-apoptotic oncogene *BCL2* is located at 18q21.3. Over-expression of *BCL2* is seen frequently in recurrent tumors (Colombel *et al.*, 1993; Krajewska *et al.*, 1996; McDonnell *et al.*, 1992), but seems not to be caused by amplification of the gene (Nupponen and Visakorpi, 1999). The role *BCL2* is suggested to play in prostate cancer is interesting. *BCL2* expression inhibits apoptosis of prostate cancer cells subjected to androgen deprivation (Geave *et al.*, 1999). If this hypothesis holds true, *BCL2* would present a very attractive therapeutic target, potentially reducing the risk of recurrent cancer.

2.3 Androgen receptor

In addition to *BCL2*, the androgen receptor gene (*AR*) has been implicated in recurrence of prostate cancer. Visakorpi *et al.* (1995a) found frequent amplification of chromosome Xq in recurrent tumors, whereas Xq is very rarely amplified in primary tumors. This group went on to confirm that the *AR* gene was the target of this amplification (Visakorpi *et al.*, 1995b). Amplification leading to over-expression of AR after androgen deprivation therapy is an understandable way of how prostate tumor cells overcome the decreased levels of circulating androgens.

An additional means of enhancing AR signaling after androgen deprivation prostate cancer cells develop are activating mutations in *AR* (Taplin *et al.*, 1995; Tilley *et al.*, 1997).

2.4 Metastasis genes

Aberrations in two genes have been associated with metastatic prostate cancer. The gene for the cell adhesion molecule *E-cadherin* (*CDH1*) on 16q has been extensively studied in prostate cancer progression. Reduced expression of *CDH1* or its accessory protein alpha-catenin are frequent events in advanced prostate cancer (Morton *et al.*, 1993; Umbas *et al.*, 1992). Although allelic loss at 16q is common in prostate cancers, reduced expression of *CDH1* seems not to be caused by this mechanism (Murant *et al.*, 2000). The *KAI1* gene at 11p11.2 shows decreased expression in metastases and suppresses metastasis in an animal model (Dong *et al.*, 1995). The down-regulation of the gene is not caused by mutation or allelic loss (Dong *et al.*, 1996), but rather by post-transcriptional events. *KAI1* expression is positively regulated by p53, leading to a hypothesis that loss of p53 function in the late stages of tumor progression causes down-regulation of *KAI1*, with subsequent metastasis (Mashimo *et al.*, 1998).

2.5 Tumor suppressor genes

The genetic regions exhibiting allelic loss or chromosomal deletions most frequently in prostate cancer are two separate sites on chromosome 8p: 8p23 and 8p12-p22 (Bova *et al.*, 1993; Cher *et al.*, 1994, 1996; Cunningham *et al.*, 1996; Joos *et al.*, 1995; Macoska *et al.*, 1995; Nupponen *et al.*, 1998; Visakorpi *et al.*, 1995). Loss of 8p appears to be an early event in prostate cancer development as also prostate interepithelial neoplasias show LOH at this location (Emmert-Buck *et al.*, 1996). However, no clear candidates for the specific genes involved have appeared, although several genes, including *NKX3.1*, *MSR*, *N33*, and *PYK2* have been actively investigated (Bookstein *et al.*, 1997; He *et al.*, 1997; Inazawa *et al.*, 1996). Other sites of loss/deletion in prostate cancer mainly occur in the late stages of cancer progression. Loss of function of such tumor suppressor genes as *TP53*, *RB1*, *INK4* (*p16*), and *PTEN/MMAC*, are seen almost exclusively in advanced cases of prostate cancer (Bookstein *et al.*, 1993; Cairns *et al.*, 1997; Gaddipati *et al.*, 1997; Heidenberg *et al.*, 1995; Jarrard *et al.*, 1997; Kubota *et al.*, 1995; Tamimi *et al.*, 1996; Tricoli *et al.*, 1996). Clearly, the initiating genes in prostate cancer development still remain to be uncovered.

3. Inherited susceptibility to prostate cancer – overview

The hereditary component of prostate cancer is believed to be at least as significant as that of either breast or colon cancer. Despite this no major susceptibility genes for prostate cancer have been identified, in contrast to the success demonstrated in the two latter tumor types. The lack of success in identifying the genes responsible for hereditary prostate cancer stems from several challenges faced by investigators studying this disease: (1) the relatively late age of onset even in the case of hereditary cases; (2) the high sporadic rate of prostate cancer; (3) the genetic heterogeneity of the disease; (4) the absence of a clear way of distinguishing between hereditary and sporadic cases.

While the actual genes responsible for prostate cancer susceptibility remain elusive, several genetic regions have been linked to hereditary prostate cancer. The first of these identified, and termed *HPC1*, was originally thought to account for up to 34% of hereditary prostate cancer. Confirmatory studies have strengthened the evidence for the presence of *HPC1*. However, it has become clear that families linked to *HPC1* account for a minority of hereditary prostate cancer families; specifically those that demonstrate early age of onset as well as a large number of affected cases. Another genetic locus with significant evidence of linkage to prostate cancer susceptibility was found on Xq27–28, adding strength to the hypothesis that a fraction of prostate cancer segregates as an X-linked trait. Two other loci on chromosome 1 have also been implicated as harboring prostate cancer susceptibility genes. The first of these, *PCAP*, is located distally from *HPC1* on 1q42. The other on 1p36, *CaPB*, was identified in families segregating brain tumors as well as prostate cancer. Finally, a susceptibility locus on chromosome 16q has been suggested based on a study of multiplex sibships.

4. Epidemiological evidence for the presence of prostate cancer susceptibility genes

The major risk factor for developing prostate cancer is age, with the incidence rate increasing from 34–440:100 000, from the age of 60 to 80 years in Caucasian men in the U.S. (Kosary *et al.*, 1995). There also exist vast differences in prostate cancer incidence among different ethnic groups. The largest incidence rate is found in African–American men, and the lowest among men in China and Japan (Muir *et al.*, 1987). Although these differences are likely to be explained to a great extent by dietary and lifestyle differences between the populations, genetic susceptibility clearly may have a role in prostate cancer development.

The first evidence that prostate cancer susceptibility may segregate as an inherited trait was described in 1960 by Woolf *et al.* (1960). This retrospective study utilized the Utah cancer registry and the extensive population registries of the Mormon Church. Here the risk of death from prostate cancer was determined to be threefold higher in first-degree male relatives of men who had died of the disease than in men with no such relatives. After this initial study, several reports have confirmed the increased risk of prostate cancer in relatives of men affected with the disease. All of these studies have consistently reported the relative risk to be two- to threefold (Cannon *et al.*, 1982; Carter *et al.*, 1990; Goldgar *et al.*, 1994; Meikle and Stanish 1982; Spitz *et al.*, 1991; Steinberg *et al.*, 1990; Whittemore *et al.*, 1995). Interestingly, the study by Whittemore *et al.* (1995) shows that, despite the vast differences in prostate cancer incidence between different ethnic groups, the increase in risk for relatives does not differ significantly among the groups. Finally, consistent with prostate cancer susceptibility being an inherited trait, the risk to male relatives increases dramatically with the more affected men there are in the family (Steinberg *et al.*, 1990).

While familial clustering described above is clearly indicative of a genetic component being present in prostate cancer development, shared environmental factors between siblings can be argued to contribute to the observed clustering as well. Twin studies are used in genetics to further dissect the role of hereditary factors as a cause of disease. Such studies are carried out by comparing the concordance of disease status between monozygotic twins (genetically identical) and dizygotic twins (half of their genes shared). Ahlbom *et al.* (1997) performed a twin study of all cancers in the Swedish population. Of all cancer types studied, including breast and colorectal cancer, the increase in risk if the affected twin was monozygotic rather than dizygotic was greatest in prostate cancer. Twin studies focusing solely on prostate cancer have generated similar results. In another Swedish twin study by Gronberg *et al.* (1994) 1% of monozygotic twins were concordant for prostate cancer compared to 0.2% of dizygotic twins. Page *et al.* (1997) using a cohort of 31 848 twins from the United States, demonstrated a significantly higher concordance rate of 27.1% in monozygotic twins than the 7.1% in dizygotic twin pairs.

5. Segregation analyses

The evidence gleaned from both the observations of familial clustering of prostate cancer as well as the twin studies strongly suggest the presence of hereditary factors in the development of prostate cancers. However, these studies do not provide evidence as to how the susceptibility to prostate cancer is inherited. To answer this question it is necessary to perform complex segregation analysis in families with multiple cases of prostate cancer. Results from such studies have given us information on the mode of inheritance of prostate cancer (dominant/recessive/X-linked), the frequency of the risk allele, and the estimated risk of developing prostate cancer in carriers of such an allele.

The first segregation study was performed by Carter *et al.* (1992) in 691 families of predominantly Caucasian origin. The families were ascertained through probands undergoing radical prostatectomy for primary clinically localized prostate cancer. The probands were also relatively young at disease onset: mean 59.3 years compared to the median age of 73.5 years at diagnosis among Caucasians in the U.S. Furthermore, the study population was gathered between the years 1982 and 1989, before the advent of PSA-testing to identify possible prostate cancer cases. Current use of PSA-testing, leading to detection of early-stage cancers, has been suggested to partly explain the dramatic increase in the prostate cancer incidence (Jacobsen *et al.*, 1995). In this study an autosomal dominant model was best fitted to explain the inheritance of prostate cancer. The frequency of the allele conferring susceptibility to prostate cancer was predicted to be 0.3%. The cumulative risk of developing prostate cancer was estimated to be 88% in carriers of the high-risk allele and 5% in non-carriers.

In a study carried out in Sweden, in which the investigators looked at all prostate cancer cases, not just those with clinically localized disease, Gronberg *et al.* (1997a) confirmed an autosomal dominant mode of inheritance. Their estimate of the frequency of the high-risk allele was substantially higher, though, at 1.67%.

Accordingly, they predicted a lower lifetime penetrance of 63%. The strength of this report is that this was effectively a population-based study, which was not biased towards less aggressive prostate tumors, as are those that are restricted to clinically localized disease. In a separate study, Gronberg et al. (1997b) have suggested that hereditary prostate cancers are more aggressive and present at a more advanced stage than sporadic tumors.

In the largest segregation study to date carried out by Schaid et al. (1998), while the results were generally consistent with an autosomal dominant model, evidence for more complex inheritance patterns was also found. The study population comprised of 4288 probands undergoing radical prostatectomy for clinically localized prostate cancer at the Mayo Clinic (Rochester, Minnesota). The frequency of the high-risk allele in this study was found to be 0.8%. However, significant evidence against the presence of Hardy-Weinberg equilibrium (HWE) was found, and under the assumption of HWE not being present, the gene frequency was predicted to be 0.6%. Even if HWE was not assumed, no single model was found to adequately explain the familial clustering of prostate cancer. The lack of an adequate model was especially apparent in the older age groups of patients. Schaid et al. (1998) also found a greater risk for developing prostate cancer in men with affected brothers compared to those with affected fathers. This observation could be due to under-reporting of prostate cancer in fathers of affected patients. It can also result from a recessive or X-linked inheritance, a possibility suggested in an earlier study by Monroe et al. (1995). In a population-based multi-ethnic cohort from Southern California, Monroe et al. reported a two-fold greater risk of developing prostate cancer in brothers versus sons of prostate cancer patients.

In conclusion, segregation analyses suggest a genetically complex inheritance pattern for prostate cancer susceptibility. In families with early age at diagnosis, the mode of inheritance is most likely autosomal dominant, and the causative allele rare and conferring a high risk. In older age groups the high number of sporadic prostate cancer cases that aggregate in families by chance make the dissection of the mode of inheritance difficult. Evidence, however, exists that a part of prostate cancer is caused by X-linked, or recessively inherited genes (*Figure 1*).

6. Linkage studies

6.1 Challenges in linkage studies for prostate cancer

For linkage studies to produce statistically significant results, an adequate number of informative meioses need to be present in the study set. An informative meiosis means that both genotype and phenotype can be unequivocally established. In a disease such as prostate cancer the late age of onset presents problems for both requirements. For example, if a man develops prostate cancer at age 70, it is likely that both of his parents are deceased and unavailable for genotyping. The man's siblings may also have passed away. The man's offspring are likely too young for prostate cancer to have arisen and cannot be used to determine phenotype.

1. Family consistent with autosomal dominant inheritance of prostate cancer susceptibility

2. Family consistent with X-linked inheritance of prostate cancer susceptibility

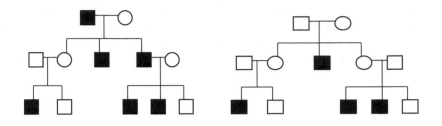

3. Family in which inheritance pattern is unclear

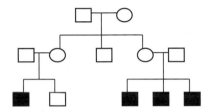

Figure 1. Working criteria for a hereditary prostate cancer family: Three or more first degree relatives diagnosed with prostate cancer; A diagnosed case of prostate cancer in each of three successive generations; Two or more 1st or 2nd degree relatives diagnosed with prostate cancer before the age of 55 years.

Incomplete penetrance further confounds the genotype–phenotype correlation. A more significant problem, however, is the high incidence of sporadic prostate cancer, especially in the older age groups. Since one in five men will develop prostate cancer during their lifetime, some families with obvious clustering of prostate cancer will have risen purely by chance. In addition, families in which an inherited susceptibility gene segregates, some prostate cancer cases can be sporadic in genesis. Without clear means of determining between the sporadic and hereditary forms of prostate cancer in such cases, linkage analysis will be greatly hindered. Finally, since it is obvious that prostate cancer susceptibility is genetically heterogenous, large numbers of families will need to be analyzed to overcome this diluting effect.

6.2 Targeted linkage versus genome wide screens

Linkage studies can be performed in two ways: as a targeted approach of a genetic region known to be of interest in the etiology of the given disease; or as a genome-wide screen, without prior predictions where in the genome putative suscepti-bility loci may reside. Genome-wide screens are performed with a set of generally anonymous genetic markers dispersed evenly throughout the human genome. Marker sets, in which the average interval between two markers is 10 cM, indi-cating ~350 markers in total, are generally used. Molecular genetic analyses of somatic genetic aberrations in prostate tumors have suggested the presence of prostate cancer susceptibility loci in a number of human chromosomes, including 8, 10, 16, and 17 (reviewed in Isaacs and Bova, 1998). Linkage analysis has not provided evidence of inherited susceptibility loci being present in these loci, apart from 16q (Cannon-Albright and Eeles, 1995; Suarez *et al.* 2000). Therefore, linkage analysis efforts in prostate cancer have concentrated on genome wide screens of large family sets, of which three have presented with evidence of linkage to date (*Table 1*).

6.3 The first prostate cancer susceptibility gene, HPC1, identified at 1q24–25

Smith *et al.* (1996) presented very significant evidence of linkage to 1q24–25 in 91 prostate cancer families collected from the United States and Sweden. The maximum 2-point LOD-score (logarithm of the odds of linkage) of 3.65 was observed at marker D1S2883 at a recombination fraction of 0.18. Significant evidence of heterogencity was present with an estimate of 34% of the families being linked to this region. Using an alpha value of 0.34 (alpha denotes the fraction of linked families), a maximum multi-point LOD score of 5.43 assuming heterogeneity was achieved. No clinical features appeared to distinguish the families that were linked from those that were not. However, two interesting observations were made on the source of the heterogeneity. The linked families had a lower age of diagnosis than those showing no linkage. Two African-American families were present in the sample set. Both families showed evidence of linkage to 1q24–25, with a combined LOD score of 1.4.

Confirmatory evidence for the presence of *HPC1* has come from three independent studies. In the study of Cooney *et al.* (1997) in a series of 59 families of mixed ethnicity an NPL (non-parametric linkage) Z score of 1.58 ($P = 0.0574$) at marker D1S466 was achieved. If stratified on the basis of young age at onset and presence of three or more cases of prostate cancer in the family the NPL Z score was 1.72 ($P = 0.0451$) at the same marker. More interestingly the study sample included six African-American families, which themselves had an NPL Z score of 1.39 ($P = 0.0848$) at marker D1S158. Hsieh *et al.* (1997) reported an NPL Z score of 1.71 ($P = 0.046$) in a set of 92 families. Stratified on the basis of age at onset, the 46 families with mean age at diagnosis of under 67 years, the NPL Z score was 2.04 ($P = 0.023$). By far the most significant confirmatory evidence for *HPC1* comes

Table 1. Prostate cancer susceptibility loci identified to date

Locus	Genetic region	Reference	Confirmatory studies
HPC1	1q24–25	Smith *et al.* (1996)	Cooney *et al.* (1997), Hsieh *et al.* (1997), Neuhausen *et al.* (1999)
HPCX	Xq27–28	Xu *et al.* (1998)	
PCaP	1q42	Berthon *et al.* (1998)	Gibbs *et al.* (1999a)
CaPB	1p36	Gibbs *et al.* (1999b)	Suarez *et al.* (2000)
Unnamed	16q23	Suarez *et al.* (2000)	

from a study by Neuhausen *et al.* (1999) in 44 large (an average of 10.7 affected individuals per pedigree) prostate cancer pedigrees from Utah. For the interval between markers D1S196 to D1S416, the authors observed a multipoint LOD score of 2.06 ($P = 0.002$). The evidence for linkage came predominantly from the quartile of families with the lowest mean age at diagnosis.

The genetic heterogeneity and the fact that *HPC1* accounts for only a fraction of hereditary prostate cancer comes apparent with the findings of three groups who found no evidence of linkage to 1q24–25. McIndoe *et al.* (1997) found no support for linkage to HPC1 using either parametric or model-free analyses in their set of 49 high-risk families. Even stratification by age at diagnosis did not increase LOD scores. As a part of a genome-wide screen Berthon *et al.* (1998) reported negative 2-point parametric LOD scores for three markers on 1q24–25. No attempts to stratify the 47 families of French and German descent were made. In the collaborative linkage effort from the United Kingdom, Quebec, and Texas, Eeles *et al.* (1998) reported negative NPL scores in the total set of 136 families. However, in the 35 families with at least four affected cases, positive NPL scores were seen.

The seemingly contradictory results from the studies described above prompted Xu *et al.* (2000) to perform a meta-analysis on 772 hereditary prostate cancer families. Six markers on 1q24–25 were analyzed, and provided a peak parametric multi-point LOD score assuming heterogeneity (HLOD) of 1.40 ($P = 0.01$) at marker D1S212. The proportion of families estimated to be linked to HPC1 was 6% (1 – LOD unit support interval 1–12%). In an effort to distinguish the families most likely to be linked the families were divided based on the presence, or lack of, male-to-male transmission of disease susceptibility. The rationale behind this division is that lack of male-to-male transmission can be viewed as indicative of X-linkage. And the evidence that a substantial share of prostate cancer is caused by a gene segregating on the X-chromosome has further strengthened with the finding of linkage to Xq27–28 (Xu *et al.*, 1998, and see below). In the 491 families, with a record of male-to-male transmission, the peak HLOD was 2.56 ($P = 0.0006$) and a corresponding alpha 0.11. If further stratified by early age at disease diagnosis, and presence of five or more affected cases, the evidence for linkage was even more significant: the peak HLOD was 2.25 ($P = 0.0001$) and alpha 0.29. The results of the meta-analysis demonstrate that *HPC1* linkage is real, but the proportion of hereditary prostate cancer families accounted for by *HPC1* seems to be less than 10%. However, in the subset of families defined by male to male transmission, early age at diagnosis, and large number of affected individuals, *HPC1* appears to play a substantial role.

6.4 More evidence for prostate cancer susceptibility to be an X-linked trait

In the segregation analysis by Schaid *et al.* (1998), an increased risk for brothers of prostate cancer cases compared to sons of cases was found. Several population-based studies describing the familial clustering of prostate cancer have reported similar results (Cerhan *et al.*, 1999; Hayes *et al.*, 1995; Monroe *et al.*, 1995; Narod *et al.*, 1995; Schuurman *et al.*, 1999; Woolf *et al.*, 1960). While these results can be caused by under-reporting of prostate cancer in previous generations; i.e. in fathers of prostate cancer cases, the results may also be interpreted as evidence of X-linked or recessive inheritance of prostate cancer susceptibility. In the initial genome-wide screen for linkage by Smith *et al.* (1996) a region on the X-chromosome was implicated. The investigators continued their efforts with a denser marker set and increasing the size of the family collection to 360 families from the United States, Sweden and Finland (Xu *et al.*, 1998). In the extended family collection a peak 2-point LOD score of 4.6 at marker DXS1113 on Xq27–28 was observed. The peak multi-point score was 3.85 between markers DXS1120 and DXS297. The locus was designated as HPCX and determined to account for ~16% of the combined sample population. As can be expected for an X-linked disease, most of the linkage evidence for HPX comes from families, in which no male-to-male transmission of prostate cancer is observed. This is in contrast with the evidence for *HPC1*, and indeed, in a heterogeneity analysis of the Johns Hopkins University prostate family collection, families linked to either *HPC1* or *HPCX* constitute two independent groups of families (Xu *et al.*, 1998).

6.5 Another susceptibility locus, designated PCaP, identified on 1q42–43

Berthon and co-workers describe a second locus on 1q located 60 cM distal to *HPC1* on 1q42–43 (Berthon *et al.*, 1998). The study population in this report comprises of 47 families of French and German descent. All of the families fulfilled the criteria suggested by Carter *et al.* (1990) for hereditary prostate cancer families. These investigators performed a genome-wide screen and achieved a peak 2-point LOD score of 2.7 at theta 0.1 maximized over three different genetic models. Significant evidence of heterogeneity was reported with *PCAP* estimated to account for 50% of the families. Gibbs *et al.* (1999a) evaluated the evidence for linkage at the *PCAP* locus in a set of 152 prostate cancer families. No significant evidence for linkage was seen in the whole set. Suggestive evidence for linkage was seen in the subset of families of at least five affected men. The *PCAP* locus seems thus to account for only a minority of prostate cancer families.

6.6 A link between prostate cancer and brain tumors

A genome-wide scan for linkage on 70 prostate cancer families revealed suggestive evidence of linkage at 1p36 (Gibbs *et al.*, 1999b). This genetic region has been of interest previously in cancer genetics due to loss of heterozygosity observed at this site in brain and other central nervous system tumors (Bello *et al* 1994a, b, 1995a, b; Kraus *et al.*, 1995; Maris *et al.*, 1995; Schleiermacher *et al.*, 1994; White *et al.*, 1995). Prompted by this knowledge, the investigators stratified their family set on the basis of brain tumors appearing in some families. In the 12 such families a 2-point LOD-score of 3.22 at theta 0.06 with marker D1S507 was observed. This observation provides the first genetic evidence of a link between prostate and brain cancers. The link has been observed before in epidemiological studies, in which the risk of brain cancer in relatives of prostate cancer patients is significantly increased (Carter *et al.*, 1993; Goldgar *et al.*, 1994; Isaacs *et al.*, 1995). Partial confirmatory evidence for the presence of a prostate-brain cancer susceptibility locus on 1p36 was presented by Suarez *et al.* (2000) in genome screen of multiplex sibships. In 13 families with both brain and prostate cancer, marginally positive Z scores (~1) were observed at three adjacent markers on 1p35.1.

6.7 A locus on 16q is suggested by a genome-wide screen of multiplex sibships

In a large study of 230 multiplex sibships, Suarez *et al.* (2000) present evidence for a susceptibility locus on chromosome 16q. Despite the large sample size, the LOD scores presented fail to gain statistical significance as required of a genome-wide scan. The highest Z score at marker D16S3096 on 16q is 3.15. However, the earlier evidence of LOH at 16q being consistently present in prostate tumors (reviewed in Isaacs and Bova, 1998) make the finding of Suarez *et al.* intriguing. The interval in which positive LOD scores were found was very broad, however, and no clear candidate region was identified. Apart from partially confirming the brain–prostate cancer link described by Gibbs *et al.* (1999; see above), Suarez *et al.* find little to no evidence of linkage in any of the other susceptibility loci suggested thus far. This despite using three means of classification (age at onset, degree of family history, presence of breast cancer in the family) to stratify the families. Nominally significant Z scores were observed additionally on chromosomes 2q, 12p, 15q, and 16p.

7. Association analyses

Association analyses are carried out by comparing the frequency of a given allele in patients versus the allele's frequency in controls (people not affected with the disease in question). While generally considered as an alternative to linkage analysis in estimating the effects of a genetic locus to disease susceptibility, the underlying principle and hence the interpretation of results, are very different for association analyses. A significant linkage result can be taken to mean that the searched for gene lies in close proximity to the anonymous marker with which the result was achieved – the two loci are 'linked'. A statistically significant result from an association analysis only means that the allele in question is present in a greater frequency in affected individuals compared to control individuals – the disease and allele are associated. No positional evidence can be gleaned from the results of a pure association analysis. However, several advantages are offered by association analyses compared to linkage analysis; and with correct interpretation of the results, such studies can be of significant value in determining genetic components of disease.

Association analyses are usually performed with known variants (functional or otherwise) of genes that can be hypothesized to be involved in the etiology of the disease in question. The identification of more and more single nucleotide polymorphisms (SNPs) and the technological advances that have made it possible to detect thousands of SNPs with ease, make association studies more and more attractive. Another advantage of association analysis is that these studies are not restricted to analyzing family-based samples. The advantages have weighed more in the reasoning of several investigative groups, and a good number of association analyses of prostate cancer have been performed.

7.1 Do the breast cancer susceptibility genes, BRCA1 and BRCA2, also confer susceptibility to prostate cancer?

Several epidemiological studies have suggested a link between prostate and breast cancer etiology (Anderson and Badzioch, 1992; Ekman *et al.*, 1997; McCahy *et al.*, 1996; Thiessen, 1974). However, in analyzing the risk of cancer at other sites in a large number of prostate cancer families, Isaacs *et al.* (1995) found only brain tumors to be present in significant excess in the families. After the identification of the *BRCA1* and *BRCA2* genes, and mutations therein, it has become possible to further dissect the role of these genes in prostate cancer susceptibility. Two alternative study approaches have resulted in seemingly contradictory results. When looking at prostate cancer patient series (either sporadic, familial, or those with an early age at diagnosis) a paucity of both *BRCA1* and *BRCA2* mutations has been observed (Langston *et al.*, 1996; Lehrer *et al.*, 1998; Nastiuk *et al.*, 1999; Wilkens *et al.*, 1999). When looking at the risk of prostate cancer in known mutation carriers, the relative risks for those men harboring *BRCA2* mutations have ranged from 4.65 to 16% (Ford *et al.*, 1994; Struewing *et al.*, 1997). The discrepancy between the two sets of results suggests that while especially *BRCA2* mutations increase the risk of prostate cancer in carriers, the effect of these mutations on prostate cancer incidence or the familial aggregation thereof is minute.

7.2 The role of the androgen receptor in hereditary prostate cancer

The androgen receptor (AR) plays a significant role in sporadic prostate cancer progression, especially in so-called hormone refractory tumors, i.e., malignancies that after an initial, favorable response to androgen deprivation therapy, relapse and become hormone refractory (reviewed in Koivisto *et al.*, 1998). The mechanism behind this phenomenon is the amplification of the *AR* gene (Visakorpi *et al.*, 1995). Due to the obvious biological significance of AR in prostate cells, this gene is also one of the most studied as a putative cause of the hereditary forms of prostate cancer. In the *AR* gene there exist two polymorphic tri-nucleotide repeats coding for poly-glutamine (CAG) and poly-glycine (GGN). Increase in androgen specific transcriptional activity is inversely related to the length of the poly-glutamine tract (Chamberlain *et al.*, 1994, Kazemi-Esfarjani *et al.*, 1995, Sobue *et al.*, 1994). Several groups have reported an association between short poly-glutamine repeat length and prostate cancer (Giovannucci *et al.*, 1997, Kantoff *et al.*, 1998, Stanford *et al.*, 1997). The study by Stanford *et al.* (1997) in addition demonstrates that men with short repeats in both tracts are at even greater risk for developing prostate cancer. Underscoring the role of androgen metabolism in prostate cancer development, polymorphisms in several other genes of this pathway, such as 3-betahydroxysteroid dehydrogenase type II and 5-alpha reductase, have been implicated in conferring increased risk for prostate cancer (Devgan *et al.*, 1997, Makridakis *et al.*, 1997).

7.3 Vitamin D receptor and prostate cancer

Vitamin D deficiency has been postulated to be a risk factor for prostate cancer, and *in vitro* evidence exists that vitamin D can suppress the growth of cultured prostate cancer cells (Braun *et al.*, 1995; Corder *et al.*, 1993; Gann *et al.*, 1996; Krill *et al.*, 1999; Lyles *et al.*, 1980; Miller *et al.*, 1995; Peehl *et al.*, 1994; Schwartz and Hulka, 1990). The 3'-untranslated region of the vitamin D receptor (*VDR*) gene contains several polymorphisms with possible functional significance. The *VDRb* allele, the putatively less active variant, has been shown to be associated with increased risk for prostate cancer, as well as more advanced disease (Ingles *et al.*, 1997, Taylor *et al.*, 1996). These results have not been corroborated (Kibel *et al.*, 1998, Ma *et al.*, 1998) though, and recent research has shown that the afore-mentioned polymorphisms may lack effect on VDR function (Durrin *et al.*, 1999; Gross *et al.*, 1998).

8. Possible correlations of phenotype and genotype

The tumors of several hereditary cancer syndromes differ enough from sporadic cases to be discerned on a clinicopathological basis. For example, in adenomatous polyposis coli (*APC*) hundreds of small polyps are found in the large intestines of carriers of *APC*-gene mutations. In hereditary breast cancer, the tumors of *BRCA1* mutation carriers are nearly all aggressive, grade III tumors and of negative hormone receptor status (Breast Cancer Linkage Consortium, 1997). Finding such differences between hereditary and sporadic cases of prostate cancer would greatly facilitate identification of the susceptibility genes.

In search of distinguishing features of hereditary prostate cancer Carter *et al.* (1993) examined a number of clinical features: clinical stage, preoperative PSA level, pathological stage, and prostate weight, in tumors from prostate cancer families. No differences were found in the three groups divided on the basis of the degree of hereditary prostate cancer characteristics being present in the family. The first group consisted of men who were considered as having a hereditary form of the disease: 3 or more affected in a single generation, prostate cancer occurred in three successive generations, or two cases of prostate cancer diagnosed before the age of 55. The second and third groups consisted of individuals with either no affected family members, or family members were affected, but not to the degree as in the first group.

Bilateral breast cancer is considered one of the hallmarks of the hereditary form of the disease. In contrast, prostate cancer is a multi-focal disease also in the sporadic form, and no difference in the number of tumor foci is observed between sporadic and hereditary cases (Bastacky *et al.*, 1995). The effect of family history on the progression of prostate cancer is controversial. If measured by PSA elevation after initial treatment by radical prostatectomy, no difference is noted between the hereditary and non-hereditary forms (Bova *et al.*, 1998). Kupelian *et al.* (1997), however, have reported that tumors in hereditary patients have a greater tendency to progress more rapidly than in sporadic cases. Death as a result of prostate cancer has been reported to be more frequent in affected men with a family history of the disease (Rodriguez *et al.*, 1998), although conflicting reports exist also (Gronberg *et al.*, 1998)

The genetic heterogeneity behind hereditary prostate cancer may well have been a confounding factor in these studies. Determination of definite correlations between genotype and phenotype will be possible only after the identification of the genes responsible for hereditary prostate cancer. However, studies characterizing those families that are linked to distinct genetic regions have generated interesting results. Families that are potentially linked to *HPC1* have a significantly lower age at diagnosis (63.7 versus 65.9 years) than those that show no evidence of linkage to 1q24–25 (Gronberg *et al.*, 1997b). There was an excess of higher grade cancers in the linked group and advanced stage disease was also more common in the families with haplotype sharing at the *HPC1* locus. The 1q-linked families tended also to have more affected cases. In contrast, families that are linked to Xq27–28, have a high mean age at diagnosis (over 65 years), and are more likely to have fewer men affected. In the *HPC1* families a non-significant trend towards excess of breast and colon cancer was seen. The finding of *CAPB* was based on a subset of families having both prostate and breast cancer (Gibbs *et al.*, 1999b). The elucidation of the real nature of these associations awaits the identification of the prostate susceptibility genes. Already, however, these data are being used to help in the search for these genes.

References

Agus, D.B., Scher, H.I., Higgins, B., Fox, W.D., Heller, G., Fazzari, M., Cordon-Cardo, C. and Golde, D.W. (1999) Response of prostate cancer to anti-Her-2/neu antibody in androgen-dependent and -independent human xenograft models. *Cancer Res.* 59: 4761–4764.

Ahlbom, A., Lichtenstein, P., Malmstrom, H., Feychting, M., Hemminki, K. and Pedersen, N.L. (1997) Cancer in twins: genetic and nongenetic familial risk factors. *J. Natl. Cancer Inst.* 89: 287–293.

Anderson, D.E. and Badzioch, M.D. Breast cancer risks in relatives of male breast cancer patients. *J. Natl. Cancer Inst.* 84: 1114–1117.

Bastacky, S.I., Wojno, K.J., Walsh, P.C., Carmichael, M.J. and Epstein, J.I. (1995) Pathological features of hereditary prostate cancer. *J. Urol.* 153: 987–992.

Bello, M.J., de Campos, J.M., Kusak, E.M., Vaquero, J., Sarasa, J.L., Pestana, A. and Rey, J.A. (1994a) Allelic loss at 1p is associated with tumor progression of meningiomas. *Genes Chromosomes Cancer* 9: 296–298

Bello, M.J., Leone, P.E., Nebreda, P. *et al.* (1995a) Allelic status of chromosome 1 in neoplasms of the nervous system. *Cancer Genet Cytogenet* 83: 160–164.

Bello, M.J., Leone, P.E., Vaquero, J., de Campos, J.M., Kusak, M.E., Sarasa, J.L., Pestana, A. and Rey, J.A. (1995b) Allelic loss at 1p and 19q frequently occurs in association and may represent early oncogenic events in oligodendroglial tumors. *Internat. J. Cancer* 64: 207–210

Bello, M.J., Vaquero, J., de Campos, J.M., Kusak, M.E., Sarasa, J.L., Saez-Castresana, J.L., Pestana, A. and Rey, J.A. (1994b) Molecular analysis of chromosome 1 abnormalities in human gliomas reveals frequent loss of 1p in oligodendroglial tumors. *Internat. J. Cancer* 57: 172–175

Berthon, P., Valeri, A., Cohen-Akenine, A. *et al.* (1998) Predisposing gene for early-onset prostate cancer, localized on chromosome 1q42.2–43. *Am. J. Hum. Genet.* 62: 1416–1424.

Bookstein, R., Bova, G.S., MacGrogan, D., Levy, A. and Isaacs, W.B. (1997) Tumor-suppressor genes in prostatic oncogenesis: a positional approach. *Br. J. Urol.* [Supplement] 79: 28–36.

Bookstein, R., MacGrogan, D., Hilsenbeck, S.G., Sharkey, F., Allred, D.C. (1993) p53 is mutated in a subset of advanced-stage prostate cancers. *Cancer Res.* 53: 3369–3373.

Bova, G.S., Carter, B.S., Bussemakers, M.J.G. *et al.* (1993) Homozygous deletion and frequent allelic loss of chromosome 8p22 loci in human prostate cancer. *Cancer Res.* 53: 3869–3873.

Bova, G.S., Partin, A.W., Isaacs, S.D., Carter, B.S., Beaty, T.L., Isaacs, W.B. and Walsh, P.C. (1998) Biological aggressiveness of hereditary prostate cancer: long-term evaluation following radical prostatectomy. *J. Urol.* **160**: 660–663.

Braun, M.M., Helzlsouer, K.J., Hollis, B.W. and Comstock, G.W. (1995) Prostate cancer and prediagnostic levels of serum vitamin D metabolites (Maryland, United States). *Cancer Causes Control* **6**: 235–239.

Breast Cancer Linkage Consortium. (1997) Pathology of familial breast cancer: differences between breast cancers in carriers of BRCA1 or BRCA2 mutations and sporadic cases. *Lancet* **349**: 1505–1510

Bubendorf, L., Kononen, J., Koivisto, P. *et al.* (1999) Survey of gene amplifications during prostate cancer progression by high-throughout fluorescence in situ hybridization on tissue microarrays. (Published erratum appears in Cancer Research 1999, 5: 1388). *Cancer Res.* **59**: 803–806.

Buttyan, R., Sawczuk, I.S., Benson, M.C., Siegal, J.D. and Olsson, C.A. (1987) Enhanced expression of the c-myc protooncogene in high-grade human prostate cancers. *Prostate* **11**: 327–337.

Cairns, P., Okami, K., Halachmi, S. *et al.* (1997) Frequent inactivation of PTEN/MMAC1 in primary prostate cancer. *Cancer Res.* **57**: 4997–5000.

Cannon, L., Bishop, D.T., Skolnick, M., Hunt, S., Lyon, J.L. and Smart, C.R. (1982) Genetic epidemiology of prostate cancer in the Utah Mormon genealogy. *Cancer Surveys* **1**: 47–69.

Cannon-Albright, L. and Eeles, R. (1995) Progress in prostate cancer. *Nature Genet.* **9**: 336–338

Carter, B.S., Beaty, T.H., Steinberg, G.D., Childs, B. and Walsh, P.C. (1992) Mendelian inheritance of familial prostate cancer. *Proc. Natl. Acad. Sci. USA* **89**: 3367–3371.

Carter, B.S., Bova, G.S., Beaty, T.H., Steinberg, G.D., Childs, B., Isaacs, W.B. and Waalsh, P.C. (1993) Hereditary prostate cancer: epidemiologic and clinical features. [Review]. *J. Urol.* **150**: 797–802.

Carter, B.S., Bova, S., Beaty, T.H., Steinberg, G.D., Childs, B., Isaacs, W.B. and Walsh, P.C. (1993) Hereditary prostate cancer: epidemiologic and clinical features. *J. Urol.* **150**: 797–802

Carter, B.S., Carter, H.B. and Isaacs, J.T. (1990) Epidemiologic evidence regarding predisposing factors to prostate cancer. *Prostate* **16**: 187–197.

Carter, B.S., Epstein, J.I. and Isaacs, W.B. (1990) ras gene mutations in human prostate cancer. *Cancer Res.* **50**: 6830–6832.

Cerhan, J.R., Parker, A.S., Putnam, S.D., Chiu, B.C., Lynch, C.F., Cohen, M.B., Torner, J.C. and Cantor, K.P. (1999) Family history and prostate cancer risk in a population-based cohort of Iowa men. *Cancer Epidemiol. Bio. Prevent.* **8**: 53–60.

Chamberlain, N.L., Driver, E.D. and Miesfeld, R.L. (1994) The length and location of CAG trinucleotide repeats in the androgen receptor N-terminal domain affect transactivation function. *Nucleic Acids Res.* **22**: 3181–3186.

Cher, M.L., Bova, G.S., Moore, D.H., Small, E.J., Carroll, P.R., Pin, S.S., Epstein, J.I., Isaacs, W.B. and Jensen, R.H. (1996) Genetic alterations in untreated metastases and androgen-independent prostate cancer detected by comparative genomic hybridization and allelotyping. *Cancer Res.* **56**: 3091–3102.

Cher, M.L., MacGrogan, D., Bookstein, R., Brown, J.A., Jenkins, R.B. and Jensen, R. (1994) Comparative genomic hybridization, allelic imbalance, and fluorescence in situ hybridization on chromosome 8 in prostate cancer. *Genes Chromosomes Cancer* **11**: 153–162.

Colombel, M., Symmans, F., Gil, S., O'Toole, K.M., Chopin, D., Benson, M., Olsson, C.A., Korsmeyer, S. and Buttyan, R. (1993) Detection of the apoptosis-suppressing oncoprotein bcl-2 in hormone-refractory human prostate cancers. *Am. J. Pathol.* **143**: 390–400.

Cooncy, K.A., McCarthy, J.D., Lange, E., Huang, L., Miesfeldt, S., Montie, J.E., Oesterling, J.E., Sandler, H.M. and Lange, K. (1997) Prostate cancer susceptibility locus on chromosome 1q: a confirmatory study [see comments]. *J. Natl. Cancer Inst.* **89**: 955–959.

Corder, E.H., Guess, H.A., Hulka, B.S. *et al.* (1993) Vitamin D and prostate cancer: a prediagnostic study with stored sera. *Cancer Epidemiol. Biomarkers. Prevent.* **2**: 467–472.

Craft, N., Shostak, Y., Carey, M. and Sawyers, C.L. (1999) A mechanism for hormone-independent prostate cancer through modulation of androgen receptor signaling by the HER-2/neu tyrosine kinase. *Nature Med.* **3**: 280–285.

Cunningham, J.M., Shan, A., Wick, M.J. *et al.* (1996) Allelic imbalance and microsatellite instability in prostatic adenocarcinoma. *Cancer Res.* **56**: 4475–4484.

Devgan, S.A., Henderson, B.E., Yu, M.C., Shi, C.Y., Pike, M.C., Ross, R.K., Reichardt, J.K. (1997) Genetic variation of 3 beta-hydroxysteroid dehydrogenase type II in three racial/ethnic groups: implications for prostate cancer risk. *Prostate* **33**: 9–12.

Dong, J.T., Lamb, P.W., Rinker-Schaeffer, C.W., Vukanovic, J., Ichikawa, T., Isaacs, J.T. and Barrett, J.C. (1995) KAI1, a metastasis suppressor gene for prostate cancer on human chromosome 11p11. *Science* 268: 884–886

Dong, J.T., Suzuki, H., Pin, S.S., Bova, G.S., Schalken, J.A., Isaacs, W.B., Barrett, J.C. and Isaacs, J.T. (1996) Down-regulation of the KAI1 metastasis suppressor gene during the progression of human prostatic cancer infrequently involves gene mutation allelic loss. *Cancer Res.* 56: 4387–4390

Durrin, L.K., Haile, R.W., Ingles, S.A. and Coetzee, G.A. (1999) Vitamin D receptor 3'-untranslated region polymorphisms: lack of effect on mRNA stability. *Biochimia Biophysica Acta.* 1453: 311–320.

Eeles, R.A., Durocher, F., Edwards, S. *et al.* (1998) Linkage analysis of chromosome 1q markers in 136 prostate cancer families. The Cancer Research Campaign/British Prostate Group U.K. Familial Prostate Cancer Study Collaborators. *Am. J. Hum. Genet.* 62: 653–658.

Ekman, P., Pan, Y., Li, C. and Dich, J. (1997) Environmental and genetic factors: a possible link with prostate cancer. *Br. J. Urol.* 79 Suppl 2: 35–41.

Emmert-Buck, M.R., Vocke, C.D., Pozzatti, R.O. *et al.* Allelic loss on chromosome 8p12–21 in microdissected prostatic intraepithelial neoplasia. *Cancer Res.* 55: 2959–2962.

Feuer., E.J. (1997) Lifetime probability of cancer. *J. Natl. Cancer Inst.* 89: 279.

Fleming, W.H., Hamel, A., MacDonald, R., Ramsey, E., Pettigrew, N.M., Johnston, B., Dodd, J.G. and Matusik, R.J. (1986) Expression of the c-myc proto-oncogene in human prostatic carcinoma and benign prostatic hyperplasia. *Cancer Res.* 46: 1535–1538.

Ford, D., Easton, D.F., Bishop, D.T., Narod, S.A. and Goldgar, D.E. (1994) Risks of cancer in BRCA1-mutation carriers. Breast Cancer Linkage Consortium. *Lancet* 343: 692–695.

Gaddipati, J.P., McLeod, D.G., Sesterhenn, I.A., Hussussian, C.J., Tong, Y.A., Seth, P., Dracopoli, N.C., Moul, J.W. and Srivastava, S. (1997) Mutations of the p16 gene product are rare in prostate cancer. *Prostate* 30: 188–194.

Gann, P.H., Ma, J., Hennekens, C.H., Hollis, B.W., Haddad, J.G. and Stampfer, M.J. (1996) Circulating vitamin D metabolites in relation to subsequent development of prostate cancer. *Cancer Epidemiol. Biomarkers. Prevent.* 5: 121–126.

Gibbs, M., Chakrabarti, L., Stanford, J.L. *et al.* (1999a) Analysis of chromosome 1q42.2–43 in 152 families with high risk of prostate cancer. *Am. J. Hum. Genet.* 64: 1087–1095.

Gibbs, M., Stanford, J.L., McIndoe, R.A. *et al.* (1999b) Evidence for a rare prostate cancer-susceptibility locus at chromosome 1p36. *Am. J. Hum. Genet.* 64: 776–787.

Giovannucci, E., Stampfer, M.J., Krithivas, K., Brown, M., Dahl, D., Brufsky, A., Talcott, J., Hennekens, C.H. and Kantoff, P.W. (1997) The CAG repeat within the androgen receptor gene and its relationship to prostate cancer [published erratum appears in *Proc. Natl. Acad. Sci. USA* 1997 Jul 22;94(15): 8272]. *Proc. Natl. Acad. Sci. USA* 94: 3320–3323.

Gleave, M., Tolcher, A., Miyake, H., Nelson, C., Brown, B., Beraldi, E. and Goldie, J. (1999) Progression to androgen independence is delayed by adjuvant treatment with antisense Bcl-2 oligodeoxynucleotides after castration in the LNCaP prostate tumor model. *Clin. Cancer Res.* 10: 2891–2898.

Goldgar, D.E., Easton, D.F., Cannon-Albright, L.A. and Skolnick, M.H. (1994) Systematic population-based assessment of cancer risk in first-degree relatives of cancer probands. *J. Natl. Cancer Inst.* 86: 1600–1608.

Grönberg, H., Damber, L. and Damber, J.E. (1994) Studies of genetic factors in prostate cancer in a twin population. *J. Urol.* 152: 1484–1487; discussion 1487–1489.

Grönberg, H., Damber, L., Damber, J.E. and Iselius, L. (1997a) Segregation analysis of prostate cancer in Sweden: support for dominant inheritance. *Am. J. Epidemiol.* 146: 552–557.

Grönberg, H., Damber, L., Tavelin, B. and Damber, J.E. (1998) No difference in survival between sporadic, familial and hereditary prostate cancer. *Br. J. Urol.* 82: 564–567.

Grönberg, H., Isaacs, S.D., Smith, J.R.. *et al.* (1997b) Characteristics of prostate cancer in families potentially linked to the hereditary prostate cancer 1 (HPC1) locus [see comments]. *JAMA* 278: 1251–1255.

Gross, C., Musiol, I.M., Eccleshall, T.R., Malloy, P.J. and Feldman, D. (1998) Vitamin D receptor gene polymorphisms: analysis of ligand binding and hormone responsiveness in cultured skin fibroblasts. *Biochem. Biophys. Res. Commun.* 242: 467–473.

Hayes, R.B., Liff, J.M., Pottern, L.M. *et al.* (1995) Prostate cancer risk in U.S. blacks and whites with a family history of cancer. *Internat. J. Cancer* 60: 361–364.

He, W.W., Sciavolino, P.J., Wing, J. *et al.* (1997) A novel human prostate-specific, androgen-regulated homeobox gene (NKX3.1) that maps to 8p21, a region frequently deleted in prostate cancer. *Genomics* **43**: 69–77.

Heidenberg, H.B., Sesterhenn, I.A., Gaddipati, J.P., Weghorst, C.M., Buzard, G.S., Moul, J.W. and Srivastava, S. (1995) Alteration of the tumor suppressor gene p53 in a high fraction of hormone refractory prostate cancer. *Urology* **54**: 414–421.

Henrikkson, M. and Lusher, B. (1996) Proteins of the Myc network: essential regulators of cell growth and differentiation. *Adv. Cancer Res.* **68**: 109–182.

Hsieh, C.L., Oakley-Girvan, I., Gallagher, R.P. *et al.* (1997) Re: prostate cancer susceptibility locus on chromosome 1q: a confirmatory study [letter; comment]. *J. Natl. Cancer Inst.* **89**: 1893–1894.

Hyytinen, E.R., Frierson, H.F. Jr., Boyd, J.C., Chung, L.W. and Dong, J.T. (1999) Three distinct regions of allelic loss at 13q14, 13q21–22, and 13q33 in prostate cancer. *Genes Chromosomes Cancer* **25**: 108–14.

Inazawa, J., Sasaki, H., Hagura, K., Kakazu, N, Abe, T. and Sasaki, T. (1996) Precise localization of the human gene encoding cell adhesion kinase beta (CAK beta/ PYK2) to chromosome 8 at p21.1 by fluorescence in situ hybridization. *Hum. Genet.* **98**: 508–510.

Ingles, S.A., Ross, R.K., Yu, M.C., Irvine, R.A., La Pera, G., Haile, R.W. and Coetzee, G.A. (1997) Association of prostate cancer risk with genetic polymorphisms in vitamin D receptor and androgen receptor [see comments]. *J. Natl. Cancer Inst.* **89**: 166–170.

Isaacs, S.D., Kiemeney, L.A., Baffoe-Bonnie, A., Beaty, T.H. and Walsh, P.C. (1995) Risk of cancer in relatives of prostate cancer probands. *J. Natl. Cancer Inst.* **87**: 991–996.

Isaacs, W.B. and Bova, G.S. (1998) Prostate cancer. In: Vogelstein B, Kinzler KW, (eds). *The Genetic Basis of Human Cancer.* New York: McGraw-Hill, pp 653–660.

Jacobsen, S.J., Katusic, S.K., Bergstralh, E.J., Oesterling, J.E., Ohrt, D., Klee, G.G. and Chute, C.G. (1995) Incidence of prostate cancer diagnosis in the eras before and after serum prostate-specific antigen testing. *JAMA* **274**: 1445–1449.

Jarrard, D.F., Bova, G.S., Ewing, C.M. *et al.* (1997) Deletional, mutational, and methylation analyses of CDKN2 (p16/MTS1) in primary and metastatic prostate cancer. *Genes Chromosomes Cancer* **19**: 90–96.

Joos, S., Bergerheim, U., Pan, Y., Matsuyama, H., Bentz, M., duManoir, S. and Lichter, P. (1995) Mapping of chromosomal gains and losses in prostate cancer by comparative genomic hybridization. *Genes Chromosomes Cancer* **14**: 267–276.

Kantoff, P., Giovannucci, E. and Brown, M. (1998) The androgen receptor CAG repeat polymorphism and its relationship to prostate cancer. *Biochimia Biophysica Acta.* **1378**: C1–5.

Kazemi-Esfarjani, P., Trifiro, M.A. and Pinsky, L. (1995) Evidence for a repressive function of the long polyglutamine tract in the human androgen receptor: possible pathogenetic relevance for the (CAG)n-expanded neuronopathies. *Hum. Mol. Genet.* **4**: 523–527.

Keetch, D.W., Humphrey, P.A., Smith, D.S., Stahl, D. and Catalona, W.J. (1996) Clinical and pathological features of hereditary prostate cancer. *J. Urol.* **155**: 1841–1843.

Kibel, A.S., Isaacs, S.D., Isaacs, W.B. and Bova, G.S. (1998) Vitamin D receptor polymorphisms and lethal prostate cancer. *J. Urol.* **160**: 1405–1409.

Koivisto, P., Kolmer, M., Visakorpi, T. and Kallioniemi, O.P. (1998) Androgen receptor gene and hormonal therapy failure of prostate cancer. *Am. J. Pathol.* **152**: 1–9.

Kosary, C.L., Ries, L.A.G., Miller, B.A., Hankey, B.F., Harras, A. and Edwards, B.K. (eds) (1995) SEER cancer statistics review, 1973–1992: tables and graphs. NIH publ 96–2789. National Cancer Institute, Bethesda.

Krajewska, M., Kraajewski, S., Epstein, J.I., Shabaik, A., Saauvageot, J., Song, K., Kitadaa, S. and Reed, J.C. (1996) Immunohistochemical analysis of bcl-2, bax, bcl-X, and mcl-1 expression in prostate cancers. *Am. J. Pathol.* **148**: 1567–1576.

Kraus, J.A., Koopmann, J., Kaskel, P., Maintz, D., Brandner, S., Schramm, J., Louis, D.N., Wiestler, O.D. and von Deimling A. (1995) Shared allelic losses on chromosomes 1p and 19q suggest a common origin of oligodendroglioma and oligoastrocytoma. *J. Neuropathol. Experiment. Neurol.* **54**: 91–95

Krill, D., Stoner, J., Konety, B.R., Becich, M.J. and Getzenberg, R.H. (1999) Differential effects of vitamin D on normal human prostate epithelial and stromal cells in primary culture. *Urology* 1999; **54**: 171–177.

Kubota, Y., Shuin, T., Uemura, H., Fujinami, K., Miyamoto, H., Torigoe, S., Dobashi, Y., Kitamura, H., Iwasaki, Y. and Danenberg, K. (1995) Tumor suppressor gene p53 mutations in human prostate cancer. *Prostate* 27: 18–24

Kupelian, P.A., Kupelian, V.A., Witte, J.S., Macklis, R. and Klein, E.A. (1997) Family history of prostate cancer in patients with localized prostate cancer: an independent predictor of treatment outcome. *J. Clin. Oncol.* 15: 1478–1480.

Langston, A.A., Stanford, J.L., Wicklund, K.G., Thompson, J.D., Blazej, R.G. and Ostrander, E.A. (1996) Germ-line BRCA1 mutations in selected men with prostate cancer [letter]. *Am. J. Hum. Genet.* 58: 881–884.

Lehrer, S., Fodor, F., Stock, R.G., Stone, N.N., Eng, C., Song, H.K. and McGovern, M. Absence of 185delAG mutation of the BRCA1 gene and 6174delT mutation of the BRCA2 gene in Ashkenazi Jewish men with prostate cancer. *Br. J. Cancer* 78: 771–773.

Lichtenstein, P., Holm, N.V., Verkasalo, P.K., Iliadou, A., Kaprio, J., Koskenvuo, M., Pukkala, E., Skytthe, A. and Hemminki, K. (2000) Environmental and heritable factors in the causation of cancer – Analyses of cohorts of twins from Sweden, Denmark, and Finland. *New Engl. J. Med.* 343.

Lyles, K.W., Berry, W.R., Haussler, M., Harrelson, J.M. and Drezner, M.K. (1980) Hypophosphatemic osteomalacia: association with prostatic carcinoma. *Annals Intern. Med.* 93: 275–278.

Ma, J., Stampfer, M.J., Gann, P.H., Hough, H.L., Giovannucci, E., Kelsey, K.T., Hennekens, C.H. and Pollak, M. (1998) Vitamin D receptor polymorphisms, circulating vitamin D metabolites, and risk of prostate cancer in United States physicians. *Cancer Epidemiol. Biomarkers. Prevent.* 7: 385–390.

Macoska, J.A., Trybus, T.M., Benson, P.D., Sakr, W.A., Grignon, D.J., Wojno, K.D., Pietruk, T. and Powell, I.J. (1995) Evidence for three tumor suppressor loci on chromosome 8p in human prostate cancer. *Cancer Res.* 55: 5390–5395.

Makridakis, N., Ross, R.K., Pike, M.C. *et al.* (1997) A prevalent missense substitution that modulates activity of prostatic steroid 5alpha-reductase. *Cancer Res.* 57: 1020–1022.

Maris, J.M., White, P.S., Beltinger, C.P., Sulman, E.P., Castleberry, R.P., Shuster, J.J., Look, A.T. and Brodeur, G.M. (1995) Significance of chromosome 1p loss of heterozygosity in neuroblastoma. *Cancer Res.* 55: 4664–4669

Mark, H.F., Feldman, D., Das, S., Kye, H., Mark, S., Sun, C.L. and Samy, M. (1999) Fluorescence in situ hybridization study of HER-2/neuoncogene amplification in prostate cancer. *Experiment. Mol. Pathol.* 66: 170–178.

Mashimo, T., Watabe, M., Hirota, S., Hosobe, S., Miura, K., Tegtmeyer, P.J., Rinker-Shaeffer, C.W. and Watabe, K. (1998) The expression of the KAI1 gene, a tumor metastasis suppressor, is directly activated by p53. *Proc. Natl. Acad. Sci. USA* 95: 11307–11311.

McCahy, P.J., Harris, C.A. and Neal, D.E. (1996) Breast and prostate cancer in the relatives of men with prostate cancer. *Br. J. Urol.* 78: 552–556.

McDonnell, T.J., Tronsoco, P., Brisbay, S.M., Logothetis, C., Chung, L.W., Hsieh, J.T., Tu, S.M. and Campbell, M.L. (1992) Expression of the protooncogene bcl-2 in the prostate and its association with emergence of androgen-independent prostate cancer. *Cancer Res.* 52: 6940–6944.

McIndoe, R.A., Stanford, J.L., Gibbs, M. *et al.* (1997) Linkage analysis of 49 high-risk families does not support a common familial prostate cancer-susceptibility gene at 1q24–25. *Am. J. Hum. Genet.* 61: 347–353.

Meikle, A.W. and Stanish, W.M. (1982) Familial prostatic cancer risk and low testosterone. *J. Clin. Endocrinol. Metab.* 54: 1104–1108.

Miller, G.J., Stapleton, G.E., Hedlund, T.E. and Moffat, K.A. (1995) Vitamin D receptor expression, 24-hydroxylase activity, and inhibition of growth by 1alpha,25-dihydroxyvitamin D3 in seven human prostatic carcinoma cell lines. *Clin. Cancer Res.* 1: 997–1003.

Monroe, K.R., Yu, M.C., Kolonel, L.N., Coetzee, G.A., Wilkens, L.Y., Ross, R.K. and Henderson, B.E. (1995) Evidence of an X-linked or recessive genetic component to prostate cancer risk. *Nature Med.* 1: 827–829

Morote, J., de Torres, I., Caceres, C., Vallejo, C., Schwartz, S. Jr. and Reventos. J. (1999) Prognostic value of immunohistochemical expression of the c-erbB-2 oncoprotein in metastasic prostate cancer. *Intern. J. Cancer* 84: 421–425.

Morton, R.A., Ewing, C.M., Nagafuchi, A., Tsukita, S., Isaacs, W.B. (1993) Reduction of E-cadherin levels and deletion of the alpha-catenin gene inhuman prostate cancer cells. *Cancer Res.* 53: 3585–3590.

Muir, C.S., Waterhouse, J., Mack, T., Powell, J., Whelan, S. and Cook, P.J. (eds) (1987) Cancer incidence in five continents. Vol 5. IARC sci publ 88. IARC, Lyon.

Murant, S.J., Rolley, N., Phillips, S.M., Stower, M. and Maitland, N.J. (2000) Allelic imbalance within the E-cadherin gene is an infrequent event in prostate carcinogenesis. *Genes Chromosomes Cancer* 27: 104–109.

Narod, S.A., Dupont, A., Cusan, L., Diamond, P., Gomez, J.L., Suburu, R. and Labrie, F. (1995) The impact of family history on early detection of prostate cancer [letter]. *Nature Med.* 1: 99–101.

Nastiuk, K.L., Mansukhani, M., Terry, M.B., Kularatne, P., Rubin, M.A., Melamed, J., Gammon, M.D., Ittmann, M. and Krolewski, J.J. (1999) Common mutations in BRCA1 and BRCA2 do not contribute to early prostate cancer in Jewish men. *Prostate* 40: 172–177.

Neuhausen, S.L., Farnham, J.M., Kort, E., Tavtigian, S.V., Skolnick, M.H. and Cannon-Albright, L.A. (1999) Prostate cancer susceptibility locus HPC1 in Utah high-risk pedigrees. *Hum. Mol. Genet.* 8: 2437–2442.

Nupponen, N. and Visakorpi, T. (1999) Molecular biology of progression of prostate cancer. *Euro. J. Urol.* 35: 351–354.

Nupponen, N., Kakkola, L., Koivisto, P. and Visaakorpi, T. (1998) Genetic alterations in hormone-refractory recurrent prostate carcinomas. *Am. J. Pathol.* 153: 141–148.

Page, W.F., Braun, M.M., Partin, A.W., Caporaso, N. and Walsh, P. (1997) Heredity and prostate cancer: a study of World War II veteran twins. *Prostate* 33: 240–245.

Partin, A.W., Isaacs, J.T., Treiger, B., Coffey, D.S. (1988) Early cell motility changes associated with an increase in metastatic ability in rat prostatic cancer cells transfected with the v-Harvey-ras oncogene. *Cancer Res.* 48: 50–53.

Peehl, D.M., Skowronski, R.J., Leung, G.K., Wong, S.T., Stamey, T.A. and Feldman, D. (1994) Antiproliferative effects of 1,25-dihydroxyvitamin D3 on primary cultures of human prostatic cells. *Cancer Res.* 54: 805–810.

Rodriguez, C., Calle, E.E., Tatham, L.M., Wingo, P.A., Miracle-McMahill, H.L., Thun, M.J. and Heath, C.W. Jr. (1998) Family history of breast cancer as a predictor for fatal prostate cancer. *Epidemiology* 9: 525–529.

Schaid, D.J., McDonnell, S.K., Blute, M.L. and Thibodeau, S.N. (1998) Evidence for autosomal dominant inheritance of prostate cancer. *Am. J. Hum. Genet.* 62: 1425–1438.

Schleiermacher, G., Peter, M., Michon, J., Hugot, J.P., Vielh, P., Zucker, J.M., Magdelenat, H., Thomas, G. and Delattre, O. (1994) Two distinct deleted regions on the short arm of chromosome 1 in neuroblastoma. *Genes Chromosomes Cancer* 10: 275–281

Schuurman, A.G., Zeegers, M.P., Goldbohm, R.A. and van den Brandt, P.A. (1999) A case-cohort study on prostate cancer risk in relation to family history of prostate cancer. *Epidemiol.* 10: 192–195.

Schwartz, G.G. and Hulka, B.S. (1990) Is vitamin D deficiency a risk factor for prostate cancer? (Hypothesis). *AntiCancer Res.* 10: 1307–1311.

Smith, J.R., Freije, D., Carpten, J.D. *et al.* (1996) Major susceptibility locus for prostate cancer on chromosome 1 suggested by a genome-wide search [see comments]. *Science* 274: 1371–1374.

Sobue, G., Doyu, M., Morishima, T., Mukai, E., Yasuda, T., Kachi, T. and Mitsuma, T. (1994) Aberrant androgen action and increased size of tandem CAG repeat in androgen receptor gene in X-linked recessive bulbospinal neuronopathy. *J. Neurol. Sci.* 121: 167–171.

Spitz, M.R., Currier, R.D., Fueger, J.J., Babaian, R.J. and Newell, G.R. (1991) Familial patterns of prostate cancer: a case-control analysis. *J. Urol.* 146: 1305–1307.

Stanford, J.L., Just, J.J., Gibbs, M., Wicklund, K.G., Neal, C.L., Blumenstein, B.A. and Ostrander, E.A. (1997) Polymorphic repeats in the androgen receptor gene: molecular markers of prostate cancer risk. *Cancer Res.* 57: 1194–1198.

Stearns, M.E. and McGarvey, T. (1992) Prostate cancer: therapeutic, diagnostic, and basic studies. *Lab. Invest.* 67: 540–552.

Steinberg, G.D., Carter, B.S., Beaty, T.H., Childs, B. and Walsh, P.C. (1990) Family history and the risk of prostate cancer. *Prostate* 17: 337–347.

Struewing, J.P., Hartge, P., Wacholder, S., Baker, S.M., Berlin, M., McAdams, M., Timmerman, M.M., Brody, L.C. and Tucker, M.A. (1997) The risk of cancer associated with specific mutations of BRCA1 and BRCA2 among Ashkenazi Jews. *New Engl. J. Med.* 336: 1401–1408.

Suarez, B.K., Lin, J., Burmester, J.K. *et al.* (2000) A genome screen of multiplex sibships with prostate cancer. *Am. J. Hum. Genet.* 66: 933–944.

Takahashi, S., Qian, J., Brown, J.A., Alcaraz, A., Bostwick, D.G., Lieber, M.M. and Jenkins, R.B. (1994) Potential markers of prostate cancer aggressiveness detected by fluorescence in situ hybridization in needle biopsies. *Cancer Res.* 54: 3574–3579.

Takahashi, S., Shan, A.L., Ritland, S.R., Delacey, K.A., Bostwick, D.G., Lieber, M.M., Thibodeau, S.N. and Jenkins, R.B. (1995) Frequent loss of heterozygosity at 7q31.1 in primary prostate cancer is associated with tumor aggressiveness and progression. *Cancer Res.* 55: 4114–4119.

Tamimi, Y., Bringuier, P.P., Smit, F., van Bokhoven, A., Debruyne, F.M. and Schalken, J.A. (1996) p16 mutations/deletions are not frequent events in prostate cancer. *Br. J. Cancer* 74: 120–122.

Taplin, M.E., Bubley, G.J., Shuster, T.D., Frantz, M.E., Spooner, A.E., Ogata, G.K., Keer, H.N. and Balk, S.P. (1995) Mutation of the androgen-receptor gene in metastatic androgen-independent prostate cancer. *New Engl. J. Med.* 332: 1393–1398.

Taylor, J.A., Hirvonen, A., Watson, M., Pittman, G., Mohler, J.L. and Bell, D.A. (1996) Association of prostate cancer with vitamin D receptor gene polymorphism. *Cancer Res.* 56: 4108–4110.

Thiessen, E.U. (1974) Concerning a familial association between breast cancer and both prostatic and uterine malignancies. *Cancer* 34: 1102–1107.

Tilley, W.D., Buchanan, G., Hickey, T.E. and Bentel, J.M. (1996) Mutations in the androgen receptor gene are associated with progression of human prostate cancer to androgen independence. *Clin. Cancer Res.* 2: 277–285.

Tricoli, J.V., Gumerlock, P.H., Yao, J.L., Chi, S.G., D'Souza, S.A., Nestok, B.R. and deVere White, R.W. (1996) Alterations of the retinoblastoma gene in human prostate adenocarcinoma. *Genes Chromosomes Cancer* 15: 108–114.

Umbas, R., Schalken, J.A., Aalders, T.W., Carter, B.S., Karthaus, H.F., Schaafsma, H.E., Debruyne, F.M. and Isaacs, W.B. (1992) Expression of the cellular adhesion molecule E-cadherin is reduced or absent in high-grade prostate cancer. *Cancer Res.* 52: 5104–9.

Van Den Berg, C., Guan, X.Y., Von Hoff, D. *et al.* (1995) DNA sequence amplification in human prostate cancer identified by chromosome microdissection: potential prognostic implications. *Clin. Cancer Res.* 1: 11–18.

Visakorpi, T., Hyytinen, E., Koivisto, P. *et al.* (1995b) In vivo amplification of the androgen receptor gene and progression of human prostate cancer. *Nature Genet.* 9: 401–406.

Visakorpi, T., Kallioniemi, A.H., Syvanen, A.-C., Hyytinen, E.R., Karhu, R., Tammela, T., Isola, J.J. and Kallioniemi, O.P. (1995a) Genetic changes in primary and recurrent prostate cancer by comparative genomic hybridization. *Cancer Res.* 55: 342–347.

White, P.S., Maris, J.M., Beltinger, C. *et al.* (1995) A region of consistent deletion in neuroblastoma maps within human chromosome 1p36.2–36.3. *Proc. Natl. Acad. Sci. USA* 92: 5520–5524

Whittemore, A.S., Wu, A.H., Kolonel, L.N., John, E.M., Gallagher, R.P., Howe, G.R. and West, D.W. (1995) Family history and prostate cancer risk in black, white, and Asian men in the United States and Canada. *Am. J. Epidemiol.* 141: 732–740.

Wilkens, E.P., Freije, D., Xu, J. *et al.* (1999) No evidence for a role of BRCA1 or BRCA2 mutations in Ashkenazi Jewish families with hereditary prostate cancer. *Prostate* 39: 280–284.

Woolf, C.M. (1960) An investigation of familial aspects of carcinoma of the prostate. *Cancer* 13: 739–744.

Xu, J., International Consortium for Prostate Cancer Genetics. (2000) Combined Analysis of Hereditary Prostate Cancer Linkage to 1q24–25: Results from 772 Hereditary Prostate Cancer Families from the International Consortium for Prostate Cancer Genetics. *Am. J. Hum. Genet.* 66: 945–957.

Xu, J., Meyers, D., Freije, D. *et al.* (1998) Evidence for a prostate cancer susceptibility locus on the X chromosome. *Nature Genet.* 20: 175–179.

12

Tuberous sclerosis complex

M. Nellist, S. Verhoef, D. Lindhout, D.J.J. Halley and
A.M.W. van den Ouweland

1. Clinical aspects of tuberous sclerosis complex

Tuberous sclerosis complex (TSC) is an autosomal dominant disease with an estimated prevalence of about 1: 10 000 at birth and about 1: 15 000 by 5 years of age
(Hunt and Lindenbaum, 1984). TSC belongs to the group of neuro-ectodermal
disorders, that have in common a combination of neurological abnormalities
(epilepsy, mental retardation, structural brain abnormalities), abnormalities of
the skin (disorders in pigmentation) and benign tumor growths (hamartomas).
Besides TSC, also called Morbus Bourneville Pringle, the most prevalent diseases
in this quite heterogeneous group, are M. Von Recklinghausen (neurofibromatosis type 1), Von Hippel Lindau disease and Sturge-Weber angiomatosis. The
symptoms of these conditions not only involve the derivatives of the ectodermal
germinal layer, but may affect mesodermally derived tissues as well.

The first accurate description of the brain pathology of TSC is generally given
to Bourneville, who proposed the name 'sclérose tubéreuse' for the disease
(Bourneville, 1880). Earlier, Von Recklinghausen had described congenital
tumors in the heart and 'sclcroses' in the brain of a newborn child (Von
Recklinghausen, 1862). The most important current reference book on TSC is
undoubtedly Gomez' monograph on TSC, considered a standard work for clinicians and researchers in TSC (Gomez, 1999). Recent reviews on the clinical
aspects of tuberous sclerosis, both general and neurological, are those of Weiner *et
al.* (1998), Harrison and Bolton (1997), and Griffiths and Martland (1997).

1.1 Clinical features of tuberous sclerosis complex

In principle, all organ systems, except the skeletal musculature, can be affected by
the hamartomatous growths in TSC patients. Neurological problems, predominantly epilepsy and mental retardation, represent the most severe expression of
TSC. Epilepsy occurs in about 70% of patients at some point in life and is often
difficult to control. Successful treatment of infantile spasms with vigabatrin

---

Molecular Genetics of Cancer second edition, edited by J.K. Cowell.

(Sabril®) and improved seizure control using a neurosurgical approach in cases where the source of epileptogenic activity could be unequivocally identified have been reported (Baumgartner et al., 1997). When the onset of epilepsy is before the age of 1 year, a large percentage of these patients have impaired mental function. About two thirds of the TSC patients with epilepsy have mental retardation. Overall, mental retardation is present in just over 50% of TSC patients and only occurs in the group with a history of epileptic seizures (unpublished data).

When a diagnosis of TSC is made or suspected, cranial imaging is recommended, as over 90% of patients show subependymal nodules, cortical tubers or both (Nixon et al., 1989). Currently, cranial tomography (CT) scan is considered the most reliable method to identify paraventricular subependymal nodules, which are often calcified. Magnetic resonance imaging (MRI) is preferred for the detection of (sub)cortical tubers and is particularly useful in young children with (yet) uncalcified paraventricular nodules.

In TSC, a wide variation of skin abnormalities can occur. Hardly distinguishable white spots (hypomelanotic macules), a mild shagreen (chagrin) patch or single subungual fibroma may be the only skin sign. On the other hand, facial angiofibromas, fibrous forehead plaques or ungual (nail) fibromas can be so severe, that they are seriously disfiguring and pose a problem for both patient and dermatologist or plastic surgeon. White spots or hypomelanotic macules are thought to occur in over 90% of TSC patients, and represent an early but not pathognomonic sign pointing at a possible diagnosis of TSC, especially when noted in a child with early onset epilepsy. The macules can easily be missed on routine physical examination, and are better visible when examination of the skin is carried out with the ultraviolet Wood lamp (wavelength 360 nm). Variation in pigmentation is natural and the contrast between pigmented and hypopigmented areas of skin increases with age. Investigations with a Wood lamp, therefore, are probably of less discriminative value in adult subjects.

The most predominant and characteristic ophthalmological sign in TSC is retinal hamartoma, often located in the periphery of the retina of the eye, and thus asymptomatic. About 25% of patients exhibit this sign, which can sometimes be quite important for confirmation of the diagnosis of TSC. Generally, retinal hamartomas are benign, rarely grow and do not need treatment. Other ophthalmologic lesions reported in TSC are retinal pigment alterations, strabismus and coloboma of the iris. However, some of these symptoms may represent chance findings, rather than true associations.

A distinguishing renal feature in TSC is the combination of multiple cysts and angiomyolipoma, a benign tumor of a mixed type. The presence of fat in this tumor (best demonstrated with a CT scan) distinguishes angiomyolipoma from renal cell carcinoma. Although mostly benign, malignant angiomyolipoma with invasive growth may occur in TSC patients (Al Saleem et al., 1998). Isolated angiomyolipoma (thus without systemic TSC) also occurs as a sporadic condition, with a strong preponderance of female patients (F:M ratio 5:1). In TSC, the predominance of female patients with angiomyolipoma as a clinical sign is less striking (F:M ratio of 3.6:1) (Van Baal, 1987). The other important renal lesions in TSC are the cysts, that can be as large and multiple as in adult type polycystic kidney disease (APKD), and difficult to distinguish from APKD on ultrasound and X-ray imaging. Although angiomyolipomas tend to grow, especially during childhood, and do not resolve, renal cysts can fluctuate and appear or disappear with time (Ewalt *et al.*, 1998). Pathological studies have shown no clear macroscopical difference between cysts in TSC and cysts in APKD. However, light microscopy revealed a difference in structure in the epithelium lining the cysts walls, with a predominance of hypertrophic and hyperplastic cells in TSC tissue compared to APKD (Bernstein, 1993). It is not known whether there is a difference in structure between cysts of patients with a TSC1 or TSC2 mutation, the genes independently involved in TSC (Section 2), as insufficient patients with known mutations have been studied. The incidence and severity of the renal lesions are roughly equal for patients with a small constitutional TSC1 or TSC2 mutation. In a small proportion of TSC patients, polycystic kidney disease (PKD) is present as a second genetic disease (Sampson *et al.*, 1997), and a large deletion can be demonstrated, disrupting both the *TSC2* gene and the adjacent *PKD1* gene in a 'contiguous gene deletion syndrome'. Patients with this syndrome present with enlarged kidneys in infancy or childhood and the diagnosis of polycystic kidney disease sometimes precedes the diagnosis of TSC by many years. This group of patients has a relatively poor prognosis regarding their renal function.

Cardiac rhabdomyoma is often the earliest detectable sign of TSC since it can be visualized by fetal ultrasound scanning during pregnancy. Cardiac insufficiency, due to the presence of cardiac rhabdomyomas, increases the chance of perinatal and postnatal mortality in TSC. However, since follow-up ultrasound imaging studies show that these tumors usually become smaller during postnatal development, and sometimes even disappear completely, the chance of cardiac insufficiency in later life is reduced. The cause of this age-related regression is unknown. The reported prevalence of cardiac rhabdomyoma in TSC patients varies between 30% (Gomez, 1988) and 47% (Jozwiak *et al.*, 1994).

TSC of the lungs, in the form of lymphangioleiomyomatosis (LAM), is rare and almost uniquely present in females. LAM occurs in 1–2% of women with TSC and almost exclusively above 20 years (Gomez, 1988). On chest X-ray a so-called 'honey comb' appearance of the lungs is apparent. The presence of small cyst-like lesions combined with fatty and smooth muscle tissue gives a characteristic CT scan. LAM in the general population is associated with a poor prognosis, and death from emphysema and pulmonary insufficiency usually occurs within 5 years after diagnosis. However, in our experience, patients with pulmonary TSC can have a rather slow progression of their lung disease.

On inspection of the oral cavity two types of lesion may indicate TSC (Gomez, 1988). Firstly, gingival fibroma is the buccal equivalent of facial angiofibroma, to be differentiated from gingival hypertrophy induced by phenytoin, an anti-epileptic drug that is frequently used by TSC patients with epilepsy. Secondly, on the teeth, especially on the flat surfaces of the incisors, small enamel pits or hypoplasias occur, that cause increased sensitivity to caries. These enamel pits are also observed in unaffected individuals, with a reported prevalence of 7% (Mlynarczyk, 1991).

1.2 Diagnostic criteria of tuberous sclerosis complex

The clinical spectrum of TSC is very diverse and variable, with large differences in severity of the disease both between unrelated sporadic patients and between affected members of a family. In view of the wide clinical variability of TSC, diagnostic criteria were proposed (Gomez, 1988) and revised several times (Gomez

1991, 1999; Roach *et al.*, 1992, 1998). In the last consensus report, diagnostic criteria were divided into major and minor features (*Table 1*). For a diagnosis of TSC, at least two features should be present and in the majority of patients there can be no doubt about a diagnosis of TSC, even in a sporadic patient. Most debate arises about minimally affected relatives or a person with a monosymptomatic manifestation of TSC. The relatively high frequency of somatic mosaicism in TSC (Verhoef *et al.*, 1999a), provides an explanation for the occurrence of these very mildly affected cases (see Section 2.3). When considering a diagnosis of TSC, depending on the age of onset and the type of symptoms, a number of other diagnoses needs to be considered as well. The most important differential diagnoses are presented in *Table 2*.

A genotype–phenotype correlation between TSC1- and TSC2-related disease was suggested. In a group of 150 unrelated patients a genotype–phenotype effect was reported with regard to mental impairment. The frequency of mental impairment was significantly higher for patients with a TSC2 mutation compared

Table 1. Revised diagnostic criteria for the diagnosis of TSC

Major features
- Facial angiofibromas or forehead plaque
- Non-traumatic ungual or periungual fibroma
- Hypomelanotic macules (three or more)
- Shagreen patch (connective tissue nevus)
- Multiple retinal nodular hamartomas
- Cortical tuber[a]
- Subependymal nodule
- Subependymal giant cell astrocytoma
- Cardiac rhabdomyoma, single or multiple
- Lymphangiomyomatosis[b]
- Renal angiomyolipoma[b]

Minor features
- Multiple, randomly distributed pits in dental enamel
- Hamartomatous rectal polyps[c]
- Bone cysts[d]
- Cerebral white matter radial migration lines[d,e]
- Gingival fibromas
- Non-renal hamartoma
- Retinal achromic patch[c]
- 'Confetti' skin lesions
- Multiple renal cysts[c]

Definite TSC: Two major features or one major plus two minor features
Probable TSC: One major plus one minor feature
Possible TSC: Either one major or two minor features

[a] When cerebral cortical dysplasia and cerebral white matter migration tracts occur together, they should be counted as one rather than two features of TSC.
[b] When both lymphangiomyomatosis and renal angiomyolipomas are present, other symptoms of TSC should be present before a definite diagnosis is assigned.
[c] Histological confirmation is suggested.
[d] Radiographic confirmation is sufficient.
[e] It is felt that three or more radial migration lines should constitute a major sign.

Table 2. Differential diagnosis of TSC depending on the age and site of the prime manifestation of disease.

Age	Presenting sign	Differential diagnosis
Pregnancy	Intracardiac tumor	Rhabdomyoma Myxoma

Newborn	Intracerebral calcifications	Cerebral arteriovenous malformation
		Congenital infection
		  Toxoplasmosis
		  Cytomegaly

Child	Epilepsy, moderate to severe mental retardation	Dysmorphology syndrome Congenital infection Post-anoxic encephalopathy
	Epilepsy/infantile spasms	Symptomatic epilepsy Idiopathic epilepsy
	Hypomelanosis	Hypomelanosis of Ito
	Epilepsy, mild mental retardation, MRI abnormalities	Familial periventricular nodular heterotopias (X-linked)
	Facial angiofibroma	Cylindromatosis/epitheliomas Acné rosacea Milia
(Young) adults	Renal problems	Polycystic kidney disease Malignancy (renal cell carcinoma)
	Retinal tumor	Retinoblastoma Drusen (e.g. as a result of diabetes)
	Honeycomb lung (X-ray)	Isolated lymphangiomyomatosis

to patients with a TSC1 mutation: 67% versus 31% for the group of sporadic cases (Jones *et al.*, 1997, 1999). Our data support this slightly 'milder' mental manifestation of TSC1 mutations (59% versus 34%), although the combined data are too small to draw definitive conclusions.

2. Genes involved in tuberous sclerosis complex

2.1 Identification of the TSC1 and TSC2 genes

Linkage analyses of large TSC families showed that TSC is a disorder with locus heterogeneity. Two loci are, independently, involved: TSC1 on chromosome 9q34 (Connor *et al.*, 1987; Fryer *et al.*, 1987) and TSC2, on chromosome 16p13 (Kandt *et al.*, 1992).

A TSC patient with an unbalanced translocation between the short arm of chromosome 16 and chromosome 22, resulting in loss of the tip of chromosome 16p, and another individual with a slightly smaller chromosome 16p deletion (Wilkie *et al.*, 1990) but no signs of TSC, facilitated the cloning of the *TSC2* gene. Detailed molecular analysis of the two deletions indicated that a candidate region of approximately 300 kb remained. Strong evidence for the presence of the *TSC2* gene in this region was obtained by pulsed field gel electrophoresis (PFGE) analysis using cloned fragments from the candidate interval. Using this approach, abnormal fragments in several TSC patients were detected. Screening of cDNA libraries with genomic probes from the candidate region resulted in the identification of several candidate cDNA clones. Southern blot analysis of TSC patients using these cDNA clones as probes, led to the identification of the *TSC2* gene (The European Chromosome 16 TS Consortium, 1993). The *TSC2* gene has 41 exons, some of them alternatively spliced (Maheshwar *et al.*, 1996), and encodes a 200 kDa protein called tuberin. A small region of homology with rap1GAP, a GTPase-activating protein (GAP) was identified (see Section 3.1).

The cloning of the *TSC1* gene was hampered by the fact that no recombinations or large (chromosomal) rearrangements could be identified to reduce the candidate region below approximately 1.4 Mb. A combination of techniques including exon trapping, cDNA library screening, expressed sequence tag (EST) mapping, cDNA selection and sequence analysis of the candidate region was applied. Several cDNA clones were isolated and analysed by single strand conformation polymorphism (SSCP), heteroduplex analyses and Southern blotting, resulting in the identification of the *TSC1* gene (van Slegtenhorst *et al.*, 1997). The *TSC1* gene consists of 23 exons, the second exon being alternatively spliced, and encodes an mRNA of 8.6 kb. Hamartin, the 130 kDa protein product of the *TSC1* gene, contains a single putative transmembrane domain and a predicted coiled coil region. No further homology to other known proteins or functional domains could be identified (see Section 3.2). The *TSC1* gene, like the *TSC2* gene, is widely expressed.

2.2 Loss of heterozygosity in hamartomas

In line with Knudson's hypothesis (Knudson, 1971) it was speculated that the development of hamartomas in TSC patients was the result of a complete loss of the function of either the *TSC1* or the *TSC2* gene. This complete inactivation is achieved by the presence of a germ-line mutation on one allele and inactivation of the wild-type allele at the same locus due to a large deletion or by another mechanism (Tischfield, 1997). Several studies presented loss of heterozygosity (LOH) in a wide variety of hamartomas of TSC patients, but also in more aggressive tumors including renal cell carcinomas, giant cell astrocytomas and a pancreatic tumor (Carbonara *et al.*, 1994; Henske *et al.*, 1996; Sepp *et al.*, 1996; Verhoef *et al.*, 1999b). LOH is observed in lesions from patients with a germ-line mutation either in the *TSC1* or in the *TSC2* gene, indicating that both genes act as tumor suppressors. In two cases, no LOH, but a somatic mutation was identified in the wild-type allele. In a TSC1 patient carrying the *TSC1* germ-line mutation 2105delAAAG, a renal cell carcinoma showed a somatic mutation 1957delG in the *TSC1* wild-type allele (van Slegtenhorst *et al.*, 1997). In the second case, a TSC2 patient with the germ-line mutation 17-bp tandem dup (G3611-G3627), a facial angiofibroma was found to contain a somatic mutation identified by Southern blot analysis, indicating an intragenic deletion in the wild type *TSC2* gene (Au *et al.*, 1998, 1999).

2.3 Mutation analysis of the TSC1 and the TSC2 genes

To date, several techniques for mutation analysis have been employed such as PFGE, Southern blot analysis, SSCP, denaturing gradient gel electrophoresis

(DGGE), heteroduplex analyses, protein truncation tests (PTT) and denaturing high performance liquid chromatography (DHPLC). All reported TSC mutations are collected in the 'TSC variation database' (TSC variation database, release January, 2000). There are significant differences in the types of mutations in the *TSC1* gene compared to those in the *TSC2* gene. No large rearrangements have been identified in the *TSC1* gene. Furthermore, no *TSC1* missense mutations have been detected. With the exception of rare in-frame deletions, all mutations in the *TSC1* gene are small and result in truncated forms of hamartin. In TSC2 patients small mutations include not only nonsense and splice site mutations, but also pathogenic missense mutations. In contrast to *TSC1*, large deletions occur in up to 20% of patients with *TSC2* mutations. Some TSC patients have, in addition to the typical clinical features of TSC, many renal cysts and, as a consequence, early onset end-stage renal disease. Very often a large deletion (partially) removing both the *TSC2* and the adjacent *PKD1* gene is found in these patients (Sampson *et al.*, 1997). An example of an abnormality identified by Southern blot analysis is given in *Figure 1*.

The question remains why there are no missense abnormalities and no large rearrangements in the *TSC1* gene. One explanation might be that large rearrangements are lethal. This might also be the case for missense abnormalities, although it cannot be excluded that individuals with such abnormalities do not have the clinical features typical of TSC and are, as such, not recognized as TSC patients.

About 60% of TSC patients have parents without clinical signs of TSC. This is a consequence of the high mutation rate of the TSC genes. Relatively frequent

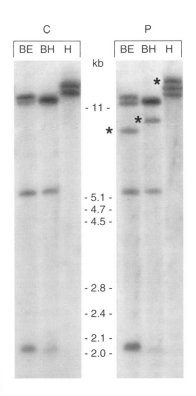

Figure 1. Southern blot analysis of genomic DNA of a control individual (C) and a TSC patient (P) using the TSC2 specific cDNA probe 4B2. The restriction enzymes used are: B, *Bam*HI; E, *Eco*RI and H, *Hind*III. Asterisks indicate the aberrant fragments resulting from a 10 kb deletion in the *TSC2* gene of the patient. Long range PCR analysis confirmed the presence of a 10 kb deletion (data not shown) (Daborah *et al.*, 2000).

germ-line and somatic mosaicism is also a result of the high mutation rate. Mutation analysis has confirmed the suspicion of germ-line mosaicism in a few families with several affected siblings and healthy parents (Rose *et al.*, 1999, Yates *et al.*, 1997). Somatic mosaicism in parents of TSC patients has been associated with (a relatively mild manifestation of) TSC (Sampson *et al.*, 1997; Verhoef *et al.*, 1995, 1999a). Most TSC families have a unique mutation and, although some recurrent mutations in both genes have been reported, no true founder mutations are known.

Renal angiomyolipomas occur in about 70% of TSC patients and in line with the idea that 'partial' TSC phenotypes can be due to TSC gene mutations, LOH studies were performed in angiomyolipomas (AML) of patients without further signs of TSC. LOH of the 16p13.3 region could be detected in 3 out of 29 AMLs (Henske *et al.*, 1995). Also, sporadic lymphangioleiomyomatosis (LAM) patients without other signs of TSC were analyzed, since about 1–2% of women with TSC have LAM and renal AMLs are present in about 50% of sporadic LAM patients. In 54% of the angiomyolipomas of the sporadic LAM patients LOH of the *TSC2* gene was observed, indicating the involvement of the *TSC2* gene in the etiology of these angiomyolipomas (Smolarek *et al.*, 1998). No LOH of the TSC1 region was found. Mutation analysis of the *TSC2* gene in 21 LAM patients did not result in the identification of germ-line mutations in the *TSC2* gene (Astrinidis *et al.*, 2000). However, in 4 angiomyolipomas and LAM tissue of sporadic LAM patients showing LOH of the TSC2 region, a somatic mutation on the other allele in the *TSC2* gene could be identified supporting a direct role of the *TSC2* gene in the pathogenesis of pulmonary LAM (Carsillo *et al.*, 2000).

2.4 Future perspectives

Since different types of mutations have been identified in the *TSC1* and *TSC2* genes, a combination of techniques must be used to search for mutations in patients. Mutations leading to a truncated protein can be detected by PTT analysis. Nucleotide substitutions underlying these mutations can be present in exon sequences, but also in introns at positions that are not being tested using SSCP, heteroduplex, DHPLC or DGGE analysis. The disadvantage of PTT analysis using RT-PCR is that the mRNA which harbors the mutation may be unstable leading to a false negative result. In addition, missense mutations cannot be detected by PTT analysis. Because the majority of the mutations in the *TSC1* gene result in a truncated protein, PTT screening is a solid technique for mutation analysis of this gene. About 20% of the *TSC2* mutations are large abnormalities and Southern blotting in combination with fluorescent *in situ* hybridization are the methods of choice to identify these mutations. Since pathogenic missense mutations have been described in the *TSC2* gene, PTT analysis as the only screening technique is not sufficient and must be combined with other methods such as SSCP, DGGE or DHPLC.

So far, only the coding regions of the two genes have been extensively tested for the presence of mutations. Some intronic mutations have been identified using PTT analysis (Mayer *et al.*, 1999), however, a systematic search for mutations in non-coding regions has not yet been conducted. Mutations may also be present in the promoter and control regions of TSC1 and TSC2. In such cases, expression assays may have to be performed to prove the effect of the sequence abnormality.

At present, using a combination of screening techniques, the mutation can be identified in about 90% of a well-defined TSC population (J. Sampson, Cardiff, pers. commun.). Therefore, it is unlikely that the existence of a third gene involved in TSC is needed to explain the current deficit of known causative mutations.

3. Functional characterization of the *TSC1* and *TSC2* gene products

Since the cloning of the *TSC1* and *TSC2* genes, research has shifted towards unravelling the functions of hamartin and tuberin, the protein products of the *TSC1* and *TSC2* genes. However, despite some significant advances, the precise functions of hamartin and tuberin remain obscure. Here we attempt to summarize what is known about these two proteins, and outline some of the goals for future investigations.

As explained in Section 2.2, LOH in TSC-associated lesions provided strong evidence that hamartin and tuberin are tumor suppressors. Tumor suppressor genes have been subdivided into those that regulate tumor growth directly, by inhibiting cell proliferation or promoting death ('gatekeepers'), and those whose inactivation causes genetic instability which in turn leads to mutations that promote growth ('caretakers'). Since genetic instability is not a feature of lesions associated with TSC, tuberin and hamartin are most likely to have direct roles in the regulation of cell proliferation. Further evidence that tuberin and hamartin do indeed function within cellular pathways controlling proliferation and growth comes from the study of a number of TSC animal models. Both *Drosophila melanogaster* (Ito and Rubin, 1999) and rat (Kobayashi *et al.*, 1995; Yeung *et al.*, 1994) strains have been identified with only one functional copy of the *TSC2* gene in the germ-line, while *TSC2* has been specifically knocked out in mice (Onda *et al.*, 1999). In *Drosophila*, a *TSC1* mutant strain, 'rocky' has also been described (Tapon and Hariharan, 2000). In each case, heterozygote animals exhibit defects in cell number, size and location, while rats and mice develop a variety of tumor-like lesions, similar to those seen in TSC. The contributions of these different animal models to our understanding of the functional significance of tuberin and hamartin are described in more detail in the following sections.

3.1 Functional characterization of tuberin

The Eker rat was first identified as a model for inherited renal cell carcinoma (Eker and Mossige, 1961). An insertion of an intracisternal A-particle element into the rat *tsc2* gene was shown to be the causative mutation (Xiao *et al.*, 1995). Re-introduction of the wild-type *TSC2* gene into renal tumor cells derived from Eker rats inhibited proliferation *in vitro* and reduced their tumorigenicity *in vivo* (Jin *et al.*, 1996). Both contact inhibition and normal cell morphology were restored. In addition, Eker rats transgenic for the wild-type *TSC2* gene did not develop the characteristic renal or associated neoplasms (Kobayashi *et al.*, 1997).

Consistent with these findings, over-expression of tuberin reduced the rate of growth of normal rat1 fibroblasts (Jin *et al.*, 1996) and supported the hypothesis that tuberin had a role in the suppression of tumorigenesis.

Tuberin is a large (200 kDa), hydrophobic protein. Alternative splicing of the TSC2 mRNA indicates that several different isoforms of tuberin are expressed, according to the tissue and stage of development (Kim *et al.*, 1995; Maheshwar *et al.*, 1996) (*Figure 2*). In addition, multiple putative phosphorylation and N-linked glycosylation sites have been identified. The functional significance of these sites and of the different isoforms is not yet known, however. The most striking result of the analysis of the predicted amino acid sequence of tuberin was the identification of a region of homology between tuberin and the GTPase activating protein (GAP) for rap1, a member of the ras family of small GTPases (The European Chromosome 16 Tuberous Sclerosis Consortium, 1993) (*Figure 2*). Hydrolysis of bound GTP to GDP by rap1 converts active rap1-GTP to inactive rap1-GDP. Since rap1GAP accelerates this reaction, it promotes the inactivation of rap1 and therefore down-regulates rap1-mediated stimulation of proliferation. Confirmation of the importance of the tuberin rap1GAP-related domain came from several sources. First, it was reported that tuberin has rap1 GAP activity (Wienecke *et al.*, 1995). Second the

(a)

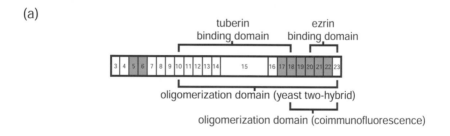

(b)

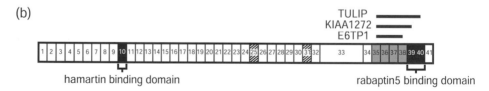

Figure 2. Structural and functional domains of hamartin and tuberin. Proteins are not drawn to scale and the relative sizes of the exons are approximate. (A) Schematic representation of hamartin, divided according to the protein coding exons (3–23). The N-terminal transmembrane domain (encoded by exons 5–6) and C-terminal coiled coil domain (encoded by exons 17–22) are indicated by the gray shading. The positions of the proposed tuberin-binding, ezrin-binding and oligomerization domains are also indicated. (B) Schematic representation of tuberin, showing all the exons (1–41) coding for protein, including the alternatively spliced exons 25 and 31 (cross-hatching). Filled exons 10 and 39–40 represent the putative hamartin-binding and rabaptin5-binding domains, respectively. Gray shading (exons 35–38) indicates the extent of the homology with rap1 GAP. The horizontal black lines indicate the regions of homology between tuberin and tulip (accession number AL050050), KIAA1272 (accession number AB033098) and E6TP1 (Gao *et al.*, 1999).

GAP-related domain is relatively highly conserved in all the TSC2 homologs identified to date across different species (rat, mouse, pufferfish, fruitfly, fission yeast). Third, several different missense mutations have been identified within this domain in TSC patients (Maheshwar et al., 1997). Finally, when this domain was over-expressed in Eker rat tumor cells, both colony formation and in vivo tumorigenicity were reduced (Jin et al., 1996). The role of rap1 in the regulation of cell growth, proliferation and differentiation is a subject of considerable recent interest (Bos, 1998), however it is not yet clear whether the putative tuberin rap1GAP activity is of any in vivo significance. Some amino acid residues necessary for rap1GAP activity and specificity are not all conserved in tuberin, perhaps explaining the relatively low activation of rap1 GTPase activity. One possibility is that tuberin may represent a new subfamily of GAP proteins for an as yet unidentified substrate or group of substrates since several proteins with homology to the tuberin rap1GAP-related domain have been identified recently (Gao et al., 1999; GenBank accession numbers AL050050, AB033098). In some cases, the homology between these proteins and tuberin extends further than the rap1GAP-related domain to encompass a region reported to be involved in binding between tuberin and the rab5 effector, rabaptin5 (Xiao et al., 1997) (Figures 2 and 3). In addition to binding rabaptin5, tuberin was reported to have rab5GAP activity. Rab5 is another ras-like GTPase and belongs to the large family of rab proteins that regulate discrete steps of the membrane transport pathway. Rab5 acts as a timer for endosome fusion (Rybin et al., 1996) and consistent with this, the rate of fluid-phase endocytosis was increased in cells derived from homozygote Eker embryos (tsc2 –/–) relative to those from wild-type embryos (tsc2 +/+). Confirmation of a role for tuberin in regulating endocytosis would be particularly exciting, as it would help strengthen the claim for the putative tuberin rab5GAP activity and suggest a possible link between tumorigenesis and membrane transport processes.

Using another approach, tuberin was shown to modulate steroid hormone receptor-mediated transcription (Henry et al., 1998). Although this finding has also not yet been confirmed independently, it is interesting because the growth of some TSC-associated tumors, and uterine and renal tumors in the Eker rat, may be influenced by steroid hormones (Fuchs-Young et al., 1996; Wolf et al., 1998).

Cell growth and proliferation is regulated by controlled passage through the cell cycle. Disruption of this process can lead to the unregulated, proliferative growth typical of tumorigenesis. Cell cycle progression can be visualized directly during development of the Drosophila eye and sophisticated genetic screens have been developed to isolate cell cycle mutants affecting this process. Recently, mutations producing enlarged cells in homozygous clones were identified and shown to be allelic with the existing Drosophila gigas mutant (Ito and Rubin, 1999). In addition to the enlarged cellular phenotype, axons from mutant gigas neurons have been shown to project beyond their normal targets and thereby modify behavior (Canal et al., 1998). The combination of large, abnormal cells and aberrant behavioral responses are a characteristic of TSC, and cloning of the gigas gene showed that it encoded a protein homologous to tuberin (26% identical amino acids; 46% similar). Analysis of the gigas phenotype provided some important clues to the possible biological role(s) of tuberin. First, like the Eker mutation, gigas is homozygous lethal and therefore essential for normal development. Second, gigas

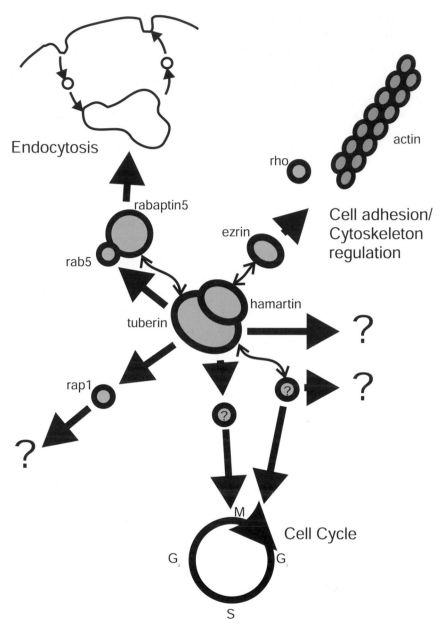

Figure 3. Possible cellular functions of the tuberin–hamartin complex. Tuberin and hamartin are shown as a complex. Shaded forms represent different proteins (not drawn to scale). Double-headed arrows indicate direct interactions between tuberin and/or hamartin and other proteins. Single-headed, straight arrows indicate the direct cellular targets of the tuberin–hamartin complex as well as the targets of the associated proteins. The proposed roles of tuberin and hamartin in endocytosis, cell adhesion and the cell cycle are indicated. Large question marks indicate that the exact function of the complex has not yet been determined precisely. Possible interactions between the tuberin–hamartin complex and additional, as yet unidentified proteins, are also indicated. Unidentified proteins are marked with a small question mark.

eye disc cells show abnormal cell cycle progression. Incorporation of 5-bromo-2-deoxyuridine into newly synthesized DNA showed that gigas cells enter S phase, while a lack of phosphorylated histone H3 indicated that they did not go through M phase. Analysis of the DNA content of gigas mutants confirmed that the cells undergo endoreplication without cell division. Consistent with this observation, sustained DNA synthesis has also been observed in cardiomyocytes derived from homozygous Eker rat embryos (Pajak et al., 1997). Taken together, these results suggest that tuberin may help regulate the transition from S phase to M phase during the cell cycle and that mutant cells may go through S phase repeatedly, replicating their DNA, without entering M phase.

Additional evidence that tuberin regulates the cell cycle came from experiments on murine fibroblasts in culture. Inhibition of tuberin expression by antisense oligonucleotides both induced cells to pass through the G_1/S transition and prevented them from entering the quiescent state (G_0) (Soucek et al., 1997). Similar treatment in logarithmically growing fibroblasts decreased the proportion of cells in G_1 while G_0 arrest by serum withdrawal was prevented and quiescent G_0-arrested fibroblasts re-entered the cell cycle. The duration of G_1 was also shortened in Eker-derived tsc2 –/– fibroblasts, relative to wild-type controls (Soucek et al., 1998b).

In related experiments, an increase in the expression of tuberin was observed after the induction of neuronal differentiation in neuroblastoma cell lines (Soucek et al., 1998a). Likewise, suppression of tuberin expression with antisense oligonucleotides inhibited the differentiation of the same cell lines. Similar effects were not observed upon differentiation of other (non-neuronal) cells although a separate report suggests that tuberin may be required for cell cycle withdrawal and terminal differentiation in cardiomyocytes (Pajak et al., 1997).

How tuberin exerts all these effects on the cell cycle is not yet clear (*Figure 3*). In the gigas mutant there is an accumulation of the G_2 cyclins (cyclins A and B) (Ito and Rubin, 1999). Conversely, in Eker tsc2 –/– fibroblasts, the levels of cyclin A are unaffected while there is mislocation of the cyclin-dependent kinase inhibitor p27 to the cytoplasm, where it is rapidly degraded (Soucek et al., 1998b).

3.2 Functional characterization of hamartin

Hamartin, the 130 kDa product of the *TSC1* gene is a hydrophilic protein (van Slegtenhorst et al., 1997). In contrast to tuberin, hamartin lacks clear regions of homology with other known proteins and there are few clues to its possible cellular function. Two interesting features have been identified, however: a putative transmembrane domain close to the N-terminus, and a large coiled coil domain close to the C-terminus (*Figure 2*). The coiled coil domain is related to similar sequences in many other, principally structural, proteins; notably myosin and proteins of the cornified cell envelope of dermal epithelium.

Exogenous expression of hamartin in mammalian cells produced a punctate labeling pattern (van Slegtenhorst et al., 1998). Subsequent expression of various hamartin fragments indicated that the coiled coil domain was necessary for this distinctive pattern. Biochemical fractionation and yeast two-hybrid experiments indicated that the punctate labeling most likely represented coiled coil-mediated aggregation (Nellist et al., 1999) (*Figure 2*). Yeast two-hybrid experiments also provided the first insights into the possible function of hamartin. A screen performed using the N-terminal domain of ezrin identified multiple cDNAs encoding the hamartin coiled coil domain. Ezrin belongs, together with radixin and moesin, to the ERM family of actin-binding proteins, thought to be important regulators of cell adhesion and migration. Binding between hamartin and the N-terminal of ezrin was confirmed *in vitro* and by co-immunoprecipitation. Consistent with an interaction between hamartin and ezrin, expression of hamartin induced the formation of focal adhesion plaques and actin filaments, while local inactivation of hamartin by chromophore-assisted laser irradiation resulted in a loss of cell-substrate adhesion (Lamb *et al.*, 2000). The authors proposed that hamartin promotes the formation of focal adhesion plaques through activation of the rho GTPase signaling pathway and presented evidence that binding between ERM proteins and hamartin may be required for this activation. In light of its large coiled coil domain, a structural function for hamartin, maybe anchoring other cellular components of the cytoskeleton or endosomes, would not be too surprising. Therefore, it is of interest to investigate whether tuberin also has a role in rho-mediated signal transduction (*Figure 3*).

3.3 Tuberin and hamartin form a complex

Consistent clinical differences between TSC patients carrying a *TSC1* mutation and those harboring *TSC2* mutations have been difficult to establish (see Section 1.2), implying that both tuberin and hamartin may be closely involved in the regulation of the same cellular process(es). Evidence that tuberin and hamartin might act in tandem was obtained with the discovery that the two proteins interact directly as part of a protein complex (Plank *et al.*, 1998; van Slegtenhorst *et al.*, 1998) (*Figure 3*). Biochemical fractionation experiments indicated that the tuberin-hamartin complex may also contain additional protein components, and that tuberin acts as a chaperone, inhibiting hamartin aggregation and maintaining the complex in a soluble and therefore presumably active form (Nellist *et al.*, 1999).

3.4 Future perspectives

The discovery that tuberin and hamartin form a protein complex is particularly relevant to the characterization of their function(s). All the experiments investigating the function of these two proteins have, until now, studied either tuberin or hamartin in isolation. The function of the tuberin–hamartin complex as a single entity has not yet been addressed.

Interestingly, although no TSC1 or TSC2 homologs have been identified in baker's yeast (*Saccharomyces cerevisiae*), fission yeast (*Schizosaccharomyces pombe*) does express two proteins with significant homology to both tuberin and hamartin (GenBank accession numbers AL109832 and Q09778 respectively). It will be interesting to investigate the effect of inactivation of one or both of these proteins on the life cycle of *S. pombe* and to determine whether any resulting mutant phenotype can be corrected by re-expression of the human equivalents (tuberin and hamartin). The presence of tuberin and hamartin homologs in fission yeast implies that both proteins may have a role in basic cellular processes that are necessary for the propagation of fission yeast as well as higher eukaryotes.

Considerable work remains to be done before the cellular functions of tuberin and hamartin can be properly defined. Identifying the roles of these two large, novel proteins in both normal and disease states is a major challenge for those involved in TSC research and may have a significant impact on more general studies of cell growth and proliferation.

4. Genetic counseling for TSC

As TSC is a complex condition, counseling should be provided by clinicians with ample experience of TSC. Issues to be covered include the clinical diversity of the condition, its inheritance, the possibility of DNA-diagnostic investigations and the chance of finding a mutation, discussion on 'de novo' versus familial TSC and information on the probabilities of somatic or germ-line mosaicism.

Pediatricians, pediatric neurologists and clinical geneticists can provide much input. They can pass relevant information about TSC to the primary care worker, the general practitioner.

As manifestations of TSC may be subtle and remain unnoticed unless specifically looked for, extensive physical examination of apparently unaffected parents and sibs of a TSC patient is necessary. The examinations as advised in most consensus publications are presented in *Table 3* (Roach *et al.*, 1998). It is debatable to what extent clinical screening is necessary. According to the literature about 98% of carriers of TSC will be identified by the combination of skin investigation, CT scan of the brain, and ophthalmologic evaluation. Renal ultrasound scanning has clinical implications if an abnormality is found, and is therefore also recommended. Once a possible sign of TSC is found, complete clinical screening should be conducted.

Table 3. Recommended investigations for first degree relatives of TSC patients

Organ	Comments
Skin	With an emphasis on the possible presence of white patches, ungual fibromas and facial angiofibroma
Central nervous system	Neurological examination, neuro-imaging by CT scan or MRI scan of the brain, in order to detect subependymal calcified nodules, cortical tubers and radial migration lines. MRI scanning is more suited to detect white matter migratory lesions and cortical tubers, whilst CT scan is a more reliable tool to detect calcified paraventricular (subependymal) nodules. As these are probably more specific at the adult age, CT scanning is the preferred method of investigation in adults, MRI scanning would be more suitable at younger age, as calcification of TSC lesions generally occurs after the first years of life
Eyes	Ophthalmologic screening, focused on peripherally located retinal hamartomas in particular. Dilatation of the pupils prior to the retinal examination is therefore recommended
Abdomen	Abdominal ultrasound to screen for renal cysts and/or angiomyolipomas, or cysts or angiomyolipomas of the liver or pancreas
Heart	Cardiac ultrasound for intramural rhabdomyomas
Mouth	Oral examination with a focus on enamel pits and/or gingival fibromas

Between one third and one half of TSC patients represent familial cases. Up to two thirds of the cases occur outside a TSC family context and are considered to be due to 'new mutations'.

In families where a causative mutation has been identified, DNA tests can replace clinical examinations for relatives. However, establishing germ-line or somatic mosaicism limits the generations to be tested. Siblings of an individual with proven mosaicism are no longer at risk to carry the familial mutation. A normal DNA result in both parents of a patient with a known mutation is no proof of a '*de novo*' mutation in the index case, since parental mosaicism may be confined to tissues that were not examined, or the degree of mosaicism might be below the detection limit. Indeed, in some families with a child with an apparently '*de novo*' mutation, a second child with TSC is born. Using linkage analysis and/or direct mutation detection, (probable) germ-line mosaicism in one of the parents may be demonstrated (Rose *et al.*, 1999; Verhoef *et al.*, 1999a; Yates *et al.*, 1997). In such cases clinical examination remains necessary to distinguish between probable germ-line mosaicism and somatic mosaicism in the parents. It is, therefore, wise to discuss the possibility of prenatal DNA testing in subsequent pregnancies with unaffected parents of an affected child in which the causative mutation is identified.

Certainly at the time of reproductive decisions, formal genetic counseling is recommended, in view of the obvious complexity of the genetics of this condition.

Acknowledgments

The authors would like to thank Ruud Koppenol for help with the figures and Noortman b.v. (Maastricht, The Netherlands), the National Tuberous Sclerosis Association (NTSA) and the Dutch Epilepsiefonds for financial support.

References

Al-Saleem, T., Wessner, L.L., Scheithauer, B.W. *et al.* (1998) Malignant tumors of the kidney, brain, and soft tissues in children and young adults with the tuberous sclerosis complex. *Cancer* **83**: 2208–2216.

Astrinidis, A., Khare, L., Carsillo, T., Smolarek, T., Au, K-S., Northrup, H. and Henske, E.P. (2000) Mutational analysis of the tuberous sclerosis gene TSC2 in patients with pulmonary lymphangioleiomyomatosis. *J. Med. Genet.* **37**: 55–57.

Au, K-S., Rodriguez, J.A., Finch, J.L., Volcik, K.A., Roach, E.S., Delgado, M.R., Rodriguez Jr, E. and Northrup, H. (1998) Germ-line mutational analysis of the TSC2 gene in 90 tuberous-sclerosis patients. *Am. J. Hum. Genet.* **62**: 286–294.

Au, K-S., Hebert, A.A., Roach E.S. and Northrup, H. (1999) Complete inactivation of the TSC2 gene leads to formation of hamartomas. *Am. J. Hum. Genet.* **65**: 1790–1795.

Baumgartner, J.E., Wheless, J.W., Kulkarni, S., Northrup, H., Au, K-S., Smith, A. and Brookshire, B. (1997) On the surgical treatment of refractory epilepsy in tuberous sclerosis complex. *Pediatr. Neurosurg.* **27**: 311–318.

Bernstein, J. (1993) Renal cystic disease in the tuberous sclerosis complex. *Pediatr. Nephrol.* **7**: 490–495.

Bos, J.L. (1998) All in the family? New insights and questions regarding interconnectivity of Ras, Rap1 and Ral. *EMBO J.* **17**: 6776–6782.

Bourneville, D.M. (1880) Sclérose tubéreuse des circonvolutions cérébrales: idiotie et épilepsie hémiplégique. *Arch Neurol (Paris)* **1**: 81–91.

Canal, I., Acebes, A. and Ferrus, A. (1998) Single neuron mosaics of the drosophila gigas mutant project beyond normal targets and modify behavior. *J. Neurosci.* **18**: 999–1008.

Carbonara, C., Longa, L., Grosso, E., Borrone C., Garrè, M.G., Brisigotti, M. and Migone, N. (1994) 9q34 loss of heterozygosity in a tuberous sclerosis astrocytoma suggests a growth suppressor-like activity also for the TSC1 gene. *Hum. Mol. Genet.* **3**: 1829–1832.

Carsillo, T., Astrinidis, A. and Henske, E.P. (2000) Mutations in the tuberous sclerosis complex gene *TSC2* are a cause of sporadic pulmonary lymphangioleiomyomatosis. *Proc. Natl Acad. Sci. USA* **97**: 6085–6090.

Connor, J.M., Pirrit, L.A., Yates, J.R.W., Fryer, A.E. and Ferguson-Smith, M.A. (1987) Linkage of the tuberous sclerosis locus to a DNA polymorphism detected by v-abl. *J. Med. Genet.* **24**: 544–546.

Daborah, S.L., Nieto, A.A., Franz, D., Jozwiak, S., van den Ouweland, A. and Kwiatkowski, D.J. (2000) Characterisation of 6 large deletions in TSC2 identified using long range PCR suggests diverse mechanisms including Alu-mediated recombinations. *J. Med. Genet.* **37**: 877–882.

Eker, R. and Mossige, J.A. (1961) A dominant gene for renal adenomas in the rats. *Nature* **189**: 858–859.

Ewalt, D.H., Sheffield, E., Sparagana, S.P., Delgado, M.R. and Roach, E.S. (1998) Renal lesion growth in children with tuberous sclerosis complex. *J. Urol.* **160**: 141–145.

Fryer, A.E., Chalmers, A., Connors, J.M., Fraser, I., Povey, S., Yates, A.D., Yates, J.R. and Osborne J.P. (1987) Evidence that the gene for tuberous sclerosis is on chromosome 9. *Lancet* **i**: 659–661.

Fuchs-Young, R., Howe, S., Hale, L., Miles, R. and Walker, C. (1996) Inhibition of estrogen-stimulated growth of uterine leiomyomas by selective estrogen receptor modulators. *Mol. Carcinog.* **17**: 151–159.

Gao, Q., Srinivasan, S., Boyer, S.N., Wazer, D.E. and Band, V. (1999) The E6 oncoproteins of high-risk papillomaviruses bind to a novel putative GAP protein, E6TP1, and target it for degradation. *Mol. Cell Biol.* **19**: 733–744.

Gomez, M.R. (ed) (1988) *Tuberous sclerosis*, 2nd edn. Raven Press, New York.

Gomez, M.R. (1991) Phenotypes of the tuberous sclerosis complex with a revision of diagnostic criteria. *Ann. NY. Acad. Sci.* **615**: 1–7.

Gomez, M.R. (ed) (1999) *Tuberous Sclerosis Complex*, 3rd Edn. Oxford University Press, New York.

Griffiths, P.D. and Martland, T.R. (1997) Tuberous sclerosis complex: the role of neuroradiology. *Neuropediatrics* **28**: 244–252.

Harrison, J.E. and Bolton, P.F. (1997) Annotation: Tuberous sclerosis. *J. Child Psychol. Psychiat.* **38**: 603–614.

Henry, K.W., Yuan, X., Koszewski, N.J., Onda, H., Kwiatkowski, D.J. and Noonan, D.J. (1998) Tuberous sclerosis gene 2 product modulates transcription mediated by steroid hormone receptor family members. *J. Biol. Chem.* **273**: 20535–20539.

Henske, E.P., Neumann, H.P.H., Scheithauer, B.W., Herbst, E.W., Short, M.P. and Kwiatkowski, D.J. (1995) Loss of heterozygosity in the tuberous sclerosis (TSC2) region of chromosome band 16p13 occurs in sporadic as well as TSC-associated renal angiomyolipomas. *Genes Chromosomes Cancer.* **13**: 295–298.

Henske, E.P., Scheithauer, B.W., Short, M.P., Wollmann, R., Nahmias, J., Hornigold, N., van Slegtenhorst, M., Welsh, C.T. and Kwiatkowski, D.J. (1996) Allelic loss is frequent in tuberous sclerosis kidney lesions but rare in brain lesions. *Am. J. Hum. Genet.* **59**: 400–406.

Hunt, A. and Lindenbaum, R. (1984) Tuberous sclerosis: a new estimate of prevalence within the Oxford region. *J. Med. Genet.* **21**: 272–277.

Ito, N. and Rubin, G.M. (1999) Gigas, a Drosophila homolog of tuberous sclerosis gene product-2, regulates the cell cycle. *Cell* **96**: 529–539.

Jin, F., Wienecke, R., Xiao, G-H., Maize, J.C., DeClue, J.E. and Yeung, R.S. (1996) Suppression of tumorigenicity by the wild-type tuberous sclerosis 2 (Tsc2) gene and its C-terminal region. *Proc. Natl Acad. Sci. USA* **93**: 9154–9159.

Jones, A.C., Daniells, C.E., Snell, R.G., Tachataki, M., Idziaszczyk, S.A., Krawczak M., Sampson, J.R. and Cheadle, J.P. (1997) Molecular genetic and phenotypic analysis reveals differences between TSC1 and TSC2 associated familial and sporadic tuberous sclerosis. *Hum. Mol. Genet.* **6**: 2155–2161.

Jones, A.C., Shyamsundar, M.M., Thomas, M.W., Maynard, J., Idziaszczyk, S., Tomkins, S., Sampson, J.R. and Cheadle, J.P. (1999) Comprehensive mutation analysis of TSC1 and TSC2 – and phenotypic correlations in 150 families with tuberous sclerosis. *Am. J. Hum. Genet.* **64**: 1305–1315.

Jozwiak, S., Kawalec, W., Dluzewska, J., Daszkowska, J., Mirkowicz-Malek, M. and Michalowicz, R. (1994) Cardiac tumors in tuberous sclerosis: their incidence and course. *Eur. J. Pediatr.* **153**: 155–157.

Kandt, R.S., Haines, J.L., Smith, M. *et al.* (1992) Linkage of an important gene locus for tuberous sclerosis to a chromosome 16 marker for polycystic kidney disease. *Nature Genet.* **2**: 37–41.

Kim, K.K., Pajak, L., Wang, H. and Field, L.J. (1995) Cloning, developmental expression, and evidence for alternative splicing of the murine tuberous sclerosis (TSC2) gene product. *Cell. Mol. Biol. Res.* **41**: 515–526.

Knudson, A.G. (1971) Mutation and cancer: statistical study of retinoblastoma. *Proc. Natl Acad. Sci. USA* **68**: 820–823.

Kobayashi, T., Hirayama, Y., Kobayashi, E., Kubo, Y and Hino, O. (1995) A germline insertion in the tuberous sclerosis (Tsc2) gene gives rise to the Eker rat model of dominantly inherited cancer. *Nature Genet.* **9**: 70–74.

Kobayashi, T., Mitani, H., Takahashi, R., Hirabayashi, M., Ueda, M., Tamura, H. and Hino, O. (1997) Transgenic rescue from embryonic lethality and renal carcinogenesis in the Eker rat model by introduction of a wild-type Tsc2 gene. *Proc. Natl Acad. Sci. USA* **94**: 3990–3993.

Lamb, R.F., Roy, C., Diefenbach, T.J., Vinters, H.V., Johnson, M.W., Jay, D.G. and Hall, A. (2000) The TSC1 tumor suppressor hamartin regulates cell adhesion through ERM proteins and the GTPase Rho. *Nat. Cell Biol.* **2**: 281–287.

Maheshwar, M.M., Sanford, R., Nellist, M., Cheadle, J.P., Sgotto, B., Vaudin, M. and Sampson, J.R. (1996) Comparative analysis and genomic structure of the tuberous sclerosis 2 (TSC2) gene in human and pufferfish. *Hum. Mol. Genet.* **5**: 131–137.

Maheshwar, M.M., Cheadle, J.P., Jones, A.C., Myring, J., Fryer, A.E., Harris, P.C. and Sampson, J.R. (1997) The GAP-related domain of tuberin, the product of the TSC2 gene, is a target for missense mutations in tuberous sclerosis. *Hum. Mol. Genet.* **6**: 1991–1996.

Mayer, K., Ballhausen, W. and Rott, H-D. (1999) Mutation screening of the entire coding regions of the TSC1 and TSC2 gene with the Protein Truncation Test (PTT) identifies frequent splicing defects. *Hum. Mutat.* **14**: 401–411.

Mlynarczyk, G. (1991) Enamel pitting. A common sign of tuberous sclerosis. *Ann. N.Y. Acad. Sci.* **615**: 367–369.

Nellist, M., van Slegtenhorst, M.A., Goedbloed, M., van den Ouweland, A.M.W., Halley, D.J.J. and van der Sluijs, P. (1999) Characterization of the cytosolic tuberin-hamartin complex. Tuberin is a cytosolic chaperone for hamartin. *J. Biol. Chem.* **274**: 35647–35652.

Nixon, J.R., Houser, O.W., Gomez, M.R. and Okazaki, H. (1989) Cerebral tuberous sclerosis: MR imaging. *Radiology* 170: 869–873.

Onda, H., Lueck, A., Marks, P.W., Warren, H.B. and Kwiatkowski, D.J. (1999) Tsc2(+/−) mice develop tumors in multiple sites that express gelsolin and are influenced by genetic background. *J. Clin. Invest.* **104**: 687–695.

Pajak, L., Jin, F., Xiao, G.H., Soonpaa, M.H., Field, L.J. and Yeung, R.S. (1997) Sustained cardiomyocyte DNA synthesis in whole embryo cultures lacking the TSC2 gene product. *Am. J. Physiol.* **273**: H1619–H1627.

Plank, T.L., Yeung, R.S. and Henske, E.P. (1998) Hamartin, the product of the tuberous sclerosis 1 (TSC1) gene, interacts with tuberin and appears to be localized to cytoplasmic vesicles. *Cancer Res.* **58**: 4766–4770.

Roach, E., Smith, M., Huttenlocher, P., Bhat, M., Alcorn, D. and Hawley, L. (1992) Diagnostic criteria: tuberous sclerosis complex. *J. Child. Neurol.* **7**: 221–224.

Roach, E.S., Gomez, M.R. and Northrup, H. (1998) Tuberous sclerosis complex consensus conference: revised clinical diagnostic criteria. *J. Child Neurol.* **13**: 624–628.

Rose, V.M., Au, K-S., Pollom, G., Roach, E.S., Prashner, H.R. and Northrup, H. (1999) Germ-line mosaicism in tuberous sclerosis: how common? *Am. J. Hum. Genet.* **64**: 986–992.

Rybin, V., Ullrich, O., Rubino, M., Alexandrov, K., Simon, I., Seabra, M.C., Goody, R. and Zerial, M. (1996) GTPase activity of Rab5 acts as a timer for endocytic membrane fusion. *Nature* 383: 266–269.

Sampson, J.R., Maheswhar, M.M., Aspinwall, R. *et al.* (1997) Renal cystic disease in tuberous sclerosis: role of the polycystic kidney disease 1 gene. *Am. J. Hum. Genet.* **61**: 843–851.

Sepp, T., Yates, J.R.W. and Green, A.J. (1996) Loss of heterozygosity in tuberous sclerosis hamartomas. *J. Med. Genet.* **33**: 962–964.

Smolarek, T.A., Wessner, L.L., McCormak, F.X., Mylet, J.C., Menon, A.G. and Henske, E.P. (1998) Evidence that lymphangioleiomyomatosis is caused by TSC2 mutations: chromosome 16p13 loss of heterozygosity in angiomyolipomas and lymph nodes from women with lymphangioleiomyomatosis. *Am. J. Hum. Genet.* **62**: 810–815.

Soucek, T., Pusch, O., Wienecke, R., DeClue, J.E. and Hengstschläger, M. (1997) Role of the tuberous sclerosis gene-2 product in cell cycle control. Loss of the tuberous sclerosis gene-2 induces quiescent cells to enter S phase. *J. Biol. Chem.* **272**: 29301–29308.

Soucek, T., Holzl, G., Bernaschek, G and Hengstschläger, M. (1998a) A role of the tuberous sclerosis gene-2 product during neuronal differentiation. *Oncogene* **16**: 2197–2204.

Soucek, T., Yeung, R.S. and Hengstschläger, M. (1998b) Inactivation of the cyclin-dependent kinase inhibitor p27 upon loss of the tuberous sclerosis complex gene-2. *Proc. Natl Acad. Sci. USA* **95**: 15653–15658.

Tapon, N.A. and Hariharan, I.K. (2000) The rocky gene regulates cell size in vivo and encodes the homolog of the human TSC1 (Tuberous sclerosis complex 1) tumor-suppressor gene. *A Dros. Res. Conf.* **41**: 19.

The European Chromosome 16 Tuberous Sclerosis Consortium (1993) Identification and characterization of the tuberous sclerosis gene on chromosome 16. *Cell* **75**: 1305–1315.

Tischfield, J.A. (1997) Loss of heterozygosity or: How I learned to stop worrying and love mitotic recombination. *Am. J. Hum. Genet.* **61**: 995–999.

TSC variation database (release January, 2000): http://www.expmed.bwh.harvard.edu/tsc/

Van Baal, J.G. Thesis (1987) Amsterdam.

Van Slegtenhorst, M., de Hoogt, R., Hermans, C. *et al.* (1997) Identification of the tuberous sclerosis gene TSC1 on chromosome 9q34. *Science* **277**: 805–808.

Van Slegtenhorst, M., Nellist, M., Nagelkerken, B. *et al.* (1998) Interaction between hamartin and tuberin, the TSC1 and TSC2 gene products. *Hum. Mol. Genet.* **7**: 1053–1057.

Verhoef, S., Bakker, L., Tempelaars, A.M.P. *et al.* (1999a) High rate of mosaicism in tuberous sclerosis complex. *Am. J. Hum. Genet.* **64**: 1632–1637.

Verhoef, S., van Diemen-Steenvoorde, R., Akkersdijk, W.L. *et al.* (1999b) Malignant pancreatic tumor within the spectrum of tuberous sclerosis complex in childhood. *Eur. J. Pediatr.* **158**: 284–287.

Verhoef, S., Vrtel, R., van Essen, T., Bakker, L., Sikkens, E., Halley, D., Lindhout, D. and van den Ouweland, A. (1995) Somatic mosaicism and clinical variation in tuberous sclerosis complex. *Lancet* **345**: 202.

Von Recklinghausen, F. (1862) Ein Herz von einen Neugeborenen. *Verhandl. Gesellsch. Geburtsh.* **15**: 73–74.

Weiner, D.M., Ewalt, D.H., Roach, E.S. and Hensle, T.W. (1998) The tuberous sclerosis complex: a comprehensive review. *J. Am. Coll. Surg.* **187**: 548–561.

Wienecke, R., Konig, A. and DeClue, J.E. (1995) Identification of tuberin, the tuberous sclerosis-2 product. Tuberin possesses specific Rap1GAP activity. *J. Biol. Chem.* **270**: 16409–16414.

Wilkie, A.O.M, Buckle, V.J., Harris, P.C. *et al.* (1990) Clinical features and molecular analysis of the α thalassemia/mental retardation syndromes. I. Cases due to deletions involving chromosome band 16p12.3. *Am. J. Hum. Genet.* **46**: 1112–1126.

Wolf, D.C., Goldsworthy, T.L., Donner, E.M., Harden, R., Fitzpatrick, B and Everitt, J.I. (1998) Estrogen treatment enhances hereditary renal tumor development in Eker rats. *Carcinogenesis* **19**: 2043–2047.

Xiao, G-H., Jin, F. and Yeung, R.S. (1995) Germ-line Tsc2 mutation in a dominantly inherited cancer model defines a novel family of rat intracisternal-A particle elements. *Oncogene* **11**: 81–87.

Xiao, G-H., Shoarinejad, F., Jin, F., Golemis, E.A. and Yeung, R.S. (1997) The tuberous sclerosis 2 gene product, tuberin, functions as a Rab5 GTPase activating protein (GAP) in modulating endocytosis. *J. Biol. Chem.* **272**: 6097–6100.

Yates, J.R.W., van Bakel, I., Sepp, T., Payne, S.J., Webb, D.W., Nevin, N.C. and Green, A.J. (1997) Female germline mosaicism in tuberous sclerosis confirmed by molecular genetic analysis. *Hum. Mol. Genet.* **6**: 2265–2269.

Yeung, R.S., Xiao, G-H., Jin, F., Lee, W.C., Testa, J.R. and Knudson, A.G. (1994) Predisposition to renal carcinoma in the Eker rat is determined by germ-line mutation of the tuberous sclerosis 2 (TSC2) gene. *Proc. Natl Acad. Sci. USA* **91**: 11413–11416.

13

Retinoblastoma: patients, tumors, gene and protein contribute to the understanding of cancer

Allison S. Duckett, Lina Dagnino and Brenda L. Gallie

1. Introduction

Unraveling the elegant interplay of proteins which regulate cell cycle progression is a continuous process. An intricate balance between growth stimulatory and inhibitory signals regulates the emergence of all the complex interacting cell types that form the organism. Perturbation of this balance can allow a cell to proliferate out of control, resulting in cancer. Multiple mechanisms control differentiation and proliferation, but a fundamental cornerstone of this process is the retinoblastoma susceptibility gene, *RB1*. This critical tumor suppressor gene was initially discovered because its mutation predisposes to retinoblastoma, a hereditary cancer of the retina. Its protein product, pRB, is also integral to the normal embryonic development and the acquisition and maintenance of the terminal differentiated fate of many cell types.

Retinoblastoma is a rare childhood tumor which develops when both copies of the *RB1* gene become mutated in developing retina, so that the amount of pRB falls below a level which may be necessary to induce terminal differentiation. The activity of pRB is controlled by cyclin/cyclin-dependent kinase (cdk) complexes and cdk inhibitors; pRB in turn regulates and controls the activities of transcription factors such as E2F. Subtle and precise modifications of pRB activity are accomplished by arrays of signaling proteins and result in normal growth and differentiation.

Inactivation of the pRB regulatory pathway is an essential initiation event for retinoblastoma, predisposes to a few specific human neoplasms, such as

---

Molecular Genetics of Cancer second edition, edited by J.K. Cowell.

osteosarcoma and melanoma, but is also important for malignant progression in many other types of tumors. The ubiquitous *RB1* growth regulatory pathway is inactivated in different specific types of cancers by mutations of different components of the pathway. Why the results of disruption of this pathway are tissue-specific, and the precise mechanisms involved in retinoblastoma tumor initiation and progression is unknown, but the answer most certainly lies in understanding the normal processes of proliferation, differentiation, and programmed cell death.

2. Retinoblastoma tumors and the *RB1* gene

2.1 Chromosomal location, loss of heterozygosity and cloning of the RB1 *gene*

The etiology of retinoblastoma was first recognized by Alfred Knudson, who postulated that two mutations are required for the initiation of retinoblastoma, leading to the development of the 'two-hit' hypothesis of tumorigenesis (Knudson, 1971). Individuals who carry a mutation in one *RB1* allele are predisposed to multiple retinoblastoma tumors in both eyes, because somatic inactivation of the remaining *RB1* allele is a relatively common event. Individuals with two normal germline *RB1* genes can also develop retinoblastoma, if both alleles are damaged in a single retinal cell; since this is a very rare somatic event, it always presents as one tumor in one eye. Retinoblastoma tumors are often discovered when light reflects off the tumor filling the eye, showing a 'cat's eye' appearance, known as leukocoria (*Figure 1*).

The locus of the *RB1* gene was first suggested by cytogenetic studies of the non-tumor cells of a few individuals who had constitutional deletion of chromosome 13q14 and developed retinoblastoma. Genetic linkage to alleles of an isoenzyme on chromosome 13 in non-deletion hereditary retinoblastoma families confirmed the locus to be chromosome 13q14 (Connolly *et al.*, 1983; Sparkes *et al.*, 1980). Study of the alleles of this isoenzyme in the normal and retinoblastoma cells of affected persons yielded a surprising result: the tumor cells had lost one allele (Godbout *et al.*, 1983). Subsequently, molecular markers revealed that large chromosomal regions surrounding chromosome 13q14 were 'lost' in retinoblastoma tumor cells (Cavenee *et al.*, 1983). This previously undescribed process was called loss of heterozygosity (LOH), and searching for LOH in specific tumors became the standard way to find the genetic loci of other tumor suppressor genes.

The locus of the *RB1* gene was identified by the discovery that DNA for a randomly isolated chromosome 13 molecular marker was missing from a retinoblastoma tumor (Dryja *et al.*, 1986). A cDNA clone isolated from this region from normal cells turned out to be encoded by the *RB1* gene, which spans 180 kb (Friend *et al.*, 1986). *RB1* contains 27 exons which encode a 4.7-kb transcript, with promoter elements within 500 kb upstream (Toguchida *et al.*, 1993). The *RB1* transcript is broadly expressed in adult tissues and encodes a protein, pRB, with 928 amino acids and a predicted molecular weight of 106 kDa. Two related proteins, p107 and p130, have been cloned which share some redundancy with pRB. It is likely that pRB, p107 and p130 perform different functions at different times during the cell cycle and in different tissue and cell types throughout development (Cobrinik *et al.*, 1993). Mutations of both p107 and p130 are involved in the induction of retinoblastoma tumors in transgenic mice, however, they are unable to predispose to tumor formation when mutated alone (reviewed in Mulligan and Jacks (1998)).

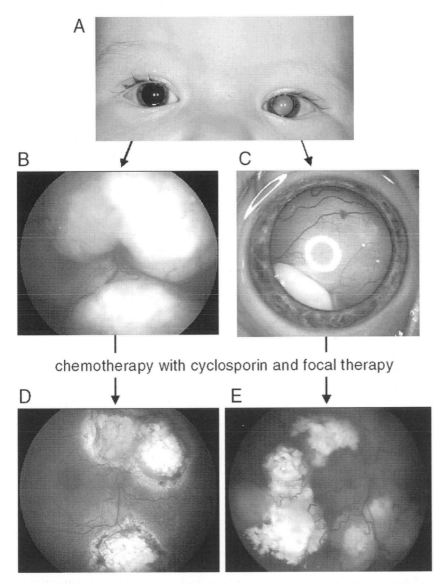

chemotherapy with cyclosporin and focal therapy

Figure 1. (A) The first sign of retinoblastoma tumor is leukocoria, caused by light reflecting off the tumor in the left eye; (B) The right eye contains Murphree Group D1 retinoblastoma; (C) The left eye contains Murphree Group D2 retinoblastoma; (D) Ten months after chemotherapy with cyclosporine and repeated focal therapy good vision is attained in the right eye with inactive tumor; (E) Despite a good regression in the left eye, recurrent tumor is treated with removal of the eye.

2.2 Mutation of the RB1 gene

Most retinoblastoma tumors lack detectable pRB consequent to *RB1* gene mutations which result in premature termination of translation (reviewed in DiCiommo *et al.*, (2000)). Mutations predisposing to retinoblastoma occur throughout the gene, are not clustered in 'hot spots' and are unique in each family, making mutation probe hybridization techniques impractical. Given the large size of the *RB1* gene, detection of mutations by direct cDNA analysis or protein truncation assays would be convenient. However, these approaches are not possible because the mutant transcripts are not detectable in normal tissue samples from patients, suggesting instability of the mRNA due to premature termination of translation when a stop codon is prematurely encountered (Dunn *et al.*, 1989). Interestingly, the *RB1* mutant mRNA can be readily found in tumor samples, which have no normal transcripts or pRB, suggesting that there is autoregulation of expression of *RB1* (Gill *et al.*, 1994).

Practical clinical mutation analysis is performed on RB probands by quantitative multiplex PCR of all exons and promoter regions, to detect changes in size and copy number. If no mutation is found by this first screen, all exons and the promoter are sequenced. With this strategy, more than 80% of *RB1* gene mutations in blood of individuals with bilateral retinoblastoma can be identified, which allows future siblings and subsequent generations to be screened for the inheritance of the mutation. Several children have been delivered prematurely because they were shown to carry an *RB1* mutation, and already had retinoblastoma tumors present at birth (Gallie *et al.*, unpublished data). Early treatment made possible by early delivery has improved the visual outcome for these children. Additionally, molecular management avoids unnecessary examinations, often under anesthesia, of children who may not be at risk for developing retinoblastoma tumors, and is five-fold less expensive for the whole family than traditional clinical screening (Noorani *et al.*, 1996).

The majority of familial *RB1* mutations result in truncated, unstable proteins which are undetectable or proteins with internal deletions of functionally critical domains. Transmission of predisposition to tumors is a dominant trait with high penetrance; the risk for retinoblastoma is greater than 90%. Low penetrance retinoblastoma families have a high frequency of unilateral tumors and unaffected carriers. The low penetrance *RB1* mutations are of two very different types (summarized in DiCiommo (2000)). First, some residual functional pRB remains when missense mutations or in-frame deletions encode nearly normal or internally truncated proteins, promoter mutations diminish transcription of an otherwise normal protein (Sakai *et al.*, 1991) or exon splice mutations variably splice correctly. Second, if total deletion of the *RB1* gene extends to include some (unknown) adjacent gene for which nullizygosity is lethal to the cell, retinal cells that suffer LOH presumably die and do not form tumors. Tumors only arise if the second *RB1* allele is a separate null mutation.

2.3 Additional chromosome abnormalities in retinoblastomas

RB1 mutations are necessary, but not sufficient for tumorigenesis in human developing retina. Additional mutations which confer selective growth advantage are required for a clone to acquire the fully malignant phenotype. The iso(6p) chromosome is almost unique to retinoblastoma (Squire *et al.*, 1984), occurs in 42% of retinoblastoma tumors and results in low level amplification of the genes present on 6p. Extra copies of chromosome 1q occur in 58% of retinoblastoma tumors, and 32% have both iso(6p) and 1q gains (D. Chen, B.L. Gallie and J. Squire, unpublished data). How these changes contribute to proliferation of embryonic retinoblasts is unknown.

Genomic amplification of *MYCN* and the adjacent DEAD box gene *DDX1* occurs in a few retinoblastoma tumors (Godbout and Squire, 1993). All retinoblastoma tumors express a moderate level of *MYCN*, without gene amplification, as does normal embryonic retina. Some retinoblastoma tumors show expression of telomerase, which seems to be an important feature of immortalization in many human tumors (Gupta *et al.*, 1996). The commonly mutated tumor suppressor gene, *TP53*, is normal in retinoblastoma tumors.

Loss of pRB can lead to proliferation, endoreduplication, or apoptosis, depending on the cell and tissue type in which pRB function is lost (Jiang *et al.*, 2000; Zacksenhaus *et al.*, 1996). It is unclear why pRB-deficiency uniquely predisposes to tumor formation in the human retina. Alteration of apoptotic pathways functioning in normal and transformed retinoblasts may be a critical part of the initiation of retinoblastoma tumors. Retinal development depends on a precise balance of survival and programmed cell death. Many cell types in the developing nervous system require the expression of trophic factors, such as nerve growth factor (NGF) to promote their survival. Failure of programmed cell death due to a third mutation may allow retinal cells to escape the usual consequence of loss of pRB and result in retinoblastoma.

2.4 RB1 *mutations and other cancers*

In 3–5% of individuals with *RB1* mutations, other types of tumors arise later in life. Most common are osteosarcomas, melanomas, and brain tumors (Eng *et al.*, 1993). External beam radiation to cure retinoblastoma, the conventional management for retinoblastoma tumors since the 1960s, increases the risk of a second malignancy by the third decade of life to at least 30%. Radiation at less than 1 year of age has a higher risk of inducing second tumors and additionally causes significant cosmetic deformity (Abramson and Frank, 1998).

It is interesting that the incidence of leukemia and other cancers in patients having germline *RB1* mutations is not elevated above the normal population, despite its broad pattern of expression (Phillips *et al.*, 1992). The *rb1*[-/-] knock-out mice reveal the reason: the common consequence of loss of *RB* in most tissues is programmed cell death, or apoptosis. In *RB1*[+/-] humans, the loss of an occasional cell by apoptosis would result in no observable effect. *RB1* mutations have been detected in almost all human tumor types, but particularly small cell lung, breast, prostate and bladder carcinomas, and osteosarcomas, and less commonly in other tumor types, such as leukemia (reviewed by Sellers and Kaelin, 1997).

3. Therapy of retinoblastoma

The safest way to treat retinoblastoma is removal of the affected eye(s). Since the early 1960s external beam radiation has been successful in saving eyes with medium sized tumors. Results with large retinoblastoma were poor. Side-effects however are severe: cosmetic and functional (cataract) deformity and a high incidence of secondary tumors in children with *RB1* mutations who also received radiation (Draper *et al.*, 1986). By 1996 many of the major retinoblastoma treatment centers world-wide showed that systemic chemotherapy combined with vigilant focal therapy offered higher likelihood of eye salvage than the conventional external beam radiation therapy. Since retinoblastoma is rare, clinical trials to optimize therapy require international teamwork in order to accrue enough patients. By working together toward the goal of clinical trials, we have seen rapid advancement in our understanding of treatment approaches to retinoblastoma (Friedman *et al.*, 2000; Gallie *et al.*, 1996; Greenwald and Strauss, 1996; Kingston *et al.*, 1996; Murphree *et al.*, 1996).

Together, we have developed a new classification (the Murphree Classification) which more accurately reflects current approaches than the classical Reese–Ellsworth classification, which was originally designed to predict the likelihood of successful treatment with external beam radiotherapy. The Murphree Classification grades each affected eye for its worst prognosis tumor and takes into account location near the optic nerve or macula, which more severely affect visual prognosis than tumors in the periphery. Group A eyes have small tumors that are usually eradicated with focal therapy (laser and/or cryotherapy) with excellent visual prognosis. Group B eyes have single tumors too large for focal therapy alone, with good visual prognosis, that can be successfully treated with brachytherapy. Group C eyes require chemotherapy to shrink tumors to a size that can be treated with focal therapy. Some eyes with small tumors near the fovea and optic nerve become Group C because chemotherapy offers the prospect of improved visual outcome by reducing the extent of focal therapy required.

Group D includes the eyes with large tumors that have already at diagnosis severely disrupted the eye (retinal detachment, vitreous seeds, etc.) that prior to the use of chemotherapy would generally have been primarily enucleated, since the prognosis to save such eyes with external beam radiotherapy is poor. Review of the outcome of Group D eyes treated with chemotherapy while preparing for multicenter clinical trials has revealed features that distinguish Group D1 eyes with good prognosis (60 to 70%) from Group D2 eyes with very poor prognosis (0 to 30%) for success with chemotherapy and focal therapy.

Group D1 eyes have very large tumors, up to total retinal detachment, and small amounts of localized vitreous seeds. These eyes can respond dramatically to full systemic chemotherapy, and then require vigilant application of focal therapy to consolidate the response. Group D2 eyes have tumor greater than 2/3 of the volume of the eye, massive vitreous seeds or solid tumor invading anterior to the retina. These eyes may initially respond to full systemic chemotherapy but ultimately have a very poor success rate to save the eye.

We have found that the multidrug resistance protein, P-glycoprotein is increased in one-third of RB enucleated primarily and in all large retinoblastoma enucleated at failure of therapy, suggesting that this protein may be responsible for the resistance of large retinoblastoma to chemotherapy (Chan *et al.*, 1996). Likewise, the good response of retinoblastoma to carboplatin, teniposide and vincristine reported by Chan and Gallie might be due to concurrent use of high dose cyclosporine (Gallie *et al.*, 1996), a known inhibitor of P-glycoprotein *in vitro*, that competitively inhibits the efflux of drugs from cells by the P-glycoprotein multidrug resistance pump.

Intraocular retinoblastoma that failed chemotherapy despite cyclosporine expressed multidrug resistance protein (MRP) rather than P-glycoprotein, whereas retinoblastoma that failed chemotherapy prior to our use of CSA showed increased P-glycoprotein (Chan *et al.*, 1997). There is currently no effective means to block MRP. However, MRP is less frequently found at diagnosis in RB (1/18) than P-glycoprotein (8/38).

Current pilot studies are collecting preliminary data on the effect of the addition of local chemotherapy (subtenon's carboplatin) to systemic chemotherapy with cyclosporine and focal therapy for Group D2 eyes. Subconjunctival carboplatin was shown to effectively inhibit intraocular retinoblastoma development in a transgenic mouse model, with no evidence of toxicity on histopathological examination (Murray *et al.*, 1997). Intraocular levels of carboplatin have been shown to be in a high therapeutic range after local administration of small doses (20 mg) in a series of eyes subsequently enucleated for retinoblastoma. A small Phase I/II study of 13 retinoblastoma eyes treated with local carboplatin showed some intraocular effect (Abramson *et al.*, 1999). Therefore, a proposed Phase I/II protocol for Group D2 eyes will combine chemotherapy (vincristine, carboplatin, etoposide) with cyclosporine to counteract multidrug resistance and a series of four subtenon's capsule injections of carboplatin. This is the best current approach to save such severely involved eyes.

Group E eyes have signs of potential extraocular disease, such as neovascular glaucoma or suspicious invasion of optic nerve on imaging. These eyes are removed and carefully examined for evidence of extraocular disease. Group F eyes have already been breached and extraocular disease is present. Retinoblastoma in the brain still has an extemely poor prognosis for life, while bone marrow involvement has been successfully treated with combinations of chemotherapy, radiation and stem cell transplant.

4. Disruption of the RB pathway in neoplasia

4.1 Somatic *RB1 mutations in human tumors*

De-regulation of components of the cell cycle control pathway that regulate the activity of pRB occurs commonly in many tumor types. The cyclin D1 gene, which inactivates pRB when complexed with cyclin-dependent kinases (cdk4 and 6), is amplified in many esophageal, breast and squamous cell carcinomas, and overexpressed in B-cell lymphomas due to chromosomal translocations (Hall and Peters, 1996; Lukas *et al.*, 1994). The cdk inhibitors p15 and p16, which inactivate cyclin D/cdk complexes, are absent or mutated in many esophageal squamous cell carcinomas, glioblastomas, non-small cell lung cancer, bladder and pancreatic carcinomas as well as in familial melanomas (reviewed in Foulkes *et al.*, 1997). Such pRB pathway alterations maintain the normal pRB in the phosphorylated, inactive state, allowing the tumor cells to proliferate. In contrast, small cell lung carcinomas, which usually lack functional pRB because of mutations in the *RB1* gene, have normal cyclin D1 levels and abundant expression of the cdk inhibitor, p16.

Although pRB loss by mutation of *RB1* is the specific rate-limiting step for initiation of transformation in retina and certain other tissues, inactivation of the pRB pathway by other means may contribute substantially to malignant progression of other specific cell types, once they have been transformed by other genetic mutations.

4.2 Inactivation of pRB by viral oncoproteins

Transforming viruses produce proteins that inactivate pRB, in order to gain control of the cell cycle. The observation of adenovirus E1A protein binding to pRB was critical in revealing the role of pRB in the cell cycle (DeCaprio *et al.*, 1989; Dyson *et al.*, 1989; Ludlow *et al.*, 1989, 1990; Whyte *et al.*, 1988). The transforming abilities of the viral proteins (adenovirus E1A, simian virus 40 (SV40) large T antigen (Tag) and human papilloma virus E7) correlate with their pRB binding abilities, and mutant viral proteins that fail to bind pRB have also lost their transforming potential (DeCaprio *et al.*, 1988; Dyson *et al.*, 1989; Heck *et al.*, 1992; Whyte *et al.*, 1988). The best characterized clinical example of a virus inducing human cancer is the causative role of human papilloma virus (HPV) in cervical carcinomas, whereby the HPV-E7 protein inactivates pRB (Munger *et al.*, 1989). Although the Epstein-Barr virus nuclear antigen EBNA-5 can bind pRB, it does not appear to alter pRB function like the other viral proteins (Szekely *et al.*, 1995).

In transgenic mice, the outcome of expression of various viral gene products under the control of tissue-specific promoters depends on the tissue and state of the cell cycle. The tumor viruses produce proteins that bind and inactivate many cellular factors, including pRB, p53 and the other pRB family proteins p107 and p130 (reviewed in DiCiommo *et al.*, 2000). For example, TAg induced tumor formation when expressed in proliferating cells and in the retina, lens and choroid plexus, but resulted in pronounced degeneration when expressed in post-mitotic cells of the retina (al-Ubaidi *et al.*, 1992). Expression of the HPV-E7 protein (which binds pRB-like proteins) only causes retinal tumor formation in $tp53^{-/-}$ mice. In heterozygous $tp53^{+/-}$ mice expressing TAg, the retina degenerates due to apoptosis.

4.3 RB1 deficiency in mice

The critical role of pRB during development was demonstrated by the generation of *rb1*-deficient mice. $rb^{-/-}$ mouse embryos die *in utero* at 13.5–15.5 days of gestation, due to the failure of differentiation of the hematopoietic and nervous systems (Clarke *et al.*, 1992; Jacks *et al.*, 1992; Lee *et al.*, 1992). There is an increase in nucleated, immature erythrocytes in the liver and neurons that lack expression of nerve growth factor receptors in the central nervous system, both evidence of failed differentiation. These cells are then eliminated by apoptosis.

Heterozygous germline mutation of *rb1* ($rb1^{+/-}$) specifically predisposes mice to tumors in the intermediate lobe of the pituitary and in the thyroid, but not to the retinoblastoma that affects $RB1^{+/-}$ humans (Jacks *et al.*, 1992; Lee *et al.*, 1992).

This suggests that susceptibility to tumorigenesis in the absence of pRB has both tissue- and species-specific components. No pituitary tumors are observed in humans with germline *RB1* mutations, easily attributed to the vestigial nature of the intermediate lobe of the pituitary in humans. But unresolved is why mice do not develop retinoblastoma. $RB^{+/-}$ mice that are also homozygous ($tp53^{-/-}$) or heterozygous ($tp53^{+/-}$) for $tp53$ mutations develop novel tumor types such as pinealoblastoma and islet cell tumors in addition to those typical of each genotype, but still do not develop retinoblastoma (Williams *et al.*, 1994a).

The common consequence of loss of pRB is apoptosis: each tissue that normally expresses *RB1* in development shows apoptosis in $rb^{-/-}$ mouse embryos. Partial rescue of $rb^{-/-}$ mice by the expression of pRB in the central nervous system extended their survival until birth, but immediate death ensued due to inadequate intercostal and diaphragm musculature to support breathing (Zacksenhaus et al., 1996). This was due to apoptosis of myoblasts and failure of myotube population expansion with concomitant endoreduplication of DNA (Jiang *et al.*, 2000). These observations suggest that when pRB is absent at critical moments of development, the cells inappropriately enter S phase while unsupported by factors which permit completion of the cell cycle. Apoptosis is the usual consequence, but under the constraints of a myotube, excessive DNA replication occurs without apoptosis or cell division.

The requirement for pRB during development has also been examined by generating chimeric animals containing normal and $rb^{-/-}$ cells. These animals reached adulthood and developed tumors of the intermediary lobe of the pituitary and lens abnormalities reminiscent of $rb^{+/-}$ mice. Although the hematopoietic and nervous system of these chimeras were normal and included $rb^{-/-}$ cells, the fraction of $rb^{-/-}$ cells in the retina was reduced (Maandag *et al.*, 1994; Williams *et al.*, 1994b). Thus, pRB appears to be required and selected for during the normal development of some cell types, but it is dispensable in most tissues. However, chimeric mice with p107/pRB double knock-out develop neoplastic lesions in the inner nuclear layer of the retina at embryonic day 17.5 (Maandag *et al.*, 1994; Robanus-Maandag *et al.*, 1998). It is possible that the $rb^{-/-}$ cells are limited in the range of specific cell fates that they can achieve, with some retinal cell types requiring *rb* expression for their specification and differentiation. It is interesting to note that mutation of *E2F1* suppresses apoptosis and extends the survival of *rb* deficient mice, but only to 17 days of gestation, indicating that RB plays other roles in addition to E2F1 regulation during embryogenesis (Tsai *et al.*, 1998).

5. Cellular function of pRB

5.1 Cellular localization and structure of pRB

pRB is a relatively abundant nuclear phosphoprotein that is strongly tethered to the nuclear matrix in its active, hypophosphorylated state. Nuclear localization of pRB is dependent on two processes: active nuclear transport using the bipartite nuclear localization signal in exon 25, and co-localization of pRB into the nucleus with proteins to which it binds, such as the E2F transcription factors (Zacksenhaus *et al.*, 1999).

Four functional domains of pRB have been defined: amino-terminal, A and B, and carboxy-terminal. Amino-terminal deletions show loss of function, although proteins that specifically bind to this domain have yet to be identified. Best characterized are the highly conserved domains A (amino acids 394–571) and B (amino acids 649–773), separated by a 'spacer' region and euphemistically termed the 'pocket' domain (reviewed by DiCiommo *et al.* (2000)). Crystal structure of the A domain reveals tertiary structure similarity to the cyclin-box of Cyclin A and the transcription factor TFIIB. More indirect data suggest that the B domain also contains a cyclin-box fold. The A/B domains are necessary for growth suppression and define a family of growth-regulatory homologous proteins that includes pRB, p107 and p130 (*Figure 2*). Many cellular factors and the viral oncoproteins E1A and Tag bind to the A/B domains, and these domains are frequently affected by mutations or deletions in tumors (Hu *et al.*, 1990).

The carboxy-terminal domain (amino acids 803–841) is additionally required for binding to cell cycle regulatory proteins such as E2F. Regulatory proteins bind the carboxy-terminal domain of pRB, for example MDM2, which is up-regulated by p53 and initiates the ubiquitin-mediated degradation of p53. The ABL1 tyrosine kinase also binds the carboxy-terminal domain of pRB (Welch and Wang, 1993). A fragment of pRB encompassing this binding site acts as a dominant negative, blocking growth suppression by pRB. Similarly, over-expression of *ABL1* blocks growth suppression.

5.2 Regulation of pRB function by phosphorylation

pRB has several different functions that are critical for normal development. The role of pRB in negative growth regulation, inhibition of cellular transformation,

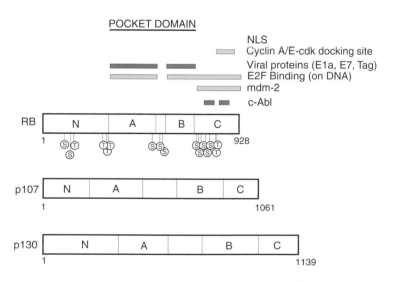

Figure 2. The RB family of proteins share three functional domains, the 'pocket' domain and the N- and C-terminal domains. The 16 possible CDK-mediated serine (S) and threonine (T) phosphorylation sites of pRB are noted. The regions encoding nuclear localization (NLS) and binding to viral and E2F proteins are indicated.

and terminal differentiation of cells is tightly regulated by phosphorylation. The importance of phosphorylation of pRB is evident: phosphorylation occurs concomitantly with or just prior to cell commitment to enter S phase and the DNA tumor virus proteins E7 and Tag, which induce cells to enter S phase, complex only with hypophosphorylated pRB (Ludlow *et al.*, 1989). The E7 and E6 proteins of the human papilloma virus types 16 and 18, which can stably associate with p53 and pRB, respectively, promote the accelerated degradation of p53 and pRB *in vitro* (Scheffner *et al.*, 1992). Cellular transcription factors also interact specifically with hypophosphorylated pRB (see below). Cell proliferation is associated with events that direct or maintain pRB phosphorylation (Hinds *et al.*, 1992).

During G_1 phase, the extent of pRB phosphorylation is progressively increased from a hypophosphorylated to a hyperphosphorylated state. The phosphorylation state of pRB regulates progression beyond a critical point in late G_1, after which the cell will proceed to S phase and DNA replication. The extent of phosphory-lation of pRB on multiple serine and threonine residues increases as the cell proceeds through S and G_2 phases (Lees *et al.*, 1991). There are 16 potential cdk consensus sites for phosphorylation, some of which regulate different specific protein interactions. Different cdk–cyclin complexes phosphorylate distinct sites on pRB, and no single cdk–cyclin can phosphorylate all the sites. Cyclin D1-cdk4/cdk6 complexes may initially phosphorylate pRB, while cyclin E- and A-containing kinase complexes modify pRB phosphorylation in late G_1 and through S and G_2 phases. Cyclin D1-ablated mice display a dramatic reduction in cell number in the neural retina due to proliferative failure during embryonic development, as if when *RB1* is first expressed it immediately blocks proliferation (Sicinski *et al.*, 1995).

At mitosis, pRB returns to the unphosphorylated state, in part through the action of phosphatases. The protein phosphatase PP1 may both dephosphorylate pRB and maintain dephosphorylation during mitosis and G_1. In the process of apoptosis, proteases cleave pRB at specific sites; it is not known if specific pRB cleavage inactivates pRB in the cell cycle, or only occurs with apoptosis.

5.3 Cell-cycle regulatory proteins in tumorigenesis and regulation of RB1 function

The phosphorylation of the retinoblastoma protein, essential for progression in the cell cycle beyond specific checkpoints, is inhibited by cdk inhibitors (cdkis). There are two distinct families of cdkis: the INK family (p15, p16, p18 and p19) and the CKI family (p21, p27 and p57). Extracellular signals such as TGF-b   or proteins such as p53 that respond to intracellular damage, stimulate the cdkis which then bind and inactivate cdks or cyclin/cdk complexes, in turn preventing the phosphorylation of pRB and maintaining the cell in G_1 arrest (*Figure 3*).

The p16 cdki is a tumor suppressor gene that, like *RB1*, when mutated contributes to both the development of specific cancers and the progression of a broad range of human cancers. The normal p16 forms binary complexes with cdk4 and 6, inhibiting phosphorylation and inactivation of pRB. The expression of p16 is high in most cells in which pRB activity is blocked by interaction with viral proteins or by mutation, which has led to suggestions that pRB may regulate the

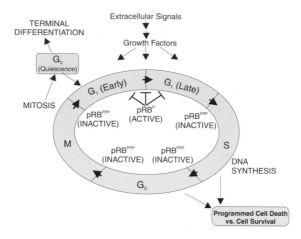

Figure 3. pRB regulates cell cycle progression. The presence of active, hypophosphorylated pRB at the G_1-S checkpoint regulates progression into S phase and subsequent DNA synthesis. Inactive, hyperphosphorylated pRB, which is present throughout the various other phases of the cell cycle (e.g., G_2, M) permits progression through the cell cycle.

expression of p16. However, we have demonstrated that p16 mRNA expression is low in retinoblastoma tumor cells, despite the fact that the loss of pRB is an early event in the transformation of these cells (Gallie *et al.*, 1999a). This indicates that the regulation of p16 mRNA expression is more complex than the direct repression by pRB that has been proposed.

The cdki p27 also plays roles in proliferation, differentiation, apoptosis, and tumorigenesis. $p27^{-/-}$ mice display increased body size and multiple organ hyperplasia, including disorganization of the retina and retinal dysplasia (Nakayama *et al.*, 1996). Studies in *Xenopus* and murine retina have shown that p27 is broadly expressed as differentiation begins, with progressive limitation to Müller glia-specific expression in the adult (Levine *et al.*, 2000; Ohnuma *et al.*, 1999). These experiments have also demonstrated a requirement for p27 during cell fate specification during retinal development that is distinct from its cell cycle regulatory kinase functions (Ohnuma *et al.*, 1999). Recent experiments have demonstrated down-regulation of p27 in retinoblastoma tumors (A.S. Duckett and B.L. Gallie, unpublished observations), however, given the multiple and distinct functions of p27, determining the significance of this alteration will be complex.

5.4 Transcriptional regulation by pRB

When pRB is active, it can activate or repress the transcriptional activity of several promoters, including regulatory genes required for cell division. Progressive phosphorylation inactivates pRB and releases pRB-bound factors (*Figure 4*). Hypo- but not hyperphosphorylated pRB binds to various cellular DNA-binding transcription factors, which allows nuclear tethering of pRB and direct transcriptional repression of downstream genes by interaction with general transcription factors (Bremner *et al.*, 1995) (reviewed in DiCiommo *et al.*, 2000).

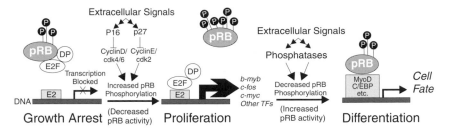

Figure 4. pRB controls cell cycle progression via transcriptional regulation. Hyperphosphorylated pRB sequesters E2F–DP complexes, preventing transcription of S phase-promoting transcription factors (TFs). Cyclin–cdk complexes increase pRB phosphorylation, physically releasing E2F–DP and permitting exit from G_1.

Several observations have placed the transcription factor E2F family as major targets of repression by pRB. Active, hypophosphorylated pRB binds to E2F proteins and suppresses transcriptional activation. Many genes encoding proteins necessary for cell proliferation contain functional E2F binding sites, whose activity is repressed by pRB. Ectopic expression of E2F-1 is sufficient to override G_1 arrest induced by pRB overexpression. Interaction of viral oncoproteins with pRB results in the release of E2F from pRB, thereby activating E2F-mediated transcription and entry into S phase. Finally, pRB can repress the activity of promoters containing E2F-binding sites, such as BMYB, the adenovirus E2 promoter, FOS and MYC. Free E2F, but not pRB-bound E2F, is degraded by ubiquitination, protecting the cell from excessive expression of cell cycle genes.

One of the proposed main targets for pRB is E2F, yet the role of E2F in the generation of retinoblastoma is not known. In the mouse, E2F-1 expression is restricted to the undifferentiated, proliferative layers of the retina, suggesting that down-regulation of E2F-1 expression may occur with retinal differentiation, and E2F-3 is expressed predominantly in the differentiating cells (Dagnino *et al.*, 1997). The expression of E2F family proteins in retinoblastoma tumors has yet to be determined.

The regulation of E2F function by pRB is complex. In simple promoters, E2F activates transcription, which is blocked by interaction with pRB. In other promoters activated by both E2F and other transcription factors (e.g. the E2F-1 promoter), the pRB–E2F complex may repress activity of adjacent promoter elements. Phosphorylation of pRB releases the inhibitory effects of the pRB–E2F complexes. Many other transcription factors bind to, or are modulated by pRB, including MyoD, the Ets factor Elf-1, PU.1 and ATF-2 (Chellappan *et al.*, 1991; Hagemeier *et al.*, 1993; Kim *et al.*, 1992; Wang *et al.*, 1993). For example, pRB enhances Sp1-mediated transactivation and appears to directly interact with ATF-2, increasing its transcriptional activity.

5.5 Role of pRB in differentiation

The observation of multiple proteins binding to pRB indicates that pRB may be highly regulated in different tissues, or have multiple functions. Induction of cancer by loss of pRB may depend on tissue- and developmental stage-specific proteins which determine the context by which the process of differentiation is dependent on pRB. For example, genes which orchestrate the morphogenesis of the retina and direct the differentiation of its layers rely on the activity of pRB to ensure terminal differentiation of specific lineages. A model for the development of retinoblastoma is shown in *Figure 5*. The undifferentiated retinoblasts do not express *RB1*; as the cells become determined to differentiate into specific cell types, the *RB1* gene is first expressed. However, certain specific retinal cell types never express *RB1*, for example rod photoreceptors (Gallie *et al.*, 1999).

In mammalian retina, just as in *Drosophila* retina, the process of differentiation involves specific deletion of excess cells by programmed cell death, while survival signals specify the future functional retinal cells, which acquire their specific destiny in part by signals from adjacent cells. If both alleles of the *RB1* gene are mutated, an increase in apoptosis is observed, indicating that the programmed cell death pathway remains functional.

However, cells destined for survival and subsequent terminal division and retinal function, would instead re-enter S phase in the absense of pRB. Many of these cells would default to apoptosis in the absense of factors supporting proliferation, as in other tissues. Linear expansion of an $RB1^{-/-}$ cell would occur, insufficient to produce a tumor, but perhaps adequate to produce a retinoma, a benign retinal manifestation of mutant *RB1* alleles. When either apoptosis or the intercellular signaling required for the induction of programmed cell death are rendered non-functional by mutation, exponential expansion of that clone could occur, resulting in a malignant retinoblastoma phenotype.

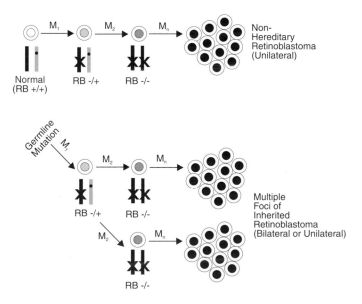

Figure 5. A model of retinoblastoma tumorigenesis in the developing retina. The occurrence of two mutations of the *RB1* gene (M_1 and M_2) in the same retinal cell is necessary for malignant transformation, but additional mutations (Mn) must accumulate for malignant progression to the retinoblastoma phenotype.

Since we and others have shown that expression of *TP53* and other genes implicated in p53-mediated apoptosis is intact in retinoblastoma tumor cells and no *TP53* mutations are found in these tumors (Gallie *et al.*, 1999a), it is likely that the defect contributing to malignant expansion is in another apoptotic pathway or a different intercellular signaling cascade. The signaling molecules which regulate differentiation and cell fate selection in *Drosophila* retina, such as the Notch receptor signaling pathway, all have mammalian homologs and are candidates to play a role in the specificity of *RB1* mutations for cancer in the retina. Disruption of tissue homeostasis in other tissues may be less deleterious due to compensatory mechanisms.

6. Summary

Retinoblastoma is a rare tumor that has yielded great insight into cancer, despite the phenotypic complexity of mutation of the *RB1* gene in various tissues in mice and humans. The capability to precisely identify the causative *RB1* mutation in each family is currently being applied with enormous implications for traditional health care, although educational and political effort is still required to get molecular diagnosis fully implemented on a global scale.

Treatment of retinoblastoma is now being evaluated in co-operative clinical trials, first addressing the role of reversal of multi-drug resistance. The multi-center clinical trial format makes possible a series of experiments to ultimately define the optimal therapy for patients world-wide.

Great advances have been made in understanding the proliferation pathway regulated by pRB, and its role in carcinogenesis. Paradoxically, the specific role of pRB in the genesis of retinoblastoma is still poorly understood. Ultimately, unraveling the mechanisms of pRB action in retinal development and differentiation will provide tools that can be exploited to suppress cell proliferation and transformation, that may be useful in the therapy and prevention of retinoblastoma and other human tumors which have acquired pRB-deficiency during their progression. Retinoblastoma has been, and will continue to be, an excellent tool to determine the earliest events in tumor initiation and malignant progression in many human cancers.

Acknowledgments

ASD is a Cambridge Commonwealth Trust Scholar and is supported by a Canadian Centennial Scholarship. LD is a Medical Research Council of Canada/Cancer Research Society Scholar. BLG is a Distinguished Scientist of the Medical Research Council of Canada. This work was supported by the National Cancer Institute of Canada with funds from the Terry Fox Run; the Medical Research Council of Canada; the Canadian Genetic Disease Network; the Retinoblastoma Family Association; and the Royal Arch Masons of Canada.

References

Abramson, D.H. and Frank, C.M. (1998) Second nonocular tumors in survivors of bilateral retinoblastoma: a possible age effect on radiation-related risk. *Ophthalmology* **105**: 573–579.

Abramson, D.H., Frank, C.M. and Dunkel, I.J. (1999) A phase I/II study of subconjunctival carboplatin for intraocular retinoblastoma. *Ophthalmology* 106: 19470–1950.

al-Ubaidi, M.R., Font, R.L., Quiambao, A.B., Keener, M.J., Liou, G.I., Overbeek, P.A. and Baehr, W. (1992) Bilateral retinal and brain tumors in transgenic mice expressing simian virus 40 large T antigen under control of the human interphotoreceptor retinoid-binding protein promoter. *J. Cell Biol.* 119: 1681–1687.

Bremner, R., Cohen, B.L., Sopta, M., Hamel, P.A., Ingles, C.J., Gallie, B.L. and Phillips, R.A. (1995) Direct transcriptional repression by pRB and its reversal by specific cyclins. *Mol. Cell Biol.* 15: 3256–3265.

Cavenee, W.K., Dryja, T.P., Phillips, R.A., Benedict, W.F., Godbout, R., Gallie, B.L., Murphree, A.L., Strong, L.C. and White, R.L. (1983) Expression of recessive alleles by chromosomal mechanisms in retinoblastoma. *Nature* 305: 779–784.

Chan, H.S.L., DeBoer, G., Thiessen, J.J. *et al.* (1996) Combining cyclosporin with chemotherapy controls intraocular retinoblastoma without requiring radiation. *Clin. Cancer Res.* 2: 1499–1508.

Chan, H.S., Lu, Y., Grogan, T.M., Haddad, G., Hipfner, D.R., Cole, S.P., Deeley, R.G., Ling, V. and Gallie, B.L. (1997) Multidrug resistance protein (MRP) expression in retinoblastoma correlates with the rare failure of chemotherapy despite cyclosporine for reversal of P-glycoprotein. *Cancer Res.* 57: 2325–2330.

Chellappan, S.P., Hiebert, S., Mudryj, M., Horowitz, J.M. and Nevins, J.R. (1991) The E2F transcription factor is a cellular target for the RB protein. *Cell* 65: 1053–1061.

Clarke, A.R., Maandag, E.R., van Roon, M., van der Lugt, N.M., van der Valk, M., Hooper, M.L., Berns, A. and te Riele, H. (1992) Requirement for a functional Rb-1 gene in murine development [see comments]. *Nature* 359: 328–330.

Cobrinik, D., Whyte, P., Peeper, D.S., Jacks, T. and Weinberg, R.A. (1993) Cell cycle-specific association of E2F with the p130 E1A-binding protein. *Genes Dev.* 7: 2392–2404.

Connolly, M.J., Payne, R.H., Johnson, G., Gallie, B.L., Alderdice, P.W., Marshall, W.H. and Lawton, R.D. (1983) Familial, EsD-linked, retinoblastoma with reduced penetrance and variable expressivity. *Hum. Genet.* 65: 122–124.

Dagnino, L., Fry, C.J., Bartley, S.M., Farnham, P., Gallie, B.L. and Phillips, R.A. (1997) Expression patterns of the E2F family of transcription factors during mouse nervous system development. *Mech. Dev.* 66: 13–25.

DeCaprio, J.A., Ludlow, J.W., Figge, J., Shew, J.-Y., Huang, C.-M., Lee, W.-H., Marsilio, E., Paucha, E. and Livingstone, D. M. (1988) SV40 large tumor antigen forms a specific complex with the product of the retinoblastoma susceptibility gene. *Cell* 54: 275–283.

DeCaprio, J.A., Ludlow, J.W., Lynch, D., Furukawa, Y., Griffin, J., Piwnica-Worms, H., Huang, C.M. and Livingston, D.M. (1989) The product of the retinoblastoma susceptibility gene has properties of a cell cycle regulatory element. *Cell* 58: 1085–1095.

DiCiommo, D., Gallie, B.L. and Bremner, R. (2000) Retinoblastoma: the disease, gene and protein provide critical leads to understand cancer. *Semin. Cancer Biol.* in press.

Draper, G.J., Sanders, B.M. and Kingston, J.E. (1986) Second primary neoplasms in patients with retinoblastoma. *Br J Cancer* 53: 661–671.

Dryja, T.P., Rapaport, J.M., Joyce, J.M. and Petersen, R.A. (1986) Molecular detection of deletions involving band q14 of chromosome 13 in retinoblastomas. *Proc. Natl Acad. Sci. USA* 83: 7391–7394.

Dunn, J.M., Phillips, R.A., Zhu, X., Becker, A. and Gallie, B.L. (1989) Mutations in the RB1 gene and their effects on transcription. *Mol. Cell Biol.* 9: 4596–4604.

Dyson, N., Howley, P.M., Munger, K. and Harlow, E. (1989) The human papilloma virus 16-E7 oncoprotein is able to bind the retinoblastoma gene product. *Science* 243: 934–936.

Eng, C., Li, F.P., Abramson, D.H., Ellsworth, R.M., Wong, F.L., Goldman, M.B., Seddon, J., Tarbell, N. and Boice, J.D. Jr. (1993) Mortality from second tumors among long-term survivors of retinoblastoma. *J. Natl Cancer Inst.* 85: 1121–1128.

Foulkes, W.D., Flanders, T.Y., Pollock, P.M. and Hayward, N.K. (1997) The CDKN2A (p16) gene and human cancer. *Mol. Med.* 3: 5–20.

Friedman, D.L., Himelstein, B., Shields, C.L., Shields, J.A., Needle, M., Miller, D., Bunin, G.R. and Meadows, A.T. (2000) Chemoreduction and local ophthalmic therapy for intraocular retinoblastoma. *J. Clin. Oncol.* 18: 12.

Friend, S.H., Bernards, R., Rogelj, S., Weinberg, R.A., Rapaport, J.M., Albert, D.M. and Dryja, T.P. (1986) A human DNA segment with properties of the gene that predisposes to retinoblastoma and osteosarcoma. *Nature* 323: 643–646.

Gallie, B.L., Budning, A., DeBoer, G., Thiessen, J., Koren, G., Verjee, Z., Ling, V. and Chan, H. (1996) Chemotherapy with focal therapy can cure intraocular retinoblastoma without radiation. *Arch. Ophthalmol.* **114**: 1321–1329.

Gallie, B.L., Campbell, C., Devlin, H., Duckett, A. and Squire, J.A. (1999) Developmental basis of retinal-specific induction of cancer by RB mutation. *Cancer Res.* **59**: 1731–1735.

Gill, R.M., Hamel, P.A., Zhe, J., Zacksenhaus, E., Gallie, B.L. and Phillips, R.A. (1994) Characterization of the human RB1 promoter and of elements involved in transcriptional regulation. *Cell Growth Differ.* **5**: 467–474.

Godbout, R., Dryja, T.P., Squire, J., Gallie, B.L. and Phillips, R.A. (1983) Somatic inactivation of genes on chromosome 13 is a common event in retinoblastoma. *Nature* **304**: 451–453.

Godbout, R. and Squire, J. (1993) Amplification of a DEAD box protein gene in retinoblastoma cell lines. *Proc. Natl Acad. Sci. USA* **90**: 7578–7582.

Greenwald, M.J. and Strauss, L.C. (1996) Treatment of intraocular retinoblastoma with carboplatin and etoposide chemotherapy. *Ophthalmology* **103**: 1989–1997.

Gupta, J., Han, L.P., Wang, P., Gallie, B.L. and Bacchetti, S. (1996) Development of retinoblastoma in the absence of telomerase activity. *J. Natl Cancer Inst.* **88**: 1152–1157.

Hagemeier, C., Bannister, A.J., Cook, A. and Kouzarides, T. (1993) The activation domain of transcription factor PU.1 binds the retinoblastoma (RB) protein and the transcription factor TFIID in vitro: RB shows sequence similarity to TFIID and TFIIB. *Proc. Natl Acad. Sci. USA* **90**: 1580–1584.

Hall, M. and Peters, G. (1996) Genetic alterations of cyclins, cyclin-dependent kinases, and Cdk inhibitors in human cancer. *Adv. Cancer Res.* **68**: 67–108.

Heck, D.V., Yee, C.L., Howley, P.M. and Munger, K. (1992) Efficiency of binding the retinoblastoma protein correlates with the transforming capacity of the E7 oncoproteins of the human papillomaviruses. *Proc. Natl Acad. Sci. USA* **89**: 4442–4446.

Hinds, P.W., Mittnacht, S., Dulic, V., Arnold, A., Reed, S.I. and Weinberg, R.A. (1992) Regulation of retinoblastoma protein functions by ectopic expression of human cyclins. *Cell* **70**: 993–1006.

Hu, Q.J., Dyson, N. and Harlow, E. (1990) The regions of the retinoblastoma protein needed for binding to adenovirus E1A or SV40 large T antigen are common sites for mutations. *EMBO J.* **9**: 1147–1155.

Jacks, T., Fazeli, A., Schmitt, E.M., Bronson, R.T., Goodell, M.A. and Weinberg, R.A. (1992) Effects of an Rb mutation in the mouse. *Science* **359**: 295–300.

Jiang, Z., Liang, P., Leng, R. *et al.* (2000) E2F1 and p53 are dispensable whereas p21Waf1/Cip1 cooperates with Rb to restrict endoreduplication and apoptosis during skeletal myogenesis. *Developmental Biology* in press.

Kim, S.-J., Onwuta, U.S., Lee, Y.I., Li, R., Botchan, M.R. and Robbins, P.D. (1992) The retinoblastoma gene product regulates Sp1 mediated transcription. *Mol. Cell Biol.* **12**: 2455–2463.

Kingston, J.E., Hungerford, J.L., Madreperla, S.A. and Plowman, P.N. (1996) Results of combined chemotherapy and radiotherapy for advanced intraocular retinoblastoma. *Arch. Ophthalmol.* **114**: 1339–1343.

Knudson, A.G.J. (1971) Mutation and cancer: statistical study of retinoblastoma. *Proc. Natl Acad. Sci. USA* **68**: 820–823.

Lee, E.Y.-H.P., Chang, C.-Y., Hu, N., Wang, Y.-C. J., Lai, C.-C., Herrup, K., Lee, W.-H. and Bradley, A. (1992) Mice deficient for Rb are nonviable and show defects in neurogenesis and haematopoiesis. *Science* **359**: 288–294.

Lees, J.A., Buchkovich, K.J., Marshak, D.R., Anderson, C.W. and Harlow, E. (1991) The retinoblastoma protein is phosphorylated on multiple sites by human cdc2. *EMBO J.* **10**: 4279–4290.

Levine, E.M., Close, J., Fero, M., Ostrovsky, A. and Reh, T.A. (2000) p27(Kip1) regulates cell cycle withdrawal of late multipotent progenitor cells in the mammalian retina. *Dev. Biol.* **219**: 299–314.

Ludlow, J.W., DeCaprio, J.A., Huang, C., Lee, W.-H., Paucha, E. and Livingston, D.M. (1989) SV40 large T antigen binds preferentially to an underphosphorylated member of the retinoblastoma susceptibility gene product family. *Cell* **56**: 57–65.

Ludlow, J.W., Shon, J., Pipas, J.M., Livingston, D.M. and DeCaprio, J.A. (1990) The retinoblastoma susceptibility gene product undergoes cell cycle-dependent dephosphorylation and binding to and release from SV40 large T. *Cell* **60**: 387–396.

Lukas, J., Jadayel, D., Bartkova, J., Nacheva, E., Dyer, M.J., Strauss, M. and Bartek, J. (1994) BCL-1/cyclin D1 oncoprotein oscillates and subverts the G1 phase control in B-cell neoplasms carrying the t(11;14) translocation. *Oncogene* **9**: 2159–2167.

Maandag, E.C., van der Valk, M., Vlaar, M., Feltkamp, C., O'Brien, J., van Roon, M., van der Lugt, N., Berns, A. and te Riele, H. (1994) Developmental rescue of an embryonic-lethal mutation in the retinoblastoma gene in chimeric mice. *EMBO J.* **13**: 4260–4268.

Mulligan, G. and Jacks, T. (1998) The retinoblastoma gene family: cousins with overlapping interests. *Trends Genet.* **14**: 223–229.

Munger, K., Werness, B.A., Dyson, N., Phelps, W.C., Harlow, E. and Howley, P.M. (1989) Complex formation of human papillomavirus E7 proteins with the retinoblastoma tumor suppressor gene product. *EMBO J.* **8**: 4099–4105.

Murphree, A.L., Villablanca, J.G., Deegan, W.F. 3rd, Sato, J., Malogolowkin, M., Fisher, A., Parker, R., Reed, E. and Gomer, C.J. (1996) Chemotherapy plus local treatment in the management of intraocular retinoblastoma. *Arch. Ophthalmol.* **114**: 1348–1356.

Murray, T.G., Cicciarelli, N., O'Brien, J.M., Hernandez, E., Mueller, R.L., Smith, B.J. and Feuer, W. (1997) Subconjunctival carboplatin therapy and cryotherapy in the treatment of transgenic murine retinoblastoma. *Arch. Ophthalmol.* **115**: 1286–1290.

Nakayama, K., Ishida, N., Shirane, M., Inomata, A., Inoue, T., Shishido, N., Horii, I., Loh, D.Y. and Nakayama, K. (1996) Mice lacking p27(Kip1) display increased body size, multiple organ hyperplasia, retinal dysplasia, and pituitary tumors. *Cell* **85**: 707–720.

Noorani, H.Z., Khan, H.N., Gallie, B.L. and Detsky, A.S. (1996) Cost comparison of molecular versus conventional screening of relatives at risk for retinoblastoma. *Am. J. Hum. Genet.* **59**: 301–307.

Ohnuma, S., Philpott, A., Wang, K., Holt, C.E. and Harris, W.A. (1999) p27Xic1, a Cdk inhibitor, promotes the determination of glial cells in Xenopus retina. *Cell* **99**: 499–510.

Phillips, R.A., Gill, R.M., Zacksenhaus, E., Bremner, R., Jiang, Z., Sopta, M., Gallie, B.L. and Hamel, P.A. (1992) Why don't germline mutations in RB1 predispose to leukemia? *Curr. Top. Microbiol. Immunol.* **182**: 485–491.

Robanus-Maandag, E., Dekker, M., van der Valk, M., Carrozza, M.L., Jeanny, J.C., Dannenberg, J.H., Berns, A. and te Riele, H. (1998) p107 is a suppressor of retinoblastoma development in pRb-deficient mice. *Genes Dev.* **12**: 1599–1609.

Sakai, T., Toguchida, J., Ohtani, N., Yandell, D.W., Rapaport, J.M. and Dryja, T.P. (1991) Allele-specific hypermethylation of the retinoblastoma tumor-suppressor gene. *Am. J. Hum. Genet.* **48**: 880–888.

Scheffner, M., Munger, K., Huibregtse, J.M. and Howley, P.M. (1992) Targeted degradation of the retinoblastoma protein by human papillomavirus E7–E6 fusion proteins. *EMBO J.* **11**: 2425–2431.

Sellers, W.R. and Kaelin, W.G. Jr. (1997) Role of the retinoblastoma protein in the pathogenesis of human cancer. *J. Clin. Oncol.* **15**: 3301–3312.

Sicinski, P., Donaher, J.L., Parker, S.B. *et al.* (1995) Cyclin D1 provides a link between development and oncogenesis in the retina and breast. *Cell* **82**: 621–630.

Sparkes, R.S., Sparkes, M.C., Wilson, M.G., Towner, J.W., Benedict, W., Murphree, A.L. and Yunis, J.J. (1980) Regional assignment of genes for human esterase D and retinoblastoma to chromosome band 13q14. *Science* **208**: 1042–1044.

Squire, J., Phillips, R.A., Boyce, S., Godbout, R., Rogers, B. and Gallie, B.L. (1984) Isochromosome 6p, a unique chromosomal abnormality in retinoblastoma: verification by standard staining techniques, new densitometric methods, and somatic cell hybridization. *Hum. Genet.* **66**: 46–53.

Szekely, L., Jiang, W.Q., Pokrovskaja, K., Wiman, K.G., Klein, G. and Ringertz, N. (1995) Reversible nucleolar translocation of Epstein-Barr virus-encoded EBNA-5 and hsp70 proteins after exposure to heat shock or cell density congestion. *J. Gen. Virol.* **76**: 2423–2432.

Toguchida, J., McGee, T.L., Paterson, J.C., Eagle, J.R., Tucker, S., Yandell, D.W. and Dryja, T.P. (1993) Complete genomic sequence of the human retinoblastoma susceptibility gene. *Genomics* **17**: 535–543.

Tsai, K.Y., Hu, Y., Macleod, K.F., Crowley, D., Yamasaki, L. and Jacks, T. (1998) Mutation of E2f-1 suppresses apoptosis and inappropriate S phase entry and extends survival of Rb-deficient mouse embryos. *Mol. Cell* **2**: 293–304.

Wang, C.-Y., Petryniak, B., Thompson, C.B., Kaelin, W.G. and Leiden, J.M. (1993) Regulation of the Ets-related transcription factor Elf-1 by binding to the retinoblastoma protein. *Science* **260**: 1330–1335.

Welch, P.J. and Wang, J.Y. (1993) A C-terminal protein-binding domain in the retinoblastoma protein regulates nuclear c-Abl tyrosine kinase in the cell cycle. *Cell* **75**: 779–790.

Whyte, P., Buchkovich, K.J., Horowitz, J.M., Friend, S.H., Raybuck, M., Weinberg, R.A. and Harlow, E. (1988) Association between an oncogene and an anti-oncogene: the adenovirus E1A proteins bind to the retinoblastoma gene product. *Nature* **334**: 124–129.

Williams, B.O., Remington, L., Albert, D.M., Mukai, S., Bronson, R.T. and Jacks, T. (1994a) Cooperative tumorigenic effects of germline mutations in Rb and p53. *Nature Genet.* **7**: 480–484.

Williams, B.O., Schmitt, E.M., Remington, L., Bronson, R.T., Albert, D.M., Weinberg, R.A. and Jacks, T. (1994b) Extensive contribution of Rb-deficient cells to adult chimeric mice with limited histopathological consequences. *EMBO J.* **13**: 4251–4259.

Zacksenhaus, E., Jiang, Z., Chung, D., Marth, J.D., Phillips, R.A. and Gallie, B.L. (1996) pRb controls proliferation, differentiation, and death of skeletal muscle cells and other lineages during embryogenesis. *Genes & Dev.* **10**: 3051–3064.

Zacksenhaus, E., Jiang, Z., Hei, Y.J., Phillips, R.A. and Gallie, B.L. (1999) Nuclear localization conferred by the pocket domain of the retinoblastoma gene product. *Biochim. Biophys. Acta* **1451**: 288–296.

14

Genetics of Wilms tumor

Mathias A.E. Frevel and Bryan R.G. Williams

1. Introduction

Wilms tumor, or nephroblastoma, is a pediatric kidney tumor that affects approximately 1/10 000 children world-wide with most cases presenting between 1 and 4 years of age (Breslow *et al.*, 1993; Matsunaga, 1981). The vast majority of cases occur sporadically, only approximately 1.5% are familial (Breslow *et al.*, 1996). A small percentage of Wilms tumors are found in association with predisposing syndromes, the WAGR, the Denys-Drash (DDS), the Beckwith-Wiedemann (BWS), and the Perlman syndrome (Beckwith, 1969; Denys *et al.*, 1967, Drash *et al.*, 1970; Grundy *et al.*, 1992; Wiedemann, 1964). The association of Wilms tumor with aniridia, a part of the WAGR phenotype that also includes genitourinary abnormalities, mental retardation and hemihypertrophy, gave the first indication for a genetic basis of the tumor (Miller *et al.*, 1964).

Wilms tumors arise from abnormally proliferating metanephric blastema, and retain a high degree of dedifferentiation. Histologically, Wilms tumors appear mostly triphasic with variable amounts of epithelial, blastemal and stromal components. A subset of tumors contain non-renal heterotypic differentiation such as muscle (Beckwith *et al.*, 1990; Yeger *et al.*, 1992). The normal kidney tissue of Wilms tumor patients often contains nephrogenic rests. These foci of abnormally persistent, embryonal cells are the precursor lesions of Wilms tumor, clonally derived from one renal stem cell. Modern treatment modalities of partial or total nephrectomy followed by chemotherapy and possibly radiation therapy have remarkably improved the survival rates of children with Wilms tumor leading to a cure rate of approximately 90%.

The statistics of familial, bilateral and unilateral Wilms tumor occurrence suggests that Wilms tumorigenesis involves two mutational events (Knudson and Strong, 1972). Chromosome 11p13 was implicated by the identification of constitutional deletions in the WAGR syndrome (Francke *et al.*, 1979; Riccardi *et al.*, 1978). Loss of heterozygosity studies showed that chromosome 11p was frequently deleted in sporadic Wilms tumors (Fearon *et al.*, 1984; Koufos *et al.*, 1984; Orkin *et al.*, 1984; Reeve *et al.*, 1984). In 1989 the Wilms tumor suppressor gene, *WT1*, was cloned from the 11p13 locus, and it was shown that inactivation of both copies of *WT1* can represent the two mutational events predicted by Knudsen and Strong to be necessary for Wilms tumorigenesis (Bonetta *et al.*, 1990; Call *et al.*, 1990;

Molecular Genetics of Cancer second edition, edited by J.K. Cowell.

Gessler *et al.*, 1990; Huang *et al.*, 1990). However, we know today that *WT1* mutations are found in only 5–10% of sporadic Wilms tumors and, hence, that *WT1* represents only one of the genes that can be involved in the onset of Wilms tumor (Coppes *et al.*, 1993; Little *et al.*, 1992; Varanasi *et al.*, 1994).

Further studies demonstrated that a second region in distal chromosome 11, comprising the 11p15 locus, is frequently deleted in Wilms tumors, and that this deletion always affects the maternal allele (Pal *et al.*, 1990; Reeve *et al.*, 1989; Schroeder *et al.*, 1987; Williams *et al.*, 1989). Prompted by the non-random LOH at 11p15, a role was proposed for genomic imprinting in the pathogenesis of Wilms tumor (Wilkins, 1988). It was suggested that a transforming gene active in Wilms tumorigenesis is transcribed from the paternal allele and genomically imprinted, hence silent on the maternal allele of chromosome 11. Subsequently, the relaxation of imprinting of the paternally transcribed fetal *insulin-like growth factor 2* gene (*IGF2*) was found in Wilms tumors which substantiated the Wilkins hypothesis and, furthermore, implied that loss of imprinting plays a role in tumorigenesis (Ogawa *et al.*, 1993b; Rainier *et al.*, 1993).

2. The 11p13 locus and the Wilms tumor suppressor gene, *WT1*

The association with the WAGR syndrome is observed in only a small subset of Wilms tumor cases (1–2%). These patients carry constitutional deletions of band 11p13. Two cloned genes from this locus are known to be responsible for parts of the WAGR phenotype. Aniridia, which is fully penetrant, has been linked to mutations in the *PAX6* gene that is involved in eye differentiation (Ton *et al.*, 1991). Wilms tumor, which develops in only 30–50% of aniridia patients, and the genito-urinary abnormalities are caused by the disruption of the *WT1* gene, a transcription factor that plays a pivotal role in both genital and renal development. For the genito-urinary abnormalities to occur a single mutant *WT1* allele is sufficient, whereas the remaining *WT1* allele has been shown to carry tumor-specific mutations in most WAGR patients conforming to the 'two-hit' hypothesis (Pritchard-Jones, 1997).

The Denys–Drash syndrome, an extremely rare sporadic syndrome of male pseudohermaphroditism, early onset nephropathy, and Wilms tumor, accounts for less than 1% of total Wilms tumor cases. In the majority of DDS cases the disease is caused by *de novo* germline mutations in *WT1* (Baird *et al.*, 1992; Bruening *et al.*, 1992; Pelletier *et al.*, 1991). However, inheritance of the most common DDS mutation from a father that was phenotypically unaffected has also been reported (Coppes *et al.*, 1992b). Over 90% of DDS mutations occur in the carboxyterminal zinc-finger region of the protein that is responsible for its DNA-binding activity. Thus, it was suggested that in DDS patients with heterozygous *WT1* mutations the constitutionally mutant WT1 protein may act in a dominant negative fashion upon dimerization, abrogating DNA binding of the wild-type protein (Hastie, 1992; Little *et al.*, 1993). However, in most tumors from DDS patients the wild-type *WT1* allele is lost suggesting that the total lack of WT1 is important for these Wilms tumors to arise. The requirement for this second genetic hit for tumor development might reflect a different mode of function of WT1 in kidney versus gonadal development.

In sporadic Wilms tumors, of which several hundred have now been analyzed, *WT1* mutations are present in less than 10% of cases (Akasaka *et al.*, 1993; Brown *et al.*, 1993; Coppes *et al.*, 1993; Cowell *et al.*, 1991; Gessler *et al.*, 1994; Haber *et al.*, 1990; Huff *et al.*, 1995; Kaneko *et al.*, 1993; Little *et al.*, 1992; Tadokoro *et al.*, 1992; Varanasi *et al.*, 1994). The genetic hit seems to occur early during kidney development since identical mutations have been detected in tumors and their precursor lesions from the same patients (Park *et al.*, 1993a). Most mutations are truncations, although some missense mutations from exon 2 to 9 are also seen. *WT1* gene mutations have been extensively reviewed by Little and Wells (Little and Wells, 1997). In tumors of predominantly stromal or heterotypic components *WT1* mutations are more frequent than in blastemal-predominant and classical triphasic tumors. Patients presenting these stromal-predominant tumors may carry a *WT1* germ-line mutation, and show tumor-specific loss of the second allele (Schumacher *et al.*, 1997). Overall, 70% of sporadic Wilms tumors with *WT1* mutations have 11p13 LOH, and for this subset loss of a functional WT1 protein may be the underlying cause of Wilms tumor development (Little and Wells, 1997). In a minority of sporadic Wilms tumors that have heterozygous *WT1* mutations the dominant negative model as proposed for DDS, or a newly acquired dominant function may account for the contribution of mutant *WT1* to Wilms tumor development.

2.1 The WT1 gene

The *WT1* gene encodes a developmentally regulated transcription factor (*Figure 1*). The gene of 10 exons, spanning approximately 50 kb, is transcribed to a 3.5 kb mRNA that undergoes alternative splicing. The WT1 protein (52–54 kD, predicted M_r 46 kD) contains a carboxyterminal DNA-binding domain with four zinc-fingers of the Cys2-His2 type, and an amino-terminal transactivation domain rich in proline, glutamine, serine, and lysin. The N-terminal part of the protein was also shown to be responsible for self-dimerization (Moffett *et al.*, 1995; Reddy *et al.*, 1995). Mutations within the zinc-finger domain found in Wilms tumor and DDS destroy the DNA-binding ability of WT1 (Haber *et al.*, 1990; Pelletier *et al.*,

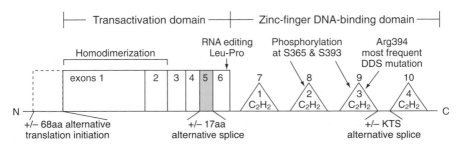

Figure 1. Scheme of WT1. The WT1 protein, encoded by exons 1–10, contains a N-terminal transactivation domain and a C-terminal DNA-binding domain with Cys2-His2-type zinc fingers 1–4.

1991), whereas the properties of the transactivation domain can be changed by mutations to confer activation instead of repression of some target promoters (Park et al., 1993b, 1993c). Two alternative splice sites give rise to four WT1 isoforms. Exon 5 can be variably spliced resulting in the presence or absence of 17 amino acids in the amino-terminal part of the protein but this appears to be specific to mammals. Usage of the second alternative splice site at the end of exon 9, that is also present in lower vertebrates, determines the presence of three amino acids, lysine-threonine-serine (KTS), between zinc-fingers 3 and 4 (Rauscher, 1993). Mutations within the +/–KTS splice junction, which upset the balance of WT1 isoforms, are the underlying cause of severe genito-urinary abnormalities seen in Frasier syndrome (Barbaux et al., 1997). RNA editing that results in a proline to leucine change at position 280, and the usage of an alternative translation start, adding 68 amino acids to WT1, increases the number of potential isoforms to 16 (Bruening and Pelletier, 1996; Sharma et al., 1994). Further complexity may be added to the regulation of WT1 at the post-transcriptional level, since phosphorylation of the WT1 protein isoforms was shown to inhibit DNA-binding and alter WT1 transcriptional activity and cellular translocation (Sakamoto et al., 1997; Ye et al., 1996).

2.2 Function of WT1 as a transcription factor

Ten years after the cloning of WT1 the true physiological targets of this transcription factor are still elusive, although many candidate genes identified in transient co-transfection experiments have been reported as potential targets of WT1. In efforts to identify the DNA-binding site of WT1, highly purified, recombinant protein (–KTS isoform) was found to bind the EGR1 consensus site 5'-GCGGGGGCG-3' (Rauscher et al., 1990). Also, the zinc-fingers 2, 3 and 4 of the WT1 DNA-binding domain have a high degree of amino acid sequence identity (61%) to the three zinc-fingers of EGR1, a member of the Early-Growth Response gene family. Subsequently, the WT1 recognition sequence for the –KTS isoform was refined to an extended motif similar to the EGR1 consensus, but of much higher affinity, and the +KTS isoform was shown to bind to TC-rich repeat sequences (Bickmore et al., 1992; Hamilton et al., 1995; Nakagama et al., 1995; Wang et al., 1993). In general, transfected WT1 has been shown to repress transcription from GC-and TC-rich promoters; a few exceptions showed promoter activation. In some cases, repression by WT1 could also be demonstrated for the endogenous gene (reviewed in Pritchard-Jones (1997)). However, the physiological relevance of these experiments is uncertain, since they overexpress individual WT1 isoforms, and can not simulate the balance of different WT1 isoforms found in vivo. Furthermore, the effects of WT1 on several test promoters are variable depending on the cell types being transfected which may be due to tissue-specific proteins that interact with and modify WT1 function (Englert et al., 1995; Madden et al., 1993; Nichols et al., 1995). Recent additions to the list of potential WT1 target genes are amphiregulin, Dax-1 and connective tissue growth factor (CTGF, Kim et al., 1999; Lee et al., 1999, Stanhope-Baker and Williams, 2000). Amphiregulin, a growth factor that can induce differentiation in an embryonic renal culture model, was induced by WT1(–KTS). Dax-1, involved in gonadal differentiation, was also activated by WT1 in partnership with the steroidogenic factor SF-1. Interestingly, the temporal and spatial expression pattern of *amphiregulin*, in the developing human kidney, and *Dax-1*, in the gonads, correlate with that of *WT1*. Although this correlation supports the physiological relevance of amphiregulin and Dax-1 induction by WT1, further experiments are needed to confirm this regulation *in vivo*. *Connective tissue growth factor* is a member of the CCN family of growth regulators and its promoter is suppressed by WT1 in both its endogenous location and in reporter constructs. Since *WT1* repression of CTGF expression is not mediated by previously identified WT1 recognition elements it may utilize a novel mechanism. Since CTGF stimulates extracellular matrix production it is possible that its over-expression due to *WT1* mutations may underlie the renal fibrosis observed in DDS.

2.3 Role of WT1 in kidney development

WT1 plays a crucial role in early urogenital development which is reflected in the failure of homozygous *wt1* mutant mice to develop kidneys and gonads (Kreidberg *et al.*, 1993). *Wt1* is expressed during embryonic development in the kidney and the gonads (*Figure 2*), as well as in the mesothelial lining of the abdomen and the spleen (Armstrong *et al.*, 1993; Pritchard-Jones *et al.*, 1990; Rackley *et al.*, 1993). In the developing kidney *WT1* is expressed specifically in the condensed mesenchyme, renal vesicles and glomerular epithelium, and the expression follows a particular pattern in which increasing levels correlate with progressing differentiation of the kidney. Levels in the kidney are highest just before, or at birth and decline rapidly after birth (Buckler *et al.*, 1991). The temporal and spatial distribution of WT1 during kidney development suggests a role for the protein in mediating mesenchyme differentiation (Armstrong *et al.*, 1993). Hence, lack of the differentiating signal initiated by WT1 may be crucial for the genesis of Wilms tumors that carry *WT1* mutations.

2.4 Expression of WT1 in other cancers

While expression of *WT1* exhibits temporal and spatial restriction in developing and mature tissues, different cancers express *WT1* at varying levels suggesting loss of control at the transcriptional level. In renal cell cancer (RCC) wild-type *WT1* is expressed in all tumors of the clear cell type regardless of stage (Campbell *et al.*, 1998). Expression of the *WT1* gene has also been reported in breast cancer and lung cancer and in cell lines derived from a variety of tumors including gastric, ovarian, colon, hepatocellular and uterine (Oji *et al.*, 1999). WT1 has been detected in normal breast tissue by immunostaining but was absent from infiltrating tumors (Silberstein *et al.*, 1997). However expression was noted in ductal carcinoma *in situ* and a correlation made between estrogen receptor expression and a lack of WT1 protein. *WT1* mRNA was detected in tumor samples but with altered ratios of splice variants perhaps underlying the altered WT1 protein expression. Interestingly, higher-grade tumors expressed both WT1 and TP53 supporting a role for WT1 in p53 stabilization (Silberstein *et al.*, 1997). Treatment of different WT1 expressing cell lines with WT1 antisense oligomers, inhibited

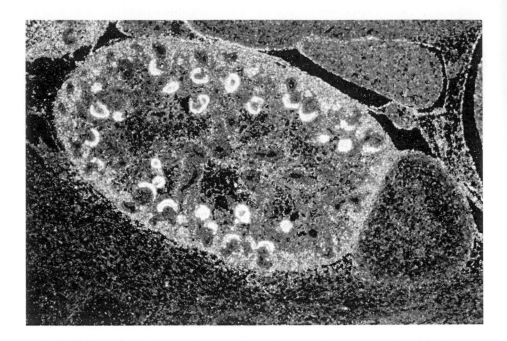

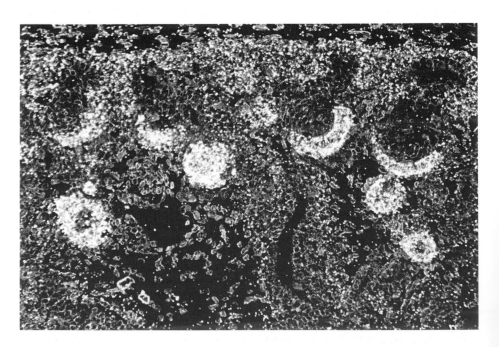

Figure 2. Expression of *WT1* during nephrogenesis. Dark field photomicrographs of sagittal sections of embryonic day 15.5 mouse kidney probed with WT13H-labeled antisense probe show (bright areas) that *WT1* is expressed at high levels in the kidney in the metanephric blastema, early condensations and podocytes.

cell growth suggesting that WT1 may play a role in facilitating the growth of solid tumors thus acting as an oncogene rather than a tumor-suppressor gene (Oji *et al.*, 1999). This may also be the case in leukemias. *WT1* is expressed in levels equivalent to fetal kidney in acute myelocytic and lymphocytic leukemia, in chronic myelocytic leukemia and in some hematopoietic progenitor cells (Pritchard-Jones and King-Underwood, 1997). The level of expression of *WT1* is highest in leukemias with immature phenotypes. *WT1* is also expressed in a high percentage of 7,12-dimethylbenz[a]anthracene (DMBA)-induced rat leukemias, where it appears to be associated with leukemiagenesis at a late stage (Osaka *et al.*, 1997).

Although wild-type WT1 is usually present in leukemias, mutations have been found in some cases, mainly acute myeloid leukemia (AML), and may confer drug resistance. In AML, the presence of *WT1* mutations has been associated with failure to achieve complete remission and a lower survival rate. AML caused by chromosomal translocations also may have underlying *WT1* mutations suggesting these lead to leukemia progression (Miyagawa *et al.*, 1999). The frequency of *WT1* gene expression in acute childhood leukemias is similar to that in adults and the expression levels are higher in AML than in ALL. Relapse cases are also associated with a higher level of *WT1* expression (Niegemann *et al.*, 1999).

WT1 mutations are found in about the same proportion of AML cases as seen in Wilms tumors. However, the *WT1* gene mutations that occur in both myeloid and biphenotypic subtypes, where they are associated with refractoriness to standard induction chemotherapy, are heterozygous, implying a dominant or dominant-negative mode of action in hematopoietic cells (King-Underwood and Pritchard-Jones, 1998). This suggests that any normal role for WT1 in hematopoiesis may be at an early progenitor stage. In accord with this, expression of *WT1* is detected in cells bearing the CD34+ phenotype but not in those cells lacking expression of CD34. Single-cell analyses revealed that expression of *WT1* occurs in the candidate stem cell-containing population of hemopoietic cells that have the phenotype CD34+ CD38–. There also appears to be differential expression of alternate transcripts of *WT1* between hemopoietic progenitor cells with the same phenotype. This limitation of expression of *WT1* to early progenitors supports a role for WT1 in hemopoietic development (Baird and Simmons, 1997).

Different cell culture systems have been used in an attempt to better define the role of WT1 in hematopoiesis and leukemiagenesis. In the murine myeloblastic leukemia cell line M1, *WT1* expression is activated 24 hours after differentiation induction by leukemia inhibitory factor (LIF). The expression of *WT1* in these cells is associated with cellular differentiation, as measured by the expression of the monocyte/macrophage marker c-fms, and the appearance of mature cells. Interestingly, the *WT1* –KTS isoform could not be ectopically expressed in M1 cells, but stable expression of the +KTS isoform led to spontaneous differentiation of the M1 myeloblasts through the monocytic differentiation pathway and concurrent expression of c-fms and the myeloid-specific cell surface marker Mac-1. Treatment of these cells with LIF results in terminal macrophage differentiation and apoptotic cell death. Thus, *WT1* gene expression not only is coincident with M1 cell monocytic differentiation *in vitro*, but a specific isoform (+KTS) can regulate this process. While these results lend further support for a role for WT1 in the control of hematopoiesis further studies including transgenic models are needed (Smith *et al.*, 1998).

In terms of leukemiagenesis, animal studies have been reported with the M1 cells. Mice inoculated with M1 cells which stably express *WT1* +KTS isoforms and undergo spontaneous monocytic differentiation without the requirement for external differentiation-inducing stimuli, show decreased tumor formation compared with control animals (Smith *et al.*, 2000). However, these results are not unexpected since the cells expressing *WT1* +KTS isoforms also exhibit apoptosis and a differentiation-promoting role for WT1 during hematopoiesis remains to be established. Expression of *WT1* is down-regulated during differentiation of leukemic cell lines such as K562 *in vitro* lending support to the notion that high *WT1* expression in hematopoietic cells is incompatible with differentiation. However, ectopic expression of the four different isoforms of WT1 in K562 does not affect the differentiation response of these cells to 12-O-tetradecanoyl-phorbol-13-acetate (TPA) suggesting that down-regulation of WT1 during induced differentiation of K562 cells is not a prerequisite for erythroid or megakaryocytic differentiation (Svedberg *et al.*, 1999).

These results are in contrast to the mechanisms of action of WT1 in human promyeloid leukemia (HL-60) cells. When WT1 or its zinc-finger domain alone was stably expressed in these cells, macrophage differentiation induced by TPA was blocked. This suggests that high level expression of *WT1* is capable of differentiation arrest of HL60 myeloid cells but not K562 erythroleukemia cells. While the reasons for this variance in results remain to be resolved, the differentiation inhibition effect appears to be mediated through the WT1 zinc-finger domain. As promoter fusion experiments suggest that the WT1 zinc-finger domains can compete with other transcription factors for common promoter elements (Deuel *et al.*, 1999).

The expression of *WT1* at high levels in AML, ALL and CML suggests it may serve as a target antigen for tumor-specific immunity. This has now been demonstrated in two recent reports. A 9-mer WT1 peptide containing a H-2Db-binding anchor motifs induced a CTL response in C57BL/6, H-2Db mice. The CTLs specifically lysed not only the peptide-pulsed target cells but also *WT1*-expressing tumor cells in an H-2Db-restricted manner. Mice immunized with the WT1 peptide rejected challenges by *WT1*-expressing tumors (Oka *et al.*, 2000). In accord with these results it has also been shown that HLA-A0201- restricted CTL, specific for WT1, kill leukemia cell lines, and inhibit colony formation by transformed CD34(+) progenitor cells isolated from patients with CML (Gao *et al.*, 2000). Taken together these results suggest WT1 is a potential target for immunotherapy. This could be through CTL-mediated purging of leukemic progenitor cells *in vitro* or for antigen-specific therapy of leukemia and other *WT1*-expressing malignancies *in vivo* (Gao *et al.*, 2000). The clinical exploitation of these findings have yet to be reported but offer potentially new avenues for immunotherapy.

3. The 11p15 locus and loss of imprinting in Wilms tumor

In some Wilms tumors allele loss of chromosome 11 is restricted to 11p15 indicating the presence of a second Wilms tumor suppressor gene, *WT2*, in distal 11p (Coppes *et al.*, 1992a; Mannens *et al.*, 1988, 1990; Reeve *et al.*, 1989; Wadey *et al.*,

1990). Further dissection using several markers within the 11p15.5–15.4 interval identified two independent regions of LOH spanning approximately 800 kb and 336 kb suggesting that two novel Wilms tumor genes may map to this region (Karnik *et al.*, 1998).

Studies on BWS patients have aided the understanding of the role of 11p15 in Wilms tumor development. BWS is a generalized overgrowth syndrome characterized by gigantism, macroglossia, visceromegaly and a number of other developmental anomalies (Beckwith, 1969; Wiedemann, 1964). In addition, individuals with BWS show a predisposition to embryonal tumors such as Wilms tumor, rhabdomyosarcoma and hepatoblastoma (Junien, 1992). In rare familial cases autosomal dominant inheritance of BWS was shown to be linked to 11p15 (Koufos *et al.*, 1989; Ping *et al.*, 1989), and in sporadic cases constitutional trisomy for 11p15 or uniparental disomy (UPD) of this region was found (Henry *et al.*, 1991; Hoovers *et al.*, 1995; Mannens *et al.*, 1994; Waziri *et al.*, 1983). In all cases the duplicated chromosome was of paternal origin pointing to the involvement of imprinted genes in the ethiology of BWS as it has been proposed for Wilms tumor (Wilkins, 1988). Consistent with this idea, there is now very strong evidence that the disruption of the parental allele-specific expression of imprinted genes, namely the *IGF2* and *H19* gene, at chromosome 11p15.5, play a role in Wilms tumorigenesis.

3.1 Loss of imprinting of IGF2

IGF2, an important autocrine growth factor, is believed to act as a tumor growth factor in Wilms tumors by way of its overexpression (Reeve *et al.*, 1985; Scott *et al.*, 1985). In the case of paternal UPD of 11p15, IGF2 overexpression is thought to result from double dose transcription of *IGF2* from two paternal alleles (*Figure 3*). A second mechanism that leads to IGF2 overexpression is loss of imprinting (LOI) that occurs in approximately one half of Wilms tumors with cytogenetically

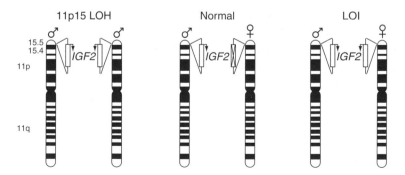

Figure 3. Chromosome 11 aberrations in Wilms tumor. In normal kidney tissue and in approximately one third of tumors chromosomes 11 are normal, with *IGF2* being transcribed from the paternal allele whereas the maternal allele is silent. Another third of cases presents with 11p loss of heterozygosity (LOH), always losing the maternal allele. Upon duplication of the paternal chromosome (partial UPD) *IGF2* is transcribed from two paternal alleles. In approximately one third of Wilms tumors with normal parental allelic contribution of 11p15 loss of imprinting (LOI) causes biallelic IGF2 expression.

normal chromosomes 11p15. In these cases the *IGF2* gene is transcribed from the paternal allele as well as from the, normally silent, maternal allele (Ogawa *et al.*, 1993a, 1993b; Rainier *et al.*, 1993).

The *IGF2* gene is one of a number of imprinted genes within a large chromosomal region of 11p15.5 that is hence referred to as an imprinted domain (Reid *et al.*, 1997). The regulation of the monoallelic expression of imprinted genes within this domain involves long ranging chromatin effects and DNA methylation, processes that we know today are interrelated. The investigation of loss of *IGF2* imprinting in Wilms tumor and in BWS has greatly contributed to the understanding of imprinting at the 11p15 locus. In the case of Wilms tumor biallelic expression of *IGF2* was shown to correlate with silencing and abnormally high DNA methylation of the *H19* gene (Moulton *et al.*, 1994; Steenman *et al.*, 1994; Taniguchi *et al.*, 1995). The *H19* gene, normally expressed from the maternal allele only, codes for an untranslated RNA of unknown function, and may itself be a growth suppressing gene (Hao *et al.*, 1993). The abnormal methylation of the maternal *H19* alleles was shown to effect an upstream region of the gene that is critical for maintaining *IGF2* and *H19* imprinting (Frevel *et al.*, 1999; Thorvaldsen *et al.*, 1998). This region contains several methylation sensitive binding sites for CTCF (CCCTC-binding factor), an eleven-zinc finger DNA-binding protein that has been shown to bind specific sites between enhancer elements and promotors to prevent gene activation (Bell *et al.*, 1999). On the maternal allele, binding of CTCF to its unmethylated sites blocks the action of downstream enhancers on *IGF2* but allows the same enhancers to promote *H19* transcription. On the paternal allele, CTCF can not bind to the methylated binding sites, leaving the downstream enhancers free to promote *IGF2* transcription (Bell and Felsenfeld, 2000; Hark *et al.*, 2000; Kanduri *et al.*, 2000; Szabo *et al.*, 2000). Aberrant methylation of *H19* was also detected in normal kidney tissue from Wilms tumor patients which implies that the epigenetic change occurred very early in kidney development (Okamoto *et al.*, 1997). Taken together, these findings suggest that hypermethylation of *H19*, either as a direct cause or in company with other undetected chromatin changes, is responsible for *H19* silencing and LOI of *IGF2* in Wilms tumors. However, other studies have found that the disruption of the monoallelic expression of *H19* and *IGF2* in Wilms tumors can also be independent events (Cui *et al.*, 1997; Ohlsson *et al.*, 1999). Finally, it is important to note that only approximately one third of Wilms tumors present maternal 11p15 LOH and another third LOI of *IGF2*, whereas one third of Wilms tumors exhibit normal monoallelic *IGF2* expression. In this latter group other genetic changes may substitute for the consequences of aberrant *IGF2* imprinting.

4. Other genetic loci associated with Wilms tumor

In addition to the two regions on chromosome 11p that have been studied in most detail at least two other loci at 16q and 7p have been firmly associated with Wilms tumor by LOH analyses. For the long arm of chromosome 16 loss of heterozygosity was found in approximately 20% of tumors (Austruy *et al.*, 1995; Coppes *et al.*, 1992a; Maw *et al.*, 1992). Two studies correlated higher rates of relapse and mortality with tumor-specific 16q LOH when compared to patients with normal chromosomes 16 (Grundy *et al.*, 1994, 1996). The adverse outcome in the cases of 16q LOH may indicate the location of a tumor suppressor gene that leads to a more aggressive tumor when inactivated.

Abnormalities of chromosome 7 have been implicated in Wilms tumorigenesis by cytogenetic studies. Trisomy of 7q and monosomy of 7p have been found as the only chromosomal aberrations in four cases (Peier *et al.*, 1995; Sawyer *et al.*, 1993; Sheng *et al.*, 1990; Wilmore *et al.*, 1994). Translocations involving the short arm of chromosome 7 with breakpoints at 7p13 and 7p15 have been reported by several groups (Hewitt *et al.*, 1991; Kaneko *et al.*, 1991; Reynolds *et al.*, 1996; Slater *et al.*, 1985; Solis *et al.*, 1988). One study detected LOH or allelic imbalance in four out of 11 cases with a shared region of allele loss over approximately 25 cM of 7p14–13 (Miozzo *et al.*, 1996). In the two most recent studies 7p LOH was found in approximately 10% of sporadic Wilms tumors, and the location of a potential tumor suppressor gene was narrowed down to band p21–15 and p22–15, respectively (Grundy *et al.*, 1998; Powlesland *et al.*, 2000). Taken together, these studies provide strong evidence for the presence of a tumor suppressor gene on the short arm of chromosome 7 that plays a role in Wilms tumor development.

A third locus, 17q12-q21, was identified by linkage analysis of a single family to harbor a susceptibility gene for Wilms tumor, termed *FWT1* for familial Wilms tumor gene 1 (Rahman *et al.*, 1996). Surprisingly, no allele loss at 17q was detected in three tumors from this family suggesting that *FWT1* is not a tumor suppressor gene. Furthermore, in a fourth tumor the *FWT1* allele that segregated with the disease was lost, leading to the speculation that a mutation in *FWT1* may contribute to the onset of some familial Wilms tumors, but not to the maintenance of the neoplastic phenotype (Rahman *et al.*, 1997). Evidence for linkage to this region was then detected in a second unrelated pedigree strengthening the hypothesis that 17q12-q21 contains a Wilms tumor susceptibility gene. However, in other families linkage to this locus was excluded, suggesting that more familial Wilms tumor genes exist (Rahman *et al.*, 1998).

The cloning of the genes on chromosome arms 7p, 16q and 17q, and potentially additional genes from 11p and other loci that may play a role in Wilms tumor development are the next steps necessary to further dissect the complex biology of this tumor.

5. Conclusions

WT1 remains the only tumor suppressor gene directly implicated in the initiation of Wilms tumorigenesis. However, the linking of *CTCF* to loss of imprinting suggests that a search for mutations in this gene may implicate it in the etiology of LOI Wilms tumors. The identification of the gene(s) mapped by LOH to 11p15.4 will allow a better understanding of the role of interstitial LOH in tumor etiology. Transcriptional profiling of Wilms tumors with different genotypes may also identify not only new targets for WT1 but also provide a more comprehensive portrait of genotype–phenotype relationships. This approach may lead to the early identification of potential anaplastic aggressive tumors and the consideration of alternatives to present treatment.

References

Akasaka, Y., Kikuchi, H., Nagai, T., Hiraoka, N., Kato, S. and Hata, J. (1993) A point mutation found in the WT1 gene in a sporadic Wilms' tumor without genitourinary abnormalities is identical with the most frequent point mutation in Denys-Drash syndrome. *FEBS Lett.* **317:** 39–43.

Armstrong, J.F., Pritchard-Jones, K., Bickmore, W.A., Hastie, N.D. and Bard, J.B. (1993) The expression of the Wilms' tumour gene, WT1, in the developing mammalian embryo. *Mech. Dev.* **40:** 85–97.

Austruy, E., Candon, S., Henry, I., Gyapay, G., Tournade, M.F., Mannens, M., Callen, D., Junien, C. and Jeanpierre, C. (1995) Characterization of regions of chromosomes 12 and 16 involved in nephroblastoma tumorigenesis. *Genes Chromosomes Cancer* **14:** 285–294.

Baird, P.N. and Simmons, P.J. (1997) Expression of the Wilms' tumor gene (WT1) in normal hemopoiesis. *Exp. Hematol.* **25:** 312–320.

Baird, P.N., Santos, A., Groves, N., Jadresic, L. and Cowell, J.K. (1992) Constitutional mutations in the WT1 gene in patients with Denys-Drash syndrome. *Hum. Mol. Genet.* **1:** 301–305.

Barbaux, S., Niaudet, P., Gubler, M.C. et al. (1997) Donor splice-site mutations in WT1 are responsible for Frasier syndrome. *Nat. Genet.* **17:** 467–470.

Beckwith, J.B. (1969) Macroglossia, omphalocele, adrenal cytomegaly, gigantism, and hyperplastic visceromegaly. *Birth Defects* **5:** 188–197.

Beckwith, J.B., Kiviat, N.B. and Bonadio, J.F. (1990) Nephrogenic rests, nephroblastomatosis, and the pathogenesis of Wilms' tumor. *Pediatr. Pathol.* **10:** 1–36.

Bell, A.C. and Felsenfeld, G. (2000) Methylation of a CTCF-dependent boundary controls imprinted expression of the Igf2 gene [see comments]. *Nature* **405:** 482–485.

Bell, A.C., West, A.G. and Felsenfeld, G. (1999) The protein CTCF is required for the enhancer blocking activity of vertebrate insulators. *Cell* **98:** 387–396.

Bickmore, W.A., Oghene, K., Little, M.H., Seawright, A., van Heyningen, V. and Hastie, N.D. (1992) Modulation of DNA binding specificity by alternative splicing of the Wilms tumor wt1 gene transcript. *Science* **257:** 235–237.

Bonetta, L., Kuehn, S.E., Huang, A. et al. (1990) Wilms tumor locus on 11p13 defined by multiple CpG island-associated transcripts. *Science* **250:** 994–997.

Breslow, N., Olshan, A., Beckwith, J.B. and Green, D.M. (1993) Epidemiology of Wilms tumor. *Med. Pediatr. Oncol.* **21:** 172–181.

Breslow, N.E., Olson, J., Moksness, J., Beckwith, J.B. and Grundy, P. (1996) Familial Wilms' tumor: a descriptive study. *Med. Pediatr. Oncol.* **27:** 398–403.

Brown, K.W., Wilmore, H.P., Watson, J.E., Mott, M.G., Berry, P.J. and Maitland, N.J. (1993) Low frequency of mutations in the WT1 coding region in Wilms' tumor. *Genes Chromosomes Cancer* **8:** 74–79.

Bruening, W. and Pelletier, J. (1996) A non-AUG translational initiation event generates novel WT1 isoforms. *J. Biol. Chem.* **271:** 8646–8654.

Bruening, W., Bardeesy, N., Silverman, B.L., Cohn, R.A., Machin, G.A., Aronson, A.J., Housman, D. and Pelletier, J. (1992) Germline intronic and exonic mutations in the Wilms' tumour gene (WT1) affecting urogenital development. *Nat. Genet.* **1:** 144–148.

Buckler, A.J., Pelletier, J., Haber, D.A., Glaser, T. and Housman, D.E. (1991) Isolation, characterization, and expression of the murine Wilms' tumor gene (WT1) during kidney development. *Mol. Cell Biol.* **11:** 1707–1712.

Call, K.M., Glaser, T., Ito, C.Y. et al. (1990) Isolation and characterization of a zinc finger polypeptide gene at the human chromosome 11 Wilms' tumor locus. *Cell* **60:** 509–520.

Campbell, C.E., Kuriyan, N.P., Rackley, R.R., Caulfield, M.J., Tubbs, R., Finke, J. and Williams, B.R. (1998) Constitutive expression of the Wilms tumor suppressor gene (WT1) in renal cell carcinoma. *Int. J. Cancer* **78:** 182–188.

Coppes, M.J., Bonetta, L., Huang, A. et al. (1992a) Loss of heterozygosity mapping in Wilms tumor indicates the involvement of three distinct regions and a limited role for nondisjunction or mitotic recombination. *Genes Chromosomes Cancer* **5:** 326–334.

Coppes, M.J., Liefers, G.J., Higuchi, M., Zinn, A.B., Balfe, J.W. and Williams, B.R. (1992b) Inherited WT1 mutation in Denys-Drash syndrome. *Cancer Res.* **52:** 6125–6128.

Coppes, M.J., Liefers, G.J., Paul, P., Yeger, H. and Williams, B.R. (1993) Homozygous somatic Wt1 point mutations in sporadic unilateral Wilms tumor. *Proc. Natl Acad. Sci. USA* **90**: 1416–1419.

Cowell, J.K., Wadey, R.B., Haber, D.A., Call, K.M., Housman, D.E. and Pritchard, J. (1991) Structural rearrangements of the WT1 gene in Wilms' tumour cells. *Oncogene* **6**: 595–599.

Cui, H., Hedborg, F., He, L., Nordenskjold, A., Sandstedt, B., Pfeifer-Ohlsson, S. and Ohlsson, R. (1997) Inactivation of H19, an imprinted and putative tumor repressor gene, is a preneoplastic event during Wilms' tumorigenesis. *Cancer Res.* **57**: 4469–4473.

Denys, P., Malvaux, P., Van Den Berghe, H., Tanghe, W. and Proesmans, W. (1967) [Association of an anatomo-pathological syndrome of male pseudohermaphroditism, Wilms' tumor, parenchymatous nephropathy and XX/XY mosaicism]. *Arch. Fr. Pediatr.* **24**: 729–739.

Deuel, T.F., Guan, L.S. and Wang, Z.Y. (1999) Wilms' tumor gene product WT1 arrests macrophage differentiation of HL-60 cells through its zinc-finger domain. *Biochem. Biophys. Res. Commun.* **254**: 192–196.

Drash, A., Sherman, F., Hartmann, W.H. and Blizzard, R.M. (1970) A syndrome of pseudohermaphroditism, Wilms' tumor, hypertension, and degenerative renal disease. *J. Pediatr.* **76**: 585–593.

Englert, C., Hou, X., Maheswaran, S., Bennett, P., Ngwu, C., Re, G.G., Garvin, A.J., Rosner, M.R. and Haber, D.A. (1995) WT1 suppresses synthesis of the epidermal growth factor receptor and induces apoptosis. *EMBO J.* **14**: 4662–4675.

Fearon, E.R., Vogelstein, B. and Feinberg, A.P. (1984) Somatic deletion and duplication of genes on chromosome 11 in Wilms' tumours. *Nature* **309**: 176–178.

Francke, U., Holmes, L.B., Atkins, L. and Riccardi, V.M. (1979) Aniridia-Wilms' tumor association: evidence for specific deletion of 11p13. *Cytogenet. Cell Genet.* **24**: 185–192.

Frevel, M.A., Sowerby, S.J., Petersen, G.B. and Reeve, A.E. (1999) Methylation sequencing analysis refines the region of H19 epimutation in Wilms tumor. *J. Biol. Chem.* **274**: 29331–29340.

Gao, L., Bellantuono, I., Elsasser, A., Marley, S.B., Gordon, M.Y., Goldman, J.M. and Stauss, H.J. (2000) Selective elimination of leukemic CD34(+) progenitor cells by cytotoxic T lymphocytes specific for WT1. *Blood* **95**: 2198–2203.

Gessler, M., Poustka, A., Cavenee, W., Neve, R.L., Orkin, S.H. and Bruns, G.A. (1990) Homozygous deletion in Wilms tumours of a zinc-finger gene identified by chromosome jumping. *Nature* **343**: 774–778.

Gessler, M., Konig, A., Arden, K. *et al.* (1994) Infrequent mutation of the WT1 gene in 77 Wilms' tumors. *Hum. Mutat.* **3**: 212–222.

Grundy, P.E., Telzerow, P.E., Breslow, N., Moksness, J., Huff, V. and Paterson, M.C. (1994) Loss of heterozygosity for chromosomes 16q and 1p in Wilms' tumors predicts an adverse outcome. *Cancer Res.* **54**: 2331–2333.

Grundy, P., Telzerow, P., Moksness, J. and Breslow, N.E. (1996) Clinicopathologic correlates of loss of heterozygosity in Wilm's tumor: a preliminary analysis. *Med. Pediatr. Oncol.* **27**: 429–433.

Grundy, R.G., Pritchard, J., Baraitser, M., Risdon, A. and Robards, M. (1992) Perlman and Wiedemann-Beckwith syndromes: two distinct conditions associated with Wilms' tumour. *Eur. J. Pediatr.* **151**: 895–898.

Grundy, R.G., Pritchard, J., Scambler, P. and Cowell, J.K. (1998) Loss of heterozygosity for the short arm of chromosome 7 in sporadic Wilms tumour. *Oncogene* **17**: 395–400.

Haber, D.A., Buckler, A.J., Glaser, T., Call, K.M., Pelletier, J., Sohn, R.L., Douglass, E.C. and Housman, D.E. (1990) An internal deletion within an 11p13 zinc finger gene contributes to the development of Wilms' tumor. *Cell* **61**: 1257–1269.

Hamilton, T.B., Barilla, K.C. and Romaniuk, P.J. (1995) High affinity binding sites for the Wilms' tumour suppressor protein WT1. *Nucleic Acids Res.* **23**: 277–284.

Hao, Y., Crenshaw, T., Moulton, T., Newcomb, E. and Tycko, B. (1993) Tumour-suppressor activity of H19 RNA. *Nature* **365**: 764–767.

Hark, A.T., Schoenherr, C.J., Katz, D.J., Ingram, R.S., Levorse, J.M. and Tilghman, S.M. (2000) CTCF mediates methylation-sensitive enhancer-blocking activity at the H19/Igf2 locus [see comments]. *Nature* **405**: 486–489.

Hastie, N.D. (1992) Dominant negative mutations in the Wilms tumour (WT1) gene cause Denys-Drash syndrome – proof that a tumour-suppressor gene plays a crucial role in normal genitourinary development. *Hum. Mol. Genet.* **1**: 293–295.

Henry, I., Bonaiti-Pellie, C., Chehensse, V., Beldjord, C., Schwartz, C., Utermann, G. and Junien, C. (1991) Uniparental paternal disomy in a genetic cancer-predisposing syndrome. *Nature* **351**: 665–667.

Hewitt, M., Lunt, P.W. and Oakhill, A. (1991) Wilms' tumour and a de novo (1;7) translocation in a child with bilateral radial aplasia. *J. Med. Genet.* **28**: 411–412.

Hoovers, J.M., Kalikin, L.M., Johnson, L.A. *et al.* (1995) Multiple genetic loci within 11p15 defined by Beckwith-Wiedemann syndrome rearrangement breakpoints and subchromosomal transferable fragments. *Proc. Natl. Acad. Sci. USA* **92**: 12456–12460.

Huang, A., Campbell, C.E., Bonetta, L. *et al.* (1990) Tissue, developmental, and tumor-specific expression of divergent transcripts in Wilms tumor. *Science* **250**: 991–994.

Huff, V., Jaffe, N., Saunders, G.F., Strong, L.C., Villalba, F. and Ruteshouser, E.C. (1995) WT1 exon 1 deletion/insertion mutations in Wilms tumor patients, associated with di- and trinucleotide repeats and deletion hotspot consensus sequences. *Am. J. Hum. Genet.* **56**: 84–90.

Junien, C. (1992) Beckwith-Wiedemann syndrome, tumourigenesis and imprinting. *Curr. Opin. Genet. Dev.* **2**: 431–438.

Kanduri, C., Holmgren, C., Pilartz, M. *et al.* (2000) The 5' flank of mouse H19 in an unusual chromatin conformation unidirectionally blocks enhancer-promoter communication. *Curr. Biol.* **10**: 449–457.

Kaneko, Y., Homma, C., Maseki, N., Sakurai, M. and Hata, J. (1991) Correlation of chromosome abnormalities with histological and clinical features in Wilms' and other childhood renal tumors. *Cancer Res.* **51**: 5937–5942.

Kaneko, Y., Takeda, O., Homma, C., Maseki, N., Miyoshi, H., Tsunematsu, Y., Williams, B.G., Saunders, G.F. and Sakurai, M. (1993) Deletion of WT1 and WIT1 genes and loss of heterozygosity on chromosome 11p in Wilms tumors in Japan. *Jpn. J. Cancer Res.* **84**: 616–624.

Karnik, P., Chen, P., Paris, M., Yeger, H. and Williams, B.R. (1998) Loss of heterozygosity at chromosome 11p15 in Wilms tumors: identification of two independent regions. *Oncogene* **17**: 237–240.

Kim, J., Prawitt, D., Bardeesy, N., Torban, E., Vicaner, C., Goodyer, P., Zabel, B. and Pelletier, J. (1999) The Wilms' tumor suppressor gene (wt1) product regulates Dax-1 gene expression during gonadal differentiation. *Mol. Cell Biol.* **19**: 2289–2299.

King-Underwood, L. and Pritchard-Jones, K. (1998) Wilms' tumor (WT1) gene mutations occur mainly in acute myeloid leukemia and may confer drug resistance. *Blood* **91**: 2961–2968.

Knudson, A.G., Jr. and Strong, L.C. (1972) Mutation and cancer: a model for Wilms' tumor of the kidney. *J. Natl Cancer Inst.* **48**: 313–324.

Koufos, A., Hansen, M.F., Lampkin, B.C., Workman, M.L., Copeland, N.G., Jenkins, N.A. and Cavenee, W.K. (1984) Loss of alleles at loci on human chromosome 11 during genesis of Wilms' tumour. *Nature* **309**: 170–172.

Koufos, A., Grundy, P., Morgan, K., Aleck, K.A., Hadro, T., Lampkin, B.C., Kalbakji, A. and Cavenee, W.K. (1989) Familial Wiedemann-Beckwith syndrome and a second Wilms tumor locus both map to 11p15.5. *Am. J. Hum. Genet.* **44**: 711–719.

Kreidberg, J.A., Sariola, H., Loring, J.M., Maeda, M., Pelletier, J., Housman, D. and Jaenisch, R. (1993) WT-1 is required for early kidney development. *Cell* **74**: 679–691.

Lee, S.B., Huang, K., Palmer, R. *et al.* (1999) The Wilms tumor suppressor WT1 encodes a transcriptional activator of amphiregulin. *Cell* **98**: 663–673.

Little, M. and Wells, C. (1997) A clinical overview of WT1 gene mutations. *Hum. Mutat.* **9**: 209–225.

Little, M.H., Prosser, J., Condie, A., Smith, P.J., Van Heyningen, V. and Hastie, N.D. (1992) Zinc finger point mutations within the WT1 gene in Wilms tumor patients. *Proc. Natl Acad. Sci. USA* **89**: 4791–4795.

Little, M.H., Williamson, K.A., Mannens, M., Kelsey, A., Gosden, C., Hastie, N.D. and van Heyningen, V. (1993) Evidence that WT1 mutations in Denys-Drash syndrome patients may act in a dominant-negative fashion. *Hum. Mol. Genet.* **2**: 259–264.

Madden, S.L., Cook, D.M. and Rauscher, F.J.d. (1993) A structure-function analysis of transcriptional repression mediated by the WT1, Wilms' tumor suppressor protein. *Oncogene* **8**: 1713–1720.

Mannens, M., Slater, R.M., Heyting, C., Bliek, J., de Kraker, J., Coad, N., de Pagter-Holthuizen, P. and Pearson, P.L. (1988) Molecular nature of genetic changes resulting in loss of heterozygosity of chromosome 11 in Wilms' tumours. *Hum. Genet.* **81**: 41–48.

Mannens, M., Devilee, P., Bliek, J., Mandjes, I., de Kraker, J., Heyting, C., Slater, R.M. and Westerveld, A. (1990) Loss of heterozygosity in Wilms' tumors, studied for six putative tumor suppressor regions, is limited to chromosome 11. *Cancer Res.* **50**: 3279–3283.

Mannens, M., Hoovers, J.M., Redeker, E. *et al.* (1994) Parental imprinting of human chromosome region 11p15.3-pter involved in the Beckwith-Wiedemann syndrome and various human neoplasia. *Eur. J. Hum. Genet.* **2**: 3–23.

Matsunaga, E. (1981) Genetics of Wilms' tumor. *Hum. Genet.* **57**: 231–246.

Maw, M.A., Grundy, P.E., Millow, L.J. *et al.* (1992) A third Wilms' tumor locus on chromosome 16q. *Cancer Res.* **52**: 3094–3098.

Miller, R.W., Fraumeni, J.F. and Manning, M.D. (1964) Association of Wilms tumour with aniridia, hemihypertrophy and other congenital malformations. *N. Engl. J. Med.* **270**: 922–927.

Miozzo, M., Perotti, D., Minoletti, F. *et al.* (1996) Mapping of a putative tumor suppressor locus to proximal 7p in Wilms tumors. *Genomics* **37**: 310–315.

Miyagawa, K., Hayashi, Y., Fukuda, T., Mitani, K., Hirai, H. and Kamiya, K. (1999) Mutations of the WT1 gene in childhood nonlymphoid hematological malignancies. *Genes Chromosomes Cancer* **25**: 176–183.

Moffett, P., Bruening, W., Nakagama, H., Bardeesy, N., Housman, D., Housman, D.E. and Pelletier, J. (1995) Antagonism of WT1 activity by protein self-association. *Proc. Natl Acad. Sci. USA* **92**: 11105–11109.

Moulton, T., Crenshaw, T., Hao, Y. *et al.* (1994) Epigenetic lesions at the H19 locus in Wilms' tumour patients. *Nat. Genet.* **7**: 440–447.

Nakagama, H., Heinrich, G., Pelletier, J. and Housman, D.E. (1995) Sequence and structural requirements for high-affinity DNA binding by the WT1 gene product. *Mol. Cell Biol.* **15**: 1489–1498.

Nichols, K.E., Re, G.G., Yan, Y.X., Garvin, A.J. and Haber, D.A. (1995) WT1 induces expression of insulin-like growth factor 2 in Wilms' tumor cells. *Cancer Res.* **55**: 4540–4543.

Niegemann, E., Wehner, S., Kornhuber, B., Schwabe, D. and Ebener, U. (1999) wt1 gene expression in childhood leukemias. *Acta Haematol.* **102**: 72–76.

Ogawa, O., Becroft, D.M., Morison, I.M., Eccles, M.R., Skeen, J.E., Mauger, D.C. and Reeve, A.E. (1993a) Constitutional relaxation of insulin-like growth factor II gene imprinting associated with Wilms' tumour and gigantism. *Nat. Genet.* **5**: 408–412.

Ogawa, O., Eccles, M.R., Szeto, J., McNoe, L.A., Yun, K., Maw, M.A., Smith, P.J. and Reeve, A.E. (1993b) Relaxation of insulin-like growth factor II gene imprinting implicated in Wilms' tumour. *Nature* **362**: 749–751.

Ohlsson, R., Flam, F., Fisher, R., Miller, S., Cui, H., Pfeifer, S. and Adam, G.I. (1999) Random monoallelic expression of the imprinted IGF2 and H19 genes in the absence of discriminative parental marks. *Dev. Genes Evol.* **209**: 113–119.

Oji, Y., Ogawa, H., Tamaki, H. *et al.* (1999) Expression of the Wilms' tumor gene WT1 in solid tumors and its involvement in tumor cell growth. *Jpn J. Cancer Res.* **90**: 194–204.

Oka, Y., Udaka, K., Tsuboi, A., Elisseeva, O.A., Ogawa, H., Aozasa, K., Kishimoto, T. and Sugiyama, H. (2000) Cancer immunotherapy targeting Wilms' tumor gene WT1 product. *J. Immunol.* **164**: 1873–1880.

Okamoto, K., Morison, I.M., Taniguchi, T. and Reeve, A.E. (1997) Epigenetic changes at the insulin-like growth factor II/H19 locus in developing kidney is an early event in Wilms tumorigenesis. *Proc. Natl Acad. Sci. USA* **94**: 5367–5371.

Orkin, S.H., Goldman, D.S. and Sallan, S.E. (1984) Development of homozygosity for chromosome 11p markers in Wilms' tumour. *Nature* **309**: 172–174.

Osaka, M., Koami, K. and Sugiyama, T. (1997) WT1 contributes to leukemogenesis: expression patterns in 7,12-dimethylbenz[a]anthracene (DMBA)-induced leukemia. *Int. J. Cancer* **72**: 696–699.

Pal, N., Wadey, R.B., Buckle, B., Yeomans, E., Pritchard, J. and Cowell, J.K. (1990) Preferential loss of maternal alleles in sporadic Wilms' tumour. *Oncogene* **5**: 1665–1668.

Park, S., Bernard, A., Bove, K.E., Sens, D.A., Hazen-Martin, D.J., Garvin, A.J. and Haber, D.A. (1993a) Inactivation of WT1 in nephrogenic rests, genetic precursors to Wilms' tumour. *Nat. Genet.* **5**: 363–367.

Park, S., Schalling, M., Bernard, A. *et al.* (1993b) The Wilms tumour gene WT1 is expressed in murine mesoderm-derived tissues and mutated in a human mesothelioma. *Nat. Genet* **4**: 415–420.

Park, S., Tomlinson, G., Nisen, P. and Haber, D.A. (1993c) Altered trans-activational properties of a mutated WT1 gene product in a WAGR-associated Wilms' tumor. *Cancer Res.* **53**: 4757–4760.

Peier, A.M., Meloni, A.M., Erling, M.A. and Sandberg, A.A. (1995) Involvement of chromosome 7 in Wilms tumor. *Cancer Genet. Cytogenet.* **79**: 92–94.

Pelletier, J., Bruening, W., Kashtan, C.E. *et al.* (1991) Germline mutations in the Wilms' tumor suppressor gene are associated with abnormal urogenital development in Denys-Drash syndrome. *Cell* 67: 437–447.

Ping, A.J., Reeve, A.E., Law, D.J., Young, M.R., Boehnke, M. and Feinberg, A.P. (1989) Genetic linkage of Beckwith-Wiedemann syndrome to 11p15. *Am. J. Hum. Genet.* 44: 720–723.

Powlesland, R.M., Charles, A.K., Malik, K.T., Reynolds, P.A., Pires, S., Boavida, M. and Brown, K.W. (2000) Loss of heterozygosity at 7p in Wilms' tumour development. *Br. J. Cancer* 82: 323–329.

Pritchard-Jones, K. (1997) Molecular genetic pathways to Wilms tumor. *Crit. Rev. Oncog.* 8: 1–27.

Pritchard-Jones, K. and King-Underwood, L. (1997) The Wilms tumour gene WT1 in leukaemia. *Leuk. Lymphoma* 27: 207–220.

Pritchard-Jones, K., Fleming, S., Davidson, D. *et al.* (1990) The candidate Wilms' tumour gene is involved in genitourinary development. *Nature* 346: 194–197.

Rackley, R.R., Flenniken, A.M., Kuriyan, N.P., Kessler, P.M., Stoler, M.H. and Williams, B.R. (1993) Expression of the Wilms' tumor suppressor gene WT1 during mouse embryogenesis. *Cell Growth Differ.* 4: 1023–1031.

Rahman, N., Arbour, L., Tonin, P., Renshaw, J., Pelletier, J., Baruchel, S., Pritchard-Jones, K., Stratton, M.R. and Narod, S.A. (1996) Evidence for a familial Wilms' tumour gene (FWT1) on chromosome 17q12- q21. *Nat. Genet.* 13: 461–463.

Rahman, N., Arbour, L., Tonin, P., Baruchel, S., Pritchard-Jones, K., Narod, S.A. and Stratton, M.R. (1997) The familial Wilms' tumour susceptibility gene, FWT1, may not be a tumour suppressor gene. *Oncogene* 14: 3099–3102.

Rahman, N., Abidi, F., Ford, D. *et al.* (1998) Confirmation of FWT1 as a Wilms' tumour susceptibility gene and phenotypic characteristics of Wilms' tumour attributable to FWT1. *Hum. Genet.* 103: 547–556.

Rainier, S., Johnson, L.A., Dobry, C.J., Ping, A.J., Grundy, P.E. and Feinberg, A.P. (1993) Relaxation of imprinted genes in human cancer. *Nature* 362: 747–749.

Rauscher, F.J.d. (1993) The WT1 Wilms tumor gene product: a developmentally regulated transcription factor in the kidney that functions as a tumor suppressor. *FASEB J.* 7: 896–903.

Rauscher, F.J.d., Morris, J.F., Tournay, O.E., Cook, D.M. and Curran, T. (1990) Binding of the Wilms' tumor locus zinc finger protein to the EGR-1 consensus sequence. *Science* 250: 1259–1262.

Reddy, J.C., Morris, J.C., Wang, J., English, M.A., Haber, D.A., Shi, Y. and Licht, J.D. (1995) WT1-mediated transcriptional activation is inhibited by dominant negative mutant proteins. *J. Biol. Chem.* 270: 10878–10884.

Reeve, A.E., Housiaux, P.J., Gardner, R.J., Chewings, W.E., Grindley, R.M. and Millow, L.J. (1984) Loss of a Harvey ras allele in sporadic Wilms' tumour. *Nature* 309: 174–176.

Reeve, A.E., Eccles, M.R., Wilkins, R.J., Bell, G.I. and Millow, L.J. (1985) Expression of insulin-like growth factor-II transcripts in Wilms' tumour. *Nature* 317: 258–260.

Reeve, A.E., Sih, S.A., Raizis, A.M. and Feinberg, A.P. (1989) Loss of allelic heterozygosity at a second locus on chromosome 11 in sporadic Wilms' tumor cells. *Mol. Cell Biol.* 9: 1799–1803.

Reid, L.H., Davies, C., Cooper, P.R. *et al.* (1997) A 1-Mb physical map and PAC contig of the imprinted domain in 11p15.5 that contains TAPA1 and the BWSCR1/WT2 region. *Genomics* 43: 366–375.

Reynolds, P.A., Powlesland, R.M., Keen, T.J., Inglehearn, C.F., Cunningham, A.F., Green, E.D. and Brown, K.W. (1996) Localization of a novel t(1;7) translocation associated with Wilms' tumor predisposition and skeletal abnormalities. *Genes Chromosomes Cancer* 17: 151–155.

Riccardi, V.M., Sujansky, E., Smith, A.C. and Francke, U. (1978) Chromosomal imbalance in the Aniridia-Wilms' tumor association: 11p interstitial deletion. *Pediatrics* 61: 604–610.

Sakamoto, Y., Yoshida, M., Semba, K. and Hunter, T. (1997) Inhibition of the DNA-binding and transcriptional repression activity of the Wilms' tumor gene product, WT1, by cAMP-dependent protein kinase-mediated phosphorylation of Ser-365 and Ser-393 in the zinc finger domain. *Oncogene* 15: 2001–2012.

Sawyer, J.R., Winkel, E.W., Redman, J.F. and Roloson, G.J. (1993) Translocation (7;7)(p13;q21) in a Wilms' tumor. *Cancer Genet. Cytogenet.* 69: 57–59.

Schroeder, W.T., Chao, L.Y., Dao, D.D., Strong, L.C., Pathak, S., Riccardi, V., Lewis, W.H. and Saunders, G.F. (1987) Nonrandom loss of maternal chromosome 11 alleles in Wilms tumors. *Am. J. Hum. Genet.* 40: 413–420.

Schumacher, V., Schneider, S., Figge, A., Wildhardt, G., Harms, D., Schmidt, D., Weirich, A., Ludwig, R. and Royer-Pokora, B. (1997) Correlation of germ-line mutations and two-hit inactivation of the WT1 gene with Wilms tumors of stromal-predominant histology. *Proc. Natl Acad. Sci. USA* **94**: 3972–3977.

Scott, J., Cowell, J., Robertson, M.E. *et al.* (1985) Insulin-like growth factor-II gene expression in Wilms' tumour and embryonic tissues. *Nature* **317**: 260–262.

Sharma, P.M., Bowman, M., Madden, S.L., Rauscher, F.J., 3rd and Sukumar, S. (1994) RNA editing in the Wilms' tumor susceptibility gene, WT1. *Genes Dev.* **8**: 720–731.

Sheng, W.W., Soukup, S., Bove, K., Gotwals, B. and Lampkin, B. (1990) Chromosome analysis of 31 Wilms' tumors. *Cancer Res.* **50**: 2786–2793.

Silberstein, G.B., Van Horn, K., Strickland, P., Roberts, C.T., Jr. and Daniel, C.W. (1997) Altered expression of the WT1 Wilms tumor suppressor gene in human breast cancer. *Proc. Natl Acad. Sci. USA* **94**: 8132–8137.

Slater, R.M., de Kraker, J., Voute, P.A. and Delemarre, J.F. (1985) A cytogenetic study of Wilms' tumor. *Cancer Genet. Cytogenet.* **14**: 95–109.

Smith, S.I., Weil, D., Johnson, G.R., Boyd, A.W. and Li, C.L. (1998) Expression of the Wilms' tumor suppressor gene, WT1, is upregulated by leukemia inhibitory factor and induces monocytic differentiation in M1 leukemic cells. *Blood* **91**: 764–773.

Smith, S.I., Down, M., Boyd, A.W. and Li, C.L. (2000) Expression of the Wilms' tumor suppressor gene, WT1, reduces the tumorigenicity of the leukemic cell line M1 in C.B-17 scid/scid mice. *Cancer Res.* **60**: 808–814.

Solis, V., Pritchard, J. and Cowell, J.K. (1988) Cytogenetic changes in Wilms' tumors. *Cancer Genet. Cytogenet.* **34**: 223–234.

Stanhope-Baker, P. and Williams, B.R.G. (2000). Identfication of connective tissue growth factor as a target of the WT1. transcriptional regulation. *J. Biol. Chem.* in press.

Steenman, M.J., Rainier, S., Dobry, C.J., Grundy, P., Horon, I.L. and Feinberg, A.P. (1994) Loss of imprinting of IGF2 is linked to reduced expression and abnormal methylation of H19 in Wilms' tumour. *Nat. Genet.* **7**: 433–439.

Svedberg, H., Chylicki, K. and Gullberg, U. (1999) Downregulation of Wilms' tumor gene (WT1) is not a prerequisite for erythroid or megakaryocytic differentiation of the leukemic cell line K562. *Exp. Hematol.* **27**: 1057–1062.

Szabo, P., Tang, S.H., Rentsendorj, A., Pfeifer, G.P. and Mann, J.R. (2000) Maternal-specific footprints at putative CTCF sites in the H19 imprinting control region give evidence for insulator function [In Process Citation]. *Curr. Biol.* **10**: 607–610.

Tadokoro, K., Fujii, H., Ohshima, A. *et al.* (1992) Intragenic homozygous deletion of the WT1 gene in Wilms' tumor. *Oncogene* **7**: 1215–1221.

Taniguchi, T., Sullivan, M.J., Ogawa, O. and Reeve, A.E. (1995) Epigenetic changes encompassing the IGF2/H19 locus associated with relaxation of IGF2 imprinting and silencing of H19 in Wilms tumor. *Proc. Natl Acad. Sci. USA* **92**: 2159–2163.

Thorvaldsen, J.L., Duran, K.L. and Bartolomei, M.S. (1998) Deletion of the H19 differentially methylated domain results in loss of imprinted expression of H19 and Igf2. *Genes Dev.* **12**: 3693–3702.

Ton, C.C., Hirvonen, H., Miwa, H. *et al.* (1991) Positional cloning and characterization of a paired box- and homeobox-containing gene from the aniridia region. *Cell* **67**: 1059–1074.

Varanasi, R., Bardeesy, N., Ghahremani, M., Petruzzi, M.J., Nowak, N., Adam, M.A., Grundy, P., Shows, T.B. and Pelletier, J. (1994) Fine structure analysis of the WT1 gene in sporadic Wilms tumors. *Proc. Natl Acad. Sci. USA* **91**: 3554–3558.

Wadey, R.B., Pal, N., Buckle, B., Yeomans, E., Pritchard, J. and Cowell, J.K. (1990) Loss of heterozygosity in Wilms' tumour involves two distinct regions of chromosome 11. *Oncogene* **5**: 901–907.

Wang, Z.Y., Qiu, Q.Q., Enger, K.T. and Deuel, T.F. (1993) A second transcriptionally active DNA-binding site for the Wilms tumor gene product, WT1. *Proc. Natl Acad. Sci. USA* **90**: 8896–8900.

Waziri, M., Patil, S.R., Hanson, J.W. and Bartley, J.A. (1983) Abnormality of chromosome 11 in patients with features of Beckwith-Wiedemann syndrome. *J. Pediatr.* **102**: 873–876.

Wiedemann, H.-R. (1964) Complexe malformatif familial avec hernie ombilicale et macroglossie – Un (syndrome Nouveau). *Journal et Genetetique Human* **13**: 223–232.

Wilkins, R.J. (1988) Genomic imprinting and carcinogenesis. *Lancet* **1**: 329–331.

Williams, J.C., Brown, K.W., Mott, M.G. and Maitland, N.J. (1989) Maternal allele loss in Wilms' tumour. *Lancet* **1**: 283–284.

Wilmore, H.P., White, G.F., Howell, R.T. and Brown, K.W. (1994) Germline and somatic abnormalities of chromosome 7 in Wilms' tumor. *Cancer Genet. Cytogenet.* **77**: 93–98.

Ye, Y., Raychaudhuri, B., Gurney, A., Campbell, C.E. and Williams, B.R. (1996) Regulation of WT1 by phosphorylation: inhibition of DNA binding, alteration of transcriptional activity and cellular translocation. *EMBO J.* **15**: 5606–5615.

Yeger, H., Cullinane, C., Flenniken, A. *et al.* (1992) Coordinate expression of Wilms' tumor genes correlates with Wilms' tumor phenotypes. *Cell Growth Differ.* **3**: 855–864.

15

Genetics of neuroblastoma

John M. Maris and Garrett M. Brodeur

1. Introduction

Neuroblastoma is the third most common pediatric cancer and is responsible for approximately 15% of all childhood cancer deaths. It is an embryonal cancer of the postganglionic sympathetic nervous system and most commonly arises in the adrenal medulla. Neuroblastoma is the most common malignant disease of infancy (Brodeur, 1998; Brodeur and Ambros, 2000; Brodeur et al., 1997) and 96% of cases occur before age ten (Brodeur et al., 1997). Although neuroblastoma is sometimes diagnosed in the perinatal period (Knudson and Strong, 1972), no environmental influences or parental exposures have been identified consistently that impact on disease occurrence (Bunin et al., 1990). Thus, the etiology of neuroblastoma remains obscure.

The clinical hallmark of neuroblastoma is heterogeneity, with the likelihood of tumor progression varying widely according to age and disease stage at diagnosis. In general, infants diagnosed before 1 year of age and/or with localized disease are curable with surgery and little or no adjuvant therapy. Some of these tumors undergo spontaneous regression (D'Angio et al., 1971) or differentiate into benign ganglioneuromas. In contrast, older children often have extensive metastases at diagnosis, and the majority of the patients die from disease progression despite intensive multimodality therapy. This clinical diversity correlates closely with several genetic features that have been identified in neuroblastomas.

Investigation of the genetics of neuroblastoma began with the karyotypic characterization of tumor-derived cell lines. The majority of neuroblastoma cell lines show deletions of the short arm of chromosome 1 (1p), which presumably represents loss of a tumor suppressor gene, as well as double minute chromatin bodies (DMs) or homogeneously staining regions (HSRs), both of which represent DNA amplification (Biedler et al., 1973; Brodeur et al., 1977). Thus, these early observations clearly demonstrated that both loss and gain of genetic material are common during neuroblastoma evolution, which is consistent with the current concepts of tumorigenesis involving both suppressor gene inactivation and oncogene activation, respectively. Therefore, this chapter will discuss genetic predisposition to

Molecular Genetics of Cancer second edition, edited by J.K. Cowell.
© 2001 BIOS Scientific Publishers Ltd, Oxford.

development of neuroblastoma, as well as the common genetic changes in neuroblastoma tumor cells. This will be followed by a review of the frequently observed alterations in gene expression that may affect their survival, differentiation and apoptosis of neuroblastoma cells.

2. Neuroblastoma predisposition

2.1 Constitutional chromosomal abnormalities and associated conditions

Germline chromosomal abnormalities have been identified rarely in children with neuroblastoma. Chromosome band 1p36 is a frequent site of somatic deletion in neuroblastoma cells, and interestingly three neuroblastoma patients have been described with germline interstitial deletions of 1p36 (Biegel et al., 1993; Fan et al., 1999; White et al., 2000). Each patient was diagnosed during infancy and had significant psychomotor retardation. The constitutional deletions overlap the location of a putative 1p36 tumor suppressor gene, suggesting that germline absence of a gene within this region may predispose to the development of neuroblastoma. In addition, constitutional balanced translocations involving 1p have been identified in two infants with neuroblastoma (Laureys et al., 1990; Mead and Cowell, 1995). These translocations also involved the long arm of chromosome 17 [t(1;17)(p36;q12–21)] in a patient with localized disease, and the long arm of chromosome 10 [t(1;10)(p22;q21)] in a patient with stage 4S disease. The NB4S gene at chromosome band 1p22 was disrupted by the latter translocation, suggesting that its inactivation may play a causal role in neuroblastoma tumorigenesis (Roberts et al., 1998). At least 14 other patients with constitutional chromosomal rearrangements and neuroblastoma have been identified (Brodeur, 1998), but the lack of a consistent pattern indicates that many of these rearrangements may be coincidental rather than causal.

In general, neuroblastoma is not associated with any congenital anomalies or genetic syndromes. One study showed a higher incidence of neurodevelopmental abnormalities in the brains of children dying of neuroblastoma, but the significance of these findings is not clear (Blatt and Hamilton, 1998). A recent review suggested a higher incidence of neuroblastoma in girls with Turner's syndrome (Blatt et al., 1997), although this has not been independently confirmed. Interestingly, another study suggested a lower incidence of neuroblastoma among patients with Down's syndrome (Sage et al., 1998).

Rarely, sporadic and familial neuroblastoma patients have been described in patients with disorders of other neural crest-derived tissues such as Hirschsprung disease, central hypoventilation (Ondine's curse) and/or neurofibromatosis type 1 (NF1) (Bower and Adkins, 1980; Gaisie, 1989; Knudson and Amsonin, 1966; Kushner et al., 1985; Maris et al., 1997a; Witzleben and Landy, 1974). These findings suggest a more global disorder of neural crest-derived cells (neurocristopathy), and genes mutated in each of the above conditions may contribute to neuroblastoma tumorigenesis in a subset of cases (Bolande, 1974; Bolande and Towler, 1970). Indeed, two studies have documented NF1 gene mutations in some neuroblastoma cell lines (Johnson *et al.*, 1993; The *et al.*, 1993), and a patient has been reported with NF1 and neuroblastoma whose tumor had a homozygous deletion of the *NF1* gene (Martinsson *et al.*, 1997). However, epidemiological evidence suggests that the association of neuroblastoma with NF1 is merely coincidental (Kushner *et al.*, 1985), and familial neuroblastoma is not linked to the *NF1* locus, even in a family in which the proband had both conditions (Maris *et al.*, 1997). Thus, germline inactivation of *NF1* does not appear to predispose to neuroblastoma, but somatically acquired inactivation may occur as a later event in tumor evolution in some cases.

RET encodes a receptor tyrosine kinase that is normally involved in neuronal differentiation. Inactivating mutations occur in a subset of familial Hirschsprung disease patients, but activating mutations lead to the multiple endocrine neoplasia type 2 syndrome. A potential involvement of *RET* in neuroblastoma tumorigenesis was suggested by mice carrying a *RET* transgene developing retroperitoneal small round blue cell tumors suggestive of neuroblastoma (Iwamoto *et al.*, 1993). However, no mutations were found in a panel of 16 neuroblastoma cell lines (Hofstra *et al.*, 1996), nor was familial neuroblastoma found to be linked to the *RET* locus at 10q11.2 (Maris *et al.*, 1996, 1997a). The other currently identified genes that are mutated in hereditary Hirschsprung disease have not yet been analyzed for somatic mutations in neuroblastomas.

2.2 Hereditary neuroblastoma

Although neuroblastoma usually occurs sporadically, 1–2% of patients have a family history of the disease (Chompret *et al.*, 1998; Knudson and Strong, 1972; Kushner *et al.*, 1986; Maris *et al.*, 1996a; Maris *et al.*, 1997b). This is similar to the other embryonal cancers of childhood, in which a subset of patients develop their cancers as a result of hereditary predisposition. Familial neuroblastoma is inherited as an autosomal dominant Mendelian trait with incomplete penetrance. In contrast to patients without hereditary predisposition, most children with familial neuroblastoma are typically diagnosed at an earlier age (usually infancy) and/or have multiple primary tumors (Knudson and Strong, 1972; Kushner *et al.*, 1986; Maris and Tonini, 2000; Maris *et al.*, 1997b). These clinical characteristics are hallmarks of the 'two-hit' model first proposed for retinoblastoma (Knudson, 1971).

It therefore appears likely that familial neuroblastoma occurs due to a germline mutation in one allele of a tumor suppressor gene (or genes). Based on review of clinical data, Knudson and Strong estimated that *de novo* germinal mutations in such a gene might initiate up to 22% of nonfamilial neuroblastomas (Knudson and Strong, 1972).

There is remarkable clinical heterogeneity among patients with hereditary neuroblastoma. Within individual families, the disease can vary from an asymptomatic clinical course with disease noted incidentally to one resulting in rapid progression and a fatal outcome (Maris *et al.*, 1997b). Thus, it has been hypothesized that the timing of inactivation of the second tumor suppressor gene allele and/or additional mutational events confer the ultimate clinical phenotype (Maris *et al.*, 1997b). The clinical heterogeneity of familial neuroblastoma may partially explain its rarity since some tumors remain occult or regress and are never detected, while others result in death prior to reproductive age.

Many candidates for the familial neuroblastoma gene locus have been proposed, including several regions containing putative tumor suppressor loci (see below). Linkage analysis at candidate loci has excluded each of these regions, including the distal short arm of chromosome 1 (Maris *et al.*, 1996a; Maris *et al.*, 1997b; Tonini *et al.*, 1997a). However, a recent genome-wide search for linkage has identified the short arm of chromosome 16 as a likely site of a hereditary neuroblastoma predisposition gene (Maris *et al.*, 2000). Studies are ongoing to determine if there are other loci harboring neuroblastoma predisposition genes.

3. Loss of genetic material and inactivation of tumor suppressor genes

3.1 Chromosome 1 alterations in neuroblastoma

Brodeur and colleagues first recognized that 1p deletions were a common karyotypic feature of neuroblastoma cell lines and advanced tumors (Brodeur *et al.*, 1977; Brodeur *et al.*, 1981). Multiple investigators have confirmed this observation, and 1p loss of heterozygosity (LOH) has been documented in 25–35% of primary tumor specimens (Caron *et al.*, 1996a; Fong *et al.*, 1989; Fong *et al.*, 1992; Gehring *et al.*, 1995; Maris *et al.*, 1995b; Maris *et al.*, 2000; Martinsson *et al.*, 1995; Takita *et al.*, 1995; White *et al.*, 1995). In addition, microcell mediated transfer of an intact human 1p to a neuroblastoma cell line resulted in decreased tumorigenicity (Bader *et al.*, 1991). It therefore appears likely that 1p harbors a neuroblastoma suppressor gene, and that this gene is inactivated in at least one-third of primary neuroblastomas.

The majority of 1p deletions in neuroblastomas are large, often resulting in loss of the majority of the short arm (Mertens *et al.*, 1997). The mechanism of deletion is complex, but frequently is the result of an unbalanced translocation with chromosome 17 (Lastowska *et al.*, 1997a; Savelyeva *et al.*, 1994; Van Roy *et al.*, 1994; Van Roy *et al.*, 1995a; Van Roy *et al.*, 1997). Virtually all neuroblastoma cell lines and primary tumors with 1p deletions reported in the literature are hemizygous for a region that includes distal 1p36.2 and all of 1p36.3 (*Figure 1*) (Cheng *et al.*, 1995; Maris and Matthay, 1999; Martinsson *et al.*, 1995; White *et al.*, 1995; White *et al.*,1997). Recent high-resolution mapping studies have confirmed the existence of a common region of deletion within chromosome subbands 1p36.2–36.3 and refined the critical region to approximately one million base pairs of DNA (Martinsson *et al.*, 1995; White *et al.*, 1995; White *et al.*, 2000).

Several studies have suggested that additional 1p tumor suppressor loci, distinct from the 1p36.2-.3 locus, contribute to neuroblastoma tumorigenesis. Schleiermacher reported three cases with interstitial deletions of 1p32 only, suggesting the possibility of an additional suppressor gene in this proximal region (Schleiermacher *et al.*, 1994). Takeda and colleagues showed that tumors with larger 1p deletions also generally had *MYCN* amplification and poor survival probability, while those with smaller deletions were more likely to have a single copy of *MYCN* and usually had a favorable clinical outcome (Takeda *et al.*, 1994).

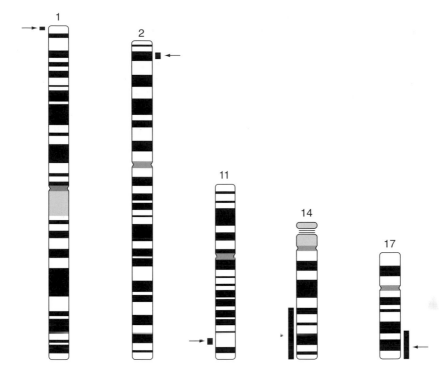

Figure 1. Regions of frequent gain or loss in human neuroblastomas. Shown are diagrammatic representations of chromosomes that show frequent loss or gain in neuroblastomas. The three most common and well-characterized regions of allelic loss involve 1p36 (35%), 11q23 (45%) and 14q32 (25%). The most frequent regions of gain involve 2p24 (amplification of the *MYCN* locus, 22%) or unbalanced gain of 17q23-qter (50–60%).

Caron and colleagues also found that tumors with *MYCN* amplification generally had 1p deletions extending proximal to 1p36, but single-copy tumors had small terminal deletions of 1p36 only (Caron *et al.*, 1993). Furthermore, the *MYCN*-amplified tumors in the latter study showed a random distribution for the parental origin of the deleted chromosome, whereas the *MYCN* single-copy tumors showed preferential deletion of the maternal homolog. These data suggest at least two tumor suppressor genes located within 1p35–36, an imprinted distal 1p36 and a nonimprinted 1p35–36.1 locus that is deleted in *MYCN*-amplified tumors. This hypothesis has not yet been independently confirmed.

Several genes have been analyzed as possible candidates for the 1p36 neuroblastoma suppressor gene. These include the *TP53* homolog *TP73* (Kaghad *et al.*, 1997) the *CDK2* homolog *CDC2L1* (p58); (Lahti *et al.*, 1994) the transcription factors *HKR3* (Maris *et al.*, 1996b), *DAN* (Shapiro *et al.*, 1993), *PAX7* (Enomoto *et al.*, 1994), *ID3* (Deed *et al.*, 1994) and *E2F2* (Saito *et al.*, 1995); the transcription elongation factor *TCEB3* (Elongin A) (Aso *et al* ., 1995); and two members of the tumor necrosis factor receptor family, *TNFR2* (Bettinger *et al.*, 1996) and *DR3* (Marsters *et al.*, 1996). However, each of these genes except *DR3* and *HKR3* are located outside the current consensus region, and no mutations have been found in the non-deleted allele of any candidate (White *et al.*, 1997). Thus, none of these candidates function as a classic tumor suppressor for neuroblastoma, but the possibility that haploinsufficiency for any candidate gene results in enhanced tumorigenicity has not been ruled out.

It is clear that 1p LOH is associated with high-risk features, such as age greater than one year at diagnosis, metastatic disease and *MYCN* amplification (Caron *et al.*, 1996; Gehring *et al.*, 1995; Maris *et al.*, 1995; Schleiermacher *et al.*, 1996). Thus, 1p LOH occurs at higher frequency in the more malignant subset of neuro-blastomas. However, there have been contrasting opinions whether or not 1p LOH can independently predict disease outcome (Caron *et al.*, 1996a; Gehring *et al.*, 1995; Maris *et al.*, 1995; Maris *et al.*, 2000; Rubie *et al.*, 1997; Schleiermacher *et al.*, 1996). A recently completed retrospective study indicated that 1p36 LOH independently predicts for a decreased event-free survival, but not overall survival (Maris *et al.*, 2000b). The apparent discrepancy among the findings of the above studies may be explained by the fact that many disease relapses in patients with nonmetastatic neuroblastoma are salvageable. In addition, 1p LOH was not independently prognostic in the high-risk group of patients, but was the only factor that was predictive of disease relapse in the low-risk cohort. Thus, 1p LOH analysis is associated with high-risk disease features, but appears to be independently prognostic for disease relapse in patients otherwise judged to be at low risk for treatment failure.

3.2 Chromosome 11 alterations in neuroblastoma

The long arm of chromosome 11 (11q) is also a frequent site of hemizygous deletion in human neuroblastomas. Indeed, 11q deletions have been noted in approximately 15–20% of reported neuroblastoma karyotypes (Mertens *et al.*, 1997), and transfer of an intact chromosome 11 into the neuroblastoma cell line NGP induced differentiation (Bader *et al.*, 1991). Constitutional rearrangements of chromosome 11q have also been observed in some neuroblastoma patients (Brodeur, 1998). Previous molecular genetic studies have demonstrated loss of 11q in about one-third of primary tumors (Brinkschmidt *et al.*, 1997; Lastowska *et al.*, 1997; Plantaz *et al.*, 1997; Srivatsan *et al.*, 1993; Takita *et al.*, 1995). A recently completed survey of 295 primary neuroblastomas has found evidence for allelic deletion in 44% of cases (Guo *et al.*, 1999; Guo *et al.*, 2000). Many of these deletions were the result of monosomy for chromosome 11, and partial deletions of 11q occurred in approximately one quarter of primary neuroblastomas. The common region of LOH mapped to 11q23.3, indicating that this is the most likely location of an 11q neuroblastoma suppressor gene (*Figure 1*) (Guo *et al.*, 1999; Guo *et al.*, 2000). In contrast to 1p LOH, there was a striking inverse correlation of 11q LOH and *MYCN* amplification. Thus, structural rearrangements resulting in partial deletion of 11q (as opposed to whole chromosome loss) are associated with high-risk disease features, but not *MYCN* amplification. Further studies will be necessary to determine if 11q LOH is an independent marker of an aggressive clinical phenotype.

3.3 Chromosome 14 alterations in neuroblastoma

Deletion of the long arm of chromosome 14 is also a common abnormality in neuroblastomas. Cytogenetic evidence for deletion is rarely observed, but allelic deletion of 14q32 loci has been reported in 18–40% of tumors studied (Fong *et al.*, 1992; Hoshi *et al.*, 2000; Srivatson *et al.*, 1993; Takayama *et al.*, 1992; Theobald *et al.*, 1999; Thompson *et al.*, 2000). Thompson and colleagues recently showed LOH in 22% of 372 primary neuroblastomas, with a common region of deletion within 14q23-qter (*Figure 1*) (Thompson *et al.*, 2000a). Two other groups have reported common regions of deletion within 14q32 of about 1 cM, although it is not clear if these regions overlap (Hoshi *et al.*, 2000; Theobald *et al.*, 1999). Clinical correlative analyses showed that 14q LOH was highly associated with 11q LOH, inversely related to *MYCN* amplification, and present in all clinical risk groups (Theobald *et al.*, 1999; Thompson *et al.*, 2000). Allelic loss at 14q32 does not appear to be independently prognostic for adverse disease outcome (Theobald *et al.*, 1999; Thompson *et al.*, 2000).

3.4 Deletions of other chromosomal regions

There are other regions of the genome that are frequently deleted in neuroblastoma cells, suggesting the existence of additional tumor suppressor genes. There have been reports of allelic loss or imbalance at chromosome arms 3p (Ejeskar *et al.*, 1998), 4p (Caron *et al.*, 1996), 5q (Meltzer *et al.*, 1996), 9p (Marshall *et al.*, 1997; Takita *et al.*, 1997) and 18q (Reale *et al.*, 1996), but these appear to occur at lower frequency than 1p or 11q LOH. These data, which are largely based on analysis of tumors with PCR-based polymorphisms, have been confirmed by recent studies utilizing comparative genomic hybridization (CGH) (Altura *et al.*, 1997; Brinkschmidt *et al.*, 1997; Lastowska *et al.*, 1997b; Plantaz *et al.*, 1997). However, the genes that are targets for these apparent non-random deletions are unknown.

3.5 Alterations of known tumor suppressor genes

No novel, neuroblastoma-specific suppressor genes have been cloned, and there is currently no evidence for consistent mutation in known tumor suppressor genes. *TP53* is the most frequently mutated gene in human cancer, but is only rarely inactivated by deletion or mutation in primary neuroblastomas (Hosoi *et al.*, 1994; Komuro *et al.*, 1993; Manhani *et al.*, 1997; Vogan *et al.*, 1993). Moll and colleagues have reported aberrant cytoplasmic localization of the p53 protein in neuroblastoma and found evidence for dysregulated G^1/S checkpoint control (Moll *et al.*, 1996). However, others have shown DNA damage to neuroblastoma cells causes normal translocation of wild-type p53 to the nucleus and induction of p21 (Caron *et al.*, 1996b). Interestingly, Keshelava and colleagues have recently demonstrated the *TP53* mutations occur in neuroblastoma cell lines established from patients at disease relapse following intensive chemotherapy, suggesting that *TP53* loss of function may confer a multi-drug resistance phenotype (Keshelava *et al.*, 2000).

CDKN2 encodes p16, another cell cycle control protein commonly inactivated in human cancers. Homozygous deletion of the p16 (and p14) locus has been observed in a few neuroblastoma cell lines and one primary tumor (Thompson *et al.*, 2000b), but inactivating mutations are rarely observed in diagnostic specimens (Bettinger *et al.*, 1995). However, one report did show lack of p16 protein expression in 61% of 74 neuroblastomas and provided evidence that hypermethylation of the p16 locus was the mechanism for functional inactivation (Takita *et al.*, 1998). The *DCC* and *DPC4* genes are located at 18q, a region that is frequently deleted in primary neuroblastomas. Although expression of *DCC* may be altered in some cell lines and tumors, no inactivating mutations of *DCC* or *DPC4* have been identified (Kong *et al.*, 1997; Reale *et al.*, 1996).

Loss of function of a wide array of DNA repair proteins can lead to a 'mutator' phenotype and DNA instability in human cancer. However, microsatellite instability is only rarely observed in primary neuroblastomas and cell lines (Hogarty *et al.*, 1998; Martinsson *et al.*, 1995; White *et al.*, 1995). Therefore, it appears unlikely that alterations in DNA repair genes play a major role in neuroblastoma tumorigenesis.

4. Gain of genetic material and oncogene activation

4.1 MYCN amplification and overexpression

DMs and HSRs are cytogenetic manifestations of gene amplification, but the presence of these findings *per se* does not indicate which gene is amplified. For neuroblastomas, the region amplified is virtually always derived from the distal short arm of chromosome 2 (*Figure 1*), and it contains the proto-oncogene *MYCN* (also known as N-*myc*) (Kohl *et al.*, 1983; Schwab *et al.*, 1983; Schwab *et al.*, 1984). Brodeur and colleagues showed that *MYCN* amplification occurs in a substantial subset of primary, untreated neuroblastomas and is highly correlated with advanced stage (Brodeur *et al.*, 1984). Seeger and co-workers then demonstrated a strong association with rapid disease progression and a poor prognosis (Seeger *et al.*, 1985). Subsequent investigations have confirmed that *MYCN* gene status is a powerful predictor of outcome, independent of clinical and other biologic features. Analysis of *MYCN* remains an essential component of evaluation for newly diagnosed neuroblastoma patients, and it serves as a paradigm for the utility of molecular biologic information in risk and treatment stratification. *MYCN* amplification is almost always present at the time of diagnosis, if it is going to occur (Brodeur *et al.*, 1987), so it appears to be an intrinsic biologic property of a subset of very aggressive tumors that frequently have a poor outcome.

The *MYCN* gene (*Table 1*) is located on the distal short arm of chromosome 2 (2p24) (Kohl *et al.*, 1983; Schwab *et al.*, 1984). Genetic material from this region is transposed to DMs during the process of amplification, and the DMs may then be linearly integrated into random chromosomal regions as HSRs (Brodeur and Seeger, 1986; Brodeur and Fong, 1989). The process of amplification usually results in 50–500 copies of the gene per cell, with correspondingly high levels of mRNA and protein expression (Seeger *et al.*, 1988). Intermediate copy level numbers (i.e. 3–10 copies) may reflect (1) a uniform but low-level amplification; (2) genetic heterogeneity, with amplification in a subset of cells (Ambros *et al.*, 2000); or (3) aneuploidy for chromosome 2, rather than amplification *per se* (Look *et al.*, 1991).

MYCN is a proto-oncogene normally expressed in the developing nervous system and selected other tissues. The *MYCN* gene product, MycN, is a nuclear phosphoprotein with a short half-life and regions of marked homology to Myc (Slamon *et al.*, 1986). Like all Myc family proteins, MycN contains an N-terminal

Table 1. Chromosomal regions and genes altered in neuroblastoma (Maris and Mathay, 1999)

Chromosomal locus	Gene name	Gene type/function[1]	Reference
1p36.2-p36.3	NB locus	Tumor suppressor	Brodeur et al., 1977
1p13	NGF	Neurotrophin; ligand for NTRK1	Kaplan et al., 1991
1q23-q31	NTRK1 (TRK-A)	Receptor tyrosine kinase	Nakagawara et al., 1993
2p12–13	MAD	May regulate MYCN	Wenzel and Schwab, 1995
2p24.1	MYCN	Proto-oncogene	Schwab et al., 1983
2p24	DDX1	RNA helicase/oncogene	George et al., 1996
3p	–	Tumor suppressor	Ejeskar et al., 1998
4p	–	Tumor suppressor	Caron et al., 1996b
7q21	PGY1 (MDR1)	Multidrug resistance	Bourhis et al., 1989
9q22.1	NTRK2 (TRK-B)	Receptor tyrosine kinase	Nakagawara et al., 1994
11p13	CD44	Integrin/metastasis suppression	Schtivelman and Bishop, 1991
11p13	BDNF	Neurotrophin; ligand for NTRK1	Nakagawara et al., 1994
11q23	–	Tumor suppressor	Guo et al., 1999, 2000
12p13	NTF3 (NT-3)	Neurotrophin; ligand for NTRK3	Yamashiro et al., 1996
14q23	MAX	Regulates MycN protein	Wenzel and Schwab, 1995
14q23-qter	–	Tumor suppressor	Thompson et al., 2000a
15q24-q25	NTRK3 (TRK-C)	Receptor tyrosine kinase	Yamashiro et al., 1996
16p13.1	MRP	Multidrug resistance	Norris et al., 1996
17q22	NME1	Nucleoside kinase/metastasis suppression	Hailat et al., 1991
17q21-qter	–	Oncogene?	Plantaz et al., 1997
18q21.1	DCC	Tumor suppressor	Reale et al., 1996
18q21.3	BCL2	Apoptosis suppression	Castle et al., 1993
19	NTF4 (NT-4)	Neurotrophin; ligand for NTRK2	Nakagawara et al., 1994

[1] Regions of allelic loss are presumed to contain tumor suppressor genes.

transactivation domain (Myc box) and a C-terminal region containing a DNA binding and dimerization motif. In order for MycN to activate transcription, it must first dimerize to a related protein called Max (Wenzel et al., 1991). Max is a ubiquitously expressed nuclear protein with a long half-life, but it lacks the amino-terminal transactivation domain that characterizes the Myc protein family. At steady-state in a quiescent (G_0) cell, Max expression is high and favors the formation of Max/Max homodimers, which repress transcription. However, with increased production of MycN, as with entrance into the cell cycle or as a result of genomic amplification, heterodimerization of MycN and Max is favored. This leads to transcriptional activation of an as yet undefined series of growth promoting genes. In addition, heterodimerization of Max with other nuclear proteins such as Mad and Mxi1 also function to repress transcription by competing with MycN for Max binding. Several genes thought to be up-regulated by Myc proteins have been identified (Gandom and Eisenman, 1997), but the essential target genes affected by MycN overexpression in neuroblastomas have not yet been defined.

MYCN functions as a classic dominant oncogene. Forced expression of *MYCN* can transform normal cells, usually in cooperation with oncogenic ras (Schwab *et al.*, 1985; Small *et al.*, 1987; Yancopoulos *et al.*, 1985). Overexpression of *MYCN* can rescue embryonic fibroblasts from senescence (Schwals and Bishop, 1988), and addition of *MYCN* antisense RNA to MycN-overexpressing neuroblastoma cell lines can decrease proliferation and/or induce differentiation (Negroni *et al.*, 1991; Schmidt *et al.*, 1994). Also, targeted overexpression of human *MYCN* in the neuroectoderm of transgenic mice consistently produces tumors resembling neuroblastoma in a dose-dependent manner (Weiss et al., 1997).

In general, there is a correlation between *MYCN* copy number and expression, and tumors with amplification usually express *MYCN* at much higher levels than are seen in tumors without amplification (Nakagawara *et al.*, 1992). Thus, *MYCN* overexpression in the context of amplification consistently identifies a subset of neuroblastomas with highly malignant behavior. However, it is controversial whether or not overexpression of *MYCN* mRNA or MycN protein has prognostic significance in nonamplified tumors (Nisen *et al.*, 1988; Seeger *et al.*, 1988; Slave *et al.*, 1990). Some neuroblastoma cell lines express high levels of *MYCN* mRNA or MycN protein without gene amplification (Seeger *et al.*, 1988; Wade *et al.*, 1993). This may be due to alterations in normal protein degradative pathways (Cohn *et al.*, 1990), or loss of *MYCN* transcriptional autoregulation (Sivak *et al.*, 1997). Some studies have suggested that MycN expression correlates inversely with survival probability (Chan *et al.*, 1997), whereas others have found no such correlation (Bordow *et al.*, 1998). Further studies in a larger cohort of consistently treated patients with standardized methods will be necessary to determine if quantitative assessment of MycN expression in tumors lacking *MYCN* amplification provides additional prognostic information.

4.2 Other genes co-amplified with MYCN

The genomic region amplified with *MYCN* is quite large, usually on the order of 500 – 1000 kilobases. Therefore, it has been postulated that additional genes located near *MYCN* may contribute to the ultimate tumor phenotype when co-amplified. High-resolution restriction mapping of the *MYCN* locus has shown that a 130 kilobase core domain is the consistent target of amplification (Reiter and Brodeur, 1996). To date, no other gene besides *MYCN* has been identified in this core domain (Reiter and Brodeur, 1998). However, the RNA helicase gene *DDX1* maps within 300 kilobases 5′ of *MYCN* and is co-amplified in approximately 40–50% of neuroblastomas with *MYCN* amplification (George *et al.*, 1996). *DDX1* can transform NIH 3T3 cells and allow for establishment of sarcomatous primary tumors in immunodeficient mice (George *et al.*, 2000). Nevertheless, *DDX1* amplification has not been identified in the absence of

MYCN amplification. Thus, *MYCN* is primarily responsible for the aggressive nature of the neuroblastomas with amplification at the 2p24.1 locus, but *DDX1* may contribute to the highly malignant nature of some *MYCN* amplified tumors.

4.3 Amplification and expression of other oncogenes

No other human oncogene has been shown to be consistently mutated, overexpressed or amplified in neuroblastoma. *NRAS* was originally identified as a transforming gene in a neuroblastoma cell line and ras cooperates with MycN to transform embryonic fibroblasts (Yancopoulos *et al.*, 1985). In addition, targeting of *HRAS* overexpression to the neuroectoderm of mice causes ganglioneuromas and occasional neuroblastomas (Sweetser *et al.*, 1997). *HRAS* overexpression has been associated with a favorable outcome in neuroblastomas (Tanaka *et al.*, 1991; Tanaka *et al.*, 1998). However, activating ras mutations in neuroblastoma primary tumors are only rarely observed (Ireland, 1989; Moley *et al.*, 1991). Coamplification of *MYCL* or *MDM2* with *MYCN* has been observed in a few neuroblastoma cell lines (Corvi *et al.*, 1995; Jinbo *et al.*, 1989; von Roy *et al.*, 1995b) and recent CGH studies have identified novel regions of amplification at 2p13–14, 2p23, 3q24–26, 4q33–35 and 6p11–22 in occasional tumors (Brinkschmidt *et al.*, 1997; Lastowska *et al.*, 1997; Plantaz *et al.*, 1997). However, the biological significance of these observations has not yet been defined.

4.4 Unbalanced gain of 17q

Recurrent abnormalities of the long arm of chromosome 17 were identified by analysis of Giemsa-banded karyotypes derived from neuroblastoma primary tumors and cell lines (Gilbert *et al.*, 1984). It has become apparent recently that gain of 17q genetic material is perhaps the most common genetic abnormality in primary neuroblastomas (*Figure 1*). Unbalanced 1;17 translocations occur frequently in primary neuroblastomas (Savelyeva *et al.*, 1994; Van Roy *et al.*, 1994) and often result in loss of distal 1p with concomitant gain of distal 17q material (Van Roy *et al.*, 1997). However, the 17q translocation breakpoints are heterogeneous and often involve other partner chromosomes, particularly 11q (Lastowska *et al.*, 1997). Recently, comparative genomic hybridization (CGH) analyses have shown that 17q21-qter gain occurs in 50–75% of primary tumors (*Figure 1*) (Altura *et al.*, 1997; Brinkschmidt *et al.*, 1997; Lastowska *et al.*, 1997; Plantaz *et al.*, 1997). Bown and colleagues demonstrated that 54% of 313 neuroblastomas analyzed at diagnosis had unbalanced 17q21-qter gain (Bown *et al.*, 1999). A 25 megabase region at 17q23.1–17qter was shown to be the smallest region of gain by FISH mapping (Meddeb *et al.*, 1996). Unbalanced 17q gain is associated with adverse prognostic features and may therefore be useful for treatment stratification (Caron, 1995; Lastowska *et al.*, 1997c). Thus, this genomic region is the likely location of a gene that contributes to neuroblastoma tumorigenesis when present in increased copy number or overexpressed.

4.5 Increased DNA content

The majority of neuroblastoma cell lines and advanced primary tumors have either a near-diploid or near-tetraploid DNA content. In contrast, favorable neuroblastomas, especially those from infants, usually have a hyperdiploid or near-triploid DNA index. The di-/tetraploid tumors are characterized by chromosomal rearrangements including amplification, deletion and unbalanced translocations. In contrast, the hyperdiploid tumors typically have whole chromosome gains with few structural rearrangements. It therefore appears that there is a fundamental difference between these types of neuroblastomas, with the di-/tetraploid tumors demonstrating genomic instability or defective repair, while the hyperdiploid tumors have a basic defect in the machinery of mitosis and chromosomal segregation (Brodeur et al., 1997). Look and colleagues first noted the association of hyperdiploidy with favorable prognostic features (Look et al., 1984), especially in children less than two years of age (Look et al., 1991). Flow cytometric analysis of tumor cell nuclei remains a relatively simple method for measuring DNA content that can lend prognostic information in this age group, although multivariate analyses will be required to show whether it is independent of more specific genetic prognostic variables (Bourhis et al., 1991; Bowman et al., 1997; Christiansen et al., 1995; Katzenstein et al., 1998).

5. Alterations in gene expression

5.1 Neurotrophin signaling pathways

Because of the central role of neurotrophin signaling in normal neuronal development, there has been considerable interest in alterations in these pathways during neuroblastoma neoplastic transformation. Neurotrophins critical to the development of the sympathetic nervous system include nerve growth factor (NGF), brain-derived neurotrophic factor (BDNF), neurotrophin-3 (NT-3) and neurotrophin-4 (NT-4). Neurotrophin signaling is mainly mediated through the Trk family of tyrosine kinases. To date, three high affinity neurotrophin receptor genes relevant to neuroblastoma have been cloned (*TRKA, TRKB* and *TRKC*), and their expression is developmentally regulated (Barbacid, 1995). TrkA serves as the high-affinity receptor for NGF, and developing neurons differentiate in response to this ligand or enter an apoptotic pathway if NGF is withdrawn. TrkB binds both BDNF and NT-4 with high-affinity and TrkC is the high affinity receptor for NT-3.

Differential expression of the neurotrophin receptors is strongly correlated with the biological and clinical features of neuroblastomas. Some primary neuroblastomas differentiate *in vitro* in the presence of NGF but die in its absence. Transfection of *TRKA* into a non-TrkA-expressing neuroblastoma cell line restores the ability to differentiate in response to NGF (Lavenius et al., 1995; Lucarelli et al., 1997). *TRKA* expression is inversely related to disease stage and *MYCN* amplification status (Nakagawara et al., 1992). Thus, high *TRKA* expression is a marker of 'favorable' neuroblastomas and is correlated with an increased probability of disease survival (Combaret et al., 1997; Kogner et al., 1993; Nakagawara et al., 1993; Suzuki et al., 1993; Svensson et al., 1997; Tanaka et al., 1995). In contrast, full length TrkB is expressed preferentially in advanced stage, *MYCN*-amplified neuroblastomas (Nakagawara et al., 1994). Many of these tumors also express *BDNF*, establishing an autocrine pathway promoting cell growth and survival (Matsumoto *et al.*, 1995; Nakagawara *et al.*, 1994). *TRKB* is either expressed in low amounts, or as the truncated isoform (lacking the tyrosine kinase domain), in biologically favorable tumors. Lastly, *TRKC* is expressed in favorable neuroblastomas, essentially all of which also express *TRKA* (Ryden *et al.*, 1996; Svensson *et al.*, 1997; Yamashiro *et al.*, 1996).

It therefore appears that the expression pattern of neurotrophin receptors at the time of malignant transformation has a critical influence on tumor behavior. An evolving model of neurotrophin ligand/receptor interactions in neuroblastomas speculates that low stage tumors, particularly those in infants, usually express high levels of TrkA (Brodeur *et al.*, 1997). If NGF is present, TrkA-expressing cells will terminally differentiate. In contrast, if NGF is limiting, TrkA expressing cells will enter a programmed cell death pathway (Nakagawara *et al.*, 1997). In contrast, *MYCN* amplified, advanced stage tumors usually express TrkB and rely on autocrine production of BDNF to maintain a growth promoting signal, even in the absence of exogenous neurotrophins.

5.2 Overexpression of multidrug resistance genes

Many neuroblastoma cell lines derived from tumors at relapse show significantly increased resistance to standard chemotherapeutic agents compared to those established at diagnosis, so acquired drug resistance is an important cause of neuroblastoma treatment failures (Keshelava *et al.*, 1997; Kuroda *et al.*, 1991). The most well-characterized genes responsible for a multidrug resistance phenotype are (1) the *PGY1* gene (previously named *MDR1*), at chromosome sub-band 7q21.1 (Bradshaw and Arceci, 1998); and (2); *MRP* encodes the multidrug resistance-associated protein. Both of these genes are efflux pumps that can render a cell resistant to natural product drugs when overexpressed (Roninson, 1991). *PGY1* expression in primary neuroblastomas increases following exposure to chemotherapy (Bates *et al.*, 1991; Bourhis *et al.*, 1989; Goldstein *et al.*, 1990). However, the contribution of *de novo PGY1* overexpression to neuroblastoma behavior has been controversial (Chan *et al.*, 1991; Dhooge *et al.*, 1997; Nakagawara *et al.*, 1990; Norris et al., 1996). Norris and colleagues, in a study of 60 untreated neuroblastomas, showed that *MRP* expression was strongly correlated with *MYCN* expression and patient survival (Norris *et al.*, 1996). These data, and the identification of a Myc protein binding site in the *MRP* promoter region (Zhu and Center, 1994), suggested a potential interaction between the MycN protein and *MRP* expression. Transfection of *MYCN* antisense RNA into a neuroblastoma cell line that expressed both *MRP* and *MYCN* caused down regulation of *MRP* mRNA to undetectable levels (Norris *et al.*, 1997). Therefore, MycN overexpression may transcriptionally activate *MRP* and lead to the drug resistance phenotype, even in untreated primary tumors.

5.3 Apoptotic signaling pathways

Neuroblastoma has the highest rate of spontaneous regression observed in human cancers. Infants with stage 4S neuroblastoma often have initial progression of metastatic disease, followed by rapid tumor involution. Indeed, spontaneous regression has been seen in tumors of infants with other stages as well. Delayed activation of developmentally programmed apoptotic pathways has been proposed as an explanation for this phenomenon (Oue *et al.*, 1996). Entrance into a programmed cell death pathway can originate with exogenous or endogenous signals. For example, NGF withdrawal is a major signal for apoptosis in the developing nervous system and mediates the elimination of redundant cells. Other cell surface proteins are involved with initiation of apoptosis in neuronal cells and neuroblastomas (reviewed by Brodeur and Castle, 2000). Members of the tumor necrosis factor receptor (TNFR) family, such as p75[LNTR] and CD95/Fas, as well as members of the retinoic acid receptor family, can mediate the induction of apoptosis in some neuroblastoma cell lines (Bunone *et al.*, 1997; Fulda *et al.*, 1997). In addition, increased CD95 expression appears to be an essential component of chemotherapy-induced apoptosis in neuroblastomas (Fulda *et al.*, 1997).

Intracellular molecules responsible for relaying the apoptotic signal include the Bcl-2 family of proteins. Apoptosis-suppressing genes such as *BCL2* and *BCLX* are highly expressed early in neuronal ontogeny. *BCL2* is highly expressed in most neuroblastoma cell lines (Dole *et al.*, 1995; Hanada *et al.*, 1993; Reed *et al.*, 1991) and primary tumors (Castle *et al.*, 1993; Hoehner *et al.*, 1995; Ikeda *et al.*, 1995; Ikegaki *et al.*, 1995; Ramani and Lu, 1994), and the level of expression is inversely related to the proportion of cells undergoing apoptosis and the degree of cellular differentiation (Hoehner *et al.*, 1997; Ikeda *et al.*, 1995; *Oue et al.*, 1996). There have been conflicting reports regarding the correlation between the level of expression of Bcl-2 in primary tumors and prognostic variables (Castle *et al.*, 1993; Ikegaki *et al.*, 1995; Mejia *et al.*, 1998; Ramani and Lu, 1994; Tonini *et al.*, 1997b), but overall the evidence suggests that there is no significant correlation. However, the Bcl-2 family of proteins may play an important role in acquired resistance to chemotherapy (Dole *et al.*, 1994; Dole *et al.*, 1995).

Finally, caspases are proteolytic enzymes responsible for the execution of the apoptotic signal. Recently, two studies have shown that increased expression of interleukin 1b-converting enzyme (ICE, caspase-1) and other caspases in primary neuroblastoma is associated with favorable biological features and improved disease outcome (Ikeda *et al.*, 1997; Nakagawara *et al.*, 1997; Posmantur *et al.*, 1997). Another recent study suggested that caspase 8 is down-regulated in most tumors and cell lines with *MYCN* amplification (Teitz *et al.*, 2000). These observations are consistent with the hypothesis that neuro-blastomas prone to undergoing apoptosis are more likely to spontaneously regress and/or respond well to cytotoxic agents, whereas caspase-deficient cells may exhibit drug resistance.

5.4 Metastasis-suppressing genes

Despite the fact that about 50% of patients have disseminated disease at diagnosis, little is known about the biology of neuroblastoma invasion and metastasis. The cell surface glycoprotein CD44 has been postulated to influence tumor cell adhesion and has been shown to be overexpressed in several murine and human cancers (Gunthert *et al.*, 1991; Tanabe *et al.*, 1993). However, the converse is true for neuroblastoma, with undetectable CD44 expression in the vast majority of neuroblastoma cell lines (Shtivelman and Bishop, 1991) and primary tumors with dissemination at diagnosis (Favrot *et al.*, 1993; Gross *et al.*, 1994). CD44 expression is inversely correlated with *MYCN* amplification and has been suggested to be independently prognostic for decreased survival probability (Combaret *et al.*, 1996; Favrot *et al.*, 1993).

The nucleoside diphosphate kinase gene family may play a role in metastasis suppression and provides another example of altered expression of a putative oncoprotein in neuroblastoma, but again in a pattern opposite that seen in other human malignancies. *NME1* (NM23-H1) encodes the nucleoside diphosphate kinase A protein (nm23A), which was originally identified on the basis of its reduced expression in murine and human metastasis systems (Bevilacqua *et al.*, 1989; Scambia *et al.*, 1996; Rosengard *et al.*, 1989). However, increased nm23A expression was noted in advanced (Stage 3 and 4) primary neuroblastomas (Hailot *et al.*, 1991; Leone *et al.*, 1993). In addition, missense mutations in a highly conserved areas of the coding region near the catalytic domain (serine 120 → glycine) were identified in a subset of these advanced tumors with increased expression (Chang *et al.*, 1994; Leone *et al.*, 1993). Overexpression of *NME1* has been correlated with a decreased probability of survival (Leone *et al.*, 1993; Takeda et al., 1996). The mutated nm23-A protein has both decreased enzymatic activity and alterations in its normal protein binding capabilities (Chang *et al.*, 1996), suggesting that overexpression of mutated nm23-A may act in a dominant-negative fashion.

Finally, activation of matrix degrading proteolytic enzymes, such as the matrix metaloproteinases (MMPs), may be required for local invasion and metastasis. *MMP9* (gelatinase B), one of best studies MMPs, appears to be selectively overexpressed in tissue extracts from Stage 4 primary neuroblastomas, as opposed to localized tumors, and *MMP9* up-regulation occurs primarily in the surrounding stromal cells (Sugiura *et al.*, 1998).

5.5 Telomerase

The integrity of chromosomal ends is maintained by the ribonucleoprotein telomerase. Increased telomerase activity is detectable in most cancer cells and appears to be a prerequisite for malignant transformation (Kim *et al.*, 1994). Hiyama and colleagues were the first to show that telomerase expression was detectable in the vast majority of neuroblastomas (96%) but not in ganglioneuromas or normal adrenal tissue (Hiyama *et al.*, 1995). In addition, very high levels of telomerase activity may correlate with adverse prognostic features and poorer survival probability (Brinkschmidt *et al.*, 1998; Hiyama *et al.*, 1997; Reynolds *et al.*, 1997). Therefore, although elevated telomerase expression may simply be a marker of escape from cellular senescence, markedly increased levels may be associated with genomic instability and an increased likelihood of additional mutational events. Semi-quantitative detection of telomerase RNA and/or protein subunit expression should be analyzed prospectively to determine if expression independently predicts for aggressive clinical behavior.

6. Genetic model of neuroblastoma development

In summary, there is increasing evidence for two or three genetic subsets of neuroblastomas that are highly predictive of clinical behavior. One recently proposed classification takes into account abnormalities of 1p, *MYCN* copy number and assessment of DNA content, and distinct genetic subsets of neuroblastomas can be identified (*Figure 2, Table 2*)(Brodeur, 1998; Brodeur and Ambros, 2000; Brodeur *et al.*, 1997).

The first group (Type 1) is characterized by mitotic dysfunction leading to a hyperdiploid or near-triploid modal karyotype, with few if any cytogenetic rearrangements. These tumors lack specific genetic changes like *MYCN* amplification or 1p LOH, and they have high *TRKA* expression. These patients are generally less than 1 year of age with localized disease and a very good prognosis. Most of the infants detected by the neuroblastoma screening studies (see below) fall into this category.

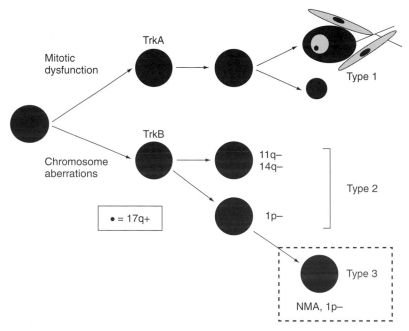

Figure 2. Genetic model of neuroblastoma development. According to this model, all neuroblastomas have a common precursor (NB) and may have a common mutation (the one responsible for familial neuroblastoma). However, a commitment is made to develop into one of three major types. The first type is characterized by mitotic dysfunction leading to a hyperdiploid or near-triploid modal karyotype (3N) with whole chromosome gains, but few if any structural cytogenetic rearrangements. These tumors usually express high levels of *TRKA*, so they are prone to either differentiation or apoptosis, depending on the presence or absence of NGF in their microenvironment. The second type generally has a near-diploid (2N) or near-tetraploid karyotype but is characterized by gross chromosomal aberrations. No consistent abnormality has been identified to date, but 17q gain is very common, and allelic loss of 1p, 11q and/or 14q is seen frequently. The third type is related to the second but is characterized by *MYCN* amplification. These tumors frequently have allelic loss of 1p. The latter tumors frequently express *TRKB* plus BDNF, probably representing an autocrine survival pathway. Thus, neuroblastoma represents fundamentally two major types and three subtypes, but they may all arise from a common precursor cell.

Table 2. Genetic/clinical subtypes of neuroblastoma (Brodeur, 1998; Brodeur and Ambros, 2000; Brodeur and Maris, 2000; Brodeur *et al.*, 1997; Maris and Matthay, 1999)

Feature	Type 1	Type 2	Type 3
MYCN	Normal	Normal	Amplified
DNA ploidy	Hyperdiploid	Near diploid	Near diploid
	Near triploid	Near tetraploid	Near tetraploid
17q gain	Rare	Common	Common
11q, 14q LOH	Rare	Common	Rare
1p LOH	Rare	Uncommon	Common
TRKA exp.	High	Low or absent	Low or absent
TRKB exp.	Truncated	Low or absent	High (full length)
TRKC exp.	High	Low or absent	Low or absent
Age	Usually < 1 y	Usually > 1 y	Usually 1–5 y
Stage	Usually 1, 2, 4	Usually 3, 4	Usually 3, 4
3-y survival	95%	~50% ~25%	

The second group is characterized by gross chromosomal aberrations and they generally have a near-diploid karyotype. No consistent abnormality has been identified to date, but 17q gain is common. Within this type, two subsets can be distinguished.

1. One subset (Type 2) is characterized by 11q deletion, 14q deletion, or other changes, but they lack *MYCN* amplification and generally lack 1p LOH. Patients with these tumors are generally older with more advanced stages of disease that is slowly progressive but often fatal.
2. The second subset (Type 3) is characterized by amplification of *MYCN*, usually with 1p36 LOH and high *TRKB* expression. These patients are generally between 1 and 5 years of age with advanced stages of disease that is rapidly progressive and frequently fatal.

It is unknown if a tumor from one type ever converts to a less favorable type, but current evidence suggests that they are genetically distinct. More precise definition of the molecular changes in neuroblastomas should result in the identification of the critical genes, proteins and pathways responsible for initiation or maintenance of the malignant state. Novel therapeutic approaches aimed at these targets should lead to the development of more effective and less toxic therapies for neuroblastoma patients.

Acknowledgments

Dr. Maris is supported in part by grant numbers R01-CA78545 and U10-CA78966 from the National Institutes of Health, Bethesda, MD. Dr. Brodeur is supported in part by grant number RO1-CA39771, PO1-NS34514 from the National Institutes of Health, Bethesda, MD, and the Audrey E. Evans Endowed Chair in Molecular Oncology. Some of this work has been published previously (Brodeur and Maris, 2000; Maris and Matthay, 1999).

References

Altura, R.A., Maris, J.M., Li, H., Boyett, J.M., Brodeur, G.M. and Look, A.T. (1997) Novel regions of chromosomal loss in familial neuroblastoma by comparative genomic hybridization. *Genes Chrom. Cancer* **19**: 176–184.

Ambros, P.F., Amann, G., Ambros, I.M. *et al.* (2000) Intratumoural heterogeneity of 1p deletions and MYCN amplification in neuroblastomas. *Med. Pediatr. Oncol.* **35**: (in press).

Aso, T., Lane, W.S., Conaway, J.W. and Conaway, R.C. (1995) Elongin (SIII): a multisubunit regulator of elongation by RNA polymerase II. *Science* **269**: 1439–1443.

Bader, S.A., Fasching, C., Brodeur, G.M. and Stanbridge, E.J. (1995) Dissociation of suppression of tumorigenicity and differentiation in vitro effected by transfer of single human chromosomes into human neuroblastoma cells. *Cell Growth Diff.* **2**: 245–255.

Bates, S.E., Shieh, C.Y. and Tsokos, M. (1991) Expression of Mdr-1/P-glycoprotein in human neuroblastoma. *Am. J. Pathol.* **139**: 305–315.

Beltinger, C.P., White, P.S., Sulman, E.P., Maris, J.M. and Brodeur, G.M. (1995) No CDKN2 mutations in neuroblastomas. Cancer Res. 55: 2053–2055.

Beltinger, C.P. White, P.S., Maris, J.M. *et al.* (1996) Physical mapping and genomic structure of the human TNFR2 gene. *Genomics* **35**: 94–100.

Bevilacqua, G., Sobel, M.E., Liotta, L.A. and Steeg, P.S. (1989) Association of low NM23 RNA levels in human primary infiltrating ductal breast carcinomas with lymph node involvement and other histopathological indicators of high metastatic potential. *Cancer Res.* **49**: 5185–5190.

Biedler, J.L., Helson, L. and Spengler, B.A. (1993) Morphology and growth, tumorigenicity, and cytogenetics of human neuroblastoma cells in continuous culture. *Cancer Res.* **33**: 2643–2652.

Biegel J.A., White, P.S., Marshall, H.N., Fujimori, M., Zackai, E.H., Scher, C.D., Brodeur, G.M. and Emanuel, B.S. (1993) Constitutional 1p36 deletion in a child with neuroblastoma. *Am. J. Hum. Genet.* **52**: 176–182.

Blatt, J. and Hamilton, R.L. (1998) Neurodevelopmental anomalies in children with neuroblastoma. *Cancer* **82**: 1603–1608.

Blatt, J., Olshan, A.F., Lee, P.A. and Ross, J.L. (1997) Neuroblastoma and related tumors in Turner's syndrome. *J. Pediatr.* **131**: 666–670.

Bolande, R.P. (1974) The neurocristopathies: A unifying concept of disease arising in neural crest maldevelopment. *Hum. Pathol.* **5**: 409–429.

Bolande, R. and Towler, W.F. (1970) A possible relationship of neuroblastoma to von Recklinghausen's disease. *Cancer* **26**: 162–175.

Bordow, S.B., Norris, M.D., Haber, P.S., Marshall, G.M. and Haber, M. (1998) Prognostic significance of MYCN oncogene expression in childhood neuroblastoma. *J. Clin. Oncol.* **16**: 3286–3294.

Bourhis, J., Benard, J., Hartmann, O., Boccon-Gibod, L., Lemerle, J. and Riou, G. (1989) Correlation of MDR1 gene expression with chemotherapy in neuroblastoma. *J. Natl Cancer Inst.* **81**: 1401–1405.

Bourhis, J., Dominici, C., McDowell, H. *et al.* (1991) N-myc genomic content and DNA ploidy in stage IVS neuroblastoma. *J. Clin. Oncol.* **9**: 1371–1375.

Bower, R.J. and Adkins, J.C. (1980) Ondine's curse and neurocristopathy. *Clin. Pediatr.* **19**: 665–668.

Bowman, L.C., Castleberry, R.P., Cantor, A. *et al.* (1997) Genetic staging of unresectable or metastatic neuroblastoma in infants: a Pediatric Oncology Group study. *J. Natl. Cancer Inst.* **89**: 373–380.

Bown, N., Cotterill, S., Lastowska, M. *et al.* (1999) Gain of chromosome arm 17q and adverse outcome in patients with neuroblastoma. *New Engl. J. Med.* **340**: 1954–1961.

Bradshaw, D.M. and Arceci, R.J. (1998) Clinical relevance of transmembrane drug efflux as a mechanism of multidrug resistance. *J. Clin. Oncol.* **16**: 3674–3690.

Brinkschmidt, C., Christiansen, H., Terpe, H.J., Simon, R., Boecker, W., Lampert, F. and Stoerkel, S. (1997) Comparative genomic hybridization (CGH) analysis of neuroblastomas – an important methodological approach in paediatric tumour pathology. *J. Pathol.* **181**: 394–400.

Brinkschmidt, C., Poremba, C., Christiansen, H, Simon, R., Schafer, K.L. Terpe, H.J., Lampert, F., Boecker, W. and Dockhorn-Dworniczak, B. (1998) Comparative genomic hybridization and telomerase activity identify two biologically different groups of 4s neuroblastomas. *Brit. J. Cancer* **77**: 2223–2229.

Brodeur, G.M. (1998) Clinical and biological aspects of neuroblastoma. In: *The Genetic Basis of Human Cancer* (eds. Vogelstein, B. and Kinzler, K.W.), pp. 691–711. McGraw-Hill, New York, NY.

Brodeur, G.M. and Ambros, P.F. (2000) Genetic and biological markers of prognosis in neuroblastoma. In: *Neuroblastoma* (eds Brodeur, G.M., Sawada, T., Tsuchida, Y. and Voute, P.A.), pp. 355–369. Elsevier Science, B.V., Amsterdam, The Netherlands.

Brodeur, G.M. and Castle, V.P. (2000) The role of apoptosis in human neuroblastomas. *Curr. Opin. Cell Biol.* (in press).

Brodeur, G.M. and Fong, C.T. (1989) Molecular biology and genetics of human neuroblastoma. *Cancer Genet. Cytogenet.* **41**: 153–174.

Brodeur, G.M. and Maris, J.M. (2000) Neuroblastoma. In: *Principles and Practice of Pediatric Oncology* (eds. Pizzo, P.A. and Poplack, D.G.), (in press). Lippincott-Raven, Philadelphia.

Brodeur, G.M. and Seeger, R.C. (1986) Gene amplification in human neuroblastomas: Basic mechanisms and clinical implications. *Cancer Genet. Cytogenet.* **19**: 101–111.

Brodeur, G.M., Sekhon, G.S. and Goldstein, M.N. (1977) Chromosomal aberrations in human neuroblastomas. *Cancer* **40**: 2256–2263.

Brodeur, G.M., Green, A.A., Hayes, F.A., Willliams, K.J., Williams, D.L. and Tsiatis, A.A. (1981) Cytogenetic features of human neuroblastomas and cell lines. *Cancer Res.* **41**: 4678–4686.

Brodeur, G., Seeger, R.C., Schwab, M., Varmus, H.E. and Bishop, J.M. (1984) Amplification of N-*myc* in untreated human neuroblastomas correlates with advanced disease stage. *Science* **224**: 1121–1124.

Brodeur, G.M., Hayes, F.A., Green, A.A., Casper, J.T., Wasson, J., Wallach, S. and Seeger, R.C. (1987) Consistent N-myc copy number in simultaneous or consecutive neuroblastoma samples from sixty individual patients. *Cancer Res.* **47**: 4248–4253.

Brodeur, G.M., Maris, J.M., Yamashiro, D.J., Hogarty, M.D. and White, P.S. (1997) Biology and genetics of human neuroblastomas. *J. Pediatr. Hematol. Oncol.* **19**: 93–101.

Bunin, G.R., Ward, E., Kramer, S., Rhee, C.A. and Meadows, A.T. (1990) Neuroblastoma and parental occupation. *Am. J. Epidemiol.* **131**: 776–780.

Bunone, G., Mariotti, A., Compagni, A., Morandi, E. and Della Valle, G. (1997) Induction of apoptosis by p75 neurotrophin receptor in human neuroblastoma cells. *Oncogene* **14**: 1463–1470.

Caron, H. (1995) Allelic loss of chromosome 1 and additional chromosome 17 material are both unfavourable prognostic markers in neuroblastoma. *Med. Pediatr. Oncol.* **24**: 215–221.

Caron, H., van Sluis, P., van Hoeve, M. *et al.* (1993) Allelic loss of chromosome 1p36 in neuroblastoma is of preferential maternal origin and correlates with N-myc amplification. *Nature Genet.* **4**: 187–190.

Caron, H., van Sluis, P., de Kraker, J. *et al.* (1996a) Allelic loss of chromosome 1p as a predictor of unfavorable outcome in patients with neuroblastoma. *N. Engl. J. Med.* **334**: 225–230.

Caron, H., van Sluis, P., Buschman, R. *et al.* (1996b) Allelic loss of the short arm of chromosome 4 in neuroblastoma suggests a novel tumour suppressor gene locus. *Hum. Genet.* **97**: 834–837.

Castle, V.P., Heidelberger, K.P., Bromberg, J., Ou, X., Dole, M. and Nunez, G. (1993) Expression of the apoptosis-suppressing protein bcl-2, in neuroblastoma is associated with unfavorable histology and N-myc amplification. *Am. J. Pathol.* **143**: 1543–1550.

Chan, H.S., Haddad, G., Thorner, P.S., DeBoer, G., Lin, Y.P., Ondrusek, N., Yeger, H. and Ling, V. (1991) P-glycoprotein expression as a predictor of the outcome of therapy for neuroblastoma. *N. Engl. J. Med.* **325**: 1608–1614.

Chan, H.S.L., Gallie, B.L., DeBoer, G., Haddad, G., Ikegaki, N., Dimitroulakos, J., Yeger, H. and Ling, V. (1997) MYCN protein expression as a predictor of neuroblastoma prognosis. *Clin. Cancer Res.* **3**: 1669–1706.

Chang, C.L., Zhu, X.X., Thoraval, D.H., Ungar, D., Rawwas, J., Hora, N., Strahler, J.H., Hanash, S.M. and Radany, E. (1994) Nm23-H1 mutation in neuroblastoma. *Nature* **370**: 335–336.

Chang, C.L., Strahler, J.R., Thoraval, D.H., Qian, M.G., Hinderer, R. and Hanash, S.M. (1996) A nucleoside diphosphate kinase A (nm23-H1) serine 120 → glycine substitution in advanced stage neuroblastoma affects enzyme stability and alters protein–protein interaction. *Oncogene* **12**: 659–667.

Cheng, N.G., Van Roy, N., Chan, A., Beitsma, M., Westerveld, A., Speleman, F. and Versteeg, R. (1995) Deletion mapping in neuroblastoma cell lines suggests two distinct tumor suppressor genes in the 1p35-36 region, only one of which is associated with N-myc amplification. *Oncogene* **10**: 291–297.

Chompret, A., de Vathaire, F., Brugieres, L., Abel, A., Raquin, M.A., Hartmann, O., Feunteun, J. and Bonaiti-Pellie, C. (1998) Excess of cancers in relatives of patients with neuroblastoma. *Med. Pediatr. Oncol.* **31**: 211A.

Christiansen, H., Sahin, K., Berthold, F., Hero, B., Terpe, H.J. and Lampert, F. (1995) Comparison of DNA aneuploidy, chromosome 1 abnormalities, MYCN amplification and CD44 expression as prognostic factors in neuroblastoma. *Eur. J. Cancer* **31A**: 541–544.

Cohn, S.L., Salwen, H., Quasney, M.W. *et al.* (1990) Prolonged N-Myc protein half-life in a neuroblastoma cell line lacking N-Myc amplification. *Oncogene* **5**: 1821–1827.

Combaret, V., Gross, N., Lasset, C., Frappaz, D., Peruisseau, G., Philip, T., Beck, D. and Favrot, M.C. (1996) Clinical relevance of CD44 cell-surface expression and N-myc gene amplification in a multicentric analysis of 121 pediatric neuroblastomas. *J. Clin. Oncol.* **14**: 25–34.

Combaret, V., Gross, N., Lasset, C. *et al.* (1997) Clinical relevance of TRKA expression on neuroblastoma: comparison with N-MYC amplification and CD44 expression. *Brit. J. Cancer* **75**: 1151–1155.

Corvi, R., Savelyeva, L., Breit, S., Wenzel, A., Handgretinger, R., Barak, J., Oren, M., Amler, L. and Schwab, M. (1995) Non-syntenic amplification of *MDM2* and *MYCN* in human neuroblastoma. *Oncogene* **10**: 1081–1086.

D'Angio, G.J., Evans, A.E. and Koop, C.E. (1971) Special pattern of widespread neuroblastoma with a favourable prognosis. *Lancet* **1**: 1046–1049.

Deed, R.W., Hirose, T., Mitchell, E.L., Santibanez-Koref, M.F. and Norton, J.D. (1994) Structural organisation and chromosomal mapping of the human ID-3 gene. *Gene* **151**: 309–314.

Dhooge, C.R., De Moerloose, B.M., Benoit, Y.C., Van Roy, N., Philippe and Laureys, G.G. (1997) Expression of the MDR 1 gene product P-glycoprotein in childhood neuroblastoma. *Cancer* **80**: 01250–1257.

Dole, M., Nunez, G., Merchant, A.K., Maybaum, J., Rode, C.K., Bloch, C.A. and Castle, V.P. (1994) Bcl-2 inhibits chemotherapy-induced apoptosis in neuroblastoma. *Cancer Res.* **54**: 3253–3259.

Dole, M.G., Jasty, R., Cooper, M.J., Thompson, C.B., Nunez, G. and Castle, V.P. (1995) Bcl-xl is expressed in neuroblastoma cells and modulates chemotherapy-induced apoptosis. *Cancer Res.* **55**: 2576–2582.

Ejeskar, K., Aburatani, H., Abrahamsson, J., Kogner, P. and Martinsson, T. (1998) Loss of heterozygosity of 3p markers in neuroblastoma tumours implicate a tumour-suppressor locus distal to the FHIT gene. *Brit. J. Cancer* **77**: 1787–1791.

Enomoto, H., Ozaki, T., Takahashi, E. *et al.* (1994) Identification of human DAN gene, mapping to the putative neuroblastoma tumor suppressor locus. *Oncogene* **9**: 2785–2791.

Fan, Y.S., Jung, J. and Hamilton, B. (1999) Small terminal deletion of 1p and duplication of 1q: cytogenetics, FISH studies, and clinical observations at newborn and at age 16 years. *Am. J. Med. Genet.* **86**: 118–123.

Favrot, M.C., Combaret, V. and Lasset, C. (1993) CD44 – a new prognostic marker for neuroblastoma. *N. Engl. J. Med.* **329**: 1965.

Fong, C.T., Dracopoli, N.C., White, P.S., Merrill, P.T., Griffith, R.C., Housman, D.E. and Brodeur, G.M. (1989) Loss of heterozygosity for the short arm of chromosome 1 in human neuroblastomas: Correlation with N-*myc* amplification. *Proc. Natl Acad. Sci. USA* **86**: 3753–3757.

Fong, C.T., White, P.S., Peterson, *et al.* (1992) Loss of heterozygosity for chromosomes 1 or 14 defines subsets of advanced neuroblastomas. *Cancer Res.* **52**: 1780–1785.

Fulda, S., Sieverts, H., Friesen, C., Herr, I. and Debatin, K.M. (1997) The CD95 (APO-1/Fas0) system mediates drug-induced apoptosis in neuroblastsoma cells. *Cancer Res.* **57**: 3823–3829.

Gaisie, G. (1989) Hirschsprungs disease, Ondine's curse, and neuroblastoma – manifestations of neurocristopathy [letter; comment]. *Pediatr. Radiol.* **20**: 136.

Grandori, C. and Eisenman, R.N. (1997) Myc target genes. *Trends Biochem. Sci.* **22**: 177–181.

Gehring, M., Berthold, F., Edler, L., Schwab, M. and Amler, L.C. (1995) The 1p deletion is not a reliable marker for the prognosis of patients with neuroblastoma. *Cancer Res.* **55**: 5366–5369.

George, R.E., Kenyon, R.M., McGuckin, A.G., Malcolm, A.J., Pearson, A.D. and Lunec, J. (1996) Investigation of co-amplification of the candidate genes ornithine decarboxylase, ribonucleotide reductase, syndecan-1 and a DEAD box gene, DDX1, with N-myc in neuroblastoma. United Kingdom Children's Cancer Study Group. *Oncogene* **12**: 1583–1587.

George, R.E., Thomas, H., McGuckin, A.G., Angus, B., Pearson, A.D.J. and Lunec, J. (2000) The DDX1 gene which is frequently co-amplified with MYCN in neuroblastoma is itself oncogenic. *Med. Pediatr. Oncol.* **35**: (in press).

Gilbert, F., Feder, M., Balaban, G., Brangman, D., Lurie, D.K., Podolsky, R., Rinaldt, V., Vinikoor, N. and Weisband, J. (1984) Human neuroblastomas and abnormalities of chromosomes 1 and 17. *Cancer Res.* 44: 5444–5449.

Goldman, S.C., Chen, C.Y., Lansing, T.J., Gilmer, T.M. and Kastan, M.B. (1996) The p53 signal transduction pathway is intact in human neuroblastoma despite cytoplasmic localization. *Am. J. Pathol.* 148: 1381–1385.

Goldstein, L.J., Fojo, A.T., Ueda, K., Crist, W., Green, A., Brodeur, G., Pastan, I. and Gottesman, M.M. (1990) Expression of the multidrug resistance, MDR1, gene in neuroblastomas. *J. Clin. Oncol.* 8: 128–136.

Gross, N., Beretta, C., Peruisseau, G., Jackson, D., Simmons, D. and Beck, D. (1994) CD44H expression by human neuroblastoma cells: relation to MYCN amplification and lineage differentation. *Cancer Res.* 54: 4238–4242.

Gunthert, U., Hofmann, M., Rudy, W. *et al.* (1991) A new variant of glycoprotein CD44 confers metastatic potential to rat carcinoma cells. *Cell* 65: 13–24.

Guo, C., White, P.S., Weiss, M.J. *et al.* (1999) Allelic deletion at 11q23 is common in MYCN single copy neuroblastomas. *Oncogene* 18: 4948–4957.

Guo, C., White, P.S., Hogarty, M.D. and Maris, J.M. (2000) Deletion of 11q23 is a frequent event in the evolution of MYCN single-copy high-risk neuroblastomas. *Med. Pediatr. Oncol.* 35: (in press).

Hailat, N., Keim, D.R., Melhem, R.F. *et al.* (1991) High levels of p19/nm23 protein in neuroblastoma are associated with advanced stage disease and with N-myc gene amplification. *J. Clin. Invest.* 88: 341–345.

Hanada, M., Krajewski, S., Tanaka, S., Cazals-Hatem, D., Spengler, B.A., Ross, R.A., Biedler, J.L. and Reed, J.C. (1993) Regulation of Bcl-2 oncoprotein levels with differentiation of human neuroblastoma cells. *Cancer Res.* 53: 4978–4986.

Hiyama, E., Hiyama, K., Yokoyama, T., Matsuura, Y., Piatyszek, M.A. and Shay, J.W. (1995) Correlating telomerase activity levels with human neuroblastoma outcomes. *Nature Med.* 1: 249–255.

Hiyama, E., Hiyama, K., Ohtsu, K., Yamaoka, H., Ichikawa, T., Shay, J.W. and Yokoyama, T. (1997) Telomerase activity in neuroblastoma: is it a prognostic indicator of clinical behaviour? *Eur. J. Cancer* 33: 1932–1936.

Hoehner, J.C., Hedborg, F., Wiklund, H.J., Olsen, L. and Pahlman, S. (1995) Cellular death in neuroblastoma: in situ correlation of apoptosis and bcl-2 expression. *Int. J. Cancer* 62: 19–24.

Hoehner, J.C., Gestblom, C., Olsen, L. and Pahlman, S. (1997) Spatial association of apoptosis-related gene expression and cellular death in clinical neuroblastoma. *Brit. J. Cancer* 75: 1185–1194.

Hofstra, R.M., Cheng, N.C., Hansen, C. *et al.* (1996) No mutations found by RET mutation scanning in sporadic and hereditary neuroblastoma. *Hum. Genet.* 97: 362–364.

Hogarty, M.D., White, P.S., Sulman, E.P. and Brodeur, G.M. (1998) Mononucleotide repeat instability is infrequent in neuroblastoma. *Cancer Genet. Cytogenet.* 106: 140–143.

Hoshi, M., Otagiri, N., Shiwaku, H.O., Asakawa, S., Shimizu, N., Kaneko, Y., Ohi, R., Hayashi, Y. and Horii, A. (2000) Detailed deletion mapping of chromosome bank 14q32 in human neuroblastoma defines a 1.1-Mb region of common allelic loss [In Process Citation]. *Br. J. Cancer* 82: 1801–1807.

Hosoi, G., Hara, J., Okamura, T., Osugi, Y., Ishihara, S., Fukuzawa, M., Okada, A., Okada, S. and Tawa, A. (1994) Low frequency of the p53 gene mutations in neuroblastoma. *Cancer* 73: 3087–3093.

Ikeda, H., Hirato, J., Akami, M., Matsuyama, S., Suzuki, N., Takahashi, A. and Kuroiwa, M. (1995) Bcl-2 oncoprotein expression and apoptosis in neuroblastoma. *J. Pediatr. Surg.* 30: 805–808.

Ikeda, H., Nakamura, Y., Hiwasa, T., Sakiyama, S., Kuida, K., Su, M.S. and Nakagawara, A. (1997) Interleukin-1 beta convertin enzyme (ICE) is preferentially expressed in neuroblastomas with favourable prognosis. *Eur. J. Cancer* 33: 2081–2083.

Ikegaki, N., Katsumata, M., Tsujimoto, Y., Nakagawara, A. and Brodeur, G.M. (1995) Relationship between bcl-2 and myc gene expression in human neuroblastoma. *Cancer Lett.* 91: 161–168.

Ireland, C.M. (1989) Activated N-Ras oncogenes in human neuroblastoma. *Cancer Res.* 49: 5530–5533.

Iwamoto, T., Taniguchi, M., Wajjwalku, W., Nakashima, I. and Takahashi, M. (1993) Neuroblastoma in a transgenic mouse carrying a metallothionein/ret fusion gene. *Brit. J. Cancer* 67: 504–507.

Jinbo, T., Iwamura, Y., Kaneko, M. and Sawaguchi, S. (1989) Coamplification of the *LMYC* and *NMYC* oncogenes in a neuroblastoma cell line. *Cancer Res.* 80: 299–303.

Johnson, M., Look, A., DeClue, J., Valentine, M. and Lowy, D. (1993) Inactivation of the NF1 gene in human melanoma and neuroblastoma cell lines without impaired regulation of GTP-Ras. *Proc. Natl Acad. Sci. USA* **90:** 5539–5543.

Kaghad, M., Bonnet, H., Yang, A. *et al.* (1997) Monoallelically expressed gene related to p53 at 1p36, a region frequently deleted in neuroblastoma and other human cancers. *Cell* **90:** 809–819.

Kaplan, D.R., Hempstead, B.L., Martin-Zanca, D., Chao, M.V. and Parada, L.F. (1991) The Trk proto-oncogene product: a signal transducing receptor for nerve growth factor. *Science* **252:** 554–558.

Katzenstein, H.M., Bowman, L.C., Brodeur, G.M. *et al.* (1998) Prognostic significance of age, MYCN oncogene amplification, tumor cell ploidy, and histology in 110 infants with stage D(S) neuroblastoma: the pediatric oncology group experience – a pediatric oncology group study. *J. Clin. Oncol.* **16:** 2007–2017.

Keshelava, N., Seeger, R.C. and Reynolds, C.P. (1997) Drug resistance in human neuroblastoma cell lines correlates with clinical therapy. *Eur. J. Cancer* **33:** 2002–2006.

Keshelava, N., Zuo, J.J., Luna, M.C., Waidyaratne, N.S., Trisch, T.J., Gomer, C.J. and Reynolds, C.P. (2000) P53 mutations and loss of function confer multi-drug resistance in neuroblastoma. *Med. Pediatr. Oncol.* **35:** (in press).

Kim, N.W., Piatyszek, M.A., Prowse, K.R. *et al.* (1994) Specific association of human telomerase activity with immortal cells and cancer. *Science* **266:** 2011–2015.

Knudson, A.G., Jr (1971) Mutation and cancer: statistical study of retinoblastoma. *Proc. Natl Acad. Sci. USA* **68:** 820–823.

Knudson, A.G., Jr. and Amromin, G.D. (1966) Neuroblastoma and ganglioneuroma in a child with multiple neurofibromatosis. Implications for the mutational origin of neuroblastoma. *Cancer* **19:** 1032–1037.

Knudson, A.G., Jr. and Strong, L.C. (1972) Mutation and cancer: neuroblastoma and pheochromocytoma. *Am. J. Hum. Genet.* **24:** 514–532.

Kogner, P., Barbany, G., Dominici, C., Castello, M.A., Raschella, G. and Persson, H. (1993) Coexpression of messenger RNA for TRK protooncogene and low affinity nerve growth factor receptor in neuroblastoma with favorable prognosis. *Cancer Res.* **53:** 2044–2050.

Kohl, N.E., Kanda, N., Schreck, R.R., Bruns, G., Latt, S.A., Gilbert, F. and Alt, F.W. (1983) Transposition and amplification of oncogene-related sequences in human neuroblastomas. *Cell* **35:** 359–367.

Komuro, H., Hayashi, Y., Kawamura, M. *et al.* (1993) Mutations of the p53 gene are involved in Ewing's sarcomas but not in neuroblastomas. *Cancer Res.* **53:** 5284–5288.

Kong, X.T., Choi, S.H., Inoue, A. *et al.* (1997) Expression and mutational analysis of the DCC, DPC4, and MADR2/JV18-1 genes in neuroblastoma. *Cancer Res.* **57:** 3772–3778.

Kuroda, H., Sugimoto, T., Ueda, K., Tsuchida, S., Horii, Y., Inazawa, J., Sato, K. and Sawada, T. (1991) Different drug sensitivity in two neuroblastoma cell lines established from the same patient before and after chemotherapy. *Int. J. Cancer* **47:** 732–737.

Kushner, B.H., Hajdu, S.I. and Helson, L. (1985) Synchronous neuroblastoma and von Recklinghausen's disease: a review of the literature. *J. Clin. Oncol.* **3:** 117–120.

Kushner, B.H., Gilbert, F. and Helson, L. (1986) Familial neuroblastoma. Case reports, literature review, and etiologic considerations. *Cancer* **57:** 1887–1893.

Lahti, J.M., Valentine, M., Xiang, J., Jones, B., Amann, J., Grenet, J., Richmond, G., Look, A.T. and Kidd, V.J. (1994) Alterations in the PITSLRE protein kinase gene complex on chromosome 1p36 in childhood neuroblastoma. *Nature Genet.* **7:** 370–375.

Lastowska, M., Roberts, P., Pearson, A.D., Lewis, I., Wolstenholme, J. and Bown, N. (1997a) Promiscuous translocations of chromosome arm 17q in human neuroblastomas. *Genes Chrom. Cancer* **19:** 143–149.

Lastowska, M., Nacheva, E., McGuckin, A., Curtis, A., Grace, C., Pearson, A. and Bown, N. (1997b) Comparative genomic hybridization study of primary neuroblastoma tumors. United Kingdom Children's Cancer Study Group. *Genes Chrom. Cancer* **18:** 162–169.

Lastowskac, M., Cotterill, S., Pearson, A.D., Roberts, P., McGuckin, A., Lewis, I. and Bown, N. (1997) Gain of chromosome arm 17q predicts unfavourable outcome in neuroblastoma patients. U.K. Children's Cancer Study Group and the U.K. Cancer Cytogenetics Group. *Eur. J. Cancer* **33:** 1627–1633.

Laureys, G., Speleman, F., Opdenakker, G., Benoit, Y. and Leroy, J. (1990) Constitutional translocation t(1;17)(p63;q12-21) in a patient with neuroblastoma. *Genes Chrom. Cancer* **2:** 252–254.

Lavenius, E., Gestblom, C., Johansson, I., Nanberg, E. and Pahlman, S. (1995) Transfection of TRK-A into human neuroblastoma cells restores their ability to differentiate in response to nerve growth factor. *Cell Growth Diff.* **6**: 727–736.

Leone, A., Seeger, R.C., Hong, C.M., Hu, Y.Y., Arboleda, M.J., Brodeur, G.M., Stram, D., Slamon, D.J. and Steeg, P.S. (1993) Evidence for nm23 RNA overexpression, DNA amplification and mutation in aggressive childhood neuroblastomas. *Oncogene* **8**: 855–865.

Look, A.T., Hayes, F.A., Nitschke, R., McWilliams, N.B. and Green, A.A. (1984) Cellular DNA content as a predictor of response to chemotherapy in infants with unresectable neuroblastoma. *N. Engl. J. Med.* **311**: 231–235.

Look, A.T., Hayes, F.A., Shuster, J.J., Douglass, E.C., Castleberry, R.P., Bowman, L.C., Smith, E.I. and Brodeur, G.M. (1991) Clinical relevance of tumor cell ploidy and N-myc gene amplification in childhood neuroblastoma: a pediatric oncology group study. *J. Clin. Oncol.* **9**: 581–591.

Lucarelli, E., Kaplan, D. and Thiele, C.J. (1997) Activation of trk-A but not trk-B signal transduction pathway inhibits growth of neuroblastoma cells. *Eur. J. Cancer* **33**: 2068–2070.

Manhani, R., Cristofani, L.M., Odone Filho, V. and Bendit, I. (1997) Concomitant p53 mutation and MYCN amplification in neuroblastoma. *Med. Pediatr. Oncol.* **29**: 206–207.

Maris, J.M. and Matthay, K.K. (1999) Molecular biology of neuroblastoma. *J. Clin. Oncol.* **17**: 2264–2279.

Maris, J.M. and Tonini, G.P. (2000) Genetics of familial neuroblastoma. In: *Neuroblastoma* (eds. Brodeur, G.M., Sawada, T., Tsuchida, Y. and Voute, P.A.), pp. 125–136. Elsevier Science B.V., Amsterdam.

Maris, J.M., White, P.S., Beltinger, C.P., Sulman, E.P., Castleberry, R.P., Shuster, J.J., Look, A.T. and Brodeur, G.M. (1995) Significance of chromosome 1p loss of heterozygosity in neuroblastoma. *Cancer Res.* **55**: 4664–4669.

Maris, J.M., Kyemba, S.M., Rebbeck, T.R. *et al.* (1996a) Familial predisposition to neuroblastoma does not map to chromosome band 1p36. *Cancer Res.* **56**: 3421–3425.

Maris, J.M., Jensen, S.J., Sulman, E.P. Beltinger, C.P., Gates, K., Allen, C., Biegel, J.A., Brodeur, G.M. and White, P.S (1996b) Cloning, chromosomal localization, physical, mapping, and genomic characterization of *HKR3*. *Genomics* **35**: 289–298.

Maris, J.M., Chatten, J., Meadows, A.T. Biegel, J.A. and Brodeur, G.M. (1999a) Familial neuroblastoma: a three-generation pedigree and a further association with Hirschsprung disease. *Med. Pediatr. Oncol.* **28**: 11–15.

Maris, J.M., Kyemba, S.M., Rebbeck, T.R., White, P.S., Sulman, E.P., Sulman, E.P., Jensen, S.J., Allen, C., Biegel, J.A. and Brodeur, G.M. (1997b) Molecular genetic analysis of familial neuroblastoma. *Eur. J. Cancer* **33**: 1923–1928.

Maris, J.M., Weiss, M.J., Guo, C., White, P.S., Hogarty, M.D., Urbanek, M., Rebbeck, T.R. and Brodeur, G.M. (2000a) Evidence for a hereditary neuroblastoma predisposition locus (HNB1) at 16p12-13. *Proc. Natl Acad. Sci. USA* **40**: (in press).

Maris, J.M., Weiss, M.J., Guo, C. *et al.* (2000b) Loss of heterozygosity at 1p36 independently predicts for disease progression, but not decreased overall survival probability, in neuroblastoma patients: A Children's Cancer Group Study. *J. Clin. Oncol.* **18**: 1888–1899.

Marshall, B., Isidro, G., Martins, A.G. and Boavida, M.G. (1997) Loss of heterozygosity at chromosome 9p21 in primary neuroblastomas: evidence for two deleted regions. *Cancer Genet. Cytogenet.* **96**: 134–139.

Marsters, S.A., Sheridan, J.P., Donahue, C.J., Pitti, R.M., Gray, C.L., Goddard, A.D., Bauer, K.D. and Ashkenazi, A. (1996) Apo-3, a new member of the tumor necrosis factor receptor family, contains a death domain and activates apoptosis and NF-kappa B. *Curr. Biol.* **6**: 1669–1676.

Martinsson, T., Sjoberg, R.M., Hedborg, F. and Kogner, P. (1995) Deletion of chromosome 1p loci and microsatellite instability in neuroblastomas analyzed with short-tandem repeat polymorphisms. *Cancer Res.* **55**: 5681–5686.

Martinsson, T., Sjoberg, R.M., Hedborg, F. and Kogner, P. (1997) Homozygous deletion of the neurofibromatosis-1 gene in the tumor of a patient with neuroblastoma. *Cancer Genet. Cytogenet.* **95**: 183–189.

Matsumoto, K., Wada, R.K., Yamashiro, J.M., Kaplan, D.R. and Thiele, C.J. (1995) Expression of brain-derived neurotrophic factor and p145TrkB affects survival, differentiation, and invasiveness of human neuroblastoma cells. *Cancer Res.* **55**: 1798–1806.

Mead, R.S. and Cowell, J.K. (1995) Molecular characterization of a (1;10)(p22;q21) constitutional translocation from a patient with neuroblastoma. *Cancer Genet. Cytogenet.* **81**: 151–157.

Meddeb, M., Danglot, G., Chudoba, I. *et al.* (1996) Additional copies of a 25 Mb chromosomal region originating from 17q23.1-17qter are present in 90% of high-grade neuroblastomas. *Genes Chrom. Cancer* **17:** 156–165.

Mejia, M.C., Navarro, S., Pellin, A., Castel, V. and Llombart-Bosch, A. (1998) Study of bcl-2 protein expression and the apoptosis phenomenon in neuroblastoma. *Anticancer Res.* **18:** 801–806.

Meltzer, S.J., O'Doherty, S.P., Frantz, C.N. *et al.* (1996) Allelic imbalance on chromosome 5q predicts long-term survival in neuroblastoma. *Brit. J. Cancer* **74:** 1855–1861.

Mertens, F., Johansson, B. and Mitelman, F. (1997) Chromosomal imbalance maps of malignant solid tumors: A cytogenetic survey of 3185 neoplasms. *Cancer Res.* **57:** 2765–2780.

Moley, J.F., Brother, M.B., Wells, S.A., Spengler, B.A., Biedler, J.L. and Brodeur, G.M. (1991) Low frequency of ras gene mutations in neuroblastomas, pheochromocytomas, and medullary thyroid cancers. *Cancer Res.* **51:** 1596–1599.

Moll, U.M., Ostermeyer, A.G., Haladay, R., Winkfield, B., Frazier, M. and Zambetti, G. (1996) Cytoplasmic sequestration of wild-type p35 protein impairs the G1 checkpoint after DNA damage. *Mol.Cell. Biol.* **16:** 1126–1137.

Nakagawara, A., Kadomatsu, K., Sato, S., Kohno, K., Takano, H., Akazawa, K., Nose, Y. and Kuwano, M. (1990) Inverse correlation between expression of multidrug resistance gene and N-Myc oncogene in human neuroblastomas. *Cancer Res.* **50:** 3043–3047.

Nakagawara, A., Arima, M., Azar, C.G., Scavarda, N.J. and Brodeur, G.M. (1992) Inverse relationship between Trk expression and N-Myc amplification in human neuroblastomas. *Cancer Res.* **52:** 1364–1368.

Nakagawara, A., Arima-Nakagawara, M., Scavarda, N.J., Azar, C.G., Cantor, A.B. and Brodeur, G.M. (1993) Association between high levels of expression of the TRK gene and favorable outcome in human neuroblastoma. *N. Engl. J. Med.* **328:** 847–854.

Nakagawara, A., Azar, C.G., Scavarda, N.J. and Brodeur, G.M. (1994) Expression and function of TRK-B and BDNF in human neuroblastomas. *Mol. Cell. Biol.* **14:** 759–767.

Nakagawara, A., Nakamura, Y., Ikeda, H., Hiwasa, T., Kuida, K., Su, M.S., Zhao, H., Cnaan, A. and Sakiyama, S. (1997) High levels of expression and nuclear localization of interleukin-1 beta converting enzyme (ICE) and CPP32 in favorable human neuroblastomas. *Cancer Res.* **57:** 4578–4584.

Negroni, A., Scarpa, S., Romeo, A., Ferrari, S., Modesti, A. and Raschella, G. (1991) Decrease of proliferation rate and induction of differentiation by a mycn antisense DNA oligomer in a human neuroblastoma cell line. *Cell Growth Diff.* **2:** 511–518.

Nisen, P.D., Waber, P.G., Rich, M.A., Pierce, S., Garvin, J.R., Jr., Gilbert, F. and Lanzkowsky, P. (1988) N-myc oncogene RNA expression in neuroblastoma. *J. Natl Cancer Inst.* **80:** 1633–1637.

Norris, M.D., Bordow, S.B., Marshall, G.M., Haber, P.S., Cohn, S.L. and Haber, M. (1996) Expression of the gene for multidrug-resistance-associated protein and outcome in patients with neuroblastoma. *N. Engl. J. Med.* **334:** 231–238.

Norris, M.D., Bordow, S.B., Haber, P.S. *et al.* (1997) Evidence that the MYCN oncogene regulates MRP gen expression in neuroblastoma. *Eur. J. Cancer* **33:** 1911–1916.

Oue, T., Fukuzawa, M., Kusafuka, T., Kohmoto, Y., Imura, K., Nagahara, S. and Okada, A. (1996) In situ detection of DNA fragmentation and expression of bcl-2 in human neuroblastoma: relation to apoptosis and spontaneous regression. *J. Pediatr. Surg.* **31:** 251–257.

Plantaz, D., Mohapatra, G., Matthay, K.K., Pellarin, M., Seeger, R.C. and Feuerstein, B.G. (1997) Gain of chromosome 17 is the most frequent abnormality detected in neuroblastoma by comparative genomic hybridization. *Am. J. Pathol.* **150:** 81–89.

Posmantur, R., McGinnis, K., Nadimpalli, R., Gilbertsen, R.B. and Wang, K.K. (1997) Characterization of CPP32-like protease activity following apoptotic challenge in SH-SY5Y neuroblastoma cells. *J. Neurochem.* **68:** 2328–2337.

Ramani, P. and Lu, Q.L. (1994) Expression of bcl-2 gene product in neuroblastoma. *J. Pathol.* **172:** 273–278.

Reale, M.A., Reyes-Mugica, M., Pierceall, W.E., Rubinstein, M.C., Hedrick, L., Cohn, S.L., Nakagawara, A., Brodeur, G.M. and Fearon, E.R. (1996) Loss of DCC expression in neuroblastoma is associated with disease dissemination. *Clin. Cancer Res.* **2:** 1097–1102.

Reed, J.C., Meister, L., Tanaka, S., Cuddy, M., Yum, S., Geyer, C. and Pleasure, D. (1991) Differential expression of Bcl2 protooncogene in neuroblastoma and other human tumor cell lines of neural origin. *Cancer Res.* **51:** 6529–6538.

Reiter, J.L. and Brodeur, G.M. (1996) High-resolution mapping of a 130-kb core region of the MYCN amplicon in neuroblastomas. *Genomics* **32**: 97–103.

Reiter, J.L. and Brodeur, G.M. (1998) MYCN is the only highly expressed gene from the core amplified domain in human neuroblastomas. *Genes Chrom. Cancer* **23**: 134–140.

Reynolds, C.P., Zuo, J.J., Kim, N.W., Wang, H., Lukens, J.N., Matthay, K.K. and Seeger, R.C. (1997) Telomerase expression in primary neuroblastomas. *Eur. J. Cancer* **33**: 1929–1931.

Roberts, T., Chernova, O. and Cowell, J.K. (1998) NB4S, a member of the TBC1 domain family genes, is truncated as a result of a constitutional t(1;10)(p22;q21) chromosome translocation in a patient with stage 4S neuroblastoma. *Hum. Mol. Genet.* **7**: 1169–1178.

Ronison, I.B. (1991) *Molecular and Cellular Biology of Multidrug Resistance in Tumor Cells*. Plenum Press: New York, NY.

Rosengard, A.M., Krutzsch, H.C., Shearn, A., Biggs, J.R., Barker, E., Margulies, I.M., King, C.R., Liotta, L.A. and Steeg, P.S. (1989) Reduced NM23/Awd protein in tumour metastasis and aberrant drosophila development. *Nature* **342**: 177–180.

Rubie, H., Delattre, O., Hartmann, O. *et al.* (1997) Loss of chromosome 1p may have a prognostic value in localised neuroblastoma: results of the French NBL 90 Study. Neuroblastoma Study Group of the Societe Francaise d'Oncologie Pediatrique (SFOP). *Eur. J. Cancer* **33**: 1917–1922.

Ryden, M., Sehgal, R., Dominici, C., Schilling, F.H., Ibanez, C.F. and Kogner, P. (1996) Expression of mRNA for the neurotrophin receptor trkC in neuroblastomas with favourable tumour stage and good prognosis. *Brit. J. Cancer* **74**: 773–779.

Saito, M., Helin, K., Valentine, M.B., Griffith, B.B., Willman, C.L., Harlow, E. and Look, A.T. (1995) Amplification of the E2F1 transcription factor gene in the HEL erythroleukemia cell line. *Genomics* **25**: 130–138.

Satge, D., Sasco, A.J., Carlsen, N.L. *et al.* (1998) A lack of neuroblastoma in Down syndrome: a study from 11 European countries. *Cancer Res.* **58**: 448–452.

Savelyeva, L., Corvi, R. and Schwab, M. (1994) Translocation involving 1p and 17q is a recurrent genetic alteration of human neuroblastoma cells. *Am. J. Hum. Genet.* **55**: 334–340.

Scambia, G., Ferrandian, G., Marone, M., Benedetti Panici, P., Giannitelli, C., Piantelli, M., Leone, A. and Mancuso, S. (1996) NM23 in ovarian cancer: correlation with clinical outcome and other clinicopathologic and biochemical prognostic parameters. *J. Clin. Oncol.* **14**: 334–342.

Schleiermacher, G., Peter, M., Michon, J., Hugot, J.P., Vielh, P., Zucker, J.M., Magdelenat, H., Thomas, G. and Delattre, O. (1994) Two distinct deleted regions on the short arm of chromosome 1 in neuroblastoma. *Genes Chrom. Cancer* **10**: 275–281.

Schleiermacher, G., Delattre, O., Peter, M. *et al.* (1996) Clinical relevance of loss of heterozygosity of the short arm of chromosome 1 in neuroblastoma: a single-institution study. *Int. J. Cancer* **69**: 73–78.

Schmidt, M.L., Salwen, H.R., Manohar, C.F., Ikegaki, N. and Cohn, S.L. (1994) The biological effects of antisense N-myc expression in human neuroblastoma. *Cell Growth Diff.* **5**: 171–178.

Schwab, M. and Bishop, J.M. (1988) Sustained expression of the human protooncogene Mycn rescues rata embryo cells from senescence. *Proc. Natl Acad. Sci. USA* **85**: 9585–9589.

Schwab, M., Alitalo, K., Klempnauer, K.H., Varmus, H.E., Bishop, J.M., Gilbert, F., Brodeur, G., Goldstein, M. and Trent, J. (1983) Amplified DNA with limited homology to myc cellular oncogene is share by human neuroblastoma cell lines and a neuroblastoma tumour. *Nature* **305**: 245–248.

Schwab, M., Varmus, H.E., Bishop, J.M., Grzeschik, K.H., Naylor, S.L., Sakaguchi, A.Y., Brodeur, G. and Trent, J. (1984) Chromosome localization in normal human cells and neuroblastomas of a gene related to c-myc. *Nature* **308**: 288–291.

Schwab, M., Varmus, H.E. and Bishop, J.M. (1985) Human N-myc gene contributes to neoplastic transformation of mammalian cells in culture. *Nature* **316**: 160–162.

Seeger, R.C., Brodeur, G.M., Sather, H., Dalton, A., Siegel, S.E., Wong, K.Y. and Hammond, D. (1985) Association of multiple copies of the N-myc oncogene with rapid progression of neuroblastomas. *N. Engl. J. Med.* **313**: 1111–1116.

Seeger, R.C., Wada, R., Brodeur, G.M., Moss, T.J., Bjork, R.L., Sousa, L. and Slamon, D.J. (1988) Expression of N-myc by neuroblastomas with one or multiple copies of the oncogene. *Prog. Clin. Biol. Res.* **271**: 41–49.

Shapiro, D.N., Sublett, J.E., Li, B., Valentine, M.B., Morris, S.W. and Noll, M. (1993) The gene for PAX7, a member of the paired-box-containing genes, is localized on human chromosome arm 1p36. *Genomics* **17**: 767–769.

Shtivelman, E. and Bishop, J.M. (1991) Expression of CD44 is repressed in neuroblastoma cells. *Mol. Cell. Biol.* **11**: 5446–5453.

Sivak, L.E., Tai, K.F., Smith, R.S., Dillon, P.A., Brodeur, G.M. and Carroll, W.L. (1997) Autoregulation of the human N-myc oncogene is disrupted in amplified but not single-copy neuroblastoma cells lines. *Oncogene* **15**: 1937–1946.

Slamon, D.J., Boone, T.C., Seeger, R.C., Keith, D.E., Chazin, V., Lee, H.C. and Souza, L.M. (1986) Identification and characterization of the protein encoded by the human N-Myc oncogene. *Science* **232**: 768–772.

Slavc, I., Ellenbogen, R., Jung, W.H., Vawter, G.F., Kretschmar, C., Grier, H. and Korf, B.R. (1990) Myc gene amplification and expression in primary human neuroblastoma. *Cancer Res.* **50**: 1459–1463.

Small, M.B., Hay, N., Schwab, M. and Bishop, J.M. (1987) Neoplastic transformation by the human gene N-myc. *Mol. Cell. Biol.* **7**: 1638–1645.

Srivatsan, E.S., Ying, K.L. and Seeger, R.C. (1993) Deletion of chromosome 11 and of 14q sequences in neuroblastoma. *Genes Chrom. Cancer* **7**: 32–37.

Sugiura, Y., Shimada, H., Seeger, R.C., Laug, W.E. and DeClerck, Y.A. (1998) Matrix metalloproteinases-2 and -9 are expressed in human neuroblastoma: contribution of stromal cells to their production and correlation with metastasis. *Cancer Res.* **58**: 2209–2216.

Suzuki, T., Bogenmann, E., Shimada, H., Stram, D. and Seeger, R.C. (1993) Lack of high-affinity nerve growth factor receptors in aggressive neuroblastomas. *J. Natl. Cancer Inst.* **85**: 377–384.

Svensson, T., Ryden, M., Schilling, F.H., Dominici, C., Sehgal, R., Ibanez, C.F. and Kogner, P. (1997) Coexpression of mRNA for the full-length neurotrophin receptor trk-C and trk-A in favourable neuroblastoma. *Eur. J. Cancer* **33**: 2058–2063.

Sweetser, D.A., Kapur, R.P., Froelick, G.J., Kafer, K.E. and Palmiter, R.D. (1997) Oncogenesis and altered differentiation induced by activated Ras in neuroblasts of transgenic mice. *Oncogene* **15**: 2783–2794.

Takayama, H., Suzuki, T., Mugishima, H. *et al.* (1992) Deletion mapping of chromosomes 14q and 1p in human neuroblastoma. *Oncogene* **7**: 1185–1189.

Takeda, O., Homma, C., Maseki, N., Sakurai, M., Kanda, N., Schwab, M., Nakamura, Y. and Kaneko, Y. (1994) There may be two tumor suppressor genes on chromosome arm 1p closely associated with biologically distinct subtypes of neuroblastoma. *Genes Chrom. Cancer* **10**: 30–39.

Takeda, O., Handa, M., Uehara, T., Maseki, N., Sakashita, A., Sakurai, M., Kanda, N., Arai, Y. and Kaneko, Y. (1996) An increased NM23H1 copy number may be a poor prognostic factor independent of LOH on 1p in neuroblastomas. *Brit. J. Cancer* **74**: 1620–1626.

Takita, J., Hayashi, Y., Kohno, T., Shiseki, M., Yamaguchi, N., Hanada, R., Yamamoto, K. and Yokota, J. (1995) Allelotype of neuroblastoma. *Oncogene* **11**: 1829–1834.

Takita, J., Hayashi, Y., Kohno, T., Yamaguchi, N., Hanada, R., Yamamoto, K. and Yokota, J. (1997) Deletion map of chromosome 9 and p16 (CDKN2A) gene alterations in neuroblastoma. *Cancer Res.* **57**: 907–912.

Takita, J., Hayashi, Y., Nakajima, T. *et al.* (1998) The p16 (CDKN2A) gene is involved in the growth of neuroblastoma cells and its expression is associated with prognosis of neuroblastoma patients. *Oncogene* **17**: 3137–3143.

Tanabe, K.K., Ellis, L.M. and Saya, H. (1993) Expression of CD44R1 adhesion molecule in colon carcinomas and metastases. *Lancet* **341**: 725–726.

Tanaka, T., Slamon, D.J., Shimada, H., Shimoda, H., Fujisawa, T., Ida, N. and Seeger, R.C. (1991) A significant association of Ha-ras p21 in neuroblastoma cells with patient prognosis. *Cancer* **68**: 1296–1302.

Tanaka, T., Hiyama, E., Sugimoto, T., Sawada, T., Tanabe, M. and Ida, N. (1995) A gene expression in neuroblastoma. The clinical significance of an immunohistochemical study. *Cancer* **76**: 1086–1095.

Tanaka, T., Sugimoto, T. and Sawada, T. (1998) Prognostic discrimination among neuroblastomas according to Ha-ras/Trk A gene expression. *Cancer* **83**: 1626–1633.

Teitz, T., Wei, T., Valentine, M.B. *et al.* (2000) Caspase 8 is deleted or silenced preferentially in childhood neuroblastomas with amplification of MYCN. *Nature Med.* **6**: 529–535.

The, I., Murthy, A., Hannigan, G., Jacoby, L., Menon, A., Gusella, J. and Bernards, A. (1993) Neurofibromatosis type 1 gene mutations in neuroblastoma. *Nat. Genet.* **3**: 62–66.

Theobald, M., Christiansen, H., Schmidt, A., Melekian, B., Wolkewitz, N., Christiansen, N.M., Brinksschmidt, C., Berthold, F. and Lampert, F. (1999) Sublocalization of putative tumor suppressor gene loci on chromosome arm 14q in neuroblastoma. *Genes Chrom. Cancer* **26**: 40–46.

Thompson, P.M., Seifried, B.A., Kyemba, S.K. *et al.* (2000a) Loss of heterozygosity for chromosome 14q in neuroblastoma. *Med. Pediatr. Oncol.* (in press).

Thompson, P.M., Maris, J.M., Hogarty, M.D., Seeger, R.C., Reynolds, C.P., Brodeur, G.M. and White, P.S. (2000b) Homozygous deletion of CDKN2A (p16INK4a/p14ARF) but not within 1p36 or at other tumor suppressor loci in neuroblastoma. *Cancer Res.* (in press).

Tonini, G.P., Lo Cunsolo, C., Cusano, R. *et al.* (1997a) Loss of heterozygosity for chromosome 1p in familial neuroblastoma. *Eur. J. Cancer* **33**: 1953–1956.

Tonini, G.P., Mazzocco, K., di Vinci, A., Geido, E., de Bernardi, B. and Giaretti, W. (1997b) Evidence of apoptosis in neuroblastoma at onset and relapse. An analysis of a large series of tumors. *J. Neuro-Oncol.* **31**: 209–215.

Van Roy, N., Laureys, G., Cheng, N.C., Willem, P., Opdenakker, G., Versteeg, R. and Speleman, F. (1994) 1;17 translocations and other chromosome 17 rearrangements in human primary neuroblastoma tumors and cell lines. *Genes Chrom. Cancer* **10**: 103–114.

Van Roy, N., Cheng, N.C., Laureys, G., Opdenakker, G., Versteeg, R. and Speleman, F. (1995a) Molecular cytogenetic analysis of 1;17 translocations in neuroblastoma. *Eur. J. Cancer* **31A**: 530–535.

Van Roy, N., Forus, A., Myklebost, O., Cheng, N.G., Versteeg, R. and Speleman, F. (1995b) Identification of two distinct chromosome 12-derived amplification units in neuroblastoma cell line NGP. *Cancer Genet. Cytogenet.* **82**: 151–154.

Van Roy, N., Laureys, G., Van Gele, M., Opdenakker, G., Miura, R., van der Drift, P., Chan, A., Versteeg, R. and Speleman, F. (1997) Analysis of 1;17 translocation breakpoints in neuroblastoma: implications for mapping of neuroblastoma genes. *Eur. J. Cancer* **33**: 1974–1978.

Vogan, K., Bernstein, M., Leclerc, J.M., Brisson, L., Brossard, J., Brodeur, G.M., Pelletier, J. and Gros, P. (1993) Absence of p53 gene mutations in primary neuroblastomas. *Cancer Res.* **53**: 5269–5273.

Wada, R.K., Seeger. R.C., Brodeur, G.M., Einhorn, P.A., Rayner, S.A., Tomayko, M.M. and Reynolds, C.P. (1993) Human neuroblastoma cell lines that express N-myc without gene amplification. *Cancer* **72**: 3346–3354.

Weiss, W.A., Aldape, K., Mohapatra, G., Feuerstein, B.G. and Bishop, J.M. (1997) Targeted expression of MYCN causes neuroblastoma in transgenic mice. *EMBO J.* **16**: 2985–2995.

Wenzel, A. and Schwab, M. (1995) The mycN/max protein complex in neuroblastoma. Short review. *Eur. J. Cancer* **31A**: 516–519.

Wenzel, A., Cziepluch, C., Hamann, U., Schurmann, J. and Schwab, M. (1991) The N-Myc oncoprotein is associated in vivo with the phosphoprotein Max (P20/22) in human neuroblastoma cells. *EMBO J.* **10**: 3703–3712.

White, P.S., Maris, J.M., Beltinger, C. *et al.* (1995) A region of consistent deletion in neuroblastoma maps within human chromosome 1p36.2-36.3. *Proc. Natl Acad. Sci. USA* **92**: 5520–5524.

White, P.S., Maris, J.M., Sulman, E.P. *et al.* (1997) Molecular analysis of the region of distal 1p commonly deleted in neuroblastoma. *Eur. J. Cancer* **33**: 1957–1961.

White, P.S., Thompson, P.M., Seifried, B.A. *et al.* (2000) Detailed molecular analysis of 1p36 in neuroblastoma. *Med. Pediatr. Oncol.* **35**: (in press).

Witzleben, C.L. and Landy, R.A. (1974) Disseminated neuroblastoma in a child with von Recklinghausen's disease. *Cancer* **34**: 786–790.

Yamashiro, D.J., Nakagawara, A., Ikegaki, N., Liu, X.G. and Brodeur, G.M. (1996) Expression of TrkC in favorable human neuroblastomas. *Oncogene* **12**: 37–41.

Yancopoulos, G.D., Nisen, P.D., Tesfaye, A., Kohl, N.E., Goldfarb, M.P. and Alt, F.W. (1985) N-myc can cooperate with ras to transform normal cells in culture. *Proc. Natl Acad. Sci. USA* **82**: 5455–5459.

Zhu, Q. and Center, M.S. (1994) Cloning and sequence analysis of the promoter region of the MRP gene of HL60 cells isolated for resistance to adriamycin. *Cancer Res.* **54**: 4488–4492.

16

Genetics of brain tumors

B.K. Ahmed Rasheed, Rodney N. Wiltshire and Sandra H. Bigner

1. Introduction

Among brain neoplasms, tumors of astrocytic origin have been studied extensively. Recent advances have uncovered many new genetic abnormalities and it is gradually becoming possible to correlate clinical and histological classes with distinct genetic abnormalities. Accumulated molecular data suggest that alterations of multiple genes in the cell cycle regulatory pathways play a major role in tumorigenesis and contribute to formation of subtypes. In this chapter we discuss the predominant genetic abnormalities associated with the common types of brain tumors.

2. Astrocytomas

Astrocytomas, tumors of one form of glial cell, can occur in all areas of the brain and spinal cord in children and adults. Although the vast majority of astrocytic neoplasms occur sporadically, they can be seen in patients with the Familial Adenomatous Polyposis syndrome, the Li-Fraumenni syndrome, and in central neurofibromatosis. The incidence of astrocytomas is approximately 7.0 per 100 000 (Velema and Percy, 1987). The World Health Organization (WHO) classification recognizes four grades of astrocytoma (Kleihues *et al.*, 1993). Grade I astrocytomas are slow-growing, non-infiltrative neoplasms, occurring mainly in children and young adults, and include juvenile pilocytic astrocytomas and gangliogliomas. Grade II astrocytomas (A) are mainly well-differentiated fibrillary astrocytomas, while the Grade III astrocytoma is a more aggressive neoplasm, the anaplastic astrocytoma (AA). The most malignant form of astrocytoma, the grade IV tumor, is the glioblastoma multiforme (GBM), which is the most common primary malignant brain tumor of adults (Burger *et al.*, 1991). Although these tumors most commonly occur in the cerebral hemispheres of older individuals, they can be seen throughout the brain and spinal cord of patients of all ages.

2.1 Chromosomal abnormalities in astrocytomas

Independent cytogenetic studies have failed to show any consistent abnormalities in juvenile pilocytic astrocytomas (Debiec-Rychter *et al.*, 1995; Neumann *et al.*,

Molecular Genetics of Cancer second edition, edited by J.K. Cowell.

1993). Allelotyping and CGH studies have also yielded similar results with no consistent loss or gain of any chromosomal regions (Blaeker et al., 1996; Schrock et al., 1996). Among low-grade adult astrocytomas, one of the consistent abnormalities reported is a gain or amplification of 8q. In CGH studies this abnormality is seen in approximately 30% (9/28) of tumors. Gain of 7q is also reported, although at a lesser frequency (Nishizaki et al., 1998; Sallinen et al., 1997; Schrock et al., 1996; Weber et al., 1996). In a comparison study by CGH of seven good prognosis vs four poor prognosis grade II astrocytomas, chromosomal gains were seen in both groups, whereas losses were frequent in tumors with poor prognosis (Sallinen et al., 1997). Loss of chromosome 10 alleles, overexpression of *PDGFA* ligand and receptors and 17p loss associated with *TP53* mutations are the common molecular abnormalities seen in grade II astrocytomas (Hermanson et al., 1992). Chromosome 10 LOH is reported at a much lower frequency (<35%) in grade II astrocytomas compared to higher grade tumors (>65%). However, the incidence of preferential loss of 10p loci, while retaining 10q loci, is higher in low grade astrocytomas (Ichimura et al., 1998; Kimmelman et al., 1996).

In anaplastic astrocytomas, in addition to a higher frequency of chromosome 10 deletion (>65%), losses of 9p, 13q and 17p are also common. The overall frequency of 9p loss is 18%, 13q loss is about 20% and 17p loss about 41%. These losses affect the known tumor suppressor genes *CDKN2A*, *RB1* and *TP53* respectively. Additionally, LOH on 19q (up to 50%) has also been reported in anaplastic astrocytomas (von Deimling et al., 1992b).

The most consistent chromosomal changes in glioblastomas (*Figure 1*) are gains of chromosome 7 (*Table 1*), seen in about 80% of tumors with abnormal stemlines, losses of chromosome 10 in 60% of tumors, losses of 9p in about a third of cases, and the presence of double minute chromosomes (Dmins) reported in up to 50% of cases (Bigner et al., 1988a; Hecht et al., 1995; Jenkins et al., 1989; Rey et al., 1987; Thiel et al., 1992). In the largest series of GBM tumors analyzed by CGH (72 tumors) the most common abnormality was 10q loss observed in 59 cases of which 47 cases lost a copy of the entire chromosome 10. Other frequent regions of loss included 9p (48 cases), 13q (35 cases), 6q (20 cases), 14q (20 cases) and 22q (18 cases). Gains of chromosome 7 regions were seen in 57 tumors. Twenty tumors showed gain of a copy of chromosome 19 and an additional 18 tumors gained portions of 19p. Gains of 20q was seen in 18 tumors and 17q in 13 tumors (Mohapatra et al., 1998). Huhn et al. (1999) have reported that chromosome 7 gains seen by CGH occur more frequently among tumors resistant to radiation therapy. Whole gains of chromosome were identified in 14 of 20 radiation resistant tumors compared to 3 of 10 in the radiation sensitive group of tumors (Huhn et al., 1999).

2.2 Alterations of PTEN

Loss of heterozygosity (LOH) analyses have confirmed losses of all or part of chromosome 10 in more than 90% of cases, in some series, and have narrowed the smallest region of overlapping deletion to 10q25 (Fults and Pedone, 1993; Rasheed et al., 1995). Most series have also identified a second region on 10p and a third site on proximal 10q has also been targeted by some observers (Ichimura et al., 1998; Karlbom et al., 1993; Kimmelman et al., 1996; Ransom et al., 1992b).

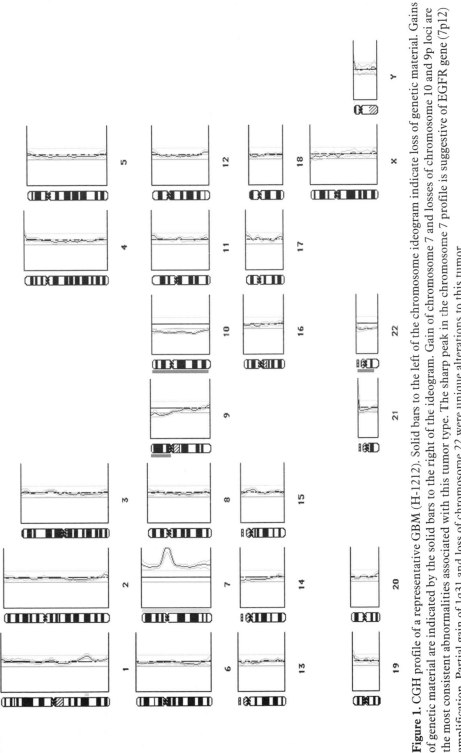

Figure 1. CGH profile of a representative GBM (H-1212). Solid bars to the left of the chromosome ideogram indicate loss of genetic material. Gains of genetic material are indicated by the solid bars to the right of the ideogram. Gain of chromosome 7 and losses of chromosome 10 and 9p loci are the most consistent abnormalities associated with this tumor type. The sharp peak in the chromosome 7 profile is suggestive of EGFR gene (7p12) amplification. Partial gain of 1q31 and loss of chromosome 22 were unique alterations to this tumor.

Table 1. Chromosomal and genetic alterations characteristic of specific types of brain tumors

Tumor type	Chromosomal or LOH abnormality	Genetic alteration
Glioblastoma	+7	Unknown
	−9p	*CDKN2A, CDKN2B*
	−10	*PTEN/MMAC1* gene mutation, *DMBT1* deletion ?
	−17p	*TP53* gene mutation
	Dmins	*EGFR* gene amplification and rearrangement
Oligodendroglioma	−1p, −19q	Unknown
Ependymoma	−22	? *NF2* gene mutation
Medulloblastoma	−17p	Unknown
	−9q	*PTCH* gene mutation
	Dmins	*MYC, NMYC* gene amplification
Meningioma	−22	*NF2* gene mutation
Schwannoma	−22	*NF2* gene mutation
Atypical teratoid/rhabdoid	−22q	*SMARCB1* deletion, mutation

Li *et al.* (1977) and Steck *et al.* (1997) have identified a gene located at 10q23 which was mutated or deleted in a subset of gliomas. This gene, called *PTEN*, for Phosphatase and Tensin homolog deleted on chromosome Ten or *MMAC1* for mutated in multiple advanced cancers, is mutated in 24–60% of glioblastomas with LOH for 10q (Fults *et al.*, 1998; Liu *et al.*, 1997; Teng *et al.*, 1997; Wang *et al.*, 1997), and in approximately 40% of prostatic and endometrial cancers (Cairns *et al.*, 1997; Maxwell *et al.*, 1998). Germline mutations of the *PTEN/MMAC1* gene are seen in Cowden disease and the Bannayan-Zonana syndrome (Marsh *et al.*, 1997, 1998). The product of this gene is a protein tyrosine phosphatase, TGFB-regulated and epithelial cell-enriched phosphatase or TEP1 (Li and Sun, 1997). Biochemical analyses have shown that PTEN can dephosphorylate the lipid second messenger Phosphatidyl inositol 3,4,5-triphosphate and act as a negative regulator of the phosphatidyl inositol 3 kinase (PI3-K) dependent pathway. Murine embryonic fibroblasts lacking *pten* have increased levels of phosphatidyl inositol triphosphate and PKB, a downstream signal target for PI3. It has also been shown that PTEN can suppress RAS transformation and there is some evidence that inactivation of PTEN and RAS activation are functionally over-lapping. It is interesting to note that *RAS* mutations are relatively rare in glioma cell lines (Tsao *et al.*, 2000). Although mutations of the *PTEN/MMAC1* gene are common in high-grade astrocytomas (glioblastomas and anaplastic astrocytomas) they are rarely seen in low-grade astrocytomas (Duerr *et al.*, 1998; Rasheed *et al.*, 1997; Schmidt *et al.*, 1999). In addition, among glioblastomas, mutations of this gene are more frequently seen in *de novo*, than in secondary tumors (Tohma *et al.*, 1998).

2.3 Other 10q genes

Although the *PTEN/MMAC1* gene is clearly implicated in a subset of gliomas, the location of this gene at 10q23, while the most frequent region of overlapping deletions in these tumors is at 10q25–6, and the observation that many astrocytomas with LOH for 10q lack mutations of this gene, has raised the possibility that another chromosome 10 gene or genes may be involved in gliomas. Candidate genes in the 10q25 region include the *MXI1*, *PAX2*, *NEURL* (human homolog of the Drosophila neuralized gene) and *DMBT1* (deleted in malignant brain tumors), located at 10q25.3–26.1 (Eagle *et al.*, 1995; Nakamura *et al.*, 1998; Stapleton *et al.*, 1993; Wechsler *et al.*, 1997). The *DMBT1* gene, which shows homology to the scavenger receptor cysteine-rich superfamily, was shown to be homozygously deleted in 9/39 glioblastomas and 2/20 medulloblastomas by Mollenhauer *et al.* and 8/21 glioblastomas in another study by Somerville *et al.* (Mollenhauer *et al.*, 1997; Somerville *et al.*, 1998). Although this gene was not expressed in 4/5 brain tumor cell lines, the lack of demonstrable point mutations in these tumors raises the possibility that this gene may not be the target of the deletions. In a study by Steck *et al.* (1999) the presence or absence of DMBT1 locus in somatic cell hybrids did not correlate with transformed phenotype. The *DMBT1* transcript was expressed in immune system cells and was similar to gp-340, a glycoprotein that is involved in the respiratory immune defense. In the normal brain, anti-DMBT1 antibodies stained a few astrocytes in the cerebellum as well as in the cerebral neocortex. Two of 10 GBMs analyzed were negative for DMBT1, the remainder showed both deficient and overexpressing cell populations. Available evidence so far suggests that DMBT1 may be involved in immune surveillance but not directly involved in glial tumorigenesis (Mollenhauer *et al.*, 2000).

TP53. LOH analyses of astrocytomas revealed that approximately 1/3 of these tumors have loss of all or part of 17p (el-Azouzi *et al.*, 1989; Frankel *et al.*, 1992; Fults *et al.*, 1989, 1992; Hermanson *et al.*, 1996; James *et al.*, 1989; Lang *et al.*, 1994a; Leenstra *et al.*, 1994; Rasheed *et al.*, 1994; Tenan *et al.*, 1994; von Deimling *et al.*, 1992a). Unlike the chromosomal deviations described above, which are seen mainly in glioblastomas, LOH for 17p occurs in astrocytomas of all grades. Point mutations of the *TP53* gene can be demonstrated in the majority of astrocytomas with 17p loss. The mutations are clustered in the same hot spots as are seen in colon, breast and lung carcinomas. The incidence of *TP53* mutations confirmed by sequence data is about 25% (73/295) in glioblastomas, 34% (49/144) in anaplastic astrocytomas and 30% (33/111) in astrocytomas (Alderson *et al.*, 1995; Chen *et al.*, 1995; Chung *et al.*, 1991; Kraus *et al.*, 1994; Kyritsis *et al.*, 1996; Lang *et al.*, 1994b; Louis, 1994; Mashiyama *et al.*, 1991). Most of the *TP53* studies have concentrated on the conserved exons 5–8, but studies which included the entire coding sequence (exons 2–11) have uncovered only a handful of mutations outside of exons 5–8. Similar to colon cancer, codons 175, 248 and 273 are frequently mutated in brain tumors, however the codon which is most frequently mutated in brain is codon 273, whereas in colon it is codon 175 (Bogler *et al.*, 1995). *TP53* mutations are associated with age of the patient. These alterations are rare among pediatric patients (Chen *et al.*, 1995; Hermanson *et al.*, 1996; Lang *et al.*, 1994b; Litofsky *et al.*, 1994; Rasheed *et al.*, 1994; Willert *et al.*, 1995) but occur in nearly 50% of tumors in the young adult, with a much lower incidence (<20%) in the patient over 50 years of age. Most of the *TP53* mutations, identified in astrocytomas are G:C→A:T transitions located at CpG sites and resemble the pattern of mutations found in colon cancer, sarcomas and lymphomas.

CDKN2A/B. Hemizygous or homozygous deletion of interferon genes was reported in glioma cell lines and in biopsies of high-grade astrocytomas (Miyakoshi *et al.*, 1990; Olopade *et al.*, 1992), but it was not clear in these early studies whether the interferon genes were the target of 9p deletions in gliomas or were simply located near the region of the target gene. Later, the *CDKN2A* and *CDKN2B* genes which are located at 9p21 were found to be homozygously deleted in various types of tumors including gliomas (Kamb *et al.*, 1994). By combining data collected on tumor biopsies in several laboratories the overall incidence of homozygous deletions is 33% (98/300) in glioblastomas; 24% in anaplastic astrocytomas (19/79); and for hemizygous deletion or LOH for 9p loci is 24% of glioblastomas and 18% of anaplastic astrocytomas. The incidence of homozygous deletions of both *CDKN2A* and *CDKN2B* is higher in xenografts, approaching 80% in some studies (Jen *et al.*, 1994). Among the 23 low-grade astrocytoma biopsies analyzed none exhibited homozygous deletion, although 5 showed LOH. Altogether there have been only three cases of mutations, all in glioblastomas (Giani and Finocchiaro, 1994; He *et al.*, 1994; Li *et al.*, 1995; Moulton *et al.*, 1995; Nishikawa *et al.*, 1995; Schmidt *et al.*, 1994; Sonoda *et al.*, 1995; Ueki *et al.*, 1994, 1996; Walker *et al.*, 1995). The high frequency of homozygous deletions on chromosome 9 and the inclusion of *CDKN2A* and *CDKN2B* gene sequences in the deleted region in most cases has led most observers to believe that *CDKN2A* and *CDKN2B* are the target suppressor genes for 9p loss in gliomas. Unlike the *TP53* gene, which usually undergoes point mutation, the most common mechanism for *CDKN2A* gene inactivation in gliomas is homozygous deletions. However, alternative mechanisms such as transcriptional silencing by hypermethylation of CpG islands may be responsible for reduced expression in some gliomas with intact *CDKN2A/CDKN2B* genes (Herman *et al.*, 1995).

2.4 Relationship between cell cycle regulators in astrocytomas

The cell cycle is regulated by hypophosphorylated RB binding to and inhibiting the transcription factor E2F, which is required for S phase entry. Specific cyclin–cdk complexes in mid to late G_1 phase hyperphosophorylate RB and release RB mediated G_1 arrest. The activity of cyclin–cdk complexes is negatively regulated by p16 the product of the *CDKN2A* gene. In addition to deletions of the *CDKN2A* and *CDKN2B* genes as discussed above, alterations of other genes involved in cell cycle regulation have been described in subsets of astrocytomas. LOH for 13q or loss of expression of the retinoblastoma (*RB1*) gene product has been described in 20–40% of glioblastomas (Fults *et al.*, 1990; He *et al.*, 1995; Henson *et al.*, 1994; James *et al.*, 1988; Ransom *et al.*, 1992b; Ueki *et al.*, 1996; Venter *et al.*, 1991), and amplification of the *CDK4* gene has been described in up to 15% of glioblastomas (Reifenberger *et al.*, 1994a; Schmidt *et al.*, 1994). Furthermore, He *et al.* (1995), Biernat *et al.* (1997) and Ueki *et al.* (1996) have shown that most glioblastomas contain only one of these three alterations: (1) *CDKN2A/CDKN2B* deletion,

(2) LOH for 13q or loss of *RB1* expression, (3) *CDK4* gene amplification or increased expression. The CDKN2A/B locus, through alternate reading frame usage codes for p14^{ARF} (encoded by exon 1B and exon 2 and 3 of *CDKN2A*), is a positive regulator of cellular p53 levels. p14^{ARF} binds to the p53/MDM2 complex and inhibits MDM2-mediated degradation of p53. Homozygous deletion of the CDKN2A/B locus can knock out both p16 as well as p14^{ARF}, thus affecting both RB1 and p53 pathways. In a large series of gliomas comprising 136 GBM, 39 AA and 15 As, Ichimura *et al.* (2000) reported an abnormality of the RB pathway in 67% of GBM and 21% of AA; and p53 pathway genes in 76% of GBM, 72% of AA and 67% of As.

2.5 EGFR amplification

The majority of glioblastomas which possess Dmins contain amplification of the *EGFR* gene (Bigner *et al.*, 1987). The *EGFR* gene has been shown to be amplified in one third to one half of glioblastomas, but only in isolated cases of anaplastic astrocytomas and rarely in other lower grade tumors. In many glioblastomas, the amplified *EFGR* gene is also rearranged (Bigner *et al.*, 1990b; Wong *et al.*, 1987). The most common class of mutants bears deletion of exons 2–7 of the gene resulting in an in-frame deletion of 801 base-pairs of coding sequence and generation of a glycine residue at the fusion point This variant receptor, designated EGFRvIII has been reported in 17–62% of glioblastomas (Ekstrand *et al.*, 1991; 1992; Humphrey *et al.*, 1990; Moscatello *et al.*, 1995; Sugawa *et al.*, 1990; Wikstrand *et al.*, 1995). The tumor cell membrane fractions, containing the mutant 140-kDa receptor, show a significant elevation in tyrosine kinase activity without its ligand (Yamazaki *et al.*, 1988). The mutant is still capable of binding with its ligand, but at a significantly reduced affinity (Batra *et al.*, 1995a).

2.6 Genetic alterations in the progression of gliomas

It has long been recognized that there are two patterns for the development of glioblastomas. The majority of these tumors occur in patients over 50 years of age, in individuals with no previous indication of a brain tumor. A second group of cases involves younger people whose glioblastomas evolve out of lower grade astrocytomas. Recent studies have provided molecular markers which in many cases distinguish between these two clinical patterns (*Table 2*). The *de novo* pathway, occurring in older patients includes tumors over 50% of which contain *EGFR* gene amplification and the majority of which lack *TP53* gene mutations. Glioblastomas evolving through progression, in contrast, seldom have *EGFR* gene amplification and more than 50% contain *TP53* gene mutations (Biernat *et al.*, 1997; Kleihues, 1998; Lang *et al.*, 1994a; Leenstra *et al.*, 1994; Rasheed *et al.*, 1994; von Deimling *et al.*, 1993; Watanabe *et al.*, 1996).

3. Oligodendrogliomas

Oligodendrogliomas are a type of glioma which occur mainly in the cerebral hemispheres of adults and are derived from the oligodendrocyte cell lineage. Well

Table 2. Relative frequency of molecular abnormalities in glioma subgroups

	High	Low
De novo glioblastoma	10q LOH *PTEN* alterations *EGFR* amplification *CDKN2A* deletions	*TP53* mutations
Progressive glioblastoma	10q LOH *TP53* mutations *PDGFRA* amplification	*PTEN* alterations *EGFR* amplification *CDKN2A* deletions
Oligodendroglioma	1p LOH 19q LOH	10q LOH *PTEN* alterations Gene amplification
Anaplastic oligodendroglioma	1p LOH 19q LOH *CDKN2A* deletions	10q LOH *PTEN* alterations Gene amplification

differentiated oligodendrogliomas which exhibit benign cytologic features are considered grade II according to the WHO classification, while anaplastic oligo-dendrogliomas which exhibit abundant mitotic activity, necrosis without pseudopalisading, glomeruloid vascular proliferation are considered grade III tumors. According to data collected by the Surveillance, Epidemiology, and End Results (SEER) project, the age adjusted incidence of oligodendroglioma is 0.33 per 100 000 (Velema and Percy, 1987).

3.1 Chromosomal abnormalities in oligodendrogliomas

Although most cytogenetic analyses of oligodendrogliomas have failed to demon-strate consistent findings, LOH studies have shown that loss of 1p and 19q occur in a substantial proportion of these tumors (*Figure 2*). Bello *et al.* (1994) reported LOH for loci on 1p in up to 100% (6/6) of oligodendrogliomas and in most (5/6) anaplastic oligodendrogliomas. Among other types of glioma only 2/11 glioblas-tomas exhibited 1p LOH suggesting that 1p LOH is characteristic of tumors of oligodendroglial origin. A high incidence of 1p loss in tumors of oligodendroglial origin has been confirmed by Reifenberger *et al.* (1994b) using LOH techniques and, by *in situ* hybridization, Hashimoto *et al.* (1995) found deletion of a 1p locus in 9/9 oligodendrogliomas (Hashimoto *et al.*, 1995; Reifenberger *et al.*, 1994b). One of the genes mapped to 1p36 is *P73* which has structural and functional simi-larity to *TP53*. No mutations were detected in the *P73* gene in a study involving 20 oligodendrogliomas with 1p LOH. A significantly favorable response to chemotherapy was reported in anaplastic oligodendrogliomas with LOH for 1p, and for both 1p and 19q. Twenty-four of 27 tumors responding to chemotherapy had 1p loss, whereas all nine tumors with no 1p LOH were chemoresistant (Cairncross *et al.*, 1998). In a related study, 7 high-grade glioma patients who had uncommonly long survival had 1p loss in their tumors. However, it may be premature to associate 1p loss to better prognosis because most of these high-grade gliomas also had some mixed oligo-astro histology (Ino *et al.*, 2000).

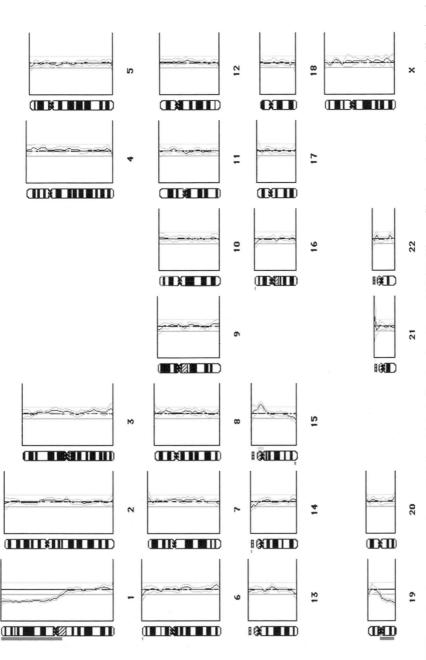

Figure 2. CGH profile of a representative well-differentiated oligodendroglioma (H-953). The profile showed a distinct pattern indicating loss of 1p and 19q as the sole abnormalities. The gain on chromosome 15 is an artifact of incomplete blocking of repetitive sequences and therefore was excluded from the analysis.

Analysis with restriction fragment length polymorphism and microsatellite markers showed loss of markers on chromosome 19 in about 63% (17/27) of grade 2 oligodendrogliomas, 75% (18/24) of grade 3 or anaplastic oligodendrogliomas, and in 48% (21/43) of mixed oligoastrocytomas. This abnormality has also been reported in about 16% (4/25) of astrocytomas, in 38% (13/34) of anaplastic astrocytomas and in 28% (37/130) of glioblastomas tested (Reifenberger et al., 1994b; Ritland et al., 1995; von Deimling et al., 1994). Loss of loci on 19q was more frequent among oligodendroglial tumors whereas astrocytic tumors mostly lost 19p alleles (Ritland et al., 1995). The 19q minimal deletion region has been mapped to a 425 kb region on 19q13.3 (Yong et al., 1995).

4. Ependymomas

Ependymomas are a category of glioma derived from the ventricular lining that can occur at many locations within the brain and spinal cord of adults and children predominantly associated with the ventricular system. Favored sites include the posterior fossa (IV ventricle) in children, the lateral ventricles in adults and the cauda equina in patients of all ages. Most lesions are classified histologically as grade II, while tumors with anaplastic features (anaplastic ependymomas, grade III) are sometimes seen. Central neurofibromatosis (NF type 2) is associated with ependymoma, most commonly of the spinal cord. SEER data indicate an age adjusted incidence of 0.18 per 100 000 (Velema and Percy, 1987).

4.1 Chromosomal abnormalities in ependymomas

The most common cytogenetic abnormality in ependymomas is loss or structural alteration of chromosome 22, seen as an isolated finding in some cases and as part of a more complex picture in others. This alteration characterizes approximately 10–20% of cases with abnormal stemlines in most cytogenetic studies and a similar incidence of chromosome 22 loss has been described in LOH studies (Agamanolis and Malone, 1995; Bijlsma et al., 1995; Bown et al., 1988; Chadduck et al., 1991; Rogatto et al., 1993; Sawyer et al., 1994; Stratton et al., 1989; Vagner-Capodano et al., 1992; Weremowicz et al., 1992; Wernicke et al., 1995). Some observers, however, have noted this abnormality in a high proportion of cases by karyotype (Chadduck et al., 1991; Rogatto et al., 1993; Wernicke et al., 1995) or by LOH analysis (James et al., 1990; Ransom et al., 1992a). Since the NF2 gene is located on 22q, it has been considered as a possible target for loss of this chromosomal region in ependymomas. However, since only one somatic mutation of this gene has been reported among the 25 ependymomas which were studied, the role of NF2 mutations in ependymomas remains speculative (Rubio et al., 1994). In addition to loss of 22q, CGH analysis has also shown non-random loss of a 6q region. Interestingly, in this study 6q was lost more often than chromsome 22. Among 23 pediatric ependymomas analyzed 6q loss was seen in five, and chromosome 22 loss in four tumors including one which exhibited both 6q and 22 loss (Reardon et al., 1999).

5. Medulloblastomas

Medulloblastomas, the most common malignant primary brain tumors of childhood, are small cell neoplasms which arise in the cerebellum. Due to the primitive morphology of the cells with lack of differentiation and their resemblance to some poorly differentiated supratentorial neoplasms, these tumors have been called 'Primitive neuroectodermal tumors (PNET)' by some observers. They are characterized by sheets of small cells with scant cytoplasm and a high mitotic rate. Medulloblastomas are usually a sporadic tumor, but can be seen in association with familial adenomatous polyposis. In the United Kingdom, the incidence of medulloblastoma has been estimated at 0.5 per 100 000 children less than 15 years (Stevens *et al.*, 1991) with an overall age-adjusted incidence reported by the Central Brain Tumor Registry of the United States (CBTRUS, 1995) of 0.2 per 100 000.

5.1 Chromosomal abnormalities in medulloblastomas

The most common specific chromosomal abnormality in medulloblastomas is loss of 17p, through formation of isochromosome 17q [i(17q)] or by unbalanced translocations (Agamanolis and Malone, 1995; Biegel *et al.*, 1989; Bigner *et al.*, 1988b; Chadduck *et al.*, 1991; Fujii *et al.*, 1994; Griffin *et al.*, 1988; Neumann *et al.*, 1993; Vagner-Capodano *et al.*, 1992). By karyotype as well as LOH studies, the incidence of this feature is approximately 30–40% (Albrecht *et al.*, 1994; Batra *et al.*, 1995b; Biegel *et al.*, 1992; Cogen *et al.*, 1990; James *et al.*, 1990; Thomas and Raffel, 1991). Allelotyping in a small number of medulloblastomas (six cases) revealed loss of 2p, 10q, 11 p or q alleles in four cases and 8p, 16q and 17p in three cases each (Blaeker *et al.*, 1996). Some of these specific chromosomal losses are also seen in CGH analysis of a larger group of primary medulloblastomas. Among 27 cases analyzed by CGH, losses of 10q (41%), 11 (41%), 16q (37%), 17p (37%) and 8p (33%) and gains of 17q (48%) and 7 (44%) were seen (Reardon *et al.*, 1997). Despite the location of the *TP53* gene on 17p, *TP53* gene mutations are uncommon in these tumors, seen in about 5% of cases (Adesina *et al.*, 1994; Badiali *et al.*, 1993; Batra *et al.*, 1995b; Biegel *et al.*, 1992; Cogen *et al.*, 1992; Ohgaki *et al.*, 1991; Raffel *et al.*, 1993; Saylors *et al.*, 1991). This observation, together with mapping of the deleted region to 17p13.1–13.3, which is distal to the *TP53* gene, suggests that another, as yet undescribed, gene is likely to be the target of 17p loss in these tumors.

5.2 Genetic alterations in medulloblastomas

Dmins are seen in about 5% of medulloblastoma biopsies, but can be identified in almost all permanent cultured cell lines and xenografts derived from these (Bigner *et al.*, 1988b, 1990a). In most samples with Dmins, amplification of the *MYC*, or less often the *NMYC* gene, can be demonstrated (Badiali *et al.*, 1991; Batra *et al.*, 1994, 1995b; Bigner *et al.*, 1990a; Friedman *et al.*, 1988; Pietsch *et al.*, 1994; Raffel *et al.*, 1990; Wasson *et al.*, 1990). The true incidence of *MYC* gene amplification is difficult to determine in these tumors because the observed incidence differs according to the method of analysis. However, a recent analysis by comparative genomic hybridization suggested that it might be as high as 18% (Schutz *et al.*, 1996).

5.3 The PTCH pathway

The gene for the Nevoid Basal cell carcinoma (Gorlin) syndrome was mapped to 9q22.3-q31 by linkage analysis. Since patients with this genetic defect are susceptible to the development of medulloblastoma, this region was investigated by LOH studies in three medulloblastomas from patients with this syndrome and 17 sporadic medulloblastomas by Schofield *et al.* (1995). They found loss of this region in both informative cases from Gorlin syndrome patients and in 3/17 (18%) sporadic tumors. All three sporadic tumors with loss were of the desmoplastic type (Schofield *et al.*, 1995). The Gorlin syndrome gene was identified as *PTCH*, the human homolog of the Drosophila patched gene by Johnson *et al.* (1996) and Hahn *et al.* (1996). Raffel *et al.* (1997) demonstrated mutations of *PTCH* in 3/5 sporadic medulloblastomas with 9q LOH. Reports from other laboratories have confirmed *PTCH* mutations in about 15% of sporadic medulloblastomas (Pietsch *et al.*, 1997; Vorechovsky *et al.*, 1997; Wolter *et al.*, 1997; Xie *et al.*, 1997). The majority of tumors containing the deletions have desmoplastic histology and exhibit LOH for 9q22. The *PTCH* gene product is a transmembrane receptor for the Sonic hedgehog (SHH) protein. This observation has prompted investigators to evaluate other members of the *PTCH* gene pathway for alterations in medulloblastoma. Reifenberger *et al.* (1998), have described one sporadic medulloblastoma with a mutation of the *SMO* gene, the product of which is known to complex with PTCH. However, Zurawel *et al.* (2000a) found no mutations in the *SHH* or the *SMO* gene in 24 tumors analyzed. Extensive screening of several other potential candidate genes in the PTCH/SMO/SHH pathway also revealed no mutations suggesting that *PTCH* is the only gene affected in this pathway in medulloblastomas. *PTCH* mutation inactivates its repressor function, increasing the expression of the transcription factor GLI1. In mice homozygous loss of *ptch* causes embryonic lethality at 9.5–10.5 days after fertilization. About 14–19% of the mice hemizygous for the *ptch* gene develop intracranial tumors resembling medulloblastoma (Goodrich *et al.*, 1997). Two different studies show that these tumors expressed normal but variable amounts of the *ptch* transcript. Despite the presence of normal *ptch* transcript the tumors showed increased levels of *gli1* mRNA and protein. These results suggested that haploinsufficiency of PTCH promotes formation of medulloblastomas in the *ptch*+/– mice (Wetmore *et al.*, 2000; Zurawel *et al.*, 2000b). In humans, *PTCH* mutations are demonstrated in about half of the tumors with 9q LOH. If haploinsufficiency operates similarly in humans, the frequency of a *PTCH* abnormality is probably higher than previously thought in medulloblastomas.

5.4 APC/β-catenin

The APC protein regulates the level of β-catenin which is important in cell–cell adhesion and an activator of the Wnt signaling pathway. Mutations in APC which lead to accumulation of β-catenin is seen in the majority of colon cancer samples.

Some colon cancer and other types of tumors have mutations in the β-catenin gene (CTNNB1) at sites that make it resistant to APC-mediated down-regulation. Medulloblastomas are often seen in Turcot syndrome, patients who carry germline mutations of the APC gene, mapped to 5q (Hamilton *et al.*, 1995). CTNNB1 has been shown to be mutated in 3/67 sporadic medulloblastomas (Zurawel *et al.*, 1998). Huang *et al.* found APC mutations in two and CTNNB1 in four of the 46 tumors examined (Huang *et al.*, 2000). APC and CTNNB1 mutations were mutually exclusive and, together, occurred in 6 of 46 (13%) of the tumors, suggesting the involvement of wingless-Wnt signaling pathway in a subset of sporadic medullo-blastomas.

6. Meningioma and schwannoma

Meningiomas are slow-growing neoplasms derived from the meningothelial cell which forms the arachnoid membrane. These tumors are composed of swirling sheets of cells with oval nuclei, often forming whorls and psammoma bodies. They are generally considered to be benign, because they can often be completely excised surgically, but occasional cases recur and can show aggressive clinical characteristics. SEER data indicate an age adjusted incidence of 0.13 per 100 000 although data reported by CBTRUS support a much higher incidence of 2.5 per 100 000 (CBTRUS, 1995; Stevens *et al.*, 1991). Schwannomas are benign neoplasms, derived from the schwann cell which forms myelin in the peripheral nervous system. They are found attached to cranial or peripheral nerves, with favored sites being the acoustic and other sensory nerves. CBTRUS data report an overall incidence of nerve sheath tumors of 0.7 per 100 000 (CBTRUS, 1995; Stevens *et al.*, 1991). Histologically they form masses of spindle-shaped cells and are usually benign. Both schwannomas, particularly bilateral lesions involving the acoustic nerves, and meningiomas are components of central neurofibromatosis or NF Type 2 and peripheral schwannomas occur in peripheral neurofibromatosis or NF Type 1.

6.1 Genetic alterations in meningiomas and schwannomas

One of the earliest chromosomal abnormalities described in a solid human tumor was monosomy for a 'G' group chromosome (Zang and Singer, 1967). With the implementation of banding techniques the missing chromosome was identified as a number 22 in the early 1970s (Mark *et al.*, 1972; Zankl and Zang, 1972). Loss or deletion of chromosome 22 is the most consistent karyotypic abnormality seen in this tumor type, occurring in about 60% of cases (Zang, 1982). LOH studies confirmed loss of 22q in 40–60% of both meningiomas and schwannomas (Cogen *et al.*, 1991; Dumanski *et al.*, 1987, 1990; Rouleau *et al.*, 1987; Seizinger *et al.*, 1987a, 1987b; Wertelecki *et al.*, 1988). This observation, taken together with the occurrence of these tumors in NF2, and linkage studies which localized the NF2 locus to 22q, raised the possibility that the *NF2* gene was the target of chromosome 22 loss in these two tumor types. Isolation of the *NF2* gene and sequencing of this gene in these tumors confirmed that 40–60% of sporadic meningiomas and schwannomas contain *NF2* gene mutations (Bijlsma *et al.*,

1994; Irving *et al.*, 1994; Jacoby *et al.*, 1994; Lekanne Deprez *et al.*, 1994; Rouleau *et al.*, 1993; Ruttledge *et al.*, 1994a, 1994b).

7. Atypical teratoid/rhabdoid tumors

Rhabdoid tumors occur in children and usually present as tumors in kidney or brain although capable of occurring in all locations of the body. In the brain these tumors may consist purely of rhabdoid cells or contain areas of rhabdoid cells adjacent to mesenchymal and epithelial tissue. Some regions of rhabdoid tumors may resemble PNET and are also designated as atypical teratoid tumors. These tumors commonly reside in the cerebellum, cerebrum, cerebellopontine angle and pineal gland. The rhabdoid tumors are extremely aggressive and usually fatal.

7.1 Genetic alterations

A commonly deleted region on 22q was identified in a series of primary renal rhabdoid tumors and cell lines (Schofield *et al.*, 1996). Subsequent studies led to the detection of mutations and identfication of *SMARCB1*, mapped to 22q11, as a candidate gene in rhabdoid tumors of the kidney and soft tissues. The *SMARCB1* gene is the human homolog of yeast *SNF5* gene. The *SNF5* gene is part of SW1/SNF5 complex and plays a role in the chromatin structure (Versteege *et al.*, 1998). Biegel *et al.* (1999) reported homozygous deletion of all exons in 2, and exon 1 in 7 of 18 CNS rhabdoid tumors analyzed. Based on the presence of the same germline mutation of *SMARCB1* in a child with CNS tumor who later developed renal tumor, it has been proposed that *SMARCB1*muations are a predisposing factor in rhabdoid tumors (Biegel *et al.*, 2000).

8. Summary

1. Gains of chromosome 7 and losses of chromosome 10, are seen in up to 80–90% of glioblastomas. Approximately 25–30% of cases with loss of chromosome 10 have mutations of the *PTEN/MMAC1* gene. About 1/3 of cases contain homozygous deletions of the *CDKN2A* gene, while some tumors with intact *CDKN2A* have loss of expression of the retinoblastoma gene or have amplification of the *CDK4* gene. Deletions of the *DMBT1* gene is present in about 20–30% of the tumors, but its role as a tumor suppressor gene remains to be established. In addition, approximately 1/2 of the cases contain amplification, often with rearrangement of the epidermal growth factor receptor (*EGFR*) gene.
2. Mutations of the *TP53* gene are more common (>50%) in low grade astrocytomas, particularly anaplastic astrocytomas, as well as low grade tumors which progress to glioblastomas.
3. Oligodendrogliomas are frequently characterized by losses of 1p and 19q. The target genes for loss of these regions remain unknown.
4. A subset of ependymomas has loss of 22q and in some series 6q. Mutation of the Neurofibromatosis Type 2 (*NF2*) gene has been described in a single case of an

ependymoma. Therefore, whether or not this gene is the target of the 22q loss in these tumors remains speculative.

5. Up to 40% of the medulloblastomas lose 17p, however, *TP53* mutations are rare in these tumors. Approximately 15% of cases have mutations of the *PTCH* gene, in association with loss of 9q. About 13% of the tumors have mutations in the *APC* or the *CTNNB1* (β-catenin) gene. The incidence of gene amplification ranged from less than 5% to 22% in medulloblastomas. The amplified gene is usually *MYC*, with a few examples of *NMYC* gene amplification.

6. Approximately 60% of meningiomas and schwannomas have loss of 22q, which is usually associated with *NF2* gene mutations.

7. Deletions or mutations of *SMARCB1* gene are seen in atypical teratoid/rhabdoid tumors.

References

Adesina, A.M., Nalbantoglu, J. and Cavenee, W.K. (1994) p53 gene mutation and mdm2 gene amplification are uncommon in medulloblastoma. *Cancer Res.* 54: 5649–5651.

Agamanolis, D.P. and Malone, J.M. (1995) Chromosomal abnormalities in 47 pediatric brain tumors. *Cancer Genet. Cytogenet.* 81: 125–134.

Albrecht, S., von Deimling, A., Pietsch, T., Giangaspero, F., Brandner, S., Kleihues, P. and Wiestler, O.D. (1994) Microsatellite analysis of loss of heterozygosity on chromosomes 9q, 11p and 17p in medulloblastomas. *Neuropathol. Appl. Neurobiol.* 20: 74–81.

Alderson, L.M., Castleberg, R.L., Harsh, G.R.T., Louis, D.N. and Henson, J.W. (1995) Human gliomas with wild-type p53 express bcl-2. *Cancer Res.* 55: 999–1001.

Badiali, M., Pession, A., Basso, G., Andreini, L., Rigobello, L., Galassi and E., Giangaspero, F. (1991) N-myc and c-myc oncogenes amplification in medulloblastomas. Evidence of particularly aggressive behavior of a tumor with c-myc amplification. *Tumori* 77: 118–121.

Badiali, M., Iolascon, A., Loda, M., Scheithauer, B.W., Basso, G., Trentini, G.P. and Giangaspero, F. (1993) p53 gene mutations in medulloblastoma. Immunohistochemistry, gel shift analysis, and sequencing. *Diagn. Mol. Pathol.* 2: 23–28.

Batra, S.K., Rasheed, B.K., Bigner, S.H. and Bigner, D.D. (1994) Oncogenes and anti-oncogenes in human central nervous system tumors. *Lab. Invest.* 71: 621–637.

Batra, S.K., Castelino-Prabhu, S., Wikstrand, C.J., Zhu, X., Humphrey, P.A., Friedman, H.S. and Bigner, D.D. (1995a) Epidermal growth factor ligand-independent, unregulated, cell-transforming potential of a naturally occurring human mutant EGFRvIII gene. *Cell Growth Differ.* 6: 1251–1259.

Batra, S.K., McLendon, R.E., Koo, J.S., Castelino-Prabhu, S., Fuchs, H.E., Krischer, J.P., Friedman, H.S., Bigner, D.D. and Bigner, S.H. (1995b) Prognostic implications of chromosome 17p deletions in human medulloblastomas. *J. Neurooncol.* 24: 39–45.

Bello, M.J., Vaquero, J., de Campos, J.M., Kusak, M.E., Sarasa, J.L., Saez-Castresana, J., Pestana, A. and Rey, J.A. (1994) Molecular analysis of chromosome 1 abnormalities in human gliomas reveals frequent loss of 1p in oligodendroglial tumors. *Int. J. Cancer* 57: 172–175.

Biegel, J.A., Rorke, L.B., Packer, R.J., Sutton, L.N., Schut, L., Bonner, K. and Emanuel, B.S. (1989) Isochromosome 17q in primitive neuroectodermal tumors of the central nervous system. *Genes Chromosomes Cancer* 1: 139–147.

Biegel, J.A., Burk, C.D., Barr, F.G. and Emanuel, B.S. (1992) Evidence for a 17p tumor related locus distinct from p53 in pediatric primitive neuroectodermal tumors. *Cancer Res.* 52: 3391–3395.

Biegel, J.A., Zhou, J.Y., Rorke, L.B., Stenstrom, C., Wainwright, L.M. and Fogelgren, B. (1999) Germ-line and acquired mutations of INI1 in atypical teratoid and rhabdoid tumors. *Cancer Res.* 59: 74–79.

Biegel, J.A., Fogelgren, B., Wainwright, L.M., Zhou, J.Y., Bevan, H. and Rorke, L.B. (2000) Germline INI1 mutation in a patient with a central nervous system atypical teratoid tumor and renal rhabdoid tumor. *Genes Chromosomes Cancer* 28: 31–37.

Biernat, W., Tohma, Y., Yonekawa, Y., Kleihues, P. and Ohgaki, H. (1997) Alterations of cell cycle regulatory genes in primary (de novo) and secondary glioblastomas. *Acta Neuropathol. (Berl)* **94**: 303–309.

Bigner, S.H., Wong, A.J., Mark, J., Muhlbaier, L.H., Kinzler, K.W., Vogelstein, B. and Bigner, D.D. (1987) Relationship between gene amplification and chromosomal deviations in malignant human gliomas. *Cancer Genet. Cytogenet.* **29**: 165–170.

Bigner, S.H., Mark, J., Burger, P.C., Mahaley, M.S. Jr., Bullard, D.E., Muhlbaier, L.H. and Bigner, D.D. (1988a) Specific chromosomal abnormalities in malignant human gliomas. *Cancer Res.* **48**: 405–411.

Bigner, S.H., Mark, J., Friedman, H.S., Biegel, J.A. and Bigner, D.D. (1988b) Structural chromosomal abnormalities in human medulloblastoma. *Cancer Genet. Cytogenet.* **30**: 91–101.

Bigner, S.H., Friedman, H.S., Vogelstein, B., Oakes, W.J. and Bigner, D.D. (1990a) Amplification of the c-myc gene in human medulloblastoma cell lines and xenografts [published erratum appears in Cancer Res 1990 Jun 15;50(12): 3809]. *Cancer Res.* **50**: 2347–2350.

Bigner, S.H., Humphrey, P.A., Wong, A.J., Vogelstein, B., Mark, J., Friedman, H.S. and Bigner, D.D. (1990b) Characterization of the epidermal growth factor receptor in human glioma cell lines and xenografts. *Cancer Res.* **50**: 8017–8022.

Bijlsma, E.K., Merel, P., Bosch, D.A., Westerveld, A., Delattre, O., Thomas, G. and Hulsebos, T.J. (1994) Analysis of mutations in the SCH gene in schwannomas. *Genes Chromosomes Cancer* **11**: 7–14.

Bijlsma, E.K., Voesten, A.M., Bijleveld, E.H., Troost, D., Westerveld, A., Merel, P., Thomas, G. and Hulsebos, T.J. (1995) Molecular analysis of genetic changes in ependymomas. *Genes Chromosomes Cancer* **13**: 272–277.

Blaeker, H., Rasheed, B.K., McLendon, R.E., Friedman, H.S., Batra, S.K., Fuchs, H.E. and Bigner, S.H. (1996) Microsatellite analysis of childhood brain tumors. *Genes Chromosomes Cancer* **15**: 54–63.

Bogler, O., Huang, H.J., Kleihues, P. and Cavenee, W.K. (1995) The p53 gene and its role in human brain tumors. *Glia* **15**: 308–327.

Bown, N.P., Pearson, A.D., Davison, E.V., Gardner-Medwin, D., Crawford, P. and Perry, R. (1988) Multiple chromosome rearrangements in a childhood ependymoma. *Cancer Genet. Cytogenet.* **36**: 25–30.

Burger, P.C., Scheithauer, B.W. and Vogel, F.S. (1991) *Surgical Pathology of the Nervous System and its Coverings.* New York: Churchill Livingstone.

Cairncross, J.G., Ueki, K., Zlatescu, M.C., et al. (1998) Specific genetic predictors of chemotherapeutic response and survival in patients with anaplastic oligodendrogliomas. *J. Natl. Cancer. Inst.* **90**: 1473–1479.

Cairns, P., Okami, K., Halachmi, S., et al. (1997) Frequent inactivation of PTEN/MMAC1 in primary prostate cancer. *Cancer Research* **57**: 4997–5000.

CBTRUS (1995): Central Brain Tumor Registry of the United States Annual Report. Chicago, Illinois.

Chadduck, W.M., Boop, F.A. and Sawyer, J.R. (1991) Cytogenetic studies of pediatric brain and spinal cord tumors. *Pediatr. Neurosurg.* **17**: 57–65.

Chen, P., Iavarone, A., Fick, J., Edwards, M., Prados, M. and Israel, M.A. (1995) Constitutional p53 mutations associated with brain tumors in young adults. *Cancer Genet. Cytogenet.* **82**: 106–115.

Chung, R., Whaley, J., Kley, N., et al. (1991) TP53 gene mutations and 17p deletions in human astrocytomas. *Genes Chromosomes Cancer* **3**: 323–331.

Cogen, P.H., Daneshvar, L., Metzger, A.K. and Edwards, M.S. (1990) Deletion mapping of the medulloblastoma locus on chromosome 17p. *Genomics* **8**: 279–285.

Cogen, P.H., Daneshvar, L., Bowcock, A.M., Metzger, A.K. and Cavalli-Sforza, L.L. (1991) Loss of heterozygosity for chromosome 22 DNA sequences in human meningioma. *Cancer Genet. Cytogenet.* **53**: 271–277.

Cogen, P.H., Daneshvar, L., Metzger, A.K., Duyk, G., Edwards, M.S. and Sheffield, V.C. (1992) Involvement of multiple chromosome 17p loci in medulloblastoma tumorigenesis. *Am. J. Hum. Genet.* **50**: 584–589.

Debiec-Rychter, M., Alwasiak, J., Liberski, P.P., Nedoszytko, B., Babinska, M., Mrozek, K., Imielinski, B., Borowska-Lehman, J. and Limon, J. (1995) Accumulation of chromosomal changes in human glioma progression. A cytogenetic study of 50 cases. *Cancer Genet. Cytogenet.* **85**: 61–67.

Duerr, E.M., Rollbrocker, B., Hayashi, Y., *et al.* (1998) PTEN mutations in gliomas and glioneuronal tumors. *Oncogene.* **16**: 2259–2264.

Dumanski, J.P., Carlbom, E., Collins, V.P. and Nordenskjold, M. (1987) Deletion mapping of a locus on human chromosome 22 involved in the oncogenesis of meningioma. *Proc. Natl Acad. Sci. USA* **84**: 9275–9279.

Dumanski, J.P., Rouleau, G.A., Nordenskjold, M. and Collins, V.P. (1990) Molecular genetic analysis of chromosome 22 in 81 cases of meningioma. *Cancer Res.* **50**: 5863–5867.

Eagle, L.R., Yin, X., Brothman, A.R., Williams, B.J., Atkin, N.B. and Prochownik, E.V. (1995) Mutation of the MXI1 gene in prostate cancer. *Nat. Genet.* **9**: 249–255.

Ekstrand, A.J., James, C.D., Cavenee, W.K., Seliger, B., Pettersson, R.F. and Collins, V.P. (1991) Genes for epidermal growth factor receptor, transforming growth factor alpha, and epidermal growth factor and their expression in human gliomas in vivo. *Cancer Res.* **51**: 2164–2172.

Ekstrand, A.J., Sugawa, N., James, C.D. and Collins, V.P. (1992) Amplified and rearranged epidermal growth factor receptor genes in human glioblastomas reveal deletions of sequences encoding portions of the N- and/or C-terminal tails. *Proc. Natl Acad. Sci. USA* **89**: 4309–4313.

el-Azouzi, M., Chung, R.Y., Farmer, G.E., *et al.* (1989) Loss of distinct regions on the short arm of chromosome 17 associated with tumorigenesis of human astrocytomas. *Proc. Natl Acad. Sci. USA* **86**: 7186–7190.

Frankel, R.H., Bayona, W., Koslow, M. and Newcomb, E.W. (1992) p53 mutations in human malignant gliomas: comparison of loss of heterozygosity with mutation frequency. *Cancer Res.* **52**: 1427–1433.

Friedman, H.S., Burger, P.C., Bigner, S.H., *et al.* (1988) Phenotypic and genotypic analysis of a human medulloblastoma cell line and transplantable xenograft (D341 Med) demonstrating amplification of c-myc. *Am. J. Pathol.* **130**: 472–484.

Fujii, Y., Hongo, T. and Hayashi, Y. (1994) Chromosome analysis of brain tumors in childhood. *Genes Chromosomes Cancer* **11**: 205–215.

Fults, D. and Pedone, C. (1993) Deletion mapping of the long arm of chromosome 10 in glioblastoma multiforme. *Genes Chromosomes Cancer* **7**: 173–177.

Fults, D., Tippets, R.H., Thomas, G.A., Nakamura, Y. and White, R. (1989) Loss of heterozygosity for loci on chromosome 17p in human malignant astrocytoma. *Cancer Res.* **49**: 6572–6577.

Fults, D., Pedone, C.A., Thomas, G.A. and White, R. (1990) Allelotype of human malignant astrocytoma. *Cancer Res.* **50**: 5784–5789.

Fults, D., Brockmeyer, D., Tullous, M.W., Pedone, C.A. and Cawthon, R.M. (1992) p53 mutation and loss of heterozygosity on chromosomes 17 and 10 during human astrocytoma progression. *Cancer Res.* **52**: 674–679.

Fults, D., Pedone, C.A., Thompson, G.E., Uchiyama, C.M., Gumpper, K.L., Iliev, D., Vinson, V.L., Tavtigian, S.V. and Perry, W.L. 3rd (1998) Microsatellite deletion mapping on chromosome 10q and mutation analysis of MMAC1, FAS, and MXI1 in human glioblastoma multiforme. *Int. J. Oncol.* **12**: 905–910.

Giani, C. and Finocchiaro, G. (1994) Mutation rate of the CDKN2 gene in malignant gliomas. *Cancer Res.* **54**: 6338–6339.

Goodrich, L.V., Milenkovic, L., Higgins, K.M. and Scott, M.P. (1997) Altered neural cell fates and medulloblastoma in mouse patched mutants. *Science* **277**: 1109–1113.

Griffin, C.A., Hawkins, A.L., Packer, R.J., Rorke, L.B. and Emanuel, B.S. (1988) Chromosome abnormalities in pediatric brain tumors. *Cancer Res.* **48**: 175–180.

Hahn, H., Wicking, C., Zaphiropoulous, P.G. *et al.* (1996) Mutations of the human homolog of Drosophila patched in the nevoid basal cell carcinoma syndrome. *Cell* **85**: 841–851.

Hamilton, S.R., Liu, B., Parsons, R.E., *et al.* (1995) The molecular basis of Turcot's syndrome [see comments]. *N. Engl. J. Med.* **332**: 839–847.

Hashimoto, N., Ichikawa, D., Arakawa, Y., *et al.* (1995) Frequent deletions of material from chromosome arm 1p in oligodendroglial tumors revealed by double-target fluorescence in situ hybridization and microsatellite analysis. *Genes Chromosomes Cancer* **14**: 295–300.

He, J., Allen, J.R., Collins, V.P., Allalunis-Turner, M.J., Godbout, R., Day, R.S. 3rd and James, C.D. (1994) CDK4 amplification is an alternative mechanism to p16 gene homozygous deletion in glioma cell lines. *Cancer Res.* **54**: 5804–5807.

He, J., Olson, J.J. and James, C.D. (1995) Lack of p16INK4 or retinoblastoma protein (pRb), or amplification-associated overexpression of cdk4 is observed in distinct subsets of malignant glial tumors and cell lines. *Cancer Res.* **55**: 4833–4836.

Hecht, B.K., Turc-Carel, C., Chatel, M., Grellier, P., Gioanni, J., Attias, R., Gaudray, P. and Hecht, F. (1995) Cytogenetics of malignant gliomas: I. The autosomes with reference to rearrangements. *Cancer Genet. Cytogenet.* **84**: 1–8.

Henson, J.W., Schnitker, B.L., Correa, K.M., von Deimling, A., Fassbender, F., Xu, H.J., Benedict, W.F., Yandell, D.W. and Louis, D.N. (1994) The retinoblastoma gene is involved in malignant progression of astrocytomas. *Ann. Neurol.* **36**: 714–721.

Herman, J.G., Merlo, A., Mao, L., Lapidus, R.G., Issa, J.P., Davidson, N.E., Sidransky, D. and Baylin, S.B. (1995) Inactivation of the CDKN2/p16/MTS1 gene is frequently associated with aberrant DNA methylation in all common human cancers. *Cancer Res.* **55**: 4525–4530.

Hermanson, M., Funa, K., Hartman, M., Claesson-Welsh, L., Heldin, C.H., Westermark, B. and Nister, M. (1992) Platelet-derived growth factor and its receptors in human glioma tissue: expression of messenger RNA and protein suggests the presence of autocrine and paracrine loops. *Cancer Res.* **52**: 3213–3219.

Hermanson, M., Funa, K., Koopmann, J., *et al.* (1996) Association of loss of heterozygosity on chromosome 17p with high platelet-derived growth factor alpha receptor expression in human malignant gliomas. *Cancer Res.* **56**: 164–171.

Huang, H., Mahler-Araujo, B.M., Sankila, A., Chimelli, L., Yonekawa, Y., Kleihues, P. and Ohgaki, H. (2000) APC mutations in sporadic medulloblastomas. *Am. J. Pathol.* **156**: 433–437.

Huhn, S.L., Mohapatra, G., Bollen, A., Lamborn, K., Prados, M.D. and Feuerstein, B.G. (1999) Chromosomal abnormalities in glioblastoma multiforme by comparative genomic hybridization: correlation with radiation treatment outcome. *Clin. Cancer Res.* **5**: 1435–1443.

Humphrey, P.A., Wong, A.J., Vogelstein, B., *et al.* (1990) Anti-synthetic peptide antibody reacting at the fusion junction of deletion-mutant epidermal growth factor receptors in human glioblastoma. *Proc. Natl Acad. Sci. USA* **87**: 4207–4211.

Ichimura, K., Schmidt, E.E., Miyakawa, A., Goike, H.M. and Collins, V.P. (1998) Distinct patterns of deletion on 10p and 10q suggest involvement of multiple tumor suppressor genes in the development of astrocytic gliomas of different malignancy grades. *Genes Chromosomes Cancer* **22**: 9–15.

Ichimura, K., Bolin, M.B., Goike, H.M., Schmidt, E.E., Moshref, A. and Collins, V.P. (2000) Deregulation of the p14ARF/MDM2/p53 pathway is a prerequisite for human astrocytic gliomas with G1-S transition control gene abnormalities. *Cancer Res.* **60**: 417–424.

Ino, Y., Zlatescu, M.C., Sasaki, H., *et al.* (2000): Long survival and therapeutic responses in patients with histologically disparate high-grade gliomas demonstrating chromosome 1p loss. *J. Neurosurg.* **92**: 983–990.

Irving, R.M., Moffat, D.A., Hardy, D.G., Barton, D.E., Xuereb, J.H. and Maher, E.R. (1994) Somatic NF2 gene mutations in familial and non-familial vestibular schwannoma. *Hum. Mol. Genet.* **3**: 347–350.

Jacoby, L.B., MacCollin, M., Louis, D.N., *et al.* (1994) Exon scanning for mutation of the NF2 gene in schwannomas. *Hum. Mol. Genet.* **3**: 413–419.

James, C.D., Carlbom, E., Dumanski, J.P., Hansen, M., Nordenskjold, M., Collins, V.P. and Cavenee, W.K. (1988): Clonal genomic alterations in glioma malignancy stages. *Cancer Res.* **48**: 5546–5551.

James, C.D., Carlbom, E., Nordenskjold, M., Collins, V.P. and Cavenee, W.K. (1989) Mitotic recombination of chromosome 17 in astrocytomas. *Proc. Natl Acad. Sci. USA* **86**: 2858–2862.

James, C.D., He, J., Carlbom, E., Mikkelsen, T., Ridderheim, P.A., Cavenee, W.K. and Collins, V.P. (1990) Loss of genetic information in central nervous system tumors common to children and young adults. *Genes Chromosomes Cancer* **2**: 94–102.

Jen, J., Harper, J.W., Bigner, S.H., Bigner, D.D., Papadopoulos, N., Markowitz, S., Willson, J.K., Kinzler, K.W. and Vogelstein, B. (1994) Deletion of p16 and p15 genes in brain tumors. *Cancer Res.* **54**: 6353–6358.

Jenkins, R.B., Kimmel, D.W., Moertel, C.A., Schultz, C.G., Scheithauer, B.W., Kelly, P.J. and Dewald, G.W. (1989) A cytogenetic study of 53 human gliomas. *Cancer Genet. Cytogenet.* **39**: 253–279.

Johnson, R.L., Rothman, A.L., Xie, J., *et al.* (1996) Human homolog of patched, a candidate gene for the basal cell nevus syndrome. *Science* **272**: 1668–1671.

Kamb, A., Gruis, N.A., Weaver-Feldhaus, J., *et al.* (1994) A cell cycle regulator potentially involved in genesis of many tumor types [see comments]. *Science* **264**: 436–440.

Karlbom, A.E., James, C.D., Boethius, J., Cavenee, W.K., Collins, V.P., Nordenskjold, M. and Larsson, C. (1993) Loss of heterozygosity in malignant gliomas involves at least three distinct regions on chromosome 10. *Hum. Genet.* **92**: 169–174.

Kimmelman, A.C., Ross, D.A. and Liang, B.C. (1996) Loss of heterozygosity of chromosome 10p in human gliomas. *Genomics* **34**: 250–254.

Kleihues, P. (1998) Subsets of glioblastoma: clinical and histological vs. genetic typing [editorial; comment]. *Brain Pathol.* **8**: 667–668.

Kleihues, P., Burger, P.C., Scheithauer and B.W. (1993) *Histological Typing of Tumours of the Central Nervous System.* 2nd ed. New York: Springer-Verlag.

Kraus, J.A., Bolln, C., Wolf, H.K., Neumann, J., Kindermann, D., Fimmers, R., Forster, F., Baumann, A. and Schlegel, U. (1994) TP53 alterations and clinical outcome in low grade astrocytomas. *Genes Chromosomes Cancer* **10**: 143–149.

Kyritsis, A.P., Xu, R., Bondy, M.L., Levin, V.A. and Bruner, J.M. (1996) Correlation of p53 immunoreactivity and sequencing in patients with glioma. *Mol. Carcinog.* **15**: 1–4.

Lang, F.F., Miller, D.C., Koslow, M. and Newcomb, E.W. (1994a) Pathways leading to glioblastoma multiforme: a molecular analysis of genetic alterations in 65 astrocytic tumors. *J. Neurosurg.* **81**: 427–436.

Lang, F.F., Miller, D.C., Pisharody, S., Koslow, M. and Newcomb, E.W. (1994b) High frequency of p53 protein accumulation without p53 gene mutation in human juvenile pilocytic, low grade and anaplastic astrocytomas. *Oncogene* **9**: 949–954.

Leenstra, S., Bijlsma, E.K., Troost, D., Oosting, J., Westerveld, A., Bosch, D.A. and Hulsebos, T.J. (1994) Allele loss on chromosomes 10 and 17p and epidermal growth factor receptor gene amplification in human malignant astrocytoma related to prognosis. *Br. J. Cancer* **70**: 684–689.

Lekanne Deprez, R.H., Bianchi, A.B., Groen, N.A., et al. (1994) Frequent NF2 gene transcript mutations in sporadic meningiomas and vestibular schwannomas. *Am. J. Hum. Genet.* **54**: 1022–1029.

Li, D.M. and Sun, H. (1997) TEP1, encoded by a candidate tumor suppressor locus, is a novel protein tyrosine phosphatase regulated by transforming growth factor beta. *Cancer Research* **57**: 2124–2129.

Li, J., Yen, C., Liaw, D., et al. (1997) PTEN, a putative protein tyrosine phosphatase gene mutated in human brain, breast, and prostate cancer [see comments]. *Science* **275**: 1943–1947.

Li, Y.J., Hoang-Xuan, K., Delattre, J.Y., Poisson, M., Thomas, G. and Hamelin, R. (1995) Frequent loss of heterozygosity on chromosome 9, and low incidence of mutations of cyclin-dependent kinase inhibitors p15 (MTS2) and p16 (MTS1) genes in gliomas. *Oncogene* **11**: 597–600.

Litofsky, N.S., Hinton, D. and Raffel, C. (1994) The lack of a role for p53 in astrocytomas in pediatric patients. *Neurosurgery* **34**: 967–972; discussion 972–973.

Liu, W., James, C.D., Frederick, L., Alderete, B.E., Jenkins, R.B. (1997) PTEN/MMAC1 mutations and EGFR amplification in glioblastomas. *Cancer Res.* **57**: 5254–5257.

Louis, D.N. (1994) The p53 gene and protein in human brain tumors. *J. Neuropathol. Exp. Neurol.* **53**: 11–21.

Mark, J., Levan, G. and Mitelman, F. (1972) Identification by fluorescence of the G chromosome lost in human meningomas. *Hereditas* **71**: 163–168.

Marsh, D.J., Dahia, P.L., Zheng, Z., Liaw, D., Parsons, R., Gorlin, R.J. and Eng, C. (1997) Germline mutations in PTEN are present in Bannayan-Zonana syndrome [letter]. *Nat. Genet.* **16**: 333–334.

Marsh, D.J., Dahia, P.L., Coulon, V., et al. (1998) Allelic imbalance, including deletion of PTEN/MMACI, at the Cowden disease locus on 10q22–23, in hamartomas from patients with Cowden syndrome and germline PTEN mutation. *Genes, Chromosomes & Cancer* **21**: 61–69.

Mashiyama, S., Murakami, Y., Yoshimoto, T., Sekiya, T. and Hayashi, K. (1991) Detection of p53 gene mutations in human brain tumors by single-strand conformation polymorphism analysis of polymerase chain reaction products. *Oncogene* **6**: 1313–1318.

Maxwell, G.L., Risinger, J.I., Gumbs, C., Shaw, H., Bentley, R.C., Barrett, J.C., Berchuck, A. and Futreal, P.A. (1998) Mutation of the PTEN tumor suppressor gene in endometrial hyperplasias. *Cancer Research* **58**: 2500–2503.

Miyakoshi, J., Dobler, K.D., Allalunis-Turner, J., et al. (1990) Absence of IFNA and IFNB genes from human malignant glioma cell lines and lack of correlation with cellular sensitivity to interferons. *Cancer Res.* **50**: 278–283.

Mohapatra, G., Bollen, A.W., Kim, D.H., Lamborn, K., Moore, D.H., Prados, M.D. and Feuerstein, B.G. (1998) Genetic analysis of glioblastoma multiforme provides evidence for subgroups within the grade. *Genes Chromosomes Cancer* **21**: 195–206.

Mollenhauer, J., Wiemann, S., Scheurlen, W., Korn, B., Hayashi, Y., Wilgenbus, K.K., von Deimling, A. and Poustka, A. (1997) DMBT1, a new member of the SRCR superfamily, on chromosome 10q25.3–26.1 is deleted in malignant brain tumours. *Nat. Genet.* **17**: 32–39.

Mollenhauer, J., Herbertz, S., Holmskov, U., *et al.* (2000) DMBT1 encodes a protein involved in the immune defense and in epithelial differentiation and is highly unstable in cancer. *Cancer Research* **60**: 1704–1710.

Moscatello, D.K., Holgado-Madruga, M., Godwin, A.K., Ramirez, G., Gunn, G., Zoltick, P.W., Biegel, J.A., Hayes, R.L. and Wong, A.J. (1995) Frequent expression of a mutant epidermal growth factor receptor in multiple human tumors. *Cancer Res.* **55**: 5536–5539.

Moulton, T., Samara, G., Chung, W.Y., Yuan, L., Desai, R., Sisti, M., Bruce, J. and Tycko, B. (1995) MTS1/p16/CDKN2 lesions in primary glioblastoma multiforme. *Am. J. Pathol.* **146**: 613–619.

Nakamura, H., Yoshida, M., Tsuiki, H., *et al.* (1998) Identification of a human homolog of the Drosophila neuralized gene within the 10q25.1 malignant astrocytoma deletion region. *Oncogene* **16**: 1009–1019.

Neumann, E., Kalousek, D.K., Norman, M.G., Steinbok, P., Cochrane, D.D. and Goddard, K. (1993) Cytogenetic analysis of 109 pediatric central nervous system tumors. *Cancer Genet. Cytogenet.* **71**: 40–49.

Nishikawa, R., Furnari, F.B., Lin, H., Arap, W., Berger, M.S., Cavenee, W.K. and Su Huang, H.J. (1995) Loss of P16INK4 expression is frequent in high grade gliomas. *Cancer Res.* **55**: 1941–1945.

Nishizaki, T., Ozaki, S., Harada, K., Ito, H., Arai, H., Beppu, T. and Sasaki, K. (1998) Investigation of genetic alterations associated with the grade of astrocytic tumor by comparative genomic hybridization. *Genes Chromosomes Cancer* **21**: 340–346.

Ohgaki, H., Eibl, R.H., Wiestler, O.D., Yasargil, M.G., Newcomb, E.W. and Kleihues, P. (1991) p53 mutations in nonastrocytic human brain tumors. *Cancer Res.* **51**: 6202–6205.

Olopade, O.I., Jenkins, R.B., Ransom, D.T., Malik, K., Pomykala, H., Nobori, T., Cowan, J.M., Rowley, J.D. and Diaz, M.O. (1992) Molecular analysis of deletions of the short arm of chromosome 9 in human gliomas. *Cancer Res.* **52**: 2523–2529.

Pietsch, T., Scharmann, T., Fonatsch, C., *et al.* (1994) Characterization of five new cell lines derived from human primitive neuroectodermal tumors of the central nervous system. *Cancer Res.* **54**: 3278–3287.

Pietsch, T., Waha, A., Koch, A., *et al.* (1997) Medulloblastomas of the desmoplastic variant carry mutations of the human homologue of Drosophila patched. *Cancer Res.* **57**: 2085–2088.

Raffel, C., Gilles, F.E. and Weinberg, K.I. (1990) Reduction to homozygosity and gene amplification in central nervous system primitive neuroectodermal tumors of childhood. *Cancer Res.* **50**: 587–591.

Raffel, C., Thomas, G.A., Tishler, D.M., Lassoff, S. and Allen, J.C. (1993) Absence of p53 mutations in childhood central nervous system primitive neuroectodermal tumors. *Neurosurgery* **33**: 301–305; discussion 305–306.

Raffel, C., Jenkins, R.B., Frederick, L., Hebrink, D., Alderete, B., Fults, D.W. and James, C.D. (1997) Sporadic medulloblastomas contain PTCH mutations. *Cancer Res.* **57**: 842–845.

Ransom, D.T., Ritland, S.R., Kimmel, D.W., Moertel, C.A., Dahl, R.J., Scheithauer, B.W., Kelly, P.J. and Jenkins, R.B. (1992a) Cytogenetic and loss of heterozygosity studies in ependymomas, pilocytic astrocytomas, and oligodendrogliomas. *Genes Chromosomes Cancer* **5**: 348–356.

Ransom, D.T., Ritland, S.R., Moertel, C.A., *et al.* (1992b) Correlation of cytogenetic analysis and loss of heterozygosity studies in human diffuse astrocytomas and mixed oligo-astrocytomas. *Genes Chromosomes Cancer* **5**: 357–374.

Rasheed, B.K., McLendon, R.E., Herndon, J.E., Friedman, H.S., Friedman, A.H., Bigner, D.D. and Bigner, S.H. (1994) Alterations of the TP53 gene in human gliomas. *Cancer Res.* **54**: 1324–1330.

Rasheed, B.K., McLendon, R.E., Friedman, H.S., Friedman, A.H., Fuchs, H.E., Bigner, D.D. and Bigner, S.H. (1995) Chromosome 10 deletion mapping in human gliomas: a common deletion region in 10q25. *Oncogene.* **10**: 2243–2246.

Rasheed, B.K., Stenzel, T.T., McLendon, R.E., Parsons, R., Friedman, A.H., Friedman, H.S., Bigner, D.D. and Bigner, S.H. (1997) PTEN gene mutations are seen in high-grade but not in low-grade gliomas. *Cancer Res.* **57**: 4187–4190.

Reardon, D.A., Michalkiewicz, E., Boyett, J.M., *et al.* (1997) Extensive genomic abnormalities in childhood medulloblastoma by comparative genomic hybridization. *Cancer Res.* 57: 4042–4047.

Reardon, D.A., Entrekin, R.E., Sublett, J., Ragsdale, S., Li, H., Boyett, J., Kepner, J.L. and Look, A.T. (1999) Chromosome arm 6q loss is the most common recurrent autosomal alteration detected in primary pediatric ependymoma. *Genes Chromosomes Cancer* 24: 230–237.

Reifenberger, G., Reifenberger, J., Ichimura, K., Meltzer, P.S. and Collins, V.P. (1994a) Amplification of multiple genes from chromosomal region 12q13–14 in human malignant gliomas: preliminary mapping of the amplicons shows preferential involvement of CDK4, SAS, and MDM2. *Cancer Res.* 54: 4299–4303.

Reifenberger, J., Reifenberger, G., Liu, L., James, C.D., Wechsler, W. and Collins, V.P. (1994b) Molecular genetic analysis of oligodendroglial tumors shows preferential allelic deletions on 19q and 1p. *Am. J. Pathol.* 145: 1175–1190.

Reifenberger, J., Wolter, M., Weber, R.G., Megahed, M., Ruzicka, T., Lichter, P. and Reifenberger, G. (1998) Missense mutations in SMOH in sporadic basal cell carcinomas of the skin and primitive neuroectodermal tumors of the central nervous system. *Cancer Res.* 58: 1798–1803.

Rey, J.A., Bello, M.J., de Campos, J.M., Kusak, M.E., Ramos, C. and Benitez, J. (1987) Chromosomal patterns in human malignant astrocytomas. *Cancer Genet. Cytogenet.* 29: 201–221.

Ritland, S.R., Ganju, V. and Jenkins, R.B. (1995) Region-specific loss of heterozygosity on chromosome 19 is related to the morphologic type of human glioma. *Genes Chromosomes Cancer* 12: 277–282.

Rogatto, S.R., Casartelli, C., Rainho, C.A. and Barbieri-Neto, J. (1993) Chromosomes in the genesis and progression of ependymomas. *Cancer Genet. Cytogenet.* 69: 146–152.

Rouleau, G.A., Wertelecki, W., Haines, J.L., *et al.* (1987) Genetic linkage of bilateral acoustic neurofibromatosis to a DNA marker on chromosome 22. *Nature* 329: 246–248.

Rouleau, G.A., Merel, P., Lutchman, M., *et al.* (1993) Alteration in a new gene encoding a putative membrane-organizing protein causes neuro-fibromatosis type 2 [see comments]. *Nature* 363: 515–521.

Rubio, M.P., Correa, K.M., Ramesh, V., MacCollin, M.M., Jacoby, L.B., von Deimling, A., Gusella, J.F. and Louis, D.N. (1994) Analysis of the neurofibromatosis 2 gene in human ependymomas and astrocytomas. *Cancer Res.* 54: 45–47.

Ruttledge, M.H., Sarrazin, J., Rangaratnam, S., *et al.* (1994a) Evidence for the complete inactivation of the NF2 gene in the majority of sporadic meningiomas. *Nat. Genet.* 6: 180–184.

Ruttledge, M.H., Xie, Y.G., Han, F.Y., Peyrard, M., Collins, V.P., Nordenskjold, M. and Dumanski, J.P. (1994b) Deletions on chromosome 22 in sporadic meningioma. *Genes Chromosomes Cancer* 10: 122–130.

Sallinen, S.L., Sallinen, P., Haapasalo, H., Kononen, J., Karhu, R., Helen, P. and Isola, J. (1997) Accumulation of genetic changes is associated with poor prognosis in grade II astrocytomas. *Am. J. Pathol.* 151: 1799–1807.

Sawyer, J.R., Sammartino, G., Husain, M., Boop, F.A. and Chadduck, W.M. (1994) Chromosome aberrations in four ependymomas. *Cancer Genet. Cytogenet.* 74: 132–138.

Saylors, R.L.d., Sidransky, D., Friedman, H.S., Bigner, S.H., Bigner, D.D., Vogelstein, B. and Brodeur, G.M. (1991) Infrequent p53 gene mutations in medulloblastomas. *Cancer Res.* 51: 4721–4723.

Schmidt, E.E., Ichimura, K., Reifenberger, G. and Collins, V.P. (1994) CDKN2 (p16/MTS1) gene deletion or CDK4 amplification occurs in the majority of glioblastomas. *Cancer Res.* 54: 6321–6324.

Schmidt, E.E., Ichimura, K., Goike, H.M., Moshref, A., Liu, L. and Collins, V.P. (1999) Mutational profile of the PTEN gene in primary human astrocytic tumors and cultivated xenografts. *J. Neuropathol. Exp. Neurol.* 58: 1170–1183.

Schofield, D., West, D.C., Anthony, D.C., Marshal, R. and Sklar, J. (1995) Correlation of loss of heterozygosity at chromosome 9q with histological subtype in medulloblastomas. *Am. J. Pathol.* 146: 472–480.

Schofield, D.E., Beckwith, J.B. and Sklar, J. (1996) Loss of heterozygosity at chromosome regions 22q11–12 and 11p15.5 in renal rhabdoid tumors. *Genes Chromosomes Cancer* 15: 10–17.

Schrock, E., Blume, C., Meffert, M.C., *et al.* (1996) Recurrent gain of chromosome arm 7q in low-grade astrocytic tumors studied by comparative genomic hybridization. *Genes Chromosomes Cancer* 15: 199–205.

Schutz, B.R., Scheurlen, W., Krauss, J., du Manoir, S., Joos, S., Bentz, M. and Lichter, P. (1996) Mapping of chromosomal gains and losses in primitive neuroectodermal tumors by comparative genomic hybridization. *Genes Chromosomes Cancer* **16**: 196–203.

Seizinger, B.R., de la Monte, S., Atkins, L., Gusella, J.F. and Martuza, R.L. (1987a) Molecular genetic approach to human meningioma: loss of genes on chromosome 22. *Proc. Natl Acad. Sci. USA* **84**: 5419–5423.

Seizinger, B.R., Rouleau, G., Ozelius, L.J., Lane, A.H., St George-Hyslop, P., Huson, S., Gusella, J.F. and Martuza, R.L. (1987b) Common pathogenetic mechanism for three tumor types in bilateral acoustic neurofibromatosis. *Science* **236**: 317–319.

Somerville, R.P., Shoshan, Y., Eng, C., Barnett, G., Miller, D. and Cowell, J.K. (1998) Molecular analysis of two putative tumour suppressor genes, PTEN and DMBT, which have been implicated in glioblastoma multiforme disease progression. *Oncogene* **17**: 1755–1757.

Sonoda, Y., Yoshimoto, T. and Sekiya, T. (1995) Homozygous deletion of the MTS1/p16 and MTS2/p15 genes and amplification of the CDK4 gene in glioma. *Oncogene* **11**: 2145–2149.

Stapleton, P., Weith, A., Urbanek, P., Kozmik, Z. and Busslinger, M. (1993) Chromosomal localization of seven PAX genes and cloning of a novel family member, PAX-9. *Nat. Genet.* **3**: 292–298.

Steck, P.A., Pershouse, M.A., Jasser, S.A., *et al.* (1997) Identification of a candidate tumour suppressor gene, MMAC1, at chromosome 10q23.3 that is mutated in multiple advanced cancers. *Nat. Genet.* **15**: 356–362.

Steck, P.A., Lin, H., Langford, L.A., Jasser, S.A., Koul, D., Yung, W.K. and Pershouse, M.A. (1999) Functional and molecular analyses of 10q deletions in human gliomas. *Genes Chromosomes Cancer* **24**: 135–143.

Stevens, M.C., Cameron, A.H., Muir, K.R., Parkes, S.E., Rcid, H. and Whitwell, H. (1991) Descriptive epidemiology of primary central nervous system tumours in children: a population-based study. *Clin. Oncol. (R. Coll. Radiol.)* **3**: 323–329.

Stratton, M.R., Darling, J., Lantos, P.L., Cooper, C.S. and Reeves, B.R. (1989) Cytogenetic abnormalities in human ependymomas. *Int. J. Cancer* **44**: 579–581.

Sugawa, N., Ekstrand, A.J., James, C.D. and Collins, V.P. (1990) Identical splicing of aberrant epidermal growth factor receptor transcripts from amplified rearranged genes in human glioblastomas. *Proc. Natl Acad. Sci. USA* **87**: 8602–8606.

Tenan, M., Colombo, B.M., Pollo, B., Cajola, L., Broggi, G. and Finocchiaro, G. (1994) p53 mutations and microsatellite analysis of loss of heterozygosity in malignant gliomas. *Cancer Genet. Cytogenet.* **74**: 139–143.

Teng, D.H., Hu, R., Lin, H., *et al.* (1997) MMAC1/PTEN mutations in primary tumor specimens and tumor cell lines. *Cancer Res.* **57**: 5221–5225.

Thiel, G., Losanowa, T., Kintzel, D., Nisch, G., Martin, H., Vorpahl, K. and Witkowski, R. (1992) Karyotypes in 90 human gliomas. *Cancer Genet. Cytogenet.* **58**: 109–120.

Thomas, G.A. and Raffel, C. (1991) Loss of heterozygosity on 6q, 16q, and 17p in human central nervous system primitive neuroectodermal tumors. *Cancer Res.* **51**: 639–643.

Tohma, Y., Gratas, C., Biernat, W., Peraud, A., Fukuda, M., Yonekawa, Y., Kleihues, P. and Ohgaki, H. (1998) PTEN (MMAC1) mutations are frequent in primary glioblastomas (de novo) but not in secondary glioblastomas. *J. Neuropathol. Exp. Neurol.* **57**: 684–689.

Tsao, H., Zhang, X., Fowlkes, K. and Haluska, F.G. (2000) Relative reciprocity of NRAS and PTEN/MMAC1 alterations in cutaneous melanoma cell lines. *Cancer Res.* **60**: 1800–1804.

Ueki, K., Rubio, M.P., Ramesh, V., Correa, K.M., Rutter, J.L., von Deimling, A., Buckler, A.J., Gusella, J.F. and Louis, D.N. (1994) MTS1/CDKN2 gene mutations are rare in primary human astrocytomas with allelic loss of chromosome 9p. *Hum. Mol. Genet.* **3**: 1841–1845.

Ueki, K., Ono, Y., Henson, J.W., Efird, J.T., von Deimling, A. and Louis, D.N. (1996) CDKN2/p16 or RB alterations occur in the majority of glioblastomas and are inversely correlated. *Cancer Res.* **56**: 150–153.

Vagner-Capodano, A.M., Gentet, J.C., Gambarelli, D., Pellissier, J.F., Gouzien, M., Lena, G., Genitori, L., Choux, M. and Raybaud, C. (1992) Cytogenetic studies in 45 pediatric brain tumors [published erratum appears in Pediatr. Hematol. Oncol. 1993 Jan-Mar;10(1): 117]. *Pediatr. Hematol. Oncol.* **9**: 223–235.

Velema, J.P. and Percy, C.L. (1987) Age curves of central nervous system tumor incidence in adults: variation of shape by histologic type. *J. Natl. Cancer Inst.* **79**: 623–629.

Venter, D.J., Bevan, K.L., Ludwig, R.L., Riley, T.E., Jat, P.S., Thomas, D.G. and Noble, M.D. (1991) Retinoblastoma gene deletions in human glioblastomas. *Oncogene* **6**: 445–448.

Versteege, I., Sevenet, N., Lange, J., Rousseau-Merck, M.F., Ambros, P., Handgretinger, R., Aurias, A. and Delattre, O. (1998) Truncating mutations of hSNF5/INI1 in aggressive paediatric cancer. *Nature* **394**: 203–206.

von Deimling, A., Eibl, R.H., Ohgaki, H., *et al.* (1992a) p53 mutations are associated with 17p allelic loss in grade II and grade III astrocytoma. *Cancer Res.* **52**: 2987–2990.

von Deimling, A., Louis, D.N., von Ammon, K., Petersen, I., Wiestler, O.D. and Seizinger, B.R. (1992b) Evidence for a tumor suppressor gene on chromosome 19q associated with human astrocytomas, oligodendrogliomas, and mixed gliomas. *Cancer Res.* **52**: 4277–4279.

von Deimling, A., von Ammon, K., Schoenfeld, D., Wiestler, O.D., Seizinger, B.R. and Louis, D.N. (1993) Subsets of glioblastoma multiforme defined by molecular genetic analysis. *Brain Pathol.* **3**: 19–26.

von Deimling, A., Nagel, J., Bender, B., Lenartz, D., Schramm, J., Louis, D.N. and Wiestler, O.D. (1994) Deletion mapping of chromosome 19 in human gliomas. *Int. J. Cancer* **57**: 676–680.

Vorechovsky, I., Tingby, O., Hartman, M., Stromberg, B., Nister, M., Collins, V.P. and Toftgard, R. (1997) Somatic mutations in the human homologue of Drosophila patched in primitive neuroectodermal tumours. *Oncogene* **15**: 361–366.

Walker, D.G., Duan, W., Popovic, E.A., Kaye, A.H., Tomlinson, F.H. and Lavin, M. (1995) Homozygous deletions of the multiple tumor suppressor gene 1 in the progression of human astrocytomas. *Cancer Res.* **55**: 20–23.

Wang, S.I., Puc, J., Li, J., Bruce, J.N., Cairns, P., Sidransky, D. and Parsons, R. (1997) Somatic mutations of PTEN in glioblastoma multiforme. *Cancer Res.* **57**: 4183–4186.

Wasson, J.C., Saylors, R.L.d., Zeltzer, P., *et al.* (1990) Oncogene amplification in pediatric brain tumors. *Cancer Res.* **50**: 2987–2990.

Watanabe, K., Tachibana, O., Sata, K., Yonekawa, Y., Kleihues, P. and Ohgaki, H. (1996): Overexpression of the EGF receptor and p53 mutations are mutually exclusive in the evolution of primary and secondary glioblastomas. *Brain Pathol.* **6**: 217–223; discussion 23–24.

Weber, R.G., Sabel, M., Reifenberger, J., Sommer, C., Oberstrass, J., Reifenberger, G., Kiessling, M. and Cremer, T. (1996) Characterization of genomic alterations associated with glioma progression by comparative genomic hybridization. *Oncogene* **13**: 983–994.

Wechsler, D.S., Shelly, C.A., Petroff, C.A. and Dang, C.V. (1997) MXI1, a putative tumor suppressor gene, suppresses growth of human glioblastoma cells. *Cancer Res.* **57**: 4905–4912.

Weremowicz, S., Kupsky, W.J., Morton, C.C. and Fletcher, J.A. (1992) Cytogenetic evidence for a chromosome 22 tumor suppressor gene in ependymoma. *Cancer Genet. Cytogenet.* **61**: 193–196.

Wernicke, C., Thiel, G., Lozanova, T., Vogel, S., Kintzel, D., Janisch, W., Lehmann, K. and Witkowski, R. (1995) Involvement of chromosome 22 in ependymomas. *Cancer Genet. Cytogenet.* **79**: 173–176.

Wertelecki, W., Rouleau, G.A., Superneau, D.W., Forehand, L.W., Williams, J.P., Haines, J.L. and Gusella, J.F. (1988) Neurofibromatosis 2: clinical and DNA linkage studies of a large kindred. *N. Engl. J. Med.* **319**: 278–283.

Wetmore, C., Eberhart, D.E. and Curran, T. (2000) The normal patched allele is expressed in medulloblastomas from mice with heterozygous germ-line mutation of patched. *Cancer Res.* **60**: 2239–2246.

Wikstrand, C.J., Hale, L.P., Batra, S.K., *et al.* (1995) Monoclonal antibodies against EGFRvIII are tumor specific and react with breast and lung carcinomas and malignant gliomas. *Cancer Res.* **55**: 3140–3148.

Willert, J.R., Daneshvar, L., Sheffield, V.C. and Cogen, P.H. (1995) Deletion of chromosome arm 17p DNA sequences in pediatric high-grade and juvenile pilocytic astrocytomas. *Genes Chromosomes Cancer* **12**: 165–172.

Wolter, M., Reifenberger, J., Sommer, C., Ruzicka, T. and Reifenberger, G. (1997) Mutations in the human homologue of the Drosophila segment polarity gene patched (PTCH) in sporadic basal cell carcinomas of the skin and primitive neuroectodermal tumors of the central nervous system. *Cancer Res.* **57**: 2581–2585.

Wong, A.J., Bigner, S.H., Bigner, D.D., Kinzler, K.W., Hamilton, S.R. and Vogelstein, B. (1987) Increased expression of the epidermal growth factor receptor gene in malignant gliomas is invariably associated with gene amplification. *Proc. Natl Acad. Sci. USA* **84**: 6899–6903.

Xie, J., Johnson, R.L., Zhang, X., *et al.* (1997) Mutations of the PATCHED gene in several types of sporadic extracutaneous tumors. *Cancer Res.* 57: 2369–2372.

Yamazaki, H., Fukui, Y., Ueyama, Y., Tamaoki, N., Kawamoto, T., Taniguchi, S. and Shibuya, M. (1988) Amplification of the structurally and functionally altered epidermal growth factor receptor gene (c-erbB) in human brain tumors. *Mol. Cell. Biol.* 8: 1816–1820.

Yong, W.H., Chou, D., Ueki, K., Harsh, G.R.t., von Deimling, A., Gusella, J.F., Mohrenweiser, H.W. and Louis, D.N. (1995) Chromosome 19q deletions in human gliomas overlap telomeric to D19S219 and may target a 425 kb region centromeric to D19S112. *J. Neuropathol. Exp. Neurol.* 54: 622–626.

Zang, K.D. (1982) Cytological and cytogenetical studies on human meningioma. *Cancer Genet. Cytogenet.* 6: 249–274.

Zang, K.D. and Singer, H. (1967) Chromosomal constitution of meningiomas. *Nature* 216: 84.

Zankl, H. and Zang, K.D. (1972) Cytological and cytogenetical studies on brain tumors. 4. Identification of the missing G chromosome in human meningiomas as no. 22 by fluorescence technique. *Humangenetik* 14: 167–169.

Zurawel, R.H., Chiappa, S.A., Allen. C. and Raffel, C. (1998) Sporadic medulloblastomas contain oncogenic beta-catenin mutations. *Cancer Res.* 58: 896–899.

Zurawel, R.H., Allen, C., Chiappa, S., Cato, W., Biegel, J., Cogen, P., de Sauvage, F., and Raffel, C. (2000a) Analysis of PTCH/SMO/SHH pathway genes in medulloblastoma. *Genes Chromosomes Cancer* 27: 44–51.

Zurawel, R.H., Allen, C., Wechsler-Reya, R., Scott, M.P., Raffel, C. (2000b) Evidence that haploinsufficiency of Ptch leads to medulloblastoma in mice. *Genes Chromosomes Cancer* 28: 77–81.

17

Molecular genetics of lung cancer

Frederic J. Kaye and Akihito Kubo

1. Introduction

Human lung cancer is a dramatic *in vivo* experiment for the model of carcinogen-induced neoplasia that is well understood by both the general public and medical establishment. This realization has finally resulted in a fall in the incidence of lung cancer cases in the United States due to a steady decrease in the prevalence of adult smokers (Greenlee *et al.*, 2000; Wingo *et al.*, 1999). In addition, by the late 1990s, more cases of lung cancer were identified in non-smoking men that were previous smokers than in current smokers. Nonetheless, lung cancer remains the most common cause of cancer deaths for both males and females (Greenlee *et al.*, 2000) and may reach over a million cases worldwide. Although recent trends have identified an increase in tobacco abuse worldwide, and among minors and young adults in the U.S., it is anticipated that overall tobacco consumption will decrease over the next century. This prediction, however, still suggests that lung cancer will continue to be a global scourge for at least 5–10 more generations. To achieve the goals of prevention, early diagnosis, and treatment, this effort will require a multi-disciplinary approach at (i) identifying the proximate carcinogens in tobacco exposure and other aerosolized and particulate toxins, (ii) understanding the cancer gene pathways that are altered by these compounds, and (iii) utilizing this knowledge for prevention and treatment strategies.

As the Human Genome Project nears completion, with the cataloguing of approximately 50 000 human genes, the research enterprise is now shifting to 'post-genomic' models. These include high throughput ventures (i) to assign all mRNA transcripts into distinct cassettes of genes that are coordinately regulated by specific internal or external signals, (ii) to identify all functionally relevant protein interactions, (iii) to design animal models for gene function, and ultimately (iv) to assign each gene to specific parallel and interconnecting enzymatic pathways that will define parameters for both growth and differentiation as well as cellular homeostasis. In the midst of this revolution, however, investigators are also continuing a painstaking effort to identify and characterize individual rate limiting and modifier genes that are selectively targeted in the development of

---

Molecular Genetics of Cancer second edition, edited by J.K. Cowell.

specific human cancer. The molecular genetics of lung cancer, however, has been particular difficult because of the difficulties to convincingly identify a familial susceptibility pattern (Tomizawa *et al.*, 1998). While age-specific analyses may help to identify these families (Gauderman and Morrison, 2000), sufficient material has not been available to date to allow for positional cloning method-ologies. As a result, progress in understanding the genetic events associated with lung cancer has relied largely on loss of heterozygosity (LOH) analyses and on cytogenetics, including modern techniques such as spectral karyotyping (SKY) (Dennis and Stock, 1999; Schrock *et al.*, 1996) or comparative genomic hybridization (CGH) (Testa *et al.*, 1997). These studies, however, have resulted in an overwhelming number of non-random loci with cytogenetic or molecular evidence for allele losses or gains that are proposed to be implicated in the initi-ation or progression of these tumors (see below). The completion of the Human Genome and Mapping Project will eventually allow for the systematic testing of all human transcripts within these loci and will validate these methods as an important tool for cancer gene discovery in lung cancer. The recent sequencing of the complete genomes of human chromosomes 21 and 22, with the mapping and cataloguing of each gene, will be an exciting test for these new opportunities (Reeves, 2000). The ultimate goal will be the ingenuity to use this information to devise effective strategies to prolong life and reduce the burden of lung cancer.

2. Histologic types of lung cancer

2.1 Small cell lung cancer

Small cell lung cancer (SCLC) is a distinct clinical, histological, genetic, and biochemical type of lung cancer that has a rapidly progressive course with a median survival of <4 months if left untreated (Ihde *et al.*, 1997). The term small cell or 'oat-cell' lung cancer is believed to be a misnomer based on the histologic appearance of small, highly condensed cells that can result from crush artifacts from the bronchoscopic biopsy and fixation. Well-preserved SCLC biopsy samples do not show this pyknotic appearance, but present as normal-sized neuroendocrine cells that cluster in a characteristic 'molding' pattern. While patients with SCLC present with a characteristic clinical and X-ray pattern, the diagnosis remains exclusively based on its histological appearance by light microscopy. Careful examination of these cells show that they contain neurose-cretory granules and express large amounts of neuroendocrine peptides, bioener-getic enzymes, and other ectopic protein products. For example, over-expression of neuron-specific enolase, the B-isoenzyme of creatine kinase, dopa-decar-boxylase, the opiomelanocortin pro-peptide, adrenocorticotrophic hormone, calcitonin, arginine vasopressin, atrial natriuretic peptide, chromogranin A, synatophysin, insulin-like growth factor, growth hormone-releasing peptide, gastrin-releasing peptide or bombesin, and IgG autoantibodies against the P/Q type of voltage-gated calcium channels, and many others (Carney *et al.*, 1985; Ihde *et al.*, 1997; Kaye *et al.*, 1987; Kiaris *et al.*, 1999; Lennon *et al.*, 1995; Moody *et al.*, 1981; Seneviratne and de Silva, 1999) have been detected. Some of these peptides, such as gastrin-releasing peptide/bombesin, insulin-like growth factor, and growth hormone releasing factor, have been suggested to function as autocrine or paracrine factors in a positive growth feedback loop (Cuttitta *et al.*, 1985; Kiaris *et al.*, 1999). Other ectopic proteins synthesized by SCLC samples are also functionally active and are responsible for paraneoplastic syndromes such as Cushing's syndrome (adrenocorticotrophic hormone), symptomatic hyponatremia (arginine vasopressin and atrial natriuretic peptide), paraneoplastic cerebellar degeneration (anti-Yo, anti-Hu, anti-Purkinje cell antibodies, and others) and/or Eaton-Lambert neuromuscular dysfunction (IgG autoantibody against the P/Q type of voltage-gated calcium channels, and others) (Lennon *et al.*, 1995; Mason *et al.*, 1997; Posner and Dalmau, 1997). These observations have resulted in several therapeutic or diagnostic clinical trials. For example, monoclonal antibodies directed against gastrin-releasing peptide have been tested as both a radiolabeled isotope marker for tumor localization and for therapeutic efficacy to disrupt a potential positive autocrine growth loop (Chaudhry *et al.*, 1999; Kelley *et al.*, 1997; Yang *et al.*, 1998). In addition, overexpression of the B isoenzyme of creatine kinase has been exploited to test if NMR analysis of phosphocreatine in lung masses can serve as a diagnostic tool (Kristjansen *et al.*, 1991) or if creatine analogs can induce growth inhibition *in vitro* and *in vivo* (Schimmel *et al.*, 1996). Since some of these ectopic peptides are similar to enzymes associated with the neural amine precursor uptake and decarboxylation (APUD) system that is scattered around the body (Erlandson and Nesland, 1994; Langley, 1994), the predominant hypothesis is that SCLC cells arise from a rare collection of normal neuroendocrine cells normally located beneath the bronchial mucosa. This implies that the distinct clinico-pathologic entity of SCLC arises from the pre-programmed pattern of differentiation of these rare neuroendocrine cells. An alternate hypothesis, however, proposes that all lung cancers arise from the same pluripotent lung stem cell (Gazdar *et al.*, 1985). This hypothesis is supported by the observation that (i) cells with features of both SCLC and non-SCLC are occasionally observed admixed within the same tumor biopsy sample, (ii) approximately 10–15% of non-SCLC exhibit neuroendocrine features and a similar number of SCLC express no neuroendocrine products (Linnoila *et al.*, 1988), and (iii) the *in vitro* demonstration that expression of an activated *HRAS* gene can induce SCLC cells to acquire morphologic and biologic features of non-SCLC tumor cells (Falco *et al.*, 1990).

2.2 Non-small cell lung cancer

Non-SCLC is a non-descriptive term that unites several histologically different types of lung cancer into one broad group that constitutes approximately 80% of all cases of lung cancer in the U.S. (Ginsberg *et al.*, 1997; Greenlee *et al.*, 2000). The major forms of non-SCLC are adenocarcinoma, bronchioloalveolar carcinoma (a type of adenocarcinoma), squamous cell carcinoma, large cell carcinoma, undifferentiated lung carcinoma, and undifferentiated carcinoma with neuroendocrine features. In addition, typical and non-typical carcinoid tumors of the lung and mesothelioma tumors of the pleural cavity are occasionally included in the category of non-SCLC, although they are better catalogued with tumors of the endocrine system and the pleura, respectively. Although each subtype of non-SCLC exhibits a markedly different pattern of cellular differentiation, their clinical presentation and response to treatment are sufficiently similar to be a useful classification for clinicians and surgeons. For example, the optimal treatment plan for all cases of early stage non-SCLC is a surgical resection with curative intent, while this option is rarely possible for patients with SCLC who are highly sensitive to their initial chemotherapy or radiotherapy regimens and often present to their physicians with extensive micrometastasis. A subtype of non-SCLC that can occasionally exhibit a unique clinical presentation is bronchioloalveolar carcinoma. This tumor, which is commonly seen in non-smokers, has a propensity for developing multifocal tumors that often behave as distinct primary tumors rather than metastatic disease. In addition, lung tumors with a similar histologic appearance can be observed as a horizontally transmitted epidemic in sheep. A pathogenic virus has been isolated from these animals, designated jaagsiekte sheep retrovirus, which has suggested a viral hypothesis for bronchioloalveolar carcinoma (Palmarini et al., 1997, 1999). No evidence for detection of a virus or retroviral reverse transcriptase activity in human tissues, however, has been confirmed to date.

Due to the increased prevalence of non-SCLC and to the frequency of surgical resections, this disease has allowed investigators to collect large numbers of primary tumor samples that can be matched with clinical information for laboratory analyses. In addition, non-SCLC arises from foci of dysplastic surface epithelial cells within the bronchial tree which has allowed investigators to collect cells and study the step-wise morphological transition through metaplasia, dysplasia, carcinoma in situ, and invasive cancer. Patients with SCLC, in contrast, rarely undergo surgery and it is very difficult to collect sufficient primary samples for genetic or biochemical analyses. Furthermore, SCLC arises as a tumor mass from underneath the mucosal epithelial surface which has prevented the identification of both early steps in tumor development and the clear isolation of the cells of origin. For these reasons, the majority of the work in the genotype analysis of primary lung cancer, in identifying prognostic factors and in assigning a multistep temporal order for gene inactivation, has been undertaken exclusively for non-SCLC samples.

3. Cytogenetics of lung cancer

Since the initial studies on a series of derived lung tumor cell lines (Whang-Peng et al., 1982), standard and molecular cytogenetic analyses continue to be the driving force in guiding investigations into the genetic etiology of lung cancer. The striking feature about the chromosomal karyotype of both SCLC and non-SCLC is the marked aneuploidy with extensive interstitial deletions, non-reciprocal translocations, and gains and losses of chromosome arms (Table 1). New techniques using spectral karyotyping (Schrock et al., 1996) and allelotyping analyses with finely mapped markers only serve to add increasing complexity to the dramatic degree of chromosomal instability and aneuploidy. Despite the almost overwhelming number of chromosomal alterations, a non-random pattern can be elucidated suggesting that many of these changes are linked with the development of these lung tumors. For example, there is compelling evidence that the

Table 1. Cytogenetic alterations in lung cancer

SCLC	non-SCLC	
1p	1p10–13	11p13–15
2q11–13; q33	1q11	11q12–23
3p12–23	2p	13p
4p	2q	13q
4q	3p14–25	14p
5q13–21; q33–35	3q23–q27	14q
6p	4p	15p
6q	4q	16q
8p21–23	5p13	17p11–13
9p	5q11–14; q21	17q11
9q	6p	18q
10q	6q15–27	19p
12p	7p12–15	19q13
13q14	7q	21p
15q	8p21–23	21q11–12
16q	9p21	
17p13	9q32	
18	10q23–q26	

Data from (Fong *et al.*, 1999; Kaye *et al.*, 1995; Kohno and Yokota, 1999; Stanton *et al.*, 2000; Testa and Siegfried, 1992; Whang-Peng *et al.*, 1991).

losses of alleles at 17p and 13q and 9p in lung cancer represent, respectively, the targeting of the *TP53*, *RB1*, and *p16/p14ARF* (*INK4*) tumor suppressor genes (*Table 2*). These findings have stimulated the search for critical genes located at other non-random sites of chromosomal gains or losses. The best example is the worldwide effort to convincingly identify the gene(s) linked to the most common deletion in lung cancer located on the short arm of chromosome 3 (see below). In addition, many of these chromosomal alterations, such as 1p, 3p, 5q, 11p and others, are also detected in many different types of cancer suggesting that the genetic alterations identified will be common to a wide range of solid and hematological malignancies. Finally, the confirmation that candidate genes at these loci are valid 'cancer genes' will continue to be a difficult task as: (i) cancer pathways, not genes, are the main target and a cohesive understanding will not be possible until the components of a specific pathway are identified and studied (Otterson *et al.*, 1994), (ii) epigenetic and gene dosage (haploinsufficiency) effects may be as important as mutational inactivation, and (iii) standard *in vivo* tumor suppression assays may not be applicable to all candidate cancer genes.

4. Recessive genes frequently targeted in lung cancer

4.1 The TP53 gene

The *TP53* gene functions as a sensor of inappropriate internal or external signals which serves to stop the proliferation of mammalian cells that are at risk for aberrant clonal expansion (see Chapter 9). This model helps explain much of the data concerning the *TP53* gene including the observations that: (i) the TP53

Table 2. Genetic alterations in lung cancer

	SCLC	non-SCLC
RB1 inactivation	90%[1]	10–15%
INK4 inactivation	5–10%	75%
TP53 mutation	90–100%	50–70%
KRAS mutation	0%	30%[2]
Telomerase activation	90–100%	75%
MYC overexpression	30%[3]	10%
PTEN mutations	10–20%	0–5%
3p allele loss[4]	100%	70–100%
Absent FHIT protein[5]	50%	75%
Elevated *BCL2*	80%	10–30%
ERBB1/EGFR[6]	20–40%	20–40%
ERBB2/neu	<5% ?	20–30%

[1] Possibly >90% since some RB1 (+) SCLC show histologic features of non-SCLC and may be misclassified (unpublished data).

[2] Prevalence may be lower in non-adenocarcinoma subtypes (Huncharek *et al.*, 1999).

[3] MYC amplification is believed to be a late event and was significantly lower in tumor samples that had not received prior cyclophosphamide/doxorubicin chemotherapy (Brennan *et al.*, 1991).

[4] At least three non-contiguous loci have been identified, but the region flanking 3p21.3 may be the most consistently deleted region.

[5] Genetic alterations are predominantly truncated, splice variants with retention of a full-length 'wild type' transcript which is associated with absent protein expression.

[6] Although *EGFR* expression is linked with the squamous cell subtype of non-SCLC, all lung cancer subtypes show some degree of overexpression by immunohistochemistry.

pathway is inactivated in essentially 100% of human tumors (Pomerantz *et al.*, 1998), (ii) inactivation of the *RB1* tumor suppressor gene leads to rapid growth arrest or apoptosis in TP53(+) cells, while it conversely leads to cell proliferation and clonal expansion in TP53(–) cells (Debbas and White, 1993) and (iii) *tp53* null mice are viable and phenotypically normal, but show a markedly increased predisposition to malignant tumors at a young age (Donehower *et al.*, 1992). In this capacity as a 'guardian of the genome' (Lane, 1992), TP53 activates its DNA binding and gene transactivation functions to induce the gene expression of a growing cassette of products (Levine, 1997; Tanaka *et al.*, 2000; Zhao *et al.*, 2000). These enzymes, in turn, function to maintain organ homeostasis by inducing a G_1/S or G_2/M cell cycle arrest or apoptosis. Understanding the signals that regulate whether the cell undergoes growth arrest, senescence, or apoptosis is still undefined but may relate to how *TP53* is initially activated. For example, different patterns of site-specific phosphorylation within the TP53 product can be detected depending on whether it is functioning to arrest or kill cells (Cuddihy *et al.*, 1999; Unger *et al.*, 1999).

Following the identification of the *TP53* gene as a frequent target for mutational inactivation in a range of different human cancers (Hollstein *et al.*, 1991), analysis of lung tumor samples showed that essentially 100% of SCLC and 50–70 % of non-SCLC had evidence for inactivating *TP53* mutations (Hensel *et al.*, 1991; Mitsudomi *et al.*, 1992; Sameshima *et al.*, 1992; Takahashi *et al.*, 1989). An important issue in understanding the role of *TP53* inactivation in human lung cancer was to define the timing of this alteration in the course of tumor progression. In contrast to studies that suggest that *TP53* mutations and/or LOH are late events in colon and ovarian cancer progression, several different investigators have demonstrated that *TP53* inactivation is observed in early, preneoplastic dysplasia of non-SCLC (Sozzi *et al.*, 1992; Sundaresan *et al.*, 1992). In addition, another study, using enhanced protein staining by immunohistochemistry as an indirect measurement for *TP53* alterations (mutant TP53 protein conformations show an increased protein half-life), demonstrated that mutant *TP53* expression was detected in 0% of normal mucosa, 8.3% of squamous metaplasia, 37.5% of mild dysplasia, 12.5% of moderate dysplasia, 93.8% of severe dysplasia, and 55% of lung carcinoma-in-situ (Bennett *et al.*, 1993). This data showed that *TP53* activation/inactivation was an early event in lung cancer and suggested that it might serve as a useful molecular marker for early diagnosis or prognosis.

Valid prognostic markers would have an impact in the clinical management of lung cancer as many patients with early stage disease relapse locally or with extensive metastasis after aggressive chest resections. In addition, while aggressive treatments with combined radiation/chemotherapy in patients with unresectable, locally advanced lung cancer prolongs median survival by only 8 weeks (as compared to radiation therapy alone), this combined treatment increases (from 5% to 18%) the percent of patients who are disease-free at 5 years (Choi *et al.*, 1997). The ability to use molecular markers in lung cancer to predict a risk–benefit ratio for individual patients, however, has been controversial. Although some studies have suggested an association between altered *TP53* expression and reduced survival (Horio *et al.*, 1993; Marchetti *et al.*, 1993; Quinlan *et al.*, 1992), the general application of this information is still undefined (Graziano *et al.*, 1999). Since it is hypothesized that 100% of all tumors will have the TP53 pathway disrupted (in order to bypass the obligatory TP53 growth arrest/apoptosis response), it is unlikely that altered TP53 function, *per se*, will have prognostic information. There is, however, considerable *in vitro* data to suggest that selected missense *TP53* mutations may confer a more deleterious, 'gain-of-function' phenotype than other types of 'loss-of-function' mutational defects, suggesting that genotype/phenotype analyses may still have a potential role in stratifying patient responses.

4.2 The RB1 gene

While the *RB1* gene was initially isolated as the susceptibility locus for the development of rare, pediatric retinal tumors (Friend *et al.*, 1986; Fung *et al.*, 1987; Lee *et al.*, 1987), functional analyses of its encoded protein product suggested that it played an essential role in the G_1 to S transition of the cell cycle within all higher eukaryotic cells. The critical role for the RB1 product as a general regulator of cell cycle control for retinal and non-retinal tissues, however, contrasts with data from human and mouse models where *RB1* inactivation preferentially targets the differentiation of discrete cell types. For example, rb –/– null mice develop normally *in utero* until day 14 of gestation and then die with abnormalities predominantly within neural and hematopoietic tissues (Jacks *et al.*, 1992). In addition, rb+/– mice develop an unusual spectrum of tumors including pituitary and medullary thyroid cancer that have origins in neural or neuroendocrine cells. These findings may be partly explained by redundancy in tumor suppressor function with other *RB1*-related family members (see below), although, it may also reflect a primary role of RB1 on the differentiation pathways of specific cell lineages.

In the case of human cancers, the neuroendocrine small cell lung cancer is the only sporadic tumor, besides pediatric retinoblastomas, that demonstrates a consistent genetic inactivation within the *RB1* gene. Interest in studying the role of the *RB1* gene in human lung cancer initially arose from cytogenetic analyses that showed many examples of deletions and non-reciprocal translocations involving the *RB1* locus at chromosomal band 13q14 (Harbour *et al.*, 1988) and from RFLP studies that demonstrated 13q allele loss in lung cancer (Yokota *et al.*, 1987). Using nucleic acid and protein analyses (Harbour *et al.*, 1988; Yokota *et al.*, 1988), investigators showed that the majority of SCLC and approximately 10–15% of non-SCLC had targeted mutations within the *RB1* gene. An analysis of 170 lung cancer samples using a sensitive protein immunoblot assay confirmed these original observations (Shimizu *et al.*, 1994).

In addition, analysis of missense or small internal deletions of the *RB* gene in lung cancer samples showed that each of these mutational events were localized to specifically disrupt the RB 'pocket' protein binding activity (Horowitz *et al.*, 1990; Kaye *et al.*, 1990) that is essential for its tumor suppressor function (see Chapter 13). Several investigators have subsequently studied the clinical implications of *RB* mutations in lung cancer. An analysis of primary lung tumors by immunohisto-chemistry through the Lung Cancer Study Group and an analysis of 171 lung cancer cell lines with matched clinical data using protein immunoblotting did not detect a correlation of RB1 status with either time to relapse or survival (Reissmann *et al.*, 1993; Shimizu *et al.*, 1994). Other investigators, however, have shown that altered RB1 protein immunostaining was associated with a worse prognosis in early stage lung tumors (Xu *et al.*, 1994). These analyses, however, are limited by the poor overall survival of most patients with lung cancer and by the small numbers of non-SCLC with absent/mutant *RB1* and the small number of SCLC with wild-type *RB1*. In addition, many immunohistochemical studies overestimate the incidence of *RB1* inactivation in non-SCLC, which is best estimated at approximately 15%.

Although the *RB1* gene pathway is only one of many targets for mutation in lung cancers, the ectopic expression of RB protein in these tumor cells results in colony suppression *in vitro* and tumor suppression *in vivo* (Kratzke *et al.*, 1993; Ookawa *et al.*, 1993). This tumor suppression was incomplete, however, with the eventual growth of small xenograft tumors that still expressed the exogenous wild-type *RB1* (Kratzke *et al.*, 1993). These findings emphasize the different mechanisms that are available to counteract RB1 function. For example, some investigators observed that a small amount of extracellular matrix (matrigel) inoc-ulated with the tumor cells can completely abrogate RB1 suppression activity, perhaps by activating selected cyclin partners which shift RB1 toward an inactive hyperphosphorylated form (Kratzke *et al.*, 1993). Regardless of the mechanism, preclinical studies suggest that introduction of *RB1* alone into *RB1*(-) tumor cells may not be an effective tool for clinical trials.

4.3 Other RB-related family members

The recognition that the viral transforming proteins from adenovirus, simian virus 40 (SV40), and human papillomavirus (HPV) could precipitate the RB1 product led to a model where RB tumor suppressor function is mediated through a 'pocket' protein-binding activity (Weinberg, 1995). In addition to RB1, however, several other protein species are co-precipitated by the same viral transforming proteins (Whyte *et al.*, 1989). Two of these unknown species were initially designated as p107 and p130 according to their apparent molecular weights on SDS-PAGE gels. The subsequent isolation of these genes revealed that they were highly related to RB1, especially in the central domains that are responsible for generating the 'pocket' protein binding function (Ewen *et al.*, 1991; Li *et al.*, 1993; Mayol *et al.*, 1993). The roles of the *RB-related 1* gene (*RBL1/p107*) and the *RB-related 2* gene (*RBL2/p130*), however, are still undefined (Kaye, 1998). Ectopic expression of all three *RB1*-related gene members results in growth suppression of mammalian cells and it has been suggested that inactivation of *RBL1* and *RBL2* may be required *in vitro* to manifest the fully transformed phenotype even in *RB1–/–* cells. In contrast to the *RB1* gene, however, mutational inactivation of these *RB1*-related genes has not, until recently, been observed in human tumor samples. In 1997, investigators reported that 1/19 SCLC cell lines showed inactivation of the *RBL2/p130* gene (Helin *et al.*, 1997). Subsequently other investigators reported that approximately 30% of non-SCLC had altered expression of *RBL2* by immunohistochemistry which was associated with a worse clinical outcome (Baldi *et al.*, 1997). In addition, these authors have reported an unusual clustering of mutations in a wide range of human tumors predominantly within the carboxy-terminal exons of *RBL2* (Cinti *et al.*, 2000; Claudio *et al.*, 2000a). For example, they reported that 11/14 lung cancer samples showed multiple missense and frameshift *RBL2* mutations (Claudio *et al.*, 2000b). In contrast, however, we have not identified mutational inactivation in either the *RBL1/p107* or the *RBL2/p130* gene in a large collection of SCLC and non-SCLC tumor cell lines (Modi *et al.*, 2000). Therefore, while simultaneous inactivation of multiple RB-related family members in lung cancer is an important hypothesis, the role of *RBL1* and *RBL2* is still undefined.

4.4 The p16/ARF locus

The identification of the *CDKN2a/p16* (*INK4*) gene as a tumor suppressor gene was initially controversial because missense mutations were rarely observed in primary tumor samples and homozygous deletions appeared to develop during adaptation to cell culture *in vitro* (Bonetta, 1994; Cairns *et al.*, 1994; Zhang *et al.*, 1994). A key observation, however, was the striking inverse correlation between RB and p16 inactivation in both SCLC and non-SCLC samples (Otterson *et al.*, 1994). This data demonstrated that *RB1* and *INK4* were part of a single pathway where either component (but not both) needed to be inactivated in essentially 100% of human lung tumors. In this model, INK4 normally functions to regulate RB1 phosphorylation by members of the cyclin dependent kinase (cdk): cyclin family. The inactivation of *INK4*, therefore, is predicted to result in constitutive hyperphosphorylation of RB1 which allows for deregulated progression through the cell cycle (Otterson *et al.*, 1994). The propensity for SCLC tumors to target the *RB1* gene for inactivation, while non-SCLC tumors target *INK4* is still unknown.

Since *INK4* gene mutations are uncommonly observed in lung cancer samples with absent INK4 protein, several studies were undertaken to understand the mechanism of inactivation. It was demonstrated that in approximately 40% of non-SCLC the *INK4* gene undergoes gene silencing that is mediated by (or, conversely, is merely associated with) hypermethylation at CpG islands within the promoter and exon 1 of p16 (Merlo *et al.*, 1995; Otterson *et al.*, 1995). In addition, exposure of a demethylating agent called 5′ aza2′deoxycytidine (decitabine) to *INK4-deficient* tumor cells resulted in the induction of wild-type levels of INK4 protein which lasted at least 7 days following a single 24 hour exposure to decitabine (Otterson *et al.*, 1995). Since the ectopic expression of *INK4* can induce the suppression of colony formation *in vitro* (Kratzke *et al.*, 1995) and decitabine has been tested in the treatment of hematological cancers (Kantarjian *et al.*, 1997; Momparler *et al.*, 1997; Sacchi *et al.*, 1999), it was hypothesized that decitabine may also induce tumor suppression *in vivo*. Despite the fact that decitabine has a global effect on demethylating which is toxic to cells in high doses, a clinical trial is currently underway to examine the efficiency of reversing gene silencing at specific cancer gene loci *in vivo* and clinical response in lung cancer and other tumors.

In addition to the *INK4* gene, this locus contains two other relevant gene products. The first identified gene was the duplicated *P15* gene that is located adjacent to *INK4* on chromosome 9p (Larsen, 1997). Although to date there is no defined role for p15 inactivation in lung cancer, this gene is regulated by TGF-β signaling and exhibits many features in common with *INK4*. Of greater interest is another recently identified gene that is partially embedded within the p16 open reading frame designated as *p14ARF* (Sharpless and DePinho, 1999). The *P14ARF* gene functions in a feedback loop to regulate TP53 stability and serves to connect the parallel RB1 and TP53 pathways (Pomerantz *et al.*, 1998; Zhang *et al.*, 1998)). Therefore, many mutations within the *INK4/ARF* open reading frame will serve to inactivate the *RB1* and *TP53* genes simultaneously. Although *RB1* and *INK4* are rarely co-inactivated in lung cancer tissues, it is still unclear whether such a tight inverse correlation will also be observed for the *TP53* and *p14ARF* genes.

4.5 3p *lung cancer gene*

The elusive *3p* gene has been the focus of intensive investigations for two decades because of the high incidence of DNA loss in lung cancer (100% of SCLC and 50–75% of non-SCLC) and because *3p* deletions are also observed in many other tumor types including renal, breast and cervical cancer. A serious handicap to these studies, however, has been the lack of any clearly defined 'lung cancer' families to allow for positional cloning methodologies. Rudimentary mapping efforts in the past decade using (i) primarily LOH studies, (ii) occasional reports of interstitial homozygous deletions, and (iii) mini-chromosome transfer methods suggested that there may be as many as three different loci on chromosome 3p proximal to the *VHL* locus at 3p25. Many candidate genes that localize to this region and which undergo tumor specific allele loss have been studied. These include a thyroid hormone receptor, a retinoic acid receptor, a tyrosine phosphatase, a serine/threonine phosphatase, *VHL*, *FHIT*, SemaIV, an E1-like ubiquitin-activating enzyme, *DUTT1*, *BAP1*, as well as many other transcribed genes of unknown function (Kaye *et al.*, 1995; McLaughlin *et al.*, 2000; Ohta *et al.*, 1996; Wang *et al.*, 1998). Mutations or absent/decreased gene expression have been observed in each of these genes in a subset of lung cancers. In addition, the recent recognition that epigenetic silencing via tumor-specific hypermethylation is an important mechanism or marker for gene inactivation has rekindled interest to take a second look at several of these genes. Ultimately, the completion of the human genome in the next year will allow the formal testing of each gene transcript for tumor suppressor activity within the multiple consensus deletion loci of chromosome 3.

4.6 The FHIT gene

The *FHIT* gene was identified at the breakpoint of a germline t(3;8) translocation found in a three-generation family with susceptibility to renal cell cancer (Cohen *et al.*, 1979; Huebner *et al.*, 1998; Ohta *et al.*, 1996). Aberrant splicing patterns, with absent protein expression, were subsequently found in many patients with common adult malignancies, including lung cancer (Otterson *et al.*, 1998; Sozzi *et al.*, 1996, 1998). Since lung cancer samples frequently show deletions involving the 3p14.2 band, the *FHIT* gene has been proposed as the specific target for these chromosomal aberrations. Further evidence supporting *FHIT* as a tumor suppressor gene for lung cancer is the observation that reintroduction of the wild-type *FHIT* gene into FHIT(-) tumor cells resulted in suppression of tumorigenicity in a nude mouse model system (Ji *et al.*, 1999; Siprashvili *et al.*, 1997). In contrast, several observations argue against a role for FHIT as a 'classic' tumor suppressor gene (Otterson *et al.*, 1998; Wu *et al.*, 2000). The *FHIT* gene covers approximately a megabase of genomic DNA within the most active chromosomal fragile region, designated FRA3B. Accordingly, allelic disruption of the *FHIT* gene may not be so surprising and *FHIT* has been observed as the fusion partner for the *HMGIC* gene in benign tumors of the parotid gland (Geurts *et al.*, 1997). Since the *HMGIC* transcription factor gene is linked to the pathogenesis of benign solid tumors with varied fusion partners, the role of *FHIT* in this tumor is undefined. Similarly, in the renal cancer family with the t(3;8) translocation where the *FHIT* gene was originally cloned (Ohta *et al.*, 1996), the derivative chromosome of the reciprocal translocation creates a fusion between the *FHIT* gene with a novel, Patched-like gene called *TRC8* (Gemmill *et al.*, 1998). Therefore, the role of FHIT as the etiologic agent in the development of renal tumors in this family is also undefined. Inhibition of tumor suppression by ectopic expression of wild-type *FHIT* is also variable as two studies could not detect significant tumor suppression (Otterson *et al.*, 1998; Wu *et al.*, 2000), and the initial report (Siprashvili *et al.*, 1997) did not detect any difference in tumor suppressor activity between wild-type *FHIT* and a mutant that abolished enzymatic dinucleoside hydrolase activity. These observations suggest that the *FHIT* gene, which is associated with absent protein expression in the majority of lung cancer samples, may be involved in tumorigenesis in ways that are distinct from the 'classic' tumor suppressor paradigm.

4.7 The PTEN gene

The *PTEN* gene (also known as *MMAC1* and *TEP1*) was initially isolated as a gene within a common homozygous deletion at chromosome 10q23 (Li *et al.*, 1997; Steck *et al.*, 1997) and by another group who identified the gene as a downstream target for TGF-β function (Li and Sun, 1997). Although the gene encodes both a dual-specificity protein phosphatase and a lipid (PIP-3) phosphatase activity, several tumor specific mutations appear to target exclusively the lipid phosphatase activity (Myers *et al.*, 1998). In addition, the catalytic pocket for PTEN is unusually wide to accommodate the larger lipid substrate, suggesting that this is the primary role for this tumor suppressor gene (Lee *et al.*, 1999).

Although early studies suggested that lung cancers rarely (0–10% frequency) showed inactivation of *PTEN* (Forgacs *et al.*, 1998; Sakurada *et al.*, 1997), other studies showed that 20–40% of SCLC and 0–8% of non-SCLC had evidence for inactivating mutations (Kohno *et al.*, 1998; Yokomizo *et al.*, 1998). In addition, another study demonstrated that 40% of head and neck tumors and 50% of non-SCLC had hemizygous allele loss at the *PTEN* locus (Okami *et al.*, 1998). While the Knudson 2-hit theory (Knudson, 1971) requires genetic or epigenetic targeting of both alleles for recessive cancer gene inactivation, evidence from the *pten* (+/–) heterozygous mouse (Di Cristofano *et al.*, 1999; Sun *et al.*, 1999) and from a growing list of other tumor suppressor genes (*tp53*, and others) suggests that gene dosage alterations (or haploinsufficiency) may be sufficient to exhibit a deleterious phenotype. In summary, *PTEN* inactivation may be a common event in the development of lung cancer, especially SCLC.

4.8 Other predicted recessive oncogenes

As discussed above, an intensive effort has been underway for almost two decades to elucidate the identity of the 'lung cancer' recessive gene that lies within one of at least three different loci within the chromosomal band 3p. Several interesting genes continue to be isolated and it is likely that within the next year a strong candidate for the elusive *3p* gene will be identified. The recognition that 40–50% of primary SCLC tumors have deletions or expansions of $(CA)_n$ dinucleotide repeats, suggested that these tumors have defects in mismatch repair which may be due to loss of the *MLH1* gene localized at chromosome 3p (Merlo *et al.*, 1994). Other investigators, however, have detected much lower rates of microsatellite instability and no abnormalities in either the *MLH1* gene or the *MSH2* genes (Fong *et al.*, 1995; Gotoh *et al.*, 1999). Several other tumor suppressor genes have been studied in lung cancer including members of the TGFβ/SMAD family, *DCC*, and *WT1*, although the incidence of mutations was low and their significance uncertain. Of interest, components of a serine/threonine protein phosphatase have been detected in approximately 15% of lung tumors (Kohno *et al.*, 1999; Wang *et al.*, 1998). If these findings are confirmed it will emphasize the importance of cancer genes in regulating signaling pathways via both kinase and phosphatase activities. In addition, bi-allelic inactivation of another novel human gene (*RASSF1*), located at the minimal consensus deletion region on chromosome 3p21.3, was recently identified in a large subset of primary lung cancer tumors. This Ras-effector homolog was shown to suppress tumor growth *in vitro* and *in vivo* and to undergo mutational inactivation in 10% of tumors and gene silencing by DNA hypermethylation in another 40% of primary tumors (Dammann *et al.*, 2000). Whether the *RASSF1* gene will finally turn out to be the elusive chromosome 3p gene should be answered shortly.

5. Dominant oncogenes in lung cancer

5.1 RAS

The concept of 'dominant' cancer genes arose from early studies on viral tumorigenesis and DNA transfection experiments into rodent cells. This classification, however, is misleading as some genes (i.e. *TP53*) may exhibit a dominant phenotype *in vitro*, but behave predominantly as a recessive oncogene when tested *in vivo* (Hollstein *et al.*, 1991; Malkin *et al.*, 1990; Srivastava *et al.*, 1990). In addition, other oncogenes may sometimes be subject to gene dosage effects (or haploinsufficiency) where retention of the wild-type allele is not sufficient to prevent aberrant gene function (Gottlieb *et al.*, 1997). More importantly, all cancer genes belong to parallel and interconnecting tumor suppressor pathways. The activated *RAS* gene is a good example of this model where (i) it is a classic 'dominant' oncogene by both viral studies and DNA transfection assays, but may require a relative loss of wild-type expression to manifest tumorigenicity (Finney and Bishop, 1993) and (ii) also belongs to well-described signal transduction pathways that impact on both RB1 and TP53 biology (Bates *et al.*, 1998; Leone *et al.*, 1997).

The *RAS* genes (predominantly *HRAS*, *KRAS* and *NRAS* family members) function as cytoplasmic switches that transduce external growth signals into a cascade toward the nucleus to regulate gene transcription. Somatic mutations at a few defined sites within the *RAS* gene appear to lock the protein in a conformation that constitutively signals growth stimulation (Barbacid, 1987). Analysis of lung tumor samples demonstrated that 20–30% of non-SCLC exhibited activating *KRAS* mutations (lower incidence in non-adenocarcinoma subtypes) while 0% of SCLC had *KRAS* mutations (Mitsudomi *et al.*, 1991b; Rodenhuis and Slebos, 1990; Slebos *et al.*, 1991). There is also considerable heterogeneity in other tumor types. For example, breast cancers rarely contain *RAS* mutations (*HRAS*), while 50% of colon adenocarcinomas and almost 90% of pancreatic adenocarcinomas carry activating *KRAS* alterations. In the case of lung cancer, *RAS* mutations have consistently been observed as late events in tumor progression (Sugio *et al.*, 1994; Zhang *et al.*, 1996) which correlates with observations in colon cancer where early activation of *RAS* is directly selected against (Jen *et al.*, 1994). In addition, multiple studies have demonstrated that *RAS* mutations in lung cancer are associated with increased tobacco use and with a shortened survival (Mitsudomi *et al.*, 1991a; 1991; Slebos *et al.*, 1990, 1991).

5.2 MYC

Early cytogenetic studies on SCLC cell lines showed homogeneously staining regions and double minute chromosomal fragments that were subsequently shown to represent regions of DNA amplification of the *MYC*, *MYCN*, and *MYCL* oncogenes (Kaye *et al.*, 1988; Little *et al.*, 1983; Nau *et al.*, 1985, 1986). While approximately 25–30% of SCLC cell lines showed amplification of one member of the *MYC* family, co-amplification of different *MYC*-related genes in the same cell line was rarely observed. In addition, the incidence of *MYC* amplification in primary tumor samples was lower (5–18%) suggesting that this occurred as a late event (Richardson and Johnson, 1993; Wong *et al.*, 1986). Since amplification of *MYCN* is associated with a worse prognosis in pediatric neuroblastoma (Brodeur *et al.*, 1984), several studies have also reported a small decrease in survival in lung tumors with DNA amplification of *MyYC* but not with *MYCN* or *MYCL* (Brennan *et al.*, 1991). The molecular consequences of *Myc* gene deregulation in lung cancer, however, is poorly understood. Although, activating missense mutations within the N-terminal transactivation domain of *MYC* have been observed in several Burkitts lymphoma samples (Bhatia *et al.*, 1993), similar structural mutations have not been reported for lung cancer. Instead, an intra-chromosomal rearrangement fusing *MYCL* with the adjacent *RLF* gene was observed in six independent SCLC samples with *MYCL* amplification (Kim *et al.*, 1998; Makela *et al.*, 1991). In addition, an anti-sense *MYCL* gene was fused with a novel cyclophilin-like gene in one SCLC sample (Kim *et al.*, 1998). The role of these chimeric mRNAs in the development or progression of these lung tumors is still unknown, however, RLF has been shown to be a zinc binding protein that can affect embryonal development in rodents when fused to *MYCL* (MacLean-Hunter *et al.*, 1994; Makela *et al.*, 1995).

5.3 SV40 infection

Human papillomavirus (HPV) is clearly established as a causative agent for the development of cervical cancer and selected anal and tonsillar tumors (zur Hausen, 2000). Since the transforming HPV proteins, E6 and E7, can sequester and/or degrade the TP53 and RB1 products, it was observed that HPV(+) cervical tumors retain wild-type *RB* and *p53* genes while HPV(–) tumors show mutational inactivation of *RB1* and *TP53* (Scheffner *et al.*, 1991). Using the analogy of HPV-induced cancers, several investigators have detected DNA sequences corresponding to the SV40 large T antigen (Tag) in several human tumors including a subset of lung tumors and the majority of pleural mesotheliomas (Butel and Lednicky, 1999; Carbone *et al.*, 1994; Waheed *et al.*, 1999). Since the Tag oncoprotein also binds and inactivates RB1 and TP53 function, these observations raised the question of whether the SV40 DNA tumor virus may be pathogenic in these mesotheliomas as well. The copy number of Tag sequences in these samples, however, is low since a two-step polymerase chain reaction (PCR) methodology was required and other investigators could not confirm these findings (Strickler *et al.*, 1996, 1998). More importantly, since the RB1 and TP53 pathways are already inactivated by loss of INK4 function in 100% of mesotheliomas (Kratzke *et al.*, 1995) and deletions within the *INK4/ARF* locus occur in the majority of primary mesothelioma tumors (Cheng *et al.*, 1994), the need for Tag expression is unclear. Despite these concerns, a preclinical rationale for Tag expression was suggested when the N-terminal 'J domain' of Tag was shown to be essential for inducing a fully transformed phenotype, even in *RB1* –/– cell lines (Christensen and Imperiale, 1995; Zalvide *et al.*, 1998). The possibility of a role for SV40 in the etiology of certain human chest tumors is, therefore, of public health importance.

5.4 Other predicted dominant oncogenes

Activation of many other cancer genes has been reported in lung cancer. The most compelling data is for activation of telomerase activity in 100% of SCLC and 85% of non-SCLC (Albanell *et al.*, 1997; Hiyama *et al.*, 1995). Activation of telomerase (mediated largely by induction of the telomerase holoenzyme) (Prescott and Blackburn, 1999) is a critical step in bypassing cell crisis and/or senescence to confer immortality and appears as an early event in the tumorigenic process (see Chapter 21). Of interest, a telomerase repressor locus for breast cancer has been proposed within the lung cancer 'minimal deletion' region of chromosome 3p (Cuthbert *et al.*, 1999; Shay, 1999).

Two different pathways to modulate programmed cell death events have been proposed in lung cancer. The first pathway involves the overexpression of Fas ligand, without the induction of the Fas receptor, in many lung cancer samples. This observation has been hypothesized to generate a paracrine mechanism to bypass T-cell anti-tumor responses by inducing the apoptosis of T-cells adjacent to the tumor (Niehans *et al.*, 1997). In addition, elevated expression of the anti-apoptotic *BCL2* gene is observed in approximately 80% of SCLC and 10–30% of non-SCLC, where it is suggested to confer a slightly improved response to therapy and survival (Apolinario *et al.*, 1997; Kaiser *et al.*, 1996; Pezzella *et al.*, 1993). The role for *FAS* and *BCL2* expression in lung cancer, however, is still undefined.

Activation of *ERBB1/EGFR*, *ERBB2/neu*, *RAF1*, HGF, IGF-like1, IGF-like2, *KIT*, E-cadherin, and *MYB* have also been reported in subsets of lung cancer samples (Bongiorno *et al.*, 1995; Cline and Battifora, 1987; Hida *et al.*, 1994; Kiefer *et al.*, 1987; Rachwal *et al.*, 1995; Yu *et al.*, 1994). The best studied of these have been members of the ERBB family where *EGFR* is overexpressed predominantly in squamous cell carcinomas, while *ERBB2* levels are predominantly elevated in the adenocarcinoma subtype of non-SCLC. The generation of monoclonal antibodies directed against these receptor tyrosine kinases and the synthesis of small molecules to block receptor activation has generated considerable interest for preclinical and *in vivo* clinical trials.

6. Therapeutic considerations

6.1 Tobacco cessation

The most tangible therapeutic option for lung cancer involves controlling the exposure of the general population to tobacco smoke. As a consequence, the National Cancer Institute has recently established an 'extraordinary opportunity' grant mechanism for implementing a national program to control tobacco addiction that will include genetic epidemiology, animal models, and community-based tobacco cessation clinics. These efforts should also take advantage of ongoing international ventures to study single nucleotide polymorphisms (SNPs) in defined ethnic groups. Ultimately, this latter approach may help map loci which are involved in either susceptibility to tobacco-related cancers or susceptibility to addictive behaviors. This work will also complement current efforts to identify and understand the role of activated carcinogens in tobacco smoke (Hecht, 1999). For example, a characteristic mutational pattern may serve as a fingerprint for specific mutagens in the environment (Harris and Hollstein, 1993) where mutational hot spots within selected oncogenes may be targeted by specific polycylic aromatic hydrocarbons (Smith et al., 2000). In addition, an intuitive working model for the etiology of carcinogen-associated lung cancer proposes that relative risks of lung cancer may vary with the ability of the host to either activate or clear the proximate carcinogens in lung tissues or to repair DNA damage (Shields and Harris, 2000; Spitz et al., 1999). Accordingly, an extensive literature exists where genetic polymorphisms within selected candidate genes have been tested for their impact on both carcinogen metabolism and lung cancer risk. The cytochrome P450 *CYP1A1* gene (activates several major classes of tobacco carcinogens) and the *GSTM1* gene (detoxifies carcinogens) have been the best studied polymorphic gene products to date, although many other genes have also been tested (Bartsch et al., 2000; Shields and Harris, 2000; Spitz et al., 1999). An underlying premise to these studies was the observation that only 10% of modest-to-heavy tobacco abusers (>25 cigarettes per day) developed lung cancer (Doll and Hill, 1950). A recent examination of cohorts from 1950 and 1990 in the United Kingdom, however, concluded that this earlier study underestimated the cumulative risk of developing lung cancer, presumably because many of the earlier subjects from 1950 had initiated their smoking habits after the age of 20 years. In fact, the more current 1990 cohort demonstrated a significantly higher risk of lung cancer at all levels of cigarette abuse, including a 25% cumulative risk of lung cancer in the '>25 cigarettes per day' category (Peto et al., 2000). Fortunately, the prevalence of cigarette abuse has decreased substantially in the United Kingdom (and to less of an extent in the United States) over the past 30 years, and the mortality from lung cancer in younger males (age 35–44 years) has dropped even greater than expected, perhaps suggesting a reduction in the hazard of smoking that may be associated with changes in tobacco constituents or other environmental exposures. These persistent, large fluctuations in lung cancer incidence will continue to make the isolation of relatively minor, low-penetrant susceptiblity loci a difficult goal.

6.2 Molecular based strategies

A large number of molecular-based diagnostic and therapeutic strategies are also underway. These strategies focus on cancer gene pathways that are known to undergo mutational inactivation in lung cancer (*Figure 1*) or on altered expression of non-specific pathways that facilitate cellular proliferation such as angiogenesis or tumor invasion through the extracellular matrix. For example, the *TP53* gene has been an attractive target for lung cancer since pre-clinical studies have shown that

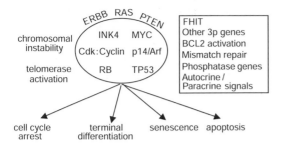

Figure 1. Genes commonly altered in lung cancer. A direct role for the genes enclosed within the rectangle is still under investigation.

introduction of an ectopic wild-type gene efficiently induces growth arrest/death (Takahashi *et al.*, 1992). Consequently, adenoviral-TP53 vectors have been tested in clinical trials that were focused primarily on treating local bronchial lesions (Nemunaitis *et al.*, 2000; Roth *et al.*, 1999). Another strategy directed at TP53 function takes advantage of the observation that certain viruses (ONYX-015) exhibit increased cytopathogenicity in tumor cells with altered TP53 function (Heise *et al.*, 1997; You *et al.*, 2000). Other investigators are working to identify small molecule inhibitors of the cyclin-regulated cell cycle pathways which would be predicted to restrain cell growth by locking the RB1 pathway in an 'inactive' hypophosphorylated form (Chen *et al.*, 1999; Kubo *et al.*, 1999; Senderowicz and Sausville, 2000). In contrast, alternate strategies exploit known biochemical properties of oncogenes, such as blocking the ability of the *RAS* oncogene to attach to the plasma membrane which is required for growth proliferation signals (Rowinsky *et al.*, 1999). Finally, clinical trials with synthetic antisense molecules, humanized monoclonal antibodies, or blocking small molecules which are directed against growth factors and/or receptors (including ERBB family members), angiogenesis pathways, or components of the apoptotic cascade are already underway with promising results in lung cancer. A challenge for many of these novel treatments, however, is to develop practical biological assays of effectiveness, especially for agents such as extracellular matrix metalloprotease inhibitors which are predicted to be cytostatic rather than cytocidal (Cockett *et al.*, 1998; Nelson *et al.*, 2000).

7. Conclusions

Knowledge about the initiation and progression of lung tumors continues to grow at a rapid pace. The completion of the Human Genome Sequence project, the careful assignment of the specific gene pathways that are altered in lung cancer, the generation of relevant animal models for SCLC and non-SCLC subtypes, and the development of large nucleotide databases (such as SNPs databases) will ultimately help health professionals focus on the optimal strategies for prevention, early detection, and treatment and may also help patients understand the biology of tobacco addiction. In the meantime patients with established lung cancer and high-risk cancer-free patients should be encouraged to enter onto clinical trials. Although there is a plethora of genetic biomarkers that are currently being tested as surrogates for prevention therapies in high-risk patients, as diagnostic markers in early detection, and as prognostic markers in patients with lung cancer, none of these to date has a proven role in clinical management. The challenge for the future will be to incorporate this new genetic information in a logical algorithm for clinical management.

References

Albanell, J., Lonardo, F., Rusch, V. *et al.* (1997) High telomerase activity in primary lung cancers: association with increased cell proliferation rates and advanced pathologic stage. *J. Natl Cancer Inst.* **89:** 1609–1615.

Apolinario, R.M., van der Valk, P., de Jong, J.S. *et al.* (1997) Prognostic value of the expression of p53, bcl-2, and bax oncoproteins, and neovascularization in patients with radically resected non-small-cell lung cancer. *J. Clin. Oncol.* **15:** 2456–2466.

Baldi, A., Esposito, V., De Luca, A., Fu, Y., Meoli, I., Giordano, G.G., Caputi, M., Baldi, F. and Giordano, A. (1997) Differential expression of Rb2/p130 and p107 in normal human tissues and in primary lung cancer. *Clin. Cancer Res.* **3:** 1691–1697.

Barbacid, M. (1987) ras genes. *Annu. Rev. Biochem.* **56:** 779–827.

Bartsch, H., Nair, U., Risch, A., Rojas, M., Wikman, H. and Alexandrov, K. (2000) Genetic polymorphism of CYP genes, alone or in combination, as a risk modifier of tobacco-related cancers. *Cancer Epidemiol. Biomarkers Prev.* **9:** 3–28.

Bates, S., Phillips, A.C., Clark, P.A., Stott, F., Peters, G., Ludwig, R.L. and Vousden, K.H. (1998) p14ARF links the tumour suppressors RB and p53 [letter]. *Nature* **395:** 124–125.

Bennett, W.P., Colby, T.V., Travis, W.D. *et al.* (1993) p53 protein accumulates frequently in early bronchial neoplasia. *Cancer Res.* **53:** 4817–4822.

Bhatia, K., Huppi, K., Spangler, G., Siwarski, D., Iyer, R. and Magrath, I. (1993) Point mutations in the c-Myc transactivation domain are common in Burkitt's lymphoma and mouse plasmacytomas. *Nat. Genet.* **5:** 56–61.

Bonetta, L. (1994) Tumor-suppressor genes. Open questions on p16 [news; comment]. *Nature* **370:** 180.

Bongiorno, P.F., al-Kasspooles, M., Lee, S.W., Rachwal, W.J., Moore, J.H., Whyte, R.I., Orringer, M.B. and Beer, D.G. (1995) E-cadherin expression in primary and metastatic thoracic neoplasms and in Barrett's oesophagus. *Br. J. Cancer* **71:** 166–172.

Brennan, J., O'Connor, T., Makuch, R.W. *et al.* (1991) myc family DNA amplification in 107 tumors and tumor cell lines from patients with small cell lung cancer treated with different combination chemotherapy regimens. *Cancer Res.* **51:** 1708–1712.

Brodeur, G.M., Seeger, R.C., Schwab, M., Varmus, H.E. and Bishop, J.M. (1984) Amplification of N-myc in untreated human neuroblastomas correlates with advanced disease stage. *Science* **224:** 1121–1124.

Butel, J.S. and Lednicky, J.A. (1999) Cell and molecular biology of simian virus 40: implications for human infections and disease. *J. Natl Cancer Inst.* **91:** 119–134.

Cairns, P., Mao, L., Merlo, A. *et al.* (1994) Rates of p16 (MTS1) mutations in primary tumors with 9p loss [letter]. *Science* **265:** 415–417.

Carbone, M., Pass, H.I., Rizzo, P., Marinetti, M., Di Muzio, M., Mew, D.J., Levine, A.S. and Procopio, A. (1994) Simian virus 40-like DNA sequences in human pleural mesothelioma. *Oncogene* **9:** 1781–1790.

Carney, D.N., Gazdar, A.F., Bepler, G., Guccion, J.G., Marangos, P.J., Moody, T.W., Zweig, M.H. and Minna, J.D. (1985) Establishment and identification of small cell lung cancer cell lines having classic and variant features. *Cancer Res.* **45:** 2913–2923.

Chaudhry, A., Carrasquillo, J.A., Avis, I.L. *et al.* (1999) Phase I and imaging trial of a monoclonal antibody directed against gastrin-releasing peptide in patients with lung cancer. *Clin. Cancer Res.* **5:** 3385–3393.

Chen, Y.N., Sharma, S.K., Ramsey, T.M., Jiang, L., Martin, M.S., Baker, K., Adams, P.D., Bair, K.W. and Kaelin, W.G., Jr. (1999) Selective killing of transformed cells by cyclin/cyclin-dependent kinase 2 antagonists [see comments]. *Proc. Natl Acad. Sci. USA* **96:** 4325–4329.

Cheng, J.Q., Jhanwar, S.C., Klein, W.M. *et al.* (1994) p16 alterations and deletion mapping of 9p21-p22 in malignant mesothelioma. *Cancer Res.* **54:** 5547–5551.

Choi, N.C., Carey, R.W., Daly, W., Mathisen, D., Wain, J., Wright, C., Lynch, T., Grossbard, M. and Grillo, H. (1997) Potential impact on survival of improved tumor downstaging and resection rate by preoperative twice-daily radiation and concurrent chemotherapy in stage IIIA non-small-cell lung cancer. *J. Clin. Oncol.* **15**: 712–722.

Christensen, J.B. and Imperiale, M.J. (1995) Inactivation of the retinoblastoma susceptibility protein is not sufficient for the transforming function of the conserved region 2-like domain of simian virus 40 large T antigen. *J. Virol.* **69**: 3945–3948.

Cinti, C., Leoncini, L., Nyongo, A. *et al.* (2000) Genetic alterations of the retinoblastoma-related gene RB2/p130 identify different pathogenetic mechanisms in and among Burkitt's lymphoma subtypes. *Am. J. Pathol.* **156**: 751–760.

Claudio, P.P., Howard, C.M., Fu, Y., Cinti, C., Califano, L., Micheli, P., Mercer, E.W., Caputi, M. and Giordano, A. (2000a) Mutations in the retinoblastoma-related gene RB2/p130 in primary nasopharyngeal carcinoma [In Process Citation]. *Cancer Res.* **60**: 8–12.

Claudio, P.P., Howard, C.M., Pacilio, C. *et al.* (2000b) Mutations in the retinoblastoma-related gene RB2/p130 in lung tumors and suppression of tumor growth in vivo by retrovirus-mediated gene transfer. *Cancer Res.* **60**: 372–382.

Cline, M.J. and Battifora, H. (1987) Abnormalities of protooncogenes in non-small cell lung cancer. Correlations with tumor type and clinical characteristics [published erratum appears in Cancer 1988 Mar 1;61(5): 1064]. *Cancer* **60**: 2669–2674.

Cockett, M.I., Murphy, G., Birch, M.L., O'Connell, J.P., Crabbe, T., Millican, A.T., Hart, I.R. and Docherty, A.J. (1998) Matrix metalloproteinases and metastatic cancer. *Biochem. Soc. Symp.* **63**: 295–313.

Cohen, A.J., Li, F.P., Berg, S., Marchetto, D.J., Tsai, S., Jacobs, S.C. and Brown, R.S. (1979) Hereditary renal-cell carcinoma associated with a chromosomal translocation. *N. Engl. J. Med.* **301**: 592–595.

Cuddihy, A.R., Li, S., Tam, N.W., Wong, A.H., Taya, Y., Abraham, N., Bell, J.C. and Koromilas, A.E. (1999) Double-stranded-RNA-activated protein kinase PKR enhances transcriptional activation by tumor suppressor p53. *Mol. Cell. Biol.* **19**: 2475–2484.

Cuthbert, A.P., Bond, J., Trott, D.A. *et al.* (1999) Telomerase repressor sequences on chromosome 3 and induction of permanent growth arrest in human breast cancer cells [see comments]. *J. Natl Cancer Inst.* **91**: 37–45.

Cuttitta, F., Carney, D.N., Mulshine, J., Moody, T.W., Fedorko, J., Fischler, A. and Minna, J.D. (1985) Bombesin-like peptides can function as autocrine growth factors in human small-cell lung cancer. *Nature* **316**: 823–826.

Dammann, R., Li, C., Yoon, J.H., Chin, P.L., Bates, S. and Pfeifer, G.P. (2000) Epigenetic inactivation of a RAS association domain family protein from the lung tumour suppressor locus 3p21.3 [In Process Citation]. *Nat. Genet.* **25**: 315–319.

Debbas, M. and White, E. (1993) Wild-type p53 mediates apoptosis by E1A, which is inhibited by E1B. *Genes Dev.* **7**: 546–554.

Dennis, T.R. and Stock, A.D. (1999) A molecular cytogenetic study of chromosome 3 rearrangements in small cell lung cancer: consistent involvement of chromosome band 3q13.2. *Cancer Genet. Cytogenet.* **113**: 134–140.

Di Cristofano, A., Kotsi, P., Peng, Y.F., Cordon-Cardo, C., Elkon, K.B. and Pandolfi, P.P. (1999) Impaired Fas response and autoimmunity in Pten+/− mice. *Science* **285**: 2122–2125.

Doll, R. and Hill, A.B. (1950) Smoking and carcinoma of the lung. Preliminary report. *BMJ* **2**: 739–748.

Donehower, L.A., Harvey, M., Slagle, B.L., McArthur, M.J., Montgomery, C., Jr., Butel, J.S. and Bradley, A. (1992) Mice deficient for p53 are developmentally normal but susceptible to spontaneous tumours. *Nature* **356**: 215–221.

Erlandson, R.A. and Nesland, J.M. (1994) Tumors of the endocrine/neuroendocrine system: an overview. *Ultrastruct. Pathol.* **18**: 149–170.

Ewen, M.E., Xing, Y.G., Lawrence, J.B. and Livingston, D.M. (1991) Molecular cloning, chromosomal mapping, and expression of the cDNA for p107, a retinoblastoma gene product-related protein. *Cell* **66**: 1155–1164.

Falco, J.P., Baylin, S.B., Lupu, R., Borges, M., Nelkin, B.D., Jasti, R.K., Davidson, N.E. and Mabry, M. (1990) v-rasH induces non-small cell phenotype, with associated growth factors and receptors, in a small cell lung cancer cell line. *J. Clin. Invest.* **85**: 1740–1745.

Finney, R.E. and Bishop, J.M. (1993) Predisposition to neoplastic transformation caused by gene replacement of H-ras1. *Science* **260**: 1524–1527.

Fong, K.M., Zimmerman, P.V. and Smith, P.J. (1995) Microsatellite instability and other molecular abnormalities in non-small cell lung cancer. *Cancer Res* **55**: 28–30.

Fong, K.M., Sekido, Y. and Minna, J.D. (1999) Molecular pathogenesis of lung cancer. *J. Thorac. Cardiovasc. Surg.* **118**: 1136–1152.

Forgacs, E., Biesterveld, E.J., Sekido, Y. *et al.* (1998) Mutation analysis of the PTEN/MMAC1 gene in lung cancer. *Oncogene* **17**: 1557–1565.

Friend, S.H., Bernards, R., Rogelj, S., Weinberg, R.A., Rapaport, J.M., Albert, D.M. and Dryja, T.P. (1986) A human DNA segment with properties of the gene that predisposes to retinoblastoma and osteosarcoma. *Nature* **323**: 643–646.

Fung, Y.K., Murphree, A.L., T'Ang, A. *et al.* (1987) Structural evidence for the authenticity of the human retinoblastoma gene. Inactivation of the retinoblastoma susceptibility gene in human breast cancers. *Science* **236**: 218–221.

Gauderman, W.J. and Morrison, J.L. (2000) Evidence for age-specific genetic relative risks in lung cancer. *Am. J. Epidemiol.* **151**: 41–49.

Gazdar, A.F., Bunn, P.A., Jr., Minna, J.D. and Baylin, S.B. (1985) Origin of human small cell lung cancer [letter]. *Science* **229**: 679–680.

Gemmill, R.M., West, J.D., Boldog, F., Tanaka, N., Robinson, L.J., Smith, D.I., Li, F. and Drabkin, H.A. (1998) The hereditary renal cell carcinoma 3;8 translocation fuses FHIT to a patched-related gene, TRC8. *Proc. Natl Acad. Sci. USA* **95**: 9572–9577.

Geurts, J.M., Schoenmakers, E.F., Roijer, E., Stenman, G. and Van de Ven, W.J. (1997) Expression of reciprocal hybrid transcripts of HMGIC and FHIT in a pleomorphic adenoma of the parotid gland. *Cancer Res.* **57**: 13–17.

Ginsberg, R., Vokes, E. & Raben, A. (1997) *Non-small cell lung cancer*, 5th edn. Lippincott-Raven, Philadelphia/New York.

Gotoh, K., Yatabe, Y., Sugiura, T., Takagi, K., Ogawa, M., Takahashi, T. and Mitsudomi, T. (1999) Frameshift mutations in TGFbetaRII, IGFIIR, BAX, hMSH3 and hMSH6 are absent in lung cancers. *Carcinogenesis* **20**: 499–502.

Gottlieb, E., Haffner, R., King, A., Asher, G., Gruss, P., Lonai, P. and Oren, M. (1997) Transgenic mouse model for studying the transcriptional activity of the p53 protein: age- and tissue-dependent changes in radiation-induced activation during embryogenesis. *EMBO J.* **16**: 1381–1390.

Graziano, S.L., Gamble, G.P., Newman, N.B., Abbott, L.Z., Rooney, M., Mookherjee, S., Lamb, M.L., Kohman, L.J. and Poiesz, B.J. (1999) Prognostic significance of K-ras codon 12 mutations in patients with resected stage I and II non-small-cell lung cancer. *J. Clin. Oncol.* **17**: 668–675.

Greenlee, R.T., Murray, T., Bolden, S. and Wingo, P.A. (2000) Cancer statistics, 2000. *CA Cancer J. Clin.* **50**: 7–33.

Harbour, J.W., Lai, S.L., Whang-Peng, J., Gazdar, A.F., Minna, J.D. and Kaye, F.J. (1988) Abnormalities in structure and expression of the human retinoblastoma gene in SCLC. *Science* **241**: 353–357.

Harris, C.C. and Hollstein, M. (1993) Clinical implications of the p53 tumor-suppressor gene. *N. Engl. J. Med.* **329**: 1318–1327.

Hecht, S.S. (1999) Tobacco smoke carcinogens and lung cancer. *J. Natl Cancer Inst.* **91**: 1194–1210.

Heise, C., Sampson-Johannes, A., Williams, A., McCormick, F., Von Hoff, D.D. and Kirn, D.H. (1997) ONYX-015, an E1B gene-attenuated adenovirus, causes tumor-specific cytolysis and antitumoral efficacy that can be augmented by standard chemotherapeutic agents [see comments]. *Nat. Med.* **3**: 639–645.

Helin, K., Holm, K., Niebuhr, A., Eiberg, H., Tommerup, N., Hougaard, S., Poulsen, H.S., Spang-Thomsen, M. and Norgaard, P. (1997) Loss of the retinoblastoma protein-related p130 protein in small cell lung carcinoma. *Proc. Natl Acad. Sci. USA* **94**: 6933–6938.

Hensel, C.H., Xiang, R.H., Sakaguchi, A.Y. and Naylor, S.L. (1991) Use of the single strand conformation polymorphism technique and PCR to detect p53 gene mutations in small cell lung cancer. *Oncogene* **6**: 1067–1071.

Hida, T., Ueda, R., Sekido, Y., Hibi, K., Matsuda, R., Ariyoshi, Y., Sugiura, T., Takahashi, T. and Takahashi, T. (1994) Ectopic expression of c-kit in small-cell lung cancer. *Int. J. Cancer Suppl.* **8**: 108–109.

Hiyama, K., Hiyama, E., Ishioka, S., Yamakido, M., Inai, K., Gazdar, A.F., Piatyszek, M.A. and Shay, J.W. (1995) Telomerase activity in small-cell and non-small-cell lung cancers [see comments]. *J. Natl Cancer Inst.* **87**: 895–902.

Hollstein, M., Sidransky, D., Vogelstein, B. and Harris, C.C. (1991) p53 mutations in human cancers. *Science* **253**: 49–53.

Horio, Y., Takahashi, T., Kuroishi, T. *et al.* (1993) Prognostic significance of p53 mutations and 3p deletions in primary resected non-small cell lung cancer. *Cancer Res.* **53**: 1–4.

Horowitz, J.M., Park, S.H., Bogenmann, E., Cheng, J.C., Yandell, D.W., Kaye, F.J., Minna, J.D., Dryja, T.P. and Weinberg, R.A. (1990) Frequent inactivation of the retinoblastoma anti-oncogene is restricted to a subset of human tumor cells. *Proc. Natl Acad. Sci. USA* **87**: 2775–2779.

Huebner, K., Druck, T., Siprashvili, Z., Croce, C.M., Kovatich, A. and McCue, P.A. (1998) The role of deletions at the FRA3B/FHIT locus in carcinogenesis. *Recent Results Cancer Res.* **154**: 200–215.

Huncharek, M., Muscat, J. and Geschwind, J.F. (1999) K-ras oncogene mutation as a prognostic marker in non-small cell lung cancer: a combined analysis of 881 cases. *Carcinogenesis* **20**: 1507–1510.

Ihde, D.C., Pass, H.I. and Glatstein, E. (1997) *Small cell lung cancer*, 5th edn. Lippincott-Raven, Philadelphia/New York.

Jacks, T., Fazeli, A., Schmitt, E.M., Bronson, R.T., Goodell, M.A. and Weinberg, R.A. (1992) Effects of an Rb mutation in the mouse. *Nature* **359**: 295–300.

Jen, J., Powell, S.M., Papadopoulos, N., Smith, K.J., Hamilton, S.R., Vogelstein, B. and Kinzler, K.W. (1994) Molecular determinants of dysplasia in colorectal lesions. *Cancer Res.* **54**: 5523–5526.

Ji, L., Fang, B., Yen, N., Fong, K., Minna, J.D. and Roth, J.A. (1999) Induction of apoptosis and inhibition of tumorigenicity and tumor growth by adenovirus vector-mediated fragile histidine triad (FHIT) gene overexpression. *Cancer Res.* **59**: 3333–3339.

Kaiser, U., Schilli, M., Haag, U., Neumann, K., Kreipe, H., Kogan, E. and Havemann, K. (1996) Expression of bcl-2-protein in small cell lung cancer. *Lung Cancer* **15**: 31 40.

Kantarjian, H.M., O'Brien, S.M., Keating, M. *et al.* (1997) Results of decitabine therapy in the accelerated and blastic phases of chronic myelogenous leukemia. *Leukemia* **11**: 1617–1620.

Kaye, F.J. (1998) The retinoblastoma-like protein family: still in the shadow of the RB gene? [editorial; comment]. *J. Natl Cancer Inst.* **90**: 1418–1419.

Kaye, F.J., McBride, O.W., Battey, J.F., Gazdar, A.F. and Sausville, E.A. (1987) Human creatine kinase-B complementary DNA. Nucleotide sequence, gene expression in lung cancer, and chromosomal assignment to two distinct loci. *J. Clin. Invest.* **79**: 1412–1420.

Kaye, F., Battey, J., Nau, M., Brooks, B., Seifter, E., De Greve, J., Birrer, M., Sausville, E. and Minna, J. (1988) Structure and expression of the human L-myc gene reveal a complex pattern of alternative mRNA processing. *Mol. Cell Biol.* **8**: 186–195.

Kaye, F.J., Kim, Y.W. and Otterson, G.A. (1995) *Molecular biology of lung cancer*. BIOS Scientific Publishers Ltd, Oxford.

Kaye, F.J., Kratzke, R.A., Gerster, J.L. and Horowitz, J.M. (1990) A single amino acid substitution results in a retinoblastoma protein defective in phosphorylation and oncoprotein binding. *Proc. Natl Acad. Sci. USA* **87**: 6922–6926.

Kelley, M.J., Linnoila, R.I., Avis, I.L., Georgiadis, M.S., Cuttitta, F., Mulshine, J.L. and Johnson, B.E. (1997) Antitumor activity of a monoclonal antibody directed against gastrin-releasing peptide in patients with small cell lung cancer. *Chest* **112**: 256–261.

Kiaris, H., Schally, A.V., Varga, J.L., Groot, K. and Armatis, P. (1999) Growth hormone-releasing hormone: an autocrine growth factor for small cell lung carcinoma [see comments]. *Proc. Natl Acad. Sci. USA* **96**: 14894–14898.

Kiefer, P.E., Bepler, G., Kubasch, M. and Havemann, K. (1987) Amplification and expression of protooncogenes in human small cell lung cancer cell lines. *Cancer Res.* **47**: 6236–6242.

Kim, J.O., Nau, M.M., Allikian, K.A., Makela, T.P., Alitalo, K., Johnson, B.E. and Kelley, M.J. (1998) Co-amplification of a novel cyclophilin-like gene (PPIE) with L-myc in small cell lung cancer cell lines. *Oncogene* **17**: 1019–1026.

Knudson, A., Jr. (1971) Mutation and cancer: statistical study of retinoblastoma. *Proc. Natl Acad. Sci. USA* **68**: 820–823.

Kohno, T. and Yokota, J. (1999) How many tumor suppressor genes are involved in human lung carcinogenesis? *Carcinogenesis* **20**: 1403–1410.

Kohno, T., Takahashi, M., Manda, R. and Yokota, J. (1998) Inactivation of the PTEN/MMAC1/TEP1 gene in human lung cancers. *Genes Chromosomes Cancer* **22**: 152–156.

Kohno, T., Takakura, S., Yamada, T., Okamoto, A., Tanaka, T. and Yokota, J. (1999) Alterations of the PPP1R3 gene in human cancer. *Cancer Res.* **59**: 4170–4174.

Kratzke, R.A., Shimizu, E., Geradts, J., Gerster, J.L., Segal, S., Otterson, G.A. and Kaye, F.J. (1993) RB-mediated tumor suppression of a lung cancer cell line is abrogated by an extract enriched in extracellular matrix. *Cell Growth Differ.* **4**: 629–635.

Kratzke, R.A., Otterson, G.A., Lincoln, C.E., Ewing, S., Oie, H., Geradts, J. and Kaye, F.J. (1995) Immunohistochemical analysis of the p16INK4 cyclin-dependent kinase inhibitor in malignant mesothelioma. *J. Natl Cancer Inst.* **87**: 1870–1875.

Kristjansen, P.E., Spang-Thomsen, M. and Quistorff, B. (1991) Different energy metabolism in two human small cell lung cancer subpopulations examined by 31P magnetic resonance spectroscopy and biochemical analysis in vivo and in vitro. *Cancer Res.* **51**: 5160–5164.

Kubo, A., Nakagawa, K., Varma, R.K. *et al.* (1999) The p16 status of tumor cell lines identifies small molecule inhibitors specific for cyclin-dependent kinase 4. *Clin. Cancer Res.* **5**: 4279–4286.

Lane, D.P. (1992) Cancer. p53, guardian of the genome. *Nature* **358**: 15–16.

Langley, K. (1994) The neuroendocrine concept today. *Ann. NY Acad. Sci.* **733**: 1–17.

Larsen, C.J. (1997) Contribution of the dual coding capacity of the p16INK4a/MTS1/CDKN2 locus to human malignancies. *Prog. Cell Cycle Res.* **3**: 109–124.

Lee, J.O., Yang, H., Georgescu, M.M., Di Cristofano, A., Maehama, T., Shi, Y., Dixon, J.E., Pandolfi, P. and Pavletich, N.P. (1999) Crystal structure of the PTEN tumor suppressor: implications for its phosphoinositide phosphatase activity and membrane association. *Cell* **99**: 323–334.

Lee, W.H., Shew, J.Y., Hong, F.D., Sery, T.W., Donoso, L.A., Young, L.J., Bookstein, R. and Lee, E.Y. (1987) The retinoblastoma susceptibility gene encodes a nuclear phosphoprotein associated with DNA binding activity. *Nature* **329**: 642–645.

Lennon, V.A., Kryzer, T.J., Griesmann, G.E., O'Suilleabhain, P.E., Windebank, A.J., Woppmann, A., Miljanich, G.P. and Lambert, E.H. (1995) Calcium-channel antibodies in the Lambert-Eaton syndrome and other paraneoplastic syndromes. *N. Engl. J. Med.* **332**: 1467–1474.

Leone, G., DeGregori, J., Sears, R., Jakoi, L. and Nevins, J.R. (1997) Myc and Ras collaborate in inducing accumulation of active cyclin E/Cdk2 and E2F [published erratum appears in Nature 1997 Jun 26;387(6636): 932]. *Nature* **387**: 422–426.

Levine, A.J. (1997) p53, the cellular gatekeeper for growth and division. *Cell* **88**: 323–331.

Li, D.M. and Sun, H. (1997) TEP1, encoded by a candidate tumor suppressor locus, is a novel protein tyrosine phosphatase regulated by transforming growth factor beta. *Cancer Res.* **57**: 2124–2129.

Li, J., Yen, C., Liaw, D. *et al.* (1997) PTEN, a putative protein tyrosine phosphatase gene mutated in human brain, breast, and prostate cancer [see comments]. *Science* **275**: 1943–1947.

Li, Y., Graham, C., Lacy, S., Duncan, A.M. and Whyte, P. (1993) The adenovirus E1A-associated 130-kD protein is encoded by a member of the retinoblastoma gene family and physically interacts with cyclins A and E. *Genes Dev.* **7**: 2366–2377.

Linnoila, R.I., Mulshine, J.L., Steinberg, S.M., Funa, K., Matthews, M.J., Cotelingam, J.D. and Gazdar, A.F. (1988) Neuroendocrine differentiation in endocrine and nonendocrine lung carcinomas. *Am. J. Clin. Pathol.* **90**: 641–652.

Little, C.D., Nau, M.M., Carney, D.N., Gazdar, A.F. and Minna, J.D. (1983) Amplification and expression of the c-myc oncogene in human lung cancer cell lines. *Nature* **306**: 194–196.

MacLean-Hunter, S., Makela, T.P., Grzeschiczek, A., Alitalo, K. and Moroy, T. (1994) Expression of a rlf/L-myc minigene inhibits differentiation of embryonic stem cells and embroid body formation. *Oncogene* **9**: 3509–3517.

Makela, T.P., Kere, J., Winqvist, R. and Alitalo, K. (1991) Intrachromosomal rearrangements fusing L-myc and rlf in small-cell lung cancer. *Mol. Cell Biol.* **11**: 4015–4021.

Makela, T.P., Hellsten, E., Vesa, J., Hirvonen, H., Palotie, A., Peltonen, L. and Alitalo, K. (1995) The rearranged L-myc fusion gene (RLF) encodes a Zn-15 related zinc finger protein. *Oncogene* **11**: 2699–2704.

Malkin, D., Li, F.P., Strong, L.C. *et al.* (1990) Germ line p53 mutations in a familial syndrome of breast cancer, sarcomas, and other neoplasms [see comments]. *Science* **250**: 1233–1238.

Marchetti, A., Buttitta, F., Merlo, G. *et al.* (1993) p53 alterations in non-small cell lung cancers correlate with metastatic involvement of hilar and mediastinal lymph nodes. *Cancer Res.* **53**: 2846–2851.

Mason, W.P., Graus, F., Lang, B. *et al.* (1997) Small-cell lung cancer, paraneoplastic cerebellar degeneration and the Lambert-Eaton myasthenic syndrome. *Brain* **120**: 1279–1300.

Mayol, X., Grana, X., Baldi, A., Sang, N., Hu, Q. and Giordano, A. (1993) Cloning of a new member of the retinoblastoma gene family (pRb2) which binds to the E1A transforming domain. *Oncogene* 8: 2561–2566.

McLaughlin, P.M., Helfrich, W., Kok, K., Mulder, M., Hu, S.W., Brinker, M.G., Ruiters, M.H., de Leij, L.F. and Buys, C.H. (2000) The ubiquitin-activating enzyme E1-like protein in lung cancer cell lines. *Int. J. Cancer* 85: 871–876.

Merlo, A., Mabry, M., Gabrielson, E., Vollmer, R., Baylin, S.B. and Sidransky, D. (1994) Frequent microsatellite instability in primary small cell lung cancer. *Cancer Res.* 54: 2098–2101.

Merlo, A., Herman, J.G., Mao, L., Lee, D.J., Gabrielson, E., Burger, P.C., Baylin, S.B. and Sidransky, D. (1995) 5' CpG island methylation is associated with transcriptional silencing of the tumour suppressor p16/CDKN2/MTS1 in human cancers. *Nat. Medicine* 1: 686–692.

Mitsudomi, T., Steinberg, S.M., Oie, H.K., Mulshine, J.L., Phelps, R., Viallet, J., Pass, H., Minna, J.D. and Gazdar, A.F. (1991a) ras gene mutations in non-small cell lung cancers are associated with shortened survival irrespective of treatment intent. *Cancer Res.* 51: 4999–5002.

Mitsudomi, T., Viallet, J., Mulshine, J.L., Linnoila, R.I., Minna, J.D. and Gazdar, A.F. (1991b) Mutations of ras genes distinguish a subset of non-small-cell lung cancer cell lines from small-cell lung cancer cell lines. *Oncogene* 6: 1353–1362.

Mitsudomi, T., Steinberg, S.M., Nau, M.M. *et al.* (1992) p53 gene mutations in non-small-cell lung cancer cell lines and their correlation with the presence of ras mutations and clinical features. *Oncogene* 7: 171–180.

Modi, S., Kubo, A., Ole, H., Coxon, A.B., Rehmatulla, A. and Kaye, F.J. (2000) Protein expression of the RB-related gene family and SV40 large T antigen in mesothelioma and lung cancer. *Oncogene* 19: 4632–4639.

Momparler, R.L., Bouffard, D.Y., Momparler, L.F., Dionne, J., Belanger, K. and Ayoub, J. (1997) Pilot phase I-II study on 5-aza-2'-deoxycytidine (Decitabine) in patients with metastatic lung cancer. *Anticancer Drugs* 8: 358–368.

Moody, T.W., Pert, C.B., Gazdar, A.F., Carney, D.N. and Minna, J.D. (1981) High levels of intracellular bombesin characterize human small-cell lung carcinoma. *Science* 214: 1246–1248.

Myers, M.P., Pass, I., Batty, I.H., Van der Kaay, J., Stolarov, J.P., Hemmings, B.A., Wigler, M.H., Downes, C.P. and Tonks, N.K. (1998) The lipid phosphatase activity of PTEN is critical for its tumor suppressor function. *Proc. Natl Acad. Sci. USA* 95: 13513–13518.

Nau, M.M., Brooks, B.J., Battey, J. *et al.* (1985) L-myc, a new myc-related gene amplified and expressed in human small cell lung cancer. *Nature* 318: 69–73.

Nau, M.M., Brooks, B., Jr., Carney, D.N., Gazdar, A.F., Battey, J.F., Sausville, E.A. and Minna, J.D. (1986) Human small-cell lung cancers show amplification and expression of the N-myc gene. *Proc. Natl Acad. Sci. USA* 83: 1092–1096.

Nelson, A.R., Fingleton, B., Rothenberg, M.L. and Matrisian, L.M. (2000) Matrix metalloproteinases: biologic activity and clinical implications. *J. Clin. Oncol.* 18: 1135–1149.

Nemunaitis, J., Swisher, S.G., Timmons, T. *et al.* (2000) Adenovirus-mediated p53 gene transfer in sequence with cisplatin to tumors of patients with non-small-cell lung cancer. *J. Clin. Oncol.* 18: 609–622.

Niehans, G.A., Brunner, T., Frizelle, S.P., Liston, J.C., Salerno, C.T., Knapp, D.J., Green, D.R. and Kratzke, R.A. (1997) Human lung carcinomas express Fas ligand. *Cancer Res.* 57: 1007–1012.

Ohta, M., Inoue, H., Cotticelli, M.G. *et al.* (1996) The FHIT gene, spanning the chromosome 3p14.2 fragile site and renal carcinoma-associated t(3;8) breakpoint, is abnormal in digestive tract cancers. *Cell* 84: 587–597.

Okami, K., Wu, L., Riggins, G. *et al.* (1998) Analysis of PTEN/MMAC1 alterations in aerodigestive tract tumors. *Cancer Res.* 58: 509–511.

Ookawa, K., Shiseki, M., Takahashi, R., Yoshida, Y., Terada, M. and Yokota, J. (1993) Reconstitution of the RB gene suppresses the growth of small-cell lung carcinoma cells carrying multiple genetic alterations. *Oncogene* 8: 2175–2181.

Otterson, G.A., Kratzke, R.A., Coxon, A., Kim, Y.W. and Kaye, F.J. (1994) Absence of p16INK4 protein is restricted to the subset of lung cancer lines that retains wildtype RB. *Oncogene* 9: 3375–3378.

Otterson, G.A., Khleif, S.N., Chen, W., Coxon, A.B. and Kaye, F.J. (1995) CDKN2 gene silencing in lung cancer by DNA hypermethylation and kinetics of p16 protein induction by 5-aza 2' deoxycytidine. *Oncogene* 11: 1211–1216.

Otterson, G.A., Xiao, G.H., Geradts, J., Jin, F., Chen, W.D., Niklinska, W., Kaye, F.J. and Yeung, R.S. (1998) Protein expression and functional analysis of the FHIT gene in human tumor cells. *J. Natl Cancer Inst.* **90**: 426–432.

Palmarini, M., Fan, H. and Sharp, J.M. (1997) Sheep pulmonary adenomatosis: a unique model of retrovirus-associated lung cancer. *Trends Microbiol.* **5**: 478–483.

Palmarini, M., Sharp, J.M., de las Heras, M. and Fan, H. (1999) Jaagsiekte sheep retrovirus is necessary and sufficient to induce a contagious lung cancer in sheep. *J. Virol.* **73**: 6964–6972.

Peto, R., Darby, S., Deo, H., Silcocks, P., Whitley, E. and Doll, R. (2000) Smoking, smoking cessation, and lung cancer in the UK since 1950: combination of national statistics with two case-control studies. *BMJ* **321**: 323–329.

Pezzella, F., Turley, H., Kuzu, I., Tungekar, M.F., Dunnill, M.S., Pierce, C.B., Harris, A., Gatter, K.C. and Mason, D.Y. (1993) bcl-2 protein in non-small-cell lung carcinoma [see comments]. *N. Engl. J. Med.* **329**: 690–694.

Pomerantz, J., Schreiber-Agus, N., Liegeois, N.J. *et al.* (1998) The Ink4a tumor suppressor gene product, p19Arf, interacts with MDM2 and neutralizes MDM2's inhibition of p53. *Cell* **92**: 713–723.

Posner, J.B. and Dalmau, J. (1997) Paraneoplastic syndromes. *Curr. Opin. Immunol.* **9**: 723–729.

Prescott, J.C. and Blackburn, E.H. (1999) Telomerase: Dr Jekyll or Mr Hyde? *Curr. Opin. Genet. Dev.* **9**: 368–373.

Quinlan, D.C., Davidson, A.G., Summers, C.L., Warden, H.E. and Doshi, H.M. (1992) Accumulation of p53 protein correlates with a poor prognosis in human lung cancer. *Cancer Res.* **52**: 4828–4831.

Rachwal, W.J., Bongiorno, P.F., Orringer, M.B., Whyte, R.I., Ethier, S.P. and Beer, D.G. (1995) Expression and activation of erbB-2 and epidermal growth factor receptor in lung adenocarcinomas. *Br. J. Cancer* **72**: 56–64.

Reeves, R.H. (2000) Recounting a genetic story. *Nature* **405**: 283–284.

Reissmann, P.T., Koga, H., Takahashi, R., Figlin, R.A., Holmes, E.C., Piantadosi, S., Cordon, C.C. and Slamon, D.J. (1993) Inactivation of the retinoblastoma susceptibility gene in non-small-cell lung cancer. The Lung Cancer Study Group. *Oncogene* **8**: 1913–1919.

Richardson, G.E. and Johnson, B.E. (1993) The biology of lung cancer. *Semin. Oncol.* **20**: 105–127.

Rodenhuis, S. and Slebos, R.J. (1990) The ras oncogenes in human lung cancer. *Am. Rev. Respir. Dis.* **142**: S27–30.

Roth, J.A., Swisher, S.G. and Meyn, R.E. (1999) p53 tumor suppressor gene therapy for cancer. *Oncology (Huntingt)* **13**: 148–154.

Rowinsky, E.K., Windle, J.J. and Von Hoff, D.D. (1999) Ras protein farnesyltransferase: A strategic target for anticancer therapeutic development. *J. Clin. Oncol.* **17**: 3631–3652.

Sacchi, S., Kantarjian, H.M., O'Brien, S. *et al.* (1999) Chronic myelogenous leukemia in nonlymphoid blastic phase: analysis of the results of first salvage therapy with three different treatment approaches for 162 patients. *Cancer* **86**: 2632–2641.

Sakurada, A., Suzuki, A., Sato, M. *et al.* (1997) Infrequent genetic alterations of the PTEN/MMAC1 gene in Japanese patients with primary cancers of the breast, lung, pancreas, kidney, and ovary. *Jpn. J. Cancer Res.* **88**: 1025–1028.

Sameshima, Y., Matsuno, Y., Hirohashi, S., Shimosato, Y., Mizoguchi, H., Sugimura, T., Terada, M. and Yokota, J. (1992) Alterations of the p53 gene are common and critical events for the maintenance of malignant phenotypes in small-cell lung carcinoma. *Oncogene* **7**: 451–457.

Scheffner, M., Munger, K., Byrne, J.C. and Howley, P.M. (1991) The state of the p53 and retinoblastoma genes in human cervical carcinoma cell lines. *Proc. Natl Acad. Sci. USA* **88**: 5523–5527.

Schimmel, L., Khandekar, V.S., Martin, K.J., Riera, T., Honan, C., Shaw, D.G. and Kaddurah-Daouk, R. (1996) The synthetic phosphagen cyclocreatine phosphate inhibits the growth of a broad spectrum of solid tumors. *Anticancer Res.* **16**: 375–380.

Schrock, E., du Manoir, S., Veldman, T. *et al.* (1996) Multicolor spectral karyotyping of human chromosomes [see comments]. *Science* **273**: 494–497.

Senderowicz, A.M. and Sausville, E.A. (2000) Preclinical and clinical development of cyclin-dependent kinase modulators. *J. Natl Cancer Inst.* **92**: 376–387.

Seneviratne, U. and de Silva, R. (1999) Lambert-Eaton myasthenic syndrome. *Postgrad. Med. J.* **75**: 516–520.

Sharpless, N.E. and DePinho, R.A. (1999) The INK4A/ARF locus and its two gene products. *Curr. Opin. Genet. Dev.* **9**: 22–30.

Shay, J.W. (1999) Toward identifying a cellular determinant of telomerase repression [editorial; comment]. *J. Natl Cancer Inst.* **91**: 4–6.

Shields, P.G. and Harris, C.C. (2000) Cancer risk and low-penetrance susceptibility genes in gene-environment interactions. *J. Clin. Oncol.* **18**: 2309–2315.

Shimizu, E., Coxon, A., Otterson, G.A. *et al.* (1994) RB protein status and clinical correlation from 171 cell lines respresenting lung cancer, extrapulmonary small cell carcinoma, and mesothelioma. *Oncogene* **9**: 2441–2448.

Siprashvili, Z., Sozzi, G., Barnes, L.D. *et al.* (1997) Replacement of Fhit in cancer cells suppresses tumorigenicity. *Proc. Natl Acad. Sci. USA* **94**: 13771–13776.

Slebos, R.J., Kibbelaar, R.E., Dalesio, O. *et al.* (1990) K-ras oncogene activation as a prognostic marker in adenocarcinoma of the lung. *N. Engl. J. Med.* **323**: 561–565.

Slebos, R.J., Hruban, R.H., Dalesio, O., Mooi, W.J., Offerhaus, G.J. and Rodenhuis, S. (1991) Relationship between K-ras oncogene activation and smoking in adenocarcinoma of the human lung. *J. Natl Cancer Inst.* **83**: 1024–1027.

Smith, L.E., Denissenko, M.F., Bennett, W.P., Li, H., Amin, S., Tang, M. and Pfeifer, G.P. (2000) Targeting of lung cancer mutational hotspots by polycyclic aromatic hydrocarbons. *J. Natl Cancer Inst.* **92**: 803–811.

Sozzi, G., Miozzo, M., Donghi, R., Pilotti, S., Cariani, C.T., Pastorino, U., Della Porta, G. and Pierotti, M.A. (1992) Deletions of 17p and p53 mutations in preneoplastic lesions of the lung. *Cancer Res.* **52**: 6079–6082.

Sozzi, G., Veronese, M.L., Negrini, M. *et al.* (1996) The FHIT gene 3p14.2 is abnormal in lung cancer. *Cell* **85**: 17–26.

Sozzi, G., Pastorino, U., Moiraghi, L. *et al.* (1998) Loss of FHIT function in lung cancer and preinvasive bronchial lesions. *Cancer Res.* **58**: 5032–5037.

Spitz, M.R., Wei, Q., Li, G. and Wu, X. (1999) Genetic susceptibility to tobacco carcinogenesis [see comments]. *Cancer Invest.* **17**: 645–659.

Srivastava, S., Zou, Z.Q., Pirollo, K., Blattner, W. and Chang, E.H. (1990) Germ-line transmission of a mutated p53 gene in a cancer-prone family with Li-Fraumeni syndrome [see comments]. *Nature* **348**: 747–749.

Stanton, S.E., Shin, S.W., Johnson, B.E. and Meyerson, M. (2000) Recurrent allelic deletions of chromosome arms 15q and 16q in human small cell lung carcinomas. *Genes Chromosomes Cancer* **27**: 323–331.

Steck, P.A., Pershouse, M.A., Jasser, S.A. *et al.* (1997) Identification of a candidate tumour suppressor gene, MMAC1, at chromosome 10q23.3 that is mutated in multiple advanced cancers. *Nat. Genet.* **15**: 356–362.

Strickler, H.D., Goedert, J.J., Fleming, M., Travis, W.D., Williams, A.E., Rabkin, C.S., Daniel, R.W. and Shah, K.V. (1996) Simian virus 40 and pleural mesothelioma in humans. *Cancer Epidemiol. Biomarkers Prev.* **5**: 473–475.

Strickler, H.D., Rosenberg, P.S., Devesa, S.S., Hertel, J., Fraumeni, J.F., Jr. and Goedert, J.J. (1998) Contamination of poliovirus vaccines with simian virus 40 (1955–1963) and subsequent cancer rates [see comments]. *JAMA* **279**: 292–295.

Sugio, K., Kishimoto, Y., Virmani, A.K., Hung, J.Y. and Gazdar, A.F. (1994) K-ras mutations are a relatively late event in the pathogenesis of lung carcinomas. *Cancer Res.* **54**: 5811–5815.

Sun, H., Lesche, R., Li, D.M. *et al.* (1999) PTEN modulates cell cycle progression and cell survival by regulating phosphatidylinositol 3,4,5,-trisphosphate and Akt/protein kinase B signaling pathway. *Proc. Natl Acad. Sci. USA* **96**: 6199–6204.

Sundaresan, V., Ganly, P., Hasleton, P., Rudd, R., Sinha, G., Bleehen, N.M. and Rabbitts, P. (1992) p53 and chromosome 3 abnormalities, characteristic of malignant lung tumours, are detectable in preinvasive lesions of the bronchus. *Oncogene* **7**: 1989–1997.

Takahashi, T., Nau, M.M., Chiba, I. *et al.* (1989) p53: a frequent target for genetic abnormalities in lung cancer. *Science* **246**: 491–494.

Takahashi, T., Carbone, D., Takahashi, T., Nau, M.M., Hida, T., Linnoila, I., Ueda, R. and Minna, J.D. (1992) Wild-type but not mutant p53 suppresses the growth of human lung cancer cells bearing multiple genetic lesions. *Cancer Res.* **52**: 2340–2343.

Tanaka, H., Arakawa, H., Yamaguchi, T., Shiraishi, K., Fukuda, S., Matsui, K., Takei, Y. and Nakamura, Y. (2000) A ribonucleotide reductase gene involved in a p53-dependent cell-cycle checkpoint for DNA damage [see comments]. *Nature* **404**: 42–49.

Testa, J.R. and Siegfried, J.M. (1992) Chromosome abnormalities in human non-small cell lung cancer. *Cancer Res.* **52**: 27025–27065.

Testa, J.R., Liu, Z., Feder, M., Bell, D.W., Balsara, B., Cheng, J.Q. and Taguchi, T. (1997) Advances in the analysis of chromosome alterations in human lung carcinomas. *Cancer Genet. Cytogenet.* **95**: 20–32.

Tomizawa, Y., Adachi, J., Kohno, T., Yamaguchi, N., Saito, R. and Yokota, J. (1998) Identification and characterization of families with aggregation of lung cancer. *Jpn. J. Clin. Oncol.* **28**: 192–195.

Unger, T., Sionov, R.V., Moallem, E., Yee, C.L., Howley, P.M., Oren, M. and Haupt, Y. (1999) Mutations in serines 15 and 20 of human p53 impair its apoptotic activity. *Oncogene* **18**: 3205–3212.

Waheed, I., Guo, Z.S., Chen, G.A., Weiser, T.S., Nguyen, D.M. and Schrump, D.S. (1999) Antisense to SV40 early gene region induces growth arrest and apoptosis in T-antigen-positive human pleural mesothelioma cells. *Cancer Res.* **59**: 6068–6073.

Wang, S.S., Esplin, E.D., Li, J.L., Huang, L., Gazdar, A., Minna, J. and Evans, G.A. (1998) Alterations of the PPP2R1B gene in human lung and colon cancer. *Science* **282**: 284–287.

Weinberg, R.A. (1995) The retinoblastoma protein and cell cycle control. *Cell* **81**: 323–330.

Whang-Peng, J., Kao-Shan, C.S., Lee, E.C., Bunn, P.A., Carney, D.N., Gazdar, A.F. and Minna, J.D. (1982) Specific chromosome defect associated with human small-cell lung cancer; deletion 3p(14–23). *Science* **215**: 181–182.

Whang-Peng, J., Knutsen, T., Gazdar, A., Steinberg, S.M., Oie, H., Linnoila, I., Mulshine, J., Nau, M. and Minna, J.D. (1991) Nonrandom structural and numerical chromosome changes in non-small-cell lung cancer. *Genes Chromosomes Cancer* **3**: 168–188.

Whyte, P., Williamson, N.M. and Harlow, E. (1989) Cellular targets for transformation by the adenovirus E1A proteins. *Cell* **56**: 67–75.

Wingo, P.A., Ries, L.A., Giovino, G.A., Miller, D.S., Rosenberg, H.M., Shopland, D.R., Thun, M.J. and Edwards, B.K. (1999) Annual report to the nation on the status of cancer, 1973–1996, with a special section on lung cancer and tobacco smoking [see comments]. *J. Natl Cancer Inst.* **91**: 675–690.

Wong, A.J., Ruppert, J.M., Eggleston, J., Hamilton, S.R., Baylin, S.B. and Vogelstein, B. (1986) Gene amplification of c-myc and N-myc in small cell carcinoma of the lung. *Science* **233**: 461–464.

Wu, R., Connolly, D.C., Dunn, R.L. and Cho, K.R. (2000) Restored expression of fragile histidine triad protein and tumorigenicity of cervical carcinoma cells [In Process Citation]. *J. Natl Cancer Inst.* **92**: 338–344.

Xu, H.J., Quinlan, D.C., Davidson, A.G., Hu, S.X., Summers, C.L., Li, J. and Benedict, W.F. (1994) Altered retinoblastoma protein expression and prognosis in early-stage non-small-cell lung carcinoma. *J. Natl Cancer Inst.* **86**: 695–699.

Yang, H.K., Scott, F.M., Trepel, J.B., Battey, J.F., Johnson, B.E. and Kelley, M.J. (1998) Correlation of expression of bombesin-like peptides and receptors with growth inhibition by an anti-bombesin antibody in small-cell lung cancer cell lines. *Lung Cancer* **21**: 165–175.

Yokomizo, A., Tindall, D.J., Drabkin, H., Gemmill, R., Franklin, W., Yang, P., Sugio, K., Smith, D.I. and Liu, W. (1998) PTEN/MMAC1 mutations identified in small cell, but not in non-small cell lung cancers. *Oncogene* **17**: 475–479.

Yokota, J., Wada, M., Shimosato, Y., Terada, M. and Sugimura, T. (1987) Loss of heterozygosity on chromosomes 3, 13, and 17 in small-cell carcinoma and on chromosome 3 in adenocarcinoma of the lung. *Proc. Natl Acad. Sci. USA* **84**: 9252–9256.

Yokota, J., Akiyama, T., Fung, Y.K. *et al.* (1988) Altered expression of the retinoblastoma (RB) gene in small-cell carcinoma of the lung. *Oncogene* **3**: 471–475.

You, L., Yang, C.T. and Jablons, D.M. (2000) ONYX-015 works synergistically with chemotherapy in lung cancer cell lines and primary cultures freshly made from lung cancer patients. *Cancer Res.* **60**: 1009–1013.

Yu, D., Wang, S.S., Dulski, K.M., Tsai, C.M., Nicolson, G.L. and Hung, M.C. (1994) c-erbB-2/neu overexpression enhances metastatic potential of human lung cancer cells by induction of metastasis-associated properties. *Cancer Res.* **54**: 3260–3266.

Zalvide, J., Stubdal, H. and DeCaprio, J.A. (1998) The J domain of simian virus 40 large T antigen is required to functionally inactivate RB family proteins. *Mol. Cell Biol.* **18**: 1408–1415.

Zhang, S.Y., Klein-Szanto, A.J., Sauter, E.R., Shafarenko, M., Mitsunaga, S., Nobori, T., Carson, D.A., Ridge, J.A. and Goodrow, T.L. (1994) Higher frequency of alterations in the p16/CDKN2 gene in squamous cell carcinoma cell lines than in primary tumors of the head and neck. *Cancer Res.* **54**: 5050–5053.

Zhang, Y., Xiong, Y. and Yarbrough, W.G. (1998) ARF promotes MDM2 degradation and stabilizes p53: ARF-INK4a locus deletion impairs both the Rb and p53 tumor suppression pathways. *Cell* **92:** 725–734.

Zhang, Z., Nakamura, M., Taniguchi, E., Shan, L., Yokoi, T. and Kakudo, K. (1996) Late occurrence of K-ras gene mutations in the pathogenesis of squamous cell carcinoma of the lung: analysis in sputum [letter]. *Anal. Quant. Cytol. Histol.* **18:** 501–502.

Zhao, R., Gish, K., Murphy, M., Yin, Y., Notterman, D., Hoffman, W.H., Tom, E., Mack, D.H. and Levine, A.J. (2000) Analysis of p53-regulated gene expression patterns using oligonucleotide arrays. *Genes Dev.* **14:** 981–993.

zur Hausen, H. (2000) Papillomaviruses causing cancer: evasion from host-cell control in early events in carcinogenesis. *J. Natl Cancer Inst.* **92:** 690–699.

18

Genetics of liver cancer

A.-M. Hui, L. Sun and M. Makuuchi

1. Introduction

Hepatocellular carcinoma (HCC) is one of the most frequent malignancies worldwide, showing the highest prevalence in Asia and Africa; it is the second most common form of lethal cancer after gastric cancer in China, and the third most common after gastric and lung cancers in Japan. The incidence of HCC is generally lower in Western Europe and the United States. However, following an increase in hepatitis C virus (HCV) infection rates, it is now occurring with increasing frequency in the United States; the incidence of histologically proven HCC increased from 1.4 per 100 000 population for the period from 1976 to 1980 to 2.4 per 100 000 for the period from 1991 to 1995 (El-Serag and Mason, 1999; Ince and Wands, 1999). Although advances in the treatment of HCC have recently been achieved, the prognosis of this disease generally remains very poor. An understanding of the molecular basis of HCC is very important because novel therapeutic approaches which target these molecules may lead to improvements in HCC prevention and treatment. Identification of biomarkers related to the clinical outcome for patients with HCCs is another avenue to improve the prognosis, which may assist in the selection of therapeutic procedures. This chapter reviews the available data on the genetic and epigenetic changes involved in HCC and discusses their clinical implications.

2. Pathogenesis

2.1 Etiology

Extensive epidemiological evidence indicates that most HCCs are associated with hepatitis B virus (HBV) or HCV infection. In China and Southeast Asia, HBV is highly prevalent and accounts for more than 60% of HCC cases. In Japan, Italy and Spain, HCV infection is the major culprit and is responsible for more than 60% of HCC cases (Okuda, 1997). In this latter group of countries, HBV appears to play a relatively minor role, being involved in perhaps fewer than 25% of HCC cases. Direct evidence for the causal role of HBV in HCC comes from the fact that HBV DNA sequences can be detected in the cellular DNA in HBV-related HCCs (Brechot et al., 1980), while a segment of HBV DNA called HBx, which encodes a transcription factor, can induce liver tumors in transgenic mice (Kim *et al.*, 1991). HCV is not integrated into the host cell DNA, since it is a plus-strand RNA virus and is not reverse-transcribed to DNA.

In Africa, some areas of China and Southeast Asia, the incidence of HCC correlates with aflatoxin exposure. Aflatoxin appears to cause a specific mutation (a G to T mutation at codon 249 of the *TP53* tumor suppressor gene) (Bressac *et al.*, 1991; Hsu *et al.*, 1991; Ozturk, 1991) and probably interacts with the hepatitis virus (Bannasch *et al.*, 1995), thereby contributing to hepatocarcinogenesis.

For populations at low risk of HCC, in areas where hepatitis viruses are not epidemic, alcoholic cirrhosis is the main factor associated with this malignancy. Alcohol consumption increases the risk of HCC by a factor of four in Northern Europe and North America (Hardell *et al.*, 1984). Cirrhosis is always present in HCC patients, and once alcoholic cirrhosis occurs, cessation of drinking does not prevent HCC (Lee, 1966); therefore, in alcoholic subjects, the risk of HCC is very closely associated with cirrhosis. Alcohol may also contribute to hepatocarcinogenesis by acting as a co-carcinogen with hepatitis viruses, and may promote tumor progression by depressing human immune responses.

2.2 The multistep process of hepatocarcinogenesis

Human hepatocarcinogenesis is a multistep process. Adenomatous hyperplasia, which is generally considered to be a precancerous lesion, arises in chronically diseased liver, and subsequently progresses to the early stages of HCC (Hirohashi and Sakamoto, 1992; Sakamoto *et al.*, 1991; Takayama *et al.*, 1990). Analysis of the integration pattern of HBV DNA in adenomatous hyperplasia nodules suggests that this lesion develops due to clonal expansion from a single hepatocyte (Tsuda *et al.*, 1988). Histological features compatible with adenomatous hyperplasia are usually observed in the area surrounding an early HCC, suggesting that the cancer arises by malignant transformation of the adenomatous hyperplasia. Early HCCs are well differentiated (Edmondson's grade I; Edmondson and Steiner, 1954) and do not destroy the underlying structure of the hepatic lobules; thus, they are a type of cancer *in situ*. Early HCCs progress to advanced cancers, which are usually moderately or poorly differentiated (Edmondson's grade II or III). Some tumors show a nodule-in-nodule appearance, in which the inner nodule is Edmondson's grade II or III, while the outer portion is Edmondson's grade I. This type of tumor is considered to represent the transformation from an early to an advanced HCC, and is therefore termed early-advanced HCC or nodule-in-nodule-type HCC (Hirohashi and Sakamoto, 1992). Intrahepatic and extrahepatic metastases are thought to be later stages in the multistep process of human hepatocarcinogenesis.

The multiple stages of HCC development and progression are associated with a number of genetic or epigenetic alterations (Hirohashi, 1992; Hui and Makuuchi, 1999). Next, we discuss the molecular events involved in HCC.

3. Tumor suppressors and cyclin-dependent kinase inhibitors

Cancer has been directly associated with disruption of the cell cycle, and uncontrolled growth is a major defining hallmark of cancer cells. In mammalian cells, the transition from the G_1 to the S phase is governed by cyclin-dependent kinases (CDKs), which are activated by the binding of positive effectors, the cyclins, and inhibited by CDK inhibitors (Peter and Herskowitz, 1994; Sherr and Roberts, 1995) (see *Figure 1*). CDK inhibitors fall into two families, INK and CIP/KIP, based on their sequence homology; the former includes *INK4* (p16^{INK4}), *INK4B* (p15^{INK4B}), p18^{INK6A} and p19^{INK6B}, while the latter includes *p21* (p21$^{WAF1/CIP1}$), *p27KIP1* and *CDKN1C* (p57^{Kip2}). The major CDKs and cyclins involved in G_1–S transition are CDK2, CDK4, CDK6, cyclin D1 and cyclin E. The product of the retinoblastoma tumor suppressor gene (pRb) is a key substrate for CDKs, which bind to and negatively regulate transcription factors, such as E2F, that control S phase entry. p53 also participates in the regulation of G_1–S transition by activating *p21* through transcription and by other pathways. These modulators of G_1–S transition are extremely intriguing from the viewpoint of human carcinogenesis (Hunter and Pines, 1994; Sherr, 1996).

3.1 INK4 (p16^{INK4})

The CDK inhibitor *INK4* is also considered to be a tumor suppressor. However, in contrast to classical tumor suppressor genes, *INK4* can be inactivated by a variety of genetic or epigenetic events, including homozygous deletion (Carins *et al.*, 1995; Nobori *et al.*, 1994), mutation (Caldas *et al.*, 1994; Hussussian *et al.*, 1994), hypermethylation of the 5′ CpG island in the promotor region (Herman *et al.*, 1995; Merlo *et al.*, 1995) and post-transcriptional regulation (Hui *et al.*, 1996a; Kovar *et al.*, 1997).

To determine whether *INK4* is involved in hepatocarcinogenesis, Hui *et al.* (1996a) studied its protein by Western blotting, its messenger RNA (mRNA) by the

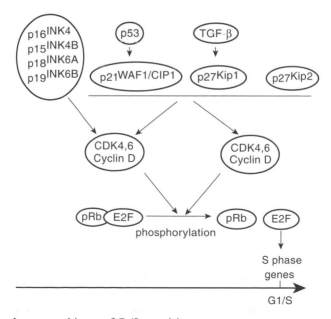

Figure 1. Regulatory machinery of G_1/S transition.

reverse-transcriptase polymerase chain reaction (RT-PCR) and Northern blotting, and its genomic status by PCR-single-strand conformation polymorphism (PCR-SSCP) and Southern blotting in both HCC cell lines and surgically resected HCC tissue. INK4 protein was absent from three of six (50%) HCC cell lines and from 11 of 32 (34%) primary tumors, but was detected in all corresponding noncancerous liver tissues, strongly suggesting an association between *INK4* inactivation and HCC. No INK4 protein was found in 2 of 8 (22%) early HCCs, and this absence rate increased to 33% for nodule-in-nodule-type HCCs (the transition between early and advanced HCC) and 40% for advanced HCCs. These findings suggest that *INK4* inactivation contributes to the early stages of HCC development and tumor progression, and provided the first evidence of the involvement of a tumor suppressor gene in the early development of HCC. Recently, Matsuda *et al.* (1999) also reported that *INK4* inactivation is associated with HCC progression.

Another novel and important finding from Hui's study (1996a) is that *INK4* seems to be inactivated at the post-transcriptional level; whereas homozygous deletions and mutations were completely undetectable, mRNA expression was very rarely absent from HCC cell lines and surgically resected tumors, including specimens that lacked INK4 protein expression. Lack of *INK4* mRNA expression was found in only one HCC cell line, and may have involved transcription loss associated with 5′ CpG island hypermethylation in the promotor region of the gene. Recently, post-transcriptional *INK4* inactivation has also been reported to occur in Ewing's tumors (Kovar *et al.*, 1997). Taken together, these studies suggest that analyzing the *INK4* gene at the level of its protein is more sensitive than analyzing it at its genomic and mRNA levels when attempting to detect *INK4* abnormalities.

Recently, many studies of *INK4* in HCC have been carried out; however, other than that of Hui *et al* (1996a), no systematic analyses at its genomic, mRNA and protein levels have been published. These studies have consistently noted the rarity of homozygous deletions and mutations of *INK4* in HCC (Biden *et al.*, 1997; Hui *et al.*, 1996a; Kita *et al.*, 1996); Lin *et al.*, 1998; Matsuda *et al.*, 1999). On the other hand, a high frequency of hypermethylation of the 5′ CpG island of the *INK4* gene, determined by methylation specific PCR (MS-PCR), has recently been reported. This abnormality was found in 62.5% of HCCs from Hong Kong (Liew *et al.*, 1999) and 45% of those from Japan (Matsuda *et al.*, 1999). In contrast, Lin *et al.* (1998), using the Southern blotting approach, detected no methylation of the *INK4* gene in HCCs from Taiwan. This disagreement is probably due to two factors, namely different etiological backgrounds and the different methods used to detect DNA methylation during the studies; however, since HCC is predominantly related to HBV in both Hong Kong and Taiwan, methodological differences seem to be the most likely reason. MS-PCR is a highly sensitive method, which can detect methylation in CpG sites even when they are located outside the sequences recognized by methylation-sensitive restriction enzymes, whereas the Southern blotting technique cannot provide information on these sites. Although promotor hypermethylation was associated with loss of INK4 protein expression as determined by immunostaining in Matsuda's study, mRNA expression was not evaluated in any of the studies. Direct evidence of an association between the *INK4* promotor methylation status and transcriptional silencing of this gene is therefore needed to elucidate the real role of DNA hypermethylation within the *INK4* gene in HCC.

Loss of INK4 protein, as determined by immunohistochemistry, has been reported to be associated with tumor cell differentiation and metastasis (Hui *et al.*, 2000a; Matsuda *et al.*, 1999); however, the impact of INK4 protein expression on the prognosis of HCC remains to be evaluated.

3.2 P21 (p21^{WAF1/CIP1})

p21$^{WAF1/CIP1}$ is a universal CDK inhibitor which is generally considered to have tumor suppressor functions. Its expression is regulated by p53-dependent and p53-independent pathways. Hui *et al.* (1997a) studied *P21* mRNA expression (by RT-PCR) and *TP53* mutational status (by PCR-SSCP and direct DNA sequencing) in HCCs from Japan. *P21* mRNA expression was reduced in 38% of these HCCs, and this reduction was associated with *TP53* mutations. Tumors with a mutant *TP53* gene expressed significantly lower levels of *P21* mRNA than tumors with the wild-type *TP53* gene and corresponding noncancerous liver tissues; there was no significant difference in *P21* mRNA expression levels between the wild-type p53 tumors and the noncancerous liver tissues. These findings suggest that *P21* expression is predominantly regulated by p53 in HCCs, and that reduced *P21* expression may contribute to hepatocarcinogenesis. Recently, the coexpression of *P21* and *TP53* has been studied immunohistochemically in 81 HCCs from Japanese patients, and *P21* expression was found to be inversely correlated with *TP53* expression in HCV-related HCCs, but not in those associated with HBV or no hepatitis virus infection (Shi *et al.*, 2000a). These results suggest that different models of *P21* regulation are involved in HCCs that differ in their viral infection status, and that reduced *P21* expression is predominantly related to *TP53* alterations in HCV-related HCCs. In support of this conclusion, Naka *et al.* (1998) reported a significant inverse correlation between the expression of P21 and p53 proteins in a series of HCCs from Japan, where HCCs are mainly associated with HCV, while Qin *et al.* (1998) demonstrated no such correlation among Chinese HCCs, the majority of which are HBV-related. However, neither of these two studies analyzed the relationship between p21$^{WAF1/CIP1}$ and p53 in separate subsets of HCCs that differed in their viral infection status.

Besides mutant *TP53*, there are two other mechanisms that probably act to down-regulate *P21* expression in HCCs (Shi *et al.*, 2000a): (1) suppression of the transcriptional activity of *P21* by HCV core protein (Ray *et al.*, 1998) and (2) degradation of p21$^{WAF1/CIP1}$ at the post-transcriptional level by ubiquitin-dependent proteolysis (Maki and Howley, 1998).

No consistent conclusions concerning an association between *P21* expression and the disease stage have been obtained from the studies published to date (Hui *et al.*, 1997a; Naka *et al.*, 1998; Qin *et al.*, 1998; Shi *et al.*, 2000a). To clarify this issue and to investigate the prognostic role of *P21* expression, the authors are currently performing a large follow-up study.

3.3 P27KIP1

P27KIP1 is another member of the CIP/KIP family of CDK inhibitors, which shares a similar 60-amino acid sequence with *P21*. Its effects in carcinogenesis seem to be related to abnormalities in its expression rather than genomic alterations. Hui *et al.* (1998a) showed that *P27KIP1* mRNA expression is decreased in 52% of HCCs, suggesting its involvement in hepatocarcinogenesis. Recently, Ito *et al.* (1999) reported that a decrease in *P27KIP1* expression, as determined by immunohisto-chemistry, is significantly associated with portal vein invasion, poor differentiation and large tumor size, and is an independent predictor of recurrence in patients with HCCs. Low-level *P27KIP1* expression has also been shown to correlate signifi-cantly with patient outcomes for other types of hepatobiliary cancers (Hui *et al.*, 1999b, 2000b).

It is interesting to draw together the results from Hui's studies (Hui *et al.*, 1996a, 1996b, 1997a, 1998a, 1998b). In this series, 88% of HCCs harbored alterations in at least one of the CDK inhibitors *INK4*, *P21* and *P27KIP1*, suggesting that CDK inhibitor dysfunctions contribute to carcinogenesis in the majority of HCCs.

3.4 RB1

An alteration in the tumor suppressor gene *RB1*, located on chromosome 13q14, is one of the most common changes seen in human cancers, and has also been reported to be involved in HCC (Hui *et al.*, 1999a; Murakami *et al.*, 1991; Nishida *et al.*, 1992; Zhang *et al.*, 1994). Loss of heterozygosity (LOH) on 13q was found in 43–47% of HCCs, and the *RB1* locus was deleted in all tumors with LOH on 13q, suggesting a close association between the Rb gene and hepatocarcinogenesis (Nishida *et al.*, 1992; Zhang *et al.*, 1994). Furthermore, LOH on 13q was signifi-cantly related to a large tumor size, poor differentiation and more advanced staging (Nishida *et al.*, 1992). Zhang *et al.* (1994) also found that allelic loss of the *RB1* gene correlated significantly with loss of pRb expression in HCCs, indicating inactivation of both alleles of this gene. However, the mutation rate in the remaining *RB1* allele appears to be low; Zhang *et al.* (1994) examined the promotor region and 27 exons of the *RB1* gene by PCR-SSCP and nucleotide sequencing methods, and a small, tumor-specific deletion within a coding region was detected in only two of the nine tumors harboring LOH at 13q. There are at least two possible reasons for the mutation rate in the *RB1* gene being lower than the incidence of allelic loss: firstly, some mutations might not be demonstrable by PCR-SSCP methods, and secondly, other mechanisms such as promotor hyperme-thylation may inactivate the remaining *RB1* allele (Sasaki *et al.*, 1991). These data also suggest that analysis at the protein level is probably more sensitive than studies at the genomic level when investigating pRb functional status.

Previous studies to examine pRb function at the protein level have focused on loss of its expression. Using an immunohistochemical technique, Hsia *et al.* (1994) reported that pRb was absent from 32% of HCCs, and this absence was associated with advanced disease stages. Recently, Hui *et al.* (1999a) found that some HCCs express significantly higher levels of pRb than corresponding noncancerous liver tissues. Using noncancerous liver tissue as a control, they classified pRb immunoreactivity into three categories based on the percentage of positive cells present: loss of expression (<1% positive cells), normal expression (1–50% positive cells) and overexpression (>50% positive cells). All noncancerous liver specimens, benign large regenerative nodules and precancerous adenomatous hyperplasia lesions showed normal pRb expression, whereas loss of pRb expression and pRb overexpression were detected in 20% (16/81 tumors) and 18% (15/81) of primary HCC lesions, respectively; the corresponding figures for metastatic HCC lesions were 39% (9/23) and 44% (10/23), approximately double the rates seen in the primaries. The rates of both pRb loss and overexpression increased significantly as the HCCs progressed from well through moderately to poorly differentiated tumors. These findings suggest that pRb overexpression may reflect an alteration in the pRb regulation pathway which results in loss of its tumor suppressor functions, and that both loss and overexpression of pRb are associated with the malignant progression of HCC and the development of metastatic disease. A study of the impact of pRb loss and overexpression on the prognosis of patients with HCCs is currently ongoing in the authors' laboratory. pRb overexpression has also consistently been found to relate to tumor progression and patient outcomes in bladder cancer (Cote *et al.*, 1998; Grossman *et al.*, 1998) and gall-bladder cancer (Shi *et al.*, 2000b).

The molecular basis of pRb overexpression is not yet fully understood. A possible mechanism involves abnormalities in the upstream genes that regulate *RB1*, such as *INK4*, which lead to pRb hyperphosphorylation and loss of its function (Benedict *et al.*, 1999). Indeed, pRb overexpression has been closely associated with loss of INK4 protein in HCCs (Hui *et al.*, 2000a). Hui *et al.* (2000a) also found that alterations in INK4 protein and pRb, alone and in combination, contribute to HCC progression.

3.5 TP53 (p53)

Mutations of *TP53* are frequently involved in a variety of cancers. The frequency and spectrum of *p53* mutations in HCCs differ with geographic regions and environmental factors. In patients from Qidong, China, and from southern Africa, where both aflatoxin exposure and HBV infection are major etiological factors underlying HCC, *TP53* mutations were detected in more than 50% of these malignancies, the majority of the mutations being a G to T substitution at codon 249 (exon 7) (Bressac *et al.*, 1991; Hsu *et al.*, 1991). In patients from Hong Kong and Taiwan, where HBV infection is an important risk factor for HCC but aflatoxin exposure is not, the overall frequency of *TP53* gene mutations ranges from 29% to 33%, and the frequency of specific codon 249 mutations is 2–13% (Lunn *et al.*, 1997; Ng *et al.*, 1994; Sheu *et al.*, 1992). In patients from Japan, in which HCC is predominantly related to HCV infection, the frequency of *TP53* mutations is around 30%, with a rate of less than 4% for codon 249 mutations (Nishida *et al.*, 1993; Oda *et al.*, 1992b). In patients from Britain, Germany, France and the United States, where populations are at lower risk of viral infection and aflatoxin exposure, the incidence of *TP53* mutations ranges from 6% to 11%, and there are few mutations at codon 249 (Bourdon *et al.*, 1995; Challen *et al.*, 1992; Kubicka *et al.*, 1995; Shieh *et al.*, 1993; Unsal et al., 1994). Overall, it seems that aflatoxin exposure is directly responsible for the mutation at codon 249, while HBV infection may enhance the likelihood of this specific mutation occurring (Yap *et al.*, 1993).

Inactivation of the *TP53* gene is generally considered to require abnormalities in both alleles. *TP53* is located at chromosome 17p, and restriction fragment length polymorphism (RFLP) analysis has revealed that almost all (94%) HCCs with *TP53* mutations show LOH at 17p (Nishida *et al.*, 1993; Oda *et al.*, 1992b). These data suggest that both alleles of the *TP53* gene are inactivated in a subset of HCCs and that *TP53* really does participate in the process of carcinogenesis.

Studies from Japan have shown that although mutations in the *TP53* gene are frequently detected in advanced HCCs, early HCCs do not exhibit such changes (Murakami *et al.*, 1991; Oda *et al.*, 1992a). Oda *et al.* (1992b) found that *TP53* gene mutations occurred more frequently in poorly differentiated HCCs (54%) than in moderately differentiated (29%) or well differentiated ones (9%). These findings suggest that *TP53* mutations contribute to an aggressive tumoral phenotype rather than acting as an inciting cause of HCC. Furthermore, in nodule-in-nodule-type HCCs, *TP53* mutations were detected in the inner nodules (the poorly or moderately differentiated HCC component) but not in the outer portions (the well differentiated HCC component), providing direct evidence that *TP53* contributes to progression from an early to an advanced HCC (Oda *et al.*, 1994).

TP53 gene mutation has been reported to be an unfavorable prognostic marker for patients with HCCs from Japan and Britain (Hayashi *et al.*, 1995; Honda *et al.*, 1998). Because of its short half-life and the minute amounts present, wild-type p53 protein is almost undetectable by immunohistochemical methods (Levine *et al.*, 1991). In contrast, mutant p53 proteins have a much longer half-life and tend to accumulate within tumor cells; therefore, if a p53 protein is detectable by immunohistochemistry, it is generally considered to be a mutant form (Bartek *et al.*, 1991). However, some exceptions have been reported (Bourdon *et al.*, 1995). Since immunohistochemical examinations are much easier to carry out than mutational analysis in clinical practice, many investigators have tried to link the expression of an altered p53 protein to clinicopathologic features of HCC. p53 protein overexpression is almost always associated with poor differentiation and a more advanced clinical stage; however, no consensus regarding the impact of *TP53* expression on clinical outcomes has yet emerged, and the majority of studies have failed to confirm that aberrant p53 protein expression is a significant predictor of patient survival (Caruso and Valentini, 1999; Hsu *et al.*, 1993; Ito *et al.*, 1999; Naka *et al.*, 1998; Ng *et al.*, 1995; Terris *et al.*, 1997).

4. Oncogenes

Mutational activation of Ras family oncogenes and amplification of the *Myc* gene have been considered to be significant events in the process of experimental hepatocarcinogenesis in many animal models (Buchmann *et al.*, 1991; Chandar *et al.*, 1989; Reynolds et al., 1985). However, activation of oncogenes occurs infrequently in human HCC, and seems to play a much less important role in human hepatocarcinogenesis than inactivation of tumor suppressors (Geissler *et al.*, 1997; Robinson, 1994). Mutations in RAS family of genes have been demonstrated in only a small subset of HCCs (Tada *et al.*, 1990; Tanaka *et al.*, 1989; Tsuda *et al.*, 1989). Overexpression of the oncogenes *MYC* and *ERBB* has also been detected in only 7% and 8% of HCCs, respectively (Collier *et al.*, 1992; Lee *et al.*, 1988).

Boix *et al.* (1993) demonstrated overexpression of the mRNA of *MET*, a receptor-type tyrosine kinase oncogene, in eight of 18 HCCs by Northern blotting analysis. Suzuki *et al.* (1994) then reported *MET* to be overexpressed at the mRNA level (as determined by Northern blotting) in 32% of 23 HCCs, and at the protein level (as determined by immunohistochemistry) in 70%. They also associated MET protein overexpression with poor to moderately differentiated tumors. The role of alterations in c-met in hepatocarcinogenesis and their clinical importance remain to be determined.

Cyclin D1 combines with and activates CDKs to promote cell progression from the G_1 to the S phase. It is generally considered to be an oncogene and is overexpressed, as a consequence of gene amplification, in many types of human cancer. Zhang *et al.* (1993) studied cyclin D1 in HCCs by Southern blotting and immunohistochemistry, and demonstrated amplification and protein overexpression in a small fraction (13%) of these tumors. Furthermore, Nishida *et al.* (1994) found that cyclin D1 was amplified in 11% of HCCs, and that this was accompanied by mRNA overexpression. Their data also suggest that cyclin D1 amplification is associated with the advanced stages of HCC. Recently, Ito *et al.* (1999) reported that cyclin D1 overexpression occurred in 32% of HCCs and was significantly associated with poor disease-free survival. In their study, cyclin E overexpression was also detected in 36% of HCCs but was not significantly related to patient outcome.

5. Allelotype

Many investigators have performed allelotype analysis on HCCs in an attempt to localize tumor suppressor loci. The combined results of the reported studies have frequently demonstrated LOH on the following chromosomes (in more than 25% of tumors): 1p, 1q, 4q, 5q, 6q, 8p, 8q, 9p, 9q, 10q, 11p, 13q, 14q, 16p, 16q, 17p, 19p and 22q. These data are summarized in *Table 1*. Some loci show variability of the frequency of LOH between studies, which may be due to the different polymorphic markers used. Some of these frequently lost regions involve known, as well as candidate, tumor suppressor genes such as the *TP53* gene (17p13.1), the *RB1* gene (13q14), *INK4* and *INK4B* genes (9p21), the insulin-like growth factor-II receptor gene (6q26-q27), the platelet-derived growth factor-receptor [beta] gene (8p21.3-p22), the hypermethylated in cancer-1 (*HIC1*) gene (17p13.3) and the E-cadherin gene (16q22.1).

Recently, Okabe *et al.* (2000) conducted a comprehensive allelotype analysis in HCC using 216 microsatellite markers situated throughout the short and long arms of all 23 chromosomes. A total of 33 loci on 4q, 6q, 8p, 8q, 9p, 9q, 13q, 16p, 16q, 17q and 19p harbored LOHs with a frequency of more than 30%. This high frequency of LOH was significantly associated with poor differentiation, vascular invasion and intrahepatic metastasis. Overall, a high rate of LOH was also correlated to HBV infection. However, there was variability among the chromosomal regions; allelic loss of 4q and 13q was associated with HBV-related HCCs, while LOH on 6q occurred more frequently in HBV-negative HCCs. These results suggest that the genetic events that occur in HCC depend on the hepatitis virus infection status.

Many investigators have studied the relationship between LOH at a specific locus and the clinicopathological features of HCC. Kuroki *et al.* (1995) reported

Table 1. Loss of chromosomal alleles in hepatocellular carcinomas

Locus	Frequency	References
1p	83%	Simon *et al.*, 1991
	33%	Konishi *et al., 1993*
	33%	Kuroki *et al.*, 1995
1q	68%	Piao *et al.*, 1998
4p11-q21	50%	Zhang *et al.*, 1990
4q	73%	Piao *et al.*, 1998
5q35-qter	50%	Ding *et al.*, 1991
6q	69%	De Souza *et al.*, 1991
8p21.3-p23.3	48%	Emai *et al.*, 1992
8p23	42%	Nagai *et al.*, 1997
8p	64%	Piao *et al.*, 1998
	60%	Pineau *et al.*, 1999
8q	77%	Piao *et al.*, 1998
8q13.3	31%	Okabe *et al.*, 2000
9p21	63%	Liew *et al.*, 1999
9p22-p23	38%	Okabe *et al.*, 2000
9q22.2	30%	Okabe *et al.*, 2000
10q	25%	Fujimori *et al.*, 1991
	25%	Takahashi *et al.*, 1992
	33%	Piao *et al.*, 1998
11p13-15	43%	Wang *et al.*, 1988
13q12	67%	Walker *et al.*, 1991
13q	40%	Piao *et al.*, 1998
14q	46%	Piao *et al.*, 1998
16q22.1-q23.2	57%	Zhang *et al.*, 1990
	52%	De Souza *et al.*, 1992
16q	59%	Piao *et al.*, 1998
16q13.3	41%	Okabe *et al.*, 2000
17q	54%	Fujimori *et al.*, 1991
	43%	Takahashi *et al.*, 1992
	46%	Piao *et al.*, 1998
17p13.3	54%	Kanai *et al.*, 1999
19p13.3	33%	Okabe *et al.*, 2000
22q11	33%	Takahashi *et al.*, 1998

that LOH on 1p frequently (31%) occurred in early or well-differentiated HCCs, suggesting that this genetic lesion participates in the early developmental stages of hepatocarcinogenesis. In contrast, Tsuda *et al.* (1990) showed that LOH on 16p frequently occurred in large, poorly differentiated and metastatic HCCs, but not in early stage HCCs, implying that loss of this region plays a role in tumor aggressiveness rather than tumor initiation.

6. DNA methylation

Alterations in DNA methylation, an epigenetic event, which modulate gene expression but do not directly change the gene sequence, play an important role in the process of human carcinogenesis. Abnormal DNA methylation is one of the most consistent changes that occurs in cancer. Alterations in the DNA methylation level within the genome of cancer cells can include coexisting global hypomethylation and regional hypermethylation in certain specific regions. DNA hypomethylation probably contributes to cancer development and progression by increasing the expression of oncogenes. DNA hypermethylation is involved in carcinogenesis through the following mechanisms. Firstly, it causes changes in the chromatin configuration, predisposing to allelic loss at a specific locus in the chromosomes. Secondly, hypermethylation of CpG islands in the promotor regions of genes leads to transcription blockade, which is in turn involved in the inactivation of some tumor suppressor genes. Finally, DNA cytosine methylation results in a specific gene mutation when 5-methylcytosine is spontaneously deaminated to a thymine. Changes in the level of DNA methylation are reversible, and reversing DNA methylation changes in cancer cells by pharmacological or molecular biological means seems to be more feasible than correcting the genetic alterations. Therefore, changes in DNA methylation are potential targets for new cancer treatments.

Many lines of evidence have indicated that DNA hypermethylation plays several important roles in human hepatocarcinogenesis (Hui and Makuuchi, 1999). The *HIC1* gene is a candidate tumor suppressor gene located at 17p13.3, near to the *TP53* locus. Recently, we studied *HIC1* mRNA expression, LOH and methylation status at the 17p13.3 (D17S5 locus) in HCC tissues, noncancerous liver tissues showing cirrhosis or chronic hepatitis and normal liver tissues (Kanai *et al.*, 1999). DNA hypermethylation at D17S5 was detected (by a Southern blotting method using methylation-sensitive restriction enzymes) in 90% of the HCC tissues and 44% of the liver tissues showing chronic hepatitis or cirrhosis, but was not demonstrated in normal liver tissues. The *HIC1* mRNA expression level was significantly lower in liver tissues with chronic hepatitis or cirrhosis than in normal liver tissues, and was decreased even more in HCC tissues. Given that both chronic hepatitis and cirrhosis can be precancerous conditions, these data suggest that DNA hypermethylation at the 17p13.3 locus and reduced *HIC1* mRNA expression may be involved in the early stages of hepatocarcinogenesis. Moreover, the *HIC1* mRNA expression levels were lower in poorly differentiated HCCs than in well or moderately differentiated ones, suggesting the possible involvement of *HIC1* in tumor progression. LOH at the 17p13.3 locus was demonstrated in 54% of the HCCs in this series. DNA hypermethylation at 17p13.3 therefore seems to contribute to hepatocarcinogenesis by decreasing *HIC1* expression and predisposing to allelic losses around this locus.

Reduced expression of E-cadherin (*CDH1*), associated with CpG island hypermethylation around the promotor region of its gene, has also been reported to be involved in human hepatocarcinogenesis (Kanai *et al.*, 1997). CpG hypermethylation was found in 67% of HCC tissues and 46% of liver tissues showing cirrhosis or chronic hepatitis, while decreased *CDH1* expression was detected in 59% of HCCs and was significantly associated with CpG hypermethylation. These data suggest that hypermethylation in the promotor region of the *CDH1* gene may participate in the progression from a precancerous condition to a cancer.

Hypermethylation of several loci on chromosome 16 has also been reported in HCCs (Kanai *et al.*, 1996). DNA hypermethylation at the TAT, D16S7 and D16S32 loci was demonstrated in 58%, 20% and 18% of HCC tissues, respectively, and was associated with advanced disease stages, vascular invasion and intrahepatic metastasis. DNA hypermethylation was also demonstrated in noncancerous liver tissues showing chronic hepatitis or cirrhosis obtained from patients with DNA hypermethylation in their HCC tissue, but not in patients without DNA hypermethylation of their HCC tissues. LOH at the TAT locus occurred in 42% of these HCC tissues, but not in any of the noncancerous liver tissues. Hypermethylation occurred more frequently than LOH at the TAT locus in HCCs (even occurring in noncancerous liver tissue showing chronic hepatitis or cirrhosis, whereas LOH did not), suggesting that DNA hypermethylation at this locus may predispose to allelic deletion. Hypomethylation at the D16S4 locus was detected in 48% of HCCs; however, its significance in hepatocarcinogenesis remains unclear.

DNA methyltransferase (DNA MTase), which transfers a methyl group from S-adenosylmethionine to the 5-position of cytidine, is responsible for changes in DNA methylation status. In order to clarify the roles of DNA MTase in human hepatocarcinogenesis, we have examined its mRNA expression in normal liver tissues, liver tissue showing chronic hepatitis or cirrhosis, HCC tissues and HCC cell lines (Sun et al., 1997). DNA MTase mRNA expression levels were significantly higher in liver tissues showing chronic hepatitis or cirrhosis than in normal liver tissues, and were higher still in HCC tissues. However, this increase in DNA MTase expression was not as well correlated with the degree of tumor cell differentiation as the hepatitis virus infection status. Nevertheless, the findings of this study suggest that elevated DNA MTase expression may be involved in the early stages of hepatocarcinogenesis.

There are several possible mechanisms by which DNA MTase may contribute to cancer development and progression. Increased DNA MTase activity leads to DNA hypermethylation at specific loci, predisposing to allelic deletions and inactivating tumor suppressor genes. In addition, DNA MTase facilitates C-to-T transition mutations at CpG sites. These mutations result from deamination of 5-methylcytosine to thymine, and are frequently observed in the TP53 gene in HCC (Oda et al., 1992b).

Since DNA MTase plays a number of significant roles in human hepatocarcinogenesis, especially in the early developmental stages, it seems an important potential target for HCC treatment and prevention. In animal models, reduction of DNA TMase levels has been proven to inhibit tumorigenicity. 5-azacytidine can inhibit the activity of MTase; however, there are at least two factors that limit its clinical applicability: firstly, global hypomethylation and regional hypermethylation at specific sites can coexist in the tumor cell genome, and secondly, 5-azacytidine is highly toxic to normal cells. Thus, the search for a biological or pharmacological weapon which can selectively modulate the methylation level in specific regions of the cancer cell genome, and which has no or only slight side-effects on normal tissue, is of key importance.

7. Tumors of multifocal origin and intrahepatic metastases

Cancer recurrence after surgery is one of the major obstacles which hinders the curability of HCC. Recurrences are frequent, with reported rates ranging from

55% to 100% (Belghiti *et al.*, 1991; Kakazu *et al.*, 1993; Yamamoto *et al.*, 1996; Zhou *et al.*, 1993). The discovery of multiple tumors at presentation is also an important factor affecting clinical outcomes. Multiple and/or recurrent HCCs can be the result of simultaneously arising multifocal tumors or of intrahepatic metastasis of a single primary tumor through the portal venous system. Discrimination between tumors of multifocal origin and metastases is important from a clinical viewpoint, since they represent different disease stages and therapeutic strategies and clinical outcomes differ between them. Compared to recurrence resulting from multifocal carcinogenesis, recurrence from intrahepatic metastasis is characterized by a short interval before recurrence and a poor prognosis. In the case of metastatic recurrence, it is also possible that cancer cells might have spread beyond the liver, thus limiting treatment options. When a recurrence results from multifocal occurrence of a new cancer, radical intervention to remove the lesions should be performed within the limits of the hepatic function reserve, and an improved outcome is usually obtained.

Pathological criteria have been used in the differential diagnosis of multifocal origin of HCCs and intrahepatic metastases (Kanai *et al.*, 1987; Tsuda *et al.*, 1992). Multiple HCCs are considered to be a result of metastasis from a single tumor (a) when they appear to be growing in contiguity with portal vein tumor thrombi or (b) when they are distributed as small satellite nodules surrounding a single main tumor. The following conditions are considered to indicate multifocal occurrence: (a) multiple early HCCs or concurrent early and advanced HCCs, (b) moderately to poorly differentiated HCCs with well differentiated components in their peripheral areas, and (c) multiple tumors with distinct differences in their histological types or in the degree of tumor cell differentiation. These pathological criteria, however, are not completely satisfactory, and differential diagnosis based on morphological findings is impossible in some patients with multiple HCCs.

In HBV-related HCCs, HBV DNA is randomly integrated into the host DNA. Thus, on Southern blotting, a restriction fragment containing integrated HBV DNA will present as a definite band because of clonal growth of a tumor cell. Since the integration patterns of HBV DNA differ between tumors with different clonal

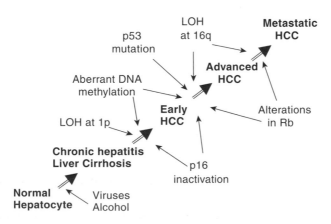

Figure 2. A molecular model of human hepatocarcinogenesis.

origins, analysis of the integration pattern has been used to determine whether multiple HCCs are of multifocal occurrence (in which case they show different integration patterns) or of metastatic origin (in which case they show an identical integration pattern) (Esumi *et al.*, 1986; Hsu *et al.*, 1990). This method is objective and accurate; however, it is only applicable to HBV-related HCCs. Patterns of chromosomal allele loss (Tsuda *et al.*, 1992) and *TP53* gene mutations (Oda *et al.*, 1992a) have also been employed as a marker during the differential diagnosis of multiple HCCs; however, these methods are only applicable to portions of HCCs with allelic loss or harboring a *TP53* mutation. The authors have shown that flow cytometric analysis of their DNA content is also useful for determining the origins of multiple HCCs, but again, this method is not totally ideal (Hui *et al.*, 1997b). Overall, none of the techniques used to date has been completely satisfactory in allowing definite identification of the clonality of multiple HCCs. Judgment of the origin of multiple and/or recurrent HCCs should therefore be made on a comprehensive basis, taking into account both pathological and genetic analyses.

8. A molecular model of human hepatocarcinogenesis

Human carcinogenesis is a multistep process in which many genetic or epigenetic alterations contribute to each stage of development and progression. A possible molecular model of hepatocarcinogenesis shown in *Figure 2*; this is a working hypothesis based on the data reviewed in this chapter. Abnormalities in DNA methylation are responsible for the early developmental stages of hepatocarcinogenesis. Increased DNA MTase expression, reduced *HIC1* expression and hypermethylation on chromosome 16 are frequently detected not only in HCCs but also in noncancerous livers showing chronic hepatitis or cirrhosis, which are widely considered to be precancerous conditions. *HIC1* inactivation and hypermethylation on chromosome 16 also contribute to later tumor progression. Inactivation of *INK4*, which occurs even in early HCCs and increases in frequency in advanced tumors, participates in hepatocarcinogenesis during both the early developmental stages and disease progression. LOH at 1p is also an early event in hepatocarcinogenesis. *TP53* mutations, abnormalities in *RB1* gene expression and LOH at chromosome 16q are observed during the later stages but not in early HCCs; these alterations seem to increase the malignancy of the tumor. Rb dysfunction and LOH at 16q are closely associated with the development of metastatic disease.

As described above, a number of genetic and epigenetic events have been reported to be involved in the development and progression of HCCs; however, the molecular mechanisms underlying the multistep process of human hepatocarcinogenesis remain incompletely understood. Differences in certain clinicopathological features between HBV-related and HCV-related HCCs imply that the development and progression of HCCs may involve different molecular mechanisms depending on the viral infection status. Unfortunately, many of the studies investigating the molecular basis of HCC have failed to provide a careful report of the association between a genetic (or epigenetic) change and the viral infection status. Comparison of the mechanisms involved in HCCs with different viral etiologies and characterization of events specific to HBV-related, HCV-related and virus-negative HCCs should be a focus for future studies. Moreover, although many genetic and epigenetic events have been shown to occur in HCCs, their impacts on clinical outcomes are not yet well clarified. New data with a bearing on this subject are necessary. An understanding of the molecular basis of carcinogenesis will provide information essential to the search for ideal targets for cancer gene therapy. In turn, the identification of such sites will mean that novel prevention strategies and therapies for HCC will be forthcoming.

References

Bannasch, P., Khoshkhou, N.I., Hacker, H.J. *et al.*, (1995) Synergistic hepatocarcinogenic effect of hepatoviral infection and dietary aflatoxin B1 in woodchucks. *Cancer Res.* 55: 3318–3330.

Bartek, J., Bartkova, J., Vojtesek, B. *et al.*, (1991) Aberrant expression of the p53 oncoprotein is a common feature of a wide spectrum of human malignancies. *Oncogene* 6: 1699–1703.

Belghiti, J., Panis, Y., Farges, O., Benhamou, J.P. and Fekete, F. (1991) Intrahepatic recurrence after resection of hepatocellular carcinoma complicating cirrhosis. *Ann. Surg.* 214: 114–117.

Benedict, W.F., Lerner, S.P., Ahou, J., Shen, X., Tokunaga, H. and Czerniak, B. (1999) Level of retinoblastoma protein expression correlates with p16 (MTS-1/INK4A/CDKN2) status in bladder cancer. *Oncogene* 18: 1197–1203.

Biden, K., Young, J., Buttenshaw, R., Searle, J., Cooksley, G., Xu, D.-B. and Leggett, B. (1997) Frequency of mutation and deletion of the tumor suppressor gene CDKN2A (MTS1/p16) in hepatocellular carcinoma from an Australian population. *Hepatology* 25: 593–597.

Boix, L., Rosa, J.L., Ventura, F., Castells, A., Bruix, J., Rodés, J. and Bartrons, R. (1994) c-met mina overexpression in human hepatocellular carcinoma. *Hepatology* 19: 88–91.

Bourdon, J.-C., D'Errico, A., Paterlini, P., Grigioni, W., May, E. and Debuire, B. (1995) p53 protein accumulation in European hepatocellular carcinoma is not always dependent on p53 gene mutation. *Gastroenterology* 108: 1176–1182.

Brechot, C., Pourcel, C., Louise, A., Rain, B. and Tiollais, P. (1980) Presence of integrated hepatitis B virus DNA sequences in cellular DNA of human hepatocellular carcinoma. *Nature* 286: 533–535.

Bressac, B., Kew, M., Wands, J. and Ozturk, M. (1991) Selective G to T mutation of p53 gene in hepatocellular carcinoma from southern Africa. *Nature* 350: 429–431.

Buchmann, A., Bauer, H.R., Mahr, J., Drinkwater, N.R., Luz, A. and Schwarz, M. (1991) Mutational activation of the c-Ha-ras gene in liver tumors of different rodent strains: correlation with susceptibility to hepatocarcinogenesis. *Proc. Natl Acad. Sci. USA* 88: 911–915.

Cairns, P., Polascik, T.J., Eby, Y. *et al.* (1995) Frequency of homozygous deletion at p16/CDKN2 in primary human tumours. *Nature Genet.* 11: 210–212.

Caldas, C., Hahn, S.A., da Costa, L.T. *et al.* (1994) Frequent somatic mutations and homozygous deletions of the p16 (MTS1) gene in pancreatic adenocarcinoma. *Nature Genet.* 8: 27–32.

Caruso, M.L. and Valentini, A.M. (1999) Overexpression of p53 in a large series of patients with hepatocellular carcinoma: a clinicopathological correlation. *Anticancer Res.* 19: 3853–3856.

Challen, C., Lunec, J., Warren, W., Collier, J. and Bassendine, M.F. (1992) Analysis of p53 tumor-suppressor gene in hepatocellular carcinomas from Britain. *Hepatology* 16: 1362–1366.

Chandar, N., Lombardi, B. and Locker, J. (1989) c-myc gene amplification during hepatocarcinogenesis by a choline-devoid diet. *Proc. Natl. Acad. Sci. USA* 86: 2703–2707.

Collier, J.D., Guo, K., Mathew, J., May, F.E., Bennett, M.K., Corbett, I.P., Bassendine, M.F. and Burt, A.D. (1992) c-erbB-2 oncogene expression in hepatocellular carcinoma and cholangiocarcinoma. *J. Hepatol.* 14: 377–380.

Cote, R.J., Dunn, M.D., Chatterjee, S.J. *et al.* (1998) Elevated and absent pRb expression is associated with bladder cancer progression and has cooperative effects with p53. *Cancer Res.* 58: 1090–1094.

De Souza, A.T., Hankins, G.R., Washington, M.K., Fine, R.L., Orton, T.C. and Jirtle, R.L. (1995) Frequent loss of heterozygosity on 6q at the mannose 6-phosphate/insulin-like growth factor II receptor locus in human hepatocellular tumors. *Oncogene* 10: 1725–1729.

Ding, S.-F., Habib, N.A., Dooley, J., Wood, C., Bowles, L. and Delhanty, J.D.A. (1991) Loss of constitutional heterozygosity on chromosome 5q in hepatocellular carcinoma without cirrhosis. *Br. J. Cancer* **64**: 1083–1087.

Edmondson, A.H. and Steiner, P.E. (1954) Primary carcinoma of the liver. A study of 100 cases among 48,900 necropsies. *Cancer* **7**: 462–503.

El-Serag, H.B. and Mason, A.C. (1999) Rising incidence of hepatocellular carcinoma in the United States. *N. Engl. J. Med.* **340**: 745–750.

Emi, M., Fujiwara, Y., Nakajima, T. *et al.* (1992) Frequent loss of heterozygosity for loci on chromosome 8p in hepatocellular carcinoma, colorectal cancer, and lung cancer. *Cancer Res.* **52**: 5368–5376.

Emi, M., Fujiwara, Y., Ohata, H., Tsuda, H., Hirohashi, S., Koike, M., Miyaki, M., Monden, M. and Nakamuya, Y. (1993) Allelic loss at chromosome band 8p21.3-p22 is associated with progression of hepatocellular carcinoma. *Genes Chrom.Cancer* **7**: 152–157.

Esumi, M., Aritaka, T., Arii, M., Suzuki, K., Tanikawa, K., Mizuo, H., Mima, T. and Shikata, T. (1986) Clonal origin of human hepatoma determined by integration of hepatitis B virus DNA. *Cancer Res.* **46**: 5767–5771.

Fujimori, M., Tokino, T., Hino, O. *et al.* (1991) Allelotype study of primary hepatocellular carcinoma. *Cancer Res.* **51**: 89–93.

Geissler, M., Gesien, A. and Wands, J.R. (1997) Molecular mechanisms of hepatocarcinognesis. In: *Liver Cancer* (eds K. Okuda and E. Tabor). Churchill Livingstone Inc., New York, pp. 59–88.

Grossman, H.B., Liebert, M., Antelo, M., Dinney, C.P.N., Hu, S.-X., Palmer, J.L. and Benedict, W.F. (1998) p53 and pRb expression predict progression in T1 bladder cancer. *Clin. Cancer Res.* **4**: 829–834.

Hardell, L., Bengtsson, N.O., Jonsson, U., Eriksson, S. and Larsson, L.G. (1984) Aetiological aspects of primary liver cancer with special regard to alcohol, organic solvents and acute intermittent porphyria – an epidemical investigation. *Br. J. Cancer* **50**: 389–397.

Hayashi, H., Sugio, K., Matsuda, T., Adachi, E., Takenaka, K. and Sugimachi, K. (1995) The clinical significance of p53 gene mutation in hepatocellular carcinomas from Japan. *Hepatology* **22**: 1702–1707.

Herman, J.G., Merlo, A., Mao, L., Lapidus, R.G., Issa, J.P.J., Davidson, N.E., Sidransky, D. and Baylin, S.B. (1995) Inactivation of the CDKN2/p16/MTS1 gene is frequently associated with aberrant DNA methylation in all common human cancers. *Cancer Res.* **55**: 4525–4530.

Hirohashi, S. (1992) Pathology and molecular mechanisms of multistage human hepatocarcinogenesis. In: *Multistage Carcinogenesis* (eds C.C. Harris, S. Hirohashi, N. Ito, H.C. Pitot, T. Sugiyama, M. Terada, J. Yokoda) Japan Sci. Soc. Press, Tokyo, pp. 87–93.

Hirohashi, S. and Sakamoto, M. (1992) Hepatocellular carcinoma in the early stage. In: *Primary Liver Cancer in Japan* (eds T. Tobe, H. Kameda, M. Okudaira, M. Ohto, Y. Endo, M. Mito, E. Okamoto, K. Tanikawa and M. Kojiro). Springer-Verlag, Tokyo, pp. 25–30.

Honda, K., Sbisa, E., Tullo, A. *et al.* (1998) p53 mutation is a poor prognostic indicator for survival in patients with hepatocellular carcinoma undergoing surgical tumor ablation. *Br. J. Cancer* **77**: 776–782.

Hsia, C.C., Di Bisceglie, A.M., Kleiner, D.E.J., Farshid, M. and Tabor, E. (1994) RB tumor suppressor gene expression in hepatocellular carcinomas from patients infected with hepatitis B virus. *J. Med. Virol.* **44**: 67–73.

Hsu, I.C., Metcalf, R.A., Sun, T., Welsh, J.A., Wang, N.J. and Harris C.C. (1991) Mutational hotspot in the p53 gene in human hepatocellular carcinomas. *Nature* **350**: 427–428.

Hsu, H.-C., Chiou, T.-J., Chen, J.-Y., Lee, C.-S., Lee, P.-H. and Peng, S.-Y. (1990) Clonality and clonal evolution of hepatocellular carcinoma with multiple nodules. *Hepatology* **13**: 923–928.

Hsu, H.-C., Tseng, H.-J., Lai, P.-L., Lee, P.-H. and Peng, S.-Y. (1993) Expression of p53 gene in 184 unifocal hepatocellular carcinomas: association with tumor growth and invasiveness. *Cancer Res.* **53**: 4691–4694.

Hui, A.-M. and Makuuchi, M. (1999) Molecular basis of multistep hepatocarcinogenesis: genetic and epigenetic events. *Scand. J. Gastroenterol.* **34**: 739–742.

Hui, A.-M., Sakamoto, M., Kanai, Y., Ino, Y., Gotoh, M., Yokota, J. and Hirohashi, S. (1996a) Inactivation of p16^{INK4} in hepatocellular carcinoma. *Hepatology* **24**: 575–579.

Hui, A.-M., Sakamoto, M., Kanai, Y., Ino, Y., Gotoh, M. and Hirohashi, S. (1996b) Cyclin-dependent kinase inhibitors and hepatocarcinogenesis. In: *Recent Advances in Gastroenterological Carcinogenesis* (eds E. Tahara, K. Sugimachi, T. Oohara). Monduzzi Editore, Bologna, pp. 481–485.

Hui, A.-M., Kanai, Y., Sakamoto, M., Tsuda, H. and Hirohashi, S. (1997a) Reduced p21$^{WAF1/CIP1}$ expression and p53 mutation in hepatocellular carcinomas. *Hepatology* 25: 575–579.

Hui, A.-M., Kawasaki, S., Imamura, H., Miyagawa, S., Ishii, K., Katsuyama, T. and Makuuchi, M. (1997b) Heterogeneity of DNA content in multiple synchronous hepatocellular carcinomas. *Br. J. Cancer* 76: 335–339.

Hui, A.-M., Sun, L., Kanai, Y., Sakamoto, M. and Hirohashi, S. (1998a) Reduced p27^{Kip1} expression in hepatocellular carcinomas. *Cancer Lett.* 132: 67–73.

Hui, A.-M., Makuuchi, M., Li, X. (1998b) Cell cycle regulators and human hepatocarcinogenesis. *Hepato-Gastroenterology* 45: 1635–1642.

Hui, A.-M., Li, X., Makuuchi, M., Takayama, T. and Kubota, K. (1999a) Overexpression and lack of retinoblastoma protein are associated with tumor progression and metastasis in hepatocellular carcinoma. *Int. J. Cancer* 84: 604–608.

Hui, A.-M., Cui, X., Makuuchi, M., Li, X., Shi, Y.-Z. and Takayama, T. (1999b) Decreased p27^{Kip1} expression and cyclin D1 overexpression, alone and in combination, influence recurrence and survival of patients with resectable extrahepatic bile duct carcinoma. *Hepatology* 30: 1167–1173.

Hui, A.-M., Shi, Y.-Z., Takayama, T. and Makuuchi, M. (2000a) Loss of p16^{INK4} protein, alone and together with loss of retinoblastinoma protein, correlate with hepatocellular carcinoma progression. *Cancer Lett.* 154: 93–99.

Hui, A.-M., Li, X., Shi, Y.-Z., Torzilli, G., Takayama, T. and Makuuchi, M. (2000b) p27^{Kip1} expression in normal epithelia, precancerous lesions and carcinomas of the gall-bladder: association with cancer progression and prognosis. *Hepatology* 31: 1068–1072.

Hunter, T. and Pines, J. (1994) Cyclins and cancer II: Cyclin D and CDK inhibitors come of age. *Cell* 79: 573–582.

Hussussian, C.J., Ptruewing, J.P., Goldstein, A.M., Higgins, P.A.T., Ally, D.S., Sheahan, M.D., Clark, Jr. W.H., Tucker, M.A. and Dracopoli, N.C. (1994) Germline p16 mutations in familial melanoma. *Nature Genet.* 8: 15–21.

Ince, N. and Wands, J.R. (1999) The increasing incidence of hepatocellular carcinoma. *N. Engl. J. Med.* 340: 798–799.

Ito, Y., Matsuura, N., Sakon, M. *et al.* (1999) Expression and prognostic roles of the G1-S modulators in hepatocellular carcinoma: p27 independently predicts the recurrence. *Hepatology* 30: 90–99.

Kakazu, T., Makuuchi, M., Kawasaki, S., Miyagawa, S., Hashikura, Y., Kosuge, T., Takayama, T. and Yamamoto, J. (1993) Repeat hepatic resection for recurrent hepatocellular carcinoma. *Hepato-Gastroenterology* 40: 337–341.

Kanai, T., Hirohashi, S., Upton, M.P. *et al.* (1987) Pathology of small hepatocellular carcinoma: a proposal for a new gross classification. *Cancer* 60: 810–819.

Kanai, Y., Ushijima, S., Tsuda, H., Sakamoto, M., Sugimura, T. and Hirohashi, S. (1996) Aberrant DNA methylation on chromosome 16 is an early event in hepatocarcinogenesis. *Jpn. J. Cancer Res.* 87: 1210–1217.

Kanai, Y., Ushijima, S., Hui, A.-M., Ochiai, A., Tsuda, H., Sakamoto, M. and Hirohashi, S. (1997) The E-cadherin gene is silenced by CpG methylation in human hepatocellular carcinomas. *Int. J. Cancer* 71: 355–359.

Kanai, Y., Hui, A.-M., Sun, L., Ushijima, S., Sakamoto, M., Tsuda, H. and Hirohashi, S. (1999) DNA hypermethylation at the D17S5 locus and reduced HIC-1 mRNA expression are associated with hepatocarcinogenesis. *Hepatology* 29: 703–709.

Kim, C.-M., Koike, K., Saito, I., Miyamura, T. and Jay, G. (1991) HBx gene of hepatitis B virus induces liver cancer in transgenic mice. *Nature* 351: 317–320.

Kita, R., Nishida, N., Fukuda, Y., Azechi, H., Matsuoka, Y., Komeda, T., Sando, T., Komeda, T. and Nakao, K. (1996) Infrequent alterations of the p16^{INK4} gene in liver cancer. *Int. J. Cancer* 67: 176–180.

Konishi, M., Kikuchi-Yanoshita, R., Tanaka, K. *et al.* (1993) Genetic changes and histopathological grades in human hepatocellular carcinomas. *Jpn J. Cancer Res.* 84: 893–899.

Kovar, H., Jug, G., Aryee, D.N.T., Zoubek, A., Ambros, P., Gruber, B., Windhager, R. and Gadner, H. (1997) Among genes involved in the RB dependent cell cycle regulatory cascade, the p16 tumor suppressor gene is frequently lost in the Ewing family of tumors. *Oncogene* 15: 2225–2232.

Kubicka, S., Trautwein, C., Schrem, H., Tillmann, H. and Manns, M. (1995) Low incidence of p53 mutations in European hepatocellular carcinomas with heterogeneous mutation as a rare events. *J. Hepatol.* 23: 412–419.

Kuroki, T., Fujiwara, Y., Tsuchiya, E., Nakamori, S., Imaoka, S., Kanematsu, T. and Nakamura, Y. (1995) Accumulation of genetic changes during development and progression of hepatocellular carcinoma: loss of heterozygosity of chromosome arm 1p occurs at an early stage of hepatocarcinogenesis. *Genes Chrom. Cancer* **13**: 163–167.

Lee, F. (1966) Cirrhosis and hepatoma in alcoholics. *Gut* **7**: 77–85.

Lee, H.S., Rajagopalan, M.S. and Vyas, G.N. (1988) A lack of direct role of hepatitis B virus in the activation of ras and c-myc oncogenes in human hepatocellular carcinogenesis. *Hepatology* **8**: 1116–1120.

Levine, A.J., Momand, J. and Finlay, C.A. (1991) The p53 tumour suppressor gene. *Nature* **351**: 453–456.

Liew, C.T., Li, H.M., Lo, K.W. *et al.* (1999) Frequent allelic loss on chromosome 9 in hepatocellular carcinoma. *Int. J. Cancer* **81**: 319–324.

Lin, Y.-W., Chen, C.-H., Huang, G.-T., Lee, P.-H., Wang, J.-T., Chen, D.-S., Lu, F.-J. and Sheu, J.-C. (1998) Infrequent mutations and no methylation of CDKN2A (p16/MTS1) and CDKN2B (p15/MTS2) in hepatocellular carcinoma in Taiwan. *Eur. J. Cancer* **34**: 1789–1795.

Lunn, R.M., Zhang, Y.J., Wang, L.Y., Chen, C.J., Lee, P.H., Lee, C.S., Tsai, W.Y. and Santella, R.M. (1997) p53 mutations, chronic hepatitis B virus infection, and aflatoxin exposure in hepatocellular carcinoma in Taiwan. *Cancer Res.* **57**: 3471–3477.

Maki, C.G. and Howley, P.M. (1997) Ubiquitination of p53 and p21 is differentially affected by ionizing and UV radiation. *Mol. Cell. Biol.* **17**: 355–363.

Matsuda, M., Ichida, T., Matsuzawa, J., Sugimura, K. and Asakura, H. (1999) p16[INK4] is inactivated by extensive CpG methylation in human hepatocellular carcinoma. *Gastroenterology* **116**: 394–400.

Merlo, A., Herman, J.G., Mao, L., Lee, D.J., Gabrielson, E., Burger, P.C., Baylin, S.B. and Sidransky, D. (1995) 5′ CpG island methylation is associated with transcriptional silencing of the tumour suppressor p16/CDKN2/MTS1 in human cancers. *Nature Med.* **1**: 686–692.

Murakami, Y., Hayashi, K., Hirohashi, S. and Sekiya, T. (1991) Aberrations of the tumor suppressor p53 and retinoblastoma genes in human hepatocellular carcinomas. *Cancer Res.* **51**: 5520–5525.

Nagai, H., Pineau, P., Tiollais, P., Buendia, M.A. and Dejean, A. (1997) Comprehensive allelotyping of human hepatocellular carcinoma. *Oncogene* **14**: 2927–2933.

Naka, T., Toyota, N., Kaneko, T. and Kaibara, N. (1998) Protein expression of p53, p21[WAF1/CIP1], and Rb as prognostic indicators in patients with surgically treated hepatocellular carcinoma. *Anticancer Res.* **18**: 555–564.

Ng, I.O.L., Chung, L.P., Tsang, S.W.Y., Lam, C.L., Lai, E.C.S., Fan, S.T. and Ng, M. (1994) p53 gene mutation spectrum in hepatocellular carcinomas in Hong Kong Chinese. *Oncogene* **9**: 985–990.

Ng, I.O.L., Lai, E.C.S., Chan, A.S.Y. and So, M.K.P. (1995) Overexpression of p53 in hepatocellular carcinoma: a clinicopathological and prognostic correlation. *J. Gastroenterol. Hepatol.* **10**: 250–255.

Nishida, N., Fukuda, Y., Kokuryu, H. *et al.* (1992) Accumulation of allelic loss on arms of chromosomes 13q, 16q and 17p in the advanced stages of human hepatocellular carcinoma. *Int. J. Cancer* **51**: 862–868.

Nishida, N., Fukuda, Y., Kokuryu, H., Toguchida, J., Yandell, D.W., Ikenaga, M., Imura, H. and Ishizaki, K. (1993) Role and mutational heterogeneity of the p53 gene in hepatocellular carcinoma. *Cancer Res.* **53**: 368–372.

Nishida, N., Fukuda, Y., Komeda, T. *et al.* (1994) Amplification and overexpression of the cyclin D1 gene in aggressive human hepatocellular carcinoma. *Cancer Res.* **54**: 3107–3110.

Nobori, T., Miura, K., Wu, D.J., Lois, A., Takabayashi, K. and Carson, D.A. (1994) Deletions of the cyclin-dependent kinase-4 inhibitor gene in multiple human cancers. *Nature* **368**: 753–756.

Oda, T., Tsuda, H., Scarpa, A., Sakamoto, M. and Hirohashi, S. (1992a) Mutation pattern of the p53 gene as a diagnostic marker for multiple hepatocellular carcinoma. *Cancer Res.* **52**: 3674–3678.

Oda, T., Tsuda, H., Scarpa, A., Sakamoto, M. and Hirohashi, S. (1992b) p53 gene mutation spectrum in hepatocellular carcinoma. *Cancer Res.* **52**: 6358–6364.

Oda, T., Tsuda, H., Sakamoto, M. and Hirohashi, S. (1994) Different mutations of the p53 gene in nodule-in-nodule hepatocellular carcinoma as an evidence for multistage progression. *Cancer Lett.* **83**: 197–200.

Okabe, H., Ikai, I., Matsuo, K. *et al.* (2000) Comprehensive allelotype study of hepatocellular carcinoma: potential differences in pathways to hepatocellular carcinoma between hepatitis B virus-positive and -negative tumors. *Hepatology* **31**: 1073–1079.

Okuda, K. (1997) Hepatitis C virus and hepatocellular carcinoma. In: *Liver Cancer* (eds K. Okuda and E. Tabor). Churchill Livingstone Inc., New York, pp. 39–50.

Ozturk, M. (1991) p53 mutation in hepatocellular carcinoma after aflatoxin exposure. *Lancet* **338:** 1356–1359.

Peter, M. and Herskowitz, I. (1994) Joining the comples: cyclin-dependent kinase inhibitory proteins and the cell cycle. *Cell* **79:** 181–184.

Piao, Z., Park, C., Park, J.H. and Kim, H. (1998) Allelotype analysis of hepatocellular carcinoma. *Int. J. Cancer* **75:** 29–33.

Pineau, P., Nagai, H., Prigent, S., Wei, Y., Gyapay, G., Weissenbach, J., Tiollais, P., Buendia, M.A. and Dejean, A. (1999) Identification of three distinct regions of allelic deletions on the short arm of chromosome 8 in hepatocellular carcinoma. *Oncogene* **18:** 3127–3134.

Qin, L.F., Ng, I.O.L., Fan, S.T. and Ng, M. (1998) p21/WAF1, p53 and PCNA expression and p53 mutation status in hepatocellular carcinoma. *Int. J. Cancer* **79:** 424–428.

Ray, R.B., Steele, R., Meyer, K. and Ray, R. (1998) Hepatitis C virus core protein represses p21WAF1/Cip1/Sid1 promoter activity. *Gene* **208:** 331–336.

Reynolds, S.H., Stowers, S.J., Patterson, R.M., Maronpot, R.R., Anronson, S.A. and Anderson, M.W. (1987) Activated oncogenes in B6C3F1 mouse liver tumors: implications for risk assessment. *Science* **237:** 1309–1316.

Robinson, W.S. (1994) Molecular events in the pathogenesis of hepadnavirus-associated hepatocellular carcinoma. *Annu. Rev. Med.* **45:** 297–323.

Sakamoto, M., Hirohashi, S. and Shimosato, Y. (1991) Early stages of multistep hepatocarcinogenesis: Adenomatous hyperplasia and early hepatocellular carcinoma. *Hum. Pathol.* **22:** 172–178.

Sasaki, T., Toguchida, J., Ohtani, N., Yandell, D.W., Rapaport, J.M. and Dryja, T.P. (1991) Allele-specific hypermethylation of the retinoblastoma tumor-suppressor gene. *Am. J. Hum. Genet.* **48:** 880–888.

Sherr, C.J. (1996) Cancer cell cycles. *Science* **274:** 1672–1677.

Sherr, C.J. and Roberts, J.M. (1995) Inhibitors of mammalian G1 cyclin-dependent kinase. *Genes. Dev.* **9:** 1149–1163.

Sheu, J.C., Huang, G.T., Lee, P.H. *et al.* (1992) Mutation of p53 gene in hepatocellular carcinoma in Taiwan. *Cancer Res.* **52:** 6098–6100.

Shi, Y.-Z., Hui, A.-M., Li, X., Takayama, T. and Makuuchi, M. (2000a) Reduced p21$^{WAF1/CIP1}$ protein expression is predominantly related to altered p53 in hepatocellular carcinomas. *Br. J. Cancer* **83:** 50–55.

Shi, Y.-Z., Hui, A.-M., Li, X., Takayama, T. and Makuuchi, M. (2000b) Overexpression of retinoblastoma protein predicts decreased survival and correlated with loss of p16^{INK4} protein in gallbladder carcinomas. *Clin. Cancer Res.* **6:** 4096–4100.

Shieh, Y.S., Nguyen, C., Vocal, M.V. and Chu, H.W. (1993) Tumor-suppressor p53 gene in hepatitis C and B virus-associated human hepatocellular carcinoma. *Int. J. Cancer* **54:** 558–562.

Simon, D., Knowles, B.B. and Weith, A. (1991) Abnormalities of chromosome 1 and loss of heterozygosity on 1p in primary hepatomas. *Oncogene* **6:** 765–770.

Sun, L., Hui, A.-M., Kanai, Y., Sakamoto, M. and Hirohashi, S. (1997) Increased DNA methyltransferase expression is associated with an early stage of human hepatocarcinogenesis. *Jpn J. Cancer Res.* **88:** 1165–1170.

Suzuki, K., Hayashi, N., Yamada, Y. *et al.* (1994) Expression of the c-met protooncogene in human hepatocellular carcinoma. *Hepatology* **20:** 1231–1236.

Takahashi, K., Kudo, J., Ishibashi, H., Hirata, Y. and Niho, Y. (1992) Frequent loss of heterozygosity on chromosome 22 in hepatocellular carcinoma. *Hepatology* **17:** 794–799.

Takayama, T., Makuuchi, M., Hirohashi, S. *et al.* (1990) Malignant transformation of adenomatous hyperplasia to hepatocellular carcinoma. *Lancet* **336:** 1150–1153.

Terris, B., Laurent-Puig, P., Belghiti, J., Degott, C., Hénin, D. and Fléjou, J.-F. (1997) Prognostic influence of clinicopathologic features, DNA-ploidy, CD44H and p53 expression in a large series of resected hepatocellular carcinoma in France. *Int. J. Cancer* **74:** 614–619.

Tsuda, H., Hirohashi, S., Shimosato, Y., Terada, M. and Hasegawa, H. (1988) Clonal origin of atypical adenomatous hyperplasia of the liver and clonal identity with hepatocellular carcinoma. *Gastroenterology* **95:** 1664–1666.

Tsuda, H., Hirohashi, S., Shimosato, Y., Ino, Y., Yoshida, T. and Terada, M. (1989) Low incidence of point mutation of c-Ki-ras and N-ras oncogenes in human hepatocellular carcinoma. *Jpn J. Cancer Res.* **80:** 196–199.

Tsuda, H., Zhang, W., Shimosato, Y., Yokota, J., Terada, M., Sugimura, T., Miyamaru, T. and Hirohashi, S. (1990) Allele loss on chromosome 16 associated with progression of human hepatocellular carcioma. *Proc. Natl Acad. Sci. USA* **87**: 6791–6794.

Tsuda, H., Oda, T., Sakamoto, M. and Hirohashi, S. (1992) Different pattern of chromosomal allele loss in multiple hepatocellular carcinomas as evidence of their multifocal origin. *Cancer Res.* **52**: 1504–1509.

Unsal, H., Yakicier, C., Marcais, C., Kew, M., Volkmann, M., Zentggraf, H., Isselbacher, K.J. and Ozturk, M. (1994) Genetic heterogeneity of hepatocellular carcinoma. *Proc. Natl Acad. Sci. USA* **91**: 822–826.

Walker, G.J., Hayward, N.K., Falvey, S. and Cooksley, W.G.E. (1991) Loss of somatic heterozygosity in hepatocellular carcinoma. *Cancer Res.* **51**: 4367–4370.

Wang, H.P. and Rogler, C.E. (1988) Deletions in human chromosome arms 11p and 13q in primary hepatocellular carcinomas. *Cytogenet. Cell Genet.* **48**: 72–78.

Yamamoto, J., Kosuge, T., Takayama, T., Shimada, K., Yamazaki, S., Ozaki, H. and Yamaguchi, N. (1996) Recurrence of hepatocellular carcinoma after surgery. *Br. J. Surg.* **83**: 1219–1222.

Yap, E.P.H., Copper, K., Maharaj, B. and McGee, J.O.D. (1993) p53 codon 249ser hot-spot mutation in HBV-negative hepatocellular carcinoma. *Lancet* **341**: 251.

Zhang, W., Hirohashi, S., Tsuda, H., Shimosato, Y., Yokota, J. and Terada, M. (1990) Frequent loss of heterozygosity on chromosomes 16 and 4 in human hepatocellular carcinoma. *Jpn J. Cancer Res.* **81**: 108–111.

Zhang, X., Xu, H.J., Murakami, Y., Sachse, R., Yashima, K., Hirohashi, S., Hu, S.X., Benedict, W.F. and Sekiya, T. (1994) Deletions of chromosome 13q, mutations in retinoblastoma 1, and retinoblastoma protein state in human hepatocellular carcinoma. *Cancer Res.* **54**: 4177–4182.

Zhang, Y.-J., Jiang, W., Chen, C.J., Lee, C.S., Kahn, S.M., Santella, R.M. and Weinstein, B. (1993) Amplification and overexpression of cyclin D1 in human hepatocellular carcinoma. *Biochem. Biophys. Res. Comm.* **196**: 1010–1016.

Zhou, X.D., Yu, Y.Q., Tang, Z.Y., Yang, B.H., Lu, J.Z., Lin, Z.Y. and Ma, Z.C. (1993) Surgical treatment of recurrent hepatocellular carcinoma. *Hepato-Gastroenterology* **40**: 333–336.

19

The genetics of bladder cancer

Margaret A. Knowles

1. Introduction

Bladder cancer is one of the most common adult malignancies world-wide. In the United Kingdom, it accounts for 12 700 new cancer cases per annum and 5300 deaths. This represents 4.2% of all new cancer cases registered and 3.3% of cancer deaths (HMSO, 1994). Similar statistics have been reported from the United States (Parker *et al.*, 1997). In men, bladder cancer is now the fourth most common cancer, representing 6% of new cancers. In women, bladder cancer represents 2% of cancers diagnosed. Peak prevalence is in the sixth and seventh decades of life and both incidence and prevalence are rising (Davies, 1982; Feldman *et al.*, 1986). In Northern Europe and North America, the majority (>90%) of bladder tumors are transitional cell carcinomas (TCC) and the remainder are squamous cell carcinoma (5%), adenocarcinoma (2%) and undifferentiated carcinoma (Koss, 1975; Mostofi *et al.*, 1973).

In regions of endemic schistosomiasis (bilharzia), bladder cancer incidence is much higher. In Egypt for example, where schistosomiasis is hyperendemic, bladder cancer is the most common cancer, representing 30% of all recorded cases (Ibrahim, 1986). Schistosomiasis-associated bladder cancer shows an earlier peak incidence, in the fourth and fifth decades of life, and there is a preponderance of squamous cell carcinoma rather than TCC (El-Bolkainy *et al.*, 1981; IARC, 1994).

Our understanding of the genetics of this disease has advanced greatly in the past decade. Current knowledge includes an extensive list of genes and mapped loci which show genetic alteration in bladder cancer, a basic understanding of the molecular pathogenesis of distinct bladder tumor types, and a repertoire of molecular alterations that are associated with specific clinical parameters. Many genetic alterations have been confirmed by more than one technique. For example, several sites of DNA deletions, denoting possible tumor suppressor gene loci have been identified by classical cytogenetics, FISH, CGH and LOH analysis. Similarly, cytogenetics, FISH, CGH, Southern blotting or quantitative PCR have identified regions of DNA overrepresentation or amplification. Several of the genes targeted by these alterations have been confirmed but others remain to be

Molecular Genetics of Cancer second edition, edited by J.K. Cowell.

identified. This review will focus on genetic changes in bladder cancer and their relationship to urothelial tumor pathogenesis and phenotype.

2. Etiology and epidemiology

The majority of patients with bladder cancer have no obvious family history. However, there is evidence for a familial association in some cases (Kantor *et al.*, 1985). An increased incidence of ureteral and renal pelvic TCC is found in families with Lynch Type II and Muir-Torre syndromes (reviewed by Kiemeney and Schoenberg, 1996) and an increased incidence of bladder TCC has been reported in retinoblastoma families (DerKinderen *et al.*, 1988). Whether familial bladder cancer exists as a separate entity from other cancer family syndromes is controversial. However, there are numerous reports of clustering of TCC of the bladder in the absence of other known familial predisposition, often with a reduced age of onset, compatible with a Mendelian pattern of autosomal dominant inheritance. For example, Lynch *et al.* reported three siblings diagnosed with bladder cancer before the age of 50 (Lynch *et al.*, 1979). There are a number of similar examples (reviewed by Kiemeney and Schoenberg, 1996). In one family (Schoenberg *et al.*, 1996), a germline translocation t(5;20)(p15;q11) was identified in the proband, suggesting possible locations for predisposing loci. In general however, the clusters of cases are relatively small and to date, no linkage studies have been attempted.

Evidence for a familial effect also comes from large epidemiological studies. For example, in a case-control study including 2982 cases of bladder cancer (Kantor *et al.*, 1985), 6% of cases compared with 4% of controls had first degree relatives with bladder cancer. These differences may be attributable to the effects of low penetrance genes such as possession of polymorphisms in xenobiotic-metabolizing enzymes. The contribution of environmental risk factors to the incidence of TCC is clear-cut and bladder cancer is the best-known example of a human cancer linked to occupational exposure to environmental carcinogens. Dyestuff production, rubber manufacturing, the textile and leather industries, printing, the metal industry and work involving exposure to petroleum products are clearly implicated (BAUS Subcommittee on Industrial Bladder Cancer, 1988; Cole *et al.*, 1972). Chemicals associated with these occupations include 4-aminobiphenyl, 2-naphthylamine, benzidine, 4,4'-methylenebis(2-chloroaniline) and o-toluidine, all considered to be human carcinogens (Tomatis *et al.*, 1978). Cigarette smoking is also a significant risk factor for TCC (Armstrong and Doll, 1974; IARC, 1986). For squamous cell carcinoma of the bladder, schistosomiasis, smoking and recurrent urinary tract infections are risk factors (Kantor *et al.*, 1984, 1988). Several other risk factors have been suggested, including use of dietary sweeteners, coffee drinking and excessive use of phenacetin-containing analgesics. Apart from use of phenacetin (McCredie *et al.*, 1983), the role of these factors remains controversial. It has been estimated that 50% of bladder tumors in men and 25% in women are smoking-attributable (IARC, 1986) and that 20–25% may be related to occupational exposure (Silverman *et al.*, 1989).

In man, interindividual variability in the metabolism of arylamines has been identified and attributed to genetic differences at one of the N-acetyltransferase loci (NAT2). A series of mutant alleles of NAT2 have been characterized which give rise to so-called 'slow' and 'fast' acetylator phenotypes. Several studies link the 'slow' acetylator phenotype which is present in 50% of individuals, to a higher risk of bladder cancer (Cartwright *et al.*, 1982; Marcus *et al.*, 2000; Risch *et al.*, 1995). It is likely that inherited variations in the ability to activate or detoxify other types of carcinogen also have considerable impact. For example, genetic defects in the carcinogen-metabolizing glutathione S-transferase genes M1 and T1 are reported to confer susceptibility to bladder cancer (Bell *et al.*, 1993; Brockmoller *et al.*, 1995; Harries *et al.*, 1997).

In schistosomiasis-associated bladder cancer, the mechanisms linking infection and cancer development are not fully understood (World Health Organisation, 1983). The infection induces chronic inflammation in the bladder due to the deposition of large numbers of eggs in the sub-epithelial tissues that eventually leads to fibrosis, stenosis and urinary retention which in turn is associated with recurrent bacterial infections. The urine of infected individuals has been shown to contain higher levels of N-nitroso compounds than that from uninfected individuals (El-Merzabani *et al.*, 1979; Tricker *et al.*, 1989). (For more detailed discussion of the epidemiology of bladder cancer, see BAUS Subcommittee on Industrial Bladder Cancer (1988); Kantor *et al.*, (1985); Thompson and Fair (1989).)

3. Chromosomal alterations in TCC

3.1 Cytogenetic changes

The likely location of many genes involved in bladder cancer development was first identified by cytogenetic studies. This literature has been reviewed by Sandberg and Berger (1994) and reveals the following common alterations: -9 or 9q-, + 7, 1p-, 1q-, 5q-, i(5p), 11p-, 6p-, 6q-, 17p-, 2q-, 3p-, +8, +11, 21q- and -Y. Notably, several studies found monosomy 9 as the sole abnormality in near diploid tumors, suggesting that this chromosome may be involved at an early stage of bladder tumor development.

3.2 Fluorescence in situ hybridization analysis

More recently, fluorescence *in situ* hybridization (FISH) has been applied to study both numerical chromosomal aberrations (using centromeric repeat probes and whole chromosome paints) and small alterations which can be detected using large insert genomic DNA clones (YACs, PACs, BACs, cosmids) as probes. One of the advantages of this technology is that metaphase spreads are not necessary for counting numerical changes and for some mapping experiments. In addition, interphase nuclei from cells in urine and from fresh, frozen and paraffin-embedded tumors have been used successfully. A variety of studies have examined numerical chromosome changes and several studies have included flow cytometric analysis of cell ploidy. Common alterations reported include monosomy 9, monosomy 15, underrepresentation of 8p, 11p and 17p, trisomies of 7 and 10 and gain of 8q. Several studies report monosomy 9 in a large proportion of tumors including near diploid tumors (e.g. Matsuyama *et al.*, 1994; Wheeless *et al.*, 1994), a large proportion of pTa non-recurrent tumors (Pycha *et al.*, 1997) and morphologically normal mucosa from tumor-bearing bladders (Hartmann *et al.*, 1999; Matsuyama *et al.*, 1994), indicating that this alteration may be involved as a very early step in tumor development. Trisomies 7 and 10 have also been reported as single alterations in diploid tumors (Matsuyama *et al.*, 1994) but are more commonly found in aneuploid tumors (Pycha *et al.*, 1997; Waldman *et al.*, 1991; Wheeless *et al.*, 1994). Similarly, loss of 8p, 11p, 17p, and gain of 8q are principally found in aneuploid, recurrent or progressive tumors (Pycha *et al.*, 1997; Sauter *et al.*, 1994, 1995; Voorter *et al.*, 1996; Wagner *et al.*, 1997; Wheeless *et al.*, 1994). FISH has also been used to detect copy number changes in specific genes including *MYC* (Sauter *et al.*, 1995), which is overrepresented or amplified, *ERBB2*, which is amplified (Sauter *et al.*, 1993), *CDKN2A* (p16) which is homozygously deleted, usually in tumors with monosomy 9 (Balázs *et al.*, 1997; Jung *et al.*, 1999; Reeder *et al.*, 1997) and TP53 which is underrepresented (Sauter *et al.*, 1994).

Taken together, classical cytogenetic studies and FISH-based numerical chromosome counts provide several important clues to the likely location of genes involved in bladder tumor development. In particular, tumor suppressor loci are implicated on chromosomes 8, 9, 11 and 17, all identified as genomic regions of underrepresentation. These will be discussed below.

3.3 Comparative genomic hybridization

Recently, comparative genomic hybridization (CGH) has provided a powerful and rapid means to identify genomic regions that are either over- or underrepresented in tumor DNA samples. CGH has been applied to bladder tumor samples in a number of studies and has identified many regions of common genetic alteration (*Table 1*). Many of the regions of underrepresentation found cytogenetically have been confirmed by CGH (e.g. 8p, 9p, 9q, 11p, 17p, Y). Various regions of overrepresentation have been reported and high level amplifications have been identified in the region of *ERBB2* (17q) and *CCND1* (11q13), regions both shown to be amplified by molecular analyses and FISH and in a number of other locations within which no oncogenes have yet been identified.

4. Oncogenes involved in TCC development

Several oncogenes are implicated in TCC development and progression (*Table 2*). These contribute to tumor development in a dominant manner via overexpression

Table 1. CGH findings in transitional cell carcinoma

Tumor stage	Losses	Gains	Amplification
Ta	9p, 9q, Y	1q, 17	11q
T1	2q, 4p, 4q, 5q, 6q, 8p, 9p, 9q, 10q, 11p, 11q, 13q, 17p, 18q, Y	1q, 3p, 3q, 5p, 6p, 8q, 10p, 17q, 20q	1q22–24, 3p22–25, 6p22, 8p 12, 8q21–22, 10p12–14, 10q22–23, 11q13, 12q12–21, 17q21, 20q13
T2–4	As for T1 + 15q	As for T1 + 7p, Xq	As for T1

Data from (Kallioniemi *et al.*, 1995; Richter *et al.*, 1997, 1998, 1999; Simon *et al.*, 1998; Voorter *et al.*, 1995; Zhao *et al.*, 1999).

Table 2. Genes mutated in transitional cell carcinoma

Oncogenes

Gene (cytogenetic location)	Alteration	Frequency/clinical association	References
FGFR3 (4p16)	Activating mutation	30–40%	Cappellen et al., 1999; Sibley et al., 2001
HRAS (11p15)	Activating mutation	1–44% overall (high grade)	Cattan et al., 2000; Fitzgerald et al., 1995; Fujita et al.,1985; Knowles and Williamson, 1993; Levesque et al., 1993; Malone et al., 1985; Ooi et al., 1994
CCND1 (11q13)	Amplification/over-expression	Amplified 10–20% all grades and stages; Recurrence; Papillary growth pattern	Proctor et al., 1991; Bringuier et al., 1994, 1996; Lee et al., 1997; Shin et al., 1997; Wagner et al., 1999
MDM2 (12q13)	Amplification/over-expression	Amplified 1–4%	Habuchi et al., 1994; Lianes et al., 1994
ERBB2 (17q21)	Amplification/over-expression	Amplified 10–14% high grade/stage; Recurrence. Survival	Coombs et al., 1991; Lonn et al., 1995; Miyamoto et al., 2000; Sauter et al., 1993; Underwood et al., 1995

Tumor suppressor genes

Gene (cytogenetic location)	Alteration	Frequency/clinical association	References
INK4A-ARF (9p21)	Homozygous deletion/methylation/mutation	20–30% high grade/stage; LOH 60% all grades/stages; Immortalization in vitro	Cairns et al., 1995; Orlow et al., 1995; Williamson et al., 1995; Yeager et al., 1998
PTCH (9q22)	Deletion/mutation	LOH 60% all grades/stages; Mutation frequency low	McGarvey et al., 1998; Simoneau et al., 1996
DBCCR1 (9q32–33)	Deletion/methylation	LOH 60% all grades/stages; No mutations detected in retained allele	Habuchi et al., 1997, 1998
TSC1 (9q34)	Deletion/mutation	LOH 50% all grades/stages; Mutation frequency low	Hornigold et al., 1999
PTEN (10q23)	Homozygous deletion/mutation	10q LOH in 35% muscle invasive; 6.6% superficial	Aveyard et al., 1999; Cairns et al., 1998
RB1 (13q14)	Deletion/mutation	10–15% overall; 37% muscle invasive	Cairns et al., 1991; Cordon-Cardo et al., 1992; Logothetis et al., 1992
TP53 (17p13)	Deletion/mutation	70% muscle invasive; High grade and stage; LOH 30% high grade/stage	Habuchi et al., 1993; Sidransky et al., 1991; Spruck et al., 1993; Williamson et al., 1994
DCC/SMAD (18q21)	Deletion	No mutation analysis to date	Brewster et al., 1994

of the normal gene product or expression of a mutant gene product with altered function. Examples of both mutated oncogenes and amplified proto-oncogenes are found in TCC and these are described below. There are also many reports of increased expression of proto-oncogenes.

4.1 FGFR3

FGFR3 encodes a member of the highly conserved family of 4 fibroblast growth factor receptor proteins. The gene is mutated in several autosomal dominant skeletal dysplasias, all of which appear to result in constitutive activation of the protein in a ligand-independent manner. In these disorders, mutation results in severe disruption in the growth of long bones compatible with a normal negative regulatory function for Fgfr in this tissue. Interestingly, studies of bladder tumors have identified similar activating mutations in 30–40% of cases of all grades and stages (Cappellen *et al.*, 1999; Sibley *et al*, in press). The *FGFR3* gene is mapped to 4p16.3 within a common region of LOH in bladder cancer (see Section 4.4). The presence of activating mutations renders this an unlikely candidate tumor suppressor but the finding of common breakpoints at 4p16.3 could possibly indicate activation in some cases via a common translocation breakpoint. We have assessed the mutation status of *FGFR3* in a series of tumors with and without 4p LOH and found no difference in mutation frequency in the two groups. It appears, therefore, that activation of *FGFR3* is unrelated to deletion of 4p and that a candidate tumor suppressor gene close to *FGFR3* remains to be identified. *FGFR3* is the most commonly mutated oncogene in TCC and further studies to elucidate its role in tumor development will be of great interest.

4.2 MYC

Copy number gains of 8q have been identified by FISH analysis (Section 2.2 above) resulting in gain of *MYC* (Sauter *et al.*, 1995). High-level amplification of *MYC* was identified in only a small number of cases (3/87). In the same study, myc protein levels were assessed and these showed no relationship to low level gains of 8q, suggesting that this does not represent a genetic mechanism for increasing *MYC* expression. This is borne out by a detailed amplicon mapping study that excluded *MYC* from the region of amplification (Bruch *et al.*, 1998). Nevertheless, over-expression of the myc protein, unrelated to gene amplification, is found in a large proportion of TCC samples including non-recurrent papillary superficial tumors (Masters *et al.*, 1988; Schmitz-Drager *et al.*, 1997).

4.3 RAS

A mutation of *HRAS* in the bladder tumor cell line T24 was the first mutation of an oncogene described in human cancer (Reddy *et al.*, 1982). Since then there have been several studies of *HRAS* mutation in TCC. The frequency of activation remains controversial. Estimates range from 1% to 44% (Cattan *et al.*, 2000; Fitzgerald *et al.*, 1995; Fujita *et al.*, 1985; Knowles and Williamson, 1993; Levesque *et al.*, 1993; Malone *et al.*, 1985; Ooi *et al.*, 1994). Mutation of *KRAS* has also been described in TCC, at low frequency (Grimmond *et al.*, 1992; Olderoy *et al.*, 1998). These studies have used a range of techniques, the most recent including highly sensitive methods to detect mutations in minor populations of cells. These latter studies have not detected a higher frequency of mutation than earlier studies and the reasons for this widely disparate set of results are not clear. Possibilities include technical problems leading to overestimation of frequency in older studies based on DNA hybridization methods or real differences between the tumor populations studied. No clear association of ras mutation with tumor phenotype has been found but *in vitro* experiments indicate that introduction of a mutant *HRAS* into human urothelial cells can induce a profound phenotypic effect, including altered responses to growth factors, cell invasion and accelerated tumor growth (Pratt *et al.*, 1992; Theodorescu *et al.*, 1991). This suggests that ras mutation has a role during tumor progression rather than initiation of TCC.

4.4 CCND1

Amplification at 11q13 has been identified in TCC both by Southern blotting (Bringuier *et al.*, 1994; Proctor *et al.*, 1991) and more recently by CGH (Richter *et al.*, 1997). The target gene driving these amplifications is believed to be *CCND1*, which encodes cyclin D1, a G1 cyclin involved in the regulation of progression from G_1 to S phase of the cycle. Amplification at 11q13 is found in >20% of tumors of all grades and stages (Proctor *et al.*, 1991) and such tumors show overexpression of cyclin D1 (Bringuier *et al.*, 1996). Protein expression studies indicate that the frequency of over-expression of cyclin D is greater than this amplification frequency but the mechanism for this has not been elucidated. Initial studies showed no obvious correlation of amplification with clinico-pathological data but a recent study reports an association of over-expression of the protein with more rapid recurrence of superficial tumors (Shin *et al.*, 1997). Immunohistochemical studies indicate that there may in fact be an inverse correlation of expression with tumor grade (Lee *et al.*, 1997; Suwa *et al.*, 1998) and a positive association with a papillary growth pattern (Lee *et al.*, 1997; Wagner *et al.*, 1999). Thus, *CCND1* is implicated early in the development of superficial papillary TCC. The absence of high levels of expression in more advanced tumors may indicate that early involvement of this protein determines subsequent tumor pathogenesis. To date, the role of this gene has not been extensively explored in TCC. Its participation in a pivotal cell cycle control mechanism (see below) provides a strong impetus for further study.

4.5 MDM2

MDM2 is a proto-oncogene that modulates p53 activity by increasing its susceptibility to proteolysis by the 26S proteosome. In turn, p53 regulates MDM2 expression, giving rise to an autoregulatory feedback loop. Amplification or over-expression of MDM2 might be expected to down-regulate p53 activity and hence provide an alternate means of p53 inactivation.

   Studies of MDM2 in TCC have shown that gene amplification is relatively rare (1–4%) (Habuchi *et al.*, 1994; Lianes *et al.*, 1994) but that over-expression of the protein is frequent (32–44%) (Lianes *et al.*, 1994; Pfister *et al.*, 2000). This appears to be more common in low-grade and -stage tumors and to be more common in p53 mutant tumors (Lianes *et al.*, 1994). Alternatively spliced MDM2 mRNAs have been identified in TCC specimens, most lacking the p53-binding domain. These were more commonly found in muscle invasive tumors and in these cases might indicate a p53-independent MDM2 transforming function (Sigalas *et al.*, 1996).

4.6 ERBB2

ERBB2 encodes a transmembrane receptor with homology to the EGF receptor. Both amplification of the *ERBB2* gene and overexpression of the erbB2 protein have been described in bladder tumors. Gene amplification, detected by Southern blotting or FISH, is more frequent in high-grade and -stage tumors (Coombs *et al.*, 1991; Sauter *et al.*, 1993) and has been associated both with tumor recurrence (Underwood *et al.*, 1995) and with decreased survival in patients with advanced TCC (Lonn *et al.*, 1995; Miyamoto *et al.*, 2000). In the latter two studies, *ERBB2* gene amplification was shown to be an independent prognostic variable. Two studies have assessed *ERBB2* amplification in recurrent tumors from the same patients and in both, amplification was found in several recurrent and/or progressive tumors from patients whose primary tumor had shown no evidence of amplification (Coombs *et al.*, 1991; Underwood *et al.*, 1995). This suggests that amplification occurs during tumor progression and taken together with the finding that muscle-invasive TCC with amplification has a worse prognosis, suggests that this confers a significant selective advantage. Several studies have assessed the levels of the erbB2 protein product by immunohistochemistry. Whilst it is clear that tumors containing an amplified *ERBB2* gene over-express the protein, it has been found that many TCCs, including a significant proportion of low-grade and -stage tumors without gene amplification, also express detectable levels of the protein. The significance of this and the potential application of erbB2 immunohistochemistry are therefore in doubt. However, the advent of accurate PCR-based methods to measure gene dosage (Heid *et al.*, 1996) may facilitate further studies at the genetic level in large tumor series.

4.7 Other potential oncogenes

CGH studies have identified a series of non-random chromosomal gains and some regions of high-level DNA amplification within which candidate genes have not yet been identified. These latter have been found predominantly in tumors of stage pT1 and above and include 1q22–24, 3p22, 6p22, 8q21–22, 10p22–23, 12q15 and 20q (*Table 1*). Rapid progress in construction of the Human Gene Map (http://www.ncbi.nlm.nih.gov/genemap/) coupled with the widespread availability of microarray technology should facilitate the identification of candidate genes within these regions.

Gains of 20q and some high-level amplifications have been consistently reported. Similar changes have been identified cytogenetically during *in vitro* transformation of human urothelial cells (Reznikoff *et al.*, 1996). *In vitro*, over-representation of 20q is associated with immortalization. Several candidate genes have been identified on 20q including *STK15/BTAK* (Zhou *et al.*, 1998) and *ZNF217* (Collins *et al.*, 1998), neither of which have yet been examined in bladder tumors. The former has been shown to induce centrosome abnormalities and aneuploidy when ectopically expressed in cultured cells. This is compatible with the observation of increasing aneuploidy in urothelial cells following immortalization.

5. Tumor suppressor genes involved in TCC development

The tumor suppressor genes now comprise a large group of genes whose inactivation via genetic or other mechanisms contributes to the development of cancer. Many of these are genes which when mutated in the germline predispose to cancer and these represent paradigms for Knudson's 'two-hit' hypothesis (Knudson, 1993). Several known tumor suppressors are mutated in sporadic cancers of the bladder (*RB1, TP53, PTEN, CDKN2A, TSC1*) (*Table 2*). Several other loci are implicated by virtue of common deletion of one allele in TCC and candidate genes remain to be identified for these. In the case of known suppressor genes, one allele is commonly deleted and the retained copy of the gene is mutated, resulting in complete abolition of gene function. Recently, however, alternative inactivation scenarios have become apparent. One allele of a tumor suppressor may be inactivated via epigenetic mechanisms, e.g. DNA hypermethylation (Jones and Laird, 1999) or loss of function of one allele alone may be sufficient to generate an altered phenotype as suggested by recent studies of heterozygous knockout mice. For example, mice with constitutional loss of one allele of *PTEN* show widespread hyperplasias (Di Cristofano *et al.*, 1998). All of these mechanisms of tumor suppressor gene inactivation are illustrated by findings in TCC.

5.1 TP53, RB, INK4A/ARF and the G$_1$ checkpoint

Regulation of progression through the cell cycle is critical for the maintenance of normal cell and tissue integrity. Appropriate control mechanisms are required to prevent progression to DNA synthesis or mitosis in the presence of various forms of cell damage. Several of the key regulators of these checkpoints are implicated as tumor suppressor genes. In G$_1$, two interrelated pathways impact on progression into S phase. These are the Rb and p53 pathways that sense and respond to DNA damage or inappropriate mitogenic signaling. These pathways are disrupted by various mechanisms in many tumor types including TCC. At the heart of these control mechanisms are the two proteins encoded by the *INK4A/ARF* locus at 9p21 (*Figure 1 a,b*). Entry into S phase is regulated by genes whose transcription is controled by the transcription factor E2F. This factor is activated by release from interaction with Rb following its phosphorylation by cyclin-dependent kinases 4 and 6 in response to activation by cyclin D1. A negative feedback loop operates via E2F up-regulation of p14ARF, which indirectly activates p53 by binding to MDM2. Similarly, p14ARF is up-regulated by oncogenic ras. The other product of the INK4A/ARF locus, p16, is a negative regulator of CDKs 4 and 6 and is also up-regulated in response to signaling via ras. p53 is up-regulated in response to

(a)

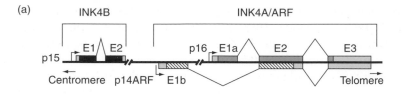

(b)

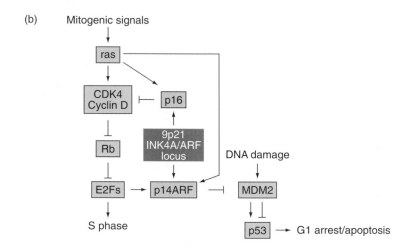

Figure 1. a. The *INK4A/ARF* and *INK4B* loci at 9p21. Transcripts for p16 and p14ARF share exons 2 but have an alternative first exon. These alternative transcripts are read in different reading frames and encode different protein products. This entire genomic region is commonly homozygously deleted in TCC. b. The p14ARF regulated interaction between p53 and /RB pathways. ARF connects p53 and Rb pathways. p16 and p14ARF are regulators of the Rb and p53 pathways respectively. p16 is a negative regulator of CDK4 activity, thus preventing phosphorylation of Rb and progression into S phase. p14ARF is upregulated by E2F and via binding to MDM2, upregulates p53, providing a negative feedback loop. All proteins shown in the diagram except E2F are genetically altered in TCC.

various signals including DNA damage, leading to cell cycle arrest or apoptosis depending on cell type. This response is tightly regulated by MDM2, which operates in a feedback loop with p53. Without doubt, there are levels of control and complexity of these pathways which remain to be elucidated, but it is already clear that these processes are tightly regulated with several failsafe mechanisms to prevent inappropriate cell division. It is not surprising therefore that several of these proteins are altered in tumor cells. In TCC, genetic alterations to all of the genes shown in *Figure 1b* apart from E2F have been described.

LOH of 17p and mutation of *TP53* on the retained allele is found in a large proportion of muscle-invasive bladder tumors (Habuchi *et al.*, 1993; Sidransky *et al.*, 1991; Spruck *et al.*, 1993; Williamson *et al.*, 1994). Similarly, *RB1* alterations are found more frequently in tumors of high grade and stage (Cairns *et al.*, 1991; Cordon-Cardo *et al.*, 1992; Ishikawa *et al.*, 1991). Several studies have shown a clear association between *TP53* and *RB1* mutation and disease progression and both are independent predictors of clinical outcome (Cordon-Cardo *et al.*, 1992; Esrig *et al.*, 1994; Lipponen, 1993; Logothetis *et al.*, 1992; Sarkis *et al.*, 1993a, 1993b; Soini *et al.*, 1993). The *INK4A/ARF* locus at 9p21 is deleted in many bladder tumors (Balázs *et al.*, 1997; Cairns *et al.*, 1991, 1994; Devlin *et al.*, 1994; Orlow *et al.*, 1995; Packenham *et al.*, 1995; Spruck *et al.*, 1994a; Williamson *et al.*, 1995). Three cell cycle regulatory proteins are encoded within a small genomic region (*Figure 1a*) and this entire region is commonly homozygously deleted in TCC, inactivating all three genes. A few tumors have been identified in which p15 is retained and the other two loci lost, suggesting that *INK4A* and *ARF* but not p15 are critical. Homozygous deletion represents an efficient mechanism of inactivation of more than one gene in close proximity. The potential prognostic implications of inactivation of one or both of these genes have not yet been studied in detail. However, transfection studies have shown that expression of p16 in TCC cell lines with deletion of the endogenous gene can cause growth arrest and inhibition of tumorigenicity (Wu *et al.*, 1996). Loss of expression of p16 is more common in cultures derived from muscle-invasive than superficial TCC and this appears to be related to ability to proliferate indefinitely *in vitro* (Yeager *et al.*, 1998).

The G_1 checkpoint can theoretically be evaded by inactivation of *RB1* or *INK4A* or by overexpression of cyclin D1. However, the failsafe mechanisms shown in *Figure 1b*, imply that complete escape may also require inactivation of either *p14ARF* or p53. Already, there is evidence that this is the case. Three studies of TCC indicate that tumors with altered expression of Rb and p53 have a worse prognosis than those with alterations to either gene alone (Cordon-Cardo *et al.*, 1997; Cote *et al.*, 1998; Grossman *et al.*, 1998). Interestingly, a study of TCC cell lines has shown that all have lesions in both Rb and p53 pathways (Markl and Jones, 1998).

5.2 Genes on the long arm of chromosome 9

LOH analysis of TCC indicates that more than 50% of tumors of all grades and stages show LOH for markers on chromosome 9 (Cairns *et al.*, 1993; Tsai *et al.*, 1990). Initial studies indicated that LOH at all loci studied was common, in accord with previous cytogenetic descriptions of monosomy 9. This led to the hypothesis that both arms of chromosome 9 may harbor tumor suppressor genes relevant to TCC development. The finding of chromosome 9 LOH at high frequency in low-grade, low-stage tumors, and more recently in morphologically normal urothelium from patients with a history of TCC (Czerniak *et al.*, 1999; Hartmann *et al.*, 1999), suggests that at least one of these genes is involved at a very early stage in tumor development. Additional evidence for this has come from studies of patients with multifocal disease, where chromosome 9 LOH showed least discordance between tumors, compatible with a role early in development (Sidransky *et al.*, 1992; Takahashi *et al.*, 1998). Identification of the target genes on chromosome 9 is therefore considered pivotal to understanding the pathogenesis of TCC.

Many large tumor series have been studied by LOH to define the location of these potential genes. This has been difficult, due to the fact that many tumors lose an entire parental homolog. Nevertheless, well-defined common regions of deletion have been defined on both chromosome arms. Deletions on 9p focus at 9p21 where the likely target genes are *CDKN2A* and p14ARF, as discussed above.

In general, fewer interstitial deletions of 9q have been identified than of 9p, and these have often been large and unhelpful in mapping. Three regions of 9q deletion have been mapped by several laboratories at 9q13–31, 9q32–33 and 9q34 (Habuchi *et al.*, 1995, 1997; Hornigold *et al.*, 1999; Simoneau *et al.*, 1996; van Tilborg *et al.*, 1999). In addition, two recent studies have identified several potential new regions of deletion (Czerniak *et al.*, 1999; Simoneau *et al.*, 1999). Simoneau *et al.* (1999) in a large series of low-grade superficial primary tumors, found a higher proportion of small interstitial regions of deletion than has previously been described. In this group of tumors, LOH on 9q (44%) exceeded that on 9p (23%). If these novel regions are confirmed, then at least five individual deletion targets must be predicted on 9q.

Three potential target genes have been identified to date. At 9q32–33, a small candidate region <850 kb in size was mapped (Habuchi *et al.*, 1997). This locus has been designated *DBC1* (*D*eleted in *B*ladder *C*ancer gene *1*). A single candidate gene, *DBCCR1*, (*D*eleted in *B*ladder *C*ancer *C*andidate *R*egion gene *1*), has been identified within the region (Habuchi *et al.*, 1998) and has been assessed by mutation, methylation and functional analyses. No small mutations were identified in the coding region of the gene on the retained allele in tumors with 9q LOH. However, a small homozygous deletion encompassing the gene has been identified in a bladder tumor (Nishiyama *et al.*, 1999) and it has been shown that the gene is transcriptionally silenced by methylation in many bladder tumor cell lines (Habuchi *et al.*, 1998). Our recent studies indicate that introduction of a *DBCCR1* cDNA clone into NIH 3T3 cells and bladder cell lines which do not express *DBCCR1* has an antiproliferative effect (Nishiyama, Gill and Knowles, unpublished). In the absence of mutation of both alleles in a relevant tumor, validation of a tumor suppressor gene is difficult and further functional studies including the generation of knockout mice will be important.

The common region of deletion at 9q34 contains the *TSC1* gene, one of the genes for the familial hamartoma syndrome tuberous sclerosis (TS). Hamartomas from TS patients commonly show LOH, indicating that this gene may act as a tumor suppressor. The *TSC1* gene was identified in 1997 (van Slegtenhorst *et al.*, 1997) and mutation analysis in TCCs with 9q34 LOH revealed a low frequency of mutation (Hornigold *et al.*, 1999). However, mutations were not found in all the tumors with LOH confined to the region of the gene, raising the possibility that loss of one allele may be sufficient or that an adjacent gene may be involved. Again, functional studies are required to confirm the authenticity of *TSC1* as a bladder tumor suppressor.

A third region of LOH defined on 9q encompasses the Gorlin syndrome locus (*PTCH*) at 9q22.3. This has been assessed as a candidate gene in two studies (McGarvey *et al.*, 1998; Simoneau *et al.*, 1996), one of which found no mutations, the other found two potential mutations. This is a much lower frequency than expected and more extensive mutation and expression analyses and functional studies are now needed.

5.3 Chromosome 10 genes

LOH of the long arm of chromosome 10 has been described in up to a third of muscle invasive tumors (Aveyard *et al.*, 1999; Cappellen *et al.*, 1997; Kagan *et al.*, 1998). The *PTEN* gene, which encodes a phosphatidylinositol phosphatase that inhibits PI3K/Akt-mediated signal transduction, has been identified within the common region of deletion. This gene is inactivated by LOH and mutation in glioblastoma, melanoma and endometrial cancer and mutations have been found in some cases of TCC. The frequency of point mutation is low and homozygous deletion may be a more common mechanism of inactivation (Aveyard *et al.*, 1999; Cairns *et al.*, 1998). Heterozygous knockout mice (PTEN +/–) have a hyper-plastic/dysplastic phenotype (Di Cristofano *et al.*, 1998), supporting the notion that *PTEN* haploinsufficiency plays a causal role in the pathogenesis of Cowden disease, Llermitte-Duclos and Banayan-Zonana syndromes in which germline mutations of the gene are found. It is possible therefore that loss of one allele is sufficient to contribute to the development of urothelial tumors.

The cell surface receptor Fas (Apo-1/CD95) which is involved in cell death signaling is also mapped close to *PTEN* within the 10q region of deletion in TCC. The role of this signaling pathway has been extensively studied in the immune system but to date there is little information on the role of fas signaling in epithelial tissues. However, a recent study of the *FAS* gene in bladder tumors has identified a number of mutations, many focused on a hotspot at nt 993 (Lee *et al.*, 1999). This is the first report of mutations of *FAS* in a non-lymphoid malignancy and as yet, the effect on urothelial cell phenotype has not been examined. Interestingly, 10q LOH was found only in 8 cases with mutation, suggesting that inactivation of both alleles was not required.

Several recent reports suggest that *PTEN* is involved in negative regulation of not only the PI3K pathway but also of growth factor signaling through inhibition of Shc and of signals generated via focal adhesions (reviewed in Besson *et al.* (1999)). There is also data linking the tumor suppressor activity of PTEN to the cell cycle machinery via Rb (Paramio *et al.*, 1999). A direct effect of *PTEN* gene transfer on CD95L-induced apoptosis has been reported in glioma cells (Wick *et al.*, 1999). Hence, some relationship between the inactivation of these two chromosome 10 genes might be expected and it will be important to examine the genetic status and function of both in a common series of TCCs.

5.4 Other potential tumor suppressor gene loci

LOH studies have identified several common regions of deletion within which candidate genes exist but have not been tested or within which no candidate genes have been identified (*Table 3*). The most commonly deleted regions include 3p (Presti *et al.*, 1991), 4p16.3 (Elder *et al.*, 1994), 4q (Polascik *et al.*, 1995), 8p (Knowles *et al.*, 1993; Ohgaki *et al.*, 1999; Takle and Knowles, 1996; Wagner *et al.*, 1997), 11p (Fearon *et al.*, 1985; Shaw and Knowles, 1995; Tsai *et al.*, 1990), 11q (Shaw and Knowles, 1995), 14q (Chang *et al.*, 1995) and 18q (Brewster *et al.*, 1994).

Deletions of 3p have been found in 48% of muscle invasive TCCs (Presti *et al.*, 1991). Three regions of deletion have been localized by deletion mapping at 3p12–14, 3p21–23 (Li *et al.*, 1996) and less frequently at 3p24.2–25 (Li *et al.*, 1996).

Table 3. Common regions of LOH where target genes have not yet been identified

Cytogenetic location	Frequency	Association with clinical parameters	References
3p	48%	Stage	(Li *et al.*, 1996; Presti *et al.*, 1991)
4p	22%	None	(Elder *et al.*, 1994; Polascik *et al.*, 1995)
4q	24%	High grade/stage	(Polascik *et al.*, 1995)
8p	23%	High grade/stage	(Choi *et al.*, 2000; Knowles *et al.*, 1993; Ohgaki *et al.*, 1999; Takle and Knowles, 1996)
9q	60%	None	(Czerniak *et al.*, 1999; Habuchi *et al.*, 1995, 1997; Simoneau *et al.*, 1999)
11p	40%	Grade	(Fearon *et al.*, 1985; Shaw and Knowles, 1995; Tsai *et al.*, 1990)
11q	15%	None	(Shaw and Knowles, 1995)
14q	10–40%	Stage	(Chang *et al.*, 1995)

No candidate genes have been assessed. It is noteworthy that a constitutional del(3)(p14-p21) has been described in a patient with bladder carcinoma (Barrios *et al.*, 1986).

Deletions have been identified in three regions of chromosome 4, at 4p16.3, 4p15 and 4q33–34 in 20–45% of tumors (Elder *et al.*, 1994; Polascik *et al.*, 1995). The region at 4p16.3 has been mapped precisely but, to date, no target gene has been identified (Bell *et al.*, 1997; Sibley, Eydmann, Shaw and Knowles, unpublished). As discussed above, this region contains *FGFR3*, which is mutated in 40% of tumors, irrespective of the presence of 4p LOH.

LOH has been reported on both the long and short arms of chromosome 5. On 5p, a region has been mapped at 5p13–12 (Bohm *et al.*, 1997, 2000). Candidate genes within this region include *DAP1*, a protein involved in mediating IFN-γ induced apoptosis and *HRAD1* the human homolog of the yeast DNA damage checkpoint gene rad1. On 5q, two regions at 5q22–23 and 5q33–34 have been mapped (von Knobloch *et al.*, 2000). No candidate genes have been identified.

Many muscle-invasive TCCs show LOH of the short arm of chromosome 8 (Knowles *et al.*, 1993, 1994). This has been confirmed by CGH studies (see above). At least two regions appear to be involved, at 8p11–12 and at 8p21–22 (Choi *et al.*, 2000; Takle and Knowles, 1996; Ohgaki *et al.*, 1999). Both have been mapped by microsatellite analysis, the more telomeric at 8p22 to within a 1 Mb interval between the markers *D8S511* and *D8S261*, a region commonly deleted in many other tumor types (Choi *et al.*, 2000; Ohgaki *et al.*, 1999). Candidate genes within this region have not yet been assessed but several candidate genes elsewhere on 8p (*POLB, PPP2CB, NAT1* and *NAT2*) have been analyzed for mutation (Eydmann and Knowles, 1997; Matsuzaki *et al.*, 1996; Takle *et al.*, unpublished). Mutations in *POLB*, at 8p11–12, were described in TCC by Matsuzaki *et al.* (1996) but not by Eydmann *et al.* (Eydmann and Knowles, 1997) and, since this gene maps centromeric to the minimum region of deletion at 8p11–12, the significance of this finding is unclear. Possibly there are two suppressor loci in close proximity.

LOH of 11p is found in approximately 40% of TCCs with a trend towards those of higher grade (Fearon *et al.*, 1985; Shaw and Knowles, 1995; Tsai *et al.*, 1990). The common region of deletion has been mapped to 11p15 (Shaw and Knowles, 1995). No candidate genes have yet been assessed. Deletions are found at lower frequency on 11q but with a stronger association with high tumor grade and stage (Shaw and Knowles, 1995).

LOH on 14q has been found in 47.2% of TCCs, more commonly in muscle-invasive tumors (Chang *et al.*, 1995). A deletion map defined two regions of deletion at 14q12 (approx. 2 cM) and 14q32.1–32.2 (approx. 3 cM). Deletions in these regions have been described in several other tumor types but no candidate genes have yet been identified.

Approximately 30% of muscle-invasive TCCs show LOH of 18q (Brewster *et al.*, 1994). The region commonly deleted includes *DCC* and *SMAD4/DPC4* but has not been mapped in detail and mutation analyses have not been reported. The importance of deletion of this region is underscored by the finding of non-random deletions of the same region during neoplastic progression of urothclial cells *in vitro* (Wu *et al.*, 1991).

6. A model for TCC progression

Most TCCs (80%) are superficial papillary tumors at presentation and only 10–20% of these lesions progress to muscle invasion, despite the frequent development of recurrences. In contrast, the 20% of tumors that are invasive at diagnosis have a much less favorable prognosis. For these, there is rarely evidence of a papillary precursor lesion and it is suggested that these tumors develop via urothelial atypia and carcinoma *in situ* (CIS). Genetic analysis of CIS has demonstrated that, although these lesions are flat and non-invasive, they share many of the genetic alterations seen in muscle-invasive TCC. Indeed, primary CIS represents an aggressive lesion with a 50% probability of progression. These clinical findings have led to the proposal of a two-pathway model for TCC development (*Figure 2*). The common development of more than one tumor either synchronously or metachronously in the same bladder complicates studies to elucidate the natural history of the disease. The relationship between these lesions is not as clear as in tissues where single lesions containing both benign and malignant pathologies provide evidence for the sequence of biological events. The sequence of events shown in *Figure 2* is based largely on histopathological evidence. Genetic information is incomplete but where known, is compatible with this. In a study of *TP53* and chromosome 9 LOH in a series of multifocal tumors, Spruck *et al.* (1994b) found a high frequency of *TP53* mutation in both CIS and dysplasia. Other studies of dysplasia and CIS confirm this (Rosin *et al.*, 1995; Wagner *et al.*, 1995). The timing and frequency of chromosome 9 alterations during the development of muscle-invasive tumors via CIS is less clear. Spruck *et al.* found 9q LOH in only 12% of dysplasias and CIS, compared with >50% of T2–T4 tumors from the same patients. Other studies have not found such a marked difference in frequency, though a slightly lower frequency of underrepresentation of chromosome 9 detected by FISH analysis has been reported in muscle-invasive compared with superficial papillary tumors (Pycha *et al.*, 1997).

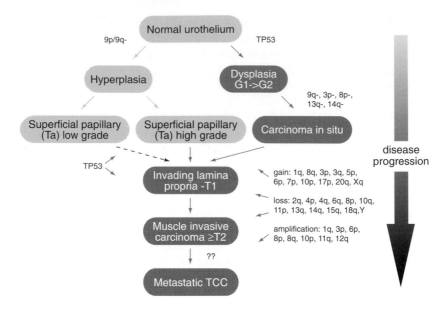

Figure 2. A model for TCC progression based on histopathological observations and genetic data. Only chromosome 9 loss is found at significant frequency in superficial papillary tumors. Many gains, losses and amplifications occur at or after Ta→T1 progression.

It is notable that, although many hundreds of superficial papillary TCC samples have been analyzed, only LOH of chromosome 9 has been found at significant frequency. All other common genetic alterations are found predominantly in muscle-invasive TCC. As discussed above, chromosome 9 alterations are likely to occur early in TCC development. There is however, a clear distinction between those tumors with LOH (often loss of an entire copy of the chromosome) and those without. It is not yet known whether these constitute phenotypically distinct groups. Two studies of chromosome 9 status and propensity to recur present conflicting results (Bartlett *et al.*, 1998; Pycha *et al.*, 1997). Prediction of tumor recurrence is an important clinical issue and much time, expenditure and patient discomfort and anxiety will be avoided if this can be predicted accurately. Similarly, markers are needed to aid diagnosis and to provide non-invasive disease monitoring in this large patient group. The lack of known genetic or gene expression changes in this large group of tumors makes them good candidates for large scale genome and expression comparisons using microarray and proteomic approaches.

TCC is commonly multifocal and there has been much discussion about the origin and clonality of these tumors. Several genetic studies have attempted to address this. These indicate that multifocal lesions usually arise from the same precursor cell (monoclonal origin) and then spread within the bladder, with subsequent independent evolution of different subclones (Lunec *et al.*, 1992; Sidransky *et al.*, 1992; Spruck *et al.*, 1994b; Takahashi *et al.*, 1998; Xu *et al.*, 1996). Occasional cases have been reported where more than one initiating event is predicted. This is not unexpected, particularly in tobacco smokers where the entire urothelium has suffered chronic exposure to carcinogens. A recent partial allelotype analysis of synchronous and metachronous multifocal, low-grade, superficial tumors, found that most tumors were almost certainly derived from a single progenitor cell. Most of these tumors were genetically stable during the course of study (up to 6 years) but some showed genetic divergence in later tumors with acquisition of LOH on 11p, 17p, 4p, 4q and 8p in addition to the LOH on 9p and/or 9q identified in the earliest tumor (Takahashi et al., 1998). It is not known whether dysplasias and CIS are more unstable genetically, but given the high frequency of *TP53* mutation in these and the large number of genetic events commonly found in invasive TCC, this seems likely. The current availability of various microdissection systems will allow a more detailed examination of the pathogenesis of TCC in the future.

7. Genetics of schistosomiasis-associated bladder cancer

Less is known about the genetics of the squamous cell carcinomas (SCC) characteristic of schistosomiasis infection despite the enormous incidence of these tumors. Schistosomiasis is endemic in 75 countries and affects more than 200 million people worldwide (WHO, 1985). There is a clear need for molecular information that could provide early detection of malignant change in infected individuals with bladder symptoms. The available genetic information is listed in *Table 4*. Several significant differences from transitional cell carcinoma samples

Table 4. Genetic alterations in schistosomiasis-associated squamous cell carcinoma of the bladder

Oncogenes		
Gene (cytogenetic location) alteration	Frequency	References
HRAS (11p15) activating mutation	14%	(Ramchuuren et al., 1995)
CCND1 (11q13) amplification	60%	(El-Rifai et al., 2000)
Tumor suppressor genes		
Gene (cytogenetic location) alteration	Frequency	References
INK4A/ARF (9p21) Homozygous deletion/LOH	65–100%	(Gonzalez-Zulueta et al., 1995; Shaw et al., 1999)
TP53 (17p13) mutation	33–86%	(Gonzalez-Zulueta et al., 1995; Habuchi et al., 1993; Ramchuuren et al., 1995; Warren et al., 1995)
LOH		
Chromosome arm	Frequency	References
3p	40%	(Shaw et al., 1999)
4p	21%	(Shaw et al., 1999)
4q	26%	(Shaw et al., 1999)
8p	37%	(Shaw et al., 1999)
11p	33%	(Shaw et al., 1999)
11q	30%	(Shaw et al., 1999)
18q	27%	(Shaw et al., 1999)

studied in the U.K. and U.S. have been found. There is a much higher frequency of involvement of the *INK4A/ARF* locus with LOH predominantly on 9p but not involving 9q and a much higher frequency of amplification in the region of the cyclin D1 gene. Other alterations are similar to those found in high-grade invasive TCC.

8. Conclusion

Knowledge of the genetic changes underlying development of TCC has increased rapidly in recent years. This is summarized in *Figure 3*. With the advent of novel technologies (e.g. 24-colour FISH, cDNA and CGH microarrays) and the near completion of the human genome sequence, it is anticipated that what is currently a long list of amplified and deleted genomic locations will shortly be transformed into a long list of genes. For the past few years many laboratories have been engaged in what has essentially been a 'fishing' and cataloging exercise. The repertoire of known genes and loci must now be examined in much greater detail both in terms of the function of the genes concerned and their impact on the clinical behavior of tumors. This may be a more demanding phase in our quest to understand urothelial transformation but undoubtedly, will be more rewarding.

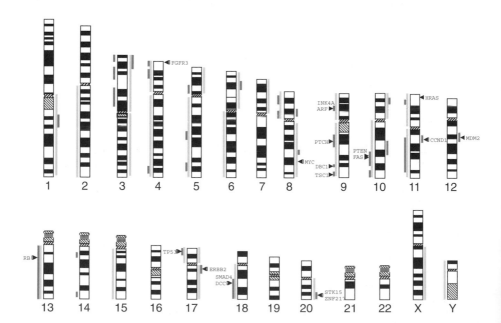

Figure 3. Summary of genetic changes identified in transitional cell carcinoma. Losses, deletions and tumor suppressor genes implicated in tumor development are shown on the left of each chromosome. Gains, amplifications and oncogenes are shown on the right. Gains and losses identified by CGH are indicated by lines in pale gray. Regions showing amplification detected by CGH or other methods and mapped regions of LOH are shown in dark gray.

References

Armstrong, B. and Doll, R. (1974) Bladder cancer mortality in England and Wales in relation to cigarette smoking and saccharin consumption. *Br. J. Prev. Soc. Med.* **28**: 233–240.

Aveyard, J.S., Skilleter, A., Habuchi, T. and Knowles, M.A. (1999) Somatic mutation of PTEN in bladder carcinoma. *Br. J. Cancer* **80**: 904–908.

Balázs, M., Carroll, P., Kerschmann, R., Sauter, G. and Waldman, F.M. (1997) Frequent homozygous deletion of cyclin-dependent kinase inhibitor 2 (MTS1, p16) in superficial bladder cancer detected by fluorescence in situ hybridization. *Genes Chromosomes Cancer* **19**: 84–89.

Barrios, L., Miro, R., Caballin, M.R., Vayreda, J., Subias, A. and Egozcue, J. (1986) Constitutional del(3)(p14-p21) in a patient with bladder carcinoma. *Cancer Genet. Cytogenet.* **21**: 171–173.

Bartlett, J.M., Watters, A.D., Ballantyne, S.A., Going, J.J., Grigor, K.M. and Cooke, T.G. (1998) Is chromosome 9 loss a marker of disease recurrence in transitional cell carcinoma of the urinary bladder? *Br. J. Cancer* **77**: 2193–2198.

BAUS Subcommittee on Industrial Bladder Cancer. (1988) Occupational bladder cancer: A guide for clinicians. *Br. J. Urol.* **61**: 183–191.

Bell, D., Taylor, J.A., Paulson, D.F., Robertson, C.N., Mohler, J.L. and Lucier, G.W. (1993) Genetic risk and carcinogen exposure: a common inherited defect of the carcinogen-metabolism gene glutathione S-transferase M1 (GSTM1) that increases susceptibility to bladder cancer. *J. Natl. Cancer Inst.* **85**: 1159–1164.

Bell, S.M., Shaw, M., Jou, Y.-S., Myers, R.M. and Knowles, M.A. (1997) Identification and characterisation of the human homologue of SH3BP2, an SH3 binding domain protein within a common region of deletion at 4p16.3 involved in bladder cancer. *Genomics* **44**: 163–170.

Besson, A., Robbins, S.M. and Yong, V.W. (1999) PTEN/MMAC1/TEP1 in signal transduction and tumorigenesis. *Eur. J. Biochem.* **263**: 605–611.

Bohm, M., Kirch, H., Otto, T., Rubben, H. and Wieland, I. (1997) Deletion analysis at the DEL-27, APC and MTS1 loci in bladder cancer: LOH at the DEL-27 locus on 5p13–12 is a prognostic marker of tumor progression. *Int. J. Cancer* **74**: 291–295.

Bohm, M., Kleine-Besten, R. and Wieland, I. (2000) Loss of heterozygosity analysis on chromosome 5p defines 5p13–12 as the critical region involved in tumor progression of bladder carcinomas. *Int. J. Cancer* **89**: 194–197.

Brewster, S.F., Gingell, J.C., Browne, S. and Brown, K.W. (1994) Loss of heterozygosity on chromosome 18q is associated with muscle-invasive transitional cell carcinoma of the bladder. *Br. J. Cancer* **70**: 697–700.

Bringuier, P.P., Tamimi, J. and Schuuring, E. (1994) Amplification of the chromosome 11q13 region in bladder tumors. *Urol. Res.* **21**: 451.

Bringuier, P.P., Tamimi, Y., Schuuring, E. and Schalken, J. (1996) Expression of cyclin D1 and EMS1 in bladder tumors: relationship with chromosome 11q13 amplification. *Oncogene* **12**: 1747–1753.

Brockmoller, J., Kerb, R., Drakoulis, N., Staffeldt, B. and Roots, I. (1995) Glutathione S-transferase M1 and its variants A and B as host factors of bladder cancer susceptibility: a case-control study. *Cancer Res.* **54**: 4103–4111.

Bruch, J., Wohr, G., Hautmann, R. *et al.* (1998) Chromosomal changes during progression of transitional cell carcinoma of the bladder and delineation of the amplified interval on chromosome arm 8q. *Genes Chromosomes Cancer* **23**: 167–174.

Cairns, P., Proctor, A.J. and Knowles, M.A. (1991) Loss of heterozygosity at the RB locus is frequent and correlates with muscle invasion in bladder carcinoma. *Oncogene* **6**: 2305–2309.

Cairns, P., Shaw, M.E. and Knowles, M.A. (1993) Initiation of bladder cancer may involve deletion of a tumor-suppressor gene on chromosome 9. *Oncogene* **8**: 1083–1085.

Cairns, P., Tokino, K., Eby, Y. and Sidransky, D. (1994) Homozygous deletions of 9p21 in primary human bladder tumors detected by comparative multiplex polymerase chain reaction. *Cancer Res.* **54**: 1422–1424.

Cairns, P., Polascik, T.J., Eby, Y. *et al.* (1995) Frequency of homozygous deletion at p16/CDKN2 in primary human tumors *Nat. Genet.* **11**: 210–212.

Cairns, P., Evron, E., Okami, K. *et al.* (1998) Point mutation and homozygous deletion of PTEN/MMAC1 in primary bladder cancers. *Oncogene* **16**: 3215–3218.

Cappellen, D., Gil Diez de Medina, S., Chopin, D., Thiery, J.P. and Radvanyi, F. (1997) Frequent loss of heterozygosity on chromosome 10q in muscle-invasive transitional cell carcinomas of the bladder. *Oncogene* **14**: 3059–3066.

Cappellen, D., De Oliveira, C., Ricol, D., de Medina, S., Bourdin, J., Sastre-Garau, X., Chopin, D., Thiery, J.P. and Radvanyi, F. (1999) Frequent activating mutations of FGFR3 in human bladder and cervix carcinomas. *Nat. Genet.* **23**: 18–20.

Cartwright, R.A., Glashan, R.W., Rogers, H.J., Ahmod, R.A., Barham-Hall, D., Higgins, E. and Kahn, M.A. (1982) The role of N-acetyltransferase in bladder carcinogenesis: a pharmacogenetic epidemiological approach to bladder cancer. *Lancet* **ii**: 842–846.

Cattan, N., Saison-Behmoaras, T., Mari, B., Mazeau, C., Amiel, J.L., Rossi, B. and Gioanni, J. (2000) Screening of human bladder carcinomas for the presence of Ha-ras codon 12 mutation. *Oncol. Rep.* **7**: 497–500.

Chang, W.Y.-H., Cairns, P., Schoenberg, M.P., Polascik, T.J. and Sidransky, D. (1995) Novel suppressor loci on chromosome 14q in primary bladder cancer. *Cancer Res.* **55**: 3246–3249.

Choi, C., Kim, M.H., Juhng, S.W. and Oh, B.R. (2000) Loss of heterozygosity at chromosome segments 8p22 and 8p11.2–21.1 in transitional-cell carcinoma of the urinary bladder. *Int. J. Cancer* **86**: 501–505.

Cole, P., Hoover, R. and Friedell, G.H. (1972) Occupation and cancer of the lower urinary tract. *Cancer* **29**: 1250–1260.

Collins, C., Rommens, J.M., Kowbel, D. *et al.* (1998) Positional cloning of ZNF217 and NABC1: genes amplified at 20q13.2 and overexpressed in breast carcinoma. *Proc. Natl Acad. Sci. USA* **95**: 8703–8708.

Coombs, L.M., Pigott, D.A., Sweeney, E., Proctor, A.J., Eydmann, M.E., Parkinson, C. and Knowles, M.A. (1991) Amplification and over-expression of c-erbB-2 in transitional cell carcinoma of the urinary bladder. *Br. J. Cancer* **63**: 601–608.

Cordon-Cardo, C., Wartinger, D., Petrylak, D., Dalbagni, G., Fair, W.R., Fuks, Z. and Reuter, V.E. (1992) Altered expression of the retinoblastoma gene product: prognostic indicator in bladder cancer. *J. Natl Cancer Inst.* **84**: 1251–1256.

Cordon-Cardo, C., Zhang, Z.-F., Dalbagni, G., Drobnjak, M., Charytonowicz, E., Hu, S.-X., Xu, H.-J., Reuter, V.E. and Benedict, W.F. (1997) Cooperative effects of p53 and pRB alterations in primary superficial bladder tumors. *Cancer Res.* **57**: 1217–1221.

Cote, R.J., Dunn, M.D., Chatterjee, S.J. *et al.* (1998) Elevated and absent pRb expression is associated with bladder cancer progression and has cooperative effects with p53. Cancer Res. **58**: 1090–1094.

Czerniak, B., Chaturvedi, V., Li, L. *et al.* (1999) Superimposed histologic and genetic mapping of chromosome 9 in progression of human urinary bladder neoplasia: implications for a genetic model of multistep urothelial carcinogenesis and early detection of urinary bladder cancer. *Oncogene* **18**: 1185–1196.

Davies, J.M. (1982). Occupational and environmental factors in bladder cancer. In: *Scientific Foundations of Urology*, (Chisholm, G. & Innes Williams, D, eds) pp. 723–727. Heinemann, London.

DerKinderen, D.J., Koten, J.W., Nagelkerke, N.J., Tan, K.E., Beemer, F.A. and Den Otter, W. (1988) Non-ocular cancer in patients with hereditary retinoblastoma and their relatives. *Int. J. Cancer* **41**: 499–504.

Devlin, J., Keen, A.J. and Knowles, M.A. (1994) Homozygous deletion mapping at 9p21 in bladder carcinoma defines a critical region within 2cM of IFNA. *Oncogene* **9**: 2757–2760.

Di Cristofano, A., Pesce, B., Cordon-Cardo, C. and Pandolfi, P.P. (1998) Pten is essential for embryonic development and tumor suppression. *Nat. Genet.* **19**: 348–355.

El-Bolkainy, M.N., Mokhtar, N.M., Ghoneim, M.A. and Hussein, M.H. (1981) The impact of schistosomiasis on the pathology of bladder carcinoma. *Cancer* **48**: 2643–2648.

Elder, P.A., Bell, S.M. and Knowles, M.A. (1994) Deletion of two regions on chromosome 4 in bladder carcinoma: definition of a critical 750kB region at 4p16.3. *Oncogene* **9**: 3433–3436.

El-Merzabani, M.M., El-Aaser, A.A. and Zakhary, N.I. (1979) A study on the aetiological factors of bilharzial bladder cancer in Egypt: 1. Nitrosamines and their precursors in urine. *Eur. J. Cancer* **15**: 287–291.

El-Rifai, W., Kamel, D., Larramendy, M.L. *et al.* (2000) DNA copy number changes in Schistosoma-associated and non-Schistosoma-associated bladder cancer. *Am. J. Pathol.* **156**: 871–878.

Esrig, D., Elmajian, D., Groshen, S. *et al.* (1994) Accumulation of nuclear p53 and tumor progression in bladder cancer. *N. Engl. J. Med.* **331**: 1259–1264.

Eydmann, M.E. and Knowles, M.A. (1997) Mutation analysis of 8p genes POLB and PPP2CB in bladder cancer. *Cancer Genet. Cytogenet.* **93**: 167–171.

Fearon, E.R., Feinberg, A.P., Hamilton, S.H. and Vogelstein, B. (1985) Loss of genes on the short arm of chromosome 11 in bladder cancer. *Nature* **318**: 377–380.

Feldman, A.R., Kessler, L., Myers, M.H. and Naughton, M.D. (1986) The prevalence of cancer: estimates based on the Connecticut Tumor Registry. *N. Engl. J. Med.* **3115**: 1394–1397.

Fitzgerald, J.M., Ramchurren, N., Rieger, K., Levesque, P., Silverman, M., Libertino, J.A. and Summerhayes, I.C. (1995) Identification of H-ras mutations in urine sediments complements cytology in the detection of bladder tumors. *J. Natl. Cancer Inst.* **87**: 129–133.

Fujita, J., Srivastava, S.K., Kraus, M.H., Rhim, J.S., Tronick, S.R. and Aaronson, S.A. (1985) Frequency of molecular alterations affecting ras protooncogenes in human urinary tract tumors. *Proc. Natl Acad. Sci. USA* **82**: 3849–3853.

Gonzalez-Zulueta, M., Shibata, A., Ohneseit, P.F. *et al.* (1995) High frequency of chromosome 9p allelic loss and *CDKN2* tumor suppressor gene alterations in squamous cell carcinoma of the bladder. *J. Natl Cancer Inst.* **87**: 1383–1393.

Grimmond, S.M., Raghavan, D. and Russell, P.J. (1992) Detection of a rare point mutation in Ki-ras of a human bladder cancer xenograft by polymerase chain reaction and direct sequencing. *Urol. Res.* **20**: 121–126.

Grossman, H.B., Liebert, M., Antelo, M., Dinney, C.P., Hu, S.X., Palmer, J.L. and Benedict, W.F. (1998) p53 and RB expression predict progression in T1 bladder cancer. *Clin. Cancer Res.* **4**: 829–834

Habuchi, T., Takahashi, R., Yamada, H. *et al.* (1993) Influence of cigarette smoking and schistosomiasis on p53 gene mutation in urothelial cancer. *Cancer Res.* **53**: 3795–3799.

Habuchi, T., Kinoshita, H., Yamada, H., Kakehi, Y., Ogawa, O., Wu, W.-J., Takahashi, R., Sugiyama, T. and Yoshida, O. (1994) Oncogene amplification in urothelial cancers with p53 gene mutation or MDM2 amplification. *J. Natl Cancer Inst.* **86**: 1331–1335.

Habuchi, T., Devlin, J., Elder, P.A. and Knowles, M.A. (1995) Detailed deletion mapping of chromosome 9q in bladder cancer: evidence for two tumor suppressor loci. *Oncogene* **11**: 1671–1674.

Habuchi, T., Yoshida, O. and Knowles, M.A. (1997) A novel candidate tumor suppressor locus at 9q32–33 in bladder cancer: localisation of the candidate region within a single 840 kb YAC. *Hum. Mol. Genet.* **6**: 913–919.

Habuchi, T., Luscombe, M., Elder, P.A. and Knowles, M.A. (1998) Structure and methylation-based silencing of a gene (DBCCR1) within a candidate bladder cancer tumor suppressor region at 9q32-q33. *Genomics* **48**: 277–288.

Harries, L.W., Stubbins, M.J., Forman, D., Howard, G.C. and Wolf, C.R. (1997) Identification of genetic polymorphisms at the glutathione S-transferase Pi locus and association with susceptibility to bladder, testicular and prostate cancer. *Carcinogenesis* **18**: 641–644.

Hartmann, A., Moser, K., Kriegmair, M., Hofstetter, A., Hofstaedter, F. and Knuechel, R. (1999) Frequent genetic alterations in simple urothelial hyperplasias of the bladder in patients with papillary urothelial carcinoma. *Am. J. Pathol.* **154**: 721–727.

Heid, C.A., Stevens, J., Livak, K.J. and Williams, P.M. (1996) Real time quantitative PCR. *Genome Res.* **6**: 986–994.

HMSO. (1994) Cancer registration statistics. England and Wales.

Hornigold, N., Devlin, J., Davies, A.M., Aveyard, J.S., Habuchi, Y. and Knowles, M.A. (1999) Mutation of the 9q34 gene *TSC1* in bladder cancer. *Oncogene* **18**: 2657–2661.

IARC (1986) *Mongraphs on the Evaluation of Carcinogenic Risk of Chemicals to Humans: Tobacco smoking.* World Health Organisation, Geneva.

IARC (1994) *Schistosomes, Liver Flukes and Helicobacter pylori.* Vol. 61. Monographs on the Evaluation of Carcinogenic Risk to Humans. International Agency for Research on Cancer, Lyon.

Ibrahim, S. (1986) Site distribution of cancer in Egypt: twelve years experience (1970–1981). In: *Cancer Prevention in Developing Countries* (Gjorgove, A. and Ismail, A., eds) pp. 45–50. Pergamon Press, Oxford.

Ishikawa, J., Xu, H.-J., Hu, S.-X., Yandell, D., Maeda, S., Kamidono, S., Benedict, W. and Takahashi, R. (1991) Inactivation of the retinoblastoma gene in human bladder and renal cell carcinomas. *Cancer Res.* **51**: 5736–5743.

Jones, P.A. and Laird, P.W. (1999) Cancer epigenetics comes of age. *Nat. Genet.* **21**: 163–167.

Jung, I., Reeder, J.E., Cox, C., Siddiqui, J.F., O'Connell, M.J., Collins, L., Yang, Z., Messing, E.M. and Wheeless, L.L. (1999) Chromosome 9 monosomy by fluorescence in situ hybridization of bladder irrigation specimens is predictive of tumor recurrence. *J. Urol.* **162**: 1900–1903.

Kagan, J., Liu, J., Stein, J.D., Wagner, S.S., Babkowski, R., Grossman, B.H. and Katz, R.L. (1998) Cluster of allele losses within a 2.5 cM region of chromosome 10 in high-grade invasive bladder cancer. *Oncogene* **16**: 909–913.

Kallioniemi, A., Kallioniemi, O.-P., Citro, G., Sauter, G., DeVries, S., Kerschmann, R., Caroll, P. and Waldman, F. (1995) Identification of gains and losses of DNA sequences in primary bladder cancer by comparative genomic hybridisation. *Genes Chromosomes Cancer* **12**: 213–219.

Kantor, A.F., Hartge, P., Hoover, R.N., Narayana, A.S., Sullivan, J.W. and Fraumeni Jr, J.F. (1984) Urinary tract infection and risk of bladder cancer. *Am. J. Epidemiol.* **119**: 510–515.

Kantor, A.F., Hartge, P., Hoover, R.N. and Fraumeni, Jr, J.F. (1985) Familial and environmental interactions in bladder cancer risk. *Int. J. Cancer* **35**: 703–706.

Kantor, A.F., Hartge, P., Hoover, R.N. and Fraumeni Jr, J.F. (1988) Epidemiological characteristics of squamous cell carcinoma and adenocarcinoma of the bladder. *Cancer Res.* **48**: 3853–3855.

Kiemeney, L.A. and Schoenberg, M. (1996) Familial transitional cell carcinoma. *J. Urol.* **156**: 867–872.

Knowles, M.A. and Williamson, M. (1993) Mutation of H-ras is infrequent in bladder cancer: confirmation by single-strand conformation polymorphism analysis, designed restriction fragment length polymorphisms, and direct sequencing. *Cancer Res.* **53**: 133–139.

Knowles, M.A., Shaw, M.E. and Proctor, A.J. (1993) Deletion mapping of chromosome 8 in cancers of the urinary bladder using restriction fragment length polymorphisms and microsatellite polymorphisms. *Oncogene* **8**: 1357–1364.

Knowles, M.A., Elder, P.A., Williamson, M., Cairns, J.P., Shaw, M.E. and Law, M.G. (1994) Allelotype of human bladder cancer. *Cancer Res.* **54**: 531–538.

Knudson, A.G. (1993) Antioncogenes and human cancer. *Proc. Natl Acad. Sci. USA* **90**: 10914–10921.

Koss, L.G. (1975). *Tumors of the Urinary Bladder. Atlas of Tumor Pathology*. Armed Forces Institute of Pathology: 2nd Series. Fascicle 11. Washington, D.C.

Lee, C.C., Yamamoto, S., Morimura, K. *et al.* (1997) Significance of cyclin D1 overexpression in transitional cell carcinomas of the urinary bladder and its correlation with histopathologic features. *Cancer* **79**: 780–789.

Lee, S.H., Shin, M.S., Park, W.S. *et al.* (1999) Alterations of Fas (APO-1/CD95) gene in transitional cell carcinomas of urinary bladder. *Cancer Res.* **59**: 3068–3072.

Levesque, P., Ramchuuren, N., Saini, K., Joyce, A., Libertino, J. and Summerhayes, I.C. (1993) Screening of human bladder tumors and urine sediments for the presence of H-*ras* mutations. *Int. J. Cancer* **55**: 785–790.

Li, M., Zhang, Z.F., Reuter, V.E. and Cordon-Cardo, C. (1996) Chromosome 3 allelic losses and microsatellite alterations in transitional cell carcinoma of the urinary bladder. *Am. J. Pathol.* **149**: 229–235.

Lianes, P., Orlow, I., Zhang, Z.-F., Oliva, M.R., Sarkis, A.S., Reuter, V.E. and Cordon-Cardo, C. (1994) Altered patterns of MDM2 and TP53 expression in human bladder cancer. *J. Natl. Cancer Inst.* **86**: 1325–1330.

Lipponen, P.K. (1993) Over-expression of p53 nuclear oncoprotein in transitional-cell bladder cancer and its prognostic value. *Int. J. Cancer* **53**: 365–370.

Logothetis, C.J., Xu, H.-J., Ro, J.Y., Hu, S.-X., Sahin, A., Ordonez, N. and Benedict, W.F. (1992) Altered expression of retinoblastoma protein and known prognostic variables in locally advanced bladder cancer. *J. Natl Cancer Inst.* **84**: 1256–1261.

Lonn, U., Lonn, S., Friberg, S., Nilsson, B., Silfversward, C. and Stenkvist, B. (1995) Prognostic value of amplification of c-erb-B2 in bladder carcinoma. *Clin. Cancer Res.* **1**: 1189–1194.

Lunec, J., Challen, C., Wright, C., Mellon, K. and Neal, D.E. (1992) c-erbB-2 amplification and identical p53 mutations in concomitant transitional carcinomas of renal pelvis and urinary bladder. *Lancet* **339**: 439–440.

Lynch, H.T., Walzak, M.P., Fried, R., Domina, A.H. and Lynch, J.F. (1979) Familial factors in bladder carcinoma. *J. Urol.* **122**: 458–461.

Malone, P.R., Visvinathan, K.V., Ponder, B.A.J., Shearer, R.J. and Summerhayes, I.C. (1985) Oncogenes and bladder cancer. *Br. J. Urol.* **57**: 664–667.

Marcus, P.M., Vineis, P. and Rothman, N. (2000) NAT2 slow acetylation and bladder cancer risk: a meta-analysis of 22 case-control studies conducted in the general population. *Pharmacogenetics* **10**: 115–122.

Markl, I.D. and Jones, P.A. (1998) Presence and location of TP53 mutation determines pattern of CDKN2A/ARF pathway inactivation in bladder cancer. *Cancer Res.* **58**: 5348–5353.

Masters, J.R., Vesey, S.G., Munn, C.F., Evan, G.I. and Watson, J.V. (1988) c-myc oncoprotein levels in bladder cancer. *Urol. Res.* **16**: 341–344.

Matsuyama, H., Bergerheim, U.S., Nilsson, I., Pan, Y., Skoog, L., Tribukait, B. and Ekman, P. (1994) Nonrandom numerical aberrations of chromosomes 7, 9, and 10 in DNA-diploid bladder cancer. *Cancer Genet. Cytogenet.* **77**: 118–124.

Matsuzaki, J., Dobashi, Y., Miyamoto, H., Ikeda, I., Fujinami, K., Shuin, T. and Kubota, Y. (1996) DNA polymerase beta gene mutations in human bladder cancer. *Mol. Carcinog.* **15**: 38–43.

McCredie, M., Stewart, J.H., Ford, J.M. and MacLennan, R.A. (1983) Phenacetin-containing analgesics and cancer of the bladder in women. *Br. J. Urol.* **55**: 220–224.

McGarvey, T.W., Maruta, Y., Tomaszewski, J.E., Linnenbach, A.J. and Malkowicz, S.B. (1998) PTCH gene mutations in invasive transitional cell carcinoma of the bladder. *Oncogene* **17**: 1167–1172.

Miyamoto, H., Kubota, Y., Noguchi, S., Takase, K., Matsuzaki, J., Moriyama, M., Takebayashi, S., Kitamura, H. and Hosaka, M. (2000) C-ERBB-2 gene amplification as a prognostic marker in human bladder cancer. *Urology* **55**: 679–683.

Mostofi, F.K., Sobin, L.H. and Torloni, H. (1973). Histologic typing of urinary bladder tumors. In: *International Histological Classification of Tumors*, Vol. 10. World Health Organisation, Geneva.

Nishiyama, H., Takahashi, T., Kakehi, Y., Habuchi, T. and Knowles, M.A. (1999) Homozygous deletion at the 9q32–33 candidate tumor suppressor locus in primary bladder cancer. *Genes Chromosomes Cancer* **26**: 171–175.

Ohgaki, K., Iida, A., Ogawa, O., Kubota, Y., Akimoto, M. and Emi, M. (1999) Localization of tumor suppressor gene associated with distant metastasis of urinary bladder cancer to a 1-Mb interval on 8p22. *Genes Chromosomes Cancer* **25**: 1–5.

Olderoy, G., Daehlin, L. and Ogreid, D. (1998) Low-frequency mutation of Ha-ras and Ki-ras oncogenes in transitional cell carcinoma of the bladder. *Anticancer Res.* **18**: 2675–2678.

Ooi, A., Herz, F., Ii, S., Cordon-Cardo, C., Fradet, Y., Mayall, B.H. and Cancer, N.M.N.f.B. (1994) Ha-ras codon 12 mutation in papillary tumors of the urinary bladder: a retrospective study. *Int. J. Oncol.* **4**: 85–90.

Orlow, I., Lacombe, L., Hannon, G.J. *et al.* (1995) Deletion of the p16 and p15 genes in human bladder tumors. *J. Natl Cancer Inst.* **87**: 1524–1529.

Packenham, J.P., Taylor, J.A., Anna, C.H., White, C.M. and Devereux, T.R. (1995) Homozygous deletions but no sequence mutations in coding regions of p15 or p16 in human primary bladder tumors. *Mol. Carcinog.* **14**: 147–151.

Paramio, J.M., Navarro, M., Segrelles, C., Gomez-Casero, E. and Jorcano, J.L. (1999) PTEN tumor suppressor is linked to the cell cycle control through the retinoblastoma protein. *Oncogene* **18**: 7462–7468.

Parker, S.L., Tong, T., Bolden, S. and Wingo, P.A. (1997) Cancer statistics, 1997. *CA Cancer J. Clin.* **47**: 5–27.

Pfister, C., Larue, H., Moore, L., Lacombe, L., Veilleux, C., Tetu, B., Meyer, F. and Fradet, Y. (2000) Tumorigenic pathways in low-stage bladder cancer based on p53, MDM2 and p21 phenotypes. *Int. J. Cancer* **89**: 100–104.

Polascik, T.J., Cairns, P., Chang, W.Y.H., Schoenberg, M.P. and Sidransky, D. (1995) Distinct regions of allelic loss on chromosome 4 in human primary bladder carcinoma. *Cancer Res.* **55**: 5396–5399.

Pratt, C.I., Kao, C., Wu, S.-Q., Gilchrist, K.W., Oyasu, R. and Reznikoff, C.A. (1992) Neoplastic progression by EJ/ras at different steps of transformation in vitro of human uroepithelial cells. *Cancer Res.* **52**: 688–695.

Presti, J.C., Jr., Reuter, V.E., Galan, T., Fair, W.R. and Cordon-Cardo, C. (1991) Molecular genetic alterations in superficial and locally advanced human bladder cancer. *Cancer Res.* **51**: 5405–5409.

Proctor, A.J., Coombs, L.M., Cairns, J.P. and Knowles, M.A. (1991) Amplification at chromosome 11q13 in transitional cell tumors of the bladder. *Oncogene* **6**: 789–795.

Pycha, A., Mian, C., Haitel, A., Hofbauer, J., Wiener, H. and Marberger, M. (1997) Fluorescence in situ hybridization identifies more aggressive types of primarily noninvasive (stage pTa) bladder cancer. *J. Urol.* **157**: 2116–2119.

Ramchuuren, N., Cooper, K. and Summerhayes, I.C. (1995) Molecular events underlying schistosomiasis-related bladder cancer. *Int. J. Cancer* **62**: 237–244.

Reddy, E.P., Reynolds, R.K., Santos, E. and Barbacid, M. (1982) A point mutation is responsible for the acquisition of transforming properties by the T24 human bladder carcinoma oncogene. *Nature* **300**: 149–152.

Reeder, J.E., Morreale, J.F., O'Connell, M.J., Stadler, W.M., Olopade, O.F., Messing, E.M. and Wheeless, L.L. (1997) Loss of the CDKN2A/p16 locus detected in bladder irrigation specimens by fluorescence in situ hybridization. *J. Urol.* **158**: 1717–1721.

Reznikoff, C.A., Belair, C.D., Yeager, T.R., Savelieva, E., Blelloch, R.H., Puthenveettil, J.A. and Cuthill, S. (1996) A molecular genetic model of human bladder cancer pathogenesis. *Semin. Oncol.* **5**: 571–584.

Richter, J., Jiang, F., Gorog, J.P., Sartorius, G., Egenter, C., Gasser, T.C., Moch, H., Mihatsch, M.J. and Sauter, G. (1997) Marked genetic differences between stage pTa and stage pT1 papillary bladder cancer detected by comparative genomic hybridization. *Cancer Res.* **57**: 2860–2864.

Richter, J., Beffa, L., Wagner, U., Schraml, P., Gasser, T.C., Moch, H., Mihatsch, M.J. and Sauter, G. (1998) Patterns of chromosomal imbalances in advanced urinary bladder cancer detected by comparative genomic hybridization. *Am. J. Pathol.* **153**: 1615–1621.

Richter, J., Wagner, U., Schraml, P. *et al.* (1999) Chromosomal imbalances are associated with a high risk of progression in early invasive (pT1) urinary bladder cancer. *Cancer Res.* **59**: 5687–5691.

Risch, A., Wallace, D.M.A., Bathers, S. and Sim, E. (1995) Slow N-acetylation genotype is a susceptibility factor in occupational and smoking related bladder cancer. *Hum. Mol. Genet.* **4**: 231–236.

Rosin, M.P., Cairns, P., Epstein, J.I., Schoenberg, M.P. and Sidransky, D. (1995) Partial allelotype of carcinoma *in situ* of the human bladder. *Cancer Res.* **55**: 5213–5216.

Sandberg, A.A. and Berger, C.S. (1994) Review of chromosome studies in urological tumors. II. Cytogenetics and molecular genetics of bladder cancer. *J. Urol.* **151**: 545–560.

Sarkis, A.S., Dalbagni, G., Cordon-Cardo, C., Zhang, Z.-F., Sheinfeld, J., Fair, W.R., Herr, H.W. and Reuter, V.E. (1993a) Nuclear overexpression of p53 protein in transitional cell bladder carcinoma: a marker for disease progression. *J. Natl Cancer Inst.* **85**: 53–59.

Sarkis, A.S., Zhang, Z.-F., Cordon-Cardo, C., Melamed, J., Dalbagni, G., Sheinfeld, J., Fair, W.R., Herr, H.W. and Reuter, V.E. (1993b) p53 nuclear overexpression and disease progression in Ta bladder carcinoma. *Int. J. Oncol.* **3**: 355–360.

Sauter, G., Moch, H., Moore, D., Carroll, P., Kerchmann, R., Chew, K., Mihatsch, M.J., Gudat, F. and Waldman, F. (1993) Heterogeneity of erbB-2 gene amplification in bladder cancer. *Cancer Res.* **53**: 2199–2203.

Sauter, G., Deng, G., Moch, H. *et al.* (1994) Physical deletion of the p53 gene in bladder cancer. Detection by fluorescence in situ hybridization. *Am. J. Pathol.* **144**: 756–766.

Sauter, G., Carroll, P., Moch, H., Kallioniemi, A., Kerschmann, R., Narayan, P., Mihatsch, M.J. and Waldman, F.M. (1995) c-myc copy number gains in bladder cancer detected by fluorescence in situ hybridization. *Am. J. Pathol.* **146**: 1131–1139.

Schmitz-Drager, B.J., Schulz, W.A., Jurgens, B., Gerharz, C.D., van Roeyen, C.R., Bultel, H., Ebert, T. and Ackermann, R. (1997) c-myc in bladder cancer. Clinical findings and analysis of mechanism. *Urol. Res.* **25**: S45–49.

Schoenberg, M., Kiemeney, L., Walsh, P.C., Griffin, C.A. and Sidransky, D. (1996) Germline translocation t(5;20)(p15;q11) and familial transitional cell carcinoma. *J. Urol.* **155**: 1035–1036.

Shaw, M.E. and Knowles, M.A. (1995) Deletion mapping of chromosome 11 in carcinoma of the bladder. *Genes Chromosomes Cancer* **13**: 1–8.

Shaw, M.E., Elder, P.A., Abbas, A. and Knowles, M.A. (1999) Partial allelotype of schistosomiasis-associated bladder cancer. *Int. J. Cancer* **80**: 656–661.

Shin, K.Y., Kong, G., Kim, W.S., Lee, T.Y., Woo, Y.N. and Lee, J.D. (1997) Overexpression of cyclin D1 correlates with early recurrence in superficial bladder cancers. *Br. J. Cancer* **75**: 1788–1792.

Sibley, K., Cuthbert-Heavens, D. and Knowles, M.A. (2001) Loss of heterozygosity at 4p16.3 and mutation of *RGFR3* in transitional cell carcinomas. *Oncogene* (in press).

Sidransky, D., von Eschenbach, A., Tsai, Y.C. *et al.* (1991) Identification of p53 gene mutations in bladder cancers and urine samples. *Science* **252**: 706–709.

Sidransky, D., Frost, P., von Eschenbach, A., Oyasu, R., Preisinger, A.C. and Vogelstein, B. (1992) Clonal origin of bladder cancer. *N. Engl. J. Med.* **326**: 737–740.

Sigalas, I., Calvert, A.H., Anderson, J.J., Neal, D.E. and Lunec, J. (1996) Alternatively spliced mdm2 transcripts with loss of p53 binding domain sequences: transforming ability and frequent detection in human cancer. *Nat. Med.* **2**: 912–917.

Silverman, D.T., Levin, L.I., Hoover, R.N. and Hortge, P. (1989) Occupational risks of bladder cancer in the United States. I. White men. *J. Natl Cancer Inst.* **81**: 1472–1480.

Simon, R., Burger, H., Brinkschmidt, C., Bocker, W., Hertle, L. and Terpe, H.J. (1998) Chromosomal aberrations associated with invasion in papillary superficial bladder cancer. *J. Pathol.* **185**: 345–351.

Simoneau, A.R., Spruck, C.H., III, Gonzalez-Zulueta, M. *et al.* (1996) Evidence for two tumor suppressor loci associated with proximal chromosome 9p to q and distal chromosome 9q in bladder cancer and the initial screening for GAS1 and PTC mutations. *Cancer Res.* **56**: 5039–5043.

Simoneau, M., Aboulkassim, T.O., LaRue, H., Rousseau, F. and Fradet, Y. (1999) Four tumor suppressor loci on chromosome 9q in bladder cancer: evidence for two novel candidate regions at 9q22.3 and 9q31. *Oncogene* **18**: 157–163.

Soini, Y., Turpeenniemi-Hujanen, T., Kamel, D., Autio-Harmainen, H., Risteli, J., Risteli, L., Nuorva, K., Paakko, P. and Vahakangas, K. (1993) p53 immunohistochemistry in transitional cell carcinoma and dysplasia of the urinary bladder correlates with disease progression. *Br. J. Cancer* **68**: 1029–1035.

Spruck, C.H., III, Rideout, W.M., III, Olumi, A.F. *et al.* (1993) Distinct pattern of p53 mutations in bladder cancer: relationship to tobacco usage. *Cancer Res.* **53**: 1162–1166.

Spruck, C.H., III, Gonzalez-Zulueta, M., Shibata, A., Simoneau, A.R., Lin, M.F., Gonzales, F., Tsai, Y.C. and Jones, P.A. (1994a) p16 gene in uncultured tumors. *Nature* **370**: 183–184.

Spruck, C.H., III, Ohnescit, P.F., Gonzalez-Zulueta, M. *et al.* (1994b) Two molecular pathways to transitional cell carcinoma of the bladder. *Cancer Res.* **54**: 784–788.

Suwa, Y., Takano, Y., Iki, M., Takeda, M., Asakura, T., Noguchi, S. and Masuda, M. (1998) Cyclin D1 protein overexpression is related to tumor differentiation, but not to tumor progression or proliferative activity, in transitional cell carcinoma of the bladder. *J. Urol.* **160**: 897–900.

Takahashi, T., Habuchi, T., Kakehi, Y., Mitsumori, K., Akao, T., Terachi, T. and Yoshida, O. (1998) Clonal and chronological genetic analysis of multifocal cancers of the bladder and upper urinary tract. *Cancer Res.* **58**: 5835–5841.

Takle, L.A. and Knowles, M.A. (1996) Deletion mapping implicates two tumor suppressor genes on chromosome 8p in the development of bladder cancer. *Oncogene* **12**: 1083–1087.

Theodorescu, D., Cornil, I., Sheehan, C., Man, M.S. and Kerbel, R.S. (1991) Ha-*ras* induction of the invasive phenotype results in up-regulation of epidermal growth factor receptors and altered responsiveness to epidermal growth factor in human papillary transitional cell carcinoma cells. *Cancer Res.* **51**: 4486–4491.

Thompson, I.M. and Fair, W.R. (1989) The epidemiology of bladder cancer. *AUA Update Ser.* **8**: 210–215.

Tomatis, L., Agthe, C., Bartsch, H., Huff, J., Montesano, R., Saracci, R., Walker, E. and Wilbourn, J. (1978) Evaluation of the carcinogenicity of chemicals: a review of the Monograph Program of the International Agency for Research on Cancer (1971 to 1977). *Cancer Res.* **38**: 877–885.

Tricker, A.R., Mostafa, M.H., Speiglhalder, B. and Preussmann, R. (1989) Urinary excretion of nitrate, nitrite and N-nitroso compounds in schistosomiasis and bladder cancer patients. *Carcinogenesis* **10**: 547–552.

Tsai, Y.C., Nichols, P.W., Hiti, A.L., Williams, Z., Skinner, D.G. and Jones, P.A. (1990) Allelic losses of chromosomes 9, 11, and 17 in human bladder cancer. *Cancer Res.* **50**: 44–47.

Underwood, M., Bartlett, J., Reeves, J., Gardiner, S., Scott, R. and Cooke, T. (1995) C-*erbB*-2 gene amplification: a molecular marker in recurrent bladder tumors? *Cancer Res.* **55**: 2422–2430.

van Slegtenhorst, M., de Hoogt, R., Hermans, C. *et al.* (1997) Identification of the tuberous sclerosis gene TSC1 on chromosome 9q34. *Science* **277**: 805–808.

van Tilborg, A.A., Groenfeld, L.E., van de Kwast, T.H. and Zwarthoff, E.C. (1999) Evidence for two candidate tumor suppressor loci on chromosome 9q in transitional cell carcinoma (TCC) of the bladder but no homozygous deletions in bladder tumor cell lines. *Br. J. Cancer* **80**: 489–494.

von Knobloch, R., Bugert, P., Jauch, A., Kalble, T. and Kovacs, G. (2000) Allelic changes at multiple regions of chromosome 5 are associated with progression of urinary bladder cancer. *J. Pathol.* **190**: 163–168.

Voorter, C., Joos, S., Bringuier, P.P. *et al.* (1995) Detection of chromosomal imbalances in transitional cell carcinoma of the bladder by comparative genomic hybridization. *Am. J. Pathol.* **146**: 1341–1354.

Voorter, C.E.M., Ummelin, M.I.J., Ramaekers, F.S.C. and Hopman, A.H.N. (1996) Loss of chromosome 11 and 11p/q imbalances in bladder cancer detected by fluorescence *in situ* hybridisation. *Int. J. Cancer* **65**: 301–307.

Wagner, U., Sauter, G., Moch, H., Novotna, H., Epper, R., Mihatsch, M.J. and Waldman, F.M. (1995) Patterns of p53, erbB-2 and EGF-r expression in premalignant lesions of the urinary bladder. *Hum. Pathol.* **26**: 970–978.

Wagner, U., Bubendorf, L., Gasser, T.C., Moch, H., Gorog, J.P., Richter, J., Mihatsch, M.J., Waldman, F.M. and Sauter, G. (1997) Chromosome 8p deletions are associated with invasive tumor growth in urinary bladder cancer. *Am. J. Pathol.* **151**: 753–759.

Wagner, U., Suess, K., Luginbuhl, T. *et al.* (1999) Cyclin D1 overexpression lacks prognostic significance in superficial urinary bladder cancer. *J. Pathol.* **188**: 44–50.

Waldman, F.M., Carroll, P.R., Kerschmann, R., Cohen, M.B., Field, F.G. and Mayall, B.H. (1991) Centromeric copy number of chromosome 7 is strongly correlated with tumor grade and labeling index in human bladder cancer. *Cancer Res.* **51**: 3807–3813.

Warren, W., Biggs, P.J., El-Baz, M., Ghoneim, M.A., Stratton, M.R. and Venitt, S. (1995) Mutations in the p53 gene in schistosomal bladder cancer: a study of 92 tumors from Egyptian patients and a comparison between spectra from schistosomal and non-schistosomal urothelial tumors. *Carcinogenesis* **16**: 1181–1189.

Wheeless, L.L., Reeder, J.E., Han, R., O'Connell, M.J., Frank, I.N., Cockett, A.T. and Hopman, A.H. (1994) Bladder irrigation specimens assayed by fluorescence in situ hybridization to interphase nuclei. *Cytometry* **17**: 319–326.

WHO. (1985). *The control of schistosomiasis. Report of a WHO Expert Committee*. Technical Report Series no 728. WHO, Geneva.

Wick, W., Furnari, F.B., Naumann, U., Cavenee, W.K. and Weller, M. (1999) PTEN gene transfer in human malignant glioma: sensitization to irradiation and CD95L-induced apoptosis. *Oncogene* **18**: 3936–3943.

Williamson, M.P., Elder, P.A. and Knowles, M.A. (1994) The spectrum of TP53 mutations in bladder carcinoma. *Genes Chromosomes Cancer* **9**: 108–118.

Williamson, M.P., Elder, P.A., Shaw, M.E., Devlin, J. and Knowles, M.A. (1995) p16 (CDKN2) is a major deletion target at 9p21 in bladder cancer. *Hum. Mol. Genet.* **4**: 1569–1577.

Wu, Q., Possati, L., Montesi, M., Gualandi, F., Rimessi, P., Morelli, C., Trabanelli, C. and Barbanti-Brodano, G. (1996) Growth arrest and suppression of tumorigenicity of bladder carcinoma cell lines induced by the *p16/CDKN2 (p16INK4A, MTS1)* gene and other loci on chromosome 9. *Int. J. Cancer* **65**: 840–846.

Wu, S.Q., Storer, B.E., Bookland, E.A., Klingelhutz, A.J., Gilchrist, K.W., Meisner, L.F., Oyasu, R. and Reznikoff, C.A. (1991) Nonrandom chromosome losses in stepwise neoplastic transformation in vitro of human uroepithelial cells. *Cancer Res.* **51**: 3323–3326.

Xu, X., Stower, M.J., Reid, I.N., Garner, R.C. and Burns, P.A. (1996) Molecular screening of multifocal transitional cell carcinoma of the bladder using p53 mutations as biomarkers. *Clin. Cancer Res.* **2**: 1795–1800.

Yeager, T.R., DeVries, S., Jarrard, D.F. *et al.* (1998) Overcoming cellular senescence in human cancer pathogenesis. *Genes Dev.* **12**: 163–174.

Zhao, J., Richter, J., Wagner, U. *et al.* (1999) Chromosomal imbalances in noninvasive papillary bladder neoplasms (pTa). *Cancer Res.* **59**: 4658–4661.

Zhou, H., Kuang, J., Zhong, L., Kuo, W.L., Gray, J.W., Sahin, A., Brinkley, B.R. and Sen, S. (1998) Tumor amplified kinase STK15/BTAK induces centrosome amplification, aneuploidy and transformation. *Nat. Genet.* **20**: 189–193.

20

Genetics of rhabdomyosarcoma

Joseph G. Pressey and Frederic G. Barr

1. Introduction

Rhabdomyosarcoma (RMS) is the most common soft-tissue sarcoma of childhood, with an annual incidence of 4 to 7 cases per million children (Kramer *et al.*, 1983). RMS comprises a heterogeneous family of tumors related to the skeletal muscle lineage. The tumors are conventionally divided into two major histologic subtypes, embryonal (ERMS) and alveolar (ARMS). ERMS typically occurs in children less than 10 years of age, most frequently in the head and neck, genitourinary tract, and the retroperitoneum. The prognosis of ERMS is generally favorable. In contrast, ARMS most commonly occurs in adolescents and young adults, typically in the trunk and extremities, and portends a poorer prognosis (Tsokos *et al.*, 1992). ARMS is associated with the hallmark 2;13 or variant 1;13 chromosomal translocations, which result in the oncogenic fusion products PAX3-FKHR and PAX7-FKHR, respectively. Most ERMS cases have allelic loss at the 11p15.5 locus, likely representing loss of function of a tumor suppressor gene, though the specific genetic alterations are not known.

RMS is also associated with less frequent, secondary genetic changes, supporting the theory of a multi-step process of tumorigenesis. Some of the secondary alterations are common to both ARMS and ERMS, whereas others are unique to only one histologic subtype, further demonstrating the genetic heterogeneity of RMS. Both primary and secondary events affect gene products that function in signal transduction or gene expression regulatory pathways. Alterations of such pathways are often also involved in inherited cancer predisposition syndromes, some of which are associated with an increased risk of RMS.

2. Chromosomal translocations in ARMS

Chromosomal analyses of RMS cases demonstrated nonrandom translocations associated with the ARMS subtype (Barr, 1997). The most prevalent finding is a translocation involving chromosomes 2 and 13, t(2;13)(q35;q14), that was

Molecular Genetics of Cancer second edition, edited by J.K. Cowell.

detected in 70% of published ARMS cases (*Figure 1*). In addition, there have been several reports of a t(1;13)(p36;q14) variant translocation. These two translocations have not been associated with any other tumor and thus appear to be specific markers for ARMS.

2.1 PAX3-FKHR and PAX7-FKHR fusions

Physical mapping and cloning studies revealed that the loci on chromosomes 2 and 1 rearranged by the t(2;13) and t(1;13) are *PAX3* and *PAX7*, respectively (Barr *et al.*, 1993; Davis *et al.*, 1994). These two genes encode highly related members of the paired box transcription factor family that are organized with N-terminal DNA binding domains consisting of paired box and homeobox motifs and C-terminal transcriptional activation domains (*Figure 2*). The chromosome 13 locus juxtaposed with both *PAX3* and *PAX7* is *FKHR* (also referred to as *FOXO1A*), which encodes a novel member of the fork head transcription factor family (Davis *et al.*, 1994; Galili *et al.*, 1993). The encoded FKHR product is organized with an N-terminal fork head DNA binding domain and a C-terminal transcriptional activation domain. The translocations break within intron 7 of *PAX3* or *PAX7* and intron 1 of *FKHR* and thus create two chimeric genes on the derivative chromosomes (Barr *et al.*, 1998; Davis *et al.*, 1995; Fitzgerald *et al.*, 2000). The *PAX3-FKHR* or *PAX7-FKHR* chimeric gene on the derivative chromosome 13 is more consistently and highly expressed, and generates chimeric transcripts consisting of 5' *PAX3* or *PAX7* exons fused to 3' *FKHR* exons. These chimeric transcripts encode fusion proteins containing the *PAX3* or *PAX7* DNA binding domain and the C-terminal *FKHR* transcriptional activation domain.

2.2 Function, expression, and localization of fusion products

A series of molecular biology studies demonstrated that these gene fusion events result in alterations at the level of protein function, gene expression, and subcellular localization. First, the PAX3-FKHR and PAX7-FKHR fusion proteins activate transcription from PAX-binding sites but are 10–100-fold more potent as

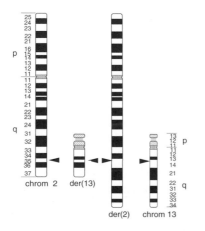

Figure 1. The t(2;13)(q35;q14) chromosomal translocation in ARMS. Schematic representation of normal and derivative chromosomes associated with the t(2;13). Translocation breakpoints are indicated by horizontal arrowheads.

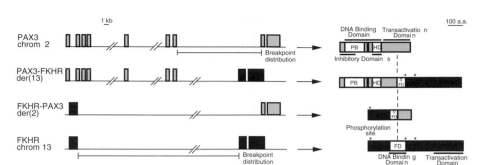

Figure 2. Chimeric genes and products generated by the 2;13 translocation in ARMS. On the left, the exons of the wild-type and fusion genes are shown as boxes above each map and the translocation breakpoint distributions are shown as line segments below the map of the wild-type genes. On the right, the protein products of the wild-type and chimeric genes are shown; the paired box, octapeptide, homeobox, and fork head domain are indicated as open boxes, and transcriptional domains are shown as solid bars. The sites phosphorylated by Akt are indicated by stars. The vertical dash line indicates the translocation fusion point.

transcriptional activators than the wild-type PAX3 and PAX7 proteins (Bennicelli *et al.*, 1996, 1999; Fredericks *et al.*, 1995). This increased function results from the insensitivity of the FKHR activation domain to inhibitory effects of N-terminal PAX3 and PAX7 domains (*Figure 2*). These N-terminal PAX3/PAX7 inhibitory domains effectively inhibit the activity of the C-terminal PAX3 or PAX7 activation domain, but have a very modest effect on the C-terminal FKHR activation domain.

The finding of greater mRNA expression of *PAX3-FKHR* and *PAX7-FKHR* relative to wild-type *PAX3* and *PAX7*, respectively, in ARMS tumors indicates that fusion gene overexpression is characteristic of ARMS and suggests that a critical level of fusion product is needed in oncogenesis (Davis and Barr, 1997). Despite the common feature of fusion gene overexpression in the two ARMS fusion subtypes, there is a striking difference in the mechanism of fusion gene overexpression between these two subtypes. In *PAX7-FKHR*-expressing tumors, the fusion gene is present in increased copy number due to *in vivo* amplification of the genomic region containing the fusion gene. In contrast, the *PAX3-FKHR* fusion gene is rarely amplified, but instead is overexpressed due to a copy number-independent increase in transcriptional rate. These findings indicate significant biological differences in the regulation of expression of these fusion genes.

The subcellular localization of the wild-type FKHR protein is regulated by an AKT-dependent signaling pathway (del Peso *et al.*, 1999). Though several of the AKT-directed phosphorylation sites are maintained in the PAX3-FKHR fusion product (*Figure 2*), the fusion protein is resistant to this regulatory pathway and shows exclusively nuclear localization. In aggregate, these findings indicate that the chromosomal changes in these tumors result in high levels of exclusively nuclear products that are potent activators of transcription from PAX3/PAX7 DNA binding sites. The end result is exaggerated activity at multiple biological levels that converges on inappropriate activation of PAX3/PAX7 target genes and ultimately contributes to tumorigenic behavior.

2.3 Phenotypic consequences of fusion gene products

Gene transfer studies support the hypothesis that the t(2;13) activates the onco-genic potential of *PAX3* by dysregulating or exaggerating its normal function in the myogenic lineage. Under conditions in which wild-type *PAX3* does not transform NIH 3T3 cells or chicken embryo fibroblasts, the *PAX3-FKHR* fusion has potent transforming activity (Lam *et al.*, 1999; Scheidler *et al.*, 1996). *PAX3-FKHR* is also more effective than *PAX3* in inhibiting myogenic differentiation of C2C12 myoblasts or MyoD-expressing 10T1/2 cells (Epstein *et al.*, 1995). Finally, experiments in which ARMS cells were treated with antisense oligonucleotides demonstrate that decreased *PAX3-FKHR* expression results in cell death, most likely due to apoptosis (Bernasconi *et al.*, 1996). Therefore, these studies indicate that the PAX3-FKHR protein impacts on cellular growth, differentiation, and apoptosis, and thus may exert an oncogenic effect through multiple pathways that exaggerate the normal role of the wild-type PAX3 protein.

3. 11p15 allelic loss in ERMS

Although no consistent chromosomal alterations have been associated with ERMS, molecular analyses of polymorphic loci have revealed frequent allelic loss on chromosome 11 (Koufos *et al.*, 1985). In these analyses, Southern blot and PCR methodology were used to screen the patient's normal cells for loci in which different alleles could be distinguished as distinct electrophoretic signals. Allelic loss is manifested by the absence of one of the two signals in the patient's tumor cells, and indicates a genetic event such as chromosome loss, deletion, or mitotic recombination that eliminates one allele and the surrounding chromosome region. The smallest region of consistent allelic loss in ERMS cases was localized to chromosomal region 11p15.5 (*Figure 3*) (Scrable *et al.*, 1987). The finding of allelic loss of chromosome 11p15.5 loci in most ERMS cases provides a consistent genetic alteration for this subtype of RMS. Though initial reports indicated that loss of heterozygosity on 11p was not observed in ARMS cases (Scrable *et al.*, 1989b), subsequent studies identified some ARMS cases with 11p allelic loss (Casola *et al.*, 1997; Visser *et al.*, 1997) and thus the specificity of this alteration for the ERMS subset is unclear.

   The presence of a consistent region of allelic loss is often indicative of the presence of a tumor suppressor gene that is inactivated in the associated malig-nancy. The localization of a putative tumor suppressor gene relevant to the ERMS tumorigenesis in chromosomal region 11p15 is supported by chromosome transfer studies in the RD ERMS cell line (Loh *et al.*, 1992). In these experiments, normal human chromosomes containing wild-type copies of putative tumor suppressor genes were transferred by microcell methodology into the RD cell line that is presumed to contain mutant copies of a tumor suppressor gene. Transfer of a normal human chromosome 11 into RD cells resulted in a significant loss of proliferative capacity, suggesting that a transferred wild-type gene restored previ-ously inactivated function. In addition, the few viable hybrids obtained after chromosome 11 transfer demonstrated rearrangements that eliminated the 11p chromosomal region. The localization of the growth-suppressing locus within the

11p region was further defined by a modification of the transfer strategy in which the subchromosomal fragments were introduced into RD cells (Koi *et al.*, 1993). The finding of growth arrest by all fragments that contained the 11p15 region confirmed the presence of a tumor suppressor gene locus in the region previously demonstrated to show allelic loss in ERMS cases (*Figure 3*).

The mechanism for inactivation of tumor suppressor genes is postulated to be a two-hit scenario in which both copies of the gene are sequentially inactivated. In this situation, a small mutation such as a point change often inactivates one of the two alleles, and the allelic loss event inactivates the second allele. Allelic loss studies of ERMS and Wilms tumor provided evidence of a different potential strategy for accomplishing the first inactivation event. Comparison of the allelic loss pattern in ERMS tumors to the allelic status of the patients' parents revealed that the ERMS tumors preferentially maintain the paternally inherited allele and lose the maternal allcle (Scrable *et al.*, 1989a). The paternal allele preference suggests the involvement of genomic imprinting, a normal epigenctic developmental process that selectively inactivates expression of alleles in a gamete-of-origin-dependent process. Studies of

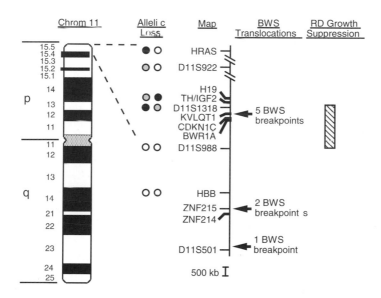

Figure 3. Mapping of the putative tumor suppressor locus in chromosomal regions 11p15.5. In the first column, a low resolution banding pattern of chromosome 11 is shown, with addition of the five sub-regions in band 11p15. In the second column, the status of these loci in two informative ERMS samples is shown such that filled circles indicate allelic loss, open circles indicate maintenance of heterozygosity, and shaded circles indicate non-informative loci (Besnard-Guerin *et al.*, 1996; Visser *et al.*, 1997). The third column shows a map of the 11p15 region derived from linkage mapping (Fain *et al.*, 1996) and physical mapping studies (Hoovers *et al.*, 1995; Hu *et al.*, 1997; Schwienbacher *et al.*, 1998; Steenman *et al.*, 2000). The fourth column presents the positions of translocation breakpoints in BWS patients (Hoovers *et al.*, 1995; Lee *et al.*, 1997; Steenman *et al.*, 2000). The fifth column delineates the position of a small subchromosomal fragment (74-1-6) that suppresses growth of RD ERMS cells (Hu *et al.*, 1997; Koi *et al.*, 1993).

the human 11p15 chromosomal region and the corresponding mouse region demonstrated the imprinting of several genes within the region; for example, *H19* is preferentially expressed from the maternally inherited alleles and *IGF2* is imprinted in the opposite direction so that the paternally inherited alleles are preferentially expressed (Tycko, 1994). Therefore, results from these allelic loss studies suggest that ERMS tumorigenesis frequently involves inactivation of an imprinted tumor suppressor by allelic loss of the active maternal allele and retention of the inactive paternal allele.

Though smaller mutations have not been generally detected in ERMS cases to implicate a specific gene, several of the genes within the 11p15.5 imprinted region represent interesting candidates (*Figure 3*). The H19 gene product is an RNA molecule that is widely expressed during fetal development in association with cell differentiation but is not apparently translated into a protein product (Tycko, 1994). A functional role is suggested by gene transfer experiments in which transfection of *H19* expression constructs into RD ERMS cells results in growth suppression (Hao *et al.*, 1993). The *CDKN1C* (*p57/KIP2*) gene is expressed during development in several tissues, including skeletal muscle, and encodes a cyclin-dependent kinase inhibitor that negatively regulates cell-cycle progression by binding to and inhibiting the activity of several G_1 cyclin–cyclin dependent kinase complexes (Matsuoka *et al.*, 1995). Both *H19* and *CDKN1C* are preferentially expressed from the maternally inherited allele (Matsuoka *et al.*, 1996; Tycko, 1994). However, *H19* is not situated within several of the small 11p15 subchromosomal fragments that suppress tumor growth (Koi *et al.*, 1993), although *CDKN1C* is contained within the region defined by the smallest 11p15 subchromosomal fragments (Hoovers *et al.*, 1995). The *BWR1A* gene in the 11p15.5 locus has also been identified as a possible candidate in RMS tumorigenesis. *BWR1A* is most highly expressed in the liver, heart, and kidney, and encodes a protein with strong homology to tetracycline resistance efflux proteins. A mutational analysis of 7 RMS cell lines detected a *BWR1A* point mutation in the TE125-T RMS cell line (Schwienbacher *et al.*, 1998).

4. Other genetic changes in RMS

Comparative genomic hybridization (CGH) studies have provided insight into chromosomal changes in RMS, highlighting genetic differences between ARMS and ERMS. The first such study analyzed 10 ERMS cases, identifying whole chromosome gains of 2, 13, 12, 8, 7, 17, 18, and 19 and losses of chromosomes 16, 10, 15, and 14 (Weber-Hall *et al.*, 1996). Only 1 ERMS was shown to have genomic amplification at 12q13–15. A later study of 12 ERMS cases from 10 patients utilizing CGH and fluorescent *in situ* hybridization (FISH) confirmed the finding of gains in chromosomes 2, 7, 8, 12, and 13, and also identified gains in 11 and 20 (Bridge *et al.*, 2000). Rather than gains in chromosome 17, the second study found chromosome 17 loss in 25% of the tumors. The latter study also found losses of chromosomal regions 9q22 and 1p35–36 in addition to the previously reported losses on chromosomes 6 and 14.

In contrast to ERMS, an initial CGH study of ARMS identified a variety of amplicons, but no consistent chromosome gains or losses. In 14 ARMS cases (10 tumors and four cell lines), amplicons were found at 12q13–15 (50%) and 2p24

(36%), and less frequently, 13q14 and 1p36 (corresponding to the PAX7-FKHR translocation), 13q31, 1q21, and 8q13–21 (Weber-Hall *et al.*, 1996). A more comprehensive CGH study of 36 ARMS cases (34 tumors and two cell lines) found regions of chromosomal loss at 16q (16%), 9q32–34 (10%), 13q14-qter (10%), and 17p (10%) among others (Gordon *et al.*, 2000). Amplification was most commonly detected at 12q13–15 (32%) and 2p24 (28%), 2q34-qter, 1p36, 15q24–26 and less frequently at previously described loci. The findings of 13q31 amplification in 19% of cases and whole chromosome 13 gain in 10% of cases suggest that increased copy number of 13q31 loci is a recurrent genetic alteration in ARMS.

The most frequent genomic amplification events in these ARMS cases involve chromosomal regions 12q13–15 and 2p24. The 12q13–15 region contains many growth-related genes encoding products such as the alpha-2 macroglobulin receptor, the zinc-finger transcription factor *GLI*, the transcription factor *CHOP*, the membrane protein SAS, the cell-cycle regulator *CDK4*, and the p53 inactivator *MDM2*. Studies demonstrated amplification of *CDK4, GLI, APR, CHOP,* and *SAS*, but not *MDM2* in the ARMS cell line RH30 (Berner *et al.*, 1996; Forus *et al.*, 1993). Chromosomal region 2p24 is well known as the site of the *MYCN* oncogene, which is amplified in a number of tumor categories, such as neuroblastoma. Similar to the CGH results, two studies utilizing Southern blot analysis found *MYCN* amplification in ARMS, but not ERMS tumors. The first study found 4 of 6 ARMS and 0 of 7 ERMS cases to have *MYCN* amplification (Dias *et al.*, 1990). The second study revealed *MYCN* amplification in 3 of 7 ARMS and 0 of 6 ERMS cases (Driman *et al.*, 1994). Neither study found *MYCN* amplification to be associated with a significantly worse clinical outcome. In a later study, FISH analysis detected *MYCN* amplification in 9 of 15 ARMS and 0 of 14 ERMS cases (Hachitanda *et al.*, 1998). The survival rate among non-amplified ARMS cases was 66% (4/6), compared to 11% (1/9) for amplified ARMS cases, suggesting that *MYCN* amplification is associated with a significantly worse survival rate.

Finally, a genome-wide screen of 33 RMS cases (27 ERMS and 6 ARMS) for loss of heterozygosity revealed several regions of allelic loss in addition to the expected 11p15 region (Visser *et al.*, 1996, 1997). Chromosomal arms showing allelic loss in at least 25% of cases were 6p, 11p, 11q, 14q, 16q, and 18p. The overall frequency of allelic loss was lower in the ARMS tumors than the ERMS tumors, supporting the view that different molecular mechanisms are involved in the development of these two RMS subtypes. Other than chromosome 11, the most frequent region of allelic loss was in 16q. Analyzing paraffin-embedded tumors, 16q loss of heterozygosity was found in 11 of 22 ERMS cases and 2 of 5 ARMS cases. These findings indicate that allelic loss on 16q occurs in a significant fraction of both ARMS and ERMS cases and thus is not specific for the ERMS category.

5. Altered pathways in sporadic RMS and inherited cancer predisposition syndromes

5.1 TP53 *tumor suppressor gene*

The *TP53* tumor suppressor gene located on chromosome 17p13.1 is the most frequently altered gene in human tumors (Hollstein *et al.*, 1991). Somatic mutations of *TP53* and dysregulation of its associated regulatory proteins have been implicated in the development of a variety of tumors, including RMS. In addition, germline mutations of TP53 are the underlying etiology of Li-Fraumeni syndrome (LFS) (Malkin *et al.*, 1990).

LFS, first described in 1969, is the familial cancer syndrome most closely linked with RMS (*Table 1*). From an initial series of 648 RMS patients, Li and Fraumeni identified four families having either sibling or cousin pairs with RMS (Li and Fraumeni, 1969). An unusually high incidence of early-onset breast cancer, sarcomas, and acute leukemias was found among family members and close relatives. Other studies confirmed the association of RMS and LFS (Birch *et al.*, 1990; Strong *et al.*, 1987). *TP53* sequence analysis in five classic LFS families led to the identification of germline mutations in LFS (Malkin *et al.*, 1990). Subsequent studies identified such mutations, most commonly point mutations in exons 5–8, in as many as 71% of classic LFS families (Varley *et al.*, 1997).

Germline *TP53* mutations have also been found in RMS patients without family histories suggestive of cancer predisposition (Diller *et al.*, 1995). Using single-strand conformation polymorphism, 3 of 33 sporadic RMS patients were found to have germline mutations. Interestingly, the study found mutations exclusively in younger patients, less than 3 years of age (3/13). In addition to germline mutations, somatic *TP53* alterations have been identified in sporadic cases of RMS. An analysis of 31 RMS cases identified 14 tumors with p53 alterations, including homozygous or hemizygous gene deletions, 17q loss of heterozygosity, or undetectable p53 mRNA or protein (Mulligan *et al.*, 1990). A smaller study found homozygous or hemizygous p53 mutations in 4 of 6 ERMS and 1 of 4 ARMS tumors or cell lines (Felix *et al.*, 1992). Other reports have shown p53 point mutations in 2 of 6 (Stratton *et al.*, 1990) and 2 of 3 (Mousses *et al.*, 1996) RMS cases analyzed. Immunohistochemical nuclear staining has been used to detect p53 accumulation in 3 of 5 ERMS and 2 of 2 ARMS cases, indirectly supporting the presence of p53 mutations (Kawai *et al.*, 1994).

Table 1. Familial cancer predisposition associated with rhabdomyosarcoma

Cancer syndrome	Locus	Gene	Characteristic malignancies	Number of RMS patients
Li-Fraumeni syndrome	17p13.1	*TP53*	Breast CA, CNS tumors, sarcomas	1–10% of RMS patients
Neurofibromatosis Type-1	17q11.2	*NFI*	Gliomas, MPNST	1–6% of RMS patients
Beckwith-Wiedemann syndrome	11p15.5	unknown	Wilms, hepatoblastoma	4 cases
Hereditary retinoblastoma	13q14	*RBI*	Retinoblastoma, osteosarcoma	5 cases
Nevoid basal cell carcinoma syndrome	9q22	*PTCH*	BCC, medulloblastoma	3 cases
Rubinstein-Taybi syndrome	16p13.3	*CREBBP*	Leukemia, CNS tumors	3 cases
Costello syndrome	unknown	unknown	RMS	4 cases

Loss of p53 tumor suppressor function in sporadic RMS may also result from overexpression of *MDM2*. Located on chromosome 12q, *MDM2* encodes a protein capable of binding and inactivating p53. *MDM2*, along with other genes at the 12q13–15 locus, may be amplified in ARMS. Using Northern blot analysis, 6 RMS-derived cell lines (3 ERMS and 3 ARMS) without *TP53* mutations, were assayed for *MDM2* gene expression (Keleti *et al.*, 1996). *MDM2* overexpression was found in 3 of 3 ARMS and 0 of 3 ERMS cell lines. Likewise, elevated *MDM2* nuclear protein was demonstrated by immunohistochemistry in each of the ARMS cell lines. The highest *MDM2* expressing cell line, RH18, was also found to have *MDM2* genomic amplification. Another immunohistochemistry study found MDM2 nuclear staining in 4 of 11 RMS cases, again suggesting a role for MDM2 in RMS tumorigenesis (Wurl *et al.*, 1996).

5.2 Neurofibromatosis type-I (NF-1) and RAS gene family

NF-1 (see Chapter 5) is a common autosomal dominant disorder with a prevalence of 1/2000 to 1/5000 (Rasmussen and Friedman, 2000). Characteristic findings include café au lait spots, axillary freckling, neurofibromas, Lisch nodules, and developmental and learning deficits. The *NF1* gene is located on chromosome 17q11.2 and codes for neurofibromin, a GTPase activating protein and negative regulator of the RAS oncogene family (Basu *et al.*, 1992; Xu *et al.*, 1990). *NF-1* is closely associated with the development of a variety of benign tumors such as peripheral neurofibromas, plexiform neurofibromas, optic tract gliomas, and pheochromocytomas. Malignancies linked to NF-1 include malignant peripheral nerve sheath tumors, RMS, and acute myelogenous leukemia (Plon and Peterson, 1997).

Large patient studies have consistently supported a link between NF-1 and RMS. The first study to review a large series of patients found 5 of 84 RMS patients with NF-1 (McKeen *et al.*, 1978). When including patients cited in the literature, the paper identified a total of 7 patients with RMS and NF-1. Four of the patients were children and three were adults. The histology of the RMS tumors was described as either ERMS or pleomorphic. A larger study found NF-1 in 5 of 249 RMS patients (2%) (Yang *et al.*, 1995). All RMS tumors were of ERMS histology. Similarly a Japanese study found 6 of 590 (1.4%) RMS patients with NF-1 in the Japanese cancer registry. Tumor histology was not specified (Matsui *et al.*, 1993). The latter two reports also noted that many of these RMS tumors in NF-1 patients occur at urogenital sites.

RAS genes, which are negatively regulated by neurofibromin, are implicated in the development of numerous malignancies (Yamamoto *et al.*, 1999). There is evidence for RAS gene family alterations in sporadic RMS. In two studies that examined a total of 17 ERMS cases, mutations of RAS genes were detected in six tumors: four in *KRAS2*, one in *NRAS*, and one in *HRAS* (Stratton *et al.*, 1989; Wilke *et al.*, 1993). No mutations were found in three ARMS cases evaluated in these studies. In addition to these tumor studies, an activating mutation in *NRAS* was detected in the RD ERMS cell line (Chardin *et al.*, 1985).

5.3 Beckwith-Wiedemann syndrome (BWS) and the 11p15.5 locus

BWS is a heterogeneous overgrowth syndrome that occurs at a frequency of 1:13 700 (Thorburn *et al.*, 1970). The syndrome is characterized by hemihypertrophy, visceromegaly, omphalocele, and macroglossia. BWS patients are at an increased risk of developing certain malignancies (7.5 to 12.5%), such as Wilms tumor, hepatoblastoma, and adrenocortical carcinoma and less commonly, neuroblastoma and RMS (Wiedemann *et al.*, 1983). The molecular basis for BWS is dysregulation or alteration of one or more imprinted genes at the 11p15 locus, which includes the growth-related gene *IGF2* and tumor suppressors *H19* and *CDKN1C* (p57/KIP2) (Li *et al.*, 1998). Allelic loss involving a similar genomic region has also been identified in other tumors associated with BWS, including Wilms tumor and hepatoblastoma (Koufos *et al.*, 1985).

There are actually few reported cases of RMS associated with BWS. Past reports include a case of an orbital RMS and two cases of ERMS arising from the bladder in young children (Aideyan and Kao, 1998; Sotelo-Avila *et al.*, 1980; Vaughan *et al.*, 1995). Another patient with BWS developed an abdominal ERMS at 22 months of age (Matsumato, 1994). In two other patients, ERMS was found in the setting of isolated hemihypertrophy, which is associated with an increased risk of malignancy and may represent a mild phenotype of BWS (Ruymann *et al.*, 1988; Samuel *et al.*, 1999).

The genetic basis of BWS is proposed to be alterations of the 11p15.5 chromosomal region, which is also the presumed locus of a tumor suppressor gene altered in a majority of ERMS cases *(Figure 3)*. Linkage analysis of families transmitting BWS localized the mutant locus to this chromosomal region (Junien, 1992; Tycko, 1994). Constitutional abnormalities of this region have also been revealed by karyotypic studies of several BWS cases, including duplications of 11p leading to extra copies of the paternally inherited region and apparently balanced translocations affecting the maternally inherited 11p15 region. In addition, polymorphic marker assays identified sporadic BWS cases without maternal contribution to the 11p region, an example of uniparental isodisomy. These chromosomal changes result in overrepresentation of the paternal alleles or disruption of the maternal alleles, and support the importance of gene imprinting in the etiology of this disease. A specific disorder in imprinting is further indicated by the finding of bialleleic expression of the normally imprinted *IGF2* gene in patients with BWS (Weksberg *et al.*, 1993). In addition to these epigenetic alterations of *IGF2*, several studies identified genetic alterations of other 11p15.5 loci: point mutations of *CDKN1C* (Hatada *et al.*, 1996), rearrangement of *KVLQT1* (Lee *et al.*, 1997), and rearrangement of more centromeric loci (Hoovers *et al.*, 1995). Clearly, BWS is a complex disease with several possible genetic alterations in the 11p15.5 chromosomal region. Some alterations lead to overexpression of the IGF2 fetal growth factor, while others serve to mutate or inactivate expression of growth suppressive genes such as *H19* and *CDKN1C*. However, the specific ERMS genetic deficits have not been identified, and whether BWS and ERMS share common genetic mutations is yet to be proven.

5.4 Hereditary retinoblastoma and the RB1 pathway

Bilateral retinoblastoma is associated with a germline mutation in the *RB1* tumor suppressor gene (see Chapter 13), which is situated in chromosomal region 13q14. Retinal tumors develop in hereditary retinoblastoma patients following a somatic mutation in the second *RB1* allele, in accord with the 'two-hit' hypothesis put forth by Knudson (Knudson, 1971). Under the control of cyclin-dependent kinase phosphorylation, the RB protein negatively regulates the G_1–S transition via binding and repression of the E2F family of transcription factors (Nevins *et al.*, 1997). Alterations in the *RB1* gene and the RB pathway are associated with the development of a variety of tumors. In addition to retinoblastoma, hereditary retinoblastoma patients are also at risk of developing other primary and treatment related cancers, typically osteosarcomas and other sarcomas (Donaldson *et al.*, 1997), including RMS. There are three cited cases of RMS in non-irradiated hereditary retinoblastoma patients (Abramson *et al.*, 1979; Levene, 1960; Meadows *et al.*, 1977). Two other cases of RMS in hereditary retinoblastoma patients occurred after radiation therapy (Meadows *et al.*, 1977).

Various sarcomas were found to have *RB1* mutations (Cance *et al.*, 1990; Weichselbaum *et al.*, 1988; Wunder *et al.*, 1991); however, no such mutations were identified in RMS. One study, utilizing Southern blot analysis, found no homozygous *RB1* deletions in three RMS cell lines (Friend *et al.*, 1987). Another study detected normal *RB1* expression by Northern blot analysis and protein immunoprecipitation in 11 human RMS cell lines, five ERMS tumors, and 13 ARMS tumors (De Chiara *et al.*, 1993). While *RB1* mutations were not detected in RMS, genetic changes were found in *CDKN2A* and *CDKN2B*, which code for *p16INK4A* and *p15INK4B*, respectively, inhibitors of cyclin-dependent kinases CDK4 and CDK6. CDK proteins inhibit RB1 function by phosphorylation, and thereby promote cell-cycle progression. Utilizing PCR and single strand conformation polymorphism analysis, homozygous deletions of *CDKN2A* and the neighboring *CDKN2B* were found in 5 of 5 RMS cell lines (2 ARMS and 3 ERMS) and 3 of 12 RMS tumors (1 ARMS and 2 ERMS) (Iolascon *et al.*, 1996). A follow-up to this study transfected *CDKN2A* into RD, a *p16INK4A*-deficient ERMS cell line. Expression of *CDKN2A* led to reduced CDK6 kinase activity, induced G_1 growth arrest, and was associated with the acquisition of a more differentiated morphology (Urashima *et al.*, 1999). Similarly, another study found CDK4 activity to be elevated in the RD cell line, contributing to the cells' inability to growth arrest in differentiating medium (Knudsen *et al.*, 1998).

5.5 Nevoid basal cell carcinoma syndrome (NBCCS) and the sonic hedgehog pathway

NBCCS (Gorlin syndrome) is a rare (incidence of 1: 56 000) autosomal dominant disorder associated with a predisposition for multiple basal cell carcinomas, medulloblastoma, ovarian fibrosarcoma, and RMS. A variety of developmental abnormalities are found, including enlarged occipitofrontal circumference, ocular hypertelorism, multiple odontogenic keratocysts, and rib anomalies (Gorlin, 1987). NBCCS results from a germline mutation in *PTCH*, a tumor suppressor gene in chromosomal region 9q22 (Johnson *et al.*, 1996). *PTCH* is a negative regulator of the sonic hedgehog signaling pathway, which plays an integral role in nervous system development and anterior–posterior limb patterning (Riddle *et al.*, 1993; Roessler *et al.*, 1996). Binding of sonic hedgehog to PTCH allows activation of the pathway, leading to up-regulation of the downstream transcription factor *GLI1*, located on the long arm of chromosome 12. Alterations in *PTCH* can lead to constitutive activation of sonic hedgehog signaling.

Several cases of RMS in the setting of NBCCS have been reported. One account detailed a 4-year-old female who developed a nasopharyngeal RMS (Beddis *et al.*, 1983). The patient's therapy was complicated by severe radiation toxicity. A case suggestive of NBCCS involved a 42 year-old male with a distant history of periorbital RMS for which he received radiation therapy. The patient later developed multiple basal cell carcinomas and a microcystic adnexal carcinoma within the radiation field (Antley and Smoller, 1999). A third case of RMS and NBCCS has also been reported (Schweisguth and Lemerle, 1968).

5.6 Rubinstein-Taybi syndrome (RTS) and chromosome 16 losses

RTS is a well-described human malformation syndrome characterized by mental retardation, facial anomalies, and broad thumbs (Rubinstein and Taybi, 1963). RTS has been linked to a mutation of the *CREBBP* gene on chromosome 16p13.3 (Imaizumi and Kuroki, 1991; Petrij *et al.*, 1995; Tommerup and van der Heiberg, 1991). Of 724 documented RTS cases, there are 17 malignant and 19 benign tumors reported (Miller and Rubinstein, 1995) including three nasopharyngeal RMS (Ruymann *et al.*, 1988; Siraganian *et al.*, 1989; Sobel and Woerner, 1981). The previously described CGH studies demonstrated chromosome 16 loss in 40% of ERMS, which suggests a possible genetic link between sporadic RMS and RTS (Weber-Hall *et al.*, 1996).

5.7 Costello syndrome

Since 1998, four cases of RMS have been reported in patients with Costello syndrome, which is characterized by failure to thrive, redundant and hyperpigmented skin, coarse facies, papillomata, and cardiomyopathy (Philip and Sigaudy, 1998). Three of the patients, aged 28 months to 6 years, were diagnosed with ERMS (Kerr *et al.*, 1998; Sigaudy *et al.*, 2000). The fourth patient was diagnosed with an extremity ARMS at 6 months of age (Feingold, 1999). The molecular basis for the disease is not yet known, although one Costello patient was found to have a balanced 1q;22q chromosomal translocation (Czeizel and Timar, 1995). Two sets of siblings and two cases of consanguinity have been documented, suggesting an autosomal recessive mode of inheritance. However, segregation analysis of 28 cases supported autosomal dominant heredity (Lurie, 1994).

5.8 Single case reports of RMS

Other genetic syndromes with isolated reports of RMS include Klinefelter syndrome (Ogur *et al.*, 1996), Duchenne muscular dystrophy (DMD) (Rossbach *et al.*, 1999), cardio-facio-cutaneous syndrome (CFCS) (Bisogno *et al.*, 1999), and Noonan syndrome (Khan *et al.*, 1995). Both Noonan syndrome and CFCS share some features of Costello syndrome. While Noonan syndrome has been linked to an increased risk of other malignancies, Klinefelter, CFCS, and DMD are not thought to be cancer predisposition syndromes.

6.   RMS mouse models

A variety of mouse models of RMS have been developed over the last decade (*Table 2*). Two of these are based on inherited cancer predisposition syndromes. It is anticipated that such mice will lend insight into the genetic changes underlying the development of RMS. These models may also provide tools for trials of novel therapeutic agents.

6.1 Patched knockout mouse

Two mouse models for NBCCS with engineered mutations in the ptch locus have been described. Homozygous *ptch* mutants are embryonic lethal as a result of profound neurodevelopmental defects. *Ptch* heterozygous mice are viable with only subtle developmental abnormalities, similar to human NBCCS patients. The first described *ptch* +/− mouse was generated by replacement of exons 1 and 2 with a LacZ/neomycin cassette (Goodrich *et al.*, 1997). Of 227 heterozygous mice on a mixed genetic background B6D2F1, 8 autopsied mice were found to have posterior fossa medulloblastomas, and one had a chest wall tumor arising from rib muscle. In the same study, 2 of 27 *ptch* +/− mice bred on a 129SV genetic background developed uncharacterized soft tissue tumors.

In the second *ptch* knockout mouse, exons 6 and 7 were replaced with a neomycin cassette (Hahn *et al.*, 1998). This genetic alteration was subsequently studied in CD1 and C57B1/6 genetic backgrounds. In the CD1 strain, 10 of 117 (9%) heterozygotes developed RMS, in comparison with the RMS incidence of 2% (1/53) in the C57B1/6 heterozygotes. In these heterozygotes, the RMS tumors were commonly associated with skeletal muscle in the rear thigh, lumbar region, or abdominal wall. The tumors were found to have upregulation of *gli1* expression, indicating activation of the SHH signaling pathway. The higher incidence of RMS in the latter model most likely is a function of the CD1 background. The two separate regions of the *ptch* gene targeted for inactivation, potentially leading to varying degrees of ptch function, must also be considered.

6.2 p53 knockout mouse

Mouse models of LFS, with homozygous and heterozygous inactivation of the *trp53* tumor suppressor gene, have been noted to develop a variety of tumors, including RMS. Unlike ptch homozygous knockouts, *p53* null mice are viable, with no obvious fetal defects. *p53* heterozygotes develop fewer tumors but theoretically better mimic the *p53* function and cancer susceptibility of LFS. The first such transgenic mouse was generated by introduction of mutant *trp53* genomic fragments from a murine Friend erythroleukemia cell line into the fertilized eggs of CD1 mice (Lavigueur *et al.*, 1989). While the mice retained the endogenous wild-type *trp53* genes, only mutant *p53* was detectable by Western blot analysis. Of 112 transgenic mice, 22 (20%) developed 27 tumors, including 2 RMS at 11.5 and 15 months of age. More frequent tumors included lung adenocarcinoma (10), osteosarcoma (8), and lymphoma (8). Another model, bred on the C57BL/6 and 129/SV genetic backgrounds, was generated by replacing exon 5

Table 2. Mouse models of RMS

Mouse model	Genetic alteration	Genetic background	Predominant tumor	RMS incidence	
				% mice	% tumors
NBCCS (Hahn *et al.*, 1998)	*ptch +/−* deletion of exons 6–7	CDI	RMS	10/117 (9%)	10/10 (100%)
Li-Fraumeni (Lavigueur *et al.*, 1989)	Mutant *p53* transgenic	C57BL/6	Lung CA	2/112 (2%)	2/27 (7%)
Li-Fraumeni (Jacks *et al.*, 1994)	*p53 +/−* deletion of exon 2 to intron 6	C57BL/6	Sarcomas	1/232 (0.4%)	1/44 (2%)
	p53 −/−	C57BL/6	Lymphoma	3/70 (4%)	3/56 (5%)
HGF/SF (Takayama *et al.*, 1997)	HGF/SF transgenic	FVB/N	Mammary gland CA	3/69 (4%)	N/A

with a neomycin cassette (Donehower *et al.*, 1992). In the homozygotes, 26 of 35 (74%) mice eventually acquired tumors, mostly lymphomas. Three undifferentiated sarcomas were documented. Only 2 of 96 heterozygotes from the same study formed neoplasms. A third model was generated by replacing the region extending from exon 2 to intron 6 with a neomycin resistance gene (Jacks *et al.*, 1994). Bred on the C57BL/6 background, 44 of 70 (63%) of the *p53* –/– mice developed lymphomas. The overall tumor frequency in the null mice was 80%. In addition, 44 of the 232 (19%) heterozygotes developed tumors, most commonly sarcomas. Of p53 heterozygotes, 1 of 232 developed RMS, while 3 of 70 null mice developed RMS.

6.3 Hepatocyte growth factor/scatter factor (HGF/SF) mouse

HGF/SF is a secreted protein expressed by mesenchymally derived tissues and involved in the activation of the tyrosine kinase receptor encoded by *MET*. An autocrine loop involving co-expression of HGF/SF and *MET* has been implicated in a variety of tumors. A transgenic mouse overexpressing HGF/SF on a FVB/N inbred background was developed by expressing the HGF/SF cDNA from the mouse metallothionein 1 (MT1) promoter (Takayama *et al.*, 1997). The resulting transgenic mouse developed a variety of cancers, including mammary gland tumors, melanomas, and RMS. In particular, 3 of 42 (7%) male transgenics developed RMS. In most tumors, transgenic HGF/SF and endogenous *c-met* mRNA were expressed. In addition, the met protein was associated with phosphorylated tyrosine complexes in most tumors.

7. Conclusions

The characteristic 2;13 and 1;13 chromosomal translocations are the most consistent, likely defining, genetic alterations in the development of ARMS. The resulting PAX3/PAX7-FKHR fusion proteins are potent oncogenic transcriptional activators of target genes with PAX3/PAX7 binding sites. In contrast, the most frequent genetic change in ERMS is 11p15.5 allelic loss, likely representing loss of function of a putative tumor suppressor gene at that locus. Numerous secondary genetic alterations have also been found in RMS, including loss of function of tumor suppressor genes and activation of oncogenes.

There is now a substantial body of reports of RMS developing in the setting of familial cancer syndromes. Often, the same dysregulated genes and pathways underlying cancer predisposition syndromes are found to be altered in sporadic RMS. Most cases of syndrome-associated RMS are ERMS, though prior to the advent of molecular testing, the ARMS histology may have been underrecognized. It is anticipated that continued reporting of such cases will better characterize the subsets of RMS developing in this setting. An improved molecular understanding of cancer syndromes will also shed further light on the genetic changes in sporadic RMS. Furthermore, these disorders may be translated into new mouse models that will further expand the understanding of RMS tumorigenesis.

References

Abramson, D.H., Ronner, H.J. and Ellsworth, R.M. (1979) Second tumors in nonirradiated bilateral retinoblastoma. *Am. J. Ophthalmol.* **87**: 624–627.

Aideyan, U.O. and Kao, S.C. (1998) Case report: Urinary bladder rhabdomyosarcoma associated with Beckwith-Wiedemann syndrome. *Clin. Radiol.* **53**: 457–459.

Antley, C.C. and Smoller, B.R. (1999) Microcystic adnexal carcinoma arising in the setting of previous radiation therapy. *J. Cut. Pathol.* **26**: 48–50.

Barr, F.G. (1997) Molecular genetics and pathogenesis of rhabdomyosarcoma. *J. Pediatr. Hematol. Oncol.* **19**: 483–491.

Barr, F.G., Galili, N., Holick, J., Biegel, J.A., Rovera, G. and Emanuel, B.S. (1993) Rearrangement of the PAX3 paired box gene in the paediatric solid tumour alveolar rhabdomyosarcoma. *Nat. Genet.* **3**: 113–117.

Barr, F.G., Nauta, L.E. and Hollows, J.C. (1998) Structural analysis of PAX3 genomic rearrangements in alveolar rhabdomyosarcoma. *Cancer Genet. Cytogenet.* **102**: 32–39.

Basu, T.N., Gutmann, D.H., Fletcher, J.A., Glover, T.W., Collins, F.S. and Downward, J. (1992) Aberrant regulation of ras proteins in malignant tumour cells from type 1 neurofibromatosis patients. *Nature* **356**: 713–715.

Beddis, I.R., Mott, M.G. and Bullimore, J. (1983) Case report: nasopharyngeal rhabdomyosarcoma and Gorlin's naevoid basal cell carcinoma syndrome. *Med. Pediatr. Oncol.* **11**: 178–179.

Bennicelli, J.L., Edwards, R.H. and Barr, F.G. (1996) Mechanism for transcriptional gain of function resulting from chromosomal translocation in alveolar rhabdomyosarcoma. *Proc. Natl Acad. Sci. USA* **93**: 5455–5459.

Bennicelli, J.L., Advani, S., Schafer, B.W. and Barr, F.G. (1999) PAX3 and PAX7 exhibit conserved cis-acting transcription repression domains and utilize a common gain of function mechanism in alveolar rhabdomyosarcoma. *Oncogene* **18**: 4348–4356.

Bernasconi, M., Remppis, A., Fredericks, W.J., Rauscher, F.J., 3rd and Schafer, B.W. (1996) Induction of apoptosis in rhabdomyosarcoma cells through down-regulation of PAX proteins. *Proc. Natl Acad. Sci. USA* **93**: 13164–13169.

Berner, J.M., Forus, A., Elkahloun, A., Meltzer, P.S., Fodstad, O. and Myklebost, O. (1996) Separate amplified regions encompassing CDK4 and MDM2 in human sarcomas. *Genes Chromosomes Cancer* **17**: 254–259.

Besnard-Guerin, C., Newsham, I., Winqvist, R. and Cavenee, W.K. (1996) A common region of loss of heterozygosity in Wilms' tumor and embryonal rhabdomyosarcoma distal to the D11S988 locus on chromosome 11p15.5. *Hum. Genet.* **97**: 163–170.

Birch, J.M., Hartley, A.L., Blair, V., Kelsey, A.M., Harris, M., Teare, M.D. and Jones, P.H. (1990) Cancer in the families of children with soft tissue sarcoma. *Cancer* **66**: 2239–2248.

Bisogno, G., Murgia, A., Mammi, I., Strafella, M.S. and Carli, M. (1999) Rhabdomyosarcoma in a patient with cardio-facio-cutaneous syndrome. *J. Pediatr. Hematol. Oncol.* **21**: 424–427.

Bridge, J.A., Liu, J., Weibolt, V. et al. (2000) Novel genomic imbalances in embryonal rhabdomyosarcoma revealed by comparative genomic hybridization and fluorescence in situ hybridization: an intergroup rhabdomyosarcoma study. *Genes Chromosomes Cancer* **27**: 337–344.

Cance, W.G., Brennan, M.F., Dudas, M.E., Huang, C.M. and Cordon-Cardo, C. (1990) Altered expression of the retinoblastoma gene product in human sarcomas. *N. Engl. J. Med.* **323**: 1457–1462.

Casola, S., Pedone, P.V., Cavazzana, A.O., Basso, G., Luksch, R., d'Amore, E.S., Carli, M., Bruni, C.B. and Riccio, A. (1997) Expression and parental imprinting of the H19 gene in human rhabdomyosarcoma. *Oncogene* **14**: 1503–1510.

Chardin, P., Yeramian, P., Madaule, P. and Tavitian, A. (1985) N-ras gene activation in the RD human rhabdomyosarcoma cell line. *Int. J. Cancer* **35**: 647–652.

Czeizel, A.E. and Timar, L. (1995) Hungarian case with Costello syndrome and translocation t(1,22). *Am. J. Med. Genet.* **57**: 501–503.

Davis, R.J. and Barr, F.G. (1997) Fusion genes resulting from alternative chromosomal translocations are overexpressed by gene-specific mechanisms in alveolar rhabdomyosarcoma. *Proc. Natl Acad. Sci. USA* **94**: 8047–8051.

Davis, R.J., D'Cruz, C.M., Lovell, M.A., Biegel, J.A. and Barr, F.G. (1994) Fusion of PAX7 to FKHR by the variant t(1;13)(p36;q14) translocation in alveolar rhabdomyosarcoma. *Cancer Res.* **54**: 2869–2872.

Davis, R.J., Bennicelli, J.L., Macina, R.A., Nycum, L.M., Biegel, J.A. and Barr, F.G. (1995) Structural characterization of the FKHR gene and its rearrangement in alveolar rhabdomyosarcoma. *Hum. Mol. Genet.* 4: 2355–2362.

De Chiara, A., T'Ang, A. and Triche, T.J. (1993) Expression of the retinoblastoma susceptibility gene in childhood rhabdomyosarcomas. *J. Natl Cancer Inst.* 85: 152–157.

del Peso, L., Gonzalez, V.M., Hernandez, R., Barr, F.G. and Nunez, G. (1999) Regulation of the forkhead transcription factor FKHR, but not the PAX3-FKHR fusion protein, by the serine/threonine kinase Akt. *Oncogene* 18: 7328–7333.

Dias, P., Kumar, P., Marsden, H.B., Gattamaneni, H.R., Heighway, J. and Kumar, S. (1990) N-myc gene is amplified in alveolar rhabdomyosarcomas (RMS) but not in embryonal RMS. *Int. J. Cancer* 45: 593–596.

Diller, L., Sexsmith, E., Gottlieb, A., Li, F.P. and Malkin, D. (1995) Germline p53 mutations are frequently detected in young children with rhabdomyosarcoma. *J. Clin. Invest.* 95: 1606–1611.

Donaldson, S.S., Egbert, P.R., Newsham, I. and Cavenee, W.K. (1997) Retinoblastoma. In: *Principles and Practice of Pediatric Oncology* (eds. Pizzo, P.A. and Poplack, D.G.), pp. 699–715. Lippincott-Raven, Philadelphia.

Donehower, L.A., Harvey, M., Slagle, B.L., McArthur, M.J., Montgomery, C.A., Jr., Butel, J.S. and Bradley, A. (1992) Mice deficient for p53 are developmentally normal but susceptible to spontaneous tumours. *Nature* 356: 215–221.

Driman, D., Thorner, P.S., Greenberg, M.L., Chilton-MacNeill, S. and Squire, J. (1994) MYCN gene amplification in rhabdomyosarcoma. *Cancer* 73: 2231–2237.

Epstein, J.A., Lam, P., Jepeal, L., Maas, R.L. and Shapiro, D.N. (1995) Pax3 inhibits myogenic differentiation of cultured myoblast cells. *J. Biol. Chem.* 270: 11719–11722.

Fain, P.R., Kort, E.N., Yousry, C., James, M.R. and Litt, M. (1996) A high resolution CEPH crossover mapping panel and integrated map of chromosome 11. *Hum. Mol. Genet.* 5: 1631–1636.

Feingold, M. (1999) Costello syndrome and rhabdomyosarcoma. *J. Med. Genet.* 36: 582–583.

Felix, C.A., Kappel, C.C., Mitsudomi, T., Nau, M.M., Tsokos, M., Crouch, G.D., Nisen, P.D., Winick, N.J. and Helman, L.J. (1992) Frequency and diversity of p53 mutations in childhood rhabdomyosarcoma. *Cancer Res.* 52: 2243–2247.

Fitzgerald, J.C., Scherr, A.M. and Barr, F.G. (2000) Structural analysis of PAX7 rearrangements in alveolar rhabdomyosarcoma. *Cancer Genet. Cytogenet.* 117: 37–40.

Forus, A., Florenes, V.A., Maelandsmo, G.M., Meltzer, P.S., Fodstad, O. and Myklebost, O. (1993) Mapping of amplification units in the q13–14 region of chromosome 12 in human sarcomas: some amplica do not include MDM2. *Cell Growth Differ.* 4: 1065–1070.

Fredericks, W.J., Galili, N., Mukhopadhyay, S., Rovera, G., Bennicelli, J., Barr, F.G. and Rauscher, F.J., 3rd. (1995) The PAX3-FKHR fusion protein created by the t(2;13) translocation in alveolar rhabdomyosarcomas is a more potent transcriptional activator than PAX3. *Mol. Cell Biol.* 15: 1522–1535.

Friend, S.H., Horowitz, J.M., Gerber, M.R., Wang, X.F., Bogenmann, E., Li, F.P. and Weinberg, R.A. (1987) Deletions of a DNA sequence in retinoblastomas and mesenchymal tumors: organization of the sequence and its encoded protein. *Proc. Natl Acad. Sci. USA* 84: 9059–9063.

Galili, N., Davis, R.J., Fredericks, W.J., Mukhopadhyay, S., Rauscher, F.J.d., Emanuel, B.S., Rovera, G. and Barr, F.G. (1993) Fusion of a fork head domain gene to PAX3 in the solid tumour alveolar rhabdomyosarcoma. *Nat. Genet.* 5: 230–235.

Goodrich, L.V., Milenkovic, L., Higgins, K.M. and Scott, M.P. (1997) Altered neural cell fates and medulloblastoma in mouse patched mutants. *Science* 277: 1109–1113.

Gordon, A.T., Brinkschmidt, C., Anderson, J., Coleman, N., Dockhorn-Dworniczak, B., Pritchard-Jones, K. and Shipley, J. (2000) A novel and consistent amplicon at 13q31 associated with alveolar rhabdomyosarcoma. *Genes Chromosomes Cancer* 28: 220–226.

Gorlin, R.J. (1987) Nevoid basal-cell carcinoma syndrome. *Medicine (Baltimore)* 66: 98–113.

Hachitanda, Y., Toyoshima, S., Akazawa, K. and Tsuneyoshi, M. (1998) N-myc gene amplification in rhabdomyosarcoma detected by fluorescence in situ hybridization: its correlation with histologic features. *Mod. Pathol.* 11: 1222–1227.

Hahn, H., Wojnowski, L., Zimmer, A.M., Hall, J., Miller, G. and Zimmer, A. (1998) Rhabdomyosarcomas and radiation hypersensitivity in a mouse model of Gorlin syndrome. *Nat. Med.* 4: 619–622.

Hao, Y., Crenshaw, T., Moulton, T., Newcomb, E. and Tycko, B. (1993) Tumour-suppressor activity of H19 RNA. *Nature* 365: 764–767.

Hatada, I., Ohashi, H., Fukushima, Y. *et al.* (1996) An imprinted gene p57KIP2 is mutated in Beckwith-Wiedemann syndrome. *Nat. Genet.* **14**: 171–173.

Hollstein, M., Sidransky, D., Vogelstein, B. and Harris, C.C. (1991) p53 mutations in human cancers. *Science* **253**: 49–53.

Hoovers, J.M., Kalikin, L.M., Johnson, L.A. *et al.* (1995) Multiple genetic loci within 11p15 defined by Beckwith-Wiedemann syndrome rearrangement breakpoints and subchromosomal transferable fragments. *Proc. Natl Acad. Sci. USA* **92**: 12456–12460.

Hu, R.J., Lee, M.P., Connors, T.D., Johnson, L.A., Burn, T.C., Su, K., Landes, G.M. and Feinberg, A.P. (1997) A 2.5-Mb transcript map of a tumor-suppressing subchromosomal transferable fragment from 11p15.5, and isolation and sequence analysis of three novel genes. *Genomics* **46**: 9–17.

Imaizumi, K. and Kuroki, Y. (1991) Rubinstein-Taybi syndrome with de novo reciprocal translocation t(2;16)(p13.3;p13.3). *Am. J. Med. Genet.* **38**: 636–639.

Iolascon, A., Faienza, M.F., Coppola, B., Rosolen, A., Basso, G., Della Ragione, F. and Schettini, F. (1996) Analysis of cyclin-dependent kinase inhibitor genes (CDKN2A, CDKN2B, and CDKN2C) in childhood rhabdomyosarcoma. *Genes Chromosomes Cancer* **15**: 217–222.

Jacks, T., Remington, L., Williams, B.O., Schmitt, E.M., Halachmi, S., Bronson, R.T. and Weinberg, R.A. (1994) Tumor spectrum analysis in p53-mutant mice. *Curr. Biol.* **4**: 1–7.

Johnson, R.L., Rothman, A.L., Xie, J. *et al.* (1996) Human homolog of patched, a candidate gene for the basal cell nevus syndrome. *Science* **272**: 1668–1671.

Junien, C. (1992) Beckwith-Wiedemann syndrome, tumourigenesis and imprinting. *Curr. Opin. Genet. Dev.* **2**: 431–438.

Kawai, A., Noguchi, M., Beppu, Y., Yokoyama, R., Mukai, K., Hirohashi, S., Inoue, H. and Fukuma, H. (1994) Nuclear immunoreaction of p53 protein in soft tissue sarcomas. A possible prognostic factor. *Cancer* **73**: 2499–2505.

Keleti, J., Quezado, M.M., Abaza, M.M., Raffeld, M. and Tsokos, M. (1996) The MDM2 oncoprotein is overexpressed in rhabdomyosarcoma cell lines and stabilizes wild-type p53 protein. *Am. J. Pathol.* **149**: 143–151.

Kerr, B., Eden, O.B., Dandamudi, R., Shannon, N., Quarrell, O., Emmerson, A., Ladusans, E., Gerrard, M. and Donnai, D. (1998) Costello syndrome: two cases with embryonal rhabdomyosarcoma. *J. Med. Genet.* **35**: 1036–1039.

Khan, S., McDowell, H., Upadhyaya, M. and Fryer, A. (1995) Vaginal rhabdomyosarcoma in a patient with Noonan syndrome. *J. Med. Genet.* **32**: 743–745.

Knudsen, E.S., Pazzagli, C., Born, T.L., Bertolaet, B.L., Knudsen, K.E., Arden, K.C., Henry, R.R. and Feramisco, J.R. (1998) Elevated cyclins and cyclin-dependent kinase activity in the rhabdomyosarcoma cell line RD. *Cancer Res.* **58**. 2042–2049.

Knudson, A.J. (1971) Mutation and cancer: statistical study of retinoblastoma. *Proc. Natl Acad. Sci. USA* **68**: 820–823.

Koi, M., Johnson, L.A., Kalikin, L.M., Little, P.F., Nakamura, Y. and Feinberg, A.P. (1993) Tumor cell growth arrest caused by subchromosomal transferable DNA fragments from chromosome 11. *Science* **260**: 361–364.

Koufos, A., Hansen, M.F., Copeland, N.G., Jenkins, N.A., Lampkin, B.C. and Cavenee, W.K. (1985) Loss of heterozygosity in three embryonal tumours suggests a common pathogenetic mechanism. *Nature* **316**: 330–334.

Kramer, S., Meadows, A.T., Jarrett, P. and Evans, A.E. (1983) Incidence of childhood cancer: experience of a decade in a population-based registry. *J. Natl Cancer Inst.* **70**: 49–55.

Lam, P.Y., Sublett, J.E., Hollenbach, A.D. and Roussel, M.F. (1999) The oncogenic potential of the Pax3-FKHR fusion protein requires the Pax3 homeodomain recognition helix but not the Pax3 paired-box DNA binding domain. *Mol. Cell Biol.* **19**: 594–601.

Lavigueur, A., Maltby, V., Mock, D., Rossant, J., Pawson, T. and Bernstein, A. (1989) High incidence of lung, bone, and lymphoid tumors in transgenic mice overexpressing mutant alleles of the p53 oncogene. *Mol. Cell Biol.* **9**: 3982–3991.

Lee, M.P., Hu, R.J., Johnson, L.A. and Feinberg, A.P. (1997) Human KVLQT1 gene shows tissue-specific imprinting and encompasses Beckwith-Wiedemann syndrome chromosomal rearrangements. *Nat. Genet.* **15**: 181–185.

Levene, M. (1960) Congenital retinoblastoma and sarcoma botryoides of the vagina. *Cancer* **13**: 532–537.

Li, F.P. and Fraumeni, J.F., Jr. (1969) Rhabdomyosarcoma in children: epidemiologic study and identification of a familial cancer syndrome. *J. Natl Cancer Inst.* **43**: 1365–1373.

Li, M., Squire, J.A. and Weksberg, R. (1998) Molecular genetics of Wiedemann-Beckwith syndrome. *Am. J. Med. Genet.* **79**: 253–259.

Loh, W.E., Jr., Scrable, H.J., Livanos, E., Arboleda, M.J., Cavenee, W.K., Oshimura, M. and Weissman, B.E. (1992) Human chromosome 11 contains two different growth suppressor genes for embryonal rhabdomyosarcoma. *Proc. Natl Acad. Sci. USA* **89**: 1755–1759.

Lurie, I.W. (1994) Genetics of the Costello syndrome. *Am. J. Med. Genet.* **52**: 358–359.

Malkin, D., Li, F.P., Strong, L.C. *et al.* (1990) Germ line p53 mutations in a familial syndrome of breast cancer, sarcomas, and other neoplasms. *Science* **250**: 1233–1238.

Matsui, I., Tanimura, M., Kobayashi, N., Sawada, T., Nagahara, N. and Akatsuka, J. (1993) Neurofibromatosis type 1 and childhood cancer. *Cancer* **72**: 2746–2754.

Matsumato, T.E.-i., Maeda, K., Niikawa, H. *et al.* (1994) Molecular analysis of a patient with Beckwith-Wiedemann syndrome, rhabdomyosarcoma and renal cell carcinoma. *Jpn. J. Human Genet.* **39**: 225–234.

Matsuoka, S., Edwards, M.C., Bai, C., Parker, S., Zhang, P., Baldini, A., Harper, J.W. and Elledge, S.J. (1995) p57KIP2, a structurally distinct member of the p21CIP1 Cdk inhibitor family, is a candidate tumor suppressor gene. *Genes Dev.* **9**: 650–662.

Matsuoka, S., Thompson, J.S., Edwards, M.C., Bartletta, J.M., Grundy, P., Kalikin, L.M., Harper, J.W., Elledge, S.J. and Feinberg, A.P. (1996) Imprinting of the gene encoding a human cyclin-dependent kinase inhibitor, p57KIP2, on chromosome 11p15. *Proc. Natl Acad. Sci. USA* **93**: 3026–3030.

McKeen, E.A., Bodurtha, J., Meadows, A.T., Douglass, E.C. and Mulvihill, J.J. (1978) Rhabdomyosarcoma complicating multiple neurofibromatosis. *J. Pediatr.* **93**: 992–993.

Meadows, A.T., D'Angio, G.J., Mike, V., Banfi, A., Harris, C., Jenkin, R.D. and Schwartz, A. (1977) Patterns of second malignant neoplasms in children. *Cancer* **40**: 1903–1911.

Miller, R.W. and Rubinstein, J.H. (1995) Tumors in Rubinstein-Taybi syndrome. *Am. J. Med. Genet.* **56**: 112–115.

Mousses, S., McAuley, L., Bell, R.S., Kandel, R. and Andrulis, I.L. (1996) Molecular and immunohistochemical identification of p53 alterations in bone and soft tissue sarcomas. *Mod. Pathol.* **9**: 1–6.

Mulligan, L.M., Matlashewski, G.J., Scrable, H.J. and Cavenee, W.K. (1990) Mechanisms of p53 loss in human sarcomas. *Proc. Natl Acad. Sci. USA* **87**: 5863–5867.

Nevins, J.R., Leone, G., DeGregori, J. and Jakoi, L. (1997) Role of the Rb/E2F pathway in cell growth control. *J. Cell. Physiol.* **173**: 233–236.

Ogur, G., Sengun, Z., Arel-Kilic, G., De Busscher, C., Basaran, S., Ozbek, U., Ayan, I., Sariban, E. and Vamos, E. (1996) Clinical and cytogenetic studies of two cases of Klinefelter syndrome with hereditary retinoblastoma and rhabdomyosarcoma. *Cancer Genet. Cytogenet.* **89**: 77–81.

Petrij, F., Giles, R.H., Dauwerse, H.G. *et al.* (1995) Rubinstein-Taybi syndrome caused by mutations in the transcriptional co-activator CBP. *Nature* **376**: 348–351.

Philip, N. and Sigaudy, S. (1998) Costello syndrome. *J. Med. Genet.* **35**: 238–240.

Plon, S.E. and Peterson, L.E. (1997) Childhood cancer, heredity, and the environment. In: *Principles and Practice of Pediatric Oncology* (eds. Pizzo, P.A. and Poplack, D.G.), pp. 11–36. Lippincott-Raven, Philadelphia.

Rasmussen, S.A. and Friedman, J.M. (2000) NF1 gene and neurofibromatosis 1. *Am. J. Epidemiol.* **151**: 33–40.

Riddle, R.D., Johnson, R.L., Laufer, E. and Tabin, C. (1993) Sonic hedgehog mediates the polarizing activity of the ZPA. *Cell* **75**: 1401–1416.

Roessler, E., Belloni, E., Gaudenz, K., Jay, P., Berta, P., Scherer, S.W., Tsui, L.C. and Muenke, M. (1996) Mutations in the human Sonic Hedgehog gene cause holoprosencephaly. *Nat. Genet.* **14**: 357–360.

Rossbach, H.C., Lacson, A., Grana, N.H. and Barbosa, J.L. (1999) Duchenne muscular dystrophy and concomitant metastatic alveolar rhabdomyosarcoma. *J. Pediatr. Hematol. Oncol.* **21**: 528–530.

Rubinstein, J.H. and Taybi, H. (1963) Broad thumbs and toes and facial abnormalities. A possible mental retardation syndrome. *Am. J. Dis. Child.* **105**: 588–608.

Ruymann, F.B., Maddux, H.R., Ragab, A., Soule, E.H., Palmer, N., Beltangady, M., Gehan, E.A. and Newton, W.A., Jr. (1988) Congenital anomalies associated with rhabdomyosarcoma: an autopsy study of 115 cases. A report from the Intergroup Rhabdomyosarcoma Study Committee (representing the Children's Cancer Study Group, the Pediatric Oncology Group, the United Kingdom Children's Cancer Study Group, and the Pediatric Intergroup Statistical Center). *Med. Pediatr. Oncol.* **16**: 33–39.

Samuel, D.P., Tsokos, M. and DeBaun, M.R. (1999) Hemihypertrophy and a poorly differentiated embryonal rhabdomyosarcoma of the pelvis. *Med. Pediatr. Oncol.* **32**: 38–43.

Scheidler, S., Fredericks, W.J., Rauscher, F.J., 3rd, Barr, F.G. and Vogt, P.K. (1996) The hybrid PAX3-FKHR fusion protein of alveolar rhabdomyosarcoma transforms fibroblasts in culture. *Proc. Natl Acad. Sci. USA* **93**: 9805–9809.

Schweisguth, O.G.-M. and Lemerle, J. (1968) Basal cell nevus syndrome. Association with congenital rhabdomyosarcoma. *Archives Francaises de Pediatrie* **25**: 1083–1093.

Schwienbacher, C., Sabbioni, S., Campi, M. *et al.* (1998) Transcriptional map of 170-kb region at chromosome 11p15.5: identification and mutational analysis of the BWR1A gene reveals the presence of mutations in tumor samples. *Proc. Natl Acad. Sci. USA* **95**: 3873–3878.

Scrable, H.J., Witte, D.P., Lampkin, B.C. and Cavenee, W.K. (1987) Chromosomal localization of the human rhabdomyosarcoma locus by mitotic recombination mapping. *Nature* **329**: 645–647.

Scrable, H., Cavenee, W., Ghavimi, F., Lovell, M., Morgan, K. and Sapienza, C. (1989a) A model for embryonal rhabdomyosarcoma tumorigenesis that involves genome imprinting. *Proc. Natl Acad. Sci. USA* **86**: 7480–7484.

Scrable, H., Witte, D., Shimada, H. *et al.* (1989b) Molecular differential pathology of rhabdomyosarcoma. *Genes Chromosomes Cancer* **1**: 23–35.

Sigaudy, S., Vittu, G., David, A., Vigneron, J., Lacombe, D., Moncla, A., Flori, E. and Philip, N. (2000) Costello syndrome: report of six patients including one with an embryonal rhabdomyosarcoma. *Eur. J. Pediatr.* **159**: 139–142.

Siraganian, P.A., Rubinstein, J.H. and Miller, R.W. (1989) Keloids and neoplasms in the Rubinstein-Taybi syndrome. *Med. Pediatr. Oncol.* **17**: 485–491.

Sobel, R.A. and Woerner, S. (1981) Rubinstein-Taybi syndrome and nasopharyngeal rhabdomyosarcoma. *J. Pediatr.* **99**: 1000–1001.

Sotelo-Avila, C., Gonzalez-Crussi, F. and Fowler, J.W. (1980) Complete and incomplete forms of Beckwith-Wiedemann syndrome: their oncogenic potential. *J. Pediatr.* **96**: 47–50.

Steenman, M., Westerveld, A. and Mannens, M. (2000) Genetics of Beckwith-Wiedemann syndrome-associated tumors: common genetic pathways. *Genes Chromosomes Cancer* **28**: 1–13.

Stratton, M.R., Fisher, C., Gusterson, B.A. and Cooper, C.S. (1989) Detection of point mutations in N-ras and K-ras genes of human embryonal rhabdomyosarcomas using oligonucleotide probes and the polymerase chain reaction. *Cancer Res.* **49**: 6324–6327.

Stratton, M.R., Moss, S., Warren, W. *et al.* (1990) Mutation of the p53 gene in human soft tissue sarcomas: association with abnormalities of the RB1 gene. *Oncogene* **5**: 1297–1301.

Strong, L.C., Stine, M. and Norsted, T.L. (1987) Cancer in survivors of childhood soft tissue sarcoma and their relatives. *J. Natl Cancer Inst.* **79**: 1213–1220.

Takayama, H., LaRochelle, W.J., Sharp, R., Otsuka, T., Kriebel, P., Anver, M., Aaronson, S.A. and Merlino, G. (1997) Diverse tumorigenesis associated with aberrant development in mice overexpressing hepatocyte growth factor/scatter factor. *Proc. Natl Acad. Sci. USA* **94**: 701–706.

Thorburn, M.J., Wright, E.S., Miller, C.G. and Smith-Read, E.H. (1970) Exomphalos-macroglossia-gigantism syndrome in Jamaican infants. *Am. J. Dis. Child* **119**: 316–321.

Tommerup, N.H., CB. van der Heiberg, A. (1991) Tentative assignment of a locus for Rubinstein-Taybi syndrome gene to 16p13.3 by a de novo reciprocal translocation, t(7;16)(q34;p13.3). *Cytogenet. Cell Genet.* **58**: 2002.

Tsokos, M., Webber, B.L., Parham, D.M. *et al.* (1992) Rhabdomyosarcoma. A new classification scheme related to prognosis. *Arch. Pathol. Lab. Med.* **116**: 847–855.

Tycko, B. (1994) Genomic imprinting: mechanism and role in human pathology. *Am. J. Pathol.* **144**: 431–443.

Urashima, M., Teoh, G., Akiyama, M., Yuza, Y., Anderson, K.C. and Maekawa, K. (1999) Restoration of p16INK4A protein induces myogenic differentiation in RD rhabdomyosarcoma cells. *Br. J. Cancer* **79**: 1032–1036.

Varley, J.M., McGown, G., Thorncroft, M., Santibanez-Koref, M.F., Kelsey, A.M., Tricker, K.J., Evans, D.G. and Birch, J.M. (1997) Germ-line mutations of TP53 in Li-Fraumeni families: an extended study of 39 families. *Cancer Res.* **57**: 3245–3252.

Vaughan, W.G., Sanders, D.W., Grosfeld, J.L., Plumley, D.A., Rescorla, F.J., Scherer, L.R., 3rd, West, K.W. and Breitfeld, P.P. (1995) Favorable outcome in children with Beckwith-Wiedemann syndrome and intraabdominal malignant tumors. *J. Pediatr. Surg.* **30**: 1042–1044.

Visser, M., Bras, J., Sijmons, C., Devilee, P., Wijnaendts, L.C., van der Linden, J.C., Voute, P.A. and Baas, F. (1996) Microsatellite instability in childhood rhabdomyosarcoma is locus specific and correlates with fractional allelic loss. *Proc. Natl Acad. Sci. USA* **93**: 9172–9176.

Visser, M., Sijmons, C., Bras, J., Arceci, R.J., Godfried, M., Valentijn, L.J., Voute, P.A. and Baas, F. (1997) Allelotype of pediatric rhabdomyosarcoma. *Oncogene* **15**: 1309–1314.

Weber-Hall, S., Anderson, J., McManus, A., Abe, S., Nojima, T., Pinkerton, R., Pritchard-Jones, K. and Shipley, J. (1996) Gains, losses, and amplification of genomic material in rhabdomyosarcoma analyzed by comparative genomic hybridization. *Cancer Res.* **56**: 3220–3224.

Weichselbaum, R.R., Beckett, M. and Diamond, A. (1988) Some retinoblastomas, osteosarcomas, and soft tissue sarcomas may share a common etiology. *Proc. Natl Acad. Sci. USA* **85**: 2106–2109.

Weksberg, R., Shen, D.R., Fei, Y.L., Song, Q.L. and Squire, J. (1993) Disruption of insulin-like growth factor 2 imprinting in Beckwith-Wiedemann syndrome. *Nat. Genet.* **5**: 143–150.

Wiedemann, H.R., Burgio, G.R., Aldenhoff, P., Kunze, J., Kaufmann, H.J. and Schirg, E. (1983) The proteus syndrome. Partial gigantism of the hands and/or feet, nevi, hemihypertrophy, subcutaneous tumors, macrocephaly or other skull anomalies and possible accelerated growth and visceral affections. *Eur. J. Pediatr.* **140**: 5–12.

Wilke, W., Maillet, M. and Robinson, R. (1993) H-ras-1 point mutations in soft tissue sarcomas. *Mod. Pathol.* **6**: 129–132.

Wunder, J.S., Czitrom, A.A., Kandel, R. and Andrulis, I.L. (1991) Analysis of alterations in the retinoblastoma gene and tumor grade in bone and soft-tissue sarcomas. *J. Natl Cancer Inst.* **83**: 194–200.

Wurl, P., Taubert, H., Bache, M. *et al.* (1996) Frequent occurrence of p53 mutations in rhabdomyosarcoma and leiomyosarcoma, but not in fibrosarcoma and malignant neural tumors. *Int. J. Cancer* **69**: 317–323.

Xu, G.F., O'Connell, P., Viskochil, D. *et al.* (1990) The neurofibromatosis type 1 gene encodes a protein related to GAP. *Cell* **62**: 599–608.

Yamamoto, T., Taya, S. and Kaibuchi, K. (1999) Ras-induced transformation and signaling pathway. *J. Biochem. (Tokyo)* **126**: 799–803.

Yang, P., Grufferman, S., Khoury, M.J., Schwartz, A.G., Kowalski, J., Ruymann, F.B. and Maurer, H.M. (1995) Association of childhood rhabdomyosarcoma with neurofibromatosis type I and birth defects. *Genet. Epidemiol.* **12**: 467–474.

21

Role of telomeres and telomerase in aging and cancer

Jerry W. Shay, Woodring E. Wright and Roger A. Schultz

1. Introduction

The ends of human chromosomes (telomeres) are composed of thousands of repeats of repetitive TTAGGG DNA sequences. Telomerase (TEE-LÓM-ER-ACE) is a ribonucleoprotein enzyme complex (a cellular reverse transcriptase) that maintains telomere length in cancer cells by adding TTAGGG repeats onto the telomeric ends, thus compensating for the normal erosion of telomeres that occurs in all dividing cells (for a recent review see Cech, 2000). Telomerase is expressed during early development and remains fully active in specific germline cells, but is undetectable in most normal somatic cells except for proliferative cells of renewal tissues (e.g. bone marrow cells, basal cells of the epidermis, proliferative endometrium, and intestinal crypt cells). In all non-reproductive proliferative cells progressive telomere shortening is observed (*Figure 1*), and when telomeres become sufficiently short further cells division is blocked, a process often referred to as replicative senescence (Harley *et al.*, 1990). Telomere shortening is a molecular means to count the number of times a cell has divided and may function to initially protect humans against the development of cancer by limiting the maximal number of permitted cell divisions. The telomere-telomerase hypothesis of aging and cancer is based on the findings that immortal cancer cells have engaged a mechanism to maintain stable telomere lengths, almost always by reactivating or up-regulating telomerase activity (*Figure 1*). In this review, we will consider the evidence in support of the telomere-telomerase hypothesis and speculate on some of the applications that may emerge in the future.

1.1 Why do cells age?

In 1825, Benjamin Gopertz described that after the age of 30 the likelihood of dying doubles every 7 years (Gompertz, 1825). As we age there is a gradual decline in performance of organ systems, resulting in the loss of reserve capacity, leading

Molecular Genetics of Cancer second edition, edited by J.K. Cowell.
© 2001 BIOS Scientific Publishers Ltd, Oxford.

The telomere hypothesis

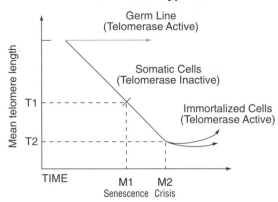

Figure 1. Telomere hypothesis of aging and cancer. In the absence of telomerase, telomeres shorten with each cell division in somatic cells. Cellular senescence (mortality stage 1 = M1, the Hayflick limit, or senescence) begins when there are on average several kilobases of telomeric TTAGGG repeats remaining. It is believed, but not proven, that the stimulus for the induction of M1 may be DNA damage signals from a rare telomere that lacks repeats. In fibroblasts, M1 requires the actions of p53 and a pRb-like activity. If these molecules or molecules in their pathway are mutated or blocked, the cells continue to divide and telomeres continue to shorten until the telomeres are so short that further replication is prevented (mortality stage 2 = M2 or crisis). In rare instances, human cells can escape M2 (giving rise to an immortal cell) and telomeres can almost always be maintained by reactivation or up-regulation of telomerase, a cellular reverse transcriptase. Most cancer cells have detectable telomerase activity and short telomeres. However, the existence of rare telomerase-negative cancers suggests that some cells may form tumors by using a DNA recombination mechanism termed ALT (*A*lternative *L*engthening of *T*elomeres) or by other pathways yet to be determined.

to an increased chance of death. In 1891, August Weisman wrote 'death takes place because a worn-out tissue cannot forever renew itself, and because a capacity for increase by means of cell division is not everlasting but finite' (Weisman, 1891). Proof that normal cells could not divide forever and thus could not infinitely renew tissues came in 1961, when Hayflick demonstrated that normal human cells had a limited replicative lifespan and became senescent after approximately 50–70 doublings (Hayflick and Moorhead, 1961). The rationale for the evolution of this limited lifespan is that it serves as an anti-tumor mechanism. Most tumor cells require 4–6 mutations to become malignant. Each mutation probably uses up at least 30 doublings (both to attain an adequate population size for the next mutation and for the usual necessity of eliminating wild-type alleles for recessive mutations). Replicative senescence would thus inhibit the progression of pre-malignant cells after they had accumulated one or two mutations. However, the benefit of limiting the number of cell doublings has to be balanced against the need of the organism for cell turnover to maintain and repair its tissues. It is likely that the number of permitted doublings has evolved to provide sufficient divisions for reasonably adequate reserves during our 'historical' expected lifespan of 30–40 years, but that increases beyond that would come at the expense of increased cancer. As sanitation and medical advances have increased our expected lifespan, decreased proliferative reserves may begin to affect tissue maintenance and contribute to the physiology of aging.

1.2 Gradual shortening of telomeres coincides with the long-term aging process

Under normal conditions human tissues can function adequately for a typical life span. Today there may be an increase in aged-related cellular decline in normal people who live to an exceptionally old age, while in the past problems from a limited cellular proliferative capacity was only observed in disease states. However, in older individuals without diseases, there is an increased incidence of chronic ulcers (Vande Berg *et al.*, 1998), wearing down of the vascular endothelium leading to arteriosclerosis, proliferative decline of retinal pigmented epithelial cells that may lead to age-related blindness, immunological deficiencies which may result in an increased incidence of cancer, and several other diseases of tissues that have an intrinsic capacity for replacement. In summary, it is believed, but not yet proven, that in some aged-related disorders, telomere decline in specific tissues and organs may contribute to aging vulnerability.

1.3 Preventing the shortening of telomeres prevents cellular aging

Telomeres were first mentioned in a lecture given by Hermann Muller in 1938 (see Muller, 1962) and from the work of Barbara McClintock (1941). However, the precise role that these structures played in cell replication was unclear at that time. It was generally agreed that there needed to be a cellular mechanism to maintain telomeres in immortal single cell organisms (Blackburn and Gall, 1978) and that a similar mechanism was needed to preserve the germline. The answer to this critical question emerged in studies with *Tetrahymena thermophila* by Greider and Blackburn (1985), who discovered the ribonucleoprotein enzyme terminal telomere transferase, now called telomerase. They found that telomeres are synthesized *de novo* by telomerase, a ribonucleoprotein enzyme that extends the 3' end of telomeres and thus elongates them (*Figure 2*). This ribonucleoprotein complex contains a reverse transcriptase and RNA template for the synthesis of the repeated sequence (Cech, 2000). It was simultaneously reported that cancer cells have shorter telomeres than do adjacent normal cells (De Lange *et al.*, 1990; Hastie *et al.*, 1990) thus providing the first link for the role of telomeres in cancer biology. Telomerase was later found to occur in extracts of immortal human cell lines (Morin, 1989) and in about 90% of all human primary tumors (Kim *et al.*, 1994). The telomerase RNA component was cloned a few years ago (Feng *et al.*, 1995) and subsequently the catalytic portion of the enzyme was cloned (Nakamura *et al.*, 1997).

While there have been many studies indicating a correlation between telomere shortening and proliferative failure of human cells (*Table 1*), the evidence for a causal relationship has only recently been demonstrated (Bodnar *et al.*, 1998). The formal proof that telomeres were rate limiting for indefinite cell proliferation was shown by introduction of the telomerase catalytic protein component (hTERT)

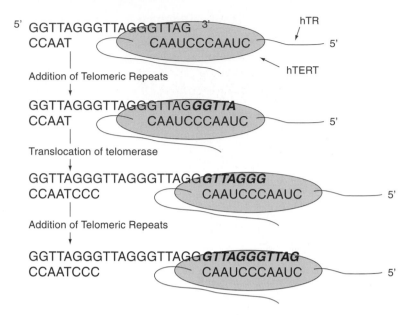

Figure 2. Reverse transcription. Telomeric sequences are synthesized by telomerase, a ribonucleoprotein enzyme (composed of both RNA and protein). Telomerase contains RNA-dependent DNA polymerase activity that uses its RNA component, hTR (complementary to the telomeric single stranded overhang) as a template in order to synthesize TTAGGG repeats (elongate) directly onto telomeric ends. The catalytic component of telomerase, hTERT, has reverse transcriptase activity. hTR and hTERT form the core of the telomerase holenzyme that binds to telomeric repeats and in combination with other cellular components adds six bases (telomeric repeat), then pauses while it repositions (translocates) the template RNA for the synthesis of the next six base pair repeat. Thus human telomerase is processive and can re-anchor its template region for the addition of more telomeric repeats. The extension of the 3′ DNA template end in turn permits additional replication of the 5′ end of the lagging strand, thus compensating for the telomere shortening that occurs in its absence.

Table 1. Evidence in support of the telomere hypothesis of cellular aging

---

- Telomeres are shorter in most somatic tissues from older individuals compared to younger individuals
- Telomeres are shorter in somatic cells than in germ line cells
- Children born with progeria (an early aging syndrome) have shorter telomeres in some tissues compared to age-matched controls
- Telomeres in normal human cells progressively shorten when grown in cell culture
- Introduction of the catalytic protein subunit of telomerase is sufficient to immortalize cells if telomeres are rate limiting for cell proliferation

---

into normal telomerase-silent human cells (*Figure 3*). This resulted in restoration of telomerase activity. Normal human cells stably expressing transfected telomerase maintained or elongated telomeres (*Figure 3*), demonstrated an indefinite life span, maintained a normal chromosome complement, and continued to grow in a normal manner. These observations provided the first direct evidence for the

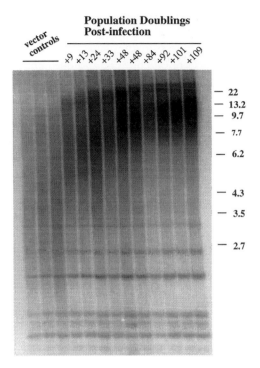

Figure 3. Telomerase elongates telomeres. The catalytic subunit of telomerase, hTERT, when introduced into normal telomerase silent fibroblasts, results in telomere elongation and prevents the occurrence of M1. Introduction of hTERT prior to M1 prevents replicative senescence while introduction of hTERT prior to M2 prevents crisis. This suggests that telomeres are involved in both M1 and M2. Telomeres are measured by digesting total genomic DNA with one or several restriction enzymes having four-base recognition sites and then running the DNA out on an agarose gel, electrophoresed, and in-gel hybridized with a TTAGGG labeled probe. This process results in genomic DNA being digested into small fragments. However, because telomeric sequences lack restriction sites, telomeres yield relatively large terminal restriction fragments (TRF) composed of 2–4 kilobase pairs of subtelomeric DNA and age-dependent amounts of telomeric repeats. Since all human chromosome ends in a typical cell are not homogeneous, the telomeres appear as smears instead of a discrete band. However, the signals in each lane can be quantitated and average telomere lengths determined.

hypothesis that telomere length determines the proliferative capacity of human cells.

1.4 Are cells expressing telomerase 'normal'?

The expression of telomerase in almost all malignancies suggests that overcoming the proliferative limits imposed by telomere shortening represents a key step in onco-genesis. Fibroblasts and epithelial cells with introduced telomerase have continued to divide for over 350 generations past the time they normally would stop dividing. Moreover, the cells also are growing and dividing in a normal manner, giving rise to normal cells without any evidence that the cells are accumulating a transformed phenotype (Jiang *et al.*, 1999; Morales *et al.*, 1999). Telomerase expression and maintenance of telomere integrity does not stimulate entry into the cell cycle, does not bypass cell cycle induced checkpoint controls, and does not lead to genomic instability. Thus telomerase alone does not directly affect the molecular processes leading to uncontrolled cell cycles, invasion or metastases. The main function for telomerase is to maintain telomeres and permit continued cell growth that is initially limited by shortened telomeres. Since telomeres normally shorten in most cells as a result of the end-replication problem (*Figure 4*), it is appropriate to ask if genetic defects in telomere metabolism may contribute to specific human diseases.

2. Chromosome instability syndromes and telomere metabolism

Heritable human disorders that exhibit chromosome instability are often associated with cancer predisposition. Historically, the identification and characterization of

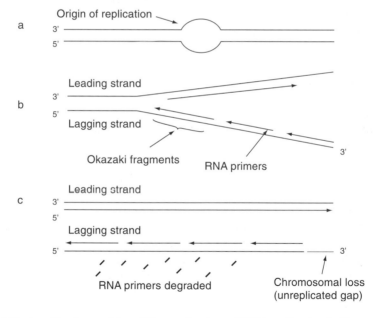

Figure 4. End replication problem. The mechanism of DNA replication in linear chromosomes is different for each of the two strands (called leading and lagging strands). The lagging strand is synthesized as discontinuous Okazaki fragments (right side of figure), whose replication is initiated with a labile RNA primer. Even if the initiation of the terminal Okazaki fragment occurs at the very end of the lagging strand of DNA, some bases will be not replicated since the terminal RNA primer cannot be replaced by DNA. Thus on the lagging strand a gap occurs at the very end that cannot be filled in by the enzymes that replicate the remainder of the DNA (this is referred to as the end replication problem). Since one strand cannot copy its end, telomere shortening occurs during progressive cell divisions. Cells that express sufficient telomerase activity are thought to extend the end of the DNA template thus compensating for the incomplete replication.

such disorders provided early documentation for the role of chromosome instability in cancer progression. While the associated instability was principally seen in the form of breaks, translocations or aneuploidy, for several of these disorders the chromosomal phenotypes have included aberrant telomere behavior. The chromosome instability disorders are reviewed here with an emphasis on the presence or lack of telomere-related defects.

2.1 Dyskeratosis congenita

Dyskeratosis congenita (DKC, DC) is characterized clinically by cutaneous pigmentation, dystrophy of the nails, leukoplakia of the oral mucosa, atresia of the lacrimal ducts, thrombocytopenia, anemia, and testicular atrophy. The most serious complications associated with the disorder are anemia and cancer, principally of the anus, mouth and skin (Milgrom et al., 1964). The most common form of DKC (DKC1) exhibits X-linked inheritance. Obligate female carriers present with clinical symptoms including skin lesions and highly skewed X inactivation is observed in white blood cells, cultured skin fibroblasts, and buccal mucosa (Devriendt et al., 1997; Ferraris et al., 1997; Vulliamy et al., 1997).

The association of pancytopenia and cancer with DKC is reminiscent of another chromosome instability syndrome, Fanconi anemia or FA, which is described in greater detail below (Bryan and Nixon, 1965; Selmanowitz and van Voolen, 1971). Although the X-linked and autosomal recessive modes of inheritance for these two disorders are distinct, the documented chromosome instability associated with FA (see below) prompted similar studies with DKC cells. DeBauche et al. (1990) found an increased frequency of chromatid breaks and chromatid gaps after X-radiation during the G_2 phase of the cell cycle and Ning et al. (1992) reported that the mean number of chromosome breaks per cell in bleomycin-treated lymphocytes was higher in patient samples than in controls. In contrast, other studies (Dokal and Luzzatto, 1994; Dokal et al., 1992) suggest that DKC lymphocytes exhibit no differences in chromosome stability when compared to controls either spontaneously or in response to clastogen exposure (bleomycin, diepoxybutane, mitomycin-c, 4-nitroquinoline-1-oxide). However, those studies did note abnormalities in the morphology (polygonal cell shape, ballooning, dendritic-like projections) and growth rate (2 times the controls) for DKC fibroblasts. Moreover, numerous unbalanced chromosomal rearrangements were noted to occur spontaneously in the DKC fibroblasts, which were also seen in bone marrow cells. The above results, although not consistent for all cell types, defined DKC as a chromosome instability syndrome.

The *DKC1* gene was linked to the factor VIII gene at Xq28 (Devriendt et al., 1997). Heiss et al. (1998) detected a 3' deletion in one DKC patient with a cDNA probe that mapped to the *DKC1* critical region. They subsequently demonstrated different missense mutations in five unrelated patients indicating that they had identified the gene responsible for X-linked DKC. The *DKC1* gene, also referred to as dyskerin, is highly conserved and homologous to the rat gene *NAP57* and the *S. cerevisiae* gene *CBF5*. The dyskerin gene contains a pseudouridine synthase motif, multiple phosphorylation sites, and a carboxy-terminal lysine-rich repeat domain. Given genetic evidence that *CBF5* is involved in the production of rRNA

(Lafontaine *et al.*, 1998), it was initially predicted that the major role of dyskerin might lie in rRNA processing and that the clinical features associated with *DKC1* might be directly related to this general defect. However, Mitchell *et al.* (1999) found that primary fibroblasts and lymphoblasts from DKC1 individuals were not deficient in general H/ACA small nucleolar RNA abundance or function. They also investigated the possible role for dyskerin in the processing of telomerase RNA, which had been shown to possess a H/ACA RNA motif. DKC1 cells were shown to exhibit a lower level of telomerase RNA, produce lower levels of telomerase activity than normal cells, and demonstrated shortened telomeres. It has been proposed that compromised telomerase function and resulting defects in telomere maintenance yield a reduced proliferative capacity in epithelial and blood cells in DKC1 patients, leading to the clinical features seen.

DKC had not been associated with telomere instability prior to the identification of a functional role for dyskerin in the production of telomerase RNA. In this regard, DKC1 represents a chromosome instability syndrome in which telomere instability was discovered *post hoc*, after the gene bearing mutations was discovered and its role in telomere function demonstrated by biochemical analysis. It is reasonable to speculate that telomere 'erosion' in a specific subset of DKC1 cells gives rise to instability through aberrant telomere associations and related translocations, resulting in the tumors seen.

2.2 Ataxia telangiectasia

Ataxia telangiectasia (AT) is an autosomal recessive chromosome instability syndrome in which patients are predisposed to development of leukemia, lymphoma and sarcoma, with other tumors that are seen less frequently. Additional clinical features include characteristic telangiectasia, café-au-lait spots, progressive neurological degeneration leading to cerebellar ataxia, immunodeficiencies associated with reduced levels of various Ig class molecules, thymic hypoplasia and gonadal atrophy. AT heterozygotes have been reported to be at risk for increased cancer (Broeks *et al.*, 2000; Swift *et al.*, 1987), although such predisposition has been disputed (Chen *et al.*, 1998; Fitzgerald *et al.*, 1997). A radiosensitivity that has come to be associated with AT was first described for patients exposed to ionizing radiation as part of their cancer therapy. A number of patients suffered severe radiation reactions and several of these patients died. Subsequently, cultured AT cells were shown to exhibit extreme radiosensitivity. (For a review of this disorder the reader is referred to Friedberg *et al.*, 1995.)

In addition to increased radiosensitivity, AT cells in culture exhibit chromosome instability in the form of increased chromosome breakage and translocations at rates 30–200 times greater than that seen in normal controls (Meyn, 1995). Such translocations commonly involve the T cell receptor and immunoglobulin heavy chain genes, providing evidence for a direct link between chromosome instability and the significant increase in lymphoreticular malignancies. In contrast, underlying causes for the pathognomonic telangiectasia and the neurological degeneration leading to ataxia in AT are less well understood.

Cells from AT patients exhibit a characteristic defect in the delay of DNA synthesis following exposure to ionizing radiation or radiomimetic drugs referred to as radioresistant DNA synthesis (Friedberg *et al.*, 1995). This defect provided early evidence that AT cells have defects in a cell-signaling checkpoint. In control cells, a checkpoint response stops DNA synthesis following exposure to agents that induce DNA strand break damage. Persistence of DNA synthesis on a damaged template likely contributes to the increase in radiosensitivity, chromosomal breaks and translocations seen in AT cells.

The gene mutated in AT cells (ATM) encodes a 350 kd protein that functions as a PI3 kinase and mediates responses to DNA damage (Savitsky *et al.*, 1995). This complex protein participates in a large and growing number of cellular functions, including the regulated activation of *TP*53 (Banin *et al.*, 1998) and *ABL* (Shafman *et al.*, 1997). The ATM gene is a member of a family of homologous genes identified in a variety of species, including MEC1 and TEL1 in *Saccharomyces cerevisiae* (Morrow *et al.*, 1995; Sanchez *et al.*, 1996), rad3+ in *S. pombe* (Bentley *et al.*, 1996) and mei-41 in *Drosophila* (Hari *et al.*, 1995). Mec1 mutants were named for a defect in 'mitotic entry checkpoint', providing additional evidence for the role of the ATM protein in damage checkpoint arrest.

Prior to the cloning of the ATM gene, it was recognized that AT cells exhibit defects in telomere biology, demonstrating aberrant telomere associations (Kojis *et al.*, 1991). After its sequence became available, interest in telomere maintenance in ATM cells was stimulated by the observation that TEL1 of the yeast *Saccharomyces cerevisiae* is the gene most closely related to ATM. Yeast strains with a mutation of TEL1 have short but stable telomeric repeats. In contrast, tlc1 mutants defective in the RNA subunit of telomerase exhibit a progressive telomere shortening and decreased viability, with survivors demonstrating variable telomere lengths maintained by Rad52p-dependent recombination events (Ritchie *et al.*, 1999). While mec1 null mutations are lethal, a viable mec1–21 mutation has been characterized and shown to be defective in telomere position effect (TPE), a pathway involved in the silencing of genes near yeast telomeres. Moreover, tel1 mec1 double mutants exhibit the senescent phenotype of tlc1 mutants and tel1 mec1 survivors also arise by Rad52p-dependent recombination (Craven and Petes, 2000). Similarly, rad3 mutants in *Saccharomyces pombe* have been shown to exhibit phenotypes including reduced telomere length and defective TPE (Matsuura *et al.*, 1999), and rad3 tel1 double mutants exhibit complete loss of telomeres and conversion of all three chromosomes into circles (Naito *et al.*, 1998). Collectively, these studies suggest a significant role for ATM and its homologs in telomere biology.

Subsequent to the early report of telomere instability in AT cells and concurrent with more recent demonstrations for the roles of yeast ATM homologs in telomere maintenance, studies have addressed the molecular nature of telomere instability in AT cells. Examination revealed that AT cell lines show higher frequencies of telomeric associations in either metaphase or interphase and that AT cells have shorter telomeres than those seen in controls (Metcalf *et al.*, 1996; Pandita *et al.*, 1995; Xi *et al.*, 1996). A correlation between telomere shortening and telomeric associations was also demonstrated. These defects were not correlated with reduced telomerase activity, as AT cells were shown to have levels of activity similar to or greater than those seen in controls (Pandita *et al.*, 1995). Similar results were found in spermatocytes, leading to the suggestion that altered telomere biology might be related to the meiotic defects associated with AT (Pandita *et al.*, 1999). Additional results suggest that the altered telomere stability in AT cells may be related to defective interactions between telomeres and the nuclear matrix and/or differences in the structure of telomeric chromatin (Smilenov *et al.*, 1999).

2.3 Nijmegan breakage syndrome

Nijmegan breakage syndrome (NBS) is a disorder characterized by many of the clinical features associated with AT, including immunological deficiencies, microcephaly, and developmental delay. Although the patients notably lack both ataxia and telangiectasia, they do exhibit the characteristic AT cellular features that include sensitivity to ionizing radiation, radioresistant DNA synthesis and chromosome instability (Friedberg *et al.*, 1995). The gene defective in NBS (p95/NBS1) has been cloned and identified as a member of a protein complex with hRad50 and hMRE11 that participates in the recognition of DNA strand break damage and subsequent signal transduction events responsible for cell cycle checkpoints (Matsuura *et al.*, 1998). A single report had described increased premature telomere shortening in NBS cells (Oexle *et al.*, 1997). Work by Lim *et al.* (2000) reveals that p95/nbs1 is phosphorylated on serine 343 in an ATM-dependent fashion following ionizing radiation. Mutation of *p95/NBS1* at this site yields a protein that does not correct the cell cycle checkpoint defect in NBS cells. These results provide a possible explanation for the close clinical and cellular relationships between these two disorders, including a possible explanation for the reported premature telomere shortening seen in both. The biochemical connection between ATM, NBS1, MRE11, RAD50, BRCA1, and p53 is becoming stronger (Wang, 2000). These genes form a link between cell cycle checkpoints activated by DNA damage (including telomeres) and DNA repair. Depending on the integrity of the genes in this pathway, many outcomes are possible. In some instances it can lead to genomic instability, driving cancer development, in other instances apoptosis, and in others senescence. Recently, links between telomeres, p95/NBS1, MRE11, and p53 and PML nuclear body structures have been reported (Lombard and Guarente, 2000; Pearson *et al.*, 2000). This network of genes may sense telomere homeostasis and thus be important in overall telomere maintenance.

2.4 Werner syndrome

Werner syndrome (WS) is a rare autosomal disorder characterized by features of premature aging. The clinical features of WS include scleroderma-like skin, arteriosclerosis, diabetes mellitus, osteoporosis, and ocular cataracts (Epstein *et al.*, 1966; Goto *et al.*, 1981). While atherosclerosis is the usual cause of death in WS patients, approximately 10% develop rare cancers including osteosarcoma and meningioma. WS cells grown in culture have been shown to exhibit a mutator phenotype, with an eight-fold higher frequency of 6-thioguanine resistant lymphocytes in patients compared to levels seen in controls (Fukuchi *et al.*, 1990). The molecular characterization of mutations arising in WS cells revealed large deletions in the HPRT gene, and such deletions were shown to involve non-homologous recombination events (Monnat et al., 1992). Chromosomal instability in WS cells is manifested by increases in translocation and cytologically visible deletions (Fukuchi et al., 1989). WS cells also exhibit a shorter life span in culture reaching population doublings only one-third those of age- and sex-matched controls. That shorter life span is associated with reduced initial telomere length (Kruk and Bohr, 1999; Tahara et al., 1997), a result consistent with other reports demonstrating shorter telomeres in cells from progeroid disorders (Allsopp et al., 1992).

The gene responsible for WS (*WRN*) has been cloned and encodes a protein with homology to the *E. coli* RecQ (Yu et al., 1996). The RecQ protein binds single-stranded DNA and has a 3'→5' helicase activity (Umezu et al., 1990). That helicase function is dependent on ATP hydrolysis and is stimulated (>100 fold) by single strand binding protein (Umezu and Nakayama, 1993). RecQ can resolve a wide variety of substrates including recombinational joint molecules generated by RecA protein (Harmon and Kowalczykowski, 1998). These properties of RecQ led to the suggestion that it may improve the fidelity of recombination by dissociating illegitimate recombination events that are initiated through microhomology. Reintroduction of the *WRN* gene into WS cells complements the premature senescence phenotype of WS cells (Hisama et al., 2000). Similarly, introduction of an exogenous *hTERT* gene into WS cells yields suppression of senescence of WS cells, as it does for most other cell types tested (Ouellette et al., 2000; Wyllie et al., 2000). Another report indicates that the introduction of *hTERT* gene into WS cells can render partial correction of the 4NQO sensitivity of those cells (Hisama et al., 2000). This suggests that the 4NQO sensitivity is intimately linked to telomere length.

In *Saccharomyces cerevisiae* the sole *recQ* homolog, Sgs1, was identified by virtue of the ability of mutants at this locus to complement the slow growth phenotype of topoisomerase 3 (TOP3) null mutants (*top3*) (Gangloff et al., 1994). *top3* mutants are viable but grow very slowly and selection for 'slow growth suppression' (Sgs) yielded sgs1. Independently, Sgs1 was identified through a yeast two-hybrid screen for proteins interacting with the c-terminal domain of topoisomerase II (Watt et al., 1995). The chromosome instability seen in sgs1 mutants is demonstrated as improper mitotic and meiotic chromosome segregation (Watt et al., 1995). While the mitotic defects have been associated with hyperrecombination of repetitive elements, no increase in meiotic recombination has been observed (Watt et al., 1996). Moreover, although sgs1 mutants display premature senescence comparable to that of WS cells, they exhibit no notable defects related to telomere biology. This may reflect fundamental differences between the aging process in higher and lower eukaryotes, with the latter more directly related to rDNA stability (Sinclair and Guarente, 1997).

2.5 Progeria

Relative to the other disorders considered in this review, Hutchinson Gilford Progeria is unique in that the disease is not associated with cancer. Indeed, to our knowledge not a single case of cancer has been reported in a progeric patient.

However, the disorder is associated with defects in telomere biology. Therefore, a better understanding of this disease is likely to provide insights into cancer and the importance and uniqueness of telomeric instability in cancer.

Progeria is characterized by premature aging very early in life, with generalized features that include short stature, alopecia, micrognathia and absence of subcutaneous fat (DeBusk, 1972). Patients rapidly develop arteriosclerosis, coronary artery disease, angina pectoris, myocardial infarction, and congestive heart failure, but no neoplasia. Skin fibroblasts from patients with Hutchinson Gilford Progeria have shorter telomeres and a limited *in vitro* life span relative to cells derived from age-matched controls (Allsopp *et al.*, 1992). Examination of telomerase expression in stimulated fresh blood lymphocytes from patients afflicted with progeria revealed robust telomerase activity, consistent with the thought that progeria is not due to a defect in telomerase activity (Ouellette *et al.*, 2000). Expression of exogenous hTERT in fibroblasts derived from patients with Hutchinson Gilford Progeria was able to circumvent telomere shortening and immortalize the cells (Ouellette *et al.*, 2000). The immortalization of progeric cells by hTERT illustrates that these cells are not deficient in the ability of telomerase to access telomeric sequences and generate long telomeres.

Normal cells demonstrate an age-related decline in DNA repair capacity and one proposed explanation for this decline is the time-dependent accumulation of DNA damage in the genome, including damage in DNA repair genes that thus reduce further repair capacity. Consistent with this notion, a second phenotype associated with Hutchinson Gilford Progeria cells is a reduced DNA repair capacity as measured by an unscheduled DNA synthesis (UDS) assay (Moriwakiet *et al.*, 1996; Wei, 1998). Reductions in DNA repair with age may be related to reversible changes in gene expression that are intimately related to telomere length. Alternatively, telomeres may serve as 'sinks' or 'storehouses' for factors required for DNA repair such that reductions in telomere length yield less available protein and measurable reductions in repair capacity. Another way to think about progeria is from a developmental perspective. It is known that telomerase is expressed for the first 3–5 months during human development and is then silenced (Wright *et al.*, 1996). It is possible that telomerase may be prematurely silenced in specific cell types in children with progeria, which would account for the shortened telomeres that are present in specific tissues. Alternatively the primary defect may have nothing to do directly with telomerase but that there is increasing cell turnover during fetal development so that some tissues may have cells with shortened telomeres. This leads to the idea that short telomeres may prevent cancer production by preventing continued cell proliferation. This would be consistent with the notion that expressing hTERT in progeric cells restores the reduced DNA repair capacity of progeric cells. This suggests in some instances that assays that measure DNA damage responses may not be directly related to the initial genetic lesion but a secondary consequence of increased cell turnover leading to shortened telomeres and poorer repair.

The lack of tumor predisposition in children with progeria is particularly striking. The fact that the progeric patients die at a young age is an insufficient argument to justify the absence of cancer. Many other genetic disorders that exhibit chromosome instability, for example xeroderma pigmentosum, Bloom syndrome and Li-Fraumeni syndrome, include very young patients with cancer. It is likely that the loss of telomeres in the absence of other genetic lesions leads to growth arrest in cells of progeric patients rather that genomic instability.

2.6 Fanconi anemia

Fanconi anemia (FA) is yet another chromosome instability syndrome associated with cancer predisposition and telomere effects. Although clinical presentation is variable, FA patients are generally born small for gestational age and present with radial aplasia, thumb abnormalities, pigmentary abnormalities, congenital heart anomalies, kidney and genital deformities, and mental retardation. Improper bone marrow function is a common feature and impacts all associated cell types lending to anemia, leukopenia and thrombocytopenia. Leukemia is common including acute myelogenous leukemia diagnosed in >15% of patients, with cancers occurring independent of the severity of other phenotypes (Giampietro *et al.*, 1993).

FA cells in culture exhibit increases in spontaneous chromosome instability and are highly susceptible to the induction of chromosome aberrations by DNA cross-linking agents, a feature used diagnostically for the disorder. The syndrome is genetically complex, represented by at least eight complementation groups. Although several of the FA genes have now been identified (Anonymous, 1996; de Winter *et al.*, 1998; Lo *et al.*, 1996; Strathdee, 1992), the underlying and unifying biochemistry in which these factors function remains to be elucidated. While chromosome instability has long been recognized as an integral part of this disorder, documentation of telomere effects is recent. Leteurtre *et al.* (1999) examined telomere length in peripheral blood lymphocyte samples representing 54 FA patients, nine heterozygotes and 51 controls. Relative to controls the telomere length was significantly shorter in the patient samples, but not in the heterozygous samples. Similar results have been reported by Kruk and Bohr (1999). It is unknown whether this telomere shortening is a direct consequence of FA gene function, or if it simply reflects increased cell turnover. Unexpectedly, Leteurtre *et al.* (1999) found telomerase enzymatic activity to be substantially higher in extracts derived from FA cells than in controls (4.8 ×), despite the shorter telomere length. This elevated activity is consistent with a variety of reports suggesting that DNA damage increases telomerase activity.

2.7 Li-Fraumeni syndrome

Li-Fraumeni syndrome (LFS) is represented as a dominant disorder in which most patients exhibit the presence of one null and one wild-type allele for the *TP53* gene. These patients exhibit a well-documented predisposition to the development of a variety of cancers at a very young age. Primary cell lines from LFS patients have been shown to possess telomeres shorter than matched controls (Kruk and Bohr, 1999). Mutations in *TP53* have been reported to increase DNA recombination rates (Mekeel *et al.*, 1997). Individuals with LFS, who are heterozygous for mutations in *TP53*, can spontaneously immortalize in cell culture (Gollahon *et al.*, 1998; Shay *et al.*, 1995). This immortalization is associated with mutation or functional loss of the other *TP53* allele with cells entering crisis. Interestingly, fibroblasts from these patients emerge from crisis in 30–40% of the time by engaging an alternative lengthening of telomeres (ALT) mechanism, perhaps because they have higher rates of recombination (Bryan *et al.*, 1997). However, epithelial cells derived from LFS patients almost always engage the telomerase activation pathway, suggesting differences between at least some epithelial cell types and fibroblasts. It has been demonstrated that LFS fibroblasts that engage the ALT mechanism contain promyelocytic leukemia (PML) bodies, whereas fibroblasts and epithelial cells that activate telomerase do not. The PML bodies also contain telomeric DNA binding proteins such as TRF1 and TRF2, replication factor A, and human RAD51 and RAD52 (Yeager *et al.*, 1999). The presence of these recombination factors in PML bodies adds support to the idea that recombination may be important in telomere metabolism in ALT pathway cells. Herbert *et al.* (1999) reported that inhibiting telomerase in LFS breast epithelial cells caused progressive telomere shortening and eventual apoptosis. To test if the ALT pathway might be engaged in such telomerase-inhibited cells, Herbert *et al.* (1999) plated ten million LFS cells during anti-telomerase induced crisis and demonstrated that there were no survivors or revertants that engaged the ALT pathway (Herbert *et al.*, 2001). In a similar study targeting telomerase using a dominant negative hTERT, no ALT pathway survivors emerged (Hahn *et al.*, 1999). Thus, it is unlikely that telomerase positive cells have an ongoing or easily engaged ALT mechanism potentially leading to drug resistance failure of clinical trials to test the efficacy of anti-telomerase cancer therapies.

2.8 Additional disorders associated with chromosome instability

A number of human disorders that have been associated with chromosome instability and cancer predisposition exhibit no evidence of defects in telomere biology. For example, xeroderma pigmentosum (XP) is characterized by photosensitivity, a very pronounced predisposition to skin cancer on the sun exposed areas of the body, and the occasional appearance of neurological defects. XP cells in culture are UV sensitive, exhibit defects in nucleotide excision repair (NER), and retain unexcised DNA damage for long periods of time (for review see Friedberg *et al.*, 1995). The observation that XP cells are defective in NER and retain bulky base damage raises questions regarding the accumulation of such damage in telomeric regions. However, to date, there have been no reports describing increased telomere shortening or abnormal telomeric associations in primary XP cultures. Bloom syndrome (BS) and Rothmund Thomson syndrome (RTS) are disorders associated with defects in DNA helicases belonging to the same family of RecQ homologs as the WRN gene (Ellis *et al.*, 1995, Kitao *et al*, 1998, 1999). Like WS, BS and RTS cells exhibit characteristic chromosome instability (Shinya *et al.*, 1993; Smith and Paterson, 1982; Vasseur, 1999). However, in contrast to WS, neither syndrome has been associated with telomere shortening. Similarly, dominant disorders such as hereditary non-polyposis colon cancer and basal cell nevus syndrome, which present with significant cancer predisposition and specific forms of chromosome instability, have not been associated with deficiencies in telomere biology.

3. Telomere shortening in mice

Compared to humans, mice have higher metabolic rates, fewer cells, longer telomeres, different patterns of gene expression, shorter life spans, and a significantly increased incidence of cancer. Since mice have shorter life spans and many fewer cells, it is unlikely that the forces of natural selection would have evolved anti-cancer protection mechanisms to the same extent as humans who are about 500–1000 times larger and require approximately 30 years to mature and nurture their offspring (Wright and Shay, 2000).

Even so, since telomere maintenance is important in human cells, it was important to determine if this was also true in intact animals. The identification of the genes encoding mouse telomerase RNA (mTR) and the mouse catalytic telomerase protein component (mTERT) made it possible to address the importance of telomerase for viability, chromosome stability, and tumorigenesis in mice (Blasco *et al.*, 1997). In initial studies, mTR 'knockout' mice are 'normal' in regards to basic physiology. Telomerase is thus not involved in some unanticipated mechanism during mouse embryogenesis and development that would make it an embryonic lethal. Telomerase knockout mice do not express detectable levels of telomerase and demonstrate progressive telomere loss and chromosome instability in later generations. Delay in a phenotype almost certainly reflects the unusually long telomere-repeat arrays in most inbred mouse species (ranging from 40–100 kb compared with just 12–15 kb for humans). The telomeric repeats in inbred strains of mice are therefore not exhausted until many (3–6) generations have elapsed (depending on the strain used). In the late generation telomerase knockout mice there is defective spermatogenesis resulting in infertility and marked impairment of proliferation in hematopoietic cells in the bone marrow and spleen. In addition, there is an increase incidence of erosive dermatitis and a marked delay in wound re-epithelialization (Greenberg *et al.*, 1999; Rudolph *et al.*, 1999, 2000). These observations indicate a role for telomerase, and hence telomeres, in the maintenance of genomic integrity and in the long-term viability of high-renewal organ systems. Interestingly, the ability of the telomerase deficient mouse to respond to physiological stresses is compromised (such as in the liver) and declines in specific organ systems indicating that they have a reduced capacity to tolerate acute stresses (a hallmark of human aging).

Another function of telomeres is to protect the chromosome ends from degradation and end–end fusions. In late generation telomerase-deficient mice there is increased genomic instability and a modest increased incidence of spontaneous cancer formation, which may be caused in part by decreased immune function due to telomere shortening. This indicates that an adequate telomere length regulation is ultimately important in the overall fitness, reserve and well-being of the mouse and warrants future studies. Knowing more about such stress responses could help explain why surgery and chemotherapy take a greater toll on older people.

4. Future applications for telomerase as a product to extend cell life span

The ability to immortalize human cells and retain normal behavior holds promise in several areas of biopharmaceutical research including drug development, screening and toxicology testing. The development of better cellular models of human disease and production of human products are among the immediate applications of this new advance. For example, in the past scientists have used primary cultures derived from patients with chromosomal instability syndromes. Since many of these cell types do not grow well, it has been difficult to compare results between laboratories. To partially solve the problem of variability of cellular reagents, there are several SV40 transformed cell lines produced from patients with chromosome instability syndromes. These cells sometimes retain the phenotype of the disease state, but in some cases they do not. With the ability to introduced hTERT and immortalize human cells, there is clear utility in developing better cell culture models of human chromosomal disorders (Shay and Wright, 2000).

The technology to introduce telomerase into cells also has the potential to produce unlimited quantities of normal human cells of virtually any tissue type and may have most immediate practical biomedical applications in the area of transplantation medicine. In the future, it may be possible to take a person's own cells, manipulate and rejuvenate them without using up their life span and then give them back to the patient. In addition, genetic engineering of telomerase-immortalized cells could lead to the development of cell-based therapies for certain genetic disorders.

Initially, this technology would be done *ex vivo*. A few cells could be removed from a person by fine needle or surgical biopsy, and the telomere clock reset in the laboratory (thus providing the cells with unlimited divisions). The cells could then be reintroduced into the person. This would have the advantage of not having to overcome the problems with immune rejection, since these are the person's own cells. In the case of a genetic disorder, the telomere clock could be reset, then the corrected gene inserted, and finally reintroduction of the cells into the patient. In the future we may not need to permanently introduce the telomerase gene; hTERT could be introduced to grow the telomeres and then be removed, or perhaps a small molecule will be discovered that can reset the telomere clock. This would avoid potential problems with having too much growth potential (and thus increased risk for cancer). However, initial results indicate that introduction of telomerase in perfectly normal cells does not increase the cancer risk very much and the cells with intact telomerase can still undergo differentiation and participate in normal tissue functions.

An example of the utility of this technology was recently demonstrated. SV40 large T-antigen maintains the differentiated phenotype of adrenal cells in culture but does not immortalize them. Introduction of hTERT prevented loss of telomeric DNA and the rapidly growing engineered bovine adrenocortical cells were then xenotransplanted into a small cylinder introduced beneath the kidney capsule of *scid* mice that had been adrenalectomized (Thomas *et al.*, 2000). While the animals without transplantation died, animals that received transplant of cells expressing both hTERT and SV40 large T-antigen survived and produced bovine cortisol to replace the mouse glucocorticoid, corticosterone. The tissue formed in the xenotransplanted animals were chimeras of normal-appearing bovine adrenal cortex cells together with mouse endothelial cells. The tissue was well vascularized and did not overgrow the polycarbonate cylinder. The proliferation rate in tissues formed by these transplanted bovine adrenal cells was low and there were no indications of malignant transformation. These experiments dramatically show that an endocrine tissue could be derived from a previously cultured somatic cell type that expressed both T-antigen and hTERT.

Genetic engineering of telomerase-immortalized human cells could also lead to the development of cell-based therapies for certain genetic disorders such as muscular dystrophy. Other areas of cell engineering that may be possible in the future include having an unlimited supply of skin cells for grafts for burn patients, for treating chronic (pressure or diabetic) ulcers, or to generate products for cosmetic applications (e.g. aging skin). Improving general immunity for older patients or those with blood disorders such as patients with AIDS or patients with bone-marrow exhaustion in myeloproliferative diseases are also possible chronic disorders that may be treated. Producing an unlimited supply of pancreatic islet cells that are glucose responsive for the treatment of diabetes, telomere modifications of the endothelial lining of blood vessels to prevent arteriosclerosis, bone marrow stem cells for blood transplants, osteoprogenitor cells for rebuilding bone, hepatocytes to alleviate cirrhotic pathology (due to viral hepatitis), and in retinal cells for the treatment of macular degeneration (a leading cause of age-related blindness) are other areas being considered.

An alternative to introducing genetically engineered cells, may be the production of proteins from normal human cell cultures which are then given back to a person. The use of engineered cells for *in vitro* markets and as alternatives to animal testing are also areas under experimental study. However, it will first be necessary to develop safety and efficacy standards, quality assurances and control, and pre-clinical and clinical evaluations. The results so far are important as they document that the introduction or activation of telomerase in normal cells is likely to have many applications and a major impact for the future of medicine.

5. Telomerase in cancer diagnostics

Since telomerase activity is present in most human tumors, but not in tissues adjacent to the tumors (Kim *et al.*, 1994; Shay, 1998; Shay and Bacchetti, 1997), much research is currently focused on the development of methods for the accurate diagnosis of cancer. Telomerase activity is detected in premalignant specimens (*in situ* lung, cervical, and breast cancers), while colon and pancreatic cancer have detectable telomerase activity at later (carcinoma) stages. The ability to use almost any clinical specimen and to demonstrate telomerase may allow the detection of cancers at an earlier stage. Telomerase activity is detected in lung cells in cancer patients obtained by bronchial alveolar lavage. In addition, fine needle aspirations (breast, liver and prostate cancer), washes (bladder and colon), and sedimented cells from urine (bladder and prostate) provide minimally invasive sources of cells to detect telomerase activity and are likely to have immediate diagnostic utility, for instance in monitoring of minimal residual disease. In an effort to improve the diagnostic value of telomerase determinations, *in situ* hybridization methods for the demonstration of the telomerase integral RNA on archival paraffin-embedded clinical specimens appears to distinguish cancer from normal cells, correlates well with telomerase activity, and thus may provide added value to telomerase activity assays. In addition, the presence or absence of telomerase may have prognostic value and help risk-stratify patients into those with favorable outcomes (to avoid unnecessary treatments for patients with low or no detectable telomerase) and those with high telomerase activity and with unfavorable outcomes (to help oncologists manage patient treatments more effectively).

6. Telomerase in cancer therapeutics

There is mounting interest in determining if telomerase inhibition (*Figure 5*) may have utility in cancer therapeutics (*Table 2*). There are several issues that need to be considered with the use of such therapy, for example the affects of inhibitors on telomerase-expressing stem cells (*Figure 6*). In the most primitive stem cells of renewal tissues (e.g. crypts of the intestine, bone marrow cells, resting lymphocytes, basal layer of the epidermis) telomerase activity is low while in the proliferative decendents of these cells telomerase activity is increased. Thus, there are telomerase-competent cells that have low activity when quiescent (not dividing) and increased activity when proliferating (dividing). However, these telomerase competent stem cells do not fully maintain telomere length (*Figure 6*) since such cells obtained from older individuals have shorter telomeres than those derived from younger individuals. In germline (reproductive) cells and tumor cells, telomerase fully maintains telomere length in contrast to stem cells (with regulated telomerase activity) and most somatic cells (with no detectable telomerase activity) in which telomeres progressively shorten with increased age (*Figure 6*).

Table 2. Telomerase/telomere inhibitor approaches

- Targeting small molecules
  - Screening combinatorial libraries
  - Screening compound collections
- Targeting the functional telomerase (template) RNA component (hTR)
  - Oligonucleotides – anti-sense hTR template
  - Hammerhead ribozymes – anti-sense hTR
  - 2–5A-anti-telomerase RNA oligonucleotides
- Targeting the catalytic protein component (hTERT)
  - Dominant-negative hTERT
  - hTERT promoter/suicide gene
  - Immunotherapy
  - Ribozyme cleavage of hTERT mRNA
  - Nucleoside analogs, reverse transcriptase inhibitors
- Targeting telomeres
  - G-quadruplex interacting compounds
  - Telomere-associated and binding proteins
- Targeting interactions of hTR, hTERT, and telomeres

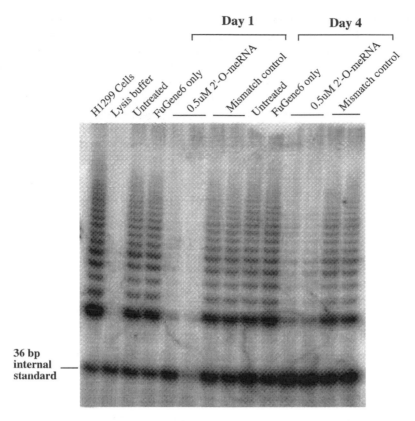

Figure 5. Anti-sense oligonucleotide based telomerase inhibition. Human cells treated with a 13-mer 2′OmeRNA oligonucleotides complementary to the telomerase RNA template region results in telomerase inhibition. Controls such as the mismatch oligonucleotides do not result in telomerase inhibition. Telomerase inhibition illustrated in this figure was accompanied by progressive telomere shortening and after a period of time that was related to the initial telomere length, cells underwent a growth arrest followed by increases in apoptosis (Herbert *et al.*, 1999). The TRAP (telomeric repeat amplification protocol) is a method developed to determine the presence or absence of telomerase activity. In this assay, if a cellular extract has telomerase activity, it can synthesize telomeric repeats onto the telomerase substrate forward primer. After this elongation step, the signal is amplified by RT-PCR using a downstream primer. Since telomerase is a processive enzyme, there are varying number of telomeric repeats added to individual primers during the elongation step resulting in a 6 base pair ladder. This assay contain an internal 36 base pair standard that can be used to help quantitate the amount of telomerase present.

It is believed that telomerase inhibition therapy should be less toxic than conventional chemotherapy, which affects all proliferating cells, including stem cells. The rate of division of the most primitive stem cells is so much slower than that of most cancer cells that the amount of telomere shortening in the stem cells should be relatively small. In addition, most tumor cells have much shorter telomeres when compared to other cell types. One research strategy is to inhibit the activity of telomerase, forcing immortal cells into a normal pattern of

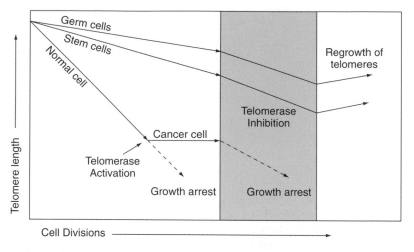

Figure 6. Model for telomerase repression in cancer. In the presence of a telomerase inhibitor, it is believed that telomerase-expressing germline cells, stem cells of renewal tissues and cancer cells would show progressive telomere shortening. Since germline cells and stem cells have longer telomeres than cancer cells, there may be a 'window of opportunity' to cause cancer cells to growth arrest and die while not irreversibily damaging normal stem cells that are generally telomerase silent or express telomerase at much reduced levels. It is believed, but not proven, that after the period of anti-telomerase treatment, germline and stem cell telomeres may return to their original length.

permanent growth arrest (senescence) or death (apoptosis). Following conventional treatments (surgery, radiotherapy, chemotherapy) anti-telomerase agents would be given to limit the proliferative capacity of the rare surviving tumor cells in the hope that this would prevent cancer recurrence. This treatment would be selective in that only cells with an activated telomerase would be affected.

   While there have been several anti-telomerase strategies proposed such as cytotoxic T cell immunity against telomerase or using gene transfer methods using the hTERT promoter with a suicide gene so that all telomerase-expressing cells will be killed, it is important to know that some of these approaches are likely to have more side effects than others (*Table 2*). Ideally, a telomerase inhibitor would be provided for a short period of time to eliminate the tumor cells population without inducing long-term damage to more primitive stem cells of renewal tissues. In previous studies, it has been demonstrated that treatment of cells for a long period of time with telomerase inhibitors results in telomere shortening, but when the inhibitor is removed, the cells eventually return to their normal telomere length. Thus, short-term reduction of telomere length in stem cells may be reversible (*Figure 6*). One important concern with this proposed treatment regimen is the prolonged time potentially required for a telomerase inhibitor to be effective. Since the mode of action of telomerase inhibitors may require telomeric shortening before inhibition of cell growth or induction of apoptosis, there may be a significant delay in efficacy. Thus methods may have to be devised to increase the rate of telomere shortening when telomerase inhibitors are used therapeutically. Telomerase inhibitors will likely be used together with or following conventional therapies, so that once the bulk of the tumor mass is eliminated anti-telomerase therapy might prevent the large number of cell divisions required for the re-growth of rare resistant cancer cells. They may also be used in early-stage cancer to prevent overgrowth of metastatic cells, as well as in high-risk patients with inherited susceptibility to cancer syndromes to prevent the emergence of telomerase-expressing cells (chemoprevention; Herbert *et al.*, 2001).

7. Conclusions

While concerns about long-term effects on cancer incidence of immortalized cells for tissue engineering are legitimate, early studies suggest that immortalization or reversible immortalization of specific cell types which can be thoroughly charac-terized prior to transplantation may have manageable risks. There is mounting evidence that cellular senescence acts as a 'cancer brake' because it takes many divi-sions to accumulate all the changes needed to become a cancer cell. In addition to the accumulation of several mutations in oncogenes and tumor suppressor genes, almost all cancer cells are immortal and, thus, have overcome the normal cellular signals that prevent continued division. Young normal cells can divide many times, but these cells are not cancer cells since they have not accumulated all the other changes needed to make a cell malignant. In most instances a cell becomes senescent before it can become a cancer cell. A central goal is to find out how to make our cancer cells mortal and die without affecting our normal cells, and our healthy but aging cells rejuvenated without increasing the risk of developing cancer. Inhibition of telomerase in cancer cells may be a viable target for anti-cancer therapeutics while expression of telomerase in normal cells may have important biopharmaceutical and medical applications. In summary, telomerase may be both an important target for cancer and for the treatment of age-related disease.

References

Allsopp, R.C., Vaziri, H., Patterson, C. *et al.* (1992) Telomere length predicts replicative capacity of human fibroblasts. *Proc. Natl Acad. Sci. USA* **89**(21): 10114–10118.

Anonymous (1996) Positional cloning of the Fanconi anaemia group A gene. The Fanconi anaemia/breast cancer consortium. *Nat. Genet.* **14**(3): 324–328.

Banin, S., Moyal, L., Shieh, S. *et al.* (1998) Enhanced phosphorylation of p53 by ATM in response to DNA damage. *Science* **281**(5383): 1674–1677.

Bentley, N.J., Holtzman, D.A., Flaggs, G. *et al.* (1996) The *Schizosaccharomyces pombe* rad3 checkpoint gene. *EMBO J.* **15**(23): 6641–6651.

Blackburn, E.H. and Gall, J.G. (1978) A tandemly repeated sequence at the termini of the extrachromosomal ribosomal RNA genes in Tetrahymena. *J. Mol. Biol.* **120**: 33–53.

Blasco, M. A. H. W. Lee, M. P. Hande, E. *et al.* (1997) Telomere shortening and tumor formation by mouse cells lacking telomerase RNA. *Cell* **91**: 25–34.

Bodnar, A.G., Ouellette, M., Frolkis, M. *et al.* (1998) Extension of life-span by introduction of telomerase into normal human cells. *Science* **279**: 349–352.

Broeks, A., Urbanus, J.H., Floore, A.N. *et al.* (2000) ATM-heterozygous germline mutations contribute to breast cancer-susceptibility. *Am. J. Hum. Genet.* **66**(2): 494–500.

Bryan, H. and Nixon, R. (1965) Dyskeratosis congenita and familial pancytopenia. *J. Am. Med. Assoc.* **192**: 203–208.

Bryan T.M., Englezou A., Dalla-Pozza L. *et al.* (1997). Evidence for an alternative mechanism for maintaining telomere length in human tumors and tumor-derived cell lines [see comments]. *Nat. Med.* **3**(11): 1271–1274.

Cech, T.R. (2000) Life at the end of the chromosome: telomeres and telomerase. *Angew. Chem. Int. Ed.* **39**: 34–43.

Chen, J., Birkholtz, G.G., Lindblom, P. *et al.* (1998) The role of ataxia-telangiectasia heterozygotes in familial breast cancer. *Cancer Res.* **58**(7): 1376–1379.

Craven, R.J. and Petes, T.D. (2000) Involvement of the checkpoint protein Mec1p in silencing of gene expression at telomeres in *Saccharomyces cerevisiae*. *Mol. Cell Biol.* **20**(7): 2378–2384.

de Winter, J.P., Waisfisz, Q., Rooimans, M.A. *et al.* (1998) The Fanconi anaemia group G gene FANCG is identical with XRCC9. *Nature Genet.* **20**(3): 281–283.

DeBauche, D.M., Pai, G.S. and Stanley, W.S. (1990) Enhanced G2 chromatid radiosensitivity in dyskeratosis congenita fibroblasts. *Am. J. Hum. Genet.* **46**(2): 350–357.

DeBusk, F.L. (1972) The Hutchinson-Gilford progeria syndrome. Report of 4 cases and review of the literature. *J. Pediat.* **80**(4): 697–724.

DeLange T., Shiue, L., Myers, M.R. *et al.* (1990) Structure and variability of human chromosome ends. *Mol. Cell. Biol.* **10**: 518–527.

Devriendt, K., Matthijs, G., Legius, E. *et al.* (1997) Skewed X-chromosome inactivation in female carriers of dyskeratosis congenita. *Am. J. Hum. Genet.* **60**(3): 581–587.

Dokal, I. and Luzzatto, L. (1994) Dyskeratosis congenita is a chromosomal instability disorder. *Leukemia and Lymphoma* **15**(1–2): 1–7.

Dokal, I., Bungey, J., Williamson, P. *et al.* (1992) Dyskeratosis congenita fibroblasts are abnormal and have unbalanced chromosomal rearrangements. *Blood* **80**(12): 3090–3096.

Ellis, N.A., Groden, J., Ye, T.Z. *et al.* (1995) The Bloom's syndrome gene product is homologous to RecQ helicases. *Cell* **83**(4): 655–666.

Epstein, C.J., Martin, G.M., Schultz A.L. *et al.* (1966) Werner's syndrome a review of its symptomatology, natural history, pathologic features, genetics and relationship to the natural aging process. *Medicine* **45**(3): 177–221.

Feng, F., Funk, W.D., Wang, S.S. *et al.* (1995) The RNA component of human telomerase. *Science* **269**: 1236–1241.

Ferraris, A.M., Forni, G.L., Mangerini R. *et al.* (1997) Nonrandom X-chromosome inactivation in hemopoietic cells from carriers of dyskeratosis congenita [letter]. *Am. J. Hum. Genet.* **61**(2): 458–461.

FitzGerald, M.G., Bean, J.M., Hegde, S.R. *et al.* (1997) Heterozygous ATM mutations do not contribute to early onset of breast cancer [see comments]. *Nature Genet.* **15**(3): 307–310.

Friedberg, E., Walker, G. and Siede, W. (1995) *DNA repair and mutagenesis*. Washington, D.C., ASM Press.

Fukuchi, K., Martin, G.M. and Monnat Jr., R.J. (1989) Mutator phenotype of Werner syndrome is characterized by extensive deletions [published erratum appears in Proc Natl Acad Sci U S A 1989 Oct;86(20): 7994]. *Proc. Natl Acad. Sci. USA* **86**(15): 5893–5897.

Fukuchi, K., Tanaka, K., Kumahara, Y. *et al.* (1990) Increased frequency of 6-thioguanine-resistant peripheral blood lymphocytes in Werner syndrome patients. *Hum. Genet.* **84**(3): 249–252.

Gangloff, S., McDonald, J.P., Bendixen, C. *et al.* (1994) The yeast type I topoisomerase Top3 interacts with Sgs1, a DNA helicase homolog: a potential eukaryotic reverse gyrase. *Mol. Cell. Biol.* **14**(12): 8391–8398.

Giampietro P.F., Adler-Brecher B., Verlander P.C. *et al.* (1993) The need for more accurate and timely diagnosis in Fanconi anemia: a report from the International Fanconi Anemia Registry. *Pediatrics* **91**(6): 1116–1120.

Gollahon, L. S., Kraus, E., Wu, T-A. *et al.* 1998. Telomerase activity status during spontaneous immortalization of Li-Fraumeni skin fibroblasts. *Oncogene* **17**: 709–718.

Gompertz, B. (1825) On the nature and function expressivity of the law of human mortality and on a new mode of determining life contingencies. *Phil. Trans. R. Soc. Lond.* **115**: 513–585.

Goto, M., Tanimoto, K., Horiuchi, Y. *et al.* (1981) Family analysis of Werner's syndrome: a survey of 42 Japanese families with a review of the literature. *Clin. Genet.* **19**(1): 8–15.

Greenberg, R.A., Chin, L., Femino, A. *et al.* (1999) Short dysfunctional telomeres impair tumorigenesis in the INK4a(delta2/3) cancer-prone mouse. *Cell* **97**: 515–525.

Greider C.W. and Blackburn E.H. (1985) Identification of a specific telomere terminal transferase enzyme with two kinds of primer specificity. *Cell* **51**: 405–413.

Hahn W.C., Counter C.M., Lundberg A.S. *et al.* (1999) Creation of human tumour cells with defined genetic elements. *Nature* **400** (6743): 464–468.

Hande, M.P., Balajee, A.S. and Natarajan, A.T. (1997) Induction of telomerase activity by UV-irradiation in Chinese hamster cells. *Oncogene* **15**(14): 1747–1752.

Hande, M.P., Lansdorp, P.M. and Natarajan, A.T. (1998) Induction of telomerase activity by in vivo X-irradiation of mouse splenocytes and its possible role in chromosome healing. *Mutat. Res.* **404**(1-2): 205–214.

Hari, K.L., Santerre, A., Sekelsky, J.J. *et al.* (1995) The mei-41 gene of *D. melanogaster* is a structural and functional homolog of the human ataxia telangiectasia gene. *Cell* **82**(5): 815–821.

Harley, C.B., Fletcher, A.B. and Greider, C.W. (1990) Telomeres shorten during aging. *Nature* **345**: 458–460.

Harmon, F.G. and Kowalczykowski, S.C. (1998) RecQ helicase, in concert with RecA and SSB proteins, initiates and disrupts DNA recombination. *Genes Develop.* **12**(8): 1134–1144.

Hastie, N.D., Dempster, M., Dunlop, M.G. *et al.* (1990) Telomere reduction in human colorectal carcinoma and with ageing. *Nature* **346**: 866–868.

Hayflick, L. and Moorhead, P.S. (1961) The limited in vitro lifetime of human diploid cell strains. *Exp. Cell Res.* **25**: 585–621.

Heiss, N.S., Knight, S.W., Vulliamy, T.J. *et al.* (1998) X-linked dyskeratosis congenita is caused by mutations in a highly conserved gene with putative nucleolar functions [see comments]. *Nature Genet.* **19**(1): 32–38.

Herbert, B-S, Pitts, A. E., Baker, S. I. *et al.* (1999) Inhibition of human telomerase in immortal human cells leads to progressive telomere shortening and cell death. *Proc. Natl Acad. Sci.* **96**: 14276–14281.

Herbert, B-S, Wright, A.C., Passons, C.M., Ali, I., Wright, W.E., Kopelovich, L. and Shay, J.W. (2001) Inhibition of the spontaneous immortalization of breast epithelial cells from individuals predisposed to breast cancer: Effects of chemopreventive and anti-telomerase agents. *J. Natl. Can. Inst.*, **93**: 39–45.

Hisama, F.M., Chen, Y.H., Meyn, M.S. *et al.* (2000) WRN or telomerase constructs reverse 4-nitroquinoline 1-oxide sensitivity in transformed Werner syndrome fibroblasts. *Cancer Res.* **60**(9): 2372–2376.

Hreidarsson, S., Kristjansson, K., Johannesson, G. *et al.* (1988) A syndrome of progressive pancytopenia with microcephaly, cerebellar hypoplasia and growth failure. *Acta Paediat. Scand.* **77**(5): 773–775.

Jiang, X.R. Jimenez, G., Chang, E. *et al.* (1999) Telomerase expression in human somatic cells does not induce changes associated with a transformed phenotype. *Nature Genet.* **21**: 111–114.

Kim, N.-W., Piatyszek, M.A., Prowse, K.R. *et al.* (1994) Specific association of human telomerase activity with immortal cells and cancer. *Science* **266**: 2011–2015.

Kitao, S., Ohsugi, I., Ichikawa, K. *et al.* (1998) Cloning of two new human helicase genes of the RecQ family: biological significance of multiple species in higher eukaryotes. *Genomics* **54**(3): 443–452.

Kitao, S., Shimamoto, A., Goto, M. *et al.* (1999) Mutations in RECQL4 cause a subset of cases of Rothmund-Thomson syndrome. *Nature Genet.* **22**(1): 82–84.

Kojis, T.L., Gatti, R.A. and Sparkes, R.S. (1991) The cytogenetics of ataxia telangiectasia. *Cancer Genet. Cytogenet.* **56**(2): 143–156.

Kruk, P.A. and Bohr, V.A. (1999) Telomeric length in individuals and cell lines with altered p53 status. *Radiat. Oncol. Invest.* **7**(1): 13–21.

Lafontaine, D.L.J., Bousquet-Antonelli, C., Henry, Y. *et al.* (1998) The box H + ACA snoRNAs carry Cbf5p, the putative rRNA pseudouridine synthase. *Genes Develop.* **12**(4): 527–537.

Leteurtre, F., Li, X., Guardiola, P. *et al.* (1999) Accelerated telomere shortening and telomerase activation in Fanconi's anaemia. *Br. J. Haematol.* **105**(4): 883–893.

Lim, D.S., Kim, S.T., Xu, B. *et al.* (2000) ATM phosphorylates p95/nbs1 in an S-phase checkpoint pathway. *Nature* **404**(6778): 613–617.

Lo Ten Foe, J.R., Rooimans, M.A., Bosnoyan-Collins, L. *et al.* (1996) Expression cloning of a cDNA for the major Fanconi anaemia gene, FAA. *Nature Genet.* **14**(3): 320–323.

Lombard, D.B. and Guarente, L. (2000). Nijmegen breakage syndrome disease protein and MRE11 at PML nuclear bodies and meiotic telomeres. *Cancer Res.* **60**: 2331–2334.

Matsuura, A., Naito, T. and Ishikawa, F. (1999) Genetic control of telomere integrity in *Schizosaccharomyces pombe*: rad3(+) and tel1(+) are parts of two regulatory networks independent of the downstream protein kinases chk1(+) and cds1(+). *Genetics* **152**(4): 1501–1512.

Matsuura, S., Tauchi, H., Nakamura, A. *et al.* (1998) Positional cloning of the gene for Nijmegen breakage syndrome. *Nature Genet.* **19**(2): 179–181.

McClintock, B. (1941) The stability of broken ends of chromosomes in *Zea mays*. *Genetics* **26**: 234–282.

Mekeel, K.L., Tang, W., Kachnic, L.A. *et al.* (1997). Inactivation of p53 results in high rates of homologous recombination. *Oncogene* **14**(15): 1847–1857.

Meyn, M.S. (1995) Ataxia-telangiectasia and cellular responses to DNA damage. *Cancer Res.* **55**(24): 5991–6001.

Metcalfe, J.A., Parkhill, J., Campbell, L. *et al.* (1996) Accelerated telomere shortening in ataxia telangiectasis. *Nature Genet.* **18**: 350–353.

Milgrom, H., Stroll, H. and Crissey, J. (1964) Dyskeratosis congenita: a case with new features. *Arch. Dermatol.* **89**: 345–349.

Mitchell, J.R., Wood, E. and Collins, K. (1999) A telomerase component is defective in the human disease dyskeratosis congenita. *Nature* **402**(6761): 551–555.

Monnat, R.J., Jr., Hackmann, A.F. and Chiaverotti, T.A. (1992) Nucleotide sequence analysis of human hypoxanthine phosphoribosyltransferase (HPRT) gene deletions. *Genomics* **13**(3): 777–787.

Morales, C.P., Holt, S.E., Ouellette, M.M. *et al.* (1999) Lack of cancer-associated changes in human fibroblasts immortalized with telomerase. *Nature Genet.* **21**: 115–118.

Morin, G.B. (1989) The human telomere terminal transferase enzyme is a ribonucleoprotein that synthesizes TTAGGG repeats. *Cell* **59**: 521–529.

Moriwaki, S., Ray, S., Tarone, R. *et al.* (1996) The effect of donor age on the processing of UV-damaged DNA by cultured human cells: reduced DNA repair capacity and increased DNA mutability. *Mutat. Res.* **364**(2): 117–123.

Morrow, D.M., Tagle, D.A., Shiloh, Y. *et al.* (1995) TEL1, an *S. cerevisiae* homolog of the human gene mutated in ataxia telangiectasia, is functionally related to the yeast checkpoint gene MEC1. *Cell* **82**(5): 831–840.

Muller, H.J. (1962) The remaking of chromosomes. In: *Studies of genetics: The selected papers of H.J. Muller*, pp 384–408, Indiana University Press, Bloomington.

Naito, T., Matsuura, A. and Ishikawa, F. (1998) Circular chromosome formation in a fission yeast mutant defective in two ATM homologues. *Nat. Genet.* **20**(2): 203–206.

Nakamura, T.M., Morin, G.B., Chapman, K.B. *et al.* (1997) Telomerase catalytic subunit homologs from fission yeast and humans. *Science* **277**: 955–959.

Ning, Y., Yongshan, Y., Pai, G.S. *et al.* (1992) Heterozygote detection through bleomycin-induced G2 chromatid breakage in dyskeratosis congenita families. *Cancer Genet. Cytogenet.* **60**(1): 31–34.

Oexle, K., Zwirner, A., Freudenberg, K. *et al.* (1997) Examination of telomere lengths in muscle tissue casts doubt on replicative aging as cause of progression in Duchenne muscular dystrophy. *Pediat. Res.* **42**(2): 226–231.

Ouellette, M.M., McDaniel, L.D., Wright, W.E. *et al.* (2000) The establishment of telomerase-immortalized cell lines representing human chromosome instability syndromes. *Hum. Mol. Genet.* **9**(3): 403–411.

Pandita, T.K., Pathak, S. and Geard, C.R. (1995) Chromosome end associations, telomeres and telomerase activity in ataxia telangiectasia cells. *Cytogenet. Cell Genet.* **71**(1): 86–93.

Pandita, T.K., Westphal, C.H., Anger, M., Sawant, S.G., Geard, C.R. Pandita, R.K. and Scherthan, H. (1999) Atm inactivation results in aberrant telomere clustering during meiotic prophase. *Mol. Cell Biol.* **19**(7): 5096–5105.

Pearson, M., Carbone, R., Sebastiani, C. *et al.* (2000). PML regulates p53 acetylation and premature senescence induced by oncogenic Ras. *Nature* **406**: 207–210.

Ritchie, K.B., Mallory, J.C. and Petes, T.D. (1999) Interactions of TLC1 (which encodes the RNA subunit of telomerase), TEL1, and MEC1 in regulating telomere length in the yeast *Saccharomyces cerevisiae*. *Mol. Cell. Biol.* **19**(9): 6065–6075.

Rudolph, K.L., Chang, S., Lee, H.W. *et al.* (1999) Longevity, stress response, and cancer in aging telomerase-deficient mice. *Cell* **96**: (5) 701–712.

Rudolph, K.L., Chang, S., Millard, M. *et al.* (2000) Inhibition of experimental cirrhosis in mice by telomerase gene therapy. *Science* **287**: 1253–1258.

Sanchez, Y., Desany, B.A., Jones, W.J. *et al.* (1996) Regulation of RAD53 by the ATM-like kinases MEC1 and TEL1 in yeast cell cycle checkpoint pathways [see comments]. *Science* **271**(5247): 357–360.

Sato, N., Mizumoto, K., Kusumoto, M., Nishio, S., Maehara, N., Urashima, T., Ogawa, T. and Tanaka, M. (2000) Up-regulation of telomerase activity in human pancreatic cancer cells after exposure to etoposide. *Br. J. Cancer* **82**(11): 1819–1826.

Savitsky, K., Bar-Shira, A., Gilad, S. *et al.* (1995) A single ataxia telangiectasia gene with a product similar to PI-3 kinase [see comments]. *Science* **268**(5218): 1749–1753.

Selmanowitz, V.J., Van Voolen, G.A. and Steier, W. (1971) Fanconi's anemia and dyskeratosis congenita. *JAMA* 216(12): 2015.

Shafei-Benaissa, E., Huret, J.L., Larregue, M. *et al.* (1994) Checks for chromosomal instability in Gorlin and non-Gorlin basal-cell carcinoma patients. *Mutat. Res.* **308**(1): 1–9.

Shafman, T., Khanna, K.K., Kedar, P. *et al.* (1997) Interaction between ATM protein and c-Abl in response to DNA damage [see comments]. *Nature* **387**(6632): 520–523.

Shay, J.W. (1998) Telomerase in cancer: Diagnostic, prognostic and therapeutic implications. *Cancer J. Scientific Amer.* **4**: 26–34.

Shay, J.W. and Bacchetti, S. (1997) A survey of telomerase in human cancer. *Eur. J. Cell Biol.* **33**: 787–791.

Shay, J.W. and Wright W.E. (2000) The use of telomerized cells for tissue engineering. *Nature Biotech.* **18**: 22–23.

Shay, J.W., Tomlinson, G., Piatyszek, M.A. *et al.* (1995) Spontaneous in vitro immortalization of breast epithelial cells from a Li-Fraumeni patient. *Mol. Cell. Biol.* **15**: 425–432.

Shinya, A., Nishigori, C., Moriwaki, S. *et al.* (1993) A case of Rothmund-Thomson syndrome with reduced DNA repair capacity [see comments]. *Arch. Dermatol.* **129**(3): 332–336.

Sinclair D.A. and Guarente L. (1997) Extrachromosomal rDNA circles – a cause of aging in yeast. *Cell* **91**(7): 1033–1042.

Smilenov, L.B., Dhar, S. and Pandita, T.K. (1999) Altered telomere nuclear matrix interactions and nucleosomal periodicity in ataxia telangiectasia cells before and after ionizing radiation treatment. *Mol. Cell. Biol.* **19**(10): 6963–6971.

Smith, P.J. and Paterson, M.C. (1982) Enhanced radiosensitivity and defective DNA repair in cultured fibroblasts derived from Rothmund Thomson syndrome patients. *Mut. Res.* **94**(1): 213–228.

Strathdee, C.A., Gavish, H., Shannon, W.R. *et al.* (1992) Cloning of cDNAs for Fanconi's anaemia by functional complementation [published erratum appears in Nature 1992 Jul 30;358(6385): 434]. *Nature* **356**(6372): 763–767.

Swift, M., Reitnauer, P.J., Morrell, D. *et al.* (1987) Breast and other cancers in families with ataxia-telangiectasia. *N. Engl. J. Med.* **316**(21): 1289–1294.

Tahara, H., Tokutake, Y., Maeda, S. *et al.* (1997) Abnormal telomere dynamics of B-lymphoblastoid cell strains from Werner's syndrome patients transformed by Epstein-Barr virus. *Oncogene* **15**(16): 1911–1920.

Thomas M., Yang L.Q. and Hornsby P.J. (2000) Formation of functional tissue from transplanted adrenocortical cells expressing telomerase reverse transcriptase *Nat. Biotechnol.* **18**(1): 39–42.

Umezu, K. and Nakayama, H. (1993) RecQ DNA helicase of *Escherichia coli*. Characterization of the helix-unwinding activity with emphasis on the effect of single-stranded DNA-binding protein. *J. Mol. Biol.* **230**(4): 1145–1150.

Umezu, K., Nakayama, K. and Nakayama, H. (1990) *Escherichia coli* RecQ protein is a DNA helicase [published erratum appears in *Proc Natl Acad Sci USA* 1990 Nov;87(22): 9072]. *Proc. Nat. Acad. Sci. USA* **87**(14): 5363–5367.

Vande Berg J.S., Rudolph, R., Hollan, C. *et al.* (1998) Fibroblast senescence in pressure ulcers. *Wound Rep. Reg.* **6**: 38–49.

Vasseur, F., Delaporte, E., Zabot, M.T. *et al.* (1999) Excision repair defect in Rothmund Thomson syndrome. *Acta Dermato-Venereologica* **79**(2): 150–152.

Vulliamy, T.J., Knight, S.W., Dokal, I. *et al.* (1997) Skewed X-inactivation in carriers of X-linked dyskeratosis congenita. *Blood* **90**(6): 2213–2216.

Wang, J.Y. (2000) Cancer: new link in a web of human genes. *Nature* **405**: 404–405.

Wang, X., Liu, Y., Chow, L.S., Wong, S.C., Tsao, G.S., Kwong, D.L., Sham, J.S. and Nicholls, J.M. (2000) Regulation of telomerase activity by gamma-radiation in nasopharyngeal carcinoma cells. *Anticancer Res.* **20**(1A): 433–437.

Watt, P.M., Louis, E.J., Borts, R.H. *et al.* (1995) Sgs1: a eukaryotic homolog of *E. coli* RecQ that interacts with topoisomerase II in vivo and is required for faithful chromosome segregation. *Cell* **81**(2): 253–260.

Watt, P.M., Hickson, I.D., Borts, R.H. *et al.* (1996) SGS1, a homologue of the Bloom's and Werner's syndrome genes, is required for maintenance of genome stability in *Saccharomyces cerevisiae*. *Genetics* **144**(3): 935–945.

Wei, Q. (1998) Effect of aging on DNA repair and skin carcinogenesis: a minireview of population-based studies. *J. Invest. Dermatol.* **3**(1): 19–22.

Weisman, A. (1891) *Essays upon heredity and kindred biological problems*, 2nd edn, Clarendon Press, Oxford.

Wright, W.E., Piatyszek, M.A., Rainey, W.E., Byrd, W. and Shay, J.W. (1996) Telomerase activity in human germline and embryonic tissues and cells. *Dev. Genet.*, **18**: 173B179.

Wright, W.E. and Shay, J.W. (2000) Telomere dynamics in cancer progression and prevention: fundamental differences in human and mouse biology. *Nature Medicine* **6**: 849–851.

Wyllie, F. S., Jones, C.J., Skinner, J.W. *et al.* (2000) Telomerase prevents the accelerated cell ageing of Werner syndrome fibroblasts. *Nature Genet.* **24**(1): 16–17.

Xia, S.J., Shammas, M.A. and Shmookler Reis, R.J. (1996) Reduced telomere length in ataxia-telangiectasia fibroblasts. *Mutat. Res.* **364**(1): 1–11.

Yeager, T.R., Neumann, A.A., Englezou, A. *et al.* (1999) Telomerase-negative immortalized human cells contain a novel type of promyelocytic leukemia (PML) body. *Cancer Res.* **59**(17): 4175–4179.

Yu, C.E., Oshima, J., Fu, J.H. *et al.* (1996) Positional cloning of the Werner's syndrome gene [see comments]. *Science* **272**(5259): 258–262.

Index